CW00725204

SUBSEA PIPELINES AND RISERS

Elsevier Internet Homepage: http://www.elsevier.com

Consult the Elsevier homepage for full catalogue information on all books, journals and electronic products and services.

Ocean Engineering Series

WATSON
Practical Ship Design
ISBN: 008-042999-8

YOUNG
Wind Generated Ocean Waves
ISBN: 008-043317-0

BAI
Pipelines and Risers
ISBN: 008-043712-5

JENSEN
Load and Global Response of Ships
ISBN: 008-043953-5

TUCKER & PITT
Waves in Ocean Engineering
ISBN: 008-043566-1

BOSE & BROOKE
Wave Energy Conversion
ISBN: 008-044212-9

PILLAY & WANG
Technology and Safety of Marine Systems
ISBN: 008-044148-3

OCHI
Hurricane-Generated Seas
ISBN: 008-044312-5

KOBYLINSKI & KASTNER
Stability and Safety of Ships
Volume I: Regulation and Operation
ISBN: 008-043001-5

BELENKY & SEVASTIANOV
Stability and Safety of Ships
Volume II: Risk of Capsizing
ISBN: 008-044354-0

Other Titles

BAI
Marine Structural Design
ISBN: 008-043921-7

CHAKRABARTI
Handbook of Offshore Engineering (2 volume set)
ISBN: 008-044381-8

MANSOUR & ERTEKIN
Proceedings of the 15th Ship and Offshore Structures Congress
(2 volume set)
ISBN:008-044076-2

Related Journals

Free specimen copy gladly sent on request. Elsevier Ltd, The Boulevard, Langford Lane, Kidlington, Oxford, OX5 1GB, UK

Applied Ocean Research
Advances in Engineering Software
CAD
Coastal Engineering
Composite Structures
Computers and Structures
Construction and Building Materials
Engineering Failure Analysis
Engineering Fracture Mechanics
Engineering Structures
Finite Elements in Analysis and Design
International Journal of Solids and Structures
Journal of Constructional Steel Research
Marine Structures
NDT & E International
Ocean Engineering
Structural Safety
Thin-Walled Structures

To Contact the Publisher

Elsevier welcomes enquiries concerning publishing proposals: books, journal special issues, conference proceedings, etc. All formats and media can be considered. Should you have a publishing proposal you wish to discuss, please contact, without obligation, the publisher responsible for Elsevier's civil and structural engineering publishing programme:

Nick Pinfield
Publishing Editor
Elsevier Ltd
The Boulevard, Langford Lane
Kidlington, Oxford
OX5 1GB, UK

Phone: +44 1865 843352
Fax: +44 1865 843920
E.mail: n.pinfield@elsevier.com

General enquiries, including placing orders, should be directed to Elsevier's Regional Sales Offices – please access the Elsevier homepage for full contact details (homepage details at the top of this page).

SUBSEA PIPELINES AND RISERS

YONG BAI

and

QIANG BAI

2005

Amsterdam – Boston – Heidelberg – London – New York – Oxford
Paris – San Diego – San Francisco – Singapore – Sydney – Tokyo

ELSEVIER B.V.
Radarweg 29
P.O. Box 211, 1000 AE
Amsterdam, The Netherlands

ELSEVIER Inc.
525 B Street, Suite 1900
San Diego,
CA 92101-4495, USA

ELSEVIER Ltd
The Boulevard, Langford Lane
Kidlington, Oxford
OX5 1GB, UK

ELSEVIER Ltd
84 Theobalds Road
London
WC1X 8RR, UK

First edition 2005

ISBN: 0-080-4456-67

Printed and bound in Great Britain by CPI Antony Rowe, Eastbourne

FOREWORD

June 2005

Being an avid pipeline engineer for many years, it was with the keenest interest that I reviewed the contents of this new text. What I discovered was a vault of valuable and compelling information for both the seasoned pipeline engineering veteran and the newest student to the field. The subject matter is very detailed and contains the necessary depth to satisfy both the daily engineering needs as well as entry level information into developing fields.

This work extends the breadth and scope of Professor Yong Bai's previous pipeline and riser text by the addition of many new subjects and further depth within principal design topics. The pipeline and riser design, system reliability and flow assurance information that has been collected and developed within this book saves engineers countless hours of acquiring and compiling technical papers and specifications. And in the fast paced offshore pipeline engineering business, time like reserves is continually in short supply.

The offshore industry is moving forward to grasp new technology at a quickened pace due to the growing global demand for hydrocarbon energy sources and the tight project budget requirements. The rapid pace requires new personnel to gain insight into difficult issues in a reduced timeframe. A text such as this has intrinsic value because engineers and support personnel can understand the complex issues faster and more thoroughly in order to assist the project teams in a more productive entry capacity. The text naturally assists seasoned engineers in catching up on an issue or two that may have been developing while they were absorbed in more traditional design projects through the years.

New pipeline and riser technologies are being required because the new energy sources are being found in deeper water depths and in more hostile environments. Project design requirements frequently include hurricanes or typhoons, earthquakes, subsea mudslides, natural seabed erosion, liquefaction and soils transport by currents plus the industry wide goal to maintain control over our natural recourses and prevent any damage to the environment. Tsunamis are now added to this exhaustive list, highlighted by this year's major disaster.

In areas like the Gulf of Mexico, routine water depths for pipeline designs are now 3,000 feet, pipelines are already installed beyond 7,000 feet and projects are encroaching on 9,000 feet depth. Deep water takes on an entirely new meaning every few years. It is likely a surprise to the students and young pipeline engineers that so much of this industry is still being redefined and reformulated due to ever evolving challenges.

To accomplish a successful pipeline and riser design with high reliability, when faced with many difficult factors, engineering issues must be correctly and accurately addressed. This text yields a roadmap not only for the pipeline engineer but also the project managers, estimators and regulatory personnel hoping to gain an appreciation of the overall issues and directed approaches to pipeline and riser design solutions.

Many of us who have worked within the subsea pipeline and riser business are in the field because it has remained continuously challenging for decades. As many complex problems are carefully delineated and solved, more economical field development scenarios evolve which generate the need for yet newer solutions presenting technical gaps to be filled by the engineering community.

This book is a wonderful text for the uninitiated student in the offshore adventure of subsea pipelines and risers filled with all the severe weather and complex soils conditions. Containing decades of expert information and insight into this most challenging environmental arena, many years of 'on the job' experience can be gained by simply reading this text.

The information compiled and presented in this text by Professor Yong Bai and his brother Dr. Qiang Bai will certainly be a valuable reference for many years to come.

Richard D. Haun, PE
Sr. Vice President
OPE Inc.

FOREWORD to "Pipeliners and Risers" Book

June 2000

This new book provides the reader with a scope and depth of detail related to the design of offshore pipelines and risers, probably not seen before in a textbook format. With the benefit of nearly 20 years of experience, Professor Yong Bai has been able to assimilate the essence of the applied mechanics aspects of offshore pipeline system design in a form of value to students and designers alike. The text is well supported by a considerable body of reference material to which Professor Yong Bai himself has made a substantial contribution over his career. I have been in the field of pipeline engineering for the best part of 25 years and in that time have seen the processes involved becoming better and better understood. This book further adds to that understanding.

Marine pipelines for the transportation of oil and gas have become a safe and reliable part of the expanding infrastructure put in place for the development of the valuable resources below the world's sea and oceans. The design of these pipelines is a relatively young technology and involves a relatively small body of specialist engineers and researchers worldwide. In the early 1980's when Professor Yong Bai began his career in pipelines, the technology was very different than it is today, being adapted from other branches of hydrodynamics, mechanical and marine engineering using code definitions and safety factors proven in other applications but not specific to the complex hydrodynamic-structure-seabed interactions seen in the behaviour of what is outwardly a simple tubular lying on or slightly below the seabed. Those designs worked then and many of the systems installed, including major oil and gas trunklines installed in the hostile waters of the North Sea, remain in safe service today. What has happened in the intervening period is that pipeline design processes have matured and have been adapted and evolved to be fit for purpose for today's more cost effective pipelines; and will continue to evolve for future application in the inevitable move into deeper waters and more hostile environments.

An aspect of the marine pipeline industry, rarely understood by those engineers working in land based design and construction, is the more critical need for a 'right first time' approach in light of the expense and complexity of the materials and the installation facilities involved, and the inability to simply 'go back and fix it' after the fact when your pipeline is sitting in water depths well beyond diver depth and only accessible by robotic systems. Money spent on good engineering up front is money well spent indeed and again a specific fit for purpose modern approach is central to the best in class engineering practice requisite for this right first time philosophy. Professor Yong Bai has made important contributions to this coming of age of our industry and the benefit of his work and knowledge is available to those who read and use this book.

It is well recognised that the natural gas resources in the world's ocean are gaining increasing importance as an energy source to help fuel world economic growth in the established and emerging economies alike. Pipelines carry a special role in the development and production of gas reserves since, at this point in time, they provide one of the most reliable means of transportation given that fewer options are available than for the movement of hydrocarbon liquids. Add to this the growing need to provide major transportation infrastructure between gas producing regions and countries wishing to import gas, and future oil transmission systems, then the requirement for new offshore pipelines appears to be set for several years to come. Even today, plans for pipeline transportation infrastructure are in development for regions with more hostile environments and deeper waters than would have been thought achievable even ten yeas ago. The challenges are out there and the industry needs a continuous influx of young pipeline engineers ready to meet those challenges. Professor Yong Bai has given us, in this volume, an excellent source of up to date practices and knowledge to help equip those who wish to be part of the exciting future advances to come in our industry.

Dr Phillip W J Raven
Group Managing Director
J P Kenny Group of Companies

PREFACE

June 2005

It has been over five years since the senior author's "Pipelines and Risers" book appeared and more than seven years since the text was written. At the time, advanced pipeline engineering, riser engineering and flow assurance were evolving rapidly as disciplines but were also approaching a degree of maturity. Perhaps it is not surprising, then, that rather little of this book now seems out-dated.

The aim of this book is to cover the theory and applications of subsea field development, from front end engineering development (FEED), detail design and installation, to testing, inspection and monitoring.

In response to a significant development in analysis, design, testing, inspection and monitoring, riser related chapters have been significantly expanded. Methods for riser design and integrity management have been addressed in detail. A chapter has been devoted to each type of the riser systems such as steel catenary risers (SCR), top tensioned risers (TTR), drilling risers, flexible risers, hybrid risers and umbilicals.

Flow assurance has become a crucially important discipline for technical feasibility and cost effectiveness of deepwater field development. It may also govern the system selection of pipelines and risers. Hence, new chapters have been added to this edition including: subsea system engineering, hydraulics, heat transfer and thermal insulation, hydrate, wax and asphaltenes, and corrosion prevention.

In preparing this book, we are grateful to Richard Haun (Senior VP of OPE Inc.) for his encouragement. Yong is indebted to his younger colleagues Johann Melillo and Xiaolin Zhang for their assistance with the new chapters on flexible risers and corrosion control respectively. Qiang would like to express his gratitude to his friends, Dr. Ian Roberts, Manager of Multiphase Services, Scandpower Petroleum Technology Inc.; Dr. Jie Lu, Technip Offshore Inc.; and Johnny C. Wu, Manatee Inc. for reviewing the flow assurance part; Ms. Helen Gao for designing the book cover.

Special thanks to our wives, Hua Peng and Kumiko Okutani, for their understanding and support.

Yong Bai, Prof., Ph.D., P.E.
President, Grenland Advanced Engineering Inc. (GAE)
Professor, Harbin Engineering University

Qiang Bai, Ph.D., P.E.
Leader – Development and Technical Group
OPE Inc.

TABLE OF CONTENTS

PART I: Mechanical Design

PART II: Pipeline Design

PART III: Flow Assurance

PART IV: Riser Engineering

PART V: Welding and Installation

PART VI: Integrity Management

PART I: Mechanical Design

Part I

Mechanical Design

Chapter 1 Introduction

1.1 Introduction

Pipelines (and risers) are used for a number of purposes in the development of offshore hydrocarbon resources (see Figure 1.1). These include e.g.:

- Export (transportation) pipelines;
- Flowlines to transfer product from a platform to export lines;
- Water injection or chemical injection flowlines;
- Flowlines to transfer product between platforms, subsea manifolds and satellite wells;
- Pipeline bundles.

The design process for each type of lines in general terms is the same. It is this general design approach that will be discussed in this book.

Design of metallic risers is similar to pipeline design, although different analysis tools and design criteria are applied. Part IV of this book is devoted to riser design.

Finally, in Chapter 16, two pipeline design projects are used as examples demonstrating how technical development described in this book is used to achieve cost saving and safety/quality.

1.2 Design Stages and Process

1.2.1 Design Stages

The design of pipelines and risers is usually performed in three stages, namely;

- Conceptual engineering,
- Preliminary engineering or pre-engineering,
- Detail engineering.

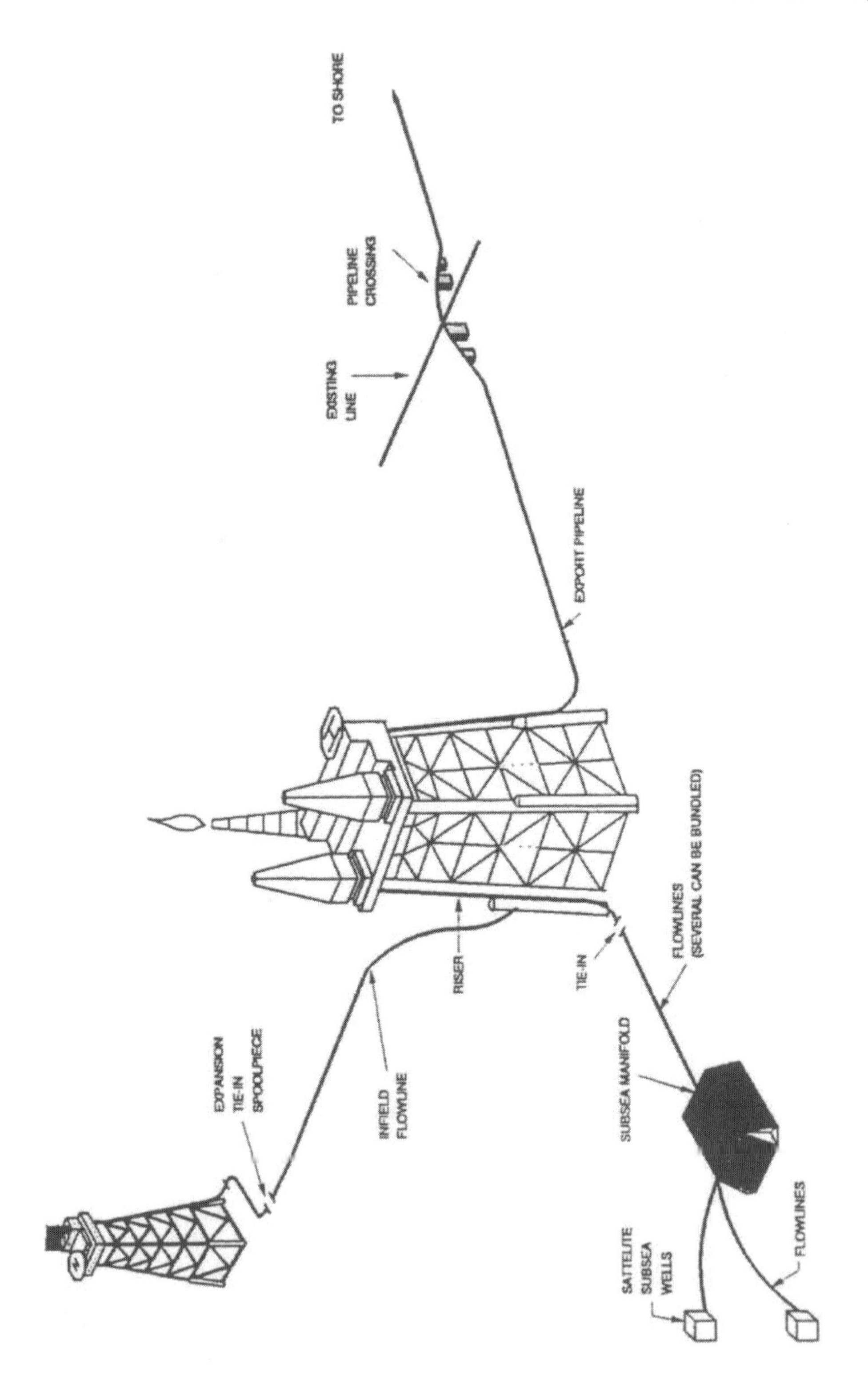

Figure 1.1 Use of flowlines offshore.

The objective and scope of each of these design stages vary depending on the operator and the size of the project. However, the primary aims are generally as follows (Langford and Kelly (1990)):

1. Conceptual Engineering

The primary objectives are normally:

- To establish technical feasibility and constraints on the system design and construction;
- To eliminate non viable options;
- To identify the required information for the forthcoming design and construction;
- To allow basic cost and scheduling exercises to be performed;
- To identify interfaces with other systems planned or currently in existence.

The value of the early engineering work is that it reveals potential difficulties and areas where more effort may be required in the data collection and design areas.

2. Preliminary engineering or basic engineering

The primary objectives are normally:

- Perform pipeline design so that system concept is fixed. This will include:

• To verify the sizing of the pipeline;
• Determining the pipeline grade and wall thickness;
• Verifying the pipeline against design and code requirements for installation, commissioning and operation;

- Prepare authority applications;

- Perform a material take off sufficient to order the linepipe (should the pipe fabrication be a long lead item, hence requiring early start-up)

The level of engineering is sometimes specified as being sufficient to detail the design for inclusion into an "Engineering, Procurement, Construction and Installation" (EPCI) tender. The EPCI contractor should then be able to perform the detailed design with the minimum number of variations as detailed in their bid.

3. Detail engineering

The detailed engineering phase is, as the description suggests, the development of the design to a point where the technical input for all procurement and construction tendering can be defined in sufficient detail.

The primary objectives can be summarized as:

- Route optimization;
- Selection of wall thickness and coating;
- Confirm code requirements on strength, Vortex-Induced Vibrations (VIV), on-bottom stability, global buckling and installation;
- Confirm the design and/or perform additional design as defined in the preliminary engineering;
- Development of the design and drawings in sufficient detail for the subsea scope. This may include pipelines, tie-ins, crossings, span corrections, risers, shore approaches, subsea structures;
- Prepare detailed alignment sheets based on most recent survey data;
- Preparation of specifications, typically covering materials, cost applications, construction activities (i.e. pipelay, survey, welding, riser installations, spoolpiece installation, subsea tie-ins, subsea structure installation) and commissioning (i.e. flooding, pigging, hydrotest, cleaning, drying);
- Prepare material take off (MTO) and compile necessary requisition information for the procurement of materials;
- Prepare design data and other information required for the certification authorities.

1.2.2 Design Process

The object of the design process for a pipeline is to determine, based on given operating parameters, the optimum pipeline size parameters. These parameters include:

- Pipeline internal diameter;
- Pipeline wall thickness;
- Grade of pipeline material;
- Type of coating-corrosion and weight (if any);
- Coating wall thickness.

The design process required to optimize the pipeline size parameters is an iterative one and is summarize in Figure 1.2. The design analysis is illustrated in Figure 1.3.

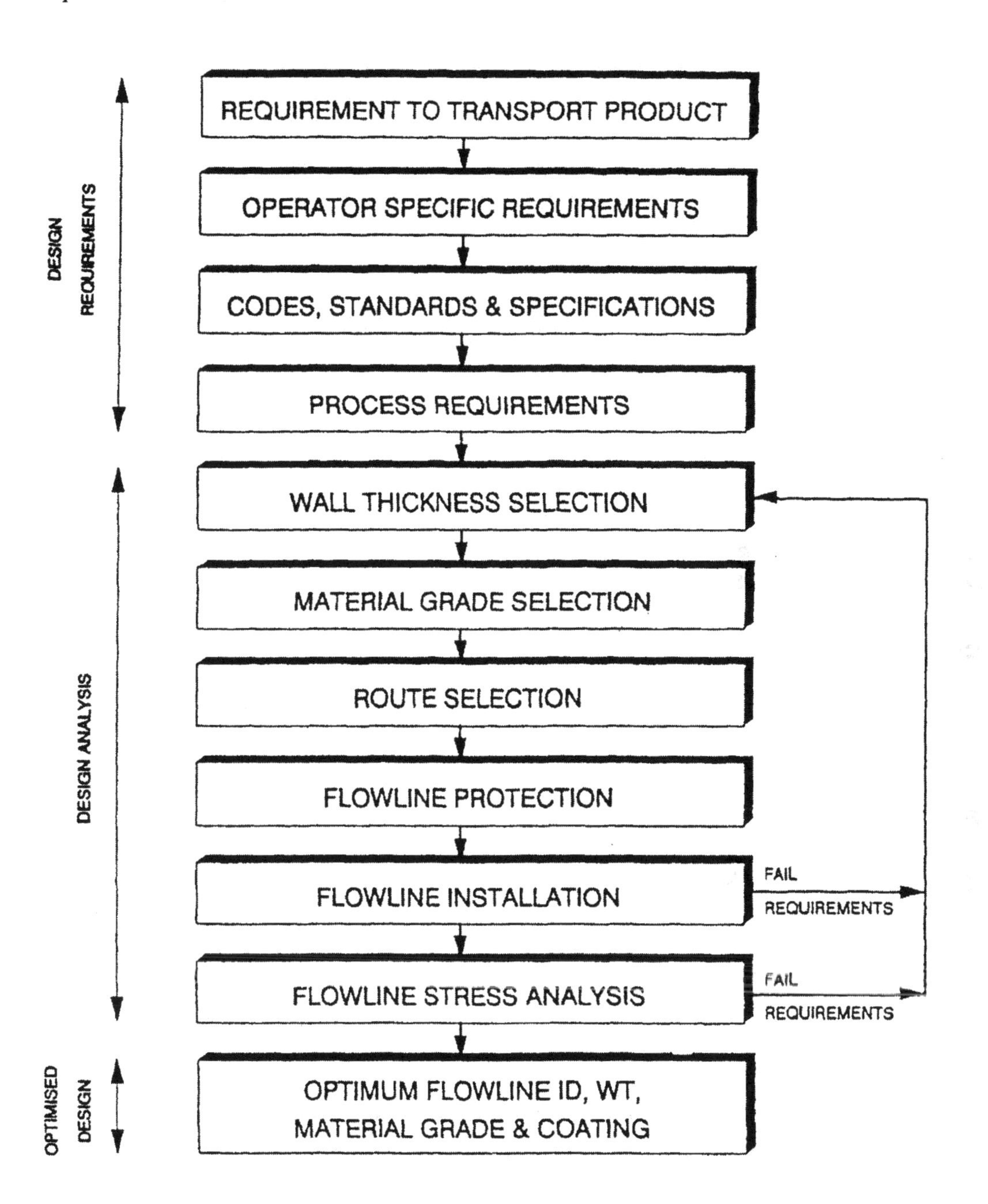

Figure 1.2 Flowline design process.

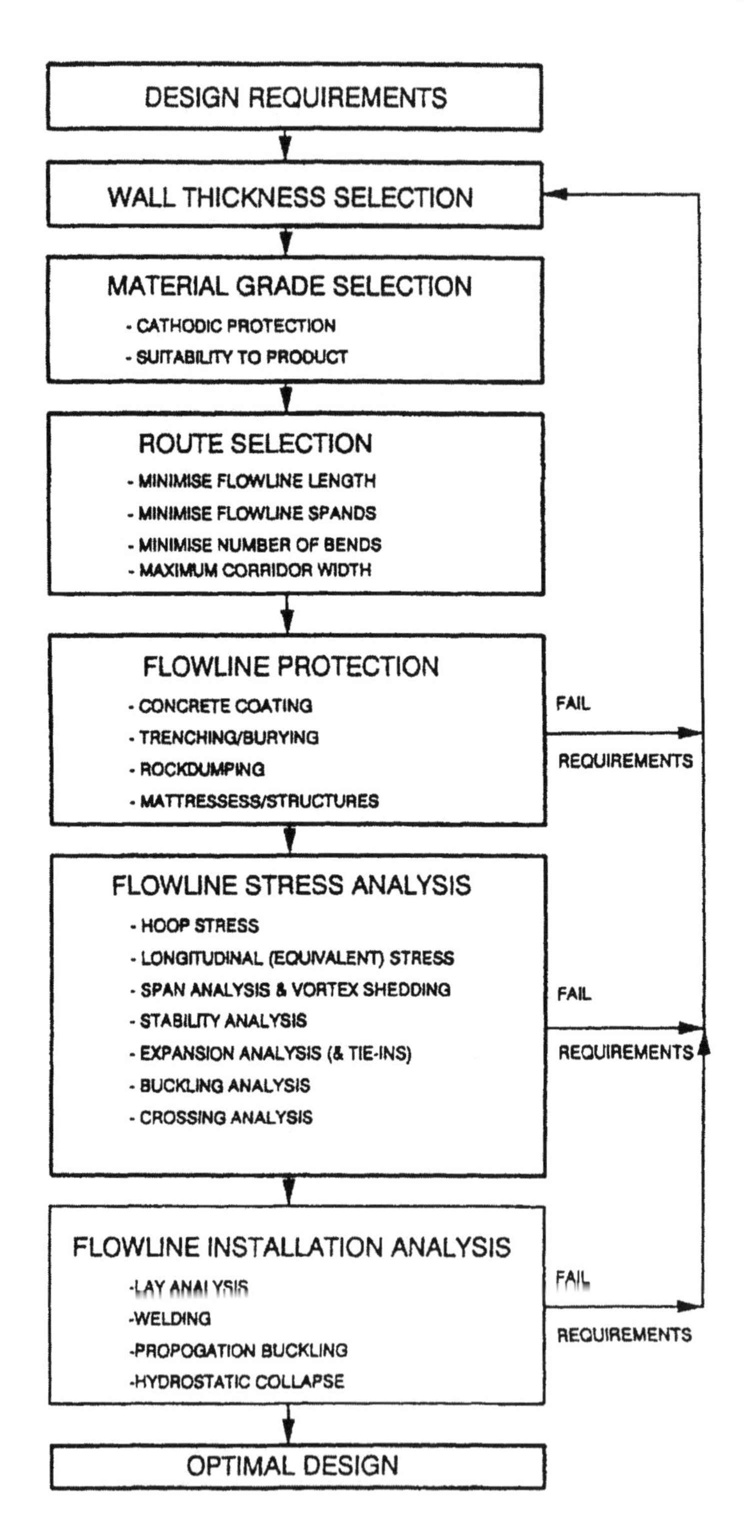

Figure 1.3 Flowline design Analysis.

Each stage in the design should be addressed whether it is conceptual, preliminary or detailed design. However, the level of analysis will vary depending on the required output. For instance, reviewing the objectives of the detailed design (Section 1.2.1), the design should be developed such that:

- Pipeline wall thickness, grade, coating and length are specified so that pipeline can be fabricated;
- Route is determined such that alignment sheets can be compiled;
- Pipeline stress analysis is performed to verify that the pipeline is within allowable stresses at all stages of installation, testing and operation. The results will also include pipeline allowable spans, tie-in details (including expansion spoolpieces), allowable testing pressures and other input into the design drawings and specifications;
- Pipeline installation analysis is performed to verify that stresses in the pipeline at all stages of installation are within allowable values. This analysis should specifically confirm if the proposed method of pipeline installation would not result in pipeline damage. The analysis will have input into the installation specifications;
- Analysis of global response;
  - Expansion, effective force and global buckling
  - Hydrodynamic response
  - Impact
- Analysis of local strength;
  - Bursting, local buckling, ratcheting
  - Corrosion defects, dent

1.3 Design Through Analysis (DTA)

A recent technical revolution in the design process has taken place in the Offshore and Marine industries. Advanced methods and analysis tools allow a more sophisticated approach to design that takes advantage of modern materials and revised design codes supporting limit state design concepts and reliability methods. The new approach is called "Design Through Analysis" where the finite element method is used to simulate global behavior of pipelines as well as local structural strength (see Bai & Damsleth (1998)). The two-step process is used in a complementary way to determine the governing limit states and to optimize a particular design.

The advantage of using advanced engineering is a substantial reduction of project CAPEX (Capital Expenditure) and OPEX (Operating Expenditure) by minimizing unnecessary conservatism in the design through a more accurate determination of the effects of local loading conditions on the structure. Rules and design codes have to cover the general design context where there are often many uncertainties in the input parameters and the application of analysis methods. Where the structure and loading conditions can be accurately modeled, realistic simulations reveal that aspects of the design codes may be overly conservative for a particular design situation. The FEM (Finite Element Methods) model simulates the true structural behavior and allows specific mitigating measures to be applied and documented.

Better quality control in pipeline production allows more accurate modeling of material while FEM analysis tools allow engineers to simulate the through-life behavior of the entire pipeline system and identify the most loaded sections or components. These are integrated into a detailed FEM model to determine the governing failure mode and limit criteria, which is compared to the design codes to determine where there is room for optimization. The uncertainties in the input data and responses can be modeled with the help of statistics to determine the probability distributions for a range of loads and effects. The reliability approach to design decisions can then be applied to optimize and document the fitness for purpose of the final product.

Engineers have long struggled with analytical methods, which only consider parts of the structural systems they are designing. How the different parts affect each other and, above all, how the structural system will respond to loading near its limiting capacity requires a non-linear model which accurately represents the loads, material and structure. The sophisticated non-linear FEM programs and high-speed computers available today allow the engineers to achieve numerical results, which agree well with observed behavior and laboratory tests.

The simulation of global response together with local strength is often necessary because design parameters and local environment are project-specific. A sub-sea pipeline is subject to loading conditions related to installation, seabed features, intervention works, testing, various operating conditions and shut-downs which prescribe a load path essential to the accurate modeling of non-linear systems involving plastic deformation and hysteresis effects. For example, simulation can verify that a pipeline system undergoing cyclic loading and displacement is self-stabilizing in a satisfactory way (shakedown) or becomes unstable needing further restraint. The simulation of pipeline behavior in a realistic environment obtained by measurement allows the engineers to identify the strength and weakness of their design to obtain safe and cost-effective solutions. Traditionally, pipeline engineers compute loads and load effects in two dimensions and either ignore or combine results to account for three-dimensional effects. This approach could lead to an overly conservative or, not so safe design. DTA has demonstrated the importance of three-dimensional (3D) FE analysis for highly loaded pipelines undergoing large thermal expansion.

Design Through Analysis (DTA) involves the following activities:

1. Perform initial design according to guidelines and codes
2. Determine global behavior by modeling complete system
3. Simulate through-life load conditions
4. Identify potential problem areas
5. Check structural failure modes and capacity by detailed FE modeling
6. Develop strategies for minimizing cost while maintaining uniform safety level
7. Perform design optimization cycles
8. Document the validity and benefits of the design
9. Provide operation and maintenance support.

In order to efficiently conduct DTA, it is necessary to develop a Pipeline Simulator System (see Chapter 1.5).

1.4 Pipeline Design Analysis

1.4.1 General

Pipeline stress analysis is performed to determine if the pipeline stresses are acceptable (in accordance with code requirements and client requirements) during pipeline installation, testing and operation. The analysis performed to verify that stresses experienced are acceptable includes:

- Hoop stress;
- Longitudinal stress; code specified
- Equivalent stress;
- Span analysis and vortex shedding;
- Stability analysis;
- Expansion analysis (tie-in design);
- Buckling analysis;
- Crossing analysis.

The first three design stages form the basis for the initial wall thickness sizing. These initial sizing calculations should also be performed in conjunction with the hydrostatic collapse/propagation buckling calculations from the installation analysis.

The methods of analyses are briefly discussed below, as an introduction to separate chapters.

1.4.2 Pipeline Stress Checks

- Hoop Stress

Hoop stress (σ_h) can be determined using the equation (see also Figure 1.4):

$$\sigma_h = (p_i - p_e)\frac{D-t}{2t} \tag{1.1}$$

where:

p_i : internal pressure

p_e : external pressure

D : outside diameter of pipeline

t : minimum wall thickness of pipeline

Depending on which code/standard, the hoop stress should not exceed a certain fraction of the Specified Minimum Yield Stress (SMYS).

- Longitudinal Stress

The longitudinal stress (σ_l) is the axial stress experienced by the pipe wall, and consists of stresses due to:

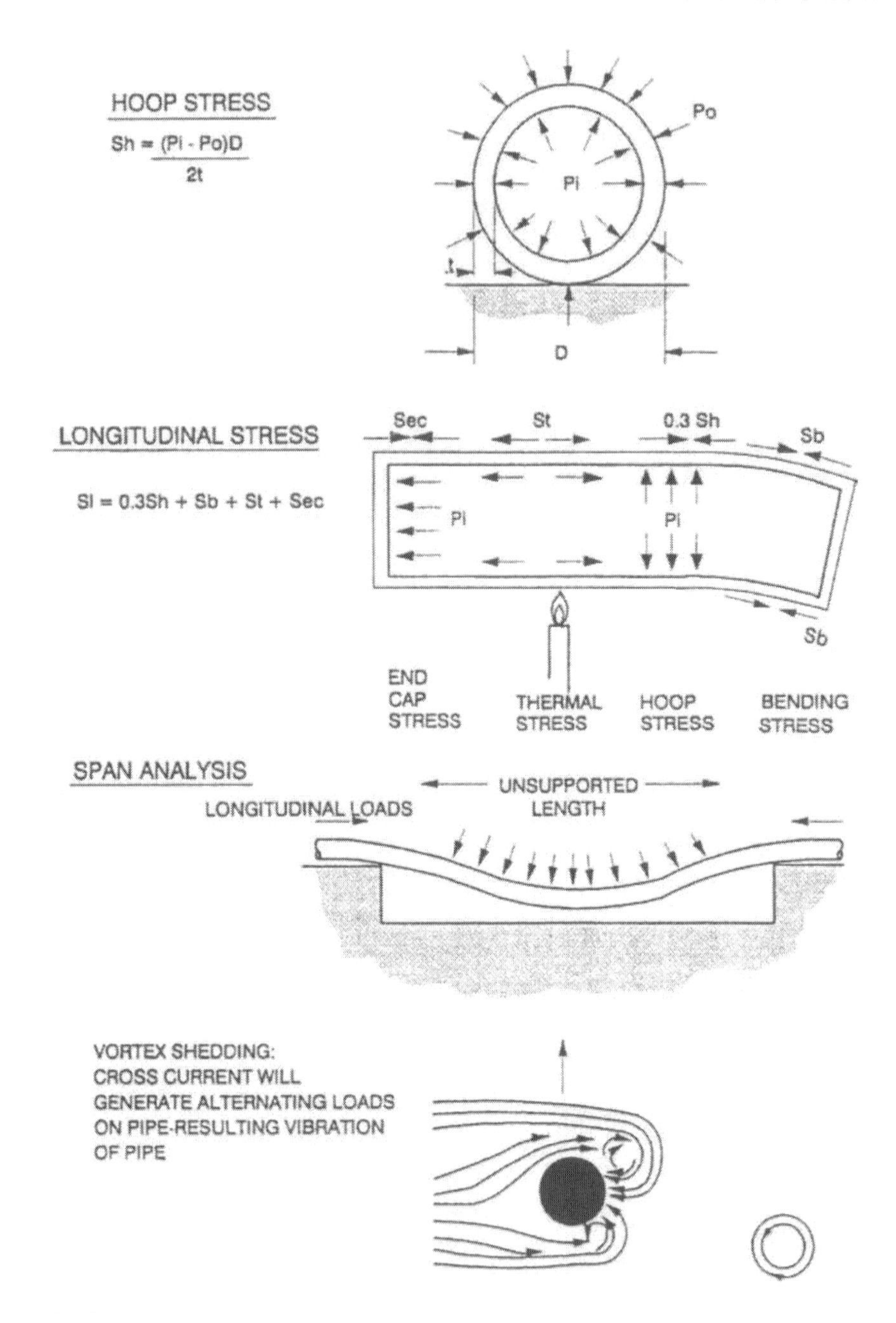

Figure 1.4 Flowline stresses and vortex shedding.

- Bending stress (σ_{lb})
- Hoop stress (σ_{lh})
- Thermal stress (σ_{lt})
- End cap force induced stress (σ_{lc})

The components of each are illustrated in Figure 1.4.

The longitudinal stress can be determined using the equation:

$$\sigma_l = 0.3\sigma_{lh} + \sigma_{lb} + \sigma_{lt} + \sigma_{lc}$$

It should be ensured that sign conventions are utilized when employing this equation (i.e. Tensile stress is positive).

- Equivalent stress

The combined stress is determined differently depending on the code/standards utilized. However, the equivalent stress (σ_e) can usually be expressed as:

$$\sigma_e = \sqrt{\sigma_h^2 + \sigma_l^2 - \sigma_h\sigma_l + 3\tau_{lh}^2} \tag{1.2}$$

where:

σ_h : hoop stress

σ_l : longitudinal stress

τ_{lh} : tangential shear stress

The components of each are illustrated in Figure 1.4.

1.4.3 Span Analysis

Over a rough seabed or on a seabed subject to scour, pipeline spanning can occur when contact between the pipeline and seabed is lost over an appreciable distance (see Figure 1.4). In such circumstances it is normal code requirements that the line is investigated for:

- Excessive yielding;
- Fatigue;
- Interference with human activities (fishing).

Due consideration to these requirements will result in the evaluation of an allowable freespan length. Should actual span lengths exceed the allowable length then correction is necessary to reduce the span for some idealized situations. This can be a very expensive exercise and, consequently, it is important that span evaluation is as accurate as possible. In many cases, a multiple span analysis has to be conducted accounting for, real seabed and in-situ structural behavior.

The flow of wave and current around a pipeline span, or any cylindrical shape, will result in the generation of sheet vortices in the wake (for turbulent flow). These vortices are shed alternately from the top and bottom of the pipe resulting in an oscillatory force being exerted on the span (see Figure 1.4).

If the frequency of shedding approaches the natural frequency of the pipeline span then severe resonance can occur. This resonance can induce fatigue failure of the pipe and cause the concrete coating to crack and possibly be lost.

The evaluation of the potential of a span to undergo resonance is based on the comparison of the shedding frequency and the natural frequency of the span. The calculation of shedding frequency is achieved using traditional mechanics although some consideration must be given to the effect of the closeness of the seabed. Simple models have, traditionally, been used to calculate the natural frequency of the span, but recent theories have shown these to be over-simplified and multiple span analysis needs to be conducted.

Another main consideration with regard to spanning is the possible interference with fishing. This is a wide subject in itself and is discussed in Chapter 12.

1.4.4 On-bottom Stability Analysis

Pipelines resting on the seabed are subject to fluid loading from both waves and steady currents. For regions of the seabed where damage may result from vertical or lateral movement of the pipeline it is a design requirement that the pipe weight is sufficient to ensure stability under the worst possible environmental conditions. In most cases this weight is provided by a concrete weight coating on the pipeline. In some circumstances the pipeline may be allowed to move laterally provided stress (or strain) limits are not exceeded. The first case is discussed briefly in this section since it is applied in the large majority of design situations. Limit-state based stability design will be discussed in Chapter 9.

The analysis of on-bottom stability is based on the simple force balance or detailed finite element analysis. The loads acting on the pipeline due to wave and current action are; the fluctuating drag, lift and inertia forces. The friction resulting from effective weight of the pipeline on the seabed to ensure stability must resist these forces. If the weight of the pipe steel and contents alone or the use of rock-berms is insufficient, then the design for stability must establish the amount of concrete coating required. In a design situation a factor of safety is required by most pipeline codes, see Figure 1.5 for component forces.

The hydrodynamic forces are derived using traditional fluid mechanics with suitable coefficient of drag, lift and diameter, roughness and local current velocities and accelerations.

The effective flow to be used in the analysis consists of two components. These are:

- The steady, current which is calculated at the position of the pipeline using boundary layer theory;
- The wave induced flow, which is calculated at the seabed using a suitable wave theory.

The selection of the flow depends on the local wave characteristics and the water depth.

The wave and current data must be related to extreme conditions. For example, the wave with a probability of occurring only once in 100 years is often used for the operational lifetime of a pipeline. A less severe wave, say 1 year or 5 years, is applied for the installation case where the pipeline is placed on the seabed in an empty condition with less submerged weight.

Friction, which depends on the seabed soils and the submerged weight of the line provide equilibrium of the pipeline. It must be remembered that this weight is reduced by the fluid lift force. The coefficient of lateral friction can vary from 0.1 to 1.0 depending on the surface of the pipeline and on the soil. Soft clays and silts provide the least friction whereas coarse sands offer greater resistance to movement.

For the pipeline to be stable on the seabed the following relationship must exist:

$$\gamma(F_D - F_I) \leq \mu(W_{sub} - F_L) \tag{1.3}$$

where:

γ : factor of safety, normally not to be taken as less than 1.1

F_D : hydrodynamic drag force per unit length (vector)

F_I : hydrodynamic inertia force per unit length (vector)

μ : lateral soil friction coefficient

W_{sub} : submerged pipe weight per unit length (vector)

F_L : hydrodynamic lift force per unit length (vector)

It can be seen that stability design is a complex procedure that relies heavily on empirical factors such as force coefficient and soil friction factors. The appropriate selection of values is strongly dependent on the experience of the engineer and the specific design conditions.

To summarize, the aim of the type of analysis described is to determine the additional weight coating required.

Should the weight of the concrete required for stability make the pipe too heavy to be installed safely then additional means of stabilization will be necessary. The two main techniques are:

- To remove the pipeline from the current forces by trenching;
- To provide additional resistance to forces by use of anchors (rock-berms) or additional weights on the pipeline.

- FLOWLINE STABILITY

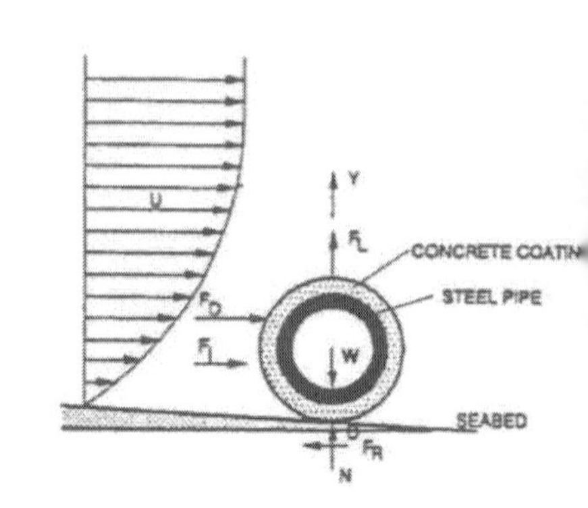

W = TOTAL SUBMERGED WEIGHT OF PIPE, INCLUDING CONCRETE COATING AND WRAP, STEEL PIPE, AND CONTENTS.

F_D = DRAG FORCE.

F_I = INERTIA FORCE.

F_L = LIFT FORCE.

N = NORMAL FORCE.

F_R = FRICTION RESISTANCE.

U = FLOW VELOCITY IN BOUNDARY LAYER.

θ = SLOPE OF SEABED.

S = FACTOR OF SAFETY.

μ = COEFFICIENT OF FRICTION.

$F_R = (W - F_L) \times \mu$

$F_R = (F_D + F_I) \times S \longrightarrow$ STABILITY

- EXPANSION ANALYSIS

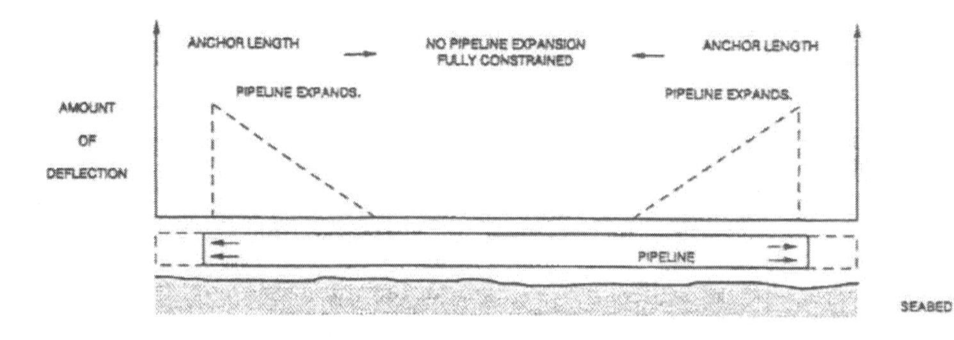

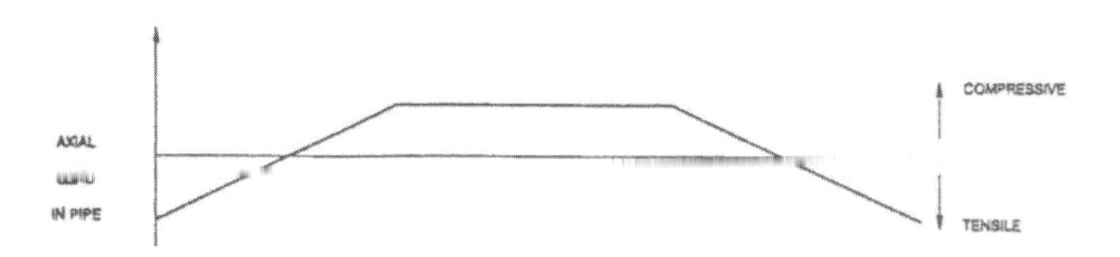

Figure 1.5 Flowline stability and expansion.

In the latter case the spacing of the anchors must be designed to eliminate the potential for sections of line between the fixed points to undergo large movements or suffer high stress levels. The safety of the line on the seabed is again the most important criterion in the stability design.

A finite element model for on-bottom stability analysis is discussed in Chapter 9.

1.4.5 Expansion Analysis

The expansion analysis determines the maximum pipeline expansion at the two termination points and the maximum associated axial load in the pipeline. Both results have significant implications in the design as:

- Axial load will determine if the line may buckle during operation, and hence additional analysis/restraint will be required;
- End expansions dictate the expansion that the tie-in spools (or other) would have to accommodate.

The degree of the expansion by the pipeline is a function of the operational parameters and the restraint on the pipeline. The line will expand up to the “anchor point”, and past this point the line does not expand (hence fully restrained). The distance between the pipeline end and this length is determined based on the operational parameters and the pipeline restraints. The less the restraint the greater the anchor length becomes and hence the greater tie-in expansion becomes (see Figure 1.5 for terminology).

1.4.6 Buckling Analysis

Buckling of a line occurs when the effective force within the line becomes so great that the line has to deflect, and so reduce these axial loads (i.e. takes a lower energy state).

As more pipelines operate at higher temperatures (over 100°C) the likelihood of buckling becomes more pertinent.

The buckling analysis will be performed to identify whether buckling is likely to occur (see Figure 1.6). If it is, then further analysis is performed to either prevent buckling or accommodate it.

A method of preventing buckling is to rock dump the pipeline. This induces even higher loads in the line but prevents it buckling. However, if the rock dump should not provide enough restraint then localized buckling may occur (i.e. upheaval buckling) which can cause failure of the line.

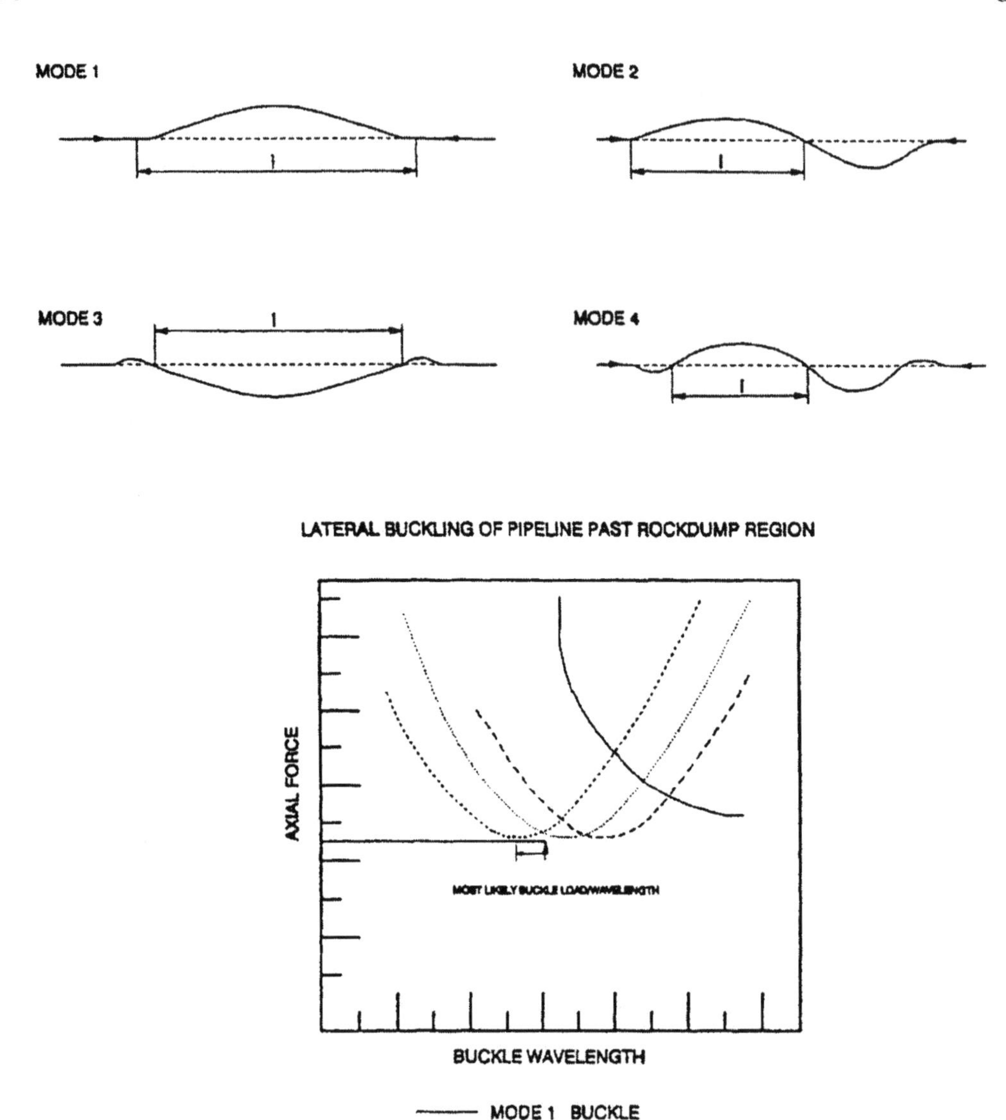

Figure 1.6 Lateral buckling of pipeline.

Another method is to accommodate the buckling problem by permitting the line to deflect (snake) on the seabed. This method is obviously cheaper than rock dumping, and results in the line experiencing lower loads. However, the analysis will probably have to be based on the limit-state design, as the pipe will have plastically deformed. This method is becoming more popular. This method can also be used with intermittent rockdumping, by permitting the line to snake and then to rockdump, this reduces the likelihood of upheaval buckling.

The methods employed in calculating upheaval and lateral buckling as well as pullover response are detailed in references, see Nystrøm et al. (1997), Tørnes et al. (1998).

1.4.7 Pipeline Installation

There are various methods of installing pipelines and risers. The methods of installation which determine the type of analysis performed are discussed as follows:

- Pipelaying by lay vessel;
- Pipelaying by reel ship;
- Pipeline installation by tow or pull method.

– Pipelaying by lay vessel

This method involves joining pipe joints on the lay vessel, where at a number of work stations welding, inspection and field joint coating take place (see Figure 1.7).

Pipelaying progresses with the lay vessel moving forward on its anchors. The pipe is placed on the seabed in a controlled S-bend shape. The curvature in the upper section, or overbend, is controlled by a supporting structure, called a stinger, fitted with rollers to minimize damage to the pipe.

The curvature in the lower portion is controlled by application of tension on the vessel using special machines.

The pipeline designer must analyze the pipelay configuration to establish that correct tension capacity and barge geometry are set up and that the pipe will not be damaged or overstressed during the lay process.

The appropriate analysis can be performed by a range of methods from simple catenary analysis to give approximate solutions, to precise analysis using finite element analysis. The main objective of the analysis is to identify stress levels in two main areas. The first is on the stinger where the pipe can undergo high bending especially at the last support. Since the curvature can now be controlled, the pipeline codes generally allow a small safety factor.

The second high stress area is in the sag bend where the pipe is subject to bending under its own weight. The curvature at the sag bend varies with pipeline lay tension, and consequently is less controllable than the overbend.

In all cases the barge geometry and tension are optimized to produce stress levels in the pipe wall within specified limits.

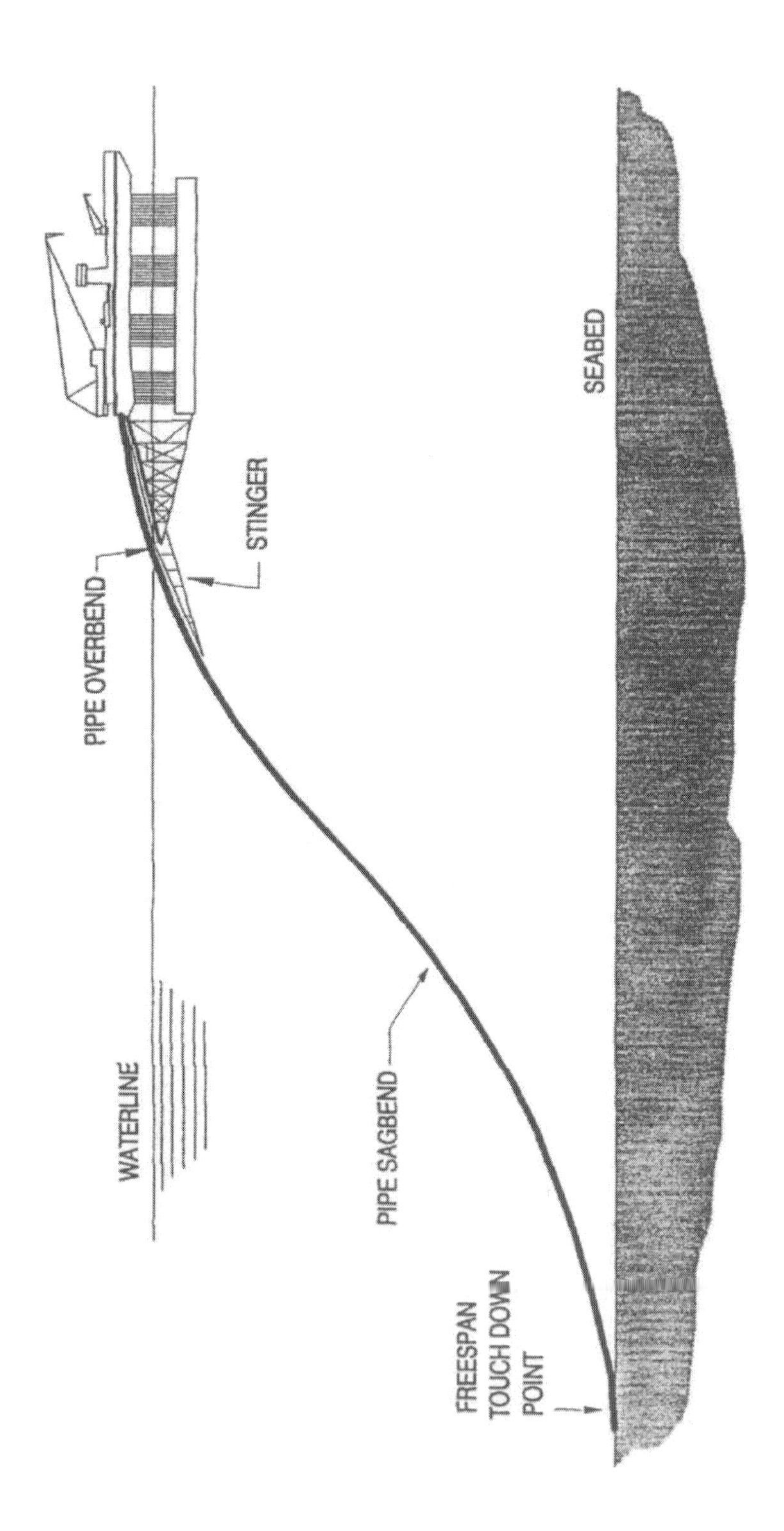

Figure 1.7 Typical pipe configuration during installation.

- Pipelaying Reelship

The pipe reeling method has been applied mainly in the North Sea, for line sizes up to 16-inch. The pipeline is made up onshore and is reeled onto a large drum on a purpose built vessel. During the reeling process the pipe undergoes plastic deformation on the drum. During installation the pipe is unreeled and straightened using a special straightened ramp. The pipe is then placed on the seabed in a similar configuration to that used by the laybarge although in most cases a steeper ramp is used and overbend curvature is eliminated.

The analysis of reeled pipelay can be carried out using the same techniques as for the laybarge. Special attention must be given to the compatibility of the reeling process with the pipeline steel grade since the welding process can cause unacceptable work hardening in higher grade steels.

A major consideration in pipeline reeling is that the plastic deformation of the pipe must be kept within limits specified by the relevant codes. Existing reelships reflect such code requirements.

- Pipeline installation by Tow or Pull

In certain circumstances a pipeline may be installed by a towing technique where long sections of line are made up onshore and towed either on the seabed or off bottom by means of an appropriate vessel (tug or pull barge). The technique has its advantages for short lines and for bundled lines where several pipelines are collected together in a carrier. In this case difficult fabrication procedures can be carried out onshore. The design procedures for towed or pulled lines are very dependent on the type of installation required. For example, it is important to control the bottom weight of a bottom towed line to minimize towing forces and at the same time give sufficient weight for stability. Thus, a high degree of weight optimization may be needed, which can involve tighter control on pipeline wall thickness tolerances than for pipelay, for example.

1.5 Pipeline Simulator

The Pipeline Simulator System comprises a new generation of pipeline modeling tools to replace in-house computer programs developed in the mid-1980s. New technology allows more accurate FEM analysis of pipeline behavior in order to optimize design and achieve cost reductions. The Simulator consists of in-place modules (global models), strength modules (local models) and LCC (life cycle cost) design modules.

The in-place modules (global models) simulate through-life behavior of pipelines, including the following design aspects:

- installation
- on-bottom stability
- expansion, upheaval and lateral buckling

- free-span VIV (Vortex Induced Vibrations)
- trawl pullover and hooking response

The in-place modules further include FEM (deterministic) and reliability (probabilistic) models. Typical reliability design is:

- calibration of safety factors used in the estimation of the appropriate cover height required to prevent upheaval buckling,
- probabilistic modeling of hydrodynamic loads and soils friction for on-bottom stability design.

The local strength modules provide tools for limit-state design to predict pipeline strength under the following failure modes (Bai and Damsleth (1997)):

- local buckling,
- bursting,
- ratcheting,
- material non-homogeneity,
- fracture and fatigue based on damage mechanics models,
- trawl impacts and dropped objects.

The local strength modules also include deterministic models and probabilistic models. Typical probabilistic models are reliability-based strength criteria, in which safety factors are calibrated using structural reliability.

The Simulator provides:

1. A through-life design approach to the pipeline model and predicted behavior.
2. Application and refinement of finite element techniques to model the behavior of pipelines in the marine environment.
3. Through life monitoring and re-assessment of pipelines in operation.

The Simulator development benefits from the experience gained in the design, development and application of the first generation engineering methodologies plus advances in PC-based computing power and software development environments.

Advanced general-purpose finite element programs (ABAQUS and ANSYS) have been applied in the practical design of pipelines as described below:

(1) *Advanced Analysis for Design*: to simulate pipeline in-place behavior during the following through-life scenarios:

- installation (Damsleth et al., 1999);
- flooding, pressure test, dewatering, filling with product;
- pressure and temperature cycling due to operation and shutdowns;
- expansion, upheaval and lateral buckling (Nystrøm et al., 1997), (Tørnes et al., 1998);
- wave and current loads;
- on-bottom stability (Ose et al., 1999);
- vortex-induced vibrations (Kristiansen et al., 1998), (Reed et al., 2000);
- trawlboard pullover and hooking (Tørnes et al., 1998);
- effects of changes to the seabed.

(2) *Numerical Tool as Alternatives to Full Scale Tests*: to develop design criteria with respect to allowable span height and energy absorption capacity requirement from consideration of protection of free-spanning pipeline against fishing gear impact loads and dropped objects loads (Tørnes et al., 1998).

Until some years ago, full-scale tests had been the only reliable method to determine strength. These tests require large amount of resources and cost. Today, many full-scale tests may be performed numerically using the finite element approach.

(3) *Numerical Structural Laboratory for Limit-state Design*: to develop design criteria with respect to structural strength and material behavior as below:

- local buckling/plastic collapse (Hauch and Bai, 1998)
- bursting strength under load-controlled and displacement controlled situations
- ratcheting of ovalisation due to cyclic loads (Kristiansen et al., 1997)
- material non-homogeneity and computational welding mechanics

(4) *Reliability-based Design*: An example of reliability-based design is to select wall-thickness, especially corrosion allowance based on reliability uncertainty analysis and LCC (Life-cycle Cost) optimization (Nødland et al., 1997a, 1997b).

(5) *Reliability-based Calibration of Safety Factors*: to select partial safety factors used in the LRFD (Load Resistance Factored Design) format by reliability-based calibrations (Bai et al., 1997), (Bai and Song, 1997).

1.6 References

1. Bai, Y. and Damsleth, P.A., (1997) "Limit-state Based Design of Offshore Pipelines", Proc. of OMAE '97.
2. Bai, Y. and Song, R., (1997) "Fracture Assessment of Dented Pipes with Cracks and Reliability-based Calibration of Safety Factors", International Journal of Pressure Vessels and Piping, Vol. 74, pp. 221-229.
3. Bai, Y., Xu, T. and Bea, R., (1997) "Reliability-based Design & Requalification criteria for Longitudinally Corroded Pipelines", Proc. of ISOPE '97.
4. Bai, Y. and Damsleth, P.A., (1998) "Design Through Analysis Applying Limit-state Concepts and Reliability Methods", Proc. of ISOPE'98. A plenary presentation at ISOPE'98.
5. Damsleth, P.A., Bai, Y., Nystrøm, P.R. and Gustafsson, C. (1999) "Deepwater Pipeline Installation with Plastic Strain", Proc. of OMAE'99.
6. Hauch, S. and Bai, Y., (1998) "Use of Finite Element Methods for the Determination of Local Buckling Strength", Proc. Of OMAE '98.
7. Kristiansen, N.Ø., Bai, Y. and Damsleth, P.A. (1997) "Ratcheting of High Pressure High Temperature Pipelines", Proc. Of OMAE '97.
8. Kristiansen, N.Ø., Tørnes, K., Nystrøm, P.R. and Damsleth, P.A. (1998) "Structural Modeling of Multi-span Pipe Configurations Subjected to Vortex Induced Vibrations", Proc. of ISOPE'98.
9. Langford, G. and Kelly, P.G., (1990) "Design, Installation and Tie-in of Flowlines", JPK Report Job No. 4680.1.
10. Nødland, S., Bai, Y. and Damsleth, P.A. (1997) "Reliability Approach to Optimize Corrosion Allowance", Proc. of Int. Conf. on Risk based & Limit-state Design & Operation of Pipelines.
11. Nødland, S., Hovdan, H. and Bai, Y., (1997). "Use of Reliability Methods to Assess the Benefit of Corrosion Allowance", Proc. of EUROCORR'97, pp.47-54 (Vol.2).
12. Nystrøm P., Tørnes K., Bai Y. and Damsleth P., (1997). "Dynamic Buckling and Cyclic Behavior of HP/HT Pipelines", Proc. of ISOPE'97.
13. Ose, B. A., Bai, Y., Nystrøm, P. R. and Damsleth, P. A. (1999) "A finite element model for In-situ Behavior of Offshore Pipelines on Uneven Seabed and its Application to On-Bottom Stability", Proc. of ISOPE'99.
14. Reid, A., Grytten, T.I. and Nystrøm, P.R., (2000) "Case Studies in Pipeline Free Span Fatigue", Proc. of ISOPE'2000.
15. Tørnes, K., Nystrøm, P., Kristiansen, N.Ø., Bai, Y. and Damsleth, P.A., (1998) "Pipeline Structural Response to Fishing Gear Pullover Loads", Proc. of ISOPE'98.

Part I

Mechanical Design

Chapter 2 Wall-thickness and Material Grade Selection

2.1 Introduction

2.1.1 General

In this section, the basis for design of wall thickness is reviewed and compared with industry practice. The codes reviewed are ABS, API, ASME B31, BS8010, DNV and ISO. Wall thickness selection is one of the most important and fundamental tasks in design of offshore pipelines. While this task involves many technical aspects related to different design scenarios, primary design loads relevant to the containment of the internal pressure are as follows:

- the differential pressure loads
- longitudinal functional loads
- external impact loads

The current design practice is to limit the hoop stress for design against the differential pressure, and to limit the equivalent stress for design against combined loads. This practice has proved to be very safe in general, except when external impact loads are critical to the integrity of the pipeline. Nevertheless, this practice has been used by the pipeline industry for decades with little change, despite significant improvements and developments in the pipeline technology, see Sotberg and Bruschi (1992) and Verley et al. (1994).

Considering the precise design and effective quality and operational control achieved by modern industry, and with the availability of new materials, it has been realized that there is a need to rationalize the wall thickness sizing practice for a safe and cost-effective design, see Jiao et al. (1996). New design codes provide guidance on application of high strength and new materials, as well as design of high pressure and high temperature pipelines.

2.1.2 Pipeline Design Codes

- ASME B31 Codes

The early history of pipeline design codes started in 1926 with the initiation of the B31 code for pressure piping followed by the well-known ASME codes B31.8 for Gas Transmission and

Distribution Piping Systems and B31.4 for Oil Transportation piping in the early 1950's. The main design principle in these two codes is that the pipeline is assessed as a pressure vessel, by limiting the hoop stress to a specific fraction of the yield stress. A brief outline of new design codes is given below:

- ISO Pipeline Code

A new pipeline code for both offshore and onshore applications is currently under development by ISO-International Standardization Organization (ISO DIS 13623, 1996). A guideline being developed as an attached document to this ISO code allows the use of structural reliability techniques by means of limit state based design procedures as those proposed by SUPERB (Jiao et al., 1996). This code and guideline represent a valuable common basis for the industry for the application of new design methods and philosophy.

- API RP1111 (1998)

The recommended practice for offshore pipelines and risers containing hydrocarbons has been updated based on limit state design concept to provide a uniform safety level. The failure mode for rupture and bursting is used as the primary design condition independent of pipe diameter, wall thickness and material grade.

- DNV Pipeline Rules

The first edition of DNV Rules for the Design, Construction and Inspection of Submarine Pipelines and Pipeline Risers was issued in 1976 and the design section was mainly based upon the ASME codes although it was written for offshore applications only.

The safety philosophy in the DNV'96 Pipeline Rules is based on that developed by the SUPERB Project. The pipeline is classified into safety classes based on location class, fluid category and potential failure consequences. Further, a limit state methodology is adopted and its basic requirement is that all relevant failure modes (limit states) are considered in design.

- ABS (2000) Guide for Building and Classing Undersea Pipelines and Risers

A new guide for building and classing undersea pipelines and risers is currently being completed. The Guide uses Working Stress Design (WSD) for the wall thickness design. The Guide optionally allows use of Limit-State Design and risk/reliability based design. It does contain new criteria for defect assessment. Criteria for other failure modes relevant for the in-place condition, installation and repair situations, as discussed by Bai and Damsleth (1997) have been evaluated/developed based on design projects, relevant JIP's and industry experience.

2.2 Material Grade Selection

2.2.1 General Principle

In this section selection of material grades for rigid pipelines and risers are discussed.

The steels applied in the offshore oil and gas industry vary from carbon steels (taken from American Petroleum Institute standards- Grade B to Grade X 70 and higher) to exotic steels (i.e. duplex). The following factors are to be considered in the selection of material grades:

- Cost;
- Resistance to corrosion effects;
- Weight requirement;
- Weldability.

The higher the grade of steel (up to exotic steels) the more expensive per volume (weight). However, as the cost of producing high grade steels has reduced, the general trend in the industry is to use these steel of higher grades. See Chapter 32. It is clear that the selection of steel grade forms a critical element of the design.

2.2.2 Fabrication, Installation and Operating Cost Considerations

The choice of material grade used for the pipelines will have cost implications on:

- Fabrication of pipeline;
- Installation;
- Operation.

<u>Fabrication</u>

The cost of steels increases for the higher grades. However, the increase in grade may permit a reduction of pipeline wall thickness. This results in the overall reduction of fabrication cost when using a high grade steel compared with a lower grade steel.

<u>Installation</u>

It is difficult to weld high grade steels, and consequently lay rate is lower compared to laying the lower grade steels. However, should the pipeline be laid in very deep water and a vessel is laying at its maximum lay tension, then the use of high grade steel may be more suitable, as the reduction in pipe weight would result in lower lay tension. In general, from an installation aspect, the lower grade steel pipelines cost less to install.

<u>Operation</u>

Depending on the product being transported in the pipeline, the pipeline may be subjected to:

- Corrosion (internal)
- Internal erosion;
- H_2S induced corrosion.

Designing for no corrosion defect may be performed by either material selection or modifying operation procedures (i.e. through use of chemical corrosion inhibitors).

2.2.3 Material Grade Optimization

Optimization of material grade is rigorously applied today based on experience gained from the past 20 years of pipeline design, and the technical advances in linepipe manufacturing and welding. The optimization is based on minimization of fabrication and installation cost while meeting operating requirements. As the selection of material grade will have a significant impact on the operating life of the pipeline, the operator is normally involved in the final selection of material grade.

2.3 Pressure Containment (hoop stress) Design

2.3.1 General

The hoop stress criterion limits the characteristic tensile hoop stress, σ_h due to a pressure differential between internal and external pressures:

$$\sigma_h < \eta_h \text{SMYS}\, k_t \tag{2.1}$$

where η_h is the design usage factor, SMYS is the Specified Minimum Yield Strength, and k_t is the material temperature derating factor. The hoop stress equation is commonly expressed in the following simple form:

$$\sigma_h = (p_i - p_e)\frac{D}{2t} \tag{2.2}$$

where p_i and p_e are the internal and external pressure respectively, D is the diameter and t is the wall thickness.

For offshore pipelines located in the off platform zone, the design (usage) factor is specified as 0.72 by all major codes. For pipelines in the near platform zone (safety zone), the usage factor is specified as 0.50 by ASME B31.8 (1992), or 0.60 by NPD (1990). The origin for design factor 0.72 can be tracked back to the (1935) B31 codes, where the working pressure was limited to 80 % of the mill test pressure which itself was calculated using Equation (2.1) with a design factor up to 0.9. The effective design factor for the working pressure was thus 0.8 x 0.9 = 0.72. Verley et al. (1994). Since the 1958 version of B31.8 codes, the factor 0.72 has been used directly to obtain the design pressure for land pipelines.

Furthermore, definition of diameter and thickness used in Eq. (2.2) varies between the codes, see Table 2.1. In recent codes, such as NPD (1990) and BS 8010 (1993), the minimum wall thickness is used rather than the nominal wall thickness while the usage factor remains unchanged. This may result in a considerably higher steel cost, indicating such codes are relatively more conservative despite of the significant improvements and developments in pipeline technology.

In most codes the maximum SMYS used in Equation (2.1) is limited to 490 MPa and the yield to tensile strength ratio to 0.85. This limits the use of high strength carbon steel such as steel grade X80 or higher. The yielding check implicitly covers other failure modes as well. To extend the material grade beyond the current limit, explicit checks for other failure modes may be necessary. See Chapter 4.

Table 2.1 Characteristic thickness and diameter used in various pipeline codes.

Code	Thickness	Diameter
ABS (2000)	Minimum	Mean
ASME B31.1 (1951)	Minimum	external – 0.8 t_{min}
ASME B31.3 (1993)	Minimum	external, mean or external - 0.8 t_{min}
ASME B31.4 (1992)	Nominal	external
ASME B31.8 (1992)	Nominal	external
BS 8010 (1993)	Minimum	external, close to mean for $D/t \leq 20$
CEN 234WG3-103 (1993)	Minimum	external
CSA-Z184-M86 (1986)	Minimum	external
Danish Guidelines	Minimum	internal
DNV (2000)	Minimum	mean
NEN 3650 (1992)	Minimum	mean
NPD (1990)	Minimum	external

2.3.2 Hoop Stress Criterion of DNV (2000)

The primary requirement of the pipe wall-thickness selection is to sustain stresses for pressure containment. The tensile hoop stress is due to the difference between internal and external pressure, and is not to exceed the permissible value as given by the following hoop stress criterion:

$$\sigma_h - (p_i - p_e)\frac{D - t_1}{2t_1} \leq \eta(SMYS - f_{y,temp}) \tag{2.3}$$

where:

- σ_h : hoop stress;
- p_i : internal pressure;
- p_e : external pressure;
- D : nominal outside diameter of pipe;
- t_1 : minimum wall thickness;
- t_1 : nominal wall-thickness – fabrication tolerance - corrosion allowance;
- SMYS : Specified Minimum Yield Stress;
- $f_{y,temp}$: derating value due to temperature.

The usage factor for pressure containment is expressed as:

$$\eta = \frac{2 \cdot \alpha_U}{\sqrt{3} \cdot \gamma_m \cdot \gamma_{sc} \cdot \gamma_{inc}} \quad (2.4)$$

where:

α_U : material strength factor;

γ_m : material resistance factor;

γ_{sc} : safety class factor;

γ_{inc} : incidental to design pressure ratio.

2.3.3 Hoop Stress Criterion of ABS (2000)

As the requirement for pressure containment, the allowable hoop stress F_h to be used in design calculations is to be determined by the following equation:

$$F_h = \eta \cdot \text{SMYS} \cdot k_T \quad (2.5)$$

where:

η : design factor (see Table 2.2, originally from B31.4 and B31.8);

SMYS : Specified Minimum Yield Strength of the material;

k_T : temperature derating factor (when temperatures is above 50°C).

Table 2.2 Design factors η for pipelines, platform piping, and risers (originally from ASME B31.4 and B31.8).

	Hoop stress	Longitudinal stress	Equivalent stress
Oil & Gas pipelines, Liquid hydrocarbon piping and risers	0.72	0.80	0.90
Gas risers on non-production platform	0.60	0.80	0.90
Gas piping, Gas risers on production platform	0.50	0.80	0.90

The hoop stress f_h in a pipe can be determined by the equation:

$$f_h = (P_i - P_e)(D - t)/(2t) \quad (2.6)$$

where:

f_h : hoop stress;

P_i : internal or external design pressure;

P_e : external design pressure;

D : nominal outside diameter of pipe;

t : minimum pipe wall thickness.

For relatively thick-walled pipes, where the ratio D/t is equal to or less than 10, a more accurate hoop stress calculation methods, resulting in a lower stress, may be used.

Design Factors and Test Pressure in the US Regulations

In the U.S.A, the production flowlines and risers are covered by 30 CFR 250, Sub part J (MMS Dept. of Interior) while export pipelines and risers are covered by 49 CFR 192 (GAS) and 49 CFR 195 (OIL) (Dept. of Transportation - DOT). CFR denotes US Code of Federal Regulations. The design factors defined in Table 2.2 are consistent with these regulations, as discussed below:

All three CFRs require hoop stress design factor 0.72 for the pipeline part. 30 CFR 250 and 49 CFR 195 require the design factor 0.60, while 49 CFR 192 requires the factor be 0.50 for risers.

Both 30 CFR 250 and 49 CFR 195 require a test pressure of 1.25 times the maximum allowable operating pressure for pipelines and risers. 49 CFR 192 requires a test pressure of 1.25 times the maximum allowable operating pressure for pipelines and 1.5 for the risers.

30 CFR 250 requires that pipelines shall not be pressure tested at a pressure which produces a stress in the pipeline in excess of 95 per cent of the Specified Minimum Yield Stress (SMYS) of the pipeline.

2.3.4 API RP1111 (1998)

Maximum Design Burst Pressure

The hydrostatic test pressure, the pipeline design pressure, and the incidental overpressure, including both internal and external pressures acting on the pipelines, shall not exceed that given by the following formulae:

$$P_t \le f_d f_e f_t P_b \tag{2.7}$$

$$P_d \le 0.80 P_t \tag{2.8}$$

$$P_a \le 0.90 P_t \tag{2.9}$$

where:

f_d : internal pressure (Burst) design factor;

: 0.90 for pipelines;

: 0.75 for pipeline risers;

f_e : weld Joint Factor, longitudinal or spiral seam welds. See ASME B31.4 or ASME B31.8. Only materials with a factor of 1.0 are acceptable;

f_t : temperature de-rating factor, as specified in ASME B31.8;

: 1.0 for temperatures less than $121^{\circ}C$;

P_a : incidental overpressure (internal minus external pressure);

P_b : specified Minimum Burst Pressure of pipe;

P_d : pipeline design pressure;

P_t : hydrostatic test pressure (internal minus external pressure).

The Specified Minimum Burst Pressure P_b is determined by the following:

$$P_b = 0.90(SMYS + SMTS)\left(\frac{t}{D-t}\right) \quad (2.10)$$

where:

D : outside diameter of pipe;

$SMYS$: Specified Minimum Yield Strength of pipe;

(See API Specification 5L, ASME B31.4, or ASME B31.8 as appropriate)

t : nominal wall thickness of pipe;

SMTS : Specified Minimum Tensile Strength of pipe.

Note: The formula for the burst pressure are for D/t>15.

Substituting the pressure test pressure into Eq. (2.8) the maximum design burst pressure:

$$P_d \le 0.80 f_d f_e f_t P_b \quad (2.11)$$

Substituting the burst pressure P_h into Eq. (2.11), the maximum design burst pressure is:

$$P_d \le 0.80 f_d f_e f_t 0.90(SMYS + SMTS)\left(\frac{t}{D-t}\right)$$

Longitudinal Load Design

The effective tension due to static primary longitudinal loads shall not exceed the value given by:

$$T_{eff} \le 0.60 T_y \quad (2.12)$$

where;

$$T_{eff} = T_a - P_i A_i + P_o A_o$$

$$T_a = \sigma_a A$$

$$T_y = SMYS \cdot A$$

$$A = A_o - A_i = \frac{\pi}{4}\left(D^2{}_o - D^2{}_i\right)$$

A : cross sectional area of pipe steel;

A_i : internal cross sectional area of the pipe;

A_o : external cross sectional area of the pipe;

P_i : internal pressure in the pipe;

P_o : external hydrostatic pressure;

SMYS : Specified Minimum Yield Stress;

T_a : Axial tension in pipe;

T_{eff} : Effective tension in pipe;

T_y : Yield tension of the pipe;

σ_a : Axial stress in the pipe wall.

Rewriting the Eq. 2.12 yields for external overpressure:

$$\sigma_{eff} = \sigma_a + \frac{\Delta Pe \cdot A_o}{A} + P_i \leq 0.60SMYS \tag{2.13}$$

where:

σ_{eff} : effective stress;

ΔPe : external over pressure.

2.4 Equivalent Stress Criterion

The equivalent stress criterion based on von Mises equivalent stress σ_{ec} may be defined as:

$$\sigma_{ec} = \sqrt{\sigma_l^2 + \sigma_h^2 - \sigma_l \sigma_h + 3\tau_c^2} < \eta_e SMYS \tag{2.14}$$

in which η_e, is the usage factor, σ_l is the characteristic longitudinal stress, σ_h is the characteristic hoop stress, and τ_c is the characteristic tangential shear stress. The Tresca yielding criterion is used in some codes.

ASME B31.8 (1992) specifies a usage factor of 0.90 for both the safety zone and the midline zone. However, this criterion is not required in situations where the pipeline experiences a predictable non-cyclic displacement of its support (e.g. fault movement or differential subsidence) or pipe sag leads to support contact as long as the consequences of yielding are not detrimental to the structural integrity of the pipeline.

BS 8010 (1993) requires a usage factor of 0.72 for risers and 0.96 for pipelines for functional and environmental (or accidental) loads respectively, and a usage factor of 1.0 for construction or hydrotest loads.

The equivalent stress equation is as following:

$$\sigma_{ec} = \sqrt{\sigma_l^2 + \sigma_h^2 - \sigma_l \sigma_h + 3\tau_c^2} \le \eta_e \cdot SMYS(T) \tag{2.15}$$

where:

σ_l : the characteristic longitudinal stress;

σ_h : the characteristic hoop stress;

$$\sigma_h = \Delta p_d \frac{D - t_2}{2t_2} \tag{2.16}$$

t_2 : $t - t_{corr}$;

Δp_d : design differential overpressure;

τ_c : the characteristic tangential shear stress;

SMYS(T) : SMYS at temperature T,

η_e : usage factor for equivalent stress.

Variations in the code requirements for combined stress criterion are evident not only in terms of the usage factor, but also with respect to applicability of the criterion. While this design format may be suitable for predominantly longitudinal stresses, it becomes irrelevant when localized stress concentration, caused by e.g. impact loads, is of concern. No explicit design criteria are currently available for design against impact loads.

For deepwater pipeline, the wall-thickness should be designed such that sufficient bending moment/strain are reserved for free-spans and external loads as discussed in Chapter 4. In addition, it is necessary to design buckle arrests to stop possible buckle propagation.

2.5 Hydrostatic Collapse

The limit external pressure 'p_l' is equal to the pipe collapse pressure and is to be calculated based on (BS8010 (1993), ABS (2000) and DNV (2000)):

$$p_l^3 - p_{el} \cdot p_l^2 - \left(p_p^2 + p_{el} \cdot p_p \cdot f_0 \cdot \frac{D}{t} \right) \cdot p_l + p_{el} \cdot p_p^2 = 0 \tag{2.17}$$

where:

$$p_{el} = \frac{2 \cdot E}{(1 - \nu^2)} \cdot \left(\frac{t}{D} \right)^3; \tag{2.18}$$

$$p_p = \eta_{fab} \cdot SMYS(T) \cdot \frac{2 \cdot t}{D}; \tag{2.19}$$

f_0 : initial out-of-roundness [1)], $(D_{max}-D_{min})/D$;

D : average diameter;

SMYS(T) : Specified Minimum Yield Strength in hoop direction;

E : Young's Module;

ν : Poisson's ratio;

η_{fab} : fabrication derating factor.

Note:

1) Out-of-roundness caused during the construction phase is to be included, but not flattening due to external water pressure or bending in as-laid position. Increased out-of-roundness due to installation and cyclic operating loads may aggravate local buckling and is to be considered. Here it is recommended that out-of-roundness, due to through life loads, be simulated using finite element analysis.

The collapse equation, Equation (2.17), is often expressed as below:

$$(p_c - p_{el}) \cdot (p_c^2 - p_p^2) = p_c \cdot p_{el} \cdot p_p \cdot f_0 \cdot \frac{D_0}{t} \tag{2.20}$$

The collapse equation can be solved by the following approach:

$p_c = y - \frac{1}{3}b$, where $P_c = P_l$ in Equation (2.17)

where:

$$b = -p_{el};$$

$$c = -\left(p_p^2 + p_p p_{el} f_0 \frac{D_o}{t}\right);$$

$$d = p_{el} p_p^2;$$

$$u = \frac{1}{3}\left(-\frac{1}{3}b^2 + c\right);$$

$$v = \frac{1}{2}\left(\frac{2}{27}b^3 - \frac{1}{3}bc + d\right);$$

$$\phi = \arccos\left(\frac{-v}{\sqrt{(-u)^3}}\right);$$

$$y = -2\sqrt{-u} \cdot \cos\left(\frac{\phi}{3} + \frac{\pi}{3}\right);$$

When calculating out of roundness, caution is required on its definition.

If D/t is less than 50, for pipes under combined bending strain and external pressure, strain criteria may be as the following, ABS (2000):

$$\left(\frac{f \cdot \varepsilon}{\varepsilon_b}\right)^{0.8} + \frac{(p_e - p_i)}{p_c} \leq 1 \quad (2.21)$$

where:

ε : bending strain in the pipe;

ε_b $= \frac{t}{2D}$;

ε_l : maximum installation bending strain or maximum in-place bending strain;

f : bending safety factor for installation bending or in-place bending;

P_c : collapse pressure.

Safety factors f should be determined by the designer with appropriate consideration of the magnitude of increases that may occur for installation bending strain or in-place bending strain. A value of 2.0 for safety factors f is recommended if no detailed information on the uncertainties of load effects is available. Safety factor f may be larger than 2.0 for cases where installation bending strain could increase significantly due to off-nominal conditions, or smaller than 2.0 for cases where bending strains are well defined (e.g. reeling) or in-place situation.

A lower safety factor may be allowed for installation phase provided that potential local buckling can be detected, repaired and buckling propagation can be stopped through use of buckle arrestors.

2.6 Wall Thickness and Length Design for Buckle Arrestors

During the installation the risk of local buckling initiating a propagating buckle will be considered to be high, hence buckle arrestors will be designed to limit the extent of the damage of a propagating buckle, see JPK (1997).

Equation used to determine whether buckle arrestor is required, may be taken as:

$$P_{pr} = 24 * SMYS * \left(\frac{t}{D}\right)^{2.5} \quad (2.22)$$

where:

P_{pr} : propagating pressure for the pipeline;

SMYS : Specified Minimum Yield Stress;

t : pipe wall thickness;

D : pipeline outer diameter.

Upon solving the following equation, feasible buckle arrestor wall thickness and length combinations is obtained. This equation is valid for thick-walled cylindrical buckle arrestors (Sriskandarajah and Mahendran, 1987).

$$\left(P_x - P_{pr}\right) = \left(P_a - P_{pr}\right) * \left[1 - \exp^{\left(-15 * \frac{t_{BA} * l_{BA}}{D_{BA}^2}\right)}\right] \tag{2.23}$$

where:

P_x : crossover pressure = $SF * P_h$; (2.24)

l_{BA} : buckle arrestor length;

SF : safety factor = 1.5;

$$P_h = \rho_w g(h_{max} + h_t + h_s) \tag{2.25}$$

where:

P_h : hydrostatic pressure;

ρ_w : seawater density;

g : gravity;

h_{max} : deepest depth with current pipeline thickness;

h_t : tidal amplitude;

h_s : storm surge.

$$P_a = 34 * SMYS * \left(\frac{t_{BA}}{D_{BA}}\right)^{2.5} \tag{2.26}$$

where:

P_a : propagating pressure for the buckle arrestor;

t_{BA} : buckle arrestor wall thickness;

D_{BA} : buckle arrestor outer diameter = $D + 2 * t_{BA}$ [m]. (2.27)

2.7 Buckle Arrestor Spacing Design

The following equations have been compiled as an approach to optimizing the buckle arrestor spacing (JPK, 1997):

$$C_{BA} = C_{Man} + C_{Matr} - C_{LP} \tag{2.28}$$

where:

C_{BA} : cost per buckle arrestor [e.g. NOK – Norwegian Kroner];

C_{Man} : assumed manufacturing cost per buckle arrestor, = 20000 [NOK];

C_{Matr} : $\rho_s * C_s * V_{BA}$; (2.29)

C_{LP} : cost of pipeline pipe saved by inserting buckle arrestor;

$$C_{LP} = L_{BA} * \left[\frac{\pi}{4} \left[(D_i + 2 * t_P)^2 - D_i^2 \right] * \rho_s * C_s \right] \text{[NOK]} \tag{2.30}$$

where:

t_P : pipeline thickness [m];

ρ_s : steel density [kg/m^3];

C_s : cost of steel (assumed) = 8 [NOK/kg];

$$V_{BA} = L_{BA} * \left[\frac{\pi}{4} * \left[(D_i + 2 * t_{BA})^2 - D_i^2 \right] \right] \tag{2.31}$$

where:

V_{BA} : volume of buckle arrestor steel [m^3];

C_P : cost of pipeline to be repaired, manufacturing assumed to be included in day rate of lay vessel [NOK];

$$C_{P\,Matr} = \rho_s * C_s * V_P \tag{2.32}$$

$$V_P = (S + 3 * h) * \left[\frac{\pi}{4} * \left[(D_i + 2 * t)^2 - D_i^2 \right] \right] \tag{2.33}$$

where:

V_P : volume of pipe to be repaired [m^3];

S : Spacing between buckle arrestors [m];

h : Greatest depth in the section considered [m];

$$C_F = 30 * (C_{LV} + C_{DSV}) \tag{2.34}$$

where:

C_F : fixed cost if repair is needed [NOK];

30 : assumed time from buckle occurs until repair [Days];
is done and regular pipe laying is started;

C_{LV} : daily rate of lay vessel (assumed) = 1.5*10^6 [NOK];

C_{DSV} : daily rate of Diving Support Vessel (DSV) (assumed) = 1.25*10^6 [NOK].

$$C_{TOTAL} = C_{BA} * X + (C_F + C_P) * M \tag{2.35}$$

where:

C_{TOTAL} : total cost of buckle arrestors, repair pipe and repair [NOK];

M : assumed probability of risk of a propagating buckle during the laying, can be between 0 and 1. It is assumed to be 1 for the first 50 km of the first lay season, after which a probability of 0.05 is assumed until the end of the first lay season.

For the second lay season a probability of 1 is assumed since the relative cost of delaying the installation of the riser is large.

$$X = \frac{L}{S} \tag{2.36}$$

where:

X : number of buckle arrestors for the pipe length considered;

L : pipe length considered [m].

2.8 References

1. ABS (2001), "Guide for Building and Classing Subsea Pipelines and Risers", American Bureau of Shipping., March 2001.
2. ASME B31.1, (1951) "Code for Pressure Piping", American Society of Mechanical Engineers.
3. ASME B31.3, (1993) "Chemical plant and petroleum refinery piping", American Society of Mechanical Engineers.
4. ASME 31.4, (1992) "Code for liquid transportation systems for hydrocarbons, liquid petroleum gas, anhydrous ammonia, and alcohol's", American Society of Mechanical Engineers.
5. ASME B31.8, (1992) "Code for Gas Transmission and Distribution Piping Systems", American Society of Mechanical Engineers (1994 Addendum).
6. API RP1111, (1998) "Design, Construction, Operation, and Maintenance of Offshore Hydrocarbon Pipelines".
7. BSI: BS 8010 (1993) "Code of Practice for Pipeline - Part 3. Pipeline Subsea: Design, Construction and Installation", British Standards Institute.
8. CEN 234WG3-103, (1993), "Pipelines for Gas Transmission", European Committee for Standardization, 1993.
9. CSA-Z184-M86, (1986), "Gas Pipeline Systems", Canadian Standard Association.
10. DNV (2000), Offshore Standard OS-F101 Submarine Pipeline Systems.
11. Jiao, G. et al. (1996), "The SUPERB Project: Wall-thickness Design Guideline for Pressure Containment of Offshore Pipelines", Proc.of OMAE'96.
12. J P Kenny A/S (1997), "Buckling Arrestor Design", Report No. D501-PK-P121-F-CE-003.
13. NEN (1992), NEN 3650, "Requirements for Steel Pipeline Transportation System", 1992.

14. NPD(1990) "Guidelines to Regulations Relating to Pipeline Systems in the Petroleum activities", 30 April 1990.
15. Sotberg, T. and Bruschi R., (1992) "Future Pipeline Design Philosophy - Framework", Int. Conf. on Offshore Mechanics and Arctic Engineering.
16. Sriskandarajah, T. and Mahendran, I. K. (1987) "Parametric Consideration of Design and Installation of Deepwater Pipelines", Brown and Root U.K. Ltd. Presented at 1987 European Seminar Offshore Oil and Gas Pipeline Technology.
17. Verley, R. et al., (1994) "Wall thickness design for high pressure offshore gas pipelines", Int. Conf. on Offshore Mechanics and Arctic Engineering.

Part I

Mechanical Design

Chapter 3 Buckling/Collapse of Deepwater Metallic Pipes

3.1 Introduction

Buckling and collapse strength of metallic pipes have been an important subject for the design of pipelines, risers and TLP tendons, as well as piping, pressure vessels, tubular structures in offshore and civil engineering.

Elastic-plastic buckling of pipes under external pressure was solved by Timoshenko as described in his book "Theory of Elastic Stability" Timoshenko and Gere (1961). In recent years, non-linear finite element analysis has been used as an accurate tool to predict buckling/collapse capacity of pipes under external pressure, bending and axial force. The finite element model has been validated against laboratory tests and applied to derive design equations. The review of the historic work and the latest research results on this topic may be found from Murphey and Langner (1985), Ellinas et al. (1986), Gresnigt (1986) and a series of journal papers by Bai et al. (1993, 1994, 1995, and 1997), Mohareb et al. (1994), Hauch and Bai (1999, 2000) and Bai et al. (1999).

NOMENCLATURE:

A : Area

D : Average diameter

d : Depth of defect/corrosion

E : Young's modulus

F : True longitudinal force

F_y : True longitudinal yield force

f_0 : Initial out-of-roundness

h : Thickness of defect wall

I : Moment of inertia

Ki : Constants

M : Moment

M_C : Moment capacity

p : Pressure

p_{burst} : Bursting pressure

p_c : Characteristic collapse pressure
p_e : External pressure
p_{el} : Elastic buckling pressure
p_i : Internal pressure
p_l : Limit pressure
p_p : Plastic buckling pressure
p_y : Yield pressure
r_{av} : Average pipe radius of non defect pipe
$SMTS$: Specified Minimum Tensile Strength
$SMYS$: Specified Minimum Yield Strength
t : Nominal wall thickness
w_i : Deflection from circular shape at point *i*
w_I : Maximum initial deflection from circular shape
β : Half of the defect/corrosion width
Δ : Anisotropy factor
$\bar{y}$: Distance from axis of bending to mass center
η : Distance from center for moment of inertia to outer fiber of wall
κ : Curvature
υ : Poisson's ratio
σ_h : Hoop stress
$\sigma_{o,l}$: Yield hoop stress
σ_{hl} : Limit hoop stress for pure pressure
σ_l : Longitudinal stress
$\sigma_{0,l}$: Yield longitudinal stress
σ_{ll} : Limit longitudinal stress for pure longitudinal force
$\sigma_{0,r}$: Yield radial stress
σ_y : Yield stress
ψ : Angle from bending plane to plastic neutral axis

3.2 Pipe Capacity under Single Load

3.2.1 General

The limit moment is highly dependent on the amount of longitudinal force and pressure loads and for cases with high external pressure also initial out-of-roundness. This is mainly due to that the deformation of the pipe caused by the additional loads either work with or against the bending induced deformation. The cross sectional deformations just before failure of pipes subjected to single loads are shown in Figure 3.1.

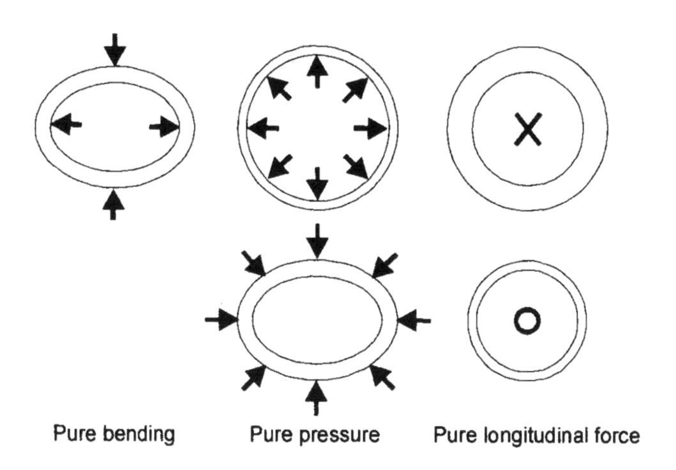

Figure 3.1 Cross sectional deformation of pipes subjected to single loads.

3.2.2 External Pressure

Initial elliptical and corrosion defects are the two major types of imperfections influencing collapse capacity of pipes. In the following, the work by Timoshenko and Gere (1961) is extended, by Bai and Hauch (1998), to account for the effect of corrosion defects.

The deviation of an initial elliptical form from a perfect circular form can be defined by a radial deflection 'w_i' that, for simplification purposes, is assumed given by the following equation:

$$w_i = w_1 \cos(2\theta) \tag{3.1}$$

in which 'w_1' is the maximum initial radial deviation from a circle and 'θ' is the central angle measured as shown in Figure 3.2.

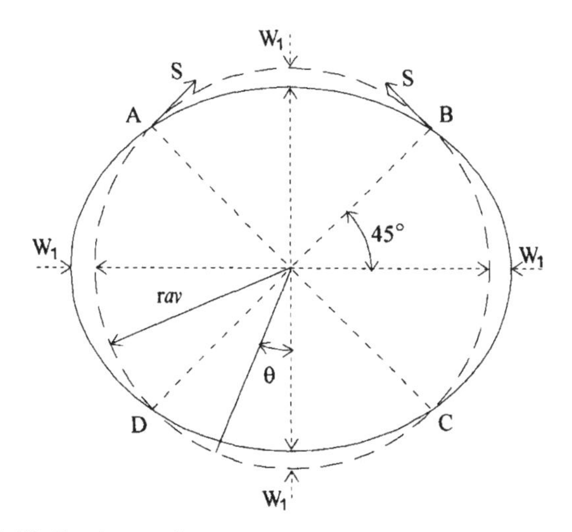

Figure 3.2 Circular and elliptic pipe section.

Under the action of external uniform pressure 'p_e', there will be an additional flattening of the pipe, and the corresponding additional radial displacement 'w' is calculated using the differential equation:

$$\frac{d^2w}{d\theta^2}+w=-\frac{12(1-\upsilon^2)Mr_{av}^2}{Et^3} \tag{3.2}$$

The decrease in the initial curvature as a consequence of the external pressure will introduce a positive bending moment in section AB and CD and a negative bending moment in section AD and BC. At points A, B, C and D the bending moment is zero, and the actions between the parts are represented by the forces 'S' tangential to the dotted circle representing the ideal circular shape.

The circle can be considered as a funicular curve for the external pressure 'p_e' and the compressive force along this curve remains constant and equal to 'S'. Thus, the bending moment at any cross section is obtained by multiplying S by the total radial displacement 'w_i + w' at the cross section. Then:

$$M = p_e r_{av}(w + w_1 \cos(2\theta)) \tag{3.3}$$

Substituting in Equation (3.3):

$$\frac{d^2w}{d\theta^2}+w=-\frac{12(1-\upsilon^2)}{Et^3}p_e r_{av}^3(w+w_1\cos(2\theta)) \tag{3.4}$$

or

$$\frac{d^2w}{d\theta^2}+w\left[1+\frac{12(1-\upsilon^2)}{Et^3}p_e r_{av}^3\right]=-\frac{12(1-\upsilon^2)}{Et^3}p_e r_{av}^3\, w_1\cos(2\theta) \tag{3.5}$$

The solution of this equation satisfying the conditions of continuity at the points A, B, C and D is

$$w=\frac{w_1 p_e}{p_{e,cr}-p_e}\cos(2\theta) \tag{3.6}$$

in which '$p_{e,cr}$' is the critical value of the uniform pressure given by equation:

$$p_{e,cr}=\frac{E}{4(1-\upsilon^2)}\left(\frac{t}{r_{av}}\right)^3 \tag{3.7}$$

$$M_{\max}=p_e r_{av}\left(w_1+\frac{w_1 p_e}{p_{e,cr}-p_e}\right)=p_e r_{av}\frac{w_1}{1-p_e/p_{e,cr}} \tag{3.8}$$

The initial yielding condition is expressed below for a rectangular cross-section with height of wall-thickness and width of unit (1):

$$\sigma_a + \sigma_b = \sigma_y \tag{3.9}$$

where 'σ_a' is the (membrane) stress induced by the external pressure and 'σ_b' is the stress induced by the bending moment. The pressure-induced stress is defined as:

$$\sigma_a = \frac{F}{A} = \frac{p_e r_{av}}{t} \tag{3.10}$$

The relationship between bending stress and moment in the elastic region is as below:

$$\sigma_b = \frac{\eta}{1+\eta / r_{av}} \frac{M_{max}}{I} = \frac{t/2}{1} \frac{M_{max}}{t^3/12} = 6\frac{M_{max}}{t^2} \tag{3.11}$$

where 'η' is the distance from the center for moment of inertia to the outer fiber, 'r_{av}' the initial curvature and 'I' the moment of inertia. From Equation (3.8) it is seen that for small values of the ratio '$p_e/p_{e,cr}$', the change in the elliptical of the pipe due to pressure can be neglected and the maximum bending moment is obtained by multiplying the compressive force '$p_e \times r_{av}$' by the initial deflection 'w_1'. When the ratio '$p/p_{e,cr}$' is not small, the change in the initial elliptical of the pipe should be considered and Equation (3.8) must be used in calculating 'M_{max}'. Thus it is found that

$$\sigma_{max} = \frac{p_e r_{av}}{t} + \frac{6 p_e r_{av}}{t^2} \frac{w_1}{1 - p_e / p_{e,cr}} \tag{3.12}$$

Assuming that this equation can be used with sufficient accuracy up to the yield point stress of the material, the following equation can be obtained:

$$\sigma_Y = \frac{p_Y r_{av}}{t} + 6 p_Y \frac{r_{av}^2}{t^2} \frac{w_1}{r_{av}} \frac{1}{1 - p_Y / p_{e,cr}} \tag{3.13}$$

from which the value of the uniform pressure, 'p_Y', at which yielding in the extreme fibers begins, can be calculated as:

$$p_Y^2 - \left[\frac{\sigma_Y t}{r_{av}} + \left(1 + 6\frac{w_1}{t}\right) p_{e,cr}\right] p_Y + \frac{\sigma_Y t}{r_{av}} p_{e,cr} = 0 \tag{3.14}$$

It should be noted that the pressure 'p_Y' determined in this manner is smaller than the pressure at which the collapsing of the pipe occurs and it becomes equal to the latter only in the case of a perfectly round pipe. Hence, by using the value of 'p_Y' calculated from Equation (3.14) as the ultimate value of pressure, the results are always on the safe side.

A corrosion defect may reduce the hoop buckling capacity of the pipe. It is here assumed that this effect can be accounted for by considering the remaining wall thickness, '$h = t–d$' (d = depth of corrosion defect) if the corrosion defect is not too wide or deep. 't' is substituted by 'h' in Equation (3.14), except for in the expression for '$p_{e,cr}$'. Buckling is an equilibrium problem and occurs when external loads are higher than or equal to internal resistance over the cross-section. The cross-section here means a rectangular one, with height of 't' or 'h' and length along the pipe longitudinal direction of (1) unit. Internal resistance is described by the cross-section with the wall-thickness of 'h' (or 't'). External loads are the moment and compression acting on the cross-section. '$p_{e,cr}$' describes the amplification of the external loads due to a combination of imperfection (i.e. w_1) and axial compression acting on the pipe-wall. The amount of amplification will not be affected by a local corrosion defect unless the defect is wide and deep. The internal resistance is reduced by the corrosion defect and therefore 'h' is used as a replacement of 't'.

Based on the above, Equation (3.14) is modified to Equation (3.15):

$$p_Y^2 - \left[\frac{\sigma_Y h}{r_{av}} + \left(1 + 6\frac{w_1}{h}\right)p_{e,cr}\right]p_Y + \frac{\sigma_Y h}{r_{av}} p_{e,cr} = 0 \tag{3.15}$$

in which '$p_{e,cr}$' is:

$$p_{e,cr} = \frac{E}{4(1-\upsilon^2)}\left(\frac{t}{r_{av}}\right)^3 \tag{3.16}$$

3.2.3 Bending Moment Capacity

The pipe cross sectional bending moment is directly proportional to the pipe curvature, see Figure 3.3. The example illustrates an initial straight pipe with low D/t (<60) subjected to a load scenario where pressure and longitudinal force are kept constant while an increasing curvature is applied.

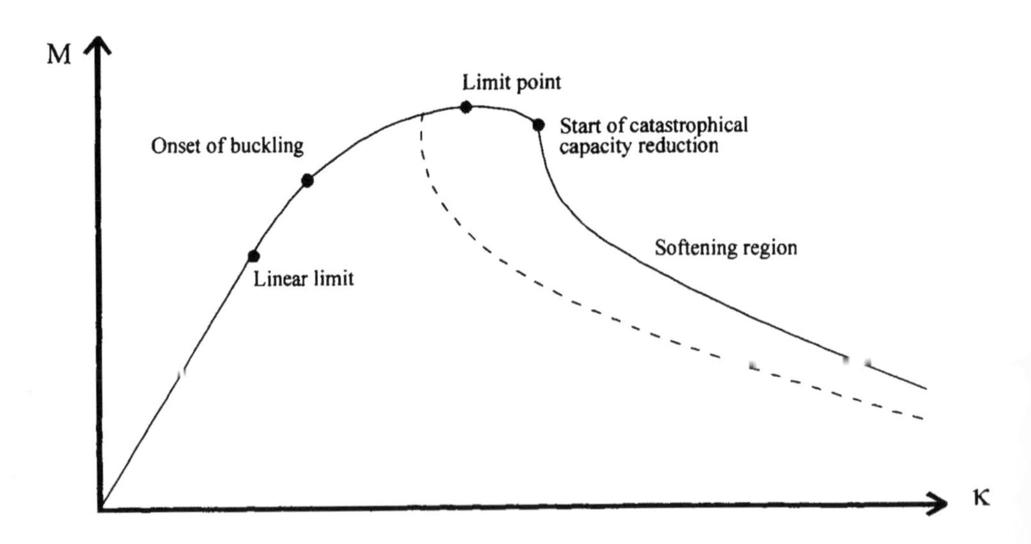

Figure 3.3 Examples of bending moment versus curvature relation.

Different significant points can be identified from the moment-curvature relationship. When applying/increasing curvature the pipe will first be subjected to global deformation inside the material's elastic range and no permanent deformation occurred. By global deformation is here meant deformation that can be looked upon as uniform over a range larger than 3-4 times the pipe diameter. After the LINEAR LIMIT of the pipe material has been reached the pipe will no longer return to its initial shape after unloading, but the deformation will still be characterized as global. If the curvature is increased further, material or geometrical imperfections will initiate ONSET OF LOCAL BUCKLING. Pipe imperfections will have an influence on at which curvature and where along the pipe the onset of local buckling will occur, but will, as long as they are small, for all practical use not influence the limit moment capacity. After the onset of local buckling has occurred, the global deformation will continue, but more and more of the applied bending energy will be accumulated in the local buckle which will continue until the LIMIT POINT is reached. At this point the maximum bending resistance of the pipe is reached and a geometrical collapse will occur if the curvature is increased. Until the point of START OF CATASTROPHIC CAPACITY REDUCTION has been reached, the geometric collapse will be "slow" and the changes in cross sectional area negligible. After this point, material softening sets in and the pipe cross section will collapse until the upper and lower pipe wall is in contact. For pipes subjected to longitudinal force and/or pressure close to the maximum capacity, START OF CATASTROPHIC CAPACITY REDUCTION occurs immediately after the LIMIT POINT. The moment curvature relation for these load conditions will be closer to that presented by the dashed line in Figure 3.3.

The moment curvature relationship provides information necessary for design against failure due to bending. Depending on the function of the pipe, any of the above-described points can be used as design limit. If the pipe is a part of a carrying structure, the elastic limit may be an obvious choice as the design limit. For pipelines and risers where the global shape is less important, this criterion will be overly conservative due to the significant remaining strength in the elastic-plastic range. Higher design strength can therefore be obtained by using design criteria based on the stress/strain levels reached at the point of onset for local buckling or at the limit point. For displacement-controlled configurations, it can even be acceptable to allow the deformation of the pipe to continue into the softening region. The rationale of this is the knowledge of the carrying capacity with high deformations combined with a precise prediction of the deformation pattern and its amplitude.

The limit bending moment for steel pipes is a function of many parameters, e.g.:

- Diameter to wall thickness ratio;
- Material stress-strain relationship;
- Material imperfections;
- Welding (Longitudinal as well as circumferential);
- Initial out-of-roundness;
- Reduction in wall thickness due to e.g. corrosion;
- Cracks (in pipe and/or welding);
- Local stress concentrations due to e.g. coating, change in wall thickness;
- Additional loads and their amplitude

3.2.4 Pure Bending

A pipe subjected to increasing pure bending will fail as a result of increased ovalization of the cross section and reduced slope in the stress-strain curve. Up to a certain level of ovalization, the decrease in moment of inertia will be counterbalanced by increased pipe wall stress due to strain hardening. When the loss in moment of inertia can no more be compensated for by the strain hardening, the moment capacity has been reached and catastrophic cross sectional collapse will occur if additional bending is applied. For low *D/t*, the failure will be initiated on the tensile side of the pipe due to stresses at the outer fibers exceeding the limiting longitudinal stress. For *D/t* higher than approximately 30-35, the hoop strength of the pipe will be so low compared to the tensile strength that the failure mode will be an inward buckling on the compressive side of the pipe. The geometrical imperfections (excluding corrosion) that are normally allowed in pipeline design will not significantly influence the moment capacity for pure bending, and the capacity can be calculated as, SUPERB (1996):

$$M_{C(F=0,\,P=0)} = \left(1.05 - 0.0015 \cdot \frac{D}{t}\right) \cdot SMYS \cdot D^2 \cdot t \tag{3.17}$$

where D is the average pipe diameter, t the wall thickness and SMYS the Specified Minimum Yield Strength. $(1.05 - 0.0015 \cdot D/t) \cdot SMYS$ represents the average longitudinal cross sectional stress at failure as a function of the diameter to wall thickness ratio.

3.2.5 Pure Internal Pressure

For pure internal pressure, the failure mode will be bursting of the cross-section, the pipe cross section expands, the pipe wall thickness decreases. The decrease in pipe wall thickness is compensated for by an increase in the hoop stress due to strain-hardening effect. At a critical pressure, the material strain hardening can no longer compensate the pipe wall thinning and the maximum internal pressure has been reached. The bursting pressure can in accordance with API (1998) be given as:

$$p_{burst} = 0.5(SMTS + SMYS) \cdot \frac{2 \cdot t}{D} \tag{3.18}$$

where, 0.5(SMTS+SMYS) is the hoop stress at failure.

3.2.6 Pure Tension

For pure tension, the failure of the pipe will be, as for bursting, the result of pipe wall thinning. When the longitudinal tensile force is increased, the pipe cross section will narrow down and the pipe wall thickness will decrease. At a critical tensile force, the cross sectional area of the pipe will be reduced until the maximum tensile stress for the pipe material is reached. The maximum tensile force can be calculated as:

$$F_l = \mathrm{SMTS} \cdot A \tag{3.19}$$

where, A is the cross sectional area and SMTS the Specified Minimum Tensile Stress.

3.2.7 Pure Compression

A pipe subjected to increasing compressive force will be subjected to Euler buckling. If the compressive force is additional increased the pipe will finally fail due to local buckling. If the pipe is restraint except from in the longitudinal direction, the maximum compressive force will be close to the tensile failure force.

$$F_l = SMTS \cdot A$$

3.3 Pipe Capacity under Couple Load

3.3.1 Combined Pressure and Axial Force

For pipes subjected to single loads, the failure is dominated by either longitudinal or hoop stresses. For the combination of pressure, longitudinal force and bending the stress level at failure will be an interaction between longitudinal and hoop stresses. This interaction can (neglecting the radial stress component and the shear stress components) be described as:

$$\frac{\sigma_l^2}{\sigma_{ll}^2} - 2\alpha \frac{\sigma_l \sigma_h}{\sigma_{ll} \sigma_{hl}} + \frac{\sigma_h^2}{\sigma_{hl}^2} = 1 \qquad (3.20)$$

where σ_l is the applied longitudinal stress, σ_h the applied hoop stress and σ_{ll} and σ_{hl} the limit stress in their respective direction. The limit stress may differ depending on if the applied load is compressive or tensile. α is a correction factor depending on the ratio between the limit stress in the longitudinal and hoop direction respectively.

For pipes under combined pressure and tension, Eq. (3.20) may be used to find the pipe strength capacity. Alternatives to Eq. (3.20) are Von Mises, Tresca's, Hill's and Tsai-Hill's yield condition. Experimental tests have been performed by e.g. Corona and Kyriakides (1988). For combined pressure and longitudinal force, the failure mode will be very similar to the ones for single loads.

In general, the ultimate strength interaction between longitudinal force and bending may be expressed by the fully plastic interaction curve for tubular cross-sections. However, if D/t is higher than 35, local buckling may occur at the compressive side, leading to a failure slightly inside the fully plastic interaction curve, Chen and Sohal (1988). When tension is dominating, the pipe capacity will be higher than the fully plastic condition due to tensile and strain-hardening effects. Based on finite element results, the critical compressive or tensile force related to bending has been found to be:

$$F_l = 0.5 \cdot (SMYS + SMTS) \cdot A \qquad (3.21)$$

where 0.5×(SMYS + SMTS) is longitudinal stress at failure.

As indicated in Figure 3.1, pressure and bending both lead to a cross sectional failure. Bending will always lead to ovalisation and finally collapse, while the pipe fails in different modes for respectively external and internal overpressure. When bending is combined with external overpressure, both loads will tend to increase the ovalisation, which leads to a rapid decrease in capacity. For bending combined with internal overpressure, the opposite is seen. Here the two failure modes work against each other and thereby "strengthen" the pipe. For high internal overpressure, the collapse will always be initiated on the tensile side of the pipe due to stresses at the outer fibers exceeding the material limit tensile stress. On the compressive side of the pipe, the high internal pressure will tend to initiate an outward buckle, which will increase the pipe diameter locally and thereby increase the moment of inertia and the bending moment capacity to the pipe. The moment capacity will therefore be expected to be higher for internal overpressure compared with a corresponding external pressure.

3.3.2 Combined External Pressure and Bending

Bai et al. (1993, 1994, 1995 and 1997) conducted a systematic study on local buckling /collapse of external pressurized pipes using the following approach:

- review experimental work
- validate finite element (ABAQUS) models by comparing numerical results with those from experimental investigation
- conduct extensive simulation of buckling behavior using validated finite element models
- develop parametric design equations accounting for major factors affecting pipe buckling and collapse
- estimate model uncertainties by comparing the developed equations with the design equations

Details of the equations are given in the above listed papers, in particular Bai et al. (1997).

The ultimate strength equation for pipes under combined external over-pressure and bending is proposed by Bai (1993) as the following function:

$$\left(\frac{M}{M_c}\right)^{\alpha} + \left(\frac{P}{P_c}\right)^{\beta} = 1 \tag{3.22}$$

In experimental tests, sets of (M/M_c) and (P/P_c) at failure are recorded. The exponents α and β can be optimized by identifying which values of α and β will provide the most stable and consistent probabilistic description of the model uncertainty in terms of mean value (preferably close to 1.0), CoV (as low as possible) and distribution type (preferably with a distinct lower bound).

Based on finite element results, Bai et al. (1993) proposed that $\alpha = \beta = 1.9$. In DNV'96 pipeline rules, a round number 2.0 is adopted.

In order to develop simple criteria for buckling/collapse of pipelines under simultaneously axial force, pressure and bending, formulation described in Section 3.4 is used (Hauch and Bai, 1999). See ABS(2000).

3.4 Pipes under Pressure Axial Force and Bending

Buckling/collapse of pipes under internal pressure, axial force and bending accounting for yield anisotropy.

In this section an analytical solution is given for the calculation of the moment capacity for a pipe under combined internal pressure, axial force and bending, with a corrosion defect symmetrical to the bending plan. The moment capacity of the pipe is here defined as the moment at which the entire cross section yields.

The solution presented in this section takes the following configurations into account: Corroded area in compression (case 1), in compression and some in tension (case 2), in tension (case 3), in tension and some in compression (case 4). The four cases are shown in Figure 3.4.

The analytical solution for the moment capacity for the four cases are presented in the following, but only case 1 is fully discussed here.

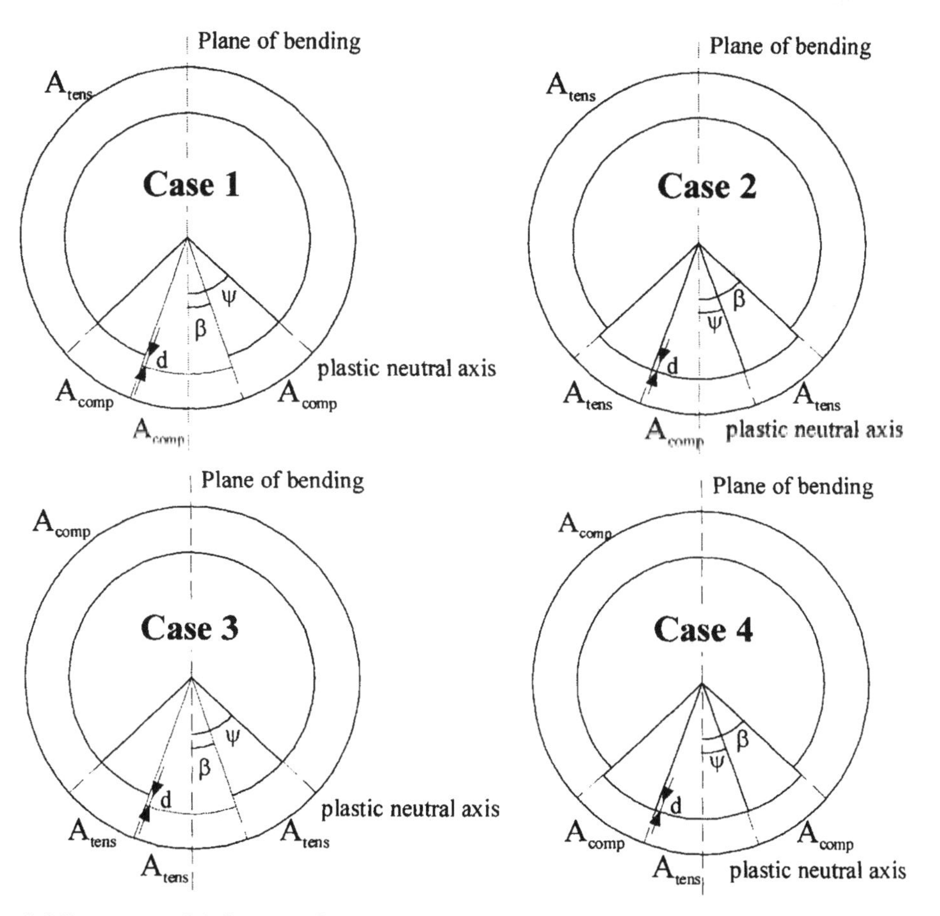

Figure 3.4 Four cases of defects and loads.

3.4.1 Case 1 – Corroded Area in Compression

To keep the complexity of the equations on a reasonable level, the following assumptions have been made:

- Diameter/wall-thickness (D/t) ratio 15-45
- No ovality and no diameter expansion, cross sections remains circular throughout deformations
- Entire cross section in yield as a consequence of applied loads
- The material model is elastic- perfectly plastic
- The defect region is symmetric around the bending plan

Initial ovality is for simplicity ignored in the solution. The rationality of this is that an initial ovality more or less will disappear when the pipe is subjected to high internal pressure under operating conditions or pressure testing conditions.

When plastic deformation is involved, the interaction between axial tension and pressure can be considered as the problem of material yielding under bi-axial loads. Neglecting all shear stress components, Hill's yield function can be expressed as a function of the longitudinal stress 'σ_l', the hoop stress 'σ_h' and the yield stress in longitudinal direction '$\sigma_{0,l}$', hoop direction '$\sigma_{0,h}$' and radial direction '$\sigma_{0,r}$', Hill, R. (1950), Kyriakides, S. et al (1988) and Madhavan R. et al. (1993):

$$\frac{\sigma_l^2}{\sigma_{0,l}^2}-\left(\frac{1}{\sigma_{0,l}^2}+\frac{1}{\sigma_{0,h}^2}-\frac{1}{\sigma_{0,r}^2}\right)\sigma_l\sigma_h+\frac{\sigma_h^2}{\sigma_{0,h}^2}=1 \tag{3.23}$$

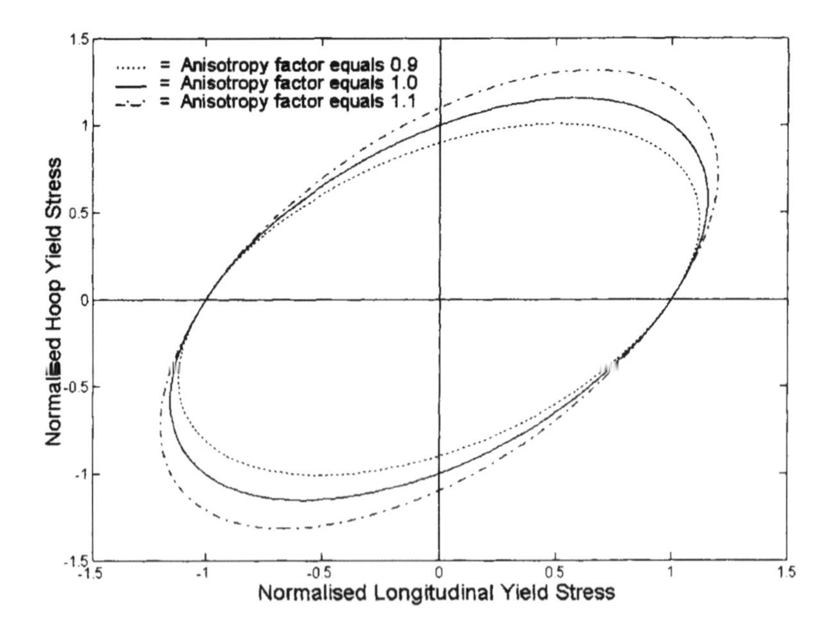

Figure 3.5 Yield surface for $\sigma_{0,h}/\sigma_{0,l}$ = $\sigma_{0,r}/\sigma_{0,l}$ = 0.9, 1.0 & 1.1.

Based on Eq. (3.23) the material yield surface will vary with $\sigma_{0,h}/\sigma_{0,l}$ and $\sigma_{0,r}/\sigma_{0,l}$ as shown in Figure 3.5.

Solving the second-degree equation, Eq. (3.23) for the longitudinal stress σ_l gives:

$$\sigma_l = \tfrac{1}{2}\Delta_1\sigma_h \pm \sigma_{0,l}\sqrt{1-\left(1-\frac{\Delta_1^2}{4\Delta_2^2}\right)\frac{\sigma_h^2}{\sigma_{0,h}^2}} \tag{3.24}$$

where:

$$\Delta_1 = \sigma_{0,l}^2\left(\frac{1}{\sigma_{0,l}^2}+\frac{1}{\sigma_{0,h}^2}-\frac{1}{\sigma_{0,r}^2}\right), \quad \Delta_2 = \frac{\sigma_{0,l}}{\sigma_{0,h}}$$

σ_{comp} is now defined as the longitudinal compressive stress in the pipe wall that would cause the pipe material to yield according to the von Mises yield criterion. The value of σ_{comp} is thereby equal to σ_l as determined above with the negative sign before the square root and σ_{tens} with the positive sign in front of the square root.

3.4.2 The Location of the Fully Plastic Neutral Axis

For a pipe with a circular and yielded cross section, the axial tension can approximately be expressed as:

$$F = A_{comp1}\sigma_{comp} + A_{comp2}\sigma_{comp} + A_{tens}\sigma_{tens} \tag{3.25}$$

$$A_{comp1} = 2(\psi - \beta)r_{av}t \tag{3.26}$$

$$A_{comp2} = 2\beta r_{av}\left(1+\frac{d}{2r_{av}}\right)t\left(1-\frac{d}{t}\right) \tag{3.27}$$

$$A_{tens} = 2(\pi - \psi)r_{av}t \tag{3.28}$$

Where A_{compl} is the compressed part of the non-defect cross section, A_{comp2} is the compressed part of the defect cross section and A_{tens} is the part of the cross section in tension, see Figure 3.6.

The plastic neutral axis is defined as the axis that divides the compressive and the tensile part of the pipe cross section, see Figures 3.4 and 3.6.

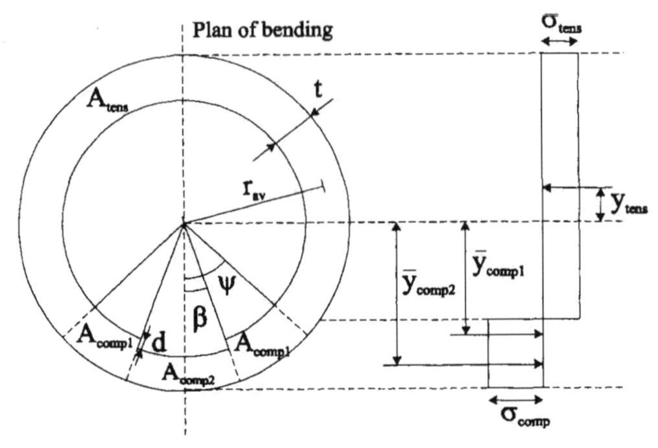

Figure 3.6 Pipe cross section and idealized stress diagram for whole plastified cross section.

Inserting the values for A_{comp1}, A_{comp2} and A_{tens} in Eq. (3.25) gives:

$$F = 2(\psi - \beta) r_{av} t \sigma_{comp} + 2\beta r_{av} \left(1 + \frac{d}{2r_{av}}\right) t \left(1 - \frac{d}{t}\right) \sigma_{comp} + 2(\pi - \psi) r_{av} t \sigma_{tens} \tag{3.29}$$

$$F = 2r_{av} t \left((\psi - k_1 \beta) \sigma_{comp} + (\pi - \psi) \sigma_{tens}\right) \tag{3.30}$$

$$k_1 = 1 - \left(1 - \frac{d}{t}\right)\left(1 + \frac{d}{2r_{av}}\right) \tag{3.31}$$

Solving for ψ:

$$\psi = \frac{F - 2r_{av} t (\pi \sigma_{tens} - k_1 \beta \sigma_{comp})}{2r_{av} t (\sigma_{comp} - \sigma_{tens})} \tag{3.32}$$

Inserting the values for σ_{tens} and σ_{comp} gives:

$$\psi = \frac{\pi + k_1 \beta}{2} - \frac{(\pi - k_1 \beta)\left(\dfrac{F}{F_Y} - \Delta \dfrac{\sigma_h}{\sigma_{0,h}}\right)}{2\sqrt{1 - (1 - \Delta^2)\left(\dfrac{\sigma_h}{\sigma_{0,h}}\right)^2}} \tag{3.33}$$

$$F_Y = 2(\pi - k_1 \beta) r_{av} t \sigma_{0,l} \tag{3.34}$$

$$\Delta = \frac{1}{2}\frac{\Delta_1}{\Delta_2} = \frac{1}{2}\frac{\sigma_{0,l}}{\sigma_{0,h}}\left(\frac{1}{\sigma_{0,l}^2} + \frac{1}{\sigma_{0,h}^2} - \frac{1}{\sigma_{0,r}^2}\right) \tag{3.35}$$

3.4.3 The Bending Moment

The bending moment capacity of the pipe can now be calculated as, see Figure 3.6:

$$M^{P}_{\sigma_h,P_e} = -\left(A_{comp1}\bar{y}_{comp1} + A_{comp2}\bar{y}_{comp2}\right)\sigma_{comp} + A_{tens}\bar{y}_{tens}\sigma_{tens} \tag{3.36}$$

Where A_{comp1}, A_{comp2} and A_{tens} are as defined above. $\bar{y}$ is the distance along the bending plan from origin to the mass center of each area.

$$\bar{y}_{comp1} = r_{av}\frac{\sin(\psi)-\sin(\beta)}{\psi-\beta} \tag{3.37}$$

$$\bar{y}_{comp2} = r_{av}\left(1+\frac{d}{2r_{av}}\right)\frac{\sin(\beta)}{\beta} \tag{3.38}$$

$$\bar{y}_{tens} = r_{av}\frac{\sin(\psi)}{\pi-\psi} \tag{3.39}$$

$$A_{comp1}\bar{y}_{comp1} = 2tr_{av}^2\left[\sin(\psi)-\sin(\beta)\right] \tag{3.40}$$

$$A_{comp2}\bar{y}_{comp2} = 2t\left(1-\frac{d}{t}\right)r_{av}^2\left(1+\frac{d}{2r_{av}}\right)^2\sin(\beta) \tag{3.41}$$

$$A_{tens}\bar{y}_{tens} = 2tr_{av}^2\sin(\psi) \tag{3.42}$$

Inserted this gives the following expression for the bending moment capacity:

$$M^{P}_{\sigma_h,F} = -2tr_{av}^2\left[\sin(\psi)-k_2\sin(\beta)\right]\sigma_{comp} + 2tr_{av}^2\sin(\psi)\sigma_{tens} \tag{3.43}$$

$$k_2 = 1-\left(1-\frac{d}{t}\right)\left(1+\frac{d}{2r_{av}}\right)^2 \tag{3.44}$$

Final expression for moment capacity (case 1)

Substituting the values for σ_{tens} and σ_{comp} into the equation gives the final expression for the bending moment capacity for case 1:

$$M_C = 2tr_{av}^2\sigma_{0,l}\left[\begin{matrix} k_2\sin(\beta)\left(\Delta\frac{\sigma_h}{\sigma_{0,h}} - \sqrt{1-\left(1-\Delta^2\right)\left(\frac{\sigma_h}{\sigma_{0,h}}\right)^2}\right) + \\ 2\sin(\psi)\sqrt{1-\left(1-\Delta^2\right)\left(\frac{\sigma_h}{\sigma_{0,h}}\right)^2} \end{matrix}\right] \tag{3.45}$$

$$\psi = \frac{\pi+k_1\beta}{2} - \frac{(\pi-k_1\beta)\left(\frac{F}{F_Y}-\Delta\frac{\sigma_h}{\sigma_{0,h}}\right)}{2\sqrt{1-\left(1-\Delta^2\right)\left(\frac{\sigma_h}{\sigma_{0,h}}\right)^2}} \tag{3.46}$$

$$F_Y = 2(\pi - k_1\beta)r_{av}t\sigma_{0,l},$$

$$k_1 = 1 - \left(1 - \frac{d}{t}\right)\left(1 + \frac{d}{2r_{av}}\right), \quad k_2 = 1 - \left(1 - \frac{d}{t}\right)\left(1 + \frac{d}{2r_{av}}\right)^2$$

The equation is valid for the following range of hoop stress and axial force:

$$-\frac{1}{\sqrt{1-\Delta^2}} \le \frac{\sigma_h}{\sigma_{0,h}} \le \frac{1}{\sqrt{1-\Delta^2}}$$

and

$$\Delta\frac{\sigma_h}{\sigma_{0,h}} - \sqrt{1-\left(1-\Delta^2\right)\left(\frac{\sigma_h}{\sigma_{0,h}}\right)^2} \le \frac{F}{F_Y} \le \Delta\frac{\sigma_h}{\sigma_{0,h}} + \frac{\pi + k_1\beta}{\pi - k_1\beta}\sqrt{1-\left(1-\Delta^2\right)\left(\frac{\sigma_h}{\sigma_{0,h}}\right)^2} \tag{3.47}$$

CASE 2 - CORROSION IN COMPRESSION AND SOME IN TENSION

Final expression for case 2:

$$M_C = 2tr_{av}^2\sigma_{Y,l}\left[\begin{array}{l} k_2\sin(\beta)\left(\Delta\frac{\sigma_h}{\sigma_{0,h}} + \sqrt{1-\left(1-\Delta^2\right)\left(\frac{\sigma_h}{\sigma_{0,h}}\right)^2}\right) + \\ 2k_4\sin(\psi)\sqrt{1-\left(1-\Delta^2\right)\left(\frac{\sigma_h}{\sigma_{0,h}}\right)^2} \end{array}\right] \tag{3.48}$$

$$k_3\psi = \frac{\pi - k_1\beta}{2} - \frac{(\pi - k_1\beta)\left(\frac{F}{F_Y} - \Delta\frac{\sigma_h}{\sigma_{0,h}}\right)}{2\sqrt{1-\left(1-\Delta^2\right)\left(\frac{\sigma_h}{\sigma_{0,h}}\right)^2}} \tag{3.49}$$

$$F_Y = 2(\pi - k_1\beta)r_{av}t\sigma_{0,l},$$

$$k_1 = 1 - \left(1 - \frac{d}{t}\right)\left(1 + \frac{d}{2r_{av}}\right), \quad k_2 = 1 - \left(1 - \frac{d}{t}\right)\left(1 + \frac{d}{2r_{av}}\right)^2$$

$$k_3 = 1 - k_1, \quad k_4 = 1 - k_2$$

The equation is valid for the following range of hoop stress and axial force:

$$-\frac{1}{\sqrt{1-\Delta^2}} \le \frac{\sigma_h}{\sigma_{0,h}} \le \frac{1}{\sqrt{1-\Delta^2}}$$

$$\Delta\frac{\sigma_h}{\sigma_{0,h}} - \frac{(2k_3 - 1)\pi + k_1\beta}{\pi - k_1\beta}\sqrt{1-\left(1-\Delta^2\right)\left(\frac{\sigma_h}{\sigma_{0,h}}\right)^2} \le \frac{F}{F_Y} \le \Delta\frac{\sigma_h}{\sigma_{0,h}} + \sqrt{1-\left(1-\Delta^2\right)\left(\frac{\sigma_h}{\sigma_{0,h}}\right)^2} \tag{3.50}$$

Case 3 - Corroded area in tension

Final expression for case 3:

$$M_C = 2tr_{av}^2\sigma_{0,l}\left[\begin{array}{l} k_2\sin(\beta)\left(\Delta\frac{\sigma_h}{\sigma_{0,h}} + \sqrt{1-\left(1-\Delta^2\right)\left(\frac{\sigma_h}{\sigma_{0,h}}\right)^2}\right) - \\ 2\sin(\psi)\sqrt{1-\left(1-\Delta^2\right)\left(\frac{\sigma_h}{\sigma_{0,h}}\right)^2} \end{array}\right] \quad (3.51)$$

$$\psi = \frac{\pi + k_1\beta}{2} + \frac{(\pi - k_1\beta)\left(\frac{F}{F_Y} - \Delta\frac{\sigma_h}{\sigma_{0,h}}\right)}{2\sqrt{1-\left(1-\Delta^2\right)\left(\frac{\sigma_h}{\sigma_{0,h}}\right)^2}} \quad (3.52)$$

$$F_Y = 2(\pi - k_1\beta)r_{av}t\sigma_{0,l}$$

$$k_1 = 1-\left(1-\frac{d}{t}\right)\left(1+\frac{d}{2r_{av}}\right), \quad k_2 = 1-\left(1-\frac{d}{t}\right)\left(1+\frac{d}{2r_{av}}\right)^2$$

The equation is valid for the following range of hoop stress and axial force:

$$-\frac{1}{\sqrt{1-\Delta^2}} \le \frac{\sigma_h}{\sigma_{Y,l}} \le \frac{1}{\sqrt{1-\Delta^2}}$$

and

$$\Delta\frac{\sigma_h}{\sigma_{0,h}} - \frac{\pi + k_1\beta}{\pi - k_1\beta}\sqrt{1-\left(1-\Delta^2\right)\left(\frac{\sigma_h}{\sigma_{0,h}}\right)^2} \le \frac{F}{F_Y} \le \Delta\frac{\sigma_h}{\sigma_{0,h}} + \sqrt{1-\left(1-\Delta^2\right)\left(\frac{\sigma_h}{\sigma_{0,h}}\right)^2} \quad (3.53)$$

Case 4 - Corroded area in tension and some in compression

Final expression for case 4:

$$M_C = 2tr_{av}^2\sigma_{0,l}\left[\begin{array}{l} k_2\sin(\beta)\left(\Delta\frac{\sigma_h}{\sigma_{0,h}} - \sqrt{1-\left(1-\Delta^2\right)\left(\frac{\sigma_h}{\sigma_{0,h}}\right)^2}\right) - \\ 2k_4\sin(\psi)\sqrt{1-\left(1-\Delta^2\right)\left(\frac{\sigma_h}{\sigma_{0,h}}\right)^2} \end{array}\right] \quad (3.54)$$

$$k_3\psi = \frac{\pi - k_1\beta}{2} + \frac{(\pi - k_1\beta)\left(\frac{F}{F_Y} - \Delta\frac{\sigma_h}{\sigma_{0,h}}\right)}{2\sqrt{1-\left(1-\Delta^2\right)\left(\frac{\sigma_h}{\sigma_{0,h}}\right)^2}} \quad (3.55)$$

$$F_Y = 2(\pi - k_1\beta)r_{av}t\sigma_{0,l}$$

$$k_1 = 1 - \left(1 - \frac{d}{t}\right)\left(1 + \frac{d}{2r_{av}}\right), \quad k_2 = 1 - \left(1 - \frac{d}{t}\right)\left(1 + \frac{d}{2r_{av}}\right)^2$$

$$k_3 = 1 - k_1, \qquad k_4 = 1 - k_2$$

The equation is valid for the following range of hoop stress and axial force:

$$-\frac{1}{\sqrt{1-\Delta^2}} \le \frac{\sigma_h}{\sigma_{0,h}} \le \frac{1}{\sqrt{1-\Delta^2}}$$

$$\Delta\frac{\sigma_h}{\sigma_{0,h}} - \sqrt{1-\left(1-\Delta^2\right)\left(\frac{\sigma_h}{\sigma_{0,h}}\right)^2} \le \frac{F}{F_Y} \le \Delta\frac{\sigma_h}{\sigma_{0,h}} + \frac{(2k_3-1)\pi + k_1\beta}{\pi - k_1\beta}\sqrt{1-\left(1-\Delta^2\right)\left(\frac{\sigma_h}{\sigma_{0,h}}\right)^2} \tag{3.56}$$

3.5 Finite Element Model

3.5.1 General

This section describes how a pipe section is modeled using the finite element method and is taken from Hauch and Bai (1999). The finite element method is a method where a physical system, such as an engineering component or structure, is divided into small sub regions/elements. Each element is an essential simple unit in space for which the behavior can be calculated by a shape function interpolated from the nodal values of the element. This in such a way that inter-element continuity tends to be maintained in the assemblage. Connecting the shape functions for each element now forms an approximating function for the entire physical system. In the finite element formulation, the principle of virtual work, together with the established shape functions are used to transform the differential equations of equilibrium into algebraic equations. In a few words, the finite element method can be defined as a Rayleigh-Ritz method in which the approximating field is interpolated in piece wise fashion from the degree of freedom that are nodal values of the field. The modeled pipe section is subject to pressure, longitudinal force and bending with the purpose to provoke structural failure of the pipe. The deformation pattern at failure will introduce both geometrical and material non-linearity. The non-linearity of the buckling/collapse phenomenon makes finite element analyses superior to analytical expressions for estimating the strength capacity.

In order to get a reliable finite element prediction of the buckling/collapse deformation behavior the following factors must be taken into account:

- A proper representation of the constitutive law of the pipe material;
- A proper representation of the boundary conditions;
- A proper application of the load sequence;
- The ability to address large deformations, large rotations, and finite strains;
- The ability to model/describe all relevant failure modes.

The material definition included in the finite element model is of high importance, since the model is subjected to deformations long into the elasto-plastic range. In the post buckling phase, strain levels between 10% and 20% is usual and the material definition should therefore at least be governing up to this level. In the present analyses, a Ramberg-Osgood stress-strain relationship has been used. For this, two points on the stress-strain curve are required along with the material Young's modules. The two points can be anywhere along the curve, and for the present model, specified minimum yield strength (SMYS) associated with a strain of 0.5% and the specified minimum tensile strength (SMTS) corresponding to approximately 20% strain has been used. The material yield limit has been defined as approximately 80% of SMYS.

The advantage in using SMYS and SMTS instead of a stress-strain curve obtained from a specific test is that the statistical uncertainty in the material stress-strain relation is accounted for. It is thereby ensured that the stress-strain curve used in a finite element analysis in general will be more conservative than that from a specific laboratory test.

To reduce computing time, symmetry of the problem has been used to reduce the finite element model to one-quarter of a pipe section. The length of the model is two times the pipe diameter, which in general will be sufficient to catch all buckling/collapse failure modes.

The general-purpose shell element used in the present model, account for finite membrane strains and allows for changes in thickness, which makes it suitable for large-strain analysis. The element definition allows for transverse shear deformation and uses thick shell theory when the shell thickness increases and discrete Kirchoff thin shell theory as the thickness decreases.

For a further discussion and verification of the used finite element model, see Bai et al (1993), Mohareb et al (1994), Bruschi et al (1995) and Hauch & Bai (1998).

3.5.2 Analytical Solution versus Finite Element Results

In the following, the above-presented equations are compared with results obtained from finite element analyses. First are the capacity equations for pipes subjected to single loads compared with finite element results for a *D/t* ratio from 10 to 60. Secondly the moment capacity equation for combined longitudinal force, pressure and bending are compared against finite element results.

3.5.3 Capacity of Pipes Subjected to Single Loads

As a verification of the finite element model, the strength capacities for single loads obtained from finite element analyses are compared against the verified analytical expressions described in the previous sections of this chapter. The strength capacity has been compared for a large range of diameter over wall thickness to demonstrate the finite element model's capability to catch the right failure mode independently of the *D/t* ratio. For all the analyses, the average diameter is 0.5088m, SMYS = 450 MPa and SMTS = 530 MPa. In Figure 3.7 the bending moment capacity found from finite element analysis has been compared against the bending moment capacity equation, Eq. (3.17). In Figure 3.8 the limit longitudinal force Eq. (3.19), in Figure 3.9 the collapse pressure Eq. (3.15) and in Figure 3.10 the bursting pressure Eq. (3.18) are compared against finite element results. The good agreement between the finite element results and analytical solutions presented in Figure 3.7-3.10 give good reasons to expect that the finite element model also will give reliable predictions for combined loads.

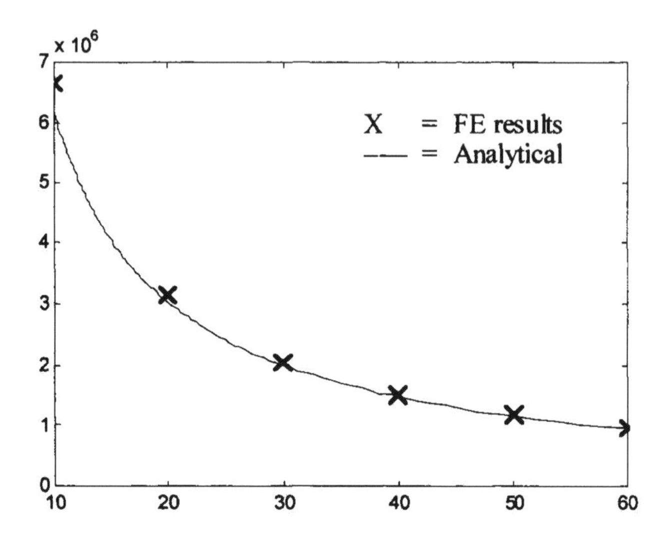

Figure 3.7 Moment capacity as a function of diameter over wall thickness for a pipe subjected to pure bending.

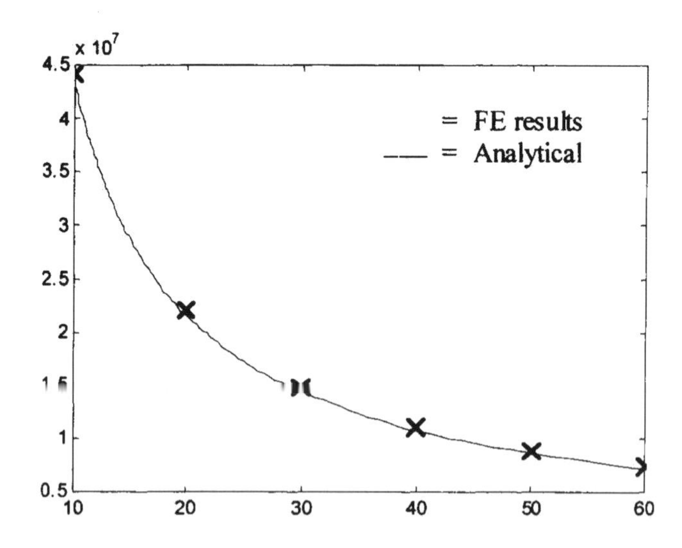

Figure 3.8 Limit longitudinal force as a function of diameter over wall thickness for a pipe subjected to pure longitudinal force.

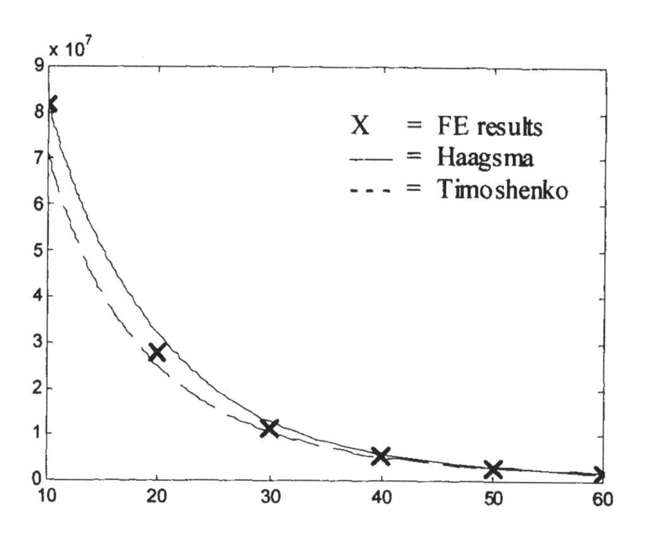

Figure 3.9 Collapse pressure as a function of diameter over wall thickness for a pipe subjected to pure external overpressure. Initial out-of-roundness f_0 equal to 1.5%.

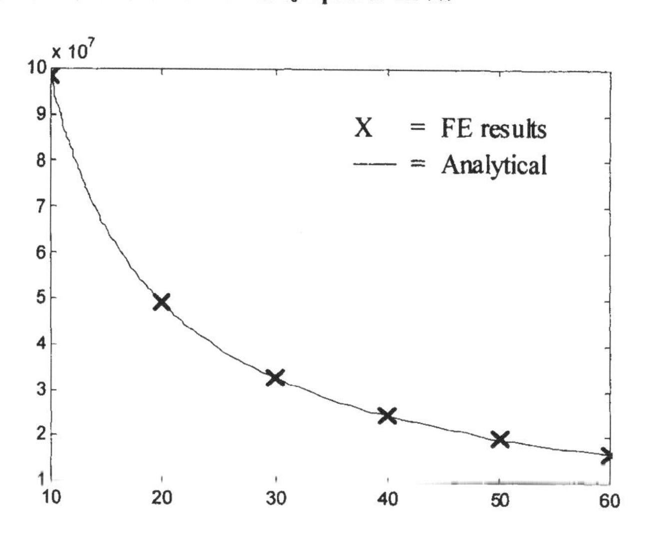

Figure 3.10 Bursting pressure as a function of diameter over wall thickness for a pipe subjected to pure internal overpressure.

3.5.4 Capacity of Pipes Subjected to Combined Loads

For the results presented in Figures 3.11-3.16 the following pipe dimensions has been used:

D/t : 35

f_0 : 1.5 %

SMYS : 450MPa

SMTS : 530MPa

α : 1/5 for external overpressure and 2/3 for internal overpressure

Figures 3.11 and 3.12 show the moment capacity surface given by Hauch and Bai (1999). In Figure 3.11 the moment capacity surface is seen from the external pressure, compressive longitudinal force side and in Figure 3.12 it is seen from above. Figures 3.7 to 3.10 have demonstrated that for single loads, the failure surface agrees well with finite element analyses for a large D/t range. To demonstrate that the failure surface also agrees with finite element analyses for combined loads, it has been cut for different fixed values of longitudinal force and pressure respectively as indicated in Figure 3.12 by the black lines. The cuts and respective finite element results are shown in Figures 3.13 to 3.16. In Figure 3.13 the moment capacity is plotted as a function of pressure. The limit pressure for external overpressure is here given by Haagsma's collapse equation (Haagsma, 1981) and the limit pressure for internal overpressure by the bursting pressure Eq. (3.18). For the non-pressurized pipe, the moment capacity is given by Eq. (3.17). In Figure 3.14, the moment capacity is plotted as a function of longitudinal force. The limit force has been given by Eq. (3.21) for both compression and tension. For a given water depth, the external pressure will be approximately constant, while the axial force may vary. Figure 3.15 shows the moment capacity as a function of longitudinal force for an external overpressure equal to 0.8 times the collapse pressure calculated by Haagsma's collapse equation. Figure 3.16 again shows the moment capacity as a function of longitudinal force, but this time for an internal overpressure equal to 0.9 times the plastic buckling pressure. Based on the results presented in Figures 3.13 to 3.16, it is concluded that the analytical deduced moment capacity and finite element results are in good agreement for the entire range of longitudinal force and pressure. The equations though tent to be a little non-conservative for external pressure very close to the collapse pressure. This is in agreement with the previous discussion about Timoshenko's and Haagsma's collapse equation.

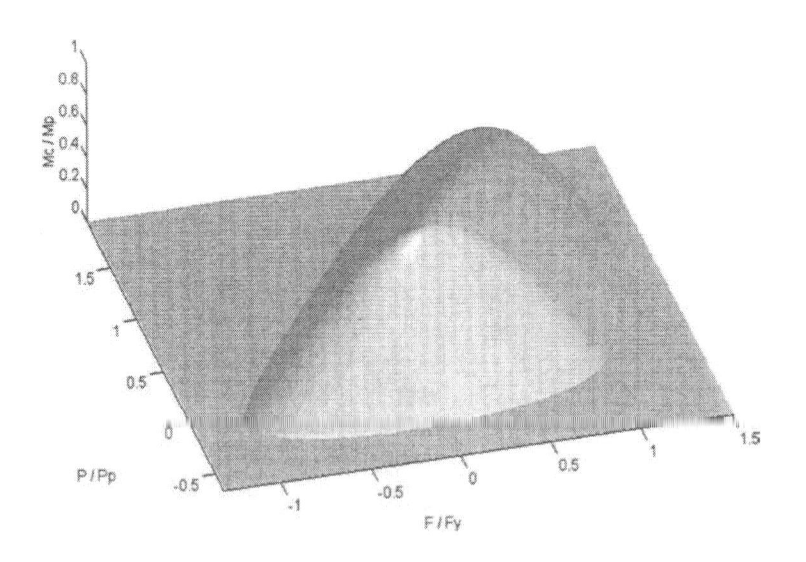

Figure 3.11 Limit bending moment surface as a function of pressure and longitudinal force.

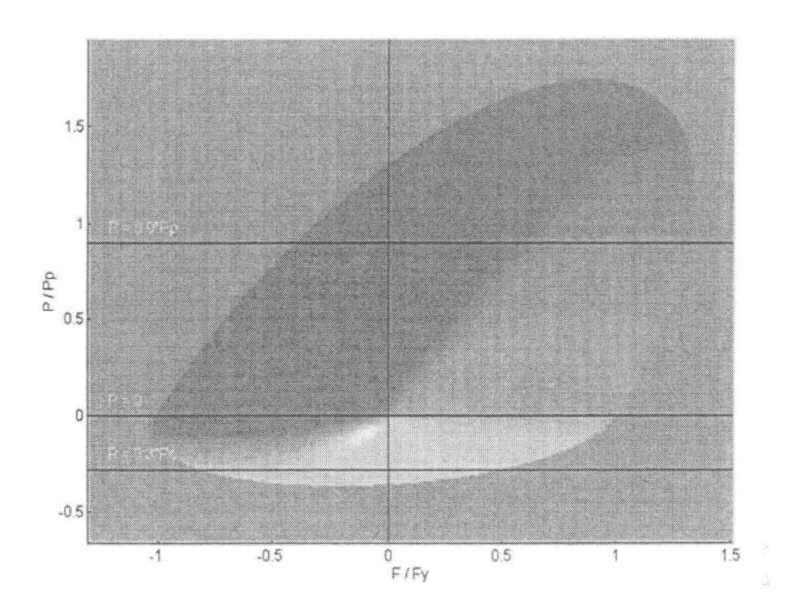

Figure 3.12 Limit bending moment surface as a function of pressure and longitudinal force including cross sections for which comparison between analytical solution and results from finite element analyses has been performed.

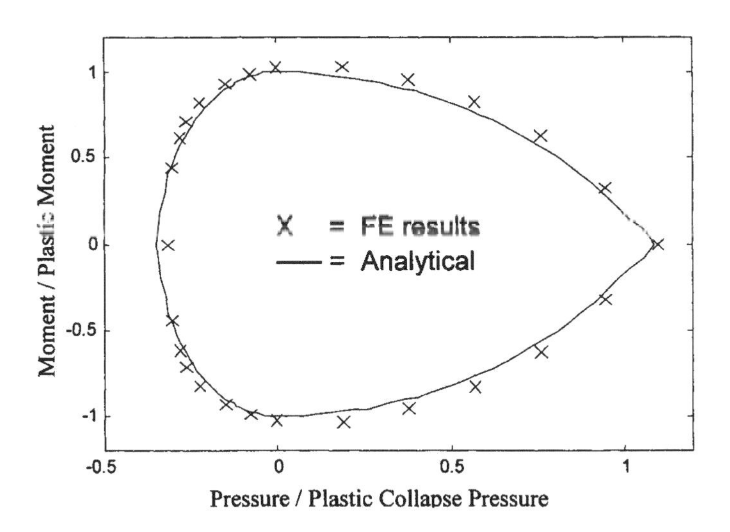

Figure 3.13 Normalized bending moment capacity as a function of pressure. No longitudinal force is applied.

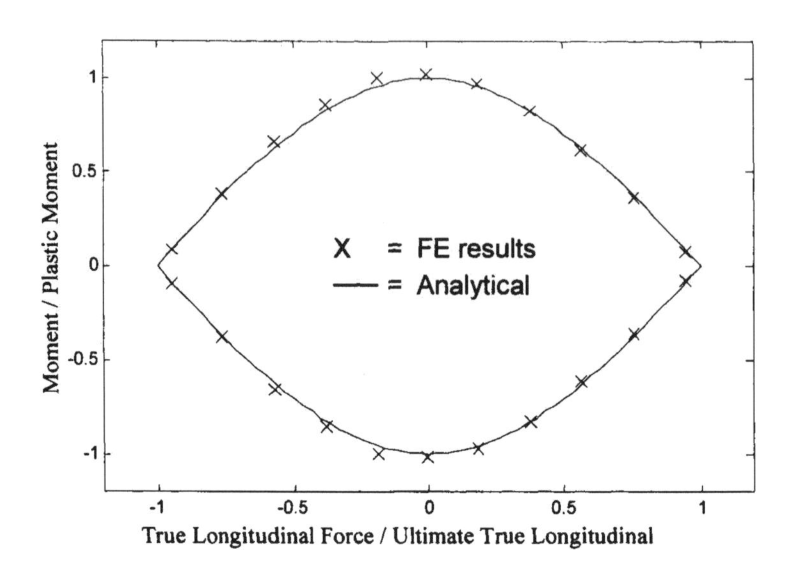

Figure 3.14 Normalized bending moment capacity as a function of longitudinal force. Pressure equal to zero.

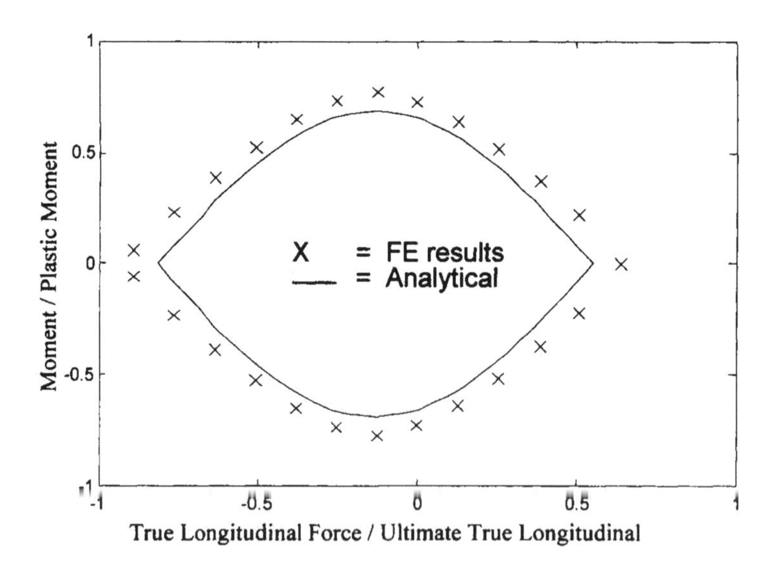

Figure 3.15 Normalized bending moment capacity as a function of longitudinal force. Pressure equal to 0.8 times Haagsma's collapse pressure.

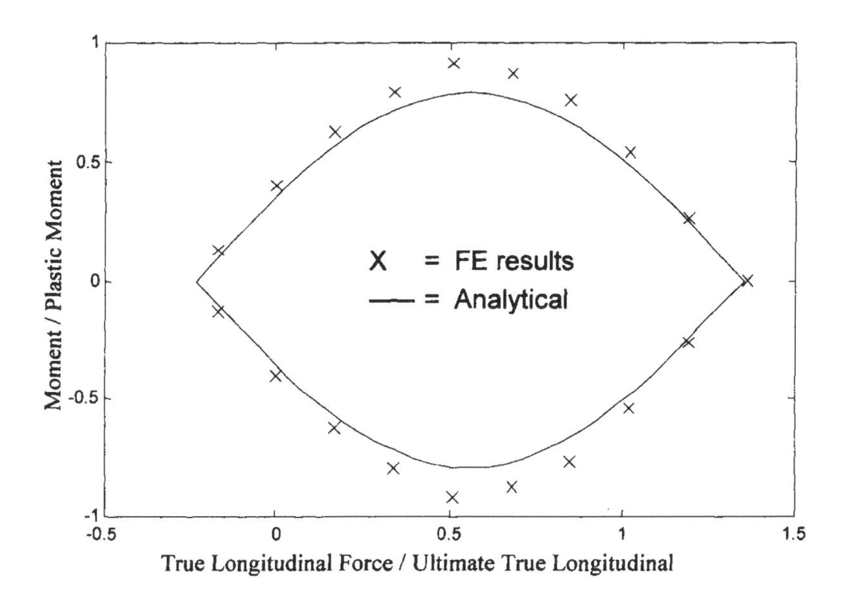

Figure 3.16 Normalized bending moment capacity as a function of longitudinal force. Pressure equal to 0.9 times the plastic buckling pressure.

3.6 References

1. ABS (2000) "Guide for Building and Classing Undersea Pipelines and Risers", American Bureau of Shipping.
2. API (1998) "Design, Construction, Operation and Maintenance of Offshore Hydrocarbon Pipelines (Limit State Design)".
3. Bai, Y., Igland, R. and Moan, T., (1993) "Tube Collapse under Combined Pressure, Tension and Bending", International Journal of Offshore and Polar Engineering, Vol. 3(2), pp. 121-129.
4. Bai, Y., Igland, R. and Moan, T., (1994) "Ultimate Limit States for Pipes under Combined Tension and Bending", International Journal of Offshore and Polar Engineering, pp. 312-319.
5. Bai, Y., Igland, R. and Moan, T., (1995) "Collapse of Thick Tubes under Combined Tension and Bending", Journal of Constructional Steel Research, pp. 233-257.
6. Bai, Y., Igland, R. and Moan, T., (1997) "Tube Collapse under Combined External Pressure, Tension and Bending", Journal of Marine Structures, Vol. 10, No.5, pp.389-410.
7. Bai, Y. and Hauch, S., (1998) "Analytical Collapse of Corroded Pipes", ISOPE '98.
8. Bai, Y., Hauch, S. and Jensen, J.C., (1999) "Local Buckling and Plastic Collapse of Corroded Pipes with Yield Anisotropy", ISOPE'99.
9. Bruschi, R., Monti, P., Bolzoni, G., Tagliaferri, R. (1995), "Finite Element Method as Numerical Laboratory for Analysing Pipeline Response under Internal Pressure, Axial Load, Bending Moment" OMAE'95.

10. BSI: BS 8010 (1993) "Code of Practice for Pipeline - Part 3. Pipeline Subsea: Design, Construction and Installation", British Standards Institute.
11. Chen, W. F., and Sohal, I. S. (1988), "Cylindrical Members in Offshore Structures" Thin-Walled Structure, Vol. 6 1988. Special Issue on Offshore Structures, Elsevier Applied Science.
12. Corona, E. and Kyriakides, A. (1988), "Collapse of Pipelines under Combined Bending and External Pressure", BOSS'88.
13. DNV (1996), "Rules for Submarine Pipeline Systems", Det Norske Veritas, December 1996.
14. Ellinas, C.P., Raven, P.W.J., Walker, A.C. and Davies, P., (1986) "Limit State Philosophy in Pipeline Design", Journal of Energy Resources Technology, Transactions of ASME, Jan. 1986.
15. Gresnigt, A.M., 1986 "Plastic Design of Buried Steel Pipelines in Settlement Areas" HERON, Delf University of Technology, Vol. 31, No. 4.
16. Haagsma, S. C., Schaap D. (1981) "Collapse Resistance of Submarine Lines Studied" Oil & Gas Journal, Feb. 1981.
17. Hauch, S. and Bai, Y. (1998), "Use of Finite Element Analysis for Local Buckling Design of Pipelines" OMAE'98.
18. Hauch, S. and Bai, Y. (1999) "Bending Moment Capacity of Pipes", OMAE'99.
19. Hauch, S. and Bai, Y. (2000) "Bending Moment Capacity of Grove Corroded Pipes", ISOPE'2000.
20. Hill, R. (1950) "The mathematical theory of plasticity" Oxford University Press, New York, ISBN 0 19 856162 8.
21. Kyriakides, S and Yeh, M. K. (1988), "Plastic Anisotropy in Drawn Metal Tubes" Journal of Engineering for Industry, Aug. 1988, Vol. 110/303.
22. Madhavan, R., Babcock, C.D. and Singer, J., (1993), "On the Collapse of Long, Thick-Walled Tubes Under External Pressure and Axial Tension", Journal of Pressure Vessel Technology, Feb. 1993, Vol. 115/15.
23. Mohareb, M. E., Elwi, A. E., Kulak, G. L. and Murray D. W. (1994), Deformational Behaviour of Line Pipe" Structural Engineering Report No. 202, University of Alberta.
24. Murphey C.E. and Langner C.G. (1985), "Ultimate Pipe Strength Under Bending, Collapse and Fatigue", OMAE'85.
25. SUPERB (1996), "Buckling and Collapse Limit State", December 1996.
26. Timoshenko, S. P. and Gere, J. M. (1961), "Theory of Elastic Stability", 3rd Edition, McGraw-Hill International Book Company.

Part I

Mechanical Design

Chapter 4 Limit-state based Strength Design

4.1 Introduction

The limit-state based strength design became crucially important when usage factors in wall-thickness design are raised from those given by traditional design codes. The discussion of a limit state design approach in this chapter is based on new Guides/Rules and a review of recent design projects and Joint Industry Projects (SUPERB, DEEPIPE, etc.), See Bai and Damsleth (1997).

This chapter presents limit-state strength criteria for pipeline design. The limit state checks conducted in the strength design are:

- Out of roundness for serviceability;
- Bursting due to internal pressure, longitudinal force and bending;
- Buckling/collapse due to pressure, longitudinal force and bending;
- Fracture of welds due to bending/tension;
- Low-cycle fatigue due to shutdowns;
- Ratcheting due to reeling and shutdowns;
- Accumulated plastic strain.

The allowable strains, equivalent stresses and bending moments are determined for the following operating scenarios.

- Empty condition;
- Water filled condition
- Pressure test condition;
- Operational conditions.

The strength criteria will be applicable for the following design situations:

- Pipeline in-place behavior;
- Trawl pullover response;
- Free-spanning pipelines;
- Pipeline dynamic free-span.

The pipeline route is divided into two zones:

- Zone 1 is the zone where no frequent human activity is anticipated along the pipeline route. For operating phases Zone 1 is classified as "Normal Safety Class"

- Zone 2 is the parts of the pipeline/riser in the near platform (manned) zone or in areas with frequent human activity. The extent of zone 2 is 500 m from the maximum facility excursion or determined based on risk analyses. For operating phases Zone 2 is classified as "High Safety Class"

For temporary (construction) phases both zones are classified as "Low Safety Class", when the pipeline does not contain any hydrocarbons.

4.2 Out of Roundness Serviceability Limit

The pipeline out of roundness is related to the maximum and minimum pipe diameters (D_{max} and D_{min}) measured from different positions around the sectional circumference and is defined according to the following equation:

$$f_0 = \frac{D_{\max} - D_{\min}}{D} \tag{4.1}$$

The out of roundness during the fabrication process is not to be more than 1.5%.

The out of roundness of the pipe may increase where the pipe is subject to reverse bending and the effect of this on subsequent straining is to be considered. For a typical pipeline the following scenarios will influence the out of roundness:

- The out of roundness may increase during the installation process where the pipe is subject to reverse inelastic bending;
- Cyclic bending may occur as a consequence of shutdowns during operation if global buckling is allowed to relieve temperature and pressure induced compressive forces.

Out of roundness due to point loads is to be checked. Critical point loads may arise at free-span shoulders, artificial supports and support settlement.

The accumulative out of roundness through life cycle is not to exceed 4%. This out of roundness requirement may be relaxed if:

The effect of out of roundness on moment capacity and strain criteria is included;
- The pigging requirements and repair systems, are met, and;
- Cyclic loads induced out-of-roundness have been considered.

Finite element analysis may be performed to calculate the increase in out of roundness during the life cycle of a pipeline. The analysis is to include fabrication tolerances and all loads applied through the pipelines life-cycle such as point loads, bending against a surface, axial load and repeated pressure, temperature and bending cycles.

4.3 Bursting

4.3.1 Hoop Stress vs. Equivalent Stress Criteria

An analytical study by Stewart (1994) and a finite element analysis have demonstrated that for a pipe under combined internal pressure and bending:

- If a pipeline section is in a displacement controlled situation, then a hoop stress criterion provides a good control of bursting.
- If a pipeline section is in a load controlled situation then an equivalent stress criterion may be applied to ensure sufficient burst strength for pipes under combined internal pressure and axial loads (the influence of bending is yet to be investigated).

For pipelines in operation, it is generally conservative to apply the equivalent stress criteria to control bursting since the dominating load is internal pressure combined with bending.

The bursting failure mode is governed by the tensile hoop stress. To ensure structural strength against bursting, the hoop stress is to fulfill the following conditions:

- yielding limit state, $\sigma_h \leq \eta_s \cdot SMYS$, where η_s is usage factor for SMYS (Specified Minimum Yield Stress)
- bursting limit state, $\sigma_h \leq \eta_u \cdot SMTS$, where η_u is usage factor for SMTS (Specified Minimum Tensile Stress)

For load-controlled situations, special consideration shall be made to bursting. It has been chosen to use an equivalent and longitudinal stress criterion according to the results from the analytical study and the finite element analysis.

4.3.2 Bursting Strength Criteria for Pipeline

The hoop stress due to pressure containment is not to exceed the following criteria:

$$(p_i - p_e)\frac{D-t}{2t} \leq \eta_S \min[SMYS(T), 0.87 * SMTS(T)] \tag{4.2}$$

where:

p_i : internal pressure

p_e : external pressure

D : nominal outside diameter of pipe

T : minimum wall thickness

SMYS (T) : specified minimum yield stress at temperature T

SMTS (T) : specified minimum tensile stress at temperature T

Temperature derating is to be accounted for if the temperature is above 50°C.

The following usage factors in Table 4.1 apply to the hoop stress equation:

Table 4.1 Usage factors for hoop stress criteria.

Material Quality Class	Usage Factor	Safety Class		
		Low	Normal	High
Class B, C [1)]	η_s	0.85	0.80	0.70
Class A[2)]	η_s	0.83	0.77	0.67

Note:

1) In order to apply these higher usage factors, the following additional requirements with respect to linepipe manufacturing must be fulfilled:

SMYS < (mean – 2*Standard Deviations) of yield stress;

SMTS < (mean – 3*Standard Deviations) of tensile stress;

t_{fab} :2* Standard Deviations of the wall-thickness.

2) After Table 2 in Section 6.4 of ISO 13623 (1997).

Stress Criteria

For internal-over pressure situations, the allowable equivalent stress and allowable longitudinal stress are η*SMYS(T) and the usage factor η given in Table 4.2 is after Table 3 in Section 6.4 of ISO13623 (1997).

Table 4.2 Equivalent stress design factor.

Load Combinations	Design factor, η
Construction and environmental loads	1.0
Functional and environmental loads	0.9
Functional, environmental and accidental loads	1.0

However, moment criteria, in Chapter 4.4 are considered to have better accuracy for strength design.

4.4 Local Buckling/Collapse

This section is based on Hauch and Bai (1999).

Local Buckling

For pipelines subjected to combined pressure, longitudinal force and bending, local buckling may occur. The failure mode may be yielding of the cross section or buckling on the compressive side of the pipe. The criteria given in this guideline may be used to calculate the maximum allowable bending moment for a given scenario. It shall be noted that the maximum allowable bending moment given in this guideline does not take fracture into account and that fracture criteria therefore may reduce the bending capacity of the pipe. This particularly applies for high-tension/high-pressure load conditions.

Load versus Displacement Controlled Situations

The local buckling check can be separated into a check for load controlled situations (bending moment) and one for displacement controlled situations (strain level). Due to the relation between applied bending moment and maximum strain in a pipe, a higher allowable strength for a given target safety level can be achieved by using a strain-based criterion than the bending moment criterion. Consequently the bending moment criterion can, conservatively be used for both load and displacement controlled situations. In this guideline only the bending moment criterion is given.

Local Buckling and Accumulated Out-of-Roundness

Increased out-of-roundness due to installation and cyclic operating loads may aggravate local buckling and is to be considered. It is recommended that out-of-roundness due to through life loads be simulated using finite element analysis.

Maximum Allowable Bending Moment

The allowable bending moment for local buckling under load controlled situations can be expressed as:

$$M_{Allowable(F,p)} = \frac{\eta_{RM}}{\gamma_c} M_p \sqrt{1-\left(1-\alpha^2\right)\left(\frac{p}{\eta_{RP} p_l}\right)^2} \cos\left(\frac{\pi}{2} \frac{\frac{\gamma_c F}{\eta_{RF} F_l} - \alpha \frac{p}{\eta_{RP} p_l}}{\sqrt{1-\left(1-\alpha^2\right)\left(\frac{p}{\eta_{RP} p_l}\right)^2}}\right) \tag{4.3}$$

where:

$M_{Allowable}$: Allowable bending moment

M_p : Plastic moment

p_l : Limit pressure

p : Pressure acting on the pipe

F_l : Limit longitudinal force

F : Longitudinal force acting on the pipe

α : Correction factor

γ_C : Condition load factor

η_R : Strength usage factor

- Correction Factor:

$$\alpha = 0.25 \frac{p_l}{F_l} \text{ for external overpressure} \tag{4.4}$$

$$\alpha = 0.25 \frac{p_l}{F_l} \text{ for internal overpressure} \tag{4.5}$$

If possible, the correction factor should be verified by finite element analyses.

- Plastic (Limit) Moment:

The limit moment may be given as:

$$M_{C(F=0,P=0)} = \left(1.05 - 0.0015 \cdot \frac{D}{t}\right) \cdot SMYS \cdot D^2 \cdot t \tag{4.6}$$

- Limit Longitudinal Force for Compression and Tension:

The limit longitudinal force may be estimated as:

$$F_l = 0.5 \cdot (SMYS + SMTS) \cdot A \tag{4.7}$$

- Limit Pressure for External Overpressure Condition:

The limit external pressure 'p_l' is to be calculated based on:

$$p_l^3 - p_{el} p_l^2 - \left(p_p^2 + p_{el} p_p f_0 \frac{D}{t}\right) p_l + p_{el} p_p^2 = 0 \tag{4.8}$$

where:

$$p_{el} \quad : \frac{2E}{(1-\nu^2)} \left(\frac{t}{D}\right)^3 \tag{4.9}$$

$$p_p \quad : \eta_{fab} SMYS \frac{2t}{D} \text{ 1)} \tag{4.10}$$

f_0 : Initial out-of-roundness [2], $(D_{max}-D_{min})/D$

E : Young's Module

υ : Poisson's ratio

Guidance note:

1. η_{fab} is 0.925 for pipes fabricated by the UO process, 0.85 for pipes fabricated by the UOE process and 1 for seamless or annealed pipes.
2. Out-of-roundness caused during the construction phase is to be included, but not flattening due to external water pressure or bending in as-laid position.

- Limit Pressure for Internal Overpressure Condition:

The limit pressure will be equal to the bursting pressure given by:

$$p_l = 0.5(SMTS + SMYS)\frac{2t}{D} \tag{4.11}$$

where:

SMYS : Specified Minimum Yield Strength in hoop direction

SMTS : Specified Minimum Tensile Strength in hoop direction

- Load and Usage Factors:

Load factor γ_C and usage factor η_R are listed in Table 4.3.

Table 4.3 Load and usage factors.

Safety factors \ Safety classes		Low	Normal	High
γ_C	Uneven seabed	1.07	1.07	1.07
	Pressure test	0.93	0.93	0.93
	Stiff supported	0.82	0.82	0.82
	Otherwise	1.00	1.00	1.00
η_{RP}	Pressure	0.95	0.93	0.90
η_{RF}	Longitudinal force	0.90	0.85	0.80
η_{RM}	Moment	0.80	0.73	0.65

Guidance notes:

- Load condition factors may be combined e.g. load condition factor for pressure test of pipelines resting on uneven seabed, 1.07×0.93 = 1.00
- Safety class is low for temporary phases. For the operating phase, safety class is normal and high for area classified as zone 1 and zone 2 respectively.

For displacement-controlled situations the following strain capacity check is given to ensure structural strength against local buckling:

$$\left(\frac{\gamma_F \cdot \gamma_c \cdot \gamma_{sncf} \cdot \gamma_D \cdot \varepsilon_{F,c} + \gamma_E \cdot \varepsilon_{E,c}}{\frac{\varepsilon_{M,c}}{\gamma_\varepsilon}}\right)^{0.8} + \frac{p_e}{\frac{p_c}{\gamma_R}} \leq 1 \tag{4.12}$$

where:

$\varepsilon_{F,c}$	: characteristic functional longitudinal strain;
$\varepsilon_{E,c}$	: characteristic environmental longitudinal strain;
$\varepsilon_{M,c}$	: characteristic buckling strain capacity;
γ_{sncf}	: strain concentration factor accounting for increased strain in the field joints due to coating stiffness discontinuities;
γ_F	: functional load factor;
γ_E	: environmental load factor;
γ_D	: dynamic load factor;
γ_C	: condition load factor;
γ_R	: strength resistance factor;
γ_ε	: strain capacity resistance factor.

Comparing to the equation established in DNV'96, Section 5 C305, two additional safety factors (γ_{sncf} and γ_D) are included. These additional safety factors are:

- γ_{sncf} for Strain Concentration Factor (SNCF).
- γ_D to take account for dynamic amplifications during a snap-through dynamic buckling (Nystrøm et al 1997).

4.5 Fracture

4.5.1 PD6493 Assessment

Fracture of the welds due to a tensile strain is normally evaluated in accordance with PD 6493 (1991).

This assessment method uses a curve (Failure Assessment Diagram) which combines the two potential failure modes: brittle fracture and plastic collapse.

Maximum weld flaws, described in Statoil R-SF-260, Pipeline Welding Specification, are to be used as the basic input for the calculations. The flaw has been assumed as maximum allowable defect due to lack of fusion between passes. The defects and material are assumed as below:

Type	: Surface flaw due to lack of fusion
Depth (a)	: 3 mm
Length (2c)	: 50 mm
CTOD	: 0.20 mm (at operating temperature)
Material	: As for Parent material

Surface flaw is chosen as the worst case scenario from acceptable flaws specified in the weld specifications. The partial safety factors recommended by PD 6493, are as below:

- For levels 2 and 3, no additional safety factors are required where worst case estimates are taken for stress level, flaw size and toughness, and all partial coefficients should be taken as unity. (Appendix A.1 of PD6493).

PD6493 FAD (Failure Assessment Diagram) gives critical stress for the given defect and material. It is necessary to convert the critical stress to critical strain and for this the Ramberg-Osgood relationship is used as defined below:

$$\varepsilon = \frac{\sigma}{E}\{1 + \frac{3}{7}(\frac{\sigma}{\sigma_{0.7}})^{n-1}\} \tag{4.13}$$

where: $\sigma_{0.7}$: 430 MPa for X65 at 20°C

n : 26 for X65 at 20°C

The allowable strain criterion used in this report is conservative due to:

- The stress-strain curves used in converting stress to strain are based on the lowest yield stress and lowest ultimate stress.
- PD6493 has been derived for load-controlled situations, but is here applied to both load and displacement-controlled situations.
- The flow stress is in PD6493 defined as the average of yield and tensile stress.

Corrosion in girth welds can significantly reduce the critical tensile strain of the girth welds if a flaw is assumed to be in the surface of the corroded weld. Fracture mechanics assessment of existing pipelines has shown that the critical strain can be between 0.1% for heavily corroded pipes and 0.5% for pipes with shallow corrosion defects. However, we shall not assume the combination of corrosion and cracks in the girth welds, although some pitting could occur in the HAZ (Heat Affect Zone). If corrosion takes place, it will occur over a certain number of years after entering into service, when the maximum strain load became lower due to reduced operating pressure and temperature, and "shakedown" of peak stress/strain levels in a number of shut downs.

In the technical report "Update of laying criteria for pipelines", Denys and Lefevre, it is stated that the failure of welds under displacement controlled situations is highly dependent on the weld matching (in particular), and on the ratio of yield to tensile strength. They gave an allowable strain of $0.61 \times 1.2/1.5 = 0.50$ % (a safety factor of 1.5 is applied to the critical strain) for defect length t/1.2 and depth 3 mm, assuming that the weld is matched and the ratio of yield to tensile strength is 0.87. They also reported that the results are very sensitive to weld matching (over-matching will increase the allowable strain considerably, and under-matching will reduce it).

The Dutch code NEN 3650 states that, normally, a tensile strain of 0.5% will not pose any problems for material and welding in accordance with their specifications. If it can be demonstrated that the ductility of the material is greater, higher strains can be tolerated accordingly.

It was stated by Canadian Standards Associations that the pipeline industry has used a longitudinal tensile strain limit of 0.5%. This limit prevents fracture initiation and plastic collapse from Circumferential weld flaws small enough to be accepted by the specification or that may have been missed by inspection. Zimmerman et al. (1992) and Price (1990) reports that the 0.5% tensile strain limit is a subjective limitation, chosen to coincide with the API yield strength specifications and does not reflect an objective failure limit.

The Troll Phase I project applied an allowable strain level of 0.4 % for a 36" gas export line, which was approved by NPD, (Koets and Guijt, 1996).

4.5.2 Plastic Collapse Assessment

It has been observed that all fracture mechanics calculations based on PD6493 lead to $S_r=1$. Where S_r is defined as:

$$S_r = \frac{\sigma_n}{\sigma_f} \tag{4.14}$$

The flow stress σ_f is according to PD6493 defined as the average of yield stress σ_y and ultimate tensile stress σ_u of the weld material. For a flat plate with surface flaw under tension, equations for the net section stress σ_n from PD6493 lead to:

$$\sigma_n = \frac{\sigma_{cr}}{1 - \left(\frac{a}{t}\right) / \left\{1 + \frac{t}{c}\right\}} \tag{4.15}$$

where:

σ_{cr} : critical stress

a : defect depth

c : half-width of the defect

t : wall-thickness of the plate

The applied PD6493 assessment criteria can then be re-expressed as:

$$\sigma_{cr} = \left[1 - \left(\frac{a}{t}\right) / \left\{1 + \frac{t}{c}\right\}\right] \frac{\sigma_y + \sigma_u}{2} \tag{4.16}$$

The above PD6493 plastic collapse equation may be valid provided that brittle fracture is not a relevant failure mode, e.g.:

- The defect depth (a) is less that 3 mm and the length (2c) is less than t (or 25 mm).
- The material CTOD is more than e.g. 0.2 mm at operating temperature.

The PD6493 plastic collapse equation can be applied to calculate allowable defect depth (a) and length (2c) for a given critical stress. In addition to PD6493 plastic collapse equation, the following plastic collapse equations are available from literature, see Bai (1993) and Denys (1992):

- CEGB R6 approach (A.G. Miller's equation);
- Willoughby's equation;
- Net section yielding collapse solution;
- CSA Z184 equation;
- Denys's equation.

Comparing with other available equations, PD6493 seems to give conservative and reasonable predictions. The PD6493 suggests that the safety factor for σ_{cr} in Eq. (4.16) is 1.1. The readers are suggested to define safety factors based on the structural reliability principles described in Chapters 37 through 39.

Chen et al. (2000) discussed formulae for plastic collapse and fracture of pipe with girth weld defects. A study of fracture criteria, conducted as part of the DEEPIPE JIP, was summarized by Igland et al. (2000).

4.6 Fatigue

4.6.1 General

Pipeline components such as risers, unsupported free spans, welds, J-lay collars, buckle arrestors, riser touchdown points and flex-joints, should be assessed for fatigue. Potential cyclic loading that can cause fatigue damage includes vortex-induced-vibrations (VIV), wave-induced hydrodynamic loads, platform movements and cyclic pressure and thermal expansion loads. The fatigue life of the component is defined as the time it takes to develop a through-wall-thickness crack of the component.

For high cycle fatigue assessment, fatigue strength is to be calculated based on laboratory tests (S-N curves) or fracture mechanics. If no detailed information is available, the F2 curve may be applied as the S-N curves for pipeline high cycle fatigue. Low cycle fatigue of girth welds may be checked based on $\Delta\varepsilon$-N curves.

The fracture mechanics approach calculates the crack growth using Paris' equation and final fracture using a recognized failure assessment diagrams (see Chapter 4.5). It may be applied to develop cracked S-N curves that are for pipes containing initial defects. If a fracture mechanics crack growth analysis is employed, the design fatigue life should be at least 10 times the service life for all components. The initial flaw size should be the maximum acceptable flaw specified for the non-destructive testing during pipe welding in question.

4.6.2 Fatigue Assessment Based on S-N Curves

The S-N curves to be used for fatigue life calculation are defined by the following formula:

$$\log N = \log a - m \cdot \log \Delta\sigma$$

where N is the allowable stress cycle numbers; a and m are parameters defining the curves, which are dependant on the material and structural detail. $\Delta\sigma$ is the stress range including the effect of stress concentration.

For the pipe wall thickness in excess of 22 mm, the S-N curve is to take the following form:

$$\log N = \log a - \frac{m}{4} \cdot \log \frac{t}{22} - m \cdot \log \Delta\sigma$$

where t is the nominal wall thickness of the pipe.

The fatigue damage may be based on the accumulation law by Palmgren-Miner:

$$D_{fat} = \sum_{i=1}^{M_c} \frac{n_i}{N_i} \leq \eta$$

where:

D_{fat} : accumulated fatigue damage

η : allowable damage ratio, to be taken as 0.1

N_i : number of cycles to failure at the i th stress range defined by S-N curve

n_i : number of stress cycles with stress range in block i

A cut-off (threshold) stress range S_o may be specified below which no significant crack growth or fatigue damage occurs. For adequately cathodic protected joints exposed to seawater, S_o is the cut-off level at $2x10^8$ cycles, see Equation (4.17).

$$S_0 = \left(\frac{2 \cdot 10^8}{C}\right)^{-\frac{1}{m}} \tag{4.17}$$

Stress ranges S smaller than S_o may be ignored when calculating the accumulated fatigue damage.

4.6.3 Fatigue Assessment Based on Δε-N Curves

The number of strain cycles to failure may be assessed according to the American Welding Society (AWS) Standards Δε-N curves, where N is a function of the range of cyclic bending strains Δε. The Δε-N curves are expressed as below:

$$\Delta\varepsilon = 0.055N^{-0.4} \qquad for\ \Delta\varepsilon \geq 0.002 \tag{4.18}$$

and

$$\Delta\varepsilon = 0.016N^{-0.25} \qquad for\ \Delta\varepsilon \leq 0.002 \tag{4.19}$$

The strain range Δε is the total amplitude of strain variations; i.e. the maximum less the minimum strains occurring in the pipe body near the weld during steady cyclic bending loads.

A study of low-cycle fatigue conducted as part of the DEEPIPE JIP was summarized by Igland et al. (2000).

4.7 Ratcheting

Ratcheting is described in general terms as signifying incremental plastic deformation under cyclic loads in pipelines subject to high pressure and high temperatures (HP/HT). The effect of ratcheting on out of roundness, local buckling and fracture is to be considered.

Two types of ratcheting are to be evaluated and the acceptance criteria are as below:

1. Ratcheting in hoop strain (the pipe expands radially) as a result of strain reversal for pipes operated at high internal pressure and high temperature. The accumulative hoop strain limit is 0.5%.
2. Ratcheting in curvature or ovalisation due to cyclic bending and external pressure. The accumulative ovalisation is not to exceed a critical value corresponding to local buckling under monotonic bending, or serviceability. The accumulative ovalisation is to be accounted for in the check of local buckling and out-of-roundness.

A simplified code check of ratcheting is that the equivalent plastic strain is not to exceed 0.1 %, based on elastic-perfectly-plastic material and assuming that the reference for zero strain is the as-built state after hydro-testing.

In case the simplified code check is violated, a finite element analysis may be applied to determine if ratcheting is a critical failure mode and quantify the amount of deformation induced by ratcheting.

4.8 Dynamic Strength Criteria

Stress criteria (i.e. allowable moments, allowable stresses etc.), or strain criteria should be specified for the dynamic stresses or strain expected during vortex induced vibrations (VIV). At the maximum amplitude of vibrations, the strength criteria defined in this Chapter should be satisfied.

4.9 Accumulated Plastic Strain

If the yield limit is exceeded, the pipe steel will accumulate plastic strain. Accumulated plastic strain may reduce the ductility and toughness of the pipe material. Special strain aging and toughness testing must then be carried out.

Accumulated plastic strain is defined as the sum of plastic strain increments irrespective of sign and direction. The plastic strain increments are to be calculated from the point where the material stress-strain curve deviates from a linear relationship, and the accumulated plastic strain are to be calculated from the time of fabrication to the end of lifetime. Limiting accumulated plastic strain is to ensure that the material properties of the pipe will not become sub-standard. This is especially relevant for the fracture toughness.

Accumulated plastic strain may also increase the hardness of the material and thus increase its susceptibility to stress corrosion cracking in the presence of H_2S. Stress corrosion cracking is also related to the stress level in the material. If the material yield limit is exceeded, the stress level will necessarily be very high. Plastic deformation of the pipe will also impose high residual stress in the material that may promote stress corrosion cracking.

The general requirement of the accumulated plastic strain is that it should be based on strain aging and toughness testing of the pipe material. It is stated that due to material considerations a permanent/plastic strain up to 2% is allowable without any testing. In practice, this is valid also for the operational case.

If the pipeline is to be exposed to more than 2% accumulated plastic strain, as is often the case for reeling installation method, the material should be strain aging tested. However, recent testing of modern pipeline steel has shown that plastic strain up to 5% or even 10% can be acceptable.

In order to have an extra safety margin, it is also desirable to have a certain ratio between the yield stress and the ultimate tensile stress. A requirement to this ratio is given in DNV'81, paragraph 5.2.6.2, where the yield stress is determined not to exceed 85% of the ultimate stress. Accumulated plastic strain will increase the yield stress of the material and also increase the yield/ultimate stresses ratio.

4.10 Strain Concentration at Field Joints Due to Coatings

It is necessary to evaluate effects of the concrete coating on strain concentrations at field joints. It is found reasonable to assume that the SNCF (Strain Concentration Factor) is 1.2. This value is mainly selected due to an allowable strain as high as 0.4% from the fracture criterion and the technical information from Ness and Verley (1996).

4.11 References

1. Bai, Y. and Damsleth, P.A. (1997) "Limit-state Based Design of Offshore Pipelines", Proc. of OMAE'97.
2. Chen, M.J., Dong, G., Jakobsen, R.A. and Bai, Y. (2000) "Assessment of Pipeline Girth Weld Defects" Proc. of ISOPE'2000.
3. Denys, R.M., (1992) "A Plastic Collapse-based Procedure for girth weld defect Acceptance" Int. Conf. on Pipeline Reliability, June 2-5, 1992, Calgary.
4. Hauch S. and Bai Y., (1999). "Bending Moment Capacity of Pipes", OMAE'99.
5. Igland, R.T., Saerik, S., Bai, Y., Berge, S., Collberg, L., Gotoh, K., Mainuon, P. and Thaulow, C. (2000) "Deepwater Pipelines and Flowlines", Proc. of OTC'2000.
6. ISO 13623 (1997) "Petroleum and Natural Gas Industries; Pipeline Transportation Systems", International Standard Organisation.
7. Koets, O.J. and Guijt J. (1996) "Troll Phase I, The Lessons Learnt", OPT'96.
8. NEN (1992), NEN 3650, "Requirements for Steel Pipeline Transportation System", 1992.
9. Ness, O.B. and Verley, R., (1996) "Strain Concentrations in Pipeline With Concrete Coating", Journal of Offshore Mechanics and Arctic Engineering, Vol.118.
10. Nystrøm P., Tørnes K., Bai Y. and Damsleth P., (1997). "Dynamic Buckling and Cyclic Behavior of HP/HT Pipelines", Proc. of ISOPE'97.
11. PD 6493, (1991) "Guidance on Methods for Assessing the Acceptability of Flows in Fusion Welded Structures".
12. Price, P. and St. J., (1990). "Canadian Standards Association Limit-states Task Force-State of Practice Review for Pipelines and Representative References".
13. Statoil Technical Specification, R-SF-260, (1991) "Pipeline Welding Specification".
14. Stewart, G. et al., (1994). "An Analytical Model to Predict the Burst Capacity of Pipelines" Proc. of OMAE'94.
15. Zimmerman, et al., (1992). "Development of Limit-states Guideline for the Pipeline Industry", OMAE '92.

PART II: Pipeline Design

Part II

Pipeline Design

Chapter 5 Soil and Pipe Interaction

5.1 Introduction

An interaction model of the contact between the pipeline and the seabed are often referred to as a pipe/soil interaction model. The pipe/soil interaction model consists of seabed stiffness and equivalent friction definition to represent the soil resistance to movement of the pipe. The equivalent friction is mainly based on coulumb friction (sand), cohesion (clay) or a combination of the two (silty-, sand- clays), the soil density and the contact pressure between the soil and pipe. It is therefore important to predict the soil contact pressure, equivalent friction and soil stiffness accurately.

5.2 Pipe Penetration in Soil

In a finite element model a non-linear pressure/penetration relationship may be used. The penetration of a statically loaded pipe into soil can be calculated as a function of pipe diameter, vertical contact pressure, soil strength parameters (undrained shear strength for clay and internal friction angle for friction materials such as sands) and submerged soil density. This penetration is to some extent complicated by the circular form of the pipeline, which leads to a combined effect of friction and bearing capacity resisting soil penetration.

In order to construct the pressure/penetration relationship mentioned above, an approach based on different methods for calculating the seabed penetration as a function of the static ground pressure has been used. Two such methods for clayey soils are the Verley and Lund (1995) method, and the buoyancy method (Håland, 1997). It is also clear that this is an approximation, since cyclic soil effects are ignored. In the following each method is given a brief description.

5.2.1 Verley and Lund Method

The Verley and Lund method is based on back calculation of pipelines with external diameter from 0.2 – 1.0 meters, resting on clays with undrained shear strength of 0.8 – 70 kPa. The method presents the following formula for calculation of pipeline penetration:

$$\frac{z}{D} = 0.0071 \cdot \left(S \cdot G^{0.3}\right)^{3.2} + 0.062 \cdot \left(S \cdot G^{0.3}\right)^{0.7} \tag{5.1}$$

where:

z : seabed penetration (m)

S : $F_c/(D \cdot s_u)$

G : $s_u/(D \cdot \gamma')$

F_c : vertical contact force (kN/m);

D : pipeline external diameter (m);

s_u : undrained shear strength (kPa);

γ' : submerged soil density (kN/m^3).

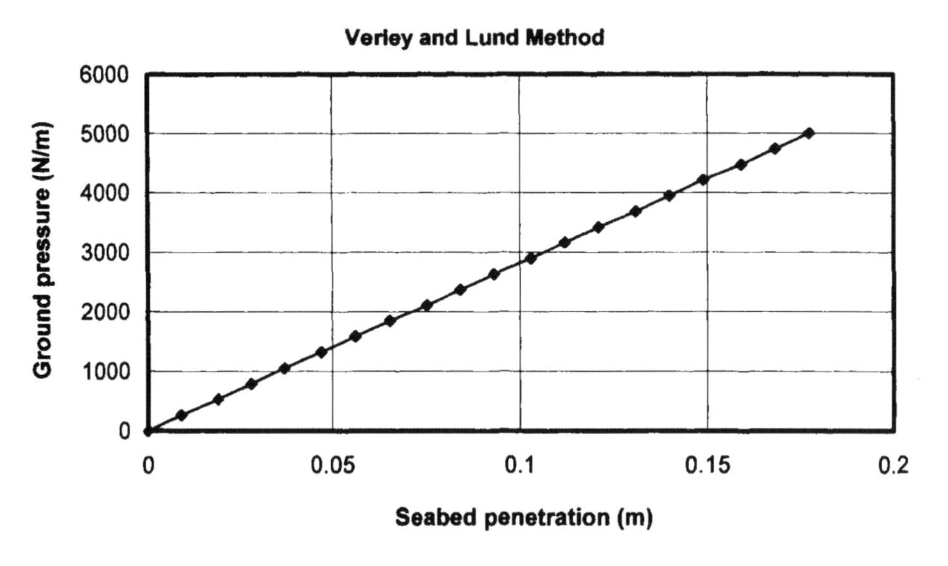

Figure 5.1 Pressure/penetration curve (Verley and Lund).

The Verley and Lund formulation is based on curve fitting to data with $S \cdot D^{0.3} < 2.5$.
For larger values the method overestimates penetration. An alternative formulation (linear), said to be valid for all values of $S \cdot D^{0.3}$, is given by:

$$\frac{z}{D} = 0.09 \cdot \left(S \cdot G^{0.3}\right) \tag{5.2}$$

5.2.2 Classical Method

The classical bearing capacity formulation was originally based on a rectangular foundation. The penetration normalized with pipe diameter is solved from the following equations:

$$\frac{z}{D} = 0.5 \cdot \left(1 - \left(1 - 4 \cdot \left(\frac{(\beta + 0.5) \cdot \psi}{2 \cdot (\beta + 0.5 \cdot \psi^3)}\right)^2\right)^{0.5}\right) \tag{5.3}$$

where:

$\psi = \alpha / (0.5 \cdot (\beta + 0.5))$;

$\alpha = F_c / (2 \cdot \gamma' \cdot D^2)$;

$\beta = (N_c \cdot s_u) / (\gamma' \cdot D)$;

N_c : bearing capacity factor, 5.14 for long foundations.

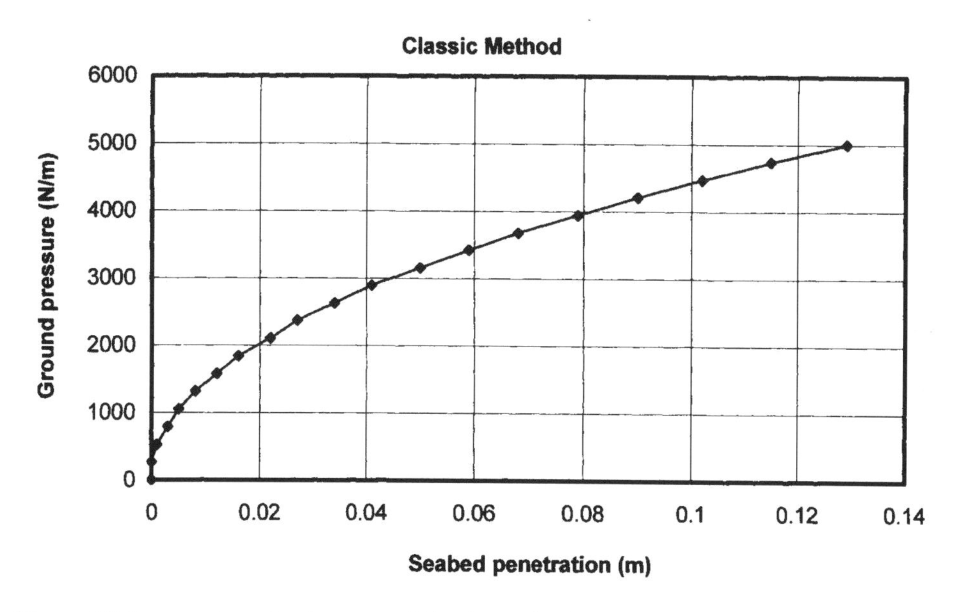

Figure 5.2 Pressure/penetration curve (Classic method).

5.2.3 Buoyancy Method

This method is intended for use with pipelines resting on very soft clays only. The buoyancy method assumes that the soil has no strength and behaves like a heavy liquid. The penetration is estimated by demanding that the soil-induced buoyancy of the pipeline is equal to the vertical contact force.

$$B = 2 \cdot \sqrt{D \cdot z - z^2}$$

$$A_s = (z / 6B) \cdot (3 \cdot z^2 + 4 \cdot B^2) \tag{5.4}$$

$$O = A_s \cdot L \cdot \gamma'$$

where:

B : width of pipeline in contact with soil;

A_s : penetrated cross sectional area of pipe;

O : buoyancy.

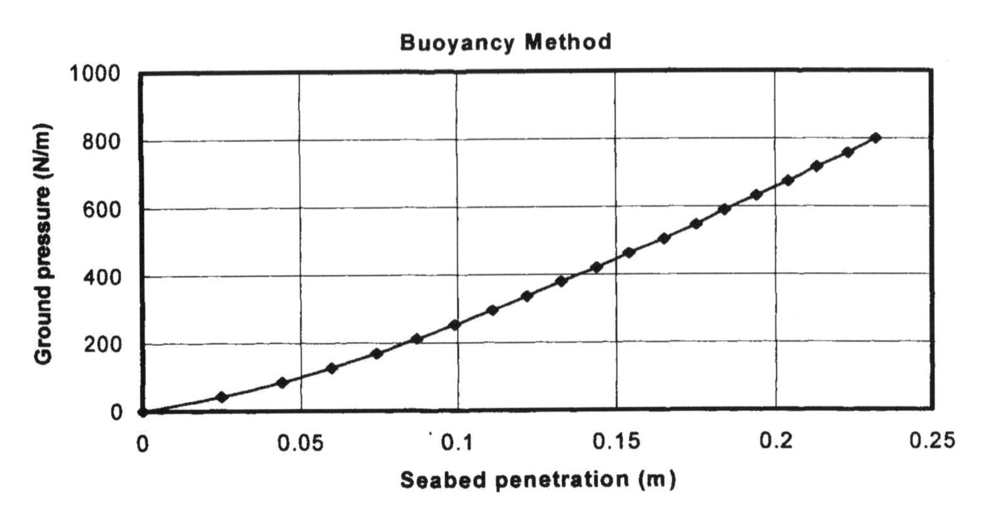

Figure 5.3 Pressure/penetration curve (Buoyancy method).

5.3 Modeling Friction and Breakout Forces

5.3.1 Anisotropic Friction

For pipelines not penetrating the seabed much, a pure Coulomb friction model can be appropriate. But, as the pipeline penetrates the seabed, the forces required moving the pipeline laterally become larger than the forces needed to move it in the longitudinal direction. This effect is due to the passive lateral soil resistance that is produced when a wedge of soil resists the pipe's motion. An anisotropic friction model that defines different friction coefficients in the lateral and longitudinal directions of the pipeline allows this effect to be investigated (Fig. 5.4).

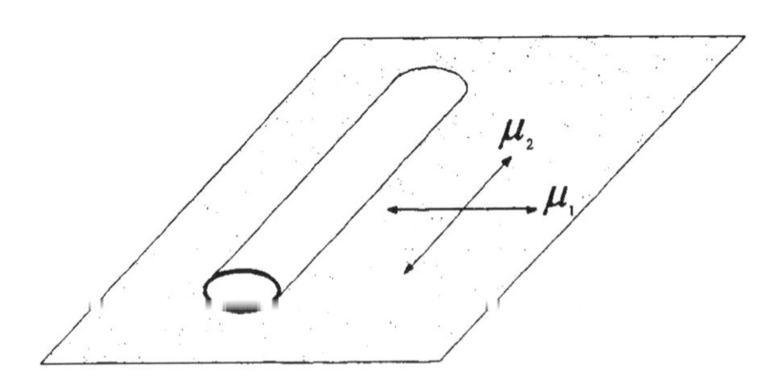

Figure 5.4 Anisotropic friction.

It may be mentioned that the torsional moment around the pipeline longitudinal axis, produced by the lateral soil-resistance force is ignored. However, the impact of this on the calculation of pipe response is believed to be negligible, unless pipeline twisting during installation is to be simulated.

5.3.2 Breakout Force

The breakout force is the maximum force needed to move the pipe from its stable position on the seabed. This force can be significantly higher than the force needed to maintain the movement after breakout due to suction and extra force needed for the pipe to "climb" out of its depression. An example curve is given in Figure 5.5.

The breakout forces, can be simulated in a finite element model, according to Brennodden (1991), which gives the following equations for the maximum breakout force in the axial and lateral direction:

Axial soil resistance (kN/m):

$$F_{a,\max} = 1.05 \cdot A_{c,calc} \cdot s_u \tag{5.5}$$

Lateral soil resistance (kN/m):

$$F_{l,\max} = 0.8 \cdot (0.2 \cdot F_c + 1.47 \cdot s_u \cdot A_{c,calc} / D) \tag{5.6}$$

where:

F_c : vertical contact force (kN/m);

$A_{c,calc}$: $2 \cdot R \cdot A\cos(1 - z / R)$ (m^2);

z : seabed penetration (m), e.g. calculated according to one of the methods in Section 5.2;

s_u : undrained shear strength (kPa).

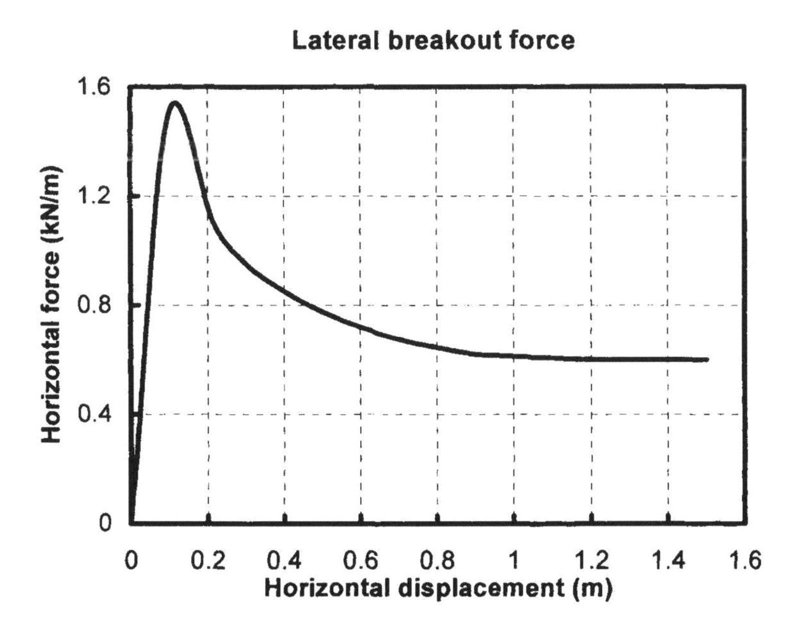

Figure 5.5 Horizontal force vs. lateral displacement.

5.4 References

1. Brennodden, H. (1991), "Troll phase I – Verification of expansion curve analysis and consolidation effects", SINTEF Geotechnical engineering.
2. Håland, G. (1997), "Penetration of large diameter pipelines", Statoil report: 97S97-8268.
3. Verley, R. and Lund, K.M., (1995) "A Soil Resistance Model for Pipelines Placed on Clay Soils", Proceedings of OMAE'95.

Part II

Pipeline Design

Chapter 6 Hydrodynamics around Pipes

6.1 Wave Simulators

Wave simulators may be established, using 2D regular long-crested and 2D random long-crested wave models. In each of the wave simulators, surface elevation, wave-induced water particle velocity and acceleration, dynamical pressure and pressure gradient, of an arbitrary point in space and time is defined mathematically. This allows the wave simulators to compute the wave kinematics during a time-domain dynamic analysis.

6.2 Choice of Wave Theory

Comprehensive studies have been conducted to identify the most suitable wave theories for representing the near-bottom kinematics due to wave action. In Dean et al. (1986) it was concluded that linear wave theory provides a good prediction of near-bottom kinematics for a wide range of relative water depth and wave steepness. One reason for this relatively good agreement is that the influence of non-linearities considered in higher order wave theories is reduced with depth below the free surface. In Kirkgoz (1986), it was also found that linear wave theory gave acceptable predictions of near seabed water particle velocities in waves close to the breaking point. It thus seems appropriate to apply linear wave theory to near seabed objects for a wide range of wave heights, periods and water depths. The calculated fluid velocities and accelerations of the surface waves, are transferred to seabed level using linear wave theory.

The 2D regular long-crested waves are useful when investigating the effects of extreme waves, while 2-D random long-crested waves are used when modeling a complete sea-state.

6.3 Mathematical Formulations Used in the Wave Simulators

6.3.1 General

Most of the theory and formulas presented in this section are available from Faltinsen (1990), Gran (1992), Hibbit et al. (1998) and Langen et al. (1997). This information has been used when programming the three wave simulators using the UWAVE subroutine in ABAQUS (Hibbit et al., 1998).

Figure 6.1 shows the parameters that are used when defining a 2D regular long-crested wave propagating in the positive x-direction.

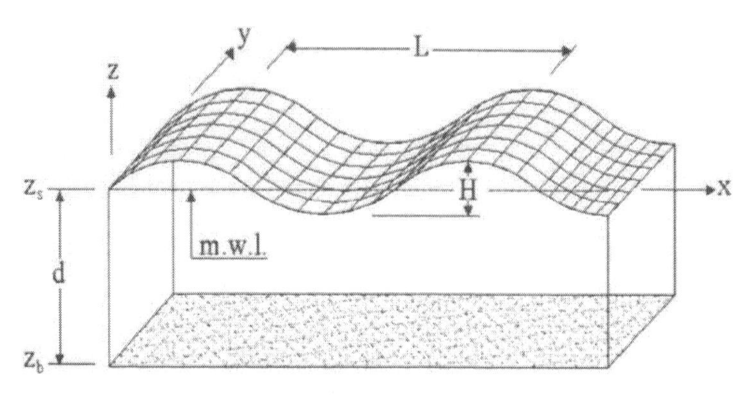

Figure 6.1 Parameters used when defining 2D regular waves.

where:

- L : wave length;
- H : wave height;
- d : $z_s - z_b$ = still water depth;
- a : wave amplitude (H/2);
- T : wave period;
- g : acceleration due to gravity;
- t : time;
- α : phase angle (radians);
- x : direction of wave propagation.

Wave frequency:

$$\omega = \frac{2\pi}{T} \tag{6.1}$$

Wave number:

$$k = \frac{2\pi}{L} \tag{6.2}$$

The dispersion relation expresses the relation between the wave period and wavelength and is given by:

$$\frac{\omega^2}{gk} = \tanh(kd) \tag{6.3}$$

6.3.2 2D Regular Long-crested Waves

The 2D regular long-crested waves (Figure 6.2) are defined by their wave amplitude and frequency, giving the following expressions for the wave kinematics:

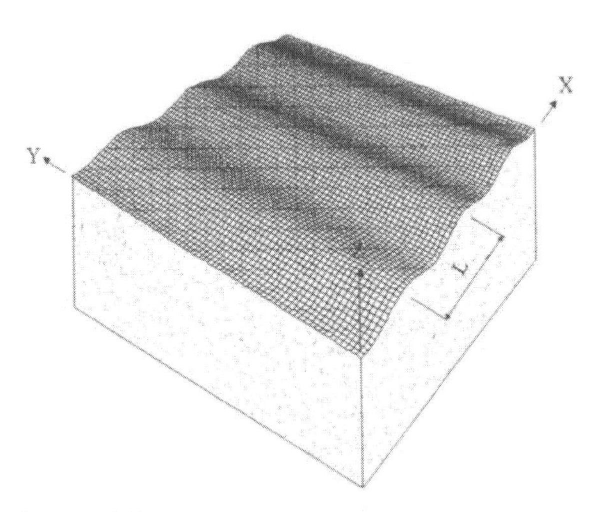

Figure 6.2 2D regular long-crested waves.

$$\eta = a \cdot \sin(\omega t - kx + \alpha) \tag{6.4}$$

Fluid velocity component in the x-direction,

$$v_x = \frac{agk}{\omega} \cdot \frac{\cosh(k(d+z))}{\cosh(kd)} \cdot \sin(\omega t - kx + \alpha) \tag{6.5}$$

Fluid velocity component in the z-direction,

$$v_z = \frac{agk}{\omega} \cdot \frac{\sinh(k(d+z))}{\cosh(kd)} \cdot \cos(\omega t - kx + \alpha) \tag{6.6}$$

Fluid acceleration component in the x-direction,

$$a_x = agk \cdot \frac{\cosh(k(d+z))}{\cosh(kd)} \cdot \cos(\omega t - kx + \alpha) \tag{6.7}$$

Fluid acceleration component in the z-direction,

$$a_z = -agk \cdot \frac{\sinh(k(d+z))}{\cosh(kd)} \cdot \sin(\omega t - kx + \alpha) \tag{6.8}$$

Dynamic pressure,

$$p_{dyn} = \rho ag \cdot \frac{\cosh(k(d+z))}{\cosh(kd)} \cdot \sin(\omega t - kx + \alpha) \tag{6.9}$$

6.3.3 2D Random Long-crested Waves

The 2D random long-crested wave (Figure 6.5) formulation is based on the use of a wave spectrum (Figure 6.3). Input of significant wave height, peak frequency, etc. (input is dependent on type of wave spectrum) defines the characteristics of the sea-state.

As an example, the JONSWAP spectrum can be defined as:

$$S(\omega) = a_p g^2 (\omega)^{-5} \exp\left(-\frac{5}{4}\left(\frac{\omega}{\omega_p}\right)^{-4}\right)\gamma^{a_p}, \quad a_p = \exp\left(\frac{-(\omega-\omega_p)^2}{2\sigma^2 \omega_p^{\ 2}}\right) \tag{6.10}$$

where:

- ω : angular frequency;
- ω_p : angular frequency of spectral peak;
- g : acceleration due to gravity;
- a_p : Phillips' constant;
- σ : spectral width parameter;
- γ : the JONSWAP peakedness parameter determined from,

$$\gamma = \begin{cases} 1, & \gamma < 1 \\ \exp\left(5.75 - 0.367 T_{peak}\sqrt{\dfrac{g}{H_s}}\right), & 1 < \gamma < 5 \\ 5, & \gamma > 5 \end{cases} \tag{6.11}$$

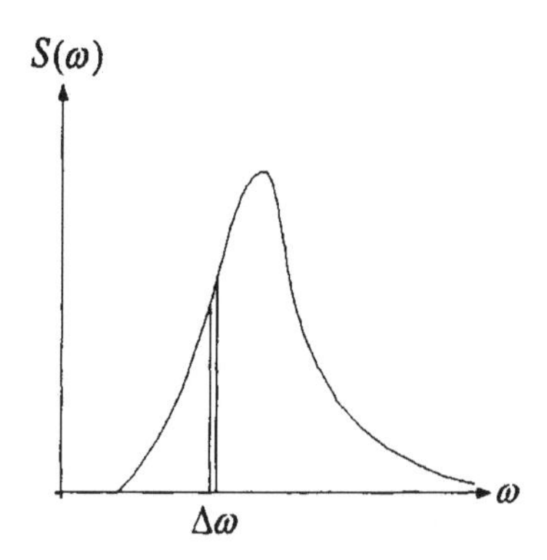

Figure 6.3 Wave spectrum.

From the wave spectrum we can find several properties. λ_n denotes the nth (stress) spectral moment defined by:

$$\lambda_n = \int_0^\infty \omega^n S_{ss}(\omega) d\omega \tag{6.12}$$

$H_{1/3}$ is the significant wave height and can be found as:

$$H_{1/3} = 4\sqrt{\lambda_0} \tag{6.13}$$

The characteristic wave period T_w may be estimated as:

$$T_w \approx 2\pi\sqrt{\frac{\lambda_0}{\lambda_2}} \tag{6.14}$$

and the spectral band width parameter ε as:

$$\varepsilon = \sqrt{1-\frac{\lambda_2^{\ 2}}{\lambda_0\lambda_4}} \tag{6.15}$$

By performing an inverse transformation, the wave amplitudes (a_i) and frequencies (ω_i) of each wave component is extracted from the wave spectrum.

Extraction of amplitudes and frequencies from the wave spectrum is for each wave component done according to:

$$a_i = \sqrt{2\cdot\Delta\omega\cdot S(\omega_i)} \tag{6.16}$$

where:

$$\omega_i = i\cdot\Delta\omega \tag{6.17}$$

$\Delta\omega$: the constant difference between successive frequencies.

$$k_i = \frac{(\omega_i)^2}{g} \text{ : the deep water dispersion relation.} \tag{6.18}$$

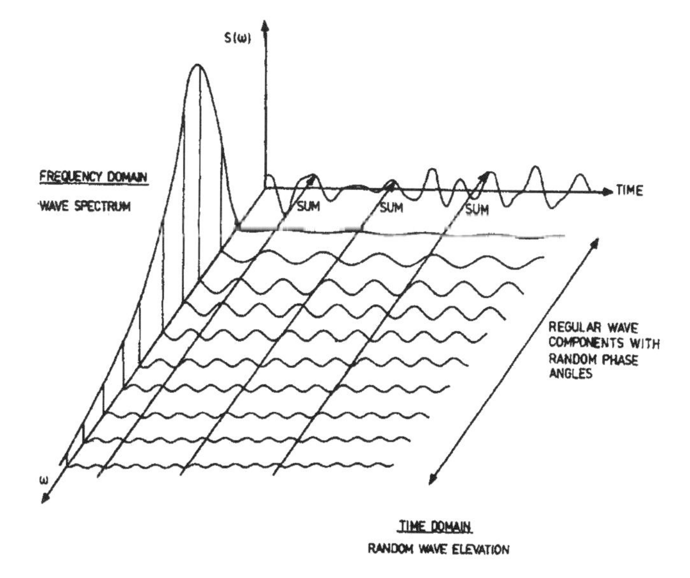

Figure 6.4 Connection between a frequency domain and a time domain representation of long-crested waves.

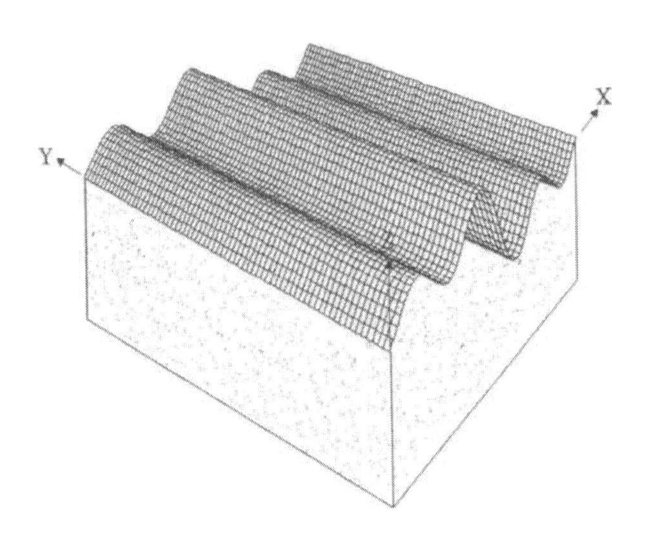

Figure 6.5 2D random long-crested waves.

Further, a random phase angle α_i, uniformly distributed between 0 and 2π is assigned to each wave component. The wave kinematics are thus represented as a sum of linear components (Figure 6.4). If "N" is the number of wave components, the sea state at a particular time and location can be represented by:

Surface elevation,

$$\eta = \sum_{i=1}^{N} a_i \cdot \sin(\omega_i t - k_i x + \alpha_i) \tag{6.19}$$

Velocity component in the x-direction,

$$v_x = g \cdot \sum_{i=1}^{N} \frac{a_i k_i}{\omega_i} \cdot \frac{\cosh(k_i(d+z))}{\cosh(k_i d)} \cdot \sin(-\omega_i t - k_i x + \alpha_i) \tag{6.20}$$

Velocity component in the z-direction,

$$v_z = g \cdot \sum_{i=1}^{N} \frac{a_i k_i}{\omega_i} \cdot \frac{\sinh(k_i(d+z))}{\cosh(k_i d)} \cdot \cos(\omega_i t - k_i x + \alpha_i) \tag{6.21}$$

Acceleration component in the x-direction,

$$a_x = g \cdot \sum_{i=1}^{N} a_i k_i \cdot \frac{\cosh(k_i(d+z))}{\cosh(k_i d)} \cdot \cos(\omega_i t - k_i x + \alpha_i) \tag{6.22}$$

Acceleration component in the z-direction,

$$a_z = -g \cdot \sum_{i=1}^{N} a_i k_i \cdot \frac{\sinh(k_i(d+z))}{\cosh(k_i d)} \cdot \sin(\omega_i t - k_i x + \alpha_i) \tag{6.23}$$

Dynamical pressure,

$$p_{dyn} = \rho g \cdot \sum_{i=1}^{N} a_i \frac{\cosh(k_i(d+z))}{\cosh(k_i d)} \cdot \sin(\omega_i t - k_i x + \alpha_i) \tag{6.24}$$

6.4 Steady Currents

For the situation where a steady current also exists the effects of the bottom boundary layer may be accounted for, and the mean current velocity over the pipe diameter may be applied in the analysis. According to DNV (1998), this has been included in the finite element model by assuming a logarithmic mean velocity profile.

$$U_c(z_D) = \frac{U(z_r)}{\ln(z_r/z_0)} \left\{ \left(\frac{e}{D}+1\right) \ln((e+D)/z_0) - \left(\frac{e}{D}\right) \ln\left(\frac{e}{z_0}\right) - 1 \right\} \tag{6.25}$$

where:

- $U(z_r)$: current velocity at reference measurement height;
- z_r : reference measurement height (usually 3m.);
- z_D : height to mid pipe (from seabed);
- z_0 : bottom roughness parameter;
- e : gap between the pipeline and the seabed;
- D : total external diameter of pipe (including any coating).

The total velocity is obtained by adding the velocities from waves and currents together:

$$v = v_{wave} + v_{current} \text{ (of a water particle)} \tag{6.26}$$

6.5 Hydrodynamic Forces

6.5.1 Hydrodynamic Drag and Inertia Forces

A pipeline section exposed to a flow will experience hydrodynamic forces, due to the combined effect of increased flow velocity above the pipe and flow separation from the pipe surface. Figure 6.6 shows the velocity distribution around the pipe. This section will explain the different components of the force vector and the expressions that are used to calculate these components.

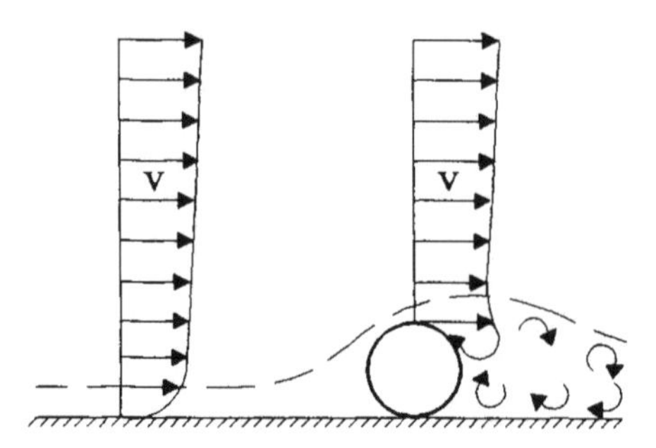

Figure 6.6 Flow field around pipe.

Pipeline Exposed to Steady Fluid Flow

Fluid drag is associated with velocities due to steady currents superposed by any waves that may be present (Figure 6.7). The expression below gives the transverse drag force component per unit length of the pipeline:

$$\text{Transverse drag force, } F_D = \frac{1}{2}\rho C_D D\, v_n |v_n| \tag{6.27}$$

where:

C_D : transverse drag coefficient.

v_n : transverse water particle velocity.

ρ : density of seawater.

D : total external diameter of pipe.

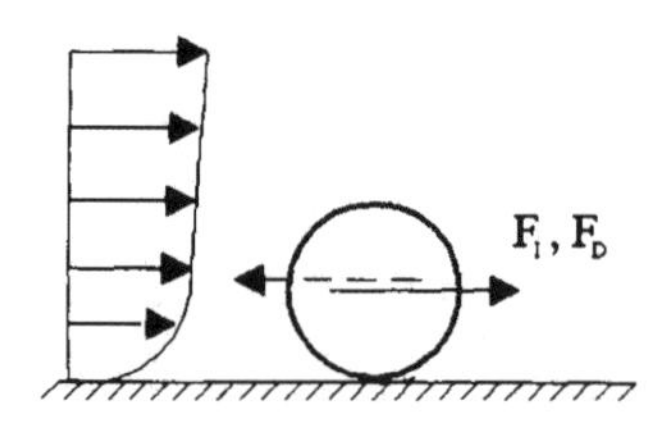

Figure 6.7 Fluid drag and inertia forces acting on a pipe section.

Pipeline Exposed to Accelerated Fluid Flow

A pipeline exposed to an accelerated fluid experiences a force proportional to the acceleration, this force is called the inertia force. The following expression gives the transverse inertia force component per unit length of a pipeline:

$$\text{Transverse inertia force, } F_I = \frac{\pi}{4}\rho D^2 C_M a_n \tag{6.28}$$

where:

C_M : (C_a+1), transverse inertia coefficient;

a_n : transverse water particle acceleration;
ρ : density of seawater;
D : Total external diameter of pipe.

The complete Morison's equation

The formula given above does not take into account that the pipe itself may have a velocity and acceleration. The inline force per unit length of a pipe is determined using the complete Morison's equation.

$$F_{IL}(t) = \frac{1}{2}\rho C_D\left(U - \frac{\partial y}{\partial t}\right)\left|U - \frac{\partial y}{\partial t}\right| + C_M\frac{\pi}{4}\rho D^2\frac{\partial U}{\partial t} - (C_M - 1)\frac{\pi}{4}\rho D^2\frac{\partial^2 y}{\partial t^2} \tag{6.29}$$

where:

ρ : sea water density;
D : outer diameter;
U : instantaneous (time dependant) flow velocity;
y : in line displacement of the pipe;
C_D : drag coefficient;
C_M : inertia coefficient;
= (C_a+1) where C_a is the added mass coefficient;
$\delta/\delta t$: differentiation with respect to time.

Drag and Inertia Coefficient Parameter Dependency

In general, the drag and inertia coefficient is given by:

$$C_D = C_D(R_e, KC, \alpha, (e/D), (k/D), (A_Z/D)) \tag{6.30}$$

$$C_M = C_M(R_e, KC, \alpha, (e/D), (A_Z/D)) \tag{6.31}$$

Reynolds number indicates the present flow regime, (i.e. laminar or turbulent) and is given as:

$$R_e = \frac{UL}{v} \tag{6.32}$$

where:

U : flow velocity;
L : characteristic length (Diameter for pipelines),;
v : cinematic viscosity.

The Keulegan-Carpenter number give information on how the flow separation around cylinders will be for ambient oscillatory planar flow ($U=U_M \sin((2\pi/T)t+\varepsilon)$)and is given as:

$$KC = \frac{U_M T}{D} \tag{6.33}$$

where:

U_M : flow velocity amplitude;

T : period;

D : diameter;

ε : phase angle;

t : time.

The current flow ratio may be applied to classify the flow regimes:

$$\alpha = \frac{U_c}{U_c + U_w} \tag{6.34}$$

where:

U_c : typical current velocity normal to pipe;

U_w : significant wave velocity normal to pipe given for each sea state (H_S, T_P, θ_W).

Note that $\alpha = 0$ corresponds to pure oscillatory flow due to waves and $\alpha = 1$ corresponds to pure (steady) current flow.

The presence of a fixed boundary near the pipe (proximity effect) has a pronounced effect on the mass coefficient. The added mass will increase as the pipe approaches a solid boundary, (see equation below).

$$C_a = 1 + \frac{1}{\left(10 \cdot \left(\frac{e}{D}\right) + 1\right)} \tag{6.35}$$

where:

e/D : gap ratio.

The natural period of the pipe oscillation will increase as the added mass increases.

The roughness number (k/D) have a large influence on the flow separation and therefore also on the drag and mass coefficient. (k=Characteristic cross-sectional dimension of the roughness on the body surface).

There is a connection between the VIV (Vortex-Induced Vibrations) and the drag force. A crude approximation can be given as:

$$C_D/C_{D0} = 1 + 2(A_Z/D) \tag{6.36}$$

where:

C_D : drag coefficient with VIV;

C_{D0} : drag coefficient with no VIV;

A_Z : cross–flow vibration amplitude.

This formula can be interpreted as saying that there is an apparent projected area D+2A_Z due to the oscillating cylinder.

6.5.2 Hydrodynamic Lift Forces

Lift force using constant lift coefficients

The lift force per unit length of a pipeline can be calculated according to:

$$\text{Vertical lift force, } F_L = \frac{1}{2}\rho D\, C_L\, v_n^{\,2} \tag{6.37}$$

where:

C_L : lift coefficient for pipe on a surface;

v_n : transverse water particle velocity (perpendicular to the direction of the lift force);

ρ : density of seawater;

D : total external diameter of pipe.

Lift force using variable lift coefficients

As can be imagined, the hydrodynamic lift coefficient (C_L) will vary as a function of the gap that might exist between the pipeline and the seabed. It can be seen from Figure 6.8 that a significant drop in the lift coefficient is present even for very small ratios of e/D. This is true both for the shear and the shear-free flow.

The lift coefficients according to Fredsøe and Sumer (1997) are given in Figure 6.8.

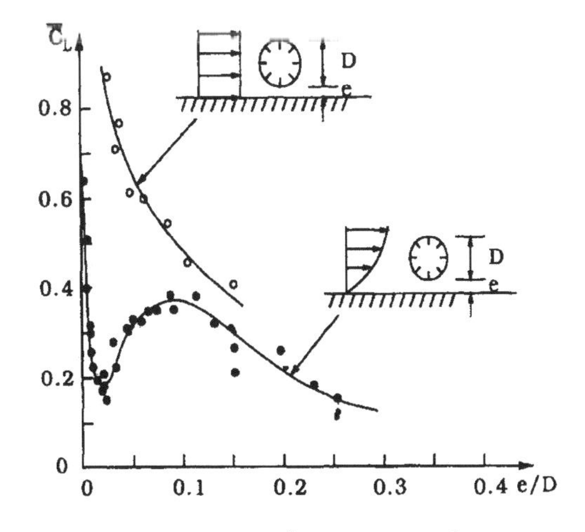

Figure 6.8 C_L in shear and shear-free flow for $10^3 < R_e < 30 \times 10^4$.

6.6 References

1. Dean, R.G., Perlin, M. (1986), "Intercomparison of Near-Bottom Kinematics by Several Wave Theories and Field and Laboratory Data", Coastal Engineering, 9.
2. DNV (1998), "Guideline No. 14 – Free-Spanning Pipelines", Det Norske Veritas.
3. Faltinsen, O.M., (1990) "Sea loads on Ships and Offshore Structures", Cambridge University Press.
4. Fredsøe, B. and Sumer, B.M., (1997) "Hydrodynamics around Cylindrical Structures", World Scientific Publishing Co.
5. Gran, S., (1992) "A Course in Ocean Engineering", Elsevier.
6. Hibbit, Karlsson and Sorensen, (1998) "ABAQUS User Manuals, Version 5.7".
7. Kirkgoz, M.S., (1986) "Particle Velocity Prediction of the Transformation Point of Plunging Breakers", Coastal Engineering, Vol. 10.
8. Langen, I., Gudmestad, O.T., Haver, S., Gilje, W. and Tjelta, T.I., (1997) "Forelesninger i Marin Teknologi", Høgskolen i Stavanger.

Part II

Pipeline Design

Chapter 7 Finite Element Analysis of In-situ Behavior

7.1 Introduction

The design of High-Pressure/High-Temperature (HP/HT) pipelines on an uneven seabed has become an important issue in the recent years. The need to gain further insight into how expansion, seabed friction and free spans influence on the pipeline behavior through selected load cases is the background for this chapter. The behavior of such pipelines is largely characterized by the tendency to undergo global buckling, either vertically if trenched or covered, or laterally if the pipeline is left fully exposed on the seabed. The main concern in the design of slender pipelines operating under HP/HT conditions is to control global buckling at some critical axial force. The large horizontal and/or vertical displacements induced by global buckling may result in high stresses and strains in the pipe wall that exceed code limits.

The simulation of the designed pipeline in a realistic three-dimensional environment obtained by measurements of the seabed topography, allows the engineers to exploit any opportunities that the pipeline behavior may offer to develop both safe and cost-effective solutions. For example, the designer can first analyze the pipeline behavior on the original seabed. If some of the load cases result in unacceptably high stress or strain, seabed modification can be simulated in the finite-element model and the analysis re-run to confirm that the modifications have lead to the desired decrease in stress or strain.

The finite element model may be a tool for analyzing the in-situ behavior of a pipeline. By the pipeline in-situ behavior it is here meant the pipeline behavior over its through-life load history. This part of the pipeline load history can consist of several sequential load cases, for example:

- Installation;
- Pressure testing (water filling and hydrotest pressure);
- Pipeline operation (content filling, design pressure, and temperature;
- Shut down/cool down cycles of pipeline;
- Upheaval and lateral buckling;
- Dynamic wave and/or current loading;
- Impact loads.

This chapter is based on a M.Sc. thesis, Ose (1998), supervised by the author and the work has been influenced by the papers presented in the conferences, Nystrøm et al. (1997), Tørnes et al. (1998) and Kristiansen et al. (1998).

7.2 Description of the Finite Element Model

In order to make a model like described above, some investigation of the problem had to be performed. This section deals with this process and describes some of the decisions that were made and problems that were to be solved during the work with this thesis.

7.2.1 Static Analysis Problems

Installation

Since the model may be used to analyze a pipeline situated on the seabed, it had to include some sort of installation process in order to find the pipeline configuration when placed on the three-dimensional seabed. This configuration would then serve as an *initial configuration* for the subsequent parts of the analysis.

Primarily it was not the behavior of the pipeline *during* the installation process that may be investigated. The important thing was to make sure that the lay-tension and lay-angle from the installation process was represented in such a way that the build-up of residual forces in the pipeline, due to friction when the pipeline lands on the seabed, was accounted for.

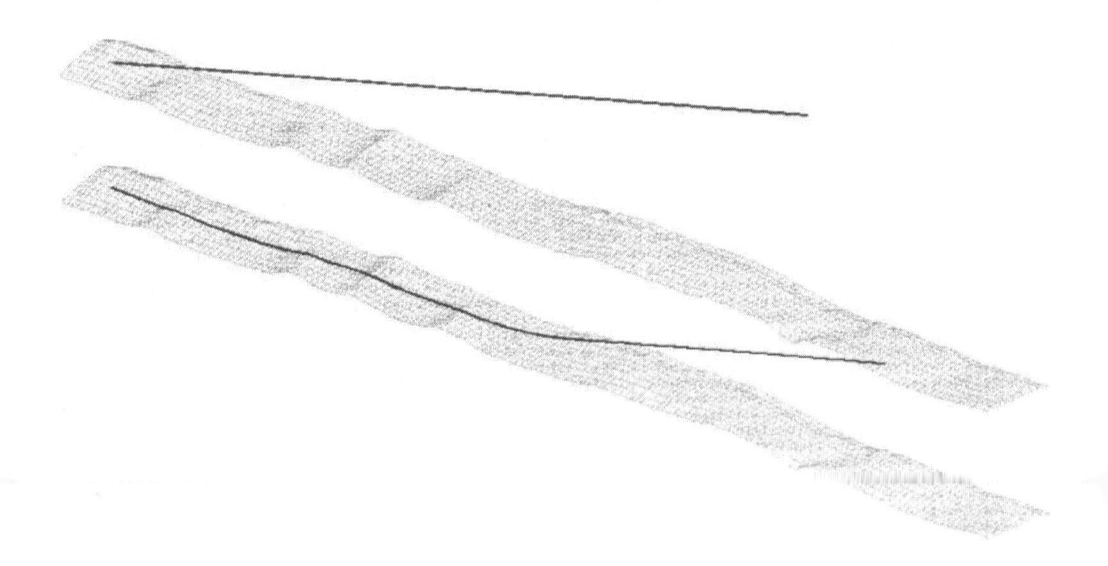

Figure 7.1 Eestablished finite-element model before and under installation.

As a result of this it was decided to make a simplified model of the installation. The model may include the possibility of applying lay tension, and, to specify the lay angel between the pipeline and the seabed to ensure good modeling of the contact forces in the touchdown zone as the pipeline lands on the seabed (Figure 7.1).

As the pipeline stretches out, a stable equilibrium between the pipeline and the seabed must be ensured. This requires a representative pipe/soil interaction model to be present. The pipe/soil interaction model will typically consist of a friction and a seabed stiffness definition. It was realized that the seabed stiffness formulation must be able to describe several pressure/penetration relationships, and that an anisotropic friction model may be used to represent the difference in frictional resistance in the longitudinal and lateral directions of the pipe.

Filling and draining of the pipeline

The filling and draining of the pipeline result in changes in the pipe weight and thus changes in the pipeline configuration. The friction force between the pipeline and the seabed is a function of the ground pressure and thus increases when the pipeline is filled.

The filling and draining of the pipeline could easily be modeled by a variation of the vertical load acting on the pipeline. But, a pipeline subjected to such load variations can in the filled condition experience large axial strains due to the change in geometry when the pipe deforms and sinks into the free-spans along the pipeline route (Figure 7.2). Due to this fact, the model to be established may use a large-displacement analysis procedure and the effect of changes in the pipe section area due to high axial straining may be accounted for. Further, the material model may be able to represent plastic behavior of the pipe section.

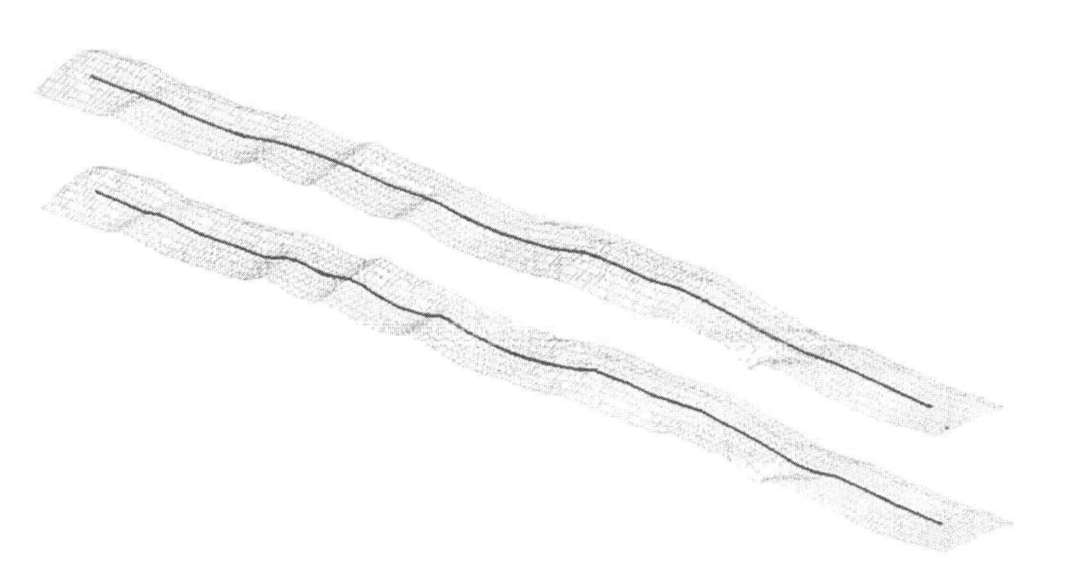

Figure 7.2 Finite-element model showing empty vs. water filled configuration.

Effects of high-pressure/high-temperature (HP/HT)

High temperatures from the contents of the pipeline causes material expansion of the pipe steel, this leads to an extension in the pipe length and the pipeline will buckle and seek new deformation paths to maintain in equilibrium (Figure 7.3). The influence of material expansion due to variation of temperature may therefore be included in the model.

Material properties such as yield stress, tensile strength and Young's modulus change with material temperature, and if necessary may be accounted for.

External hydrostatic pressure is an important factor regarding the strength capacity of deep-water pipelines. Since the model may include a fully three-dimensional seabed, the external pressure may be a function of the water depth. Internal pressure can be modeled as constant, but the possibility to account for the static head of the contents may be included.

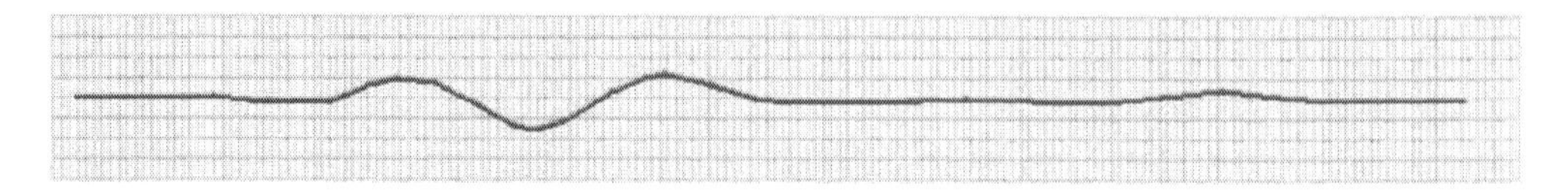

Figure 7.3 Top view of the finite-element model showing buckling due to temperature dependent material expansion (scaled displacements).

7.2.2 Dynamic Analysis Problems

Wave and current loading

Hydrodynamic forces arise from water particle velocity and acceleration. These forces can be fluctuating (caused by waves) or constant (caused by steady currents) and will result in a dynamic load pattern on the pipeline (Figure 7.4). Drag, inertia, and lift forces are of interest when analyzing the behavior of a submerged pipeline subjected to wave and/or current loading.

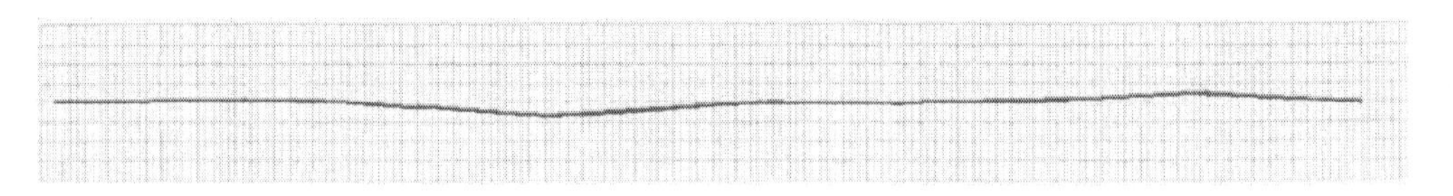

Figure 7.4 Top view of the finite-element model showing horizontal displacement when the pipeline is subjected to wave and current loading.

Because of the dynamic nature of waves, the pipeline response when subjected to this type of loading may be investigated in a dynamic analysis. Further, several wave formulations would be desirable. 2D regular or random long-crested waves and the 3D regular or random short-crested waves may be included in the finite-element model to supply the wave kinematics in a dynamic analysis.

Trawl gear pullover response

The trawl gear pullover loads may result in a dynamic plastic response. The calculation of loads and strength acceptance criteria are discussed in Chapter 34.

In a finite element analysis, implicit dynamic solution, such as that described in Chapter 7.3.2, is used to simulate the time-history of displacements, stresses and strain. Details are given in Tørnes et al. (1998).

7.3 Steps in an Analysis and Choice of Analysis Procedure

A basic concept in ABAQUS is the division of the load/problem history into steps. For each step the user chooses an analysis procedure. This means that any sequence of load history and desired type of analysis can be performed. For example in one static step the pipeline can be filled with gas, in the next static step emptied, and in the third step a dynamic analysis of the empty pipeline can be performed.

A typical load history from the established model is given as an example in Table 7.1.

Table 7.1 Typical load history in an ABAQUS analysis.

Loadstep	Action	Analysis procedure
1	Applying pipe self-weight and buoyancy.	Static
2	Applying hydrostatic external pressure.	Static
3	Applying lay tension.	Static
4	Lowering pipeline down at the seabed (see Fig. 7.1).	Static
5	Removing GAPSPHERE elements (winch).	Static
6	Modify boundary conditions.	Static
7	Water filling (see Fig. 7.2).	Static
8	Apply hydrotest pressure.	Static
9	Remove hydrotest pressure and water.	Static
10	Gas filling.	Static
11	Apply operation pressure.	Static
12	Apply operation temperature (see Fig. 7.3).	Static
13	Remove pressure and temperature.	Static
14	Apply wave and current loading (see Fig. 7.4).	Dynamic

7.3.1 The Static Analysis Procedure

The static analysis available from ABAQUS that is used in the model handles non-linearity's from large-displacements effects, material non-linearity, and boundary non-linearity's such as contact, sliding, and friction (pipe/seabed interaction). ABAQUS uses Newton's method to solve the non-linear equilibrium equations. Therefore, the solution is obtained as a series of increments with iterations to obtain equilibrium within each increment. For more information about static finite element analysis, see Cooker et al. (1991).

7.3.2 The Dynamic Analysis Procedure

A general dynamic analysis (dynamic analysis using direct integration) must be used to study the non-linear dynamic response of the pipeline. General non-linear dynamic analysis uses implicit integration of the entire model to calculate the transient dynamic response of the system. The direct integration method provided in ABAQUS called the Hilbert-Hughes-Taylor operator (which is an extension of the trapezoidal rule) is therefore used in the model. The Hilbert-Hughes-Taylor operator is implicit, the integration operator matrix must be inverted, and a set of simultaneous non-linear dynamic equilibrium equations must be solved at each time increment. This solution is done iteratively using Newton's method.

7.4 Element Types Used in the Model

Three types of elements are used in the established finite-element model (Figure 7.5). These are:

- The rigid elements of type R3D4 used to model the seabed.
- The PIPE31H pipe elements used to model the pipeline.
- The GAPSPHER elements that are used as a winch when lowering the pipeline from its initial position and down at the seabed (see Figure 7.1). These elements are removed from the model when the pipeline has landed and gained equilibrium at the seabed.

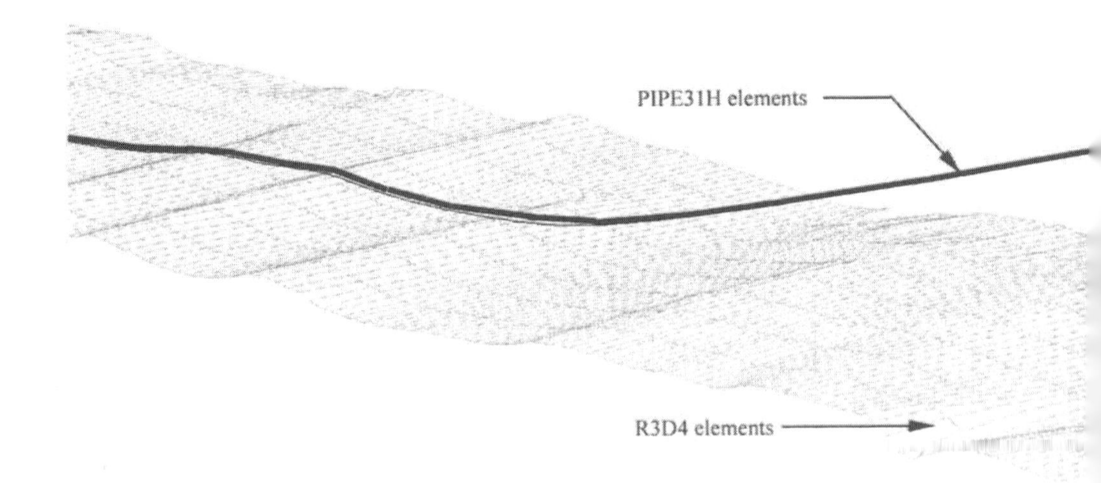

Figure 7.5 Element types used in the model.

The PIPE31H element

The 3D finite pipe element (Figure 7.6) used in the established model is the two node twelve degrees of freedom PIPE31H element. The element uses linear interpolation and therefore has a lumped mass distribution. The hybrid formulation makes the element well suited for cases with slender structures and contact problems, such as a pipe lying on the seabed.

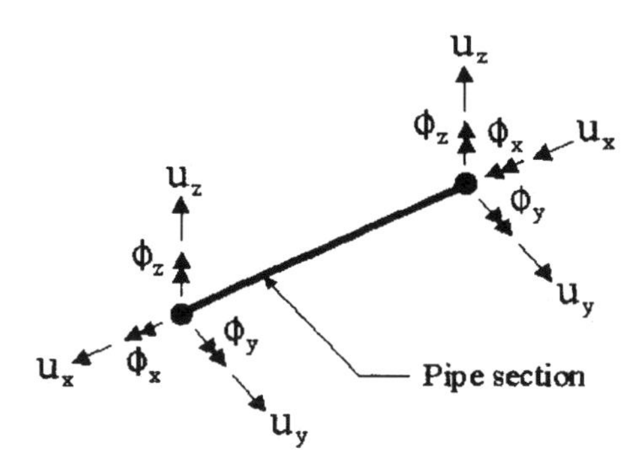

Figure 7.6 Two node twelve degrees of freedom 3D finite pipe element.

The hybrid elements are provided by ABAQUS for use in cases where it is numerically difficult to compute the axial and shear forces in the beam by the usual finite element displacement method. The problem in such cases is that slight differences in nodal positions can cause very large forces in some parts of the model, which, in turn cause large motions in other directions. The hybrid elements overcome this difficulty by using a more general formulation in which the axial and transverse shear forces in the elements are included, along with the nodal displacements and rotations, as primary variables. Although this formulation makes these elements more calculation intensive, they generally converge much faster when the pipe rotations are large and are more efficient overall in such cases.

The PIPE31H element is available with a hollow thin-walled circular section and supports the possibility for the user to specify external and/or internal pressure. The element can also account for changes in the pipe section area due to high axial straining of the pipe.

The R3D4 element

The four-node R3D4 rigid element (Figure 7.7) makes it possible to model complex surfaces with arbitrary geometry's and has been chosen when modeling the seabed topography. A very important feature of ABAQUS when modeling the seabed has been the possibility to smooth surfaces generated with the rigid elements, this leads to a much better representation of the seabed than the initial faceted surface.

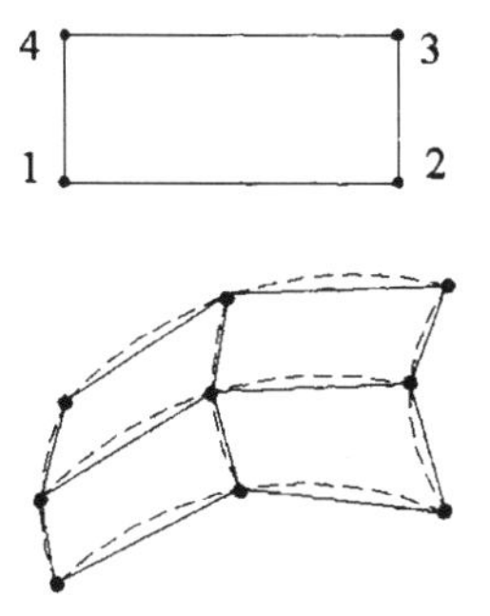

Figure 7.7 R3D4 rigid element, and example of smoothing of surface created with rigid elements.

The smoothing is done by ABAQUS creating Bèzier surfaces based on the faceted surface of the seabed formed by the rigid elements (Figure 7.7). The resulting Bèzier surfaces, unlike the faceted element surface will be smooth and have a continuous outward surface normal. The Bèzier surfaces will not match the faceted geometry of the rigid surface exactly, but the nodes of the rigid elements defining the seabed will always lie on the Bèzier surface. In addition, the user can specify the degree of smoothing in order to control the geometry of the smoothed surface.

In the established model the set of R3D4 elements defining the seabed is used as a so-called master surface for contact applications with the pipe elements. This means that a contact pair (pipe/seabed) is defined, and an interaction model is specified. This interaction model will typically consist of a seabed stiffness and friction definition.

7.5 Non-linearity and Seabed Model

The non-linear stress analysis used in the model contains up to three sources of non-linearity depending on strain level, change in geometry, and load situation:

- Material non-linearity;
- Geometric non-linearity;
- Boundary non-linearity (friction, sliding etc).

7.5.1 Material Model

The material model used is capable of representing the complete stress/strain relationship for the pipeline material, including non-linear plastic behavior (Figure 7.8).

In the elastic area the stress/strain relationship is governed by supplying the Young's modulus of the material. For the steel types commonly used as structural pipe steel, the Young's modulus will be temperature dependent. This can easily be accounted for in the model by numerically specifying the Young's modulus as a function of temperature.

The plastic behavior of the material is defined by specifying numerically the complete plastic stress/strain curve for the steel (e.g. from test data) in the material definition part of the input file. The temperature expansion coefficient of the material can also be defined as a function of temperature if necessary.

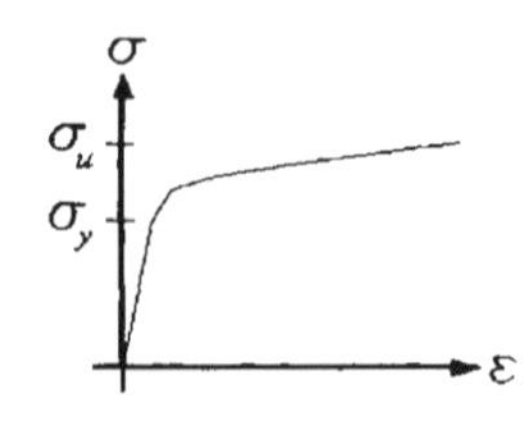

Figure 7.8 Stress/strain relationship.

7.5.2 Geometrical Non-linearity

Geometrical non-linearity is accounted for in the model. This means that strains due to change in the model geometry are calculated and that this stiffness contribution (stress stiffness) is added to the structure stiffness matrix. In addition, the instantaneous (deformed) state of the structure is always used in the next increment and updated through the calculation.

The latter feature is especially important when performing the dynamic analysis of a pipeline subjected to wave loading. By including geometrical non-linearity in the calculation, ABAQUS will use the instantaneous co-ordinates (instead of the initial) of the load integration points on the pipe elements when calculating water particle velocity and acceleration. This ensures that even if some parts of the pipeline undergoes very large lateral displacements (15-20 m.), the correct drag and inertia forces will be calculated on each of the individual pipe elements that make up the pipeline.

7.5.3 Boundary Conditions

Arbitrarily boundary conditions along the pipeline can be specified. If only a section of the total length of the pipeline is to be analyzed (e.g. between two successive rockdumpings), the user can simulate the stiffness of the rest of the pipeline with springs in each of the two pipe ends. If there are other constraints along the pipeline, these can be modeled by either fixing nodes or assigning springs to a number of nodes along the pipeline.

7.5.4 Seabed Model

The basis for constructing the 3-D seabed model is data from measurements of the seabed topography (bathymetric surveys) in the area where the pipeline is to be installed. From this information a corridor of width up to 40 m and lengths up to several kilometers is generated in the FE model to ensure a realistic environment when performing analysis of the pipeline behavior.

The seabed topography is represented with four node rigid elements that make it possible to model flat or complex surfaces with arbitrary geometrics. An advantage when modeling the three-dimensional seabed is the smoothing algorithm used by ABAQUS. The resulting smoothed surfaces; unlike the flat rigid element surfaces will have a continuous outward surface normal across element boundaries and model the seabed better. The smoothed surfaces will not match the faceted geometry of the rigid surface exactly, but the nodes of the rigid elements defining the seabed will always lie on the surface.

7.6 Validation of the Finite Element Model

A 1300-meter long pipeline section between two consecutive rock-berms was analyzed, to compare with the results of similar finite element models, Nystrøm et al. (1997), Tørnes et al. (1997). Below, the results from the water filled situation are given for the first 100 meters only, in order to get the details in the plots clear.

From the results (Fig. 7.9) it can be seen that the two in-place models (based on ABAQUS and ANSYS respectively) give close prediction of axial stress, strain, bending moment, and configuration on the seabed (Ose et al. (1999)).

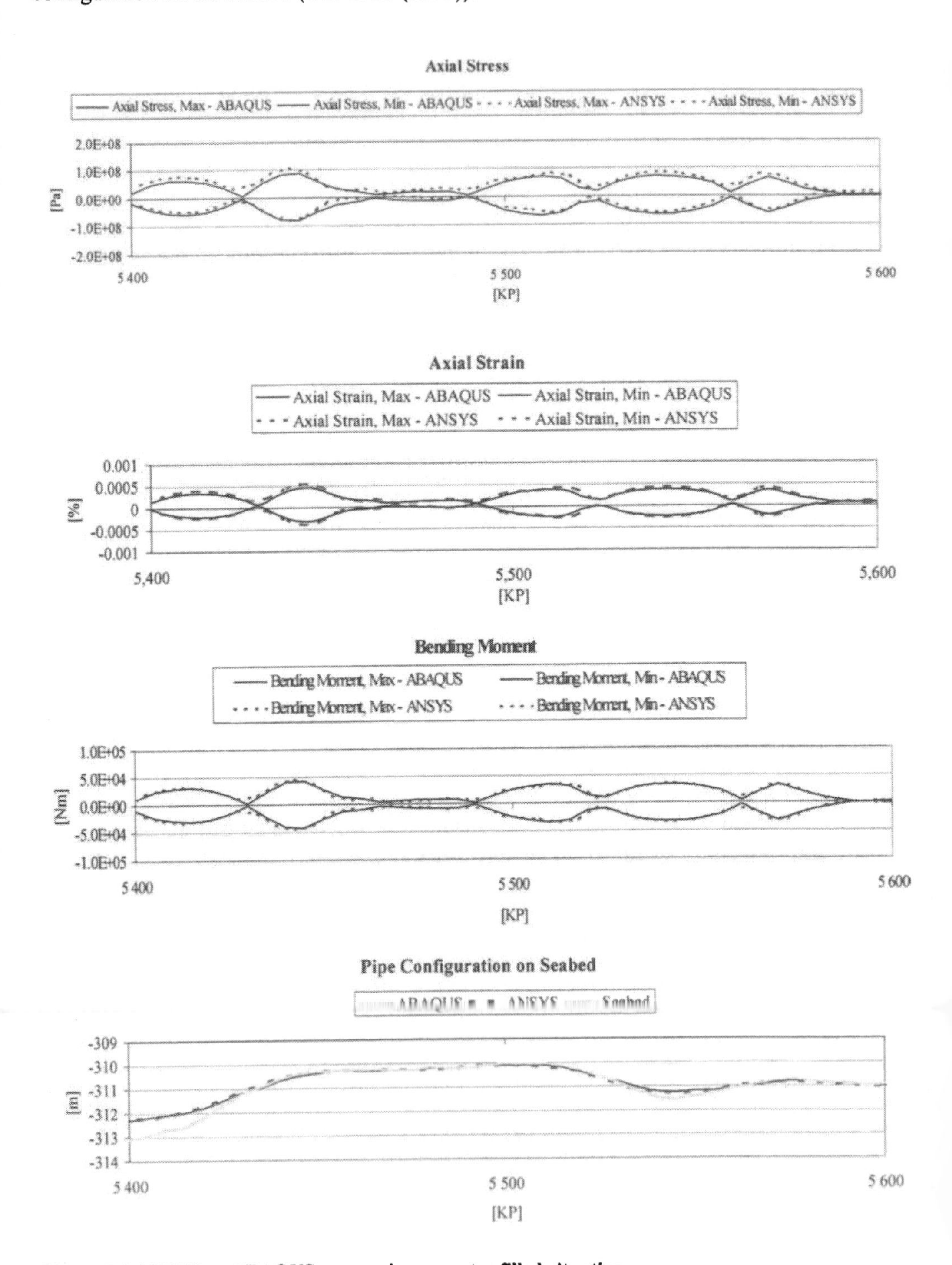

Figure 7.9 ANSYS vs ABAQUS comparison – water filled situation.

7.7 Dynamic Buckling Analysis

For a pipeline resting on a very uneven seabed, the vertical imperfections induce a more abrupt curvature than the horizontal imperfections created during laying. As a result, the critical buckling load required to lift the pipeline vertically is lower than the corresponding force needed to buckle horizontally. Although the initial movement occurs in the vertical plane, it was expected that as the pipe's contact force with the seabed diminishes the critical lateral buckling force would decrease and a lateral buckle could be initiated.

In order to investigate the pipeline behaviour on an irregular seabed as realistically as possible, a 3D pipeline - seabed FE model was developed where the 3D seabed bathymetry is imported as X,Y,Z data directly from a Digital Terrain Model (DTM) program. The 3D pipe-seabed model can then be presented as shown in Figure 7.10.

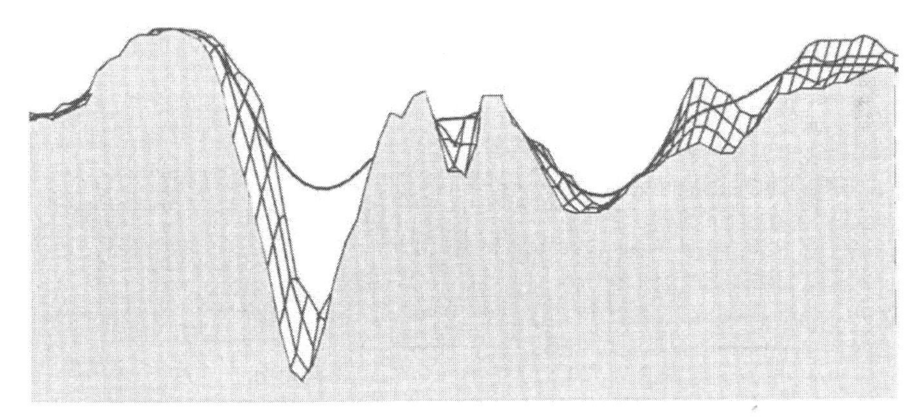

Figure 7.10 3D FEM in-place analysis model – typical details from an as-laid flowline simulation (Nystrøm et al., 1997).

The lateral buckle is a sudden loss of axial stability which results in a dynamic 'snap' movement. Nystrøm et al. (1997) investigated what actually happens with the pipe as it buckles in terms of stresses/strains, displacements, effective axial force etc. During the snap-through process, the flowline experiences acceleration and velocity leading to dynamic effects which may be significant. The transient flowline buckling behaviour was simulated using the general equilibrium equation for a dynamic system:

$$[M]\{\ddot{u}\} + [C]\{\dot{u}\} + [K]\{u\} = \{F(t)\} \tag{7.1}$$

where

$[M]$: structure mass matrix;

$[C]$: structure damping matrix;

$[K]$: structure stiffness matrix;

$\{u\}$: nodal displacement vector;

$\{\dot{u}\}$: nodal velocity vector;

$\{\ddot{u}\}$: nodal acceleration vector;

$\{F(t)\}$: time-dependent nodal force vector.

The dynamic response is caused by a change from potential energy to kinetic energy. To account for inertia effects for displacing the surrounding water, mass elements were coupled to each pipe node.

In non-linear finite element analysis, stress/strain results can be obtained from 8 positions around the pipe circumference as presented in Figure 7.11.

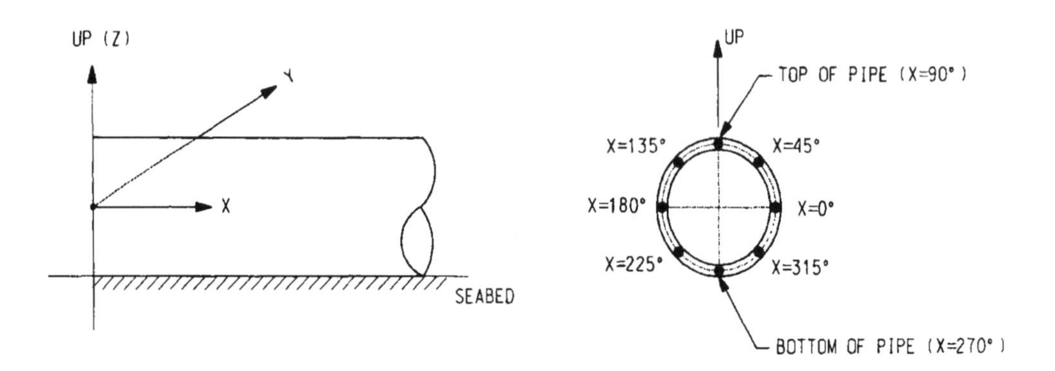

Figure 7.11 Definition of position *x* at pipe circumference.

Figure 7.12 shows the total axial strain from the 3D transient dynamic analysis at the buckle point plotted vs. time. At first, during heat up, the highest strains occur at the top of the pipe (Position 90°). However, as the pipeline has lifted off the seabed for a certain distance, the horizontal plane interacts and a lateral 'snap' buckle is initiated. The strain at Position 0° shows that at time 32 s a gradual change in the curvature is started and peaks at 37 s. The maximum total axial strain of 0.09 % occurs at Position 45° as a resultant of the bending at Positions 0°and 90°.

As the snap movement continues, the strains decrease and eventually as the 'snap' movement terminates at approx. 39 s the maximum strain at Position 45° stabilizes around 0.065 % strain. Hence, the flowline has experienced a 35 % higher total axial strain during the 'snap' movement than in the post-buckled configuration.

The strain experienced during the dynamic 'snap' is significantly higher than the post buckled static result. As the dynamic analysis results above show, the pipe buckles laterally in a continuous smooth manner which is due to the inertia effects. When the buckled pipe reaches its maximum curvature, the kinetic energy continues to displace the pipe laterally away from the buckle point in a wave form. Although this energy wave dissipates relatively quickly, it causes some relaxation of the curvature at the buckle center. As a result of this, the strains reduce as the pipe approaches a stationary condition again.

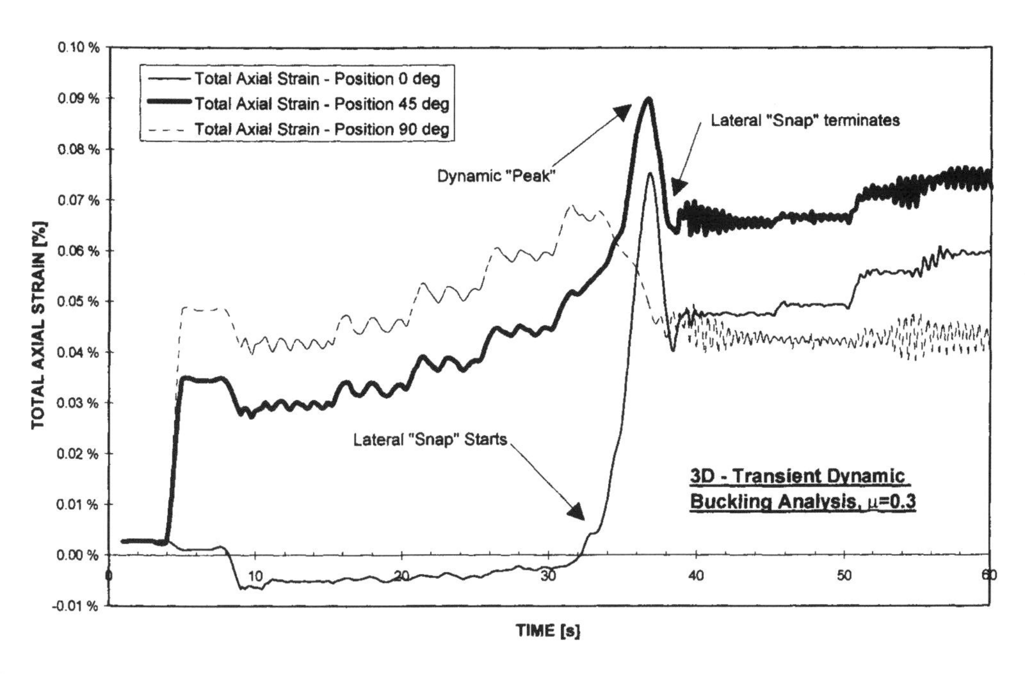

Figure 7.12 Total Axial Strain vs. Time (Nystrøm et al., 1997).

7.8 Cyclic In-place Behaviour during Shutdown Operations

Let's consider a flowline subjected to cyclic temperature and internal pressure due to shutdown events, Nystrøm et al. (1997). The cyclic capacity is defined as the maximum differential loads (temperature and pressure) between start-up and shutdown at which the structure "shakes down" to an elastic state. Some concepts can be mentioned:

a) Cycling leads to ductility exhaustion, causing fracture.
b) The pipeline remains intact, but the strains increase in each cycle until they are no longer acceptable.
c) The strains in each cycle decrease until, eventually the structure behaves purely elastic. This is called "shakedown".

The investigated flowline is assumed to be subjected to the following cyclic strains:

- 1 strain cycle during reeling-on and off the lay vessel's drum.
- 1 strain cycle during bending over the stinger and bending in the sag bend.
- 100 cycles of planned and unplanned shutdowns.

It is assumed that the shutdowns are constant, i.e. the same load range is used for the all cycles. These shutdowns are conservatively assumed to be content temperature goes from 130°C to ambient, i.e. ΔT=125°C, and internal pressure goes from 370 bar to 0 bar. A pipeline on a 3D seabed surface is analyzed having both a vertical and horizontal out-of-straightness. Internal pressure and temperature loads are applied up to full operational load. Thereafter, the internal pressure is reduced to 0 and the pipe wall temperature is gradually reduced to ambient.

The corresponding total axial strains are shown in Figure 7.13 as a function of lateral displacement. The maximum tensile and compressive strains occur during the first cycle and shake down to +0.14 % and -0.11 % respectively.

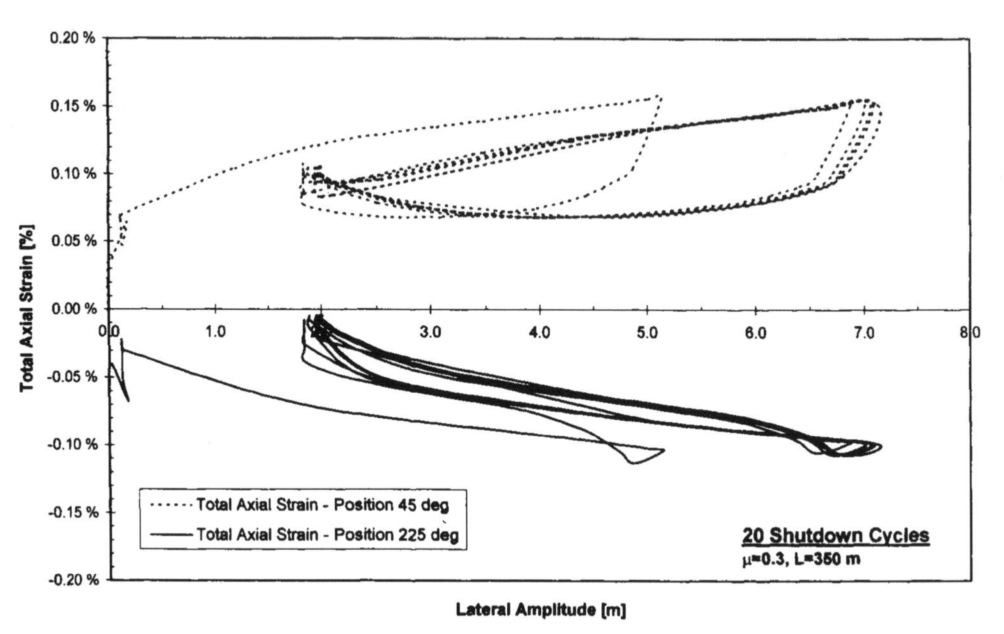

Figure 7.13 Total axial strain vs. lateral displacement (Nystrøm et al., 1997).

7.9 References

1. ANSYS Inc. (1998) "ANSYS, Ver. 5.5".
2. Cooker et al. (1991), "Concepts and Applications of Finite Element Methods".
3. Kristiansen, N.Ø., Tørnes, K., Nystrøm, P.R. and Damsleth, P.A., (1998) "Structura Modeling of Multi-span Pipe Configurations Subjected to Vortex Induced Vibrations" Proc. of ISOPE'98.
4. Nystrøm P., Tørnes K., Bai Y. and Damsleth P., (1997) "Dynamic Buckling and Cyclic Behavior of HP/HT Pipelines", Proc. of ISOPE'97.
5. Ose, B.A., (1998) "A finite element model for In-situ Behavior Simulation of Offshore Pipelines on Uneven Seabed Focusing On-Bottom Stability", A M.Sc. thesis performed under the author's supervision at Stavanger University for JP Kenny A/S, 1998.
6. Ose, B. A., Bai, Y., Nystrøm, P. R. and Damsleth, P. A., (1999) "A Finite Element Model for In-situ Behavior of Offshore Pipelines on Uneven Seabed and its Application to On-Botton Stability", Proc. of ISOPE'99.
7. Tørnes, K., Nystrøm, P. R., Damlseth, P. A., Sortland, H. (1997), "The Behavior of High Pressure, High temperature Flowlines on very Uneven Seabed", Proc. of ISOPE'97.
8. Tørnes, K., Nystrøm, P., Kristiansen, N.Ø., Bai, Y. and Damsleth, P.A., (1998) "Pipeline Structural Response to Fishing Gear Pullover Loads", Proc. of ISOPE'98.

Part II
Pipeline Design

Chapter 8 Expansion, Axial Creeping, Upheaval/Lateral Buckling

8.1 Introduction

Expansion may be induced by internal pressure and temperature. In recent years, more and more High Pressure and High Temperature (HP/HT) fields are developed using pipelines and steel catenary risers. In this design scenario, axial creeping, buckling in the form of upheaval movements and lateral movements or a combination of both, may take place.

In this Chapter, the basis for expansion analysis will be outlined first with a focus on effective force calculation. An analysis model is introduced for axial creeping of a high temperature flowline in clay soil condition. Simplified mathematical equations are presented for use in the upheaval buckling and lateral buckling designs. Finally HP/HT flowlines will be considered on uneven seabed, and their buckling behavior will be characterized as a combination of upheaval and lateral buckling modes: the flowline will uplift first and then move more laterally. There are many papers published on these subjects. Hence, we shall focus on the mathematical formulation that has actually been adopted in the current pipeline design practice. Some further readings include Ellinas et al. (1990), Kershenbaum et al. (1996), Nes et al. (1996) and etc.

8.2 Expansion

8.2.1 General Principle

A pipeline can be a single or piggybacked pipeline system, pipe-in pipe or bundled system. Its expansion is dependent on the temperature, pressure profile, pipeline self weight and friction forces. The expansion analysis will interface with:

- Tie-in Design;
- Lateral and Upheaval Buckling Assessment;
- Free-span Assessment;
- Crossing Design;
- Bottom Roughness/ Stress Assessment.

The maximum pipeline end expansion is calculated for the lower bound friction coefficient and the highest pipeline axial stresses are derived for the higher bound friction coefficient. The increased axial resistance from backfill along the route should be also accounted for. The input data for expansion analysis includes:

- Pipeline and Coatings Properties;
- Minimum Contents Weight;
- Temperature Profile & Pressure Profile;
- Geotechnical Data;
- Depth of Burial.

8.2.2 Single Flowlines

This section presents the theory behind developing the axial force profile along a flowline operating under temperature and pressure, which is being restrained by soil friction forces. The position of the virtual point is found which gives the position where the soil friction force (including backfill) equals the locked in effective force (where the axial strain is zero). The end expansion is the integration of the axial strain from this position to the end of the pipeline, and is therefore:

$$\Delta = \int_0^{VAP} (\varepsilon_{pressure} + \varepsilon_{temp} - \varepsilon_{soil})dx \tag{8.1}$$

The true axial force in the pipeline is:

$$P_{wall} = P_{eff} + \frac{\pi}{4}\Delta P(D-2t)^2 \tag{8.2}$$

The effective axial force in the partially restrained region is:

$$P_{eff} = -\mu\ \omega\ x \tag{8.3}$$

The effective axial force in the fully restrained region is:

$$P_{eff} = -\frac{\pi}{4}\Delta P(D-t)^2 + \sigma_H \upsilon A_s - E\alpha\Delta T A_s \tag{8.4}$$

where:

ε : axial strain;

P_{wall} : axial force in the pipe wall;

P_{eff} : effective axial force;

ΔP : pressure differential;

D : pipeline diameter;

t : wall thickness;

μ : axial friction coefficient;

ω : pipeline unit weight;

x : distance from free end;

σ_H : hoop stress;

υ : poisons ratio;

A_s : pipeline wall cross section area;

E : Youngs modulus;

α : thermal expansion coefficient;

ΔT : temperature differential.

8.3 Axial Creeping of Flowlines Caused by Soil Ratcheting

8.3.1 General

In recent years, continuous monitoring of some high-temperature subsea flowlines operated in the North Sea has shown that the flowlines are experiencing a gradual overall axial displacement towards the colder outlet ends. This net axial shift of the pipelines has proved to be critical for the integrity of the tie-in spools. In this Section, we shall present an analysis model for axial creeping of flowlines caused by soil ratcheting, based on Tørnes et al (2000).

8.3.2 Cyclic Soil/Pipe Interaction Model

In reality, the pipe-soil resistance in clay is far more complex than simple Coulomb friction. The frictional resistance is strongly non-linear with the maximum force (Fp) reached at a very small displacement and then gradually reduced to a residual value (Fr) at a relatively large displacement. This pipe-soil resistance is represented by curve 1 (dashed line) in Figure 8.1.

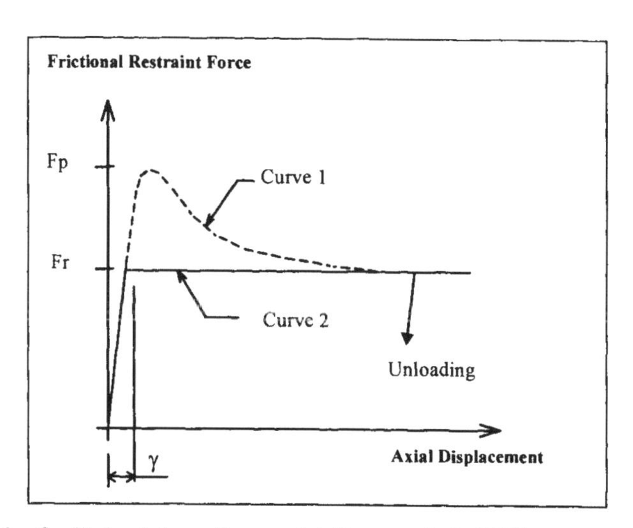

Figure 8.1 Schematic of soil/pipe interaction model (Tørnes et al., 2000).

A modified elasto-plastic Coulomb friction model as represented by curve 2 in Figure 8.1, may be used to describe the axial pipe-soil resistance in a FE model. Since the flowlines gradually creep in one direction at an approximate rate of 1m per year, this would in theory remove the peak force. Further, if the peak force is present, this would only be the case at very small displacements and thus only affect a limited part of the flowline.

The first part of the bi-linear Coulomb friction curve is linear. Initially the expanding pipeline stretches the surrounding soil and the soil resistance behaves purely elastic. When the "static" friction limit is reached, the pipe surface starts to slide relative to soil at the lower residual frictional value (plastic behaviour). The distance the pipe needs to travel before plastic sliding occur, is normally referred to as the frictional mobilisation length (γ).

8.3.3 Expansion of a "Long" Flowline with Free ends

In order to explain the creeping behaviour of the flowlines, it is therefore useful to first discuss pipeline expansion behaviour in general. First, consider an ideal pipeline with free ends laid on a perfectly flat seabed. The constant pressure and temperature applied to the pipeline, will cause it to expand and the frictional resistance of the seabed soil will be mobilised. The accumulation of frictional forces opposing the expansion will result in an effective compressive force in the pipeline which increases linearly from each pipeline end. A point is eventually reached where the strain caused by the frictional resistance exactly counterbalances the sum of the pressure and thermal strains. Beyond this point, further expansion movement is prevented and thus the pipeline is fully restrained.

This is shown in Figure 8.2 b), where the effective compressive force increases linearly from the end to a point where the pipeline becomes fully restrained. In the midsection of the pipeline, often referred to as the anchor zone, the effective compressive force will remain constant. Since a section of a "long" pipeline is fully restrained at full operating condition, it cannot undergo global axial creep.

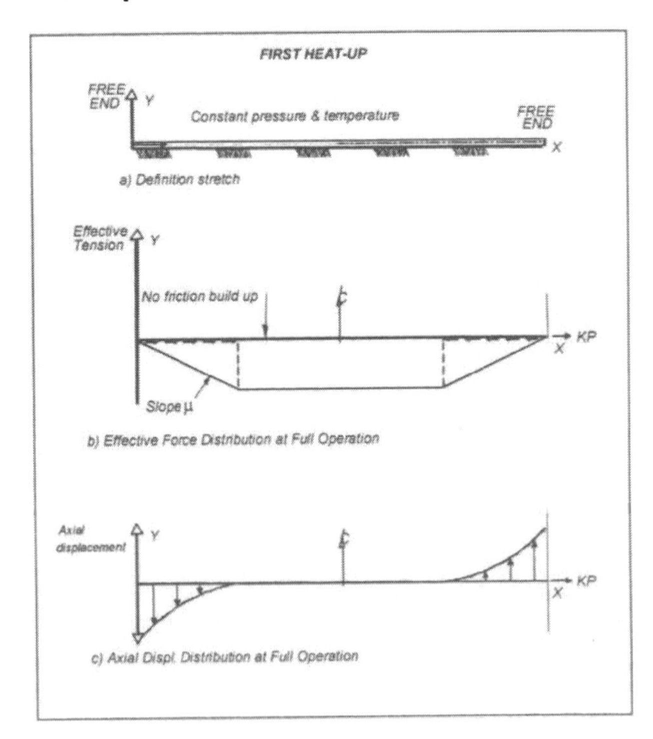

Figure 8.2 Expansion of a "long" flowline with free ends and constant pressure and temperature (Tornes et al., 2000).

8.3.4 In-situ Expansion Behavior of the Creeping Flowlines

The analysis results presented in this section were obtained by simulating the operational history of one of the production flowlines under a cyclic temperature profiles and pressure history. The effective force distribution history as the flowline is gradually heated up and cooled down during the first cycle, is presented in Figure 8.3. It shows how the negative peaks on the effective force plot will not initially be located at midline. Instead, the peak moves from the hot end towards the middle of the flowline as the heat propagates towards the cold end.

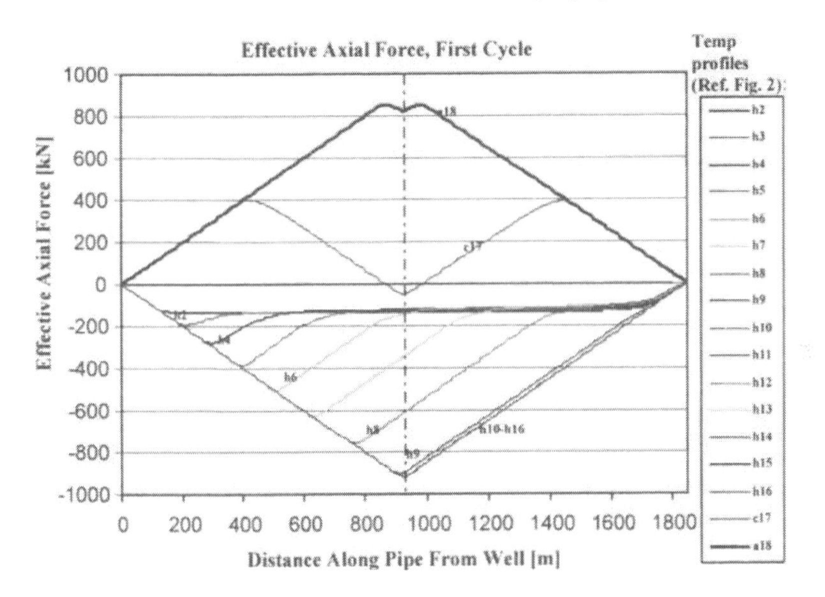

Figure 8.3 Schematic of soil/pipe interaction model (Tørnes et al., 2000).

Only at the moment friction has been mobilised along the entire length, will the maximum compressive force occur at midline. As can be seen from Figure 8.3, this occurs approximately when temperature profile number 9 is applied to the flowline. The corresponding axial displacement history from the FE analysis is shown in Figure 8.4. A schematic of a temperature profile and the corresponding effective force and axial displacement distributions at one instant during the heating up process, i.e. prior to reaching the full operating conditions, is shown in Figure 8.5.

Since the thermal strain is proportional to the temperature in the line, the tendency to expand will decrease gradually from the hot end to the point where the pipeline is at ambient temperature. As the pipeline expands towards the hot end, the accumulation of frictional force is built up away from this end of the flowline. This means that the further away from the hot end, the more resistance (higher effective force due to the accumulation of friction force) against the thermal expansion.

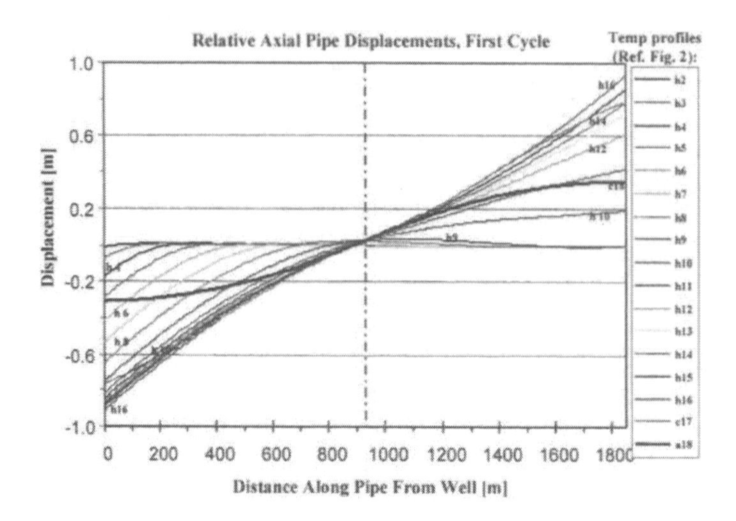

Figure 8.4 Axial pipe displacements along the flowline through the temperature cycle 1 (Tørnes et al., 2000).

The analysis results showed that this heating up process causes the entire flowline to shift towards the cold end compared to a case where it is heated up uniformly to its operating conditions (i.e. without heating it up gradually from the hot towards the cold end). Furthermore, the entire shift occurs before the maximum effective force is located at midline, i.e. before the frictional resistance is mobilised along the entire line.

The reason for this is as follows: Figure 8.5 shows how the pipeline expands in both directions from either side of the stationary point at which the maximum effective force occurs. Since this point gradually moves from the hot end towards the midline, this means that all the points on the hot half of the pipeline will first experience an expansion movement towards the cold end followed by movement in the reversed direction towards the hot end. The latter occurs as soon as the point of maximum effective force passes the point in question.

8.4 Upheaval Buckling

8.4.1 General

Global buckling is a beam-mode buckling that is characterized by an amplitude and wavelength. An upheaval buckling may take place for a buried pipeline that up-lifts due to excessive expansion.

8.4.2 Analysis of Up-lifts

A buried pipeline can form a raised loop upward out of the seabed. The buckles can overstress the pipe wall. It is assumed that the pipeline has an imperfection in the form of plastic deformations in combination with foundation imperfections. The analysis method is based on Pedersen and Jensen (1988), who presented a simplified analysis of the pipeline uplifts from the bottom of the trench. The pipeline is assumed to buckle elastically. Figure 8.6 illustrates the Pedersen and Jensen model that has been derived from an elastic beam theory for the imperfect pipeline uplifted in the x-w coordinate system.

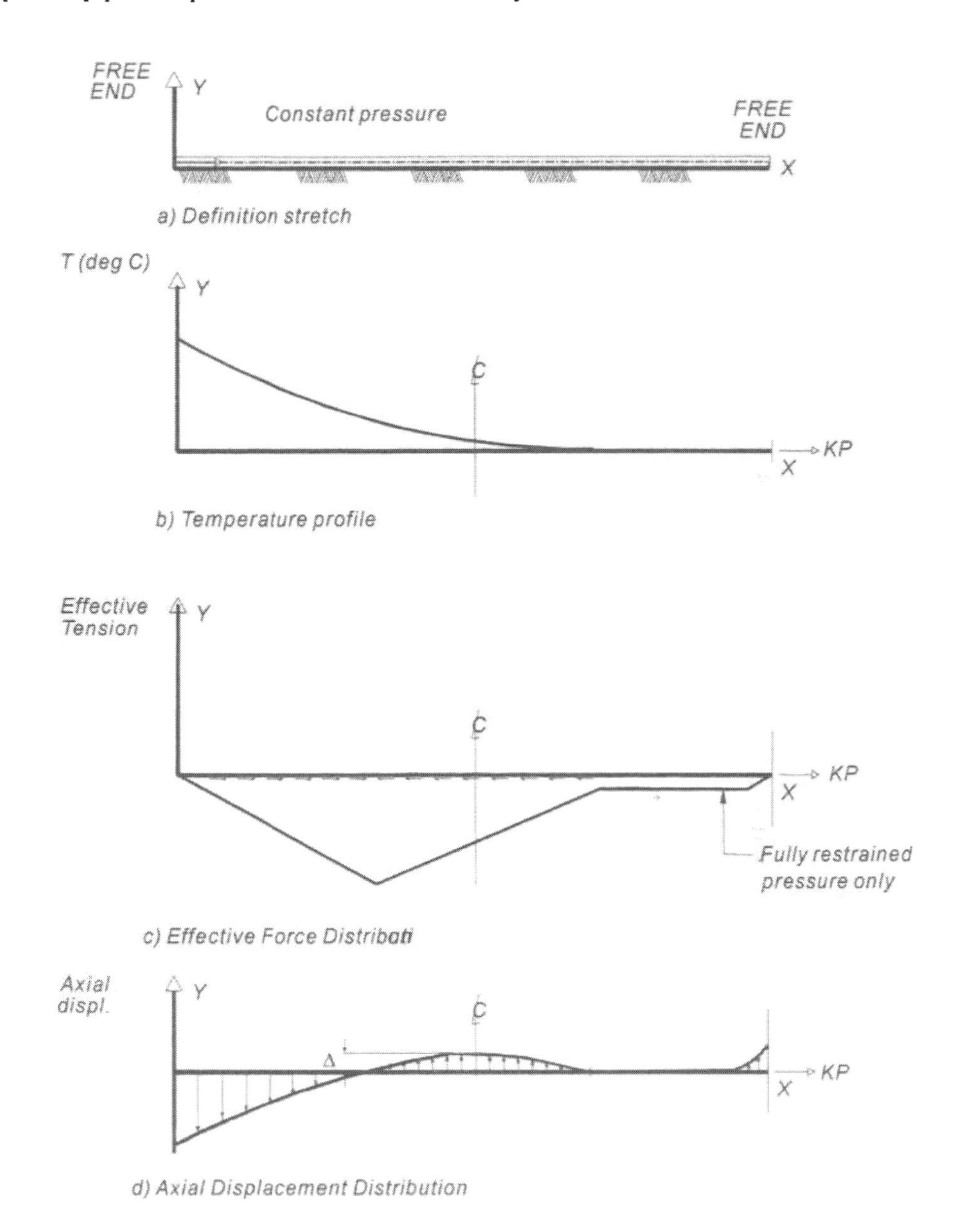

Figure 8.5 Expansion of a flowline with a free end & a temperature gradient (Tørnes et al., 2000).

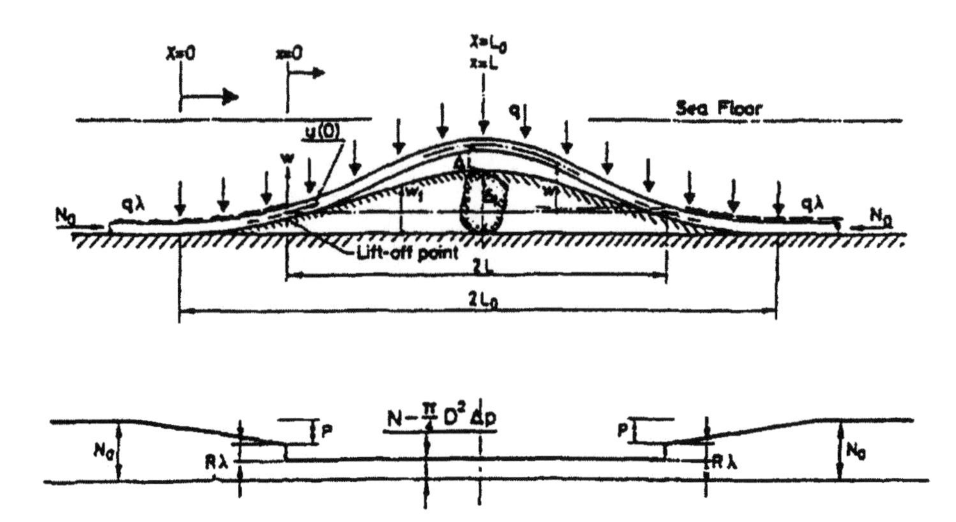

Figure 8.6 Uplifted pipelines (top), and variation in compressive axial force (bottom) (Pedersen & Jensen, 1988).

A numerical procedure for the calculation of uplift profile is:

1) Calculate the k value associated with a given free span (2L) from Equation (8.5) and then predict the axial force N using Equation (8.6), where q is the total pipe weight per unit, q_p is the weight due to pipe plastic deformation, and L_0 is half wave length, L is half uplift length, respectively .

$$(\frac{\alpha}{k^3}+\frac{\gamma}{k})\sin kL-(\frac{\beta}{k^2}-\kappa)\cos kL+\frac{1}{k^2}(-\alpha L+\beta)=0 \tag{8.5}$$

$$N=EIk^2 \tag{8.6}$$

where

$$\alpha=\frac{q+q_p}{E\cdot I},$$

$$\beta=\frac{q_p\cdot(3L-2L_0)+3\cdot q\cdot L}{3\cdot E\cdot I},$$

$$\kappa=\frac{q_f}{6E\cdot I}(L_0-L)^2 L \quad \text{for } L\le L_0,$$

or $\kappa=0$ for $L>L_0$.

2) Determine the deflection configuration from Equation (8.7);

$$w(x) = A\cos kx + B\sin kx + \frac{1}{k^2}\left(-\frac{1}{2}\alpha x^2 + \beta x + \frac{\alpha}{k^2} + \gamma\right) \tag{8.7}$$

where

$$A = -\left(\frac{\alpha}{k^4} + \frac{\gamma}{k^2}\right),$$

$$B = -\frac{\beta}{k^3} + \frac{\kappa}{k}.$$

3) Calculate the associated pressure–temperature combination from Equations (8.8) and (8.9); Figure 8.7 explains the limiting permissible temperature has been taken as the minimum temperature on the U-shaped curve in the temperature-buckling wavelength plane (Pedresen and Jensen, 1988).

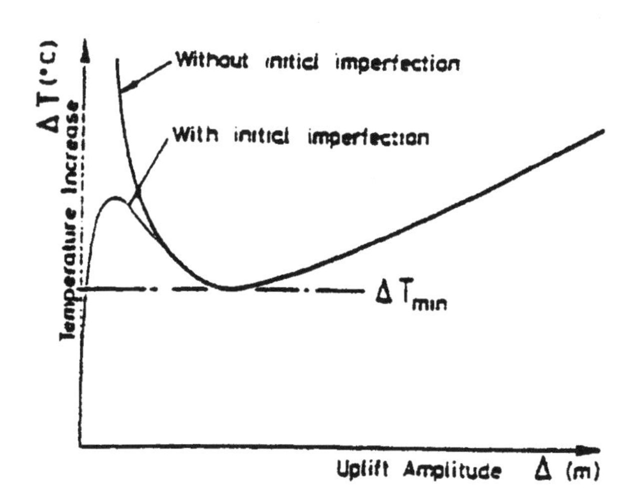

Figure 8.7 Minimum temperature increase ΔT_{min} for equilibrium curves associated with uplifted pipelines (Pedersen & Jensen, 1988).

$$N_0 = N + \left(q\lambda EA_s \int_0^L \{(w')^2 - (w_f')^2\}dx - (q\lambda L)^2\right)^{1/2} \tag{8.8}$$

$$N_0 = \bar{\alpha} EA_s \Delta T - \nu \frac{\pi}{2} D^2 \Delta P + \frac{\pi}{4} D^2 \Delta P \tag{8.9}$$

where N_0 is effective axial compression away from the buckle, N is compression in the uplift buckle, D is steel pipe outside diameter, ΔT is the temperature change, ΔP is the pressure difference between the product and the pressure at the seabed, A_s is the pipe steel cross sectional area, $\bar{\alpha}$ is thermal coefficient, ν is poisson's ratio, w' is the differential deflection, and w_f' is the differential initial deflection, respectively.

8.4.3 Upheaval Movements

To make an initial assessment of the pipelines propensity for global buckling and to derive the driving force under the design and operational conditions, a traditional upheaval buckling assessment can be performed, adopting the method given by Palmer et al (1990).

The maximum required download is found by the above relationship for a (worst case) prop support, which has the highest natural occurring curvature in the pipeline and is a factor of flowline axial force and bending stiffness:

$$\omega = 0.064\left(P_{eff} f_a\right)^2 \frac{H}{EI} \tag{8.10}$$

where

ω : total required download;

P_{eff} : effective axial force;

f_a : axial force factor;

H : prop height;

EI : flowline bending stiffness.

An analysis may be carried out to calculate the total required download for a series of prop heights, which accounts for the true axial force in the flowline by the build up soil friction forces along the route. A safety factor of 1.2 is normally applied to the axial force. The required extra download is then calculated as:

$$\omega_{net} = \omega - \omega_{PL} - \omega_{bf} \tag{8.11}$$

where:

ω_{net} : net required download (rock dump);

ω_{PL} : pipeline unit weight;

ω_{bf} : backfill unit weight.

The calculation is performed for the design and operational temperature and pressure. In the calculation, the effect of backfill is considered, for both the axial force and download components. The net required download finally can be converted into an equivalent soil or rock dump cover based on the geotechnical properties entered, as follows:

$$\omega_{net} = \gamma D_o z\left(1 + \frac{fz}{D_o}\right) \tag{8.12}$$

where:

γ : submerged weight of the soil and rocks dumped on the pipeline;

D_o : pipeline overall diameter;

z : the cover from the top of the pipe to the surface of the soil above the pipe centreline (backfill depth);

f : an uplift coefficient determined experimentally, generally about 0.7 for rock and 0.5 for sand, but occasionally much smaller in loose sand.

8.5 Lateral Buckling

8.5.1 General

Unburied pipe on the seabed will buckle laterally instead of vertically unless the lateral friction coefficients are very high. The resistance to sideways movement is the submerged weight multiplied by lateral friction coefficient. The driving force for lateral buckling is the compressive force induced by operational temperature and pressure. A lateral buckling analysis may be carried out based on Hobbs (1984), Hobbs and Liang (1989). The effective axial force is calculated for the operating condition defined within a design envelope (shown by the series of graphs) for any of the four possible modes presented.

8.5.2 Lateral Buckling of Straight Line on Flat Seabed

The parameters and equations used for the determination of the lateral buckling are presented as below. The required effective axial force to buckle can be expressed as:

$$P(z) = \frac{k_1 \cdot E \cdot I}{(L(z))^2} + k_3 \cdot \mu \cdot \omega \cdot L(z) \cdot [(1 + \frac{k_2 \cdot E \cdot A \cdot \mu_l^2 \cdot \omega \cdot (L(z))^5}{\mu (E \cdot I)^2})^{0.5} - 1) \tag{8.13}$$

The buckle amplitude is

$$y(z) = \frac{k_4 \cdot \mu_l \cdot \omega \cdot L(z)^4}{E \cdot I} \tag{8.14}$$

The force left in the buckle is

$$P_{buck}(z) = \frac{k_1 \cdot E \cdot I}{(L(z))^2} \tag{8.15}$$

The maximum moment induced in the buckle is

$$M = k_5 \cdot \mu_l \cdot \omega \cdot L(z) \tag{8.16}$$

where;

- z : the location on the pipe;
- L(z) : buckle length;
- μ_l : the lateral seabed friction coefficient;
- μ : the axial seabed friction coefficient;
- ω : pipeline submerged unit weight;
- A : cross section area of pipe;
- E : Youngs Modulus;
- K_n : buckle constant.

Values for the buckling constants are given in the following table.

Table 8.1 Buckling constants for lateral buckling mode (refer to Figure 1.6).

Buckle Mode	K_1	K_2	K_3	K_4	K_5
1	80.76	6.391e-5	0.5	2.407e-3	0.0694
2	39.48	1.743e-4	1.0	5.532e-3	0.1088
3	34.06	1.668e-4	1.294	1.032e-2	0.1434
4	28.20	2.144e-4	1.608	1.047e-2	0.1483

The maximum amount of pipe which can feed into a buckling occurring at the pipeline end is equal to the unrestrained expansion of the line. However, the force in the buckled section (P) would have to be zero. The increase in length of pipe Δl in the buckled section from the unbuckled state can be determined as:

$$\Delta l = \frac{(P_{buck} - P)\cdot L}{A\cdot E} \tag{8.17}$$

Expansion of the adjacent slipping length (L_s) as force falls from P_0 to P at the start of the buckle can be expressed as:

$$\Delta l_s = \frac{(P_{buck} - P)\cdot L_s}{A\cdot E} \tag{8.18}$$

where

$$L_s = \frac{(P_{buck} - P)}{\mu\cdot\omega} \tag{8.19}$$

Total expansion into the buckle is,

$$\Delta L = \Delta l + \Delta l_s = \frac{(P_{buck} - P)}{A\cdot E}\cdot\left(L + \frac{(P_{buck} - P)}{\mu\cdot\omega}\right) \tag{8.20}$$

For the buckling modes 2 and 4, *L* is replaced by *2L* in the above equation as they are double buckle modes. The above assessment is sufficient for conceptual design.

8.6 Interaction between Lateral and Upheaval Buckling

A finite element analysis may be applied to gain an understanding of the complex mechanism between the vertical and horizontal mode of buckling. This included a number of 2D and 3D FE simulations using a typical 3D seabed model (Nystrom et al., 1997).

3D buckling model

In the first examples, a flat seabed with one realistic vertical seabed imperfection was used (1.8 m high and 50 m long, see Figure 8.7). A 10" oil flowline with a residual lay tension of 50 kN and no internal pressure is assumed. The seabed is assumed to be flat in the lateral plane, however a pipe imperfection of W_{om}/L_o=0.001 was introduced in the pipe horizontal plane. In order to determine the effect of the restraining force, two friction coefficients were checked (0.3 and 1.0). Temperature loads were applied to the model, in addition to the residual lay tension.

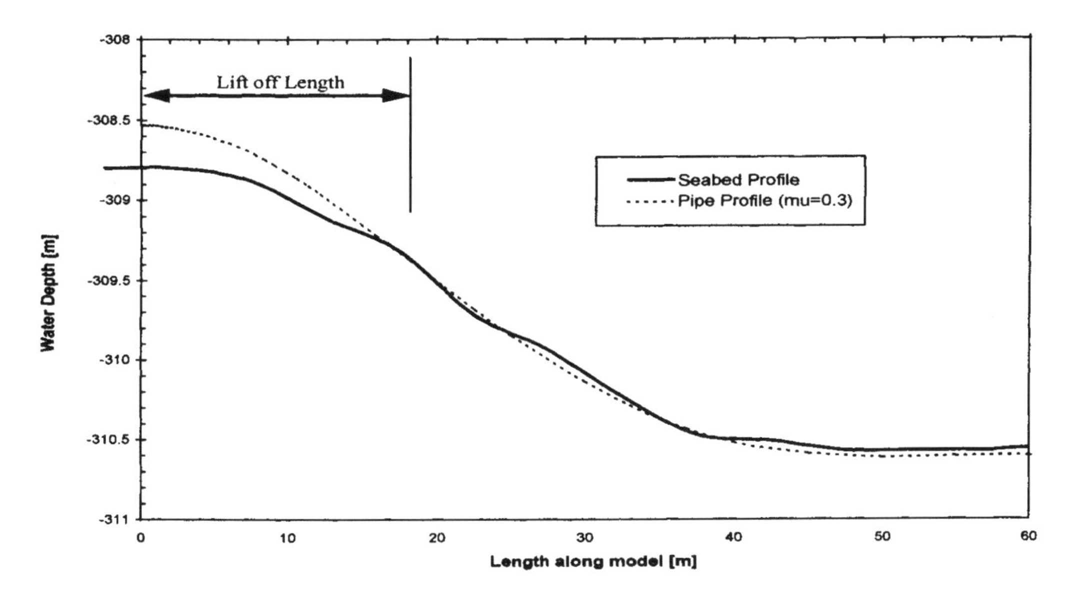

Figure 8.8 Vertical pipe-seabed profile (Half mode) (Nystrom et al., 1997).

Figure 8.8 shows that, for a friction coefficient of 0.3, the flowline starts to lift off the seabed at approximately 13°C. Little horizontal movement has yet occurred. At 16°C, the vertical displacement reaches 0.25 m but, suddenly drops down again to zero. In the same instant, the pipe 'snaps' laterally by 1.85 m to a new equilibrium configuration. As stated previously the pipe is now fallen back onto the seabed again.

The same effect can be seen for the higher friction case (μ=1.0), the differences being that the vertical movement starts at 15°C, the pipe lifts slightly higher, the lateral buckling temperature is approx. 21°C and the resulting lateral displacement is 1.4 m.

Even though the higher friction case causes the pipeline to buckle laterally at a higher temperature, Figure 8.9 shows that the flowline buckles at approximately the same effective force for both friction cases. This is an important finding for the understanding of the 3D buckling behaviour. On an uneven seabed, therefore, it appears that the critical lateral buckling force is little affected by the lateral friction resistance. This phenomenon can be explained by the fact that for a pipeline resting on an uneven seabed, the vertical imperfections can be significantly larger than the horizontal imperfections. As the flowline initially 'buckles' vertically, it looses lateral restraint (less contact, less lateral restraint) where upon it interacts with the horizontal mode and buckles laterally. Specifically, the pipe starts to lift vertically at approximately 170 kN (Ref. Figure 8.9) and starts to loose lateral frictional restraint. As a result of this reduction in lateral restraint, the lateral buckling force also decreases. The critical lateral buckling force is reached at an effective axial force of approx. 200 kN. Because the pipe now has reached the point of axial instability in the horizontal plane, the path of least resistance is chosen, and a lateral 'snap' to a new equilibrium configuration occurs. It should be noted that, in theory, not all pipes which lift off vertically will buckle laterally. A situation could occur where the flowline operational loads only are capable of lifting a section of the pipeline. In this case the vertical mode would still be governing.

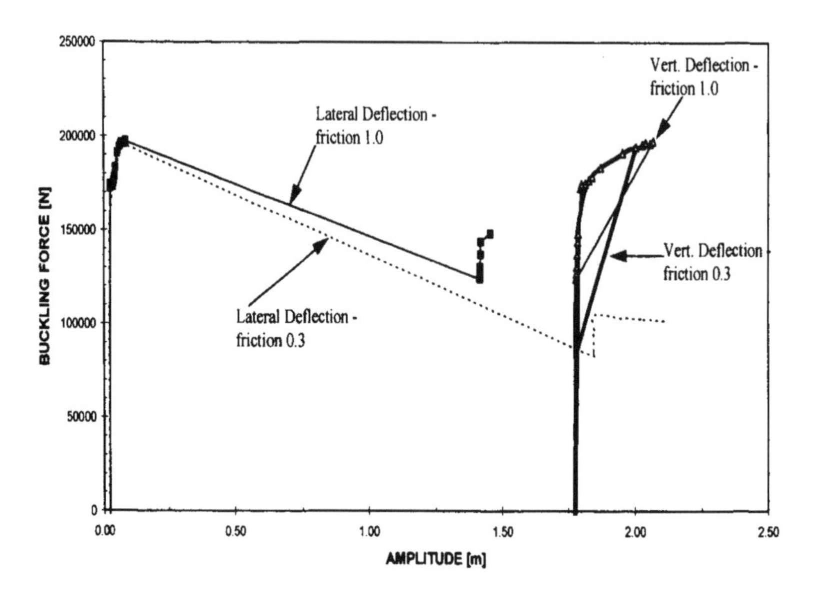

Figure 8.9 Static 3D analysis, buck force vs. deflection (Nystrom et al., 1997).

8.7 References

1. Ellinas, CP, Supple, WJ and Vastenholt, H (1990): "Prevention of Upheaval Buckling of Hot Submarine Pipelines by Means of Intermittent Rock Dumping", Proceedings 22nd Offshore Technology Conference, pp 519-528.
2. Hobbs, RE (1984): "In-Service Buckling of Heated Pipelines", Journal Transport Engineering, pp 178-179.
3. Hobbs, RE and Liang, F (1989): "Thermal Buckling of Pipelines Close to Restraints", Proceedings 8th Int. Conf on Offshore Mechanics and Arctic Eng, pp 121-127.
4. Kershenbaum, NY, Harrison, GE and Choi, HS (1996): "Subsea Pipeline Lateral Deviation Due to High Temperature Product", Proceedings 6th Int. Offshore and Polar Eng Conf, Vol 2, pp 74-79.
5. Nes, H, Sævik, S, Levold, E. and Johannesen, A., (1996): "Expansion Control Design of Large Diameter Pipelines", Proceedings 15th Int. Conf on Offshore Mechanics and Arctic Eng, Vol 5, pp 279-285.
6. Nystrøm P.R., Tørnes K., Bai Y. and Damsleth P.A., (1998): "3D Dynamic Buckling and Cyclic Behavior of HP/HT Flowlines", ISOPE-1997.
7. Pedersen, P. T. and Jensen, J.J, "Upheaval Creep of Buried Heated Pipelines with Initial Imperfections", Marine Structures, 1988.
8. Palmer, A., Ellinas, C., Richards & Guijt (1990): "Design of Submarine Pipelines against Upheaval Buckling", OTC 6335.
9. Tørnes, K., Ose, B.A., Jury, J. and Thomson, P. (2000): "Axial Creeping of High Temperature Flowlines Caused by Soil Ratcheting", OMAE-2000.

Part II

Pipeline Design

Chapter 9 On-bottom Stability

9.1 Introduction

On-bottom stability calculations are performed to establish requirements for pipeline submerged mass. The required pipeline submerged mass will have a direct impact on the required pipelay tensions, installation stresses and the pipe configuration on the seabottom. From the installation viewpoint, especially where spans are not a concern, the priority is to minimize the required pipeline submerged mass.

On-bottom stability calculations shall be performed for the operational phase and for the installation phase. For the operational phase, a combination of 100 year wave loading + 10 year current loading is to be checked, as well as 10 year wave loading + 100 year current loading. The pipeline is filled with content at the expected lowest density when considering the operational phase.

For the installation phase (temporary phase) the recurrence period may be taken as follows:

- Duration less than 3 days:
  (i) The environmental parameters for determination of environmental loads may be established based on reliable weather forecasts.
- Duration in excess of 3 days:
  (i) No danger of loss of human lives. A return period of 1 year for the relevant season may be applied.
  (ii) Danger for loss of human lives: The parameters may be defined with a 100-year seasonal return period.

However, the relevant season may not be taken less than 2 months. If the empty pipeline is left unprotected on the seabed over the winter season, combinations of 10 year current + 1 year wave, and 1 year current + 10 year wave loading will be checked. The pipeline is assumed to be air filled for the on-bottom stability analysis when considering the installation phase. For the installation condition, a minimum specific gravity of 1.1 is required.

9.2 Force Balance: the Simplified Method

The lateral pipeline stability may be assessed using two-dimensional static or three-dimensional dynamic analysis methods. The dynamic analysis methods allow limited pipe movements or check of structural strength, the acceptance criteria for dynamic analysis is explained in chapter 9.3, which is a summary of Chapter 4. The static analysis method may be expressed by Eq. (1.3) in Chapter 1.

9.3 Acceptance Criteria

9.3.1 Allowable Lateral Displacement

The selection of the allowable lateral pipeline displacement shall be based on several factors, such as:

- National regulations.
- Distance from platform or other restraint.
- Sea bed obstructions.
- Width of surveyed corridor.

If no further information is available, then the following may be used for the allowable maximum lateral displacement of the pipeline in the operational condition:

- Zone 1 (over 500 meters from an installation): 20 meters.
- Zone 2 (less than 500 meters from an installation): 0 meters.

These criteria can be relaxed or replaced if other relevant criteria (e.g. limit-state based strength criteria) are available.

9.3.2 Limit-state Strength Criteria

General

Limit-state based strength criteria have been discussed by Bai & Damsleth (1997), who have presented potential failure modes and design equations as well as design experience on detailed design projects. Details are given in Chapter 4.

9.4 Special Purpose Program for Stability Analysis

9.4.1 General

There are several analysis methods available on which to base pipeline stability design. Three different methods are used by pipeline industry:

1) Dynamic analysis
2) Generalized stability analysis
3) Simplified stability analysis

The choice of the above analysis methods is dependent on the degree of detail required in results of the design analysis.

1) Dynamic analysis involves a full dynamic simulation of a pipeline resting on the seabed, including modeling of soil resistance, hydrodynamic forces, boundary conditions and dynamic response. It may be used for detailed analysis of critical areas along a pipeline, such as pipeline crossings, riser connections etc. where a high level of detail is required on pipeline response or for reanalysis of a critical existing line.

Software: PONDUS and AGA (1993) Software

2) The Generalized stability analysis is based on a set of non-dimensional stability curves, which have been derived from a series of runs with a dynamic response model.

Software: PIPE

3) The Simplified stability analysis is based on a quasi-static balance of forces acting on the pipe, but has been calibrated with results from the generalized stability analysis. The method generally gives pipe weights that form a conservative envelope of those obtained from the generalized stability analysis.

SOFTWARE: Purpose made spreadsheets (EXCEL, LOTUS 1-2-3)

A short description of the two computer programs, PONDUS and PIPE, are given below.

9.4.2 PONDUS

PONDUS is a computer model, which computes the dynamic response of a pipeline on the seabottom due to wave and current excitation in the time domain. The response of the pipeline is non-linear due to non-linear hydrodynamic forces and non-linear interaction between the pipe and soil.

A 100 meter long pipeline section subjected to wave and current loading is modeled. The pipeline is unconstrained at its free ends to simulate an infinitely long pipeline resting on a flat seabed. The purpose of this model is to determine pipeline stability in terms of displacement, regardless of axial constraints and boundary effects. The waves are represented by a 3 hours storm with a build-up time of half an hour.

The attributes of PONDUS are summarized below:

Pipe structure:

- straight pipeline on horizontal seabottom (no free spans)
- two degree of freedom (lateral deflection & rotation about global vertical axis) at each nodal point
- variable pipe mechanical and geometric properties along the pipe
- variable end conditions (free, fixes or spring)
- constant axial force in space along the pipe

- tension effects: optional (the pipe may have an initial axial force which may increase due to lateral deflection)
- pressure effects (the pressure will contribute to the effective axial force and
- internal pressure may give tensile stress along the pipe axis)
- temperature effects (increased temperature may give compressive stress along the pipe axis)
- nodal linear springs and nodal masses may be specified
- no stiffness contribution from concrete coating

Soil force:

- simple Coulomb friction model
- comprehensive soil models for sand and clay
- soil properties may vary along the route
- soil force in pipe axis direction is not considered

Hydrodynamic force (horizontal and lift):

- several force models available
- relative velocity is considered (optional)
- regular and irregular waves with a user defined direction relative to the pipe. Time series for velocity, acceleration and coefficients along the pipe for irregular waves must be generated in separate modules (WAVESIM, PREPONDUS) and stored on file
- constant current (normal to the pipe) in time. Possible modifications due to boundary layer effects may be included in the value for current velocity

Specified force:

- a distributed force may be specified, constant along the pipe but varying in time (linear- or sine-functions)

Numerical method:

- finite element formulation with straight beam elements with two degree of freedom at each node (rotation and transverse displacement)
- small deflection theory (small rotations) for the beam elements with linear material behavior (no updating of nodal co-ordinates)
- geometric stiffness is included
- solution in time domain using the Newmark and incremental formulation
- Rayleigh damping may be specified for the pipe
- damping in the linear range of the soil may be specified
- concentrated mass formulation
- constant time step (user specified) with automatic subdivision in smaller steps in highly non-linear interval (if required)

- simple trapezoidal integration for the distributed loading along the beam elements (nodal forces only, nom moments)

9.4.3 PIPE

PIPE is based on the use of non-dimensional parameters, which allow scaling of the environmental load effects, the soil resistance and the pipeline response (lateral pipe displacement).

Three options are available for the description of the long-term wave environment:

1) scatter diagram of significant wave height, Hs and the peak period, Tp
2) analytical model for the long term distribution of Hs and Tp
3) Weibull distribution based on the definition of Hs and Tp for two return periods

Wave directionality can be specified for all options. The long-term wave elevation data are transformed to water particle velocity data. Together with the current data, these velocities form the basis for the description of the long-term hydrodynamic loading process and are used by the program for the pipeline stability design according to the specified design criteria.

Two principally different design checks are made for the stability control of the pipeline:

1) The first check is relevant for an as laid on-bottom section (not artificially trenched or buried). For a pipeline on sand soil, the design control is based on a specified permissible pipeline displacement for a given design load condition (return period). The basis for the design process is a generalized response database generated through series of pipeline response simulations with PONDUS. For the on-bottom design check on clay, a critical weight is calculated to fulfill the "no breakout criteria". The critical pipe weight has been found through series of pipeline response simulations to be the weight where the pipe is "dynamically" stable (due to penetration) for the given external load level.

2) The second design check makes an absolute static stability calculation of a pipeline trenched and or buried in the soil, sand or clay. The design check is based on static equilibrium between the hydrodynamic design loads and the soil capacity.

9.5 Use of FE Analysis for Intervention Design.

9.5.1 Design Procedure

Figure 9.1 shows the Flow-chart for seabed intervention design procedure.

9.5.2 Seabed Intervention

There are several types of seabed intervention. Examples of seabed intervention are rock dumping, trenching, burying and pre-sweeping. The purpose of seabed intervention design is to ensure that the pipeline maintains structural integrity throughout its design life. It is then a premise that a good work has been done when the design criteria is established and compared with the simulated pipeline response to a history of loads.

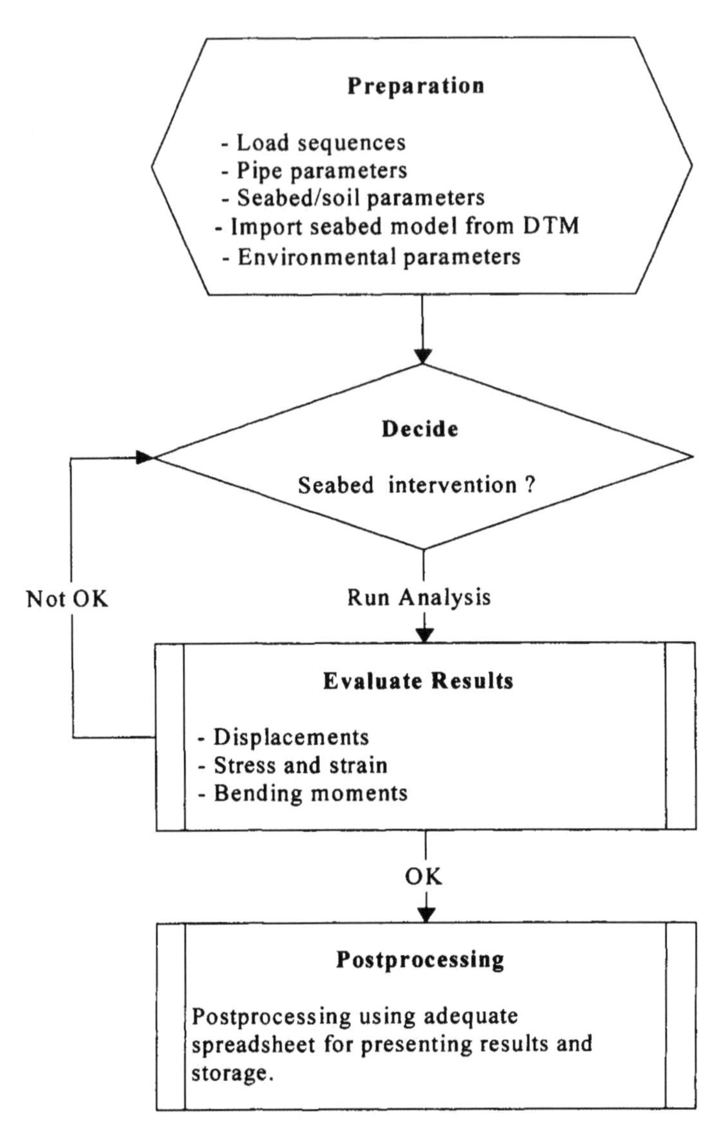

Figure 9.1 Flow-chart for seabed intervention design procedure.

The structural behavior of pipeline along its route can be analyzed using finite-element simulations of the load history from installation, flooding, hydro test, de-watering to operation. This analysis makes it possible to simulate the pipeline in-place behavior. Based on the understanding of the pipeline behavior from the analysis it is possible to select a seabed intervention design that is technically feasible and cost effective. The effect of the intervention can then be analyzed in detail for each particular location of the pipeline by finite-element simulations. The finite-element simulations are therefore a great tool/help for developing a rational intervention strategy.

This kind of simulations has also shown that the results can be quite sensitive to the shape and properties of the seabed. As a result of this the actual behavior of the pipeline can differ from the simulated behavior. Some factors are:

- Deviations between the planned route and the as-laid route.
- Actual lay tension during installation.
- Performance of seabed intervention, primarily trenching.
- Local variations in soil conditions.

It is therefore suggested to take the final decision on whether to perform seabed intervention work at some locations when as-built information becomes available.

9.5.3 Effect of Seabed Intervention

In Figure 9.2, seabed intervention in the form of trenching and rockdumping has been performed on the 3-D seabed model trying to reduce stresses and strains in the pipe from vertical loads. Results are given for maximum axial stress and bending moment, before and after intervention (Ose et al., 1999).

In Figure 9.3, seabed intervention in the form of rock dumping has been performed on the 3-D seabed model trying to reduce the lateral displacement of the pipe due to hydrodynamic loads.

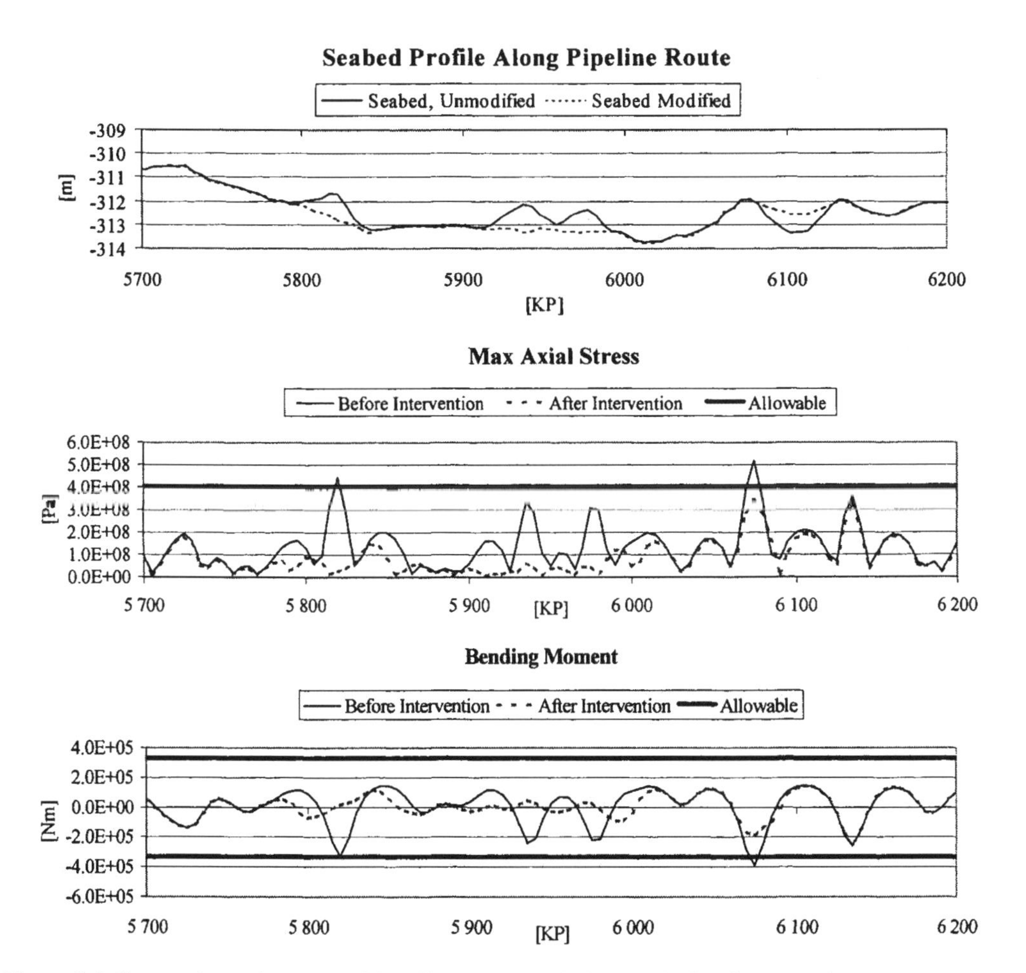

Figure 9.2 Comparison of stress and bending moment, before and after intervention.

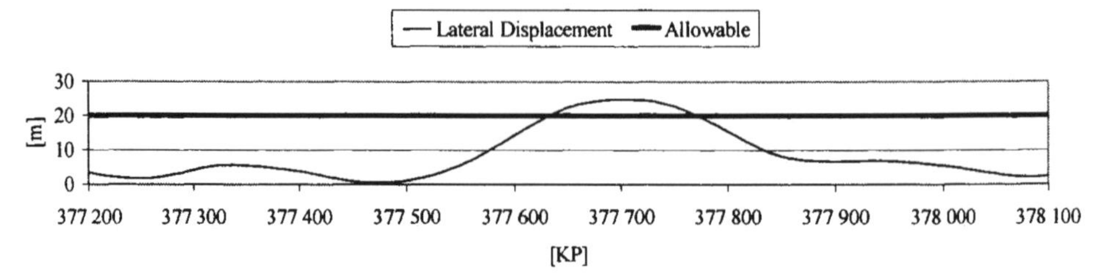

Lateral Displacement of Pipeline After Intervention

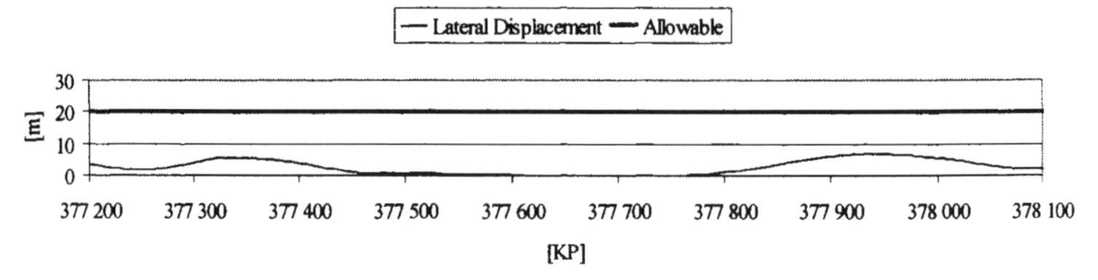

Figure 9.3 Comparison of lateral displacement of pipeline, before and after intervention.

The seabed intervention design through analysis is conducted as:

- To calculate stress, bending moment and displacements as shown for the two pipelines Figures 9.2 and 9.3.
- To compare the calculated stress, moment, and displacements with acceptance criteria.
- For the sections of pipeline where stress, moment, or displacement criteria is violated, seabed intervention is designed. The stress, moment, or displacements are then re-calculated, as shown in Figures 9.2 and 9.3, and compared with the acceptance criteria.
- This iteration is continued until acceptance criteria are fulfilled in all sections, Figure 9.1.

From the plots it can be seen that the load effects are reduced significantly as a result of the seabed intervention performed on the 3-D seabed of the analysis model.

9.6 References

1. AGA (1993) "Submarine Pipeline On-bottom Stability", Vols. 1 and 2, project PR-178-9333, American Gas Association.
2. Bai Y. and Damsleth, P.A. (1997) "Limit-state Based Design of Offshore Pipelines", Proc. of OMAE'97.
3. Ose, B. A., Bai, Y., Nystrøm, P. R. and Damsleth, P. A. (1999) "A Finite Element Model for In-situ Behavior of Offshore Pipelines on Uneven Seabed and its Application to On-Botton Stability", Proc. of ISOPE'99.
4. SINTEF PIPE Program.
5. SINTEF PONDUS Program "A Computer Program System for Pipeline Stability Design Utilizing a Pipeline Response Model".

Part II

Pipeline Design

Chapter 10 Vortex-induced Vibrations (VIV) and Fatigue

10.1 Introduction

The objective of this Chapter is to present acceptance criteria with respect to Vortex Shedding Induced Vibrations (VIV) of freespans and to outline the proposed methodology for the detailed design of pipeline systems.

Traditionally, VIV of freespans is not allowed to occur at any time during the design life of a pipeline system. In recent years a less stringent approach has become acceptable, in which VIV has been allowed provided it is demonstrated that the allowable fatigue damage is not exceeded.

Spans that are found to be critical with respect to VIV are usually corrected by placing rock berms below the pipe in order to shorten the span lengths and thus increase the natural frequency of the spans. In addition to the cost implication of placing a large number of rock berms on the seabed, the main disadvantage of this approach is that feed in of expansion into the spans will be restricted. It was demonstrated that allowing the pipeline to feed into the spans reduces the effective force, which is the prime factor in the onset of pipeline buckling. It is therefore advantageous with respect to minimizing buckling that the number of rock berm freespan supports is kept to a minimum.

Based on the above, it is proposed that the VIV criteria are as follows:

- Onset of in-line VIV is allowed during any phase of the design life provided it is demonstrated that the allowable stress and allowable fatigue damage is not exceeded.
- Onset of cross flow VIV is allowed during any phase of the design life provided it is demonstrated that the allowable stress and allowable fatigue are not exceeded.

A flowchart listing the various analysis steps to be performed during the VIV assessment are shown in Figure 10.1. (Grytten and Reid, 1999).

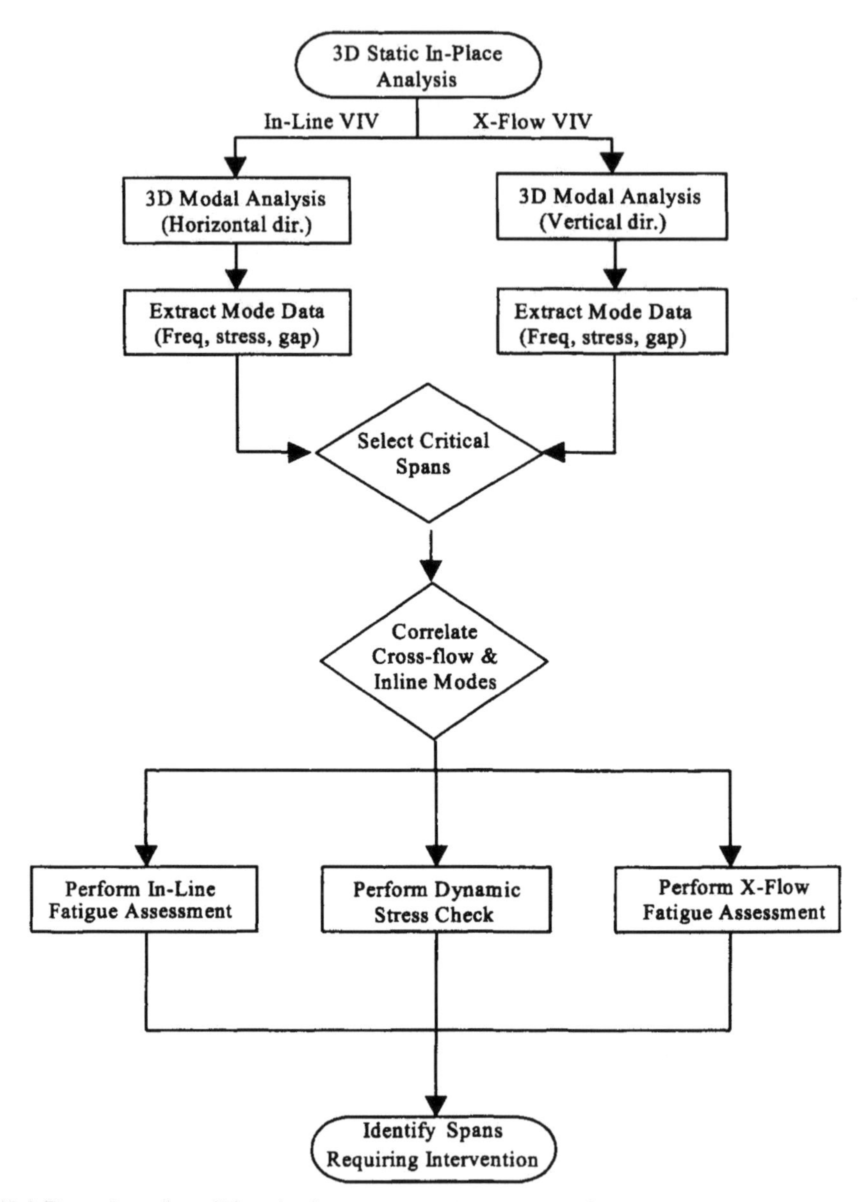

Figure 10.1 Flow chart describing the free span assessment procedure.

Design criteria applicable to different environmental conditions have been defined as follow:

1. Peak stresses or moment under extreme condition will satisfy the dynamic strength criteria given in Chapter 4.8.
2. For a) verified, a fatigue analysis will be performed.
3. The fatigue damage shall not exceed the allowable fatigue damage, η, that is normally 0.1.

Mørk et al. (1997, 1998) gave a series of papers on VIV and fatigue of free-spanning pipelines.

10.2 Free-span VIV Analysis Procedure

10.2.1 Structural Analysis

The structural properties of the given span configuration are to be characterized in terms of static and dynamic properties. The output are key parameters that can be applied in subsequent analysis involving hydrodynamic loading etc. on span.

Key parameters and relationships to be deducted are mainly:

- Relationship between loading/deflection of span and associated stresses and sectional forces/moments in pipe wall (static analysis)
- Eigenfrequencies and mode shapes of span, relationship between vibration amplitudes and stress cycles (dynamic analysis)
- Soil damping in terms of soil static and dynamic interaction with pipe.

Structural models of various complexity, analytical as well as computer based models, may be applied, ranging from simple models for simplified desk calculation to advanced finite element models for computer analysis.

Static model

Basically the static model is applied to determine stresses due to static or quasi-static loads such as deadweight of span, quasi-static wave and current loads, trawl boards, anchors. Frequently an elastic approach is selected for the pipe itself, whereas elasto-plastic soil behavior most often are adopted. This is particularly important in case of large spans supported on soft seabed.

For analysis of impact loads it is usually relevant to consider elasto-plastic behavior of the pipe as well as the soil.

Dynamic model

Basically the dynamic model is applied to determine stresses corresponding to flow induced vibrations (in conjunction with Response Amplitude Data Base/Model) for subsequent calculation of fatigue damage (in conjunction with the Fatigue Model) and for comparison with criteria for maximum allowable stresses. In-line and cross-flow vibration may be treated integrated or separately.

10.2.2 Hydrodynamic Description

Reduced Velocity

For determination of velocity ranges where VIV may occur, the reduced velocity parameter, V_R, is used, defined as:

$$V_R = \frac{U_c + U_w}{f_0 \cdot D} \tag{10.1}$$

where:

U_c : current velocity normal to pipe;

U_w : wave velocity normal to pipe;

f_0 : natural frequency of the span for a given vibration mode;

D : total outside diameter of the pipe including any coating or marine growth.

Stability Parameter

The other main parameter controlling the motions is the stability parameter, K_S, which is given as:

$$K_s = \frac{4 \cdot \pi \cdot m_e \cdot \zeta_T}{\rho \cdot D^2} \tag{10.2}$$

where:

ρ : sea water density;

ζ_T : total modal damping ratio at a given vibration.

Damping

The total damping, ζ_T, is normally considered to comprise hydrodynamic damping, soil damping and structural damping.

Hydrodynamic Damping

The hydrodynamic damping ratio accounts for the damping effect of the surrounding water. Hydrodynamic damping is proportional to the water velocity, i.e. reduces to zero as the water velocity tends towards zero. For VIV, the contribution to hydrodynamic damping within the lock-in region is set to zero since damping is already included in the response model.

Soil Damping

Soil damping ratio is the contribution of the soil to the overall damping ratio of the pipe-soil system. The soil damping is an end effect of the span therefore increasing the span length reduces the overall effect to the total damping. The soil damping is larger for the inline direction compared to the cross flow direction.

In Grytten and Reid (1999) typical values of soil damping ratios for various types of soil and span length/pipe diameter (L/D) ratios, are given. The damping values, as used in VIVA, can be interpolated for the correct span length. For continuous spans, taking the largest span length will give the most conservative value for soil damping.

It should be emphasized that the determination of pipeline soil interaction effects is encumbered with relatively large uncertainties stemming from the basic soil parameters and physical models. It is thus important that a sensitivity study is performed to investigate the effect of the above mentioned uncertainties.

Structural Damping

Structural damping ratio is the damping due to internal friction in the pipe steel material. A value of 0.005 (0.5 %) to be used if no other information is available, which is considered to be very conservative.

Effective Mass

The effective mass is defined as:

$$m_e = m(s) = m_{str} + m_c + m_a + m_{con} \tag{10.3}$$

where:

m_{str} : structural mass (including coating);

m_c : mass of content;

m_a : added mass; $m_a = \frac{\pi}{4} D^2 \cdot \rho \cdot C_a$ (10.4)

where :

C_a : added mass coefficient.

If it is assumed that the entire span is oscillating and vortex shedding occurs over the entire length, the effective mass can be defined by Equation 10.4. This assumption will contribute to a somewhat lower natural frequency and is considered to be conservative.

The Eigen period will increase as the added mass increases. The Eigen period calculation is computed during the Eigen value analysis. Secondly, Ks, the stability parameter will increase as the added mass increases. Thereby the effect of the damping will increase.

10.2.3 Soil Stiffness Analysis

Soil data is needed for setting up the structural model and for calculation of soil damping.

ASTM Unified Soil Classification System (USCS) is very convenient system for soil description in connection with pipeline projects.

Offshore sedimentation soils may convenient be labeled as either sandy soils or clayey soils. The soils parameters requested from pipeline projects are listed in Table 10.1 for sand and clay respectively.

Table 10.1 Design parameters for sandy and clayey soils.

	Sandy soil	Clayey soil
Material parameters	Gradient, Specific gravity	Liquid and plastic limits
	Void ratios in loosest and densest state	Specific gravity
		Remoulded shear strength
In-situ parameters	Void ratio and density index	Water content and liquidity index
	Bulk and dry densities	Bulk and dry densities
	Peak friction angle	Undrained shear strength
	Modulus of subgrade reaction	Drained shear strength
	Permeability	sensitivity
		consolidation parameters
		Modulus of subgrade reaction

Recommended values of some of the key parameters are listed in Table 10.2.

Table 10.2 Recommended values of key parameters and coefficients for typical offshore soils.

USCS symbol	Soil description	Submerged density (kN/m^3)	Plane Angle of Friction ϕ (deg.)	C_u (kN/m^2)
SW	Well graded sands, little or no fines	8.5-11.5	34-41	
SP	Poorly graded sands, little or no fines	7.5-10.5	34-39	
	- very loose	8.1	28	
	- medium loose	9.3	34	
	- very dense	10.6	40	
SM	Silty sands, poorly graded	8.0-11-5	31-37	
	- very loose	8.9	27	
	- medium loose	10.1	32	
	- very dense	11.4	38	
SC	Clayey sands, poorly graded	8.0-11.0	29-35	
ML	Silts and clayed silts	8.0-11.0	26-33	
CL	Clays of low to medium plasticity	8.0-11.0	N/A	
CH	Clays of high plasticity	3.0 9.0	N/A	10-100
	- very loose			10
	- medium loose			50
	- very dense			100

Table 10.3 Estimates of modulus of subgrade reaction for different types of soil.

Soil type	Subgrade reaction Ks (MPa)
Very soft clay	1-10
Soft clay	3-33
Medium clay	9-33
Hard clay	30-67
Sandy clay/ moraine clay	13-140
Loose sand	5-13
Dense sand	25-48
Silt	1-11
Rock	550-52000
Rock with marine growth	550-52000

The recommended value of modulus of subgrade reaction (K_s) are listed in Table 10.3

10.2.4 Vibration Amplitude and Stress Range Analysis

The results of the structural and environmental analysis are used as input to the calculation of the response of the free span to the environmental loads. The response may be found through the application of static or quasi-static loads or may be given directly as vibration amplitudes.

Due to the complexity of the physical processes involved, i.e. the highly non-linear nature of the fluid-elastic interaction of the vibrating span, the response of the span will generally be determined through the application of model or full scale investigation. Therefore the fluid-elastic properties of the environmental and the free span will be described by a number of governing non-dimensional parameters which are used to retrieve the relevant response data (force coefficients and oscillation amplitudes).

The response data are subsequently used to calculate:

- Stress range distribution
- Expected number of oscillations
- Fatigue damages parameter
- Maximum stress

10.2.5 Fatigue Model

To calculate the relationship between stress cycles experienced in pipe and the resulting fatigue damages, and thus the consumption of fatigue life, relationship of imperial or semi-empirical nature may be applied. This typically means a determination of the number of cycles that lead to failure for the various dynamic stress range (e.g. S-N curves) and the subsequently determination of the accumulation of the partial damages (e.g. Palmer-Miners law).

10.3 Fatigue Design Criteria

10.3.1 Accumulated Fatigue Damage

The fatigue damage shall be based on the accumulation law by Palmgren-Miner:

$$D_{fat} = \sum_{i=1}^{M_c} \frac{n_i}{N_i} \leq \eta \tag{10.5}$$

where:

- D_{fat} : accumulated fatigue damage;
- η : allowable damage ratio, normally to be taken as 0.1;
- N_i : number of cycles to failure at stress range $S(U_n)$ defined by S-N curve;
- n_i : number of equivalent stress cycles with stress range $S(U_n)$ in block i.

When several potential vibration modes may become active simultaneously at given current velocity the ode associated with the largest contribution to the fatigue damage must be applied. Formally, the fatigue damage criteria may be assessed numerically.

10.3.2 S-N Curves

When the stress range *S* (i.e. the double stress amplitude) has been established for a range of values of *Vr*, the expected fatigue damage shall be calculated by means of S-N curves.

In case Stress Concentration Factor is not applied, it is proposed that the F2 S-N curve for submerged structures in seawater is used in the detailed design, thus

- *log a* : constant equal to 11.63;
- *m* : fatigue exponent, which is equal to 3.

10.4 Response Amplitude

10.4.1 In-line VIV in Current Dominated Conditions

This section applies to current dominated situations only, i.e. for $\alpha > 0.8$ or $\alpha > 0.5$

Onset of In-line Vibrations

The onset value for the reduced velocity in the 1st instability region is given by (DNV, 1998).

First instability region:

$$V_{R\_onset} = \begin{cases} 1.0 & \text{for} \quad K_{s,d} \leq 0.4 \\ 0.6 + K_{s,d} & \text{for} \quad 0.4 \leq K_{s,d} \leq 1.6 \\ 2.2 & \text{for} \quad K_{s,d} \geq 1.6 \end{cases} \tag{10.6}$$

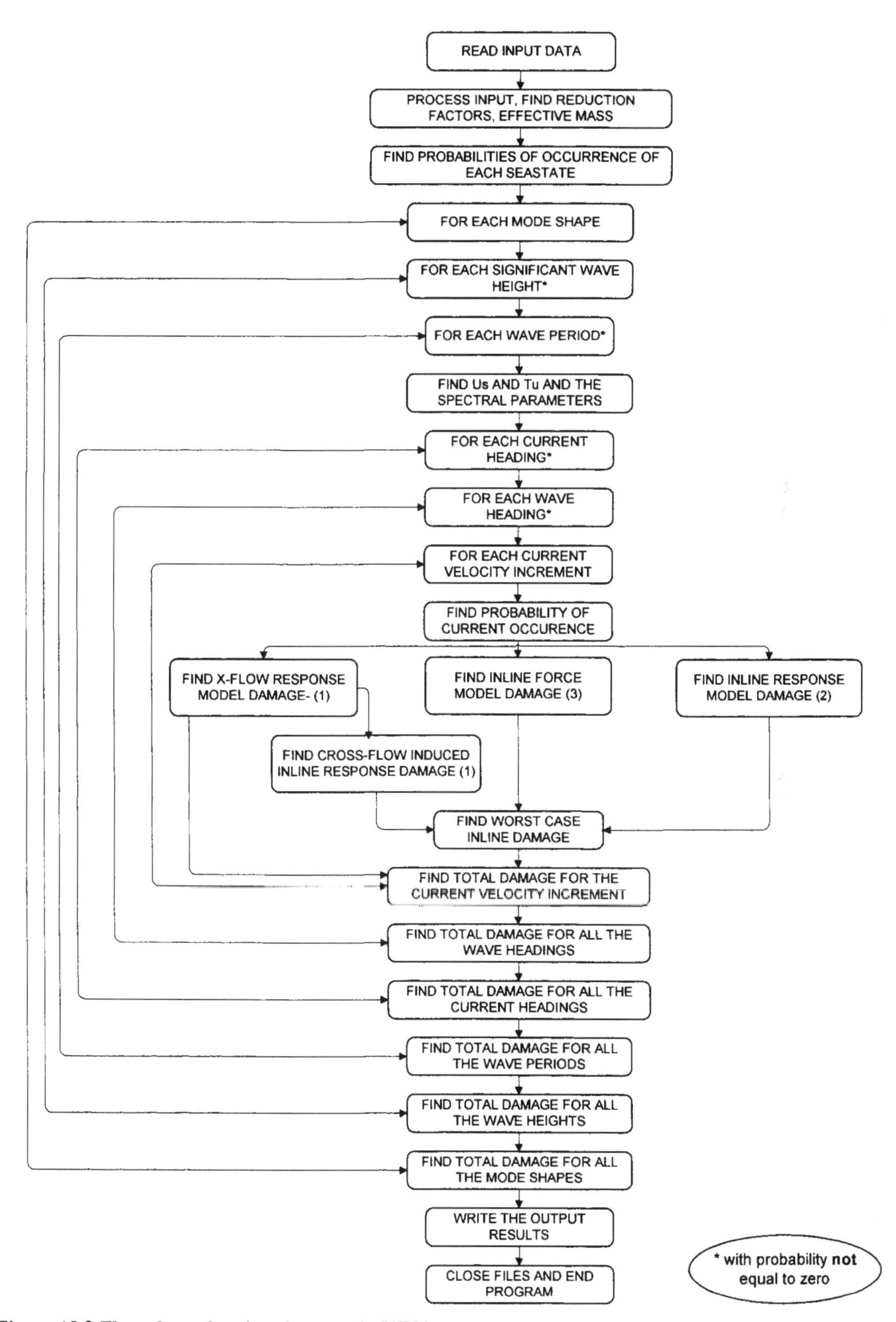

Figure 10.2 Flow chart showing the steps in VIVA.

Second instability region:

$$V_{R\_end} = \begin{cases} 4.5 - 0.8K_{s,d} & \text{for} \quad K_{s,d} \le 1.0 \\ 3.7 & \text{for} \quad K_{s,d} > 1.0 \end{cases} \tag{10.7}$$

Response

The characteristic maximum response amplitude is shown graphically in Figure 10.3 below (DNV, 1998).

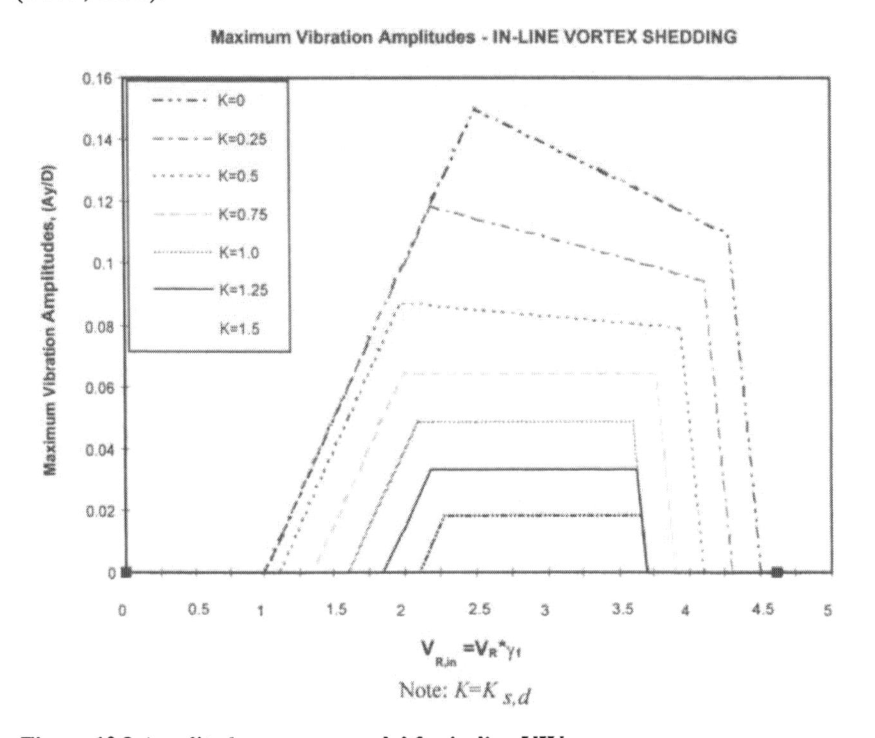

Figure 10.3 Amplitude response model for in-line VIV.

Stress Range

The in-line response of a pipeline span in current dominated conditions is associated with either alternating or symmetric vortex shedding. Contribution from the both first in-line instability region ($1.0 < V_R < 2.5$) and the second instability region ($2.5 < V_R < 4.5$) are included in this section. The stress range, S, may be approximately calculated by the in-line VIV Response Model:

$$S = 2 \cdot S_{A=1m} \cdot (A_Y / D) \cdot D \tag{10.8}$$

where: $S_{A=1m}$: unit stress amplitude (stress due to one meter in-line mode shape deflection) which is to be estimated by a dedicated FE analysis package.

(AY/D) : non-dimensional in-line VIV response amplitude.

10.4.2 Cross-flow VIV in Combined Wave and Current

This section applies to all cross-flow loads in all types of regions ($\alpha < 0.5$ & $0.5 < \alpha < 0.8$ wave dominant and $\alpha > 0.8$ current dominant).

Onset of Cross-flow Lock-on

For steady current dominated flow situations on-set of cross-flow VIV of significant amplitude occurs typically at a value of reduced velocity, V_R, between 3.0 and 5.0, whereas maximum amplitude vibrations occur at a value between 5.0 and 7.0. For wave dominated flow situations, cross flow vibrations may be initiated for V_R as low as between 2 and 3 and are in this region apparently linked to the in-line motions. For high values of V_R the motion are again de-coupled.

$$V_{R,\,onset} = \sqrt{\frac{\pi^3\left(\frac{\rho_s}{\rho} + C_m\right)}{1.5 + \left(0.27 - 0.03\cdot\left(\frac{e}{D_{pipe}}\right)\right)^2 \times \left(\pi^3\left(\frac{\rho_s}{\rho} + C_m\right) - 1.5K_S{}^2\right)}} \tag{10.9}$$

(ρ_s/ρ) is the specific mass to be taken as:

$$\left(\frac{\rho_s}{\rho}\right) = \frac{m(s)}{\frac{\pi}{4}\cdot\rho_w\cdot D_{pipe}} \tag{10.10}$$

Thus the onset of cross-flow motion will not occur if the reduced velocity is below $V_{R,\,onset}$.

Stress Range

If it is established that cross-flow VIV may occur, the span will have to be checked for fatigue damage. An important parameter is the stress range, $S(Un)$, associated with the response amplitude. The stress range may be approximately estimated as:

$$S = 2\cdot S_{FE}\cdot R\cdot f_Y\left(V_R, Kc, \alpha\right) \tag{10.11}$$

where:

- S_{FE} : stress amplitude (stress due to unit diameter mode shape deflection) to which is to be estimated by a dedicated FE analysis package;
- R : amplitude reduction factor accounting for damping and gap ratio.

The characteristic (maximum) amplitude response $f_Y(V_R,KC,\alpha)$ in combined current and wave flow may be taken from Figure 10.4 (DNV, 1998).

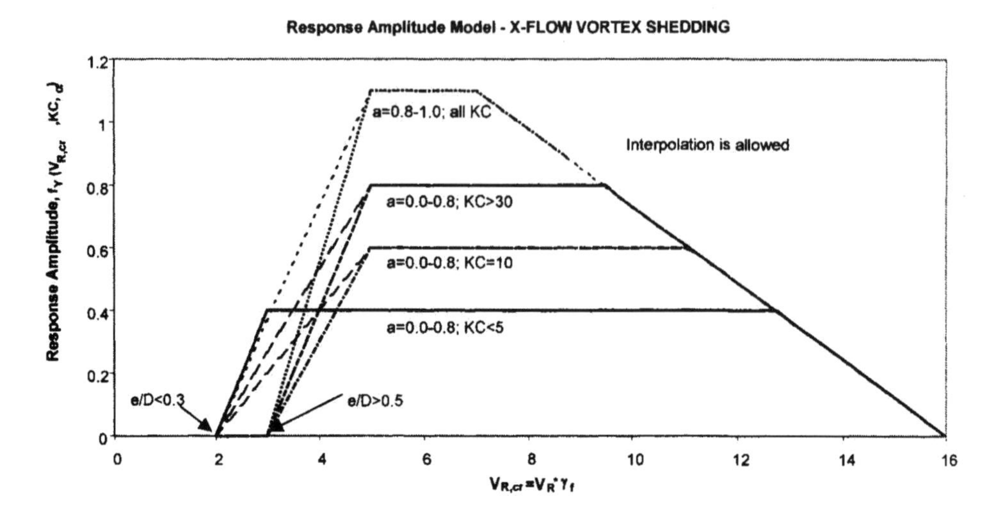

Figure 10.4 Characteristic (maximum) amplitude responses.

10.5 Modal Analysis

10.5.1 General

In order to obtain natural frequencies, modal shapes and associated normalized stress ranges for the possible modes of vibrations, a dedicated FEM analysis program should be used, and as a minimum, the following aspects have to be considered:

1. the flexural behavior of the pipeline is modeled considering both the bending stiffness and the geometrical stiffness;

2. the effective axial force which governs the bending behavior of the span is taken into account;

3. interaction between spanning pipe section and pipe lying on the seabed adjacent to the span should be considered (multispan project).

Due consideration to points 1 and 2 above is given in both the single span and multi-span modal analyses:

- In the FEM analyses, the change (increase) in the overall stiffness due to the deflected shape of the as-laid span is taken into account. This includes second order effects such as stress stiffening due to the sagging of the span.

- The axial effective force, i.e. the sum of the external forces acting on the pipe is also accounted for. It should be noted that the effective force will change considerably during the various phases of the design life.

In order to achieve this, it is important to ensure that a realistic load history is modeled prior to performing the modal analyses

10.5.2 Single Span Modal Analysis

Single span analyses are performed in order to assess the onset of in-line and cross-flow VIV as well as to calculate fatigue damage if it is found that onset of VIV may occur.

The modal analyses of a single span with simple boundary conditions will be used to assess the onset of in-line and cross flow VIV and associated fatigue damage. The justification for this is as follows:

- The multispan analysis can be carried out on the assumption that the pipeline along the routing is an input to the FEM program and the actual span length is considered. In this report the actual span lengths are unknown and the span criteria is the only interest. In such cases the boundary condition will have to be assumed. The main reason for performing a multispan analysis is to take into account the interaction between adjacent spans, i.e. several spans may respond as a system. Although this may be important for the vertical mode of vibration where the seabed between adjacent spans may form a fixed point about which the pipeline may pivot during the vibration, this effect is less important for the horizontal mode where the lateral seabed friction will oppose the movement and where there is no fixed point to which to pivot about. The VIV analysis for the actual span lengths will be considered in the inplace analysis.

- The single span model is adequate for the VIV analysis based on that the actual span lengths are unknown which is one condition to carry out the multispan analysis.

Both Fixed-Fixed and Pinned-Pinned boundary conditions have been analyzed together with a range of axial effective forces.

10.5.3 Multiple Span Modal Analysis

Two dimensional multispan analyses are performed in order to assess the onset of Cross-flow VIV as well as to calculate fatigue damage, if it is found that onset VIV may occur. The multispan analyses will take account of the interaction between adjacent spans. The analysis will not be performed in this report, but the analysis will be a part of the inplace analysis report.

The multispan analysis can be carried out on the assumption that the pipeline along the routing is an input to the FEM program and the actual span length is considered. In this report the actual span lengths are unknown and the span criteria is the only interest. The VIV analysis for the actual span lengths will be considered in the in place analysis report.

The criteria used for Cross-flow VIV is to keep the natural frequency of the spans above the VIV frequency corresponding to the onset of Cross-flow vibrations.

In order to ensure that the boundary conditions for each individual span are properly accounted for, the cross-flow vortex shedding is subject to a more rigorous modal analysis in which the longitudinal on-bottom configuration is assessed as part of the 2D static analysis.

The spans natural frequencies are checked against the corresponding frequency for onset of cross-flow VIV for the various design conditions (installation, water filled and operation). Those with a potential for experiencing significant cross-flow VIV are identified and measures to prevent or limit fatigue of the pipe will be evaluated on a case by case basis.

10.6 Example Cases

10.6.1 General

To give an insight into design against free span fatigue, a 40" pipeline has been assessed under operating conditions. For a complete analysis the damage during the empty and the water-filled conditions should be assessed and included in the overall damage accumulation.

Each case represents a typical pipeline with characteristic dimensions and flow parameters as provided below, see Table 10.4. (Reid et al., 2000).

Table 10.4 Pipeline input parameters.

Input parameters \ Pipe diameter	40 inch
Outer pipe diameter (m)	1.016
Wall thickness (m)	0.030
Corrosion coating thickness (m)	0.006
Corrosion coating density (kg/m^3)	1300
Concrete coating thickness (m)	0.03
Concrete coating density (kg/m^3)	2500
Residual lay tension (kN)	1350
Content density operating (kg/m^3)	180
Internal operating pressure (bar)	175
Operating Temperature (°C)	25
Internal hydrotest pressure (bar)	200

A typical uneven seabed has been selected in order to obtain a wide range of span lengths giving high fatigue damage. The soil is medium stiff clay. The configuration and loads from the static analysis are used as the basis for the eigen mode analysis. The modal analysis is carried out for the horizontal (in-line) and the vertical (cross-flow) directions. The in-line eigen mode values (natural frequencies) tend to be lower than the cross-flow values due to the stiffness of the span end conditions. To limit the amount of data processed for this exercise, the fatigue is assessed for the first ten modes. The eigen modes for in-line and cross-flow are presented below for a 40inch pipeline:

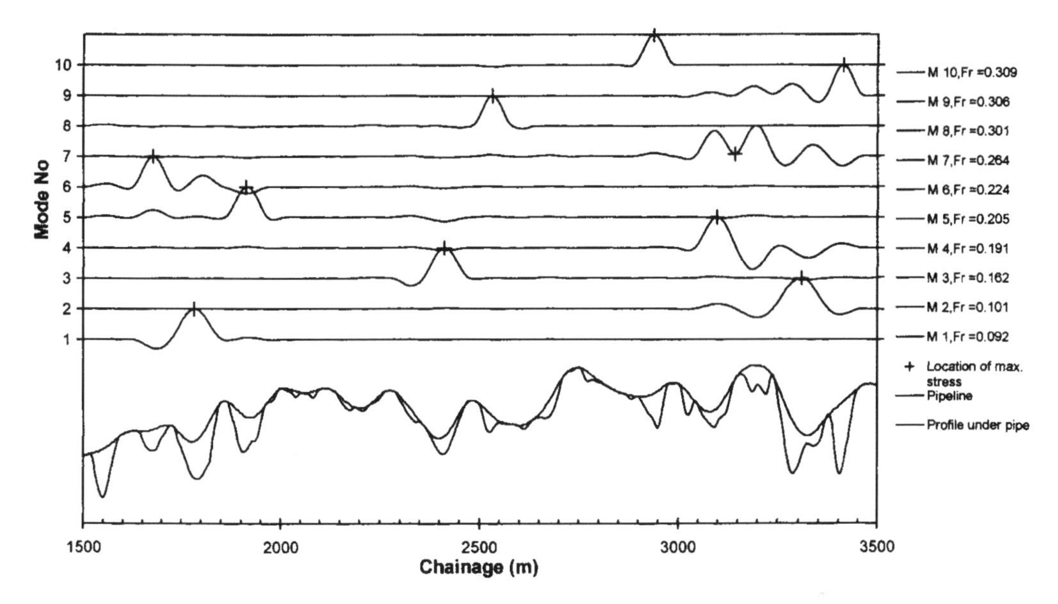

Figure 10.5 In-line modes for a 40" pipe.

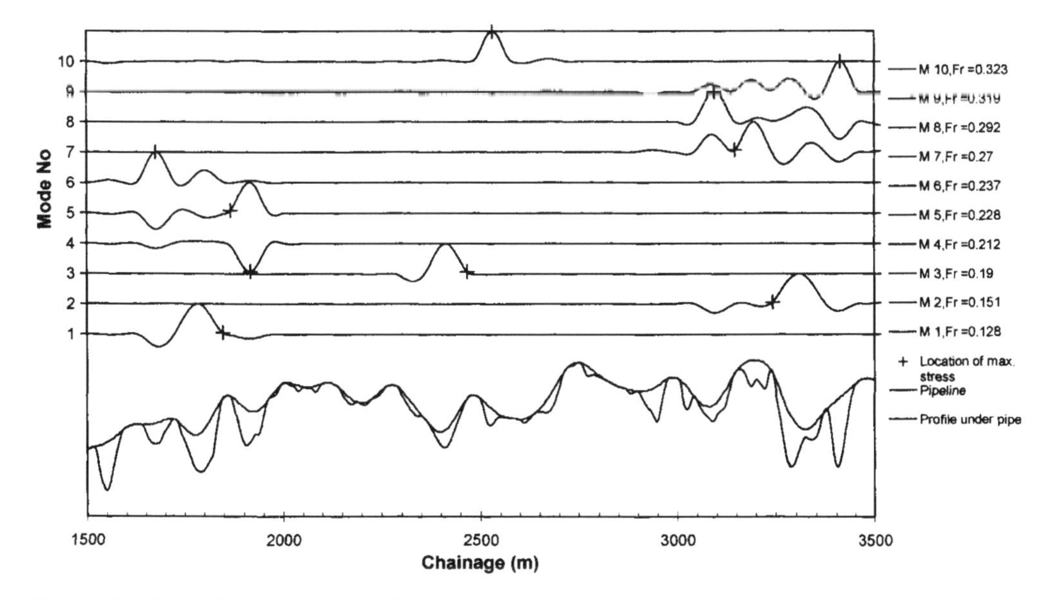

Figure 10.6 Cross-flow modes for a 40 inch pipe.

The modal analyses also generate unit stress amplitude results i.e. the stress at each node for one meter of modal deflection. This is used to find the stresses along the pipe for VIV and wave induced vibrations. The location of the maximum bending stress over the section is the point where the fatigue damage is evaluated. The maximum stress is normally located on the shoulder or at the mid-span of the dominating free span in each mode.

The cross-flow and the in-line mode shapes need to be correlated in order to take account of the cross-flow induced in-line fatigue. For the first three modes in the figures above, the in-line and the cross-flow modes are clearly linked. There are two clear cross-flow modes at the same location as the fourth in-line mode. Both the seventh and the eighth cross-flow show large excitations over the same span. The seventh mode is conservatively chosen because the natural frequency is lower. For the tenth in-line frequency the corresponding mode was found to be the twelfth cross-flow mode (not shown).

10.6.2 Fatigue Assessment

For the long-term environment description typical North Sea omni-directional wave and current distributions are applied.

- joint-frequency spectrum (Hs and Tp) of 3 hourly sea states
- 3 parameter Weibull current distribution of the 10 minute average current measurement at 3 meters above the seabed

The water depth is approximately 120 meters so the longer period waves will have an effect on the pipeline. The loads are initially considered acting at 90° to the pipe. This is a conservative assumption, which reduces the run time during the first stage of the analysis. It is used for screening purposes in order to determine which spans are critical and therefore require a more detailed assessment.

Structural damping of 1% is taken for each of the pipelines. The fatigue resistance is determined from the two-slope F1 S-N curve in seawater with cathodic protection. The damage is found for an operational life of 50 years for all expected environmental conditions. Ten percent of the total fatigue damage is allowed during the temporary phases, i.e. empty and water filled conditions and a further ten percent for the installation. The results are shown in Table 10.5 where the damage acceptance criterion for the operating condition is 0.48.

The results come from the Fortran based program VIVA (Grytten and Reid, 1999) for calculating pipeline free span fatigue.

As expected due to the rough seabed the pipes experience a high level of fatigue damage and intervention is required. There is unacceptable damage for the both the cross-flow and the in-line directions. The lowest modes show high force model damage, in particular at KP 1841 and KP 3319. Most of the spans have cross-flow damage and therefore also experience cross-flow induced in-line damage.

Table 10.5 Fatigue assessment results.

Span	40"	
KP	**In-line**	**X-flow**
1585	N/A	N/A
1718	0.55	0.55
1841	3380	16.7
1956	0.95	2.06
2436	1.61	2.71
2571	0.19	0
2704	N/A	N/A
2970	0.05	0
3136	0.69	0.02
3236	0.39	0.18
3319	943	3.71
3444	0.17	0.003

Note. Span damage given as N/A indicates that the span is not among the first ten eigen modes.

Supports are placed at the mid-spans of all the unacceptable locations. Optimization of the support location is possible to reduce the rock volume by placing the supports nearer the span shoulders. Further static and modal analyses and a fatigue re-assessment were carried out. The results are given below.

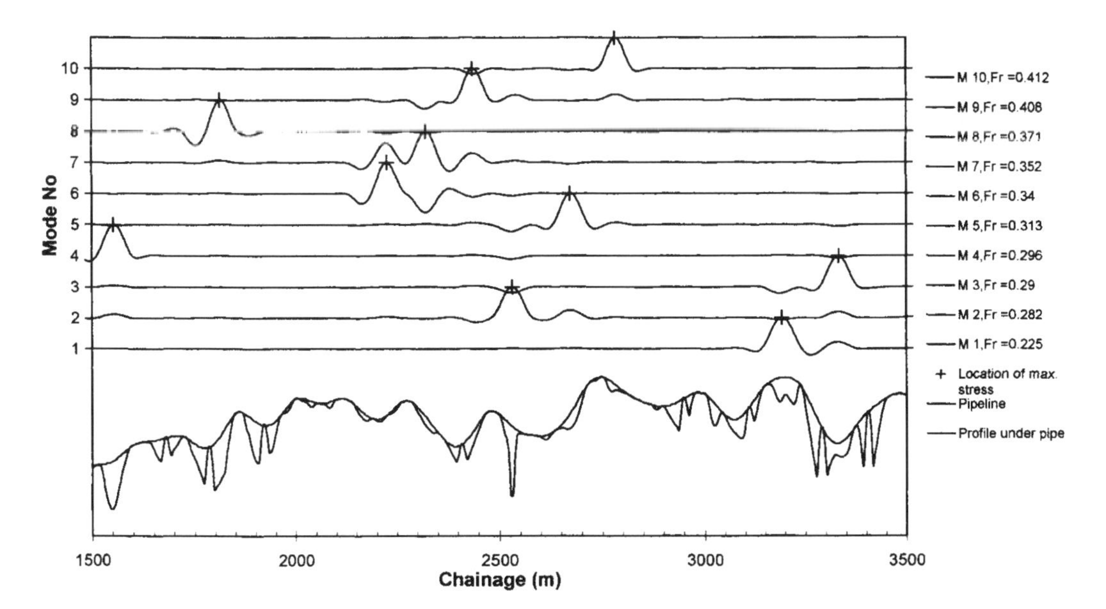

Figure 10.7 In-line mode shapes, 40 inch pipe with supports.

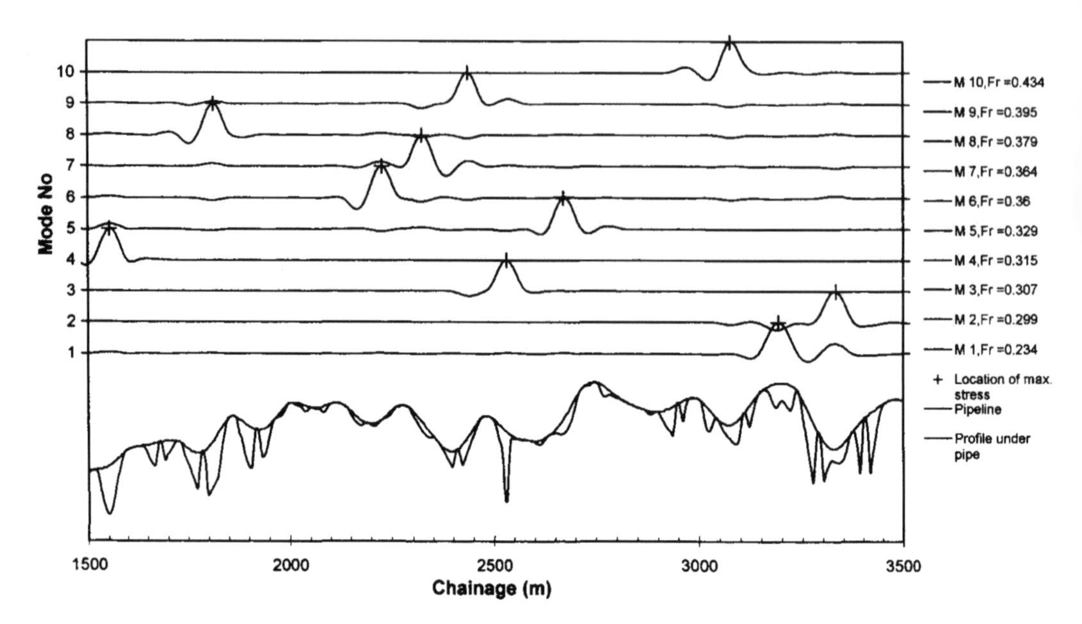

Figure 10.8 Cross-flow mode shapes, 40 inch pipe with supports.

10.7 References

1. DNV, (1998) "Guideline No. 14-Free-Spanning Pipelines", Det Norske Veritas.
2. Grytten, T.I. and Reid, A. (1999) "VIVA Guidelines Volume 2 – Theory", J P Kenny A/S.
3. Mørk, K., Vitali, L., and Verley, R. (1997) "Design Guideline for Free Spanning Pipelines," OMAE.
4. Mørk, K. and Fyrileiv, O. (1998) "Fatigue Design According to the DNV Guideline for Free Spanning Pipelines", OMAE.
5. Mørk, K., Verley, R., Bryndum, M. And Bruschi, R. (1998) "Introduction to the DNV Guideline for Free Spanning Pipelines" OMAE.
6. Reid, A., Grytten, T.I. and Nystrøm, P.R. (2000) "Case Studies in Pipeline Free Span Fatigue", Proc. of ISOPE'2000.

Part II
Pipeline Design

Chapter 11 Force Model and Wave Fatigue

11.1 Introduction

Free-spanning subsea pipelines subject to oscillating environmental loads may fatigue at the welded joints. Remedial seabed intervention by trenching and rock dumping is intended to ensure that the span lengths are acceptable, but often at great cost. Therefore, the spans have to be carefully assessed with respect to fatigue due to vortex-induced vibrations and wave-induced oscillations.

A considerable amount of work has been performed to develop methods for assessment of vortex-induced vibrations, see Tura et al. (1994) and Mørk et al. (1997). However, there is a lack of comprehensive mathematical formulations specifically dealing with wave-induced fatigue. In order to clearly present the theoretical background, a few equations available from reference books (e.g. Blevins (1994), Sumer and Fredsøe, (1997)) are presented. However, the rest of the paper is devoted to a methodology to assess wave-induced fatigue that the authors feel has not been given enough focus in the literature.

When calculating the fatigue damage due to transverse oscillations, it is first necessary to determine the stress amplitudes. This chapter describes the methodology used to calculate the stress amplitudes using a wave force model, based upon the well-known Morison Equation. Two different approaches are developed to solve the equation of the motion of the free-spanning pipeline:

- to solve the non-linear equation of motion numerically. This solution is called a 'Time Domain Solution';
- to linearize the Morison equation and solve it analytically. This is called a 'Frequency Domain Solution'.

11.2 Fatigue Analysis

11.2.1 Fatigue of Free-spanning Pipelines

- Vortex-induced Vibrations and Wave-induced Oscillation

A complete fatigue assessment of a free spanning pipeline should consider the loading due to both Vortex Induced Vibrations (VIV) and wave-induced oscillations.

While an amplitude response model may be applied when the vibrations of the free span are dominated by vortex induced resonance phenomena, a (Morison) force model is used to compute the free span response to waves through application of calibrated hydrodynamic loads.

Fatigue due to free-span oscillations is considered in two directions (see Figure 11.1):

- In-line with the wave and current direction (horizontal plane);
- Cross-flow direction (vertical plane).

- Combined In-line Fatigue

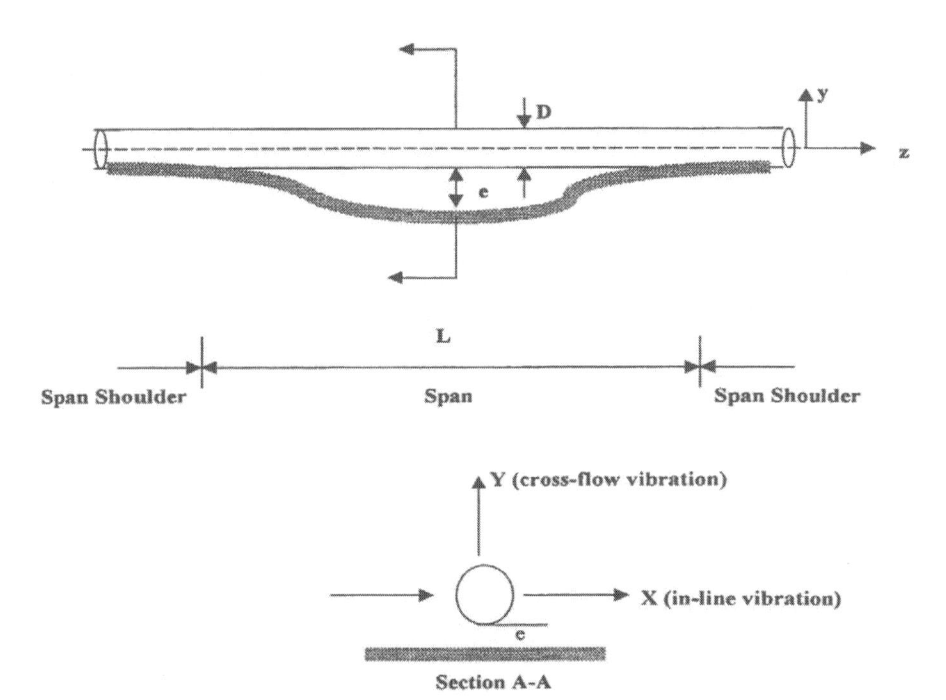

Figure 11.1 Free-spanning pipeline and its in-line and cross-flow directions.

There are three sources of in-line fatigue:

- in-line motion due to cyclic wave-induced oscillations, which may be simulated using a Force Model;
- in-line vibrations due to in-line vortex-induced resonance, which may be simulated using an Amplitude Response Model;
- in-line vibrations due to coupling from cross-flow vibrations.

Theoretically, it is necessary to add the fatigue damage due to all of the above. The accumulated fatigue is obtained by accounting for all sea-states and the joint probability of sea-state combined with current. Since vortex shedding has been thoroughly discussed in DNV (1998), this chapter shall focus on the wave-induced fatigue.

- Current Conditions

The current velocity is statistically described by a Weibull distribution as:

$$F_{U_{ref}}(U(z_{ref})) = 1 - \exp\left(-\left(\frac{U_{ref} - \gamma_{ref}}{\alpha_{ref}}\right)^{\beta_{ref}}\right)$$

Where $\gamma_{ref}, \beta_{ref}, \alpha_{ref}$ are Weibull parameters. The current velocity at a given depth $U(z_{ref})$ is transferred to current velocity at pipe level.

- Long-term Wave Statistics

Long term statistics are to be applied in the fatigue damage assessment, whereby the wave climate is represented by a scatter diagram of the joint probability of the sea state vector $\Theta = \lfloor H_s, T_p, \theta_w \rfloor$ and the wave spectrum, defined by significant wave height H_s, peak period T_p, and main wave direction θ_w.

- Short-term Wave Conditions

An irregular sea-state is assumed to be a short-term stationary process represented by a wave spectrum,

$$S_{\eta\eta}(f, \bar{\theta}) = S_{\eta\eta}(f)W(\bar{\theta})$$

The directional properties are usually modeled as:

$$W(\bar{\theta}) = \begin{Bmatrix} k_w \cos^s(\bar{\theta}) & ,|\bar{\theta}| < \frac{\pi}{2} \\ 0 & ,|\bar{\theta}| > \frac{\pi}{2} \end{Bmatrix}; k_w = \frac{1}{\sqrt{\pi}} \frac{\Gamma\left(\frac{s}{2}+1\right)}{\Gamma\left(\frac{s}{2}+\frac{1}{2}\right)}$$

The non-directional spectrum $S_{\eta\eta}(f)$ adopted in this chapter is the JONSWAP spectrum. The velocity and acceleration spectra at pipe level are derived from the directional wave spectrum through a transformation, using Airy wave theory:

$$S_{UU}(f, \bar{\theta}) = G_U^2(f) S_{\eta\eta}(\omega, \bar{\theta}), \quad S_{AA}(f, \bar{\theta}) = G_A^2(f) S_{\eta\eta}(\omega, \bar{\theta})$$

where:

$$G_U(f) = \frac{2\pi f \cosh(k(D+e))}{\sinh(kh)}, \quad G_A(f) = \frac{(2\pi f)^2 \cosh(k(D+e))}{\sinh(kh)}$$

where: f : wave frequency in (1/sec), $f = \omega/2\pi$;

k : wave number;

D : outer pipe diameter;

E : gap between seabed and pipe;

H : water depth to pipe.

11.2.2 Fatigue Damage Assessment Procedure

The following assumptions are made for the force model:

- The mass, the axial force, the stiffness and the structural damping are constant over time and along the pipe.
- The mean (main) wave direction is assumed to be perpendicular to the pipe and all the energy is assumed concentrated around the main wave direction.
- The time domain fatigue model can include a statistically distributed current velocity, or a fixed current velocity,
- The frequency domain fatigue model does not account for current.
- Given the assumptions, the fatigue damage assessment procedure may be summarized as:
- Characterization of Sea Environment: The wave environment is represented by the frequency of occurrences of various sea states, defined by the sea state vector $\Theta = \lfloor H_s, T_p, \theta_w \rfloor$ and the wave spectrum. The current is described by a Weibull distribution of current velocity.
- Dynamic response analysis: waves of appropriate frequencies, heights and directions are selected. The dynamic response and the loading of the pipeline are computed for each wave condition. The dynamic response analysis that is usually referred as the force model can be developed based on the time domain approach, hybrid time/frequency domain approach, and frequency domain approach. The results are expressed as the load or displacement transfer function per unit wave amplitude.
- Structural Analysis: Structural analysis is conducted to determine the stress transfer function per unit load or per unit displacement at each hotspot in the pipeline.
- Stress Transfer Function: The load transfer function per unit wave amplitude as a function of wave frequency is multiplied by the stress transfer function per unit load to determine the stress transfer function per unit wave amplitude as a function of wave frequency.
- Stress Concentration Factor: The geometric SCF should be considered in the fatigue assessment. The SCF is determined by Finite Element Analysis.
- Hotspot Stress Transfer Function: The stress transfer function is multiplied by the stress concentration factor to determine the hotspot stress transfer function.
- Long-term Stress Range: Based on the wave spectrum, wave scatter diagram and hotspot stress response per unit wave amplitude, the long-term stress range can be determined. This is done by multiplying the ordinate of the wave amplitude spectrum for each sea state by the ordinate squared of the hotspot stress transfer function to determine the stress spectrum. The stress range distribution is assumed to follow a Rayleigh distribution. The long-term stress range is then defined through a short-term Rayleigh distribution within each sea state for different wave directions. This summation can be further used to fit the Weibull distribution.
- S-N Classification: For each critical location considered, S-N curves will be assigned based on the structural geometry, applied loading and welding quality.
- Based on the long-term hotspot stress distribution and the S-N classification, the fatigue analysis and design of free-spanning pipeline may be conducted.

11.2.3 Fatigue Damage Acceptance Criteria

The design philosophy is that vibrations due to vortex shedding and oscillations due to wave action are allowed provided that the fatigue acceptance criteria are satisfied for the total number of stress cycles. The fatigue damage assessment is to be based on the Miner's rule:

$$D_{fat} = \sum \frac{n_i}{N_i} \leq \eta \tag{11.1}$$

where D_{fat} is the accumulated life time fatigue damage, η is the allowable damage ratio and N_i is the number of cycles to failure at stress S_i defined by the S-N curve on the form:

$$N_i = C * S_i^{-m} \tag{11.2}$$

m is a fatigue exponent and C is the characteristic fatigue strength constant. The number of cycles n_i corresponding to the stress range block S_i is given by,

$$n_i = P(\bullet) f_v T_{life} \tag{11.3}$$

$P(\bullet)$ is the probability of a combined wave and current induced flow event. f_v is the dominating vibration frequency of the considered pipe response and T_{life} is the time of exposure to fatigue load effects.

Applying the partial safety factors, the above equations my be re-expressed as

$$D_{fat} = \frac{T_{life}}{C} \sum f_v (\gamma_S S(\gamma_f, \gamma_k \ldots))^m P(\bullet) \leq \eta \tag{11.4}$$

where γ_f, γ_k and γ_s denote partial safety factors for natural frequency, damping (stability parameter) and stress range, respectively. For normal safety class, it is suggests that $\gamma_f = \gamma_k = \gamma_s = 1.3$ and $\eta = 0.6$.

11.2.4 Fatigue Damage Calculated Using Time Domain Solution

The fatigue damage may be evaluated independently for each sea-state of the scatter diagram in terms of H_s, T_p and θ_w as below.

$$D_{fat} = \frac{T_{life}}{C} \sum_{H_S T_p \theta_w} P(\bullet) \int_0^{\infty} \max\left[f_v \left(\gamma_s S\left(\gamma_f, \gamma_k \ldots\right)\right)^m\right] dF_{Uc} \tag{11.5}$$

where:

$P(\bullet)$ is the joint probability of occurrence for the given sea state in terms of significant wave height H_s, wave peak period T_p, mean wave direction.

dF_{Uc} denotes the long term distribution function for the current velocity. The notation "max" denotes that the mode associated with the largest contribution to the fatigue damage must be applied when several potential vibration modes may be active at a given current velocity.

In the Time Domain analysis, the irregular wave induced short term particle velocity at pipe level is represented by regular waves for a range of wave frequencies.

The stress range is calculated in the Time Domain Force model for each sea-state with a constant value of wave induced velocity amplitude but for a range of current velocities, from zero to a maximum value with nearly zero probability of occurrence. The calculated stress ranges are used when evaluating the integral in Equation (11.5). For each sea-state, the fatigue damage associated with each current velocity is multiplied by the probability of occurrence of the current velocity. When stress ranges for all sea-states are obtained through the force model, the fatigue damage is calculated using Equation (11.5).

11.2.5 Fatigue Damage Calculated Using Frequency Domain Solution

- Fatigue Damage for One Sea-State

For narrow banded response, the accumulated damage of a sea-state may be expressed in the continuous form:

$$D_{fat} = \int_0^\infty \frac{n(S)}{N(S)} dS$$

where $n(S)dS$ represents the number of stress ranges between S and $S+dS$. If a stationary response process of duration T_{life} is assumed, the total number of stress cycles will be:

$$N = v_0 T_{life} = \frac{T_{life}}{2\pi}\sqrt{\frac{m_2}{m_0}}$$

in which case, one can obtain:

$$n(S)dS = Np(S)dS = v_0 T_{life}\, p(S)dS$$

where $p(S)$ is the probability density function for stress range S given by:

$$p(S) = \frac{S}{4\sigma^2}\exp\left(-\frac{S^2}{8\sigma^2}\right)$$

Then, one can obtain

$$D_{fat} = v_0 T_{life}\int_0^\infty \frac{p(S)}{N(S)} dS = \frac{T_{life}}{2\pi}\sqrt{\frac{m_2}{m_0}}\,\frac{1}{m}\int \frac{S^{m+1}}{4\sigma^2}\exp\left(\frac{S^2}{8\sigma^2}\right) dS$$

using notation

$$t = \frac{S^2}{8\sigma^2}$$

and Gamma function,

$$\int_0^{\infty} e^{-t} t^{\frac{m}{2}} dt = \Gamma\left(1+\frac{m}{2}\right)$$

We may get

$$D_{fat} = \frac{T_{life}}{2\pi}\sqrt{\frac{m_2}{m_0}} \cdot m_0^{\frac{m}{2}} \left(\frac{8^{\frac{m}{2}}}{C}\right) \cdot \Gamma\left(1+\frac{m}{2}\right)$$

- <u>Fatigue Damage for All Sea-States</u>

From the damage equation for one sea-state, one may easily calculate the damage accumulated for all sea-states. If the response process is wide banded process, the Wirsching's Rain Flow Correction Factor is recommended to correct the conservatism due to narrow banded assumption (Wirsching and Light, 1980).

$$D_{fat} = \sum_{i}^{all\ sea-states} \frac{T_{Life}}{2\pi}\sqrt{\frac{m_{2i}}{m_{0i}}} \cdot m_{0i}^{\frac{m}{2}} \left(\frac{8^{\frac{m}{2}}}{C}\right) \cdot \Gamma\left(1+\frac{m}{2}\right) \cdot \lambda(\varepsilon, m)\gamma_s^m \tag{11.6}$$

where:

$\lambda(\varepsilon, m)$: Rain Flow correction factor;

$\lambda(\varepsilon, m)$: $a + (1-a)(1-\varepsilon)^b$;

a : 0,926-0,003m;

b : 1,587m-2,323;

m_{0_i} : spectral zero moment of the hotspot stress spectrum;

m_{2_i} : spectral second moment of the hotspot stress spectrum;

ε : band width of the hot spot stress spectrum.

Based on Equation (11.6), the transformation of a stress range spectrum to a fatigue damage is straightforward. Applying a spectral fatigue analysis, analytical expressions may be derived as the transfer functions from wave spectra to bottom velocity spectra, to response amplitude spectra and finally to stress range spectrum.

11.3 Force Model

11.3.1.1 The Equation of In-line Motion for a Single Span

The equation of in-line motion for a Bernoulli-Euler beam subject to wave forces represented by the Morison force, damping forces and the axial force is given by:

$$M\frac{\partial^2 z}{\partial t^2}+C\frac{\partial z}{\partial t}+EI\frac{\partial^4 z}{\partial x^4}-T\frac{\partial^2 z}{\partial x^2} = \begin{array}{l}\frac{1}{2}\rho DC_D\left(U-\frac{\partial z}{\partial t}\right)\left|U-\frac{\partial z}{\partial t}\right| \\ +C_M\frac{\pi}{4}\rho D^2\frac{\partial U}{\partial t} \\ -(C_M-1)\frac{\pi}{4}\rho D^2\frac{\partial^2 z}{\partial t^2}\end{array} \quad (11.7)$$

where:

Z	: in-line displacement of the pipe, and is a function of t and x;
X	: position along the pipe;
t	: time;
M	: mass of the pipe and the mass of fluid inside;
C	: damping parameter;
EI	: bending stiffness parameter where E is the elasticity module and I is the inertia moment for bending;
T	: effective force (T is negative if compression);
U	: time dependent instantaneous flow velocity;
ρ	: water density;
D	: pipe diameter;
C_D	: drag coefficient;
C_M	: (C_a+1), inertia coefficient and C_a is the added mass coefficient.

C_D and C_M are functions of the Keulegan-Carpenter (KC) number and the ratio between current velocity and wave velocity α. The added mass coefficient is taken from the Figure 9-1 of the DNV (1998) guideline, multiplied by a factor due to the gap between the span and the seabed.

The motion of the beam as a function of time and position along the beam is obtained by solving Equation (11.7) with appropriate boundary conditions.

Equation (11.7) is a non-linear partial differential equation that can not be solved analytically. The dependency of the position along the pipe axis can be eliminated from the equation by applying modal analysis. The modal analysis is based upon the assumption that the vibration mode shape of the beam is represented by a summation of beam eigen modes, whereby increasing the number of modes improves the accuracy. Modal analysis reduces the non-linear partial differential equation to a set of non-linear ordinary differential equations.

The non-linear ordinary differential equations can either be solved numerically or linearised and then solved analytically. The first approach is called a 'Time Domain Solution' and the latter a 'Frequency Domain Solution'.

The time domain approach demands more computing power than the frequency domain approach, but the latter will in some cases give erroneous results. In this context it shall also be mentioned that the Morison force representation is empirical, and originally intended to be used on stationary vertical piles. Since the first presentation of the formula it has been verified to cover other scenarios. The relative velocity model, used to describe the wave forces on a vibrating cylinder. The force coefficients are empirical and probably obtained from experiments with regular waves.

11.3.2 Modal Analysis

The modal analysis method reduces the partial differential equation to a set of ordinary differential equations. The key assumption is that the vibration mode of the beam can be described by a superposition of the eigen-modes. Eigen-frequencies and modes are determined from the equation of motion describing free vibrations.

$$M\frac{\partial^2 z}{\partial t^2}+EI\frac{\partial^4 z}{\partial x^4}-T\frac{\partial^2 z}{\partial x^2}=0$$

Solutions to the above equation are expressed as:

$$z(t,x)=\psi(x)\chi(t)$$

where:

$$\chi(t)=\cos(\omega t+\phi), n=1,2,3,...,$$

and

$$\psi(x)=c_1\cosh(s_1x)+c_2\sinh(s_1x)+c_3\cos(s_2x)+c_4\sin(s_2x)$$

$$s_1=\sqrt{\left(\frac{T^2}{4E^2I^2}+\frac{\rho A\omega^2}{EI}\right)^{\frac{1}{2}}+\frac{T}{2EI}}$$

$$s_2=\sqrt{\left(\frac{T^2}{4E^2I^2}+\frac{\rho A\omega^2}{EI}\right)^{\frac{1}{2}}-\frac{T}{2EI}}$$

The boundary conditions for a beam with end springs may be expressed as:

BC 1: $EI\frac{d^2\psi(0)}{dx^2}=k_{r_1}\frac{d\psi(0)}{dx}$

BC 2: $EI\frac{d^2\psi(l)}{dx^2}=-k_{r_2}\frac{d\psi(l)}{dx}$

BC 3: $T\frac{d\psi(0)}{dx}-EI\frac{d^3\psi(0)}{dx^3}=k_{t_1}\psi(0)$

$$\textbf{BC 4: } T\frac{d\psi(l)}{dx} - EI\frac{d^3\psi(l)}{dx^3} = -k_{t_2}\psi(l)$$

where:

- k_{t_1} : translational spring stiffness, left end of beam;
- k_{t_2} : translational spring stiffness, right end of beam;
- k_{r_1} : rotational spring stiffness, left end of beam;
- k_{r_2} : rotational spring stiffness, right end of beam;
- l : length of pipe.

Applying the boundary conditions in the general solutions, 4 linear equations are obtained from which ω is solved as frequency determinant. When ω is known the four coefficients except for an arbitrary factor can be determined.

The frequency determinant may be derived as:

$$\det\begin{vmatrix} s_1^2 & -\frac{k_{r_1}}{EI}s_1 & -s_2^2 & -\frac{k_{r_1}}{EI}s_2 \\ s_1^2\cosh(s_1 l) + \frac{k_{r_2}}{EI}s_1\sinh(s_1 l) & s_1^2\sinh(s_1 l) + \frac{k_{r_2}}{EI}s_1\cosh(s_1 l) & -s_2^2\cos(s_2 l) - \frac{k_{r_2}}{EI}s_2\sin(s_2 l) & -s_2^2\sin(s_2 l) + \frac{k_{r_2}}{EI}s_2\cos(s_2 l) \\ \frac{k_{t_1}}{EI} & s_1^3 + \frac{T}{EI}s_1 & \frac{k_{t_1}}{EI} & \frac{T}{EI}s_2 - s_2^3 \\ \left(s_1^3 + \frac{T}{EI}s_1\right)\sinh(s_1 l) - \frac{k_{t_2}}{EI}\cosh(s_1 l) & \left(s_1^3 + \frac{T}{EI}s_1\right)\cosh(s_1 l) - \frac{k_{t_2}}{EI}\sinh(s_1 l) & \left(s_2^3 - \frac{T}{EI}s_2\right)\sin(s_2 l) - \frac{k_{t_2}}{EI}\cos(s_2 l) & \left(\frac{T}{EI}s_2 - s_2^3\right)\cos(s_2 l) - \frac{k_{t_2}}{EI}\sin(s_2 l) \end{vmatrix} = 0$$

The solution to the original equation of motion, Equation (11.7), is assumed to be a product between the time response function and the eigen-modes as below:

$$z(x,t) = \sum_{n=1}^{m} Z_n(t)\psi_n(x) \tag{11.8}$$

11.3.3 Time Domain Solution

- <u>The Generalized Equation of Motion</u>

Inserting Equation (11.8) into Equation (11.7) gives:

$$\sum_{n=1}^{m}\left(\psi_n(x)\left[(M+M_a)\frac{d^2(Z_n(t))}{dt^2}+C\frac{d(Z_n(t))}{dt}\right]\right)$$
$$+\sum_{n=1}^{m}\left(Z_n(t)\left[EI\frac{d^4(\psi_n(x))}{dx^4}-T\frac{d^2(\psi_n(x))}{dx^2}\right]\right)=$$
$$\frac{1}{2}\rho DC_D\left(U-\sum_{n=1}^{m}\left(\psi_n(x)\frac{dZ_n(t)}{dt}\right)\right)\left|U-\sum_{n=1}^{m}\left(\psi_n(x)\frac{dZ_n(t)}{dt}\right)\right|$$
$$+C_M\frac{\pi}{4}\rho D^2\frac{dU}{dt}$$

where:

$$M_a=(C_m-1)\frac{\pi}{4}PD^2$$

Multiplying the equation through by $\psi_j(x)$ and integrating over the beam length gives:

$$\int_{x=0}^{l}\left\{\sum_{n=1}^{m}\left(\psi_n(x)\left[(M+M_a)\frac{d^2(Z_n(t))}{dt^2}+C\frac{d(Z_n(t))}{dt}\right]\right)\right\}\psi_j(x)dx$$
$$+\int_{x=0}^{l}\sum_{n=1}^{m}\left(Z_n(t)\left[EI\frac{d^4(\psi_n(x))}{dx^4}-T\frac{d^2(\psi_n(x))}{dx^2}\right]\right)\psi_j(x)dx=$$
$$\int_{x=0}^{l}\left\{\frac{1}{2}\rho DC_D\left(U-\sum_{n=1}^{m}\left(\psi_n(x)\frac{dZ_n(t)}{dt}\right)\right)\left|U-\sum_{n=1}^{m}\left(\psi_n(x)\frac{dZ_n(t)}{dt}\right)\right|\right\}\psi_j(x)dx$$
$$+\int_{x=0}^{l}\left\{C_M\frac{\pi}{4}\rho D^2\frac{dU}{dt}\right\}\psi_j(x)dx$$

Using the orthogonality properties results in:

$$\left[(M+M_a)\frac{d^2(Z_n(t))}{dt^2}+C\frac{d(Z_n(t))}{dt}\right]\int_{x=0}^{l}\psi_n^{\,2}(x)dx$$
$$+Z_n(t)\int_{x=0}^{l}\left[EI\frac{d^4(\psi_n(x))}{dx^4}-T\frac{d^2(\psi_n(x))}{dx^2}\right]\psi_n(x)dx$$
$$=\left(for\ n=1,\ldots,m\right)$$
$$\int_{x=0}^{l}\left\{\frac{1}{2}\rho DC_D\left(U-\sum_{n=1}^{m}\left(\psi_n(x)\frac{dZ_n(t)}{dt}\right)\right)\left|U-\sum_{n=1}^{m}\left(\psi_n(x)\frac{dZ_n(t)}{dt}\right)\right|\right\}\psi_n(x)dx$$
$$+C_M\frac{\pi}{4}\rho D^2\frac{dU}{dt}\int_{x=0}^{l}\psi_n(x)dx$$

The generalized equation of motion is therefore given by:

$$M_n \frac{d^2(Z_n(t))}{dt^2} + C_n \frac{d(Z_n(t))}{dt} + K_n Z_n(t) = F_n \tag{11.9}$$

where:

$$M_n = \int_{x=0}^{l} (M + M_a)\psi_n^2(x)dx = (M + M_a)\int_{x=0}^{l} \psi_n^2(x)dx$$

$$C_n = \int_{x=0}^{l} 2\zeta_n\omega_n m\psi_n^2(x)dx = 2\zeta_n\omega_n m\int_{x=0}^{l} \psi_n^2(x)dx$$

$$K_n = \int_{x=0}^{l} \left[EI\frac{d^4(\psi_n(x))}{dx^4} - T\frac{d^2(\psi_n(x))}{dx^2} \right]\psi_n(x)dx$$

$$F_n = \begin{array}{l} \displaystyle\int_{x=0}^{l} \left\{ \frac{1}{2}\rho DC_D \left(U - \sum_{n=1}^{m}\left(\psi_n(x)\frac{dZ_n(t)}{dt} \right) \right) \left| U - \sum_{n=1}^{m}\left(\psi_n(x)\frac{dZ_n(t)}{dt} \right) \right| \right\} \psi_n(x)dx \\ \displaystyle\qquad + \int_{x=0}^{l} \left\{ C_M \frac{\pi}{4}\rho D^2 \frac{dU}{dt} \right\} \psi_n(x)dx \end{array}$$

When Equation (11.9) is solved, the motion of the beam as a function of time and position along the pipe is given by Equation (11.8).

There are two ways of determining the response time-history when using the time domain model. One is to solve Equation (11.9) for a spectrum of representative regular waves; the other is to generate an irregular wave velocity time history from the wave spectrum and use this when solving Equation (11.9).

Preparation for numerical solution

The time domain approach is to construct a time history of the irregular sea surface from a wave spectrum $S_X(\omega)$. Given, such a spectrum, the velocity and acceleration of water particles given by the linear wave theory are:

$$U(t) = \sum_{i=-n}^{n} \overline{\omega}_i \sqrt{2S_{hh}(\overline{\omega})\Delta\overline{\omega}} \cos(\overline{\omega}_i t + \theta_i)$$

$$\dot{U}(t) = \sum_{i=-n}^{n} \overline{\omega}_i^2 \sqrt{2S_{hh}(\overline{\omega})\Delta\overline{\omega}} \sin(\overline{\omega}_i t + \theta_i)$$

where:

$S_{hh}(\varpi) = \frac{1}{4} S_{\eta\eta}(\varpi)$ is the wave height spectrum;

θ_i is the phase angle uniformly distributed from 0 to 2π.

Given the above equations, a time series of velocity and acceleration can be constructed. The span motion can then be analyzed in the time domain to obtain a time history of the response.

Before Equation (11.9) can be solved it is necessary to recast it, because the numerical differential equation solver used only handles first order ordinary differential equations. By introducing a new variable the equations become:

$$\frac{dZ_n(t)}{dt} = \tilde{Z}_n \tag{11.10}$$

$$\frac{d^2 Z_n(t)}{dt^2} = \frac{d\tilde{Z}_n}{dt} \tag{11.11}$$

$$\begin{aligned}(M+M_a)\frac{d\left(\tilde{Z}_n(t)\right)}{dt}\int_{x=0}^{l}\psi_n^{\,2}(x)dx = \\ \int_{x=0}^{l}\left\{\frac{1}{2}\rho DC_D\left(U-\sum_{n=1}^{m}\left(\psi_n(x)\tilde{Z}_n(t)\right)\right)\left|U-\sum_{n=1}^{m}\left(\psi_n(x)\tilde{Z}_n(t)\right)\right|\right\}\psi_n(x)dx \\ +C_M\frac{\pi}{4}\rho D^2\frac{dU}{dt}\int_{x=0}^{l}\psi_n(x)dx - C\tilde{Z}_n(t)\int_{x=0}^{l}\psi_n^{\,2}(x)dx \\ -Z_n(t)\int_{x=0}^{l}\left[EI\frac{d^4\left(\psi_n(x)\right)}{dx^4}-T\frac{d^2\left(\psi_n(x)\right)}{dx^2}\right]\psi_n(x)dx\end{aligned} \tag{11.12}$$

Equations (11.10) and (11.12) are solved to obtain $Z_n(t)$. The pipe movement is then given by Equation (11.8).

The spectrum of the pipe response is then calculated from the response time history by Fourier Transformation. The advantage of the time history simulation is that non-linearity's in the loading and response may correctly be taken into account. However, the calculation of the transfer function also involves a linearisation process that is basically only appropriate for the sea-state for which the simulation was done.

The accuracy of the solution increases when m increases. Unfortunately the number of simultaneous equations that are to be solved increases by two times m. The value of m is therefore determined from test runs.

<u>Stress calculation</u>

When the beam motion as a function of time and position along the x-axis is obtained, the stress range is given by:

$$\Delta\sigma = E\frac{\partial^2 z(x,t)}{\partial x^2}D$$

If the beam has elementary supports (pin-pin, fix-fix, pin-fix), the maximum bending moment will occur at the beam middle or ends. If the beam is supported by springs the maximum moment does not necessarily occur at these positions.

11.3.4 Frequency Domain Solution

- The generalized equation of motion

The frequency domain model presented herein is based on a linearised version of the Morison equation. In order to linearise the non-linear drag term it is assumed that $U \gg \frac{\partial z}{\partial t}$, the following linearisation is then proposed (Verley (1992)):

$$\left(U - \frac{\partial z}{\partial t}\right)\left|U - \frac{\partial z}{\partial t}\right| \cong U|U| - 2|U|\frac{\partial z}{\partial t}$$

A value for the absolute velocity being used in a statistical sense is averaged over the entire sea-state,

$$|U| = \sqrt{\frac{8}{\pi}}\sigma_U \quad , \sigma_U = RMS(U(t))$$

then

$$\left(U - \frac{\partial z}{\partial t}\right)\left|U - \frac{\partial z}{\partial t}\right| = K_L U - 2K_L\frac{\partial z}{\partial t}$$

The equation of motion can then be re-expressed as:

$$(M + M_A)\frac{\partial^2 z}{\partial t^2} + (C + 2K_D K_L)\frac{\partial z}{\partial t} + EI\frac{\partial^4 z}{\partial x^4} - T\frac{\partial^2 z}{\partial x^2} = K_D K_L U + K_M\frac{\partial U}{\partial t}$$

where:

$$K_D = \frac{1}{2}\rho D C_D$$

$$K_M = C_M\frac{\pi}{4}\rho D^2$$

$$M_A = -(C_M - 1)\frac{\pi}{4}\rho D^2\frac{\partial^2 z}{\partial t^2}$$

$$K_L = \sqrt{\frac{8}{\pi}}\sigma_U$$

The generalized equation of motion then becomes:

$$M_n \frac{d^2 Z_n(t)}{dt^2} + C_n \frac{dZ_n(t)}{dt} + K_n Z_n(t) = F_n(t) \;\; for \;\;\; n = 1,...,m$$

where:

$$M_n = \int_0^l (M + M_a)\psi^2{}_n(x)dx = (M + M_a)\int_0^l \psi^2{}_n(x)dx$$

$$C_n = \int_0^l (C + 2K_D K_L)\psi^2{}_n(x)dx = (C + 2K_D K_L)\int_0^l \psi^2{}_n(x)dx$$

$$K_n = \int_{x=0}^l \left[EI\frac{d^4(\psi_n(x))}{dx^4} - T\frac{d^2(\psi_n(x))}{dx^2} \right]\psi_n(x)dx$$

$$F_n = \left(K_D \; K_L \; U + K_M \; \frac{\partial U}{\partial A} \right) \int_0^l \psi_n(x) \; dx$$

- The Transfer Function between Wave Forces and Displacements

Using short-term wave conditions, the forcing function spectrum is given by:

$$S_{FF}(f) = K_D{}^2 K_L{}^2 S_{UU}(f) + K_M{}^2 S_{AA}(f)$$

The spectrum for the n'th generalized forcing function, $F_n(t)$, is:

$$F_n(t) = \int_0^l \psi_n(x)F(t)dx = F(t)\int_0^l \psi_n(x)dx$$

$$S_{F_nF_n}(f) = \left[\int_0^l \psi_n(x)dx \right]^2 S_{FF}(f)$$

The z_n -n'th response spectrum is:

$$S_{Z_nZ_n}(f) = \left| M_{transfer,n}(f) \right|^2 S_{F_nF_n}(f)$$

where $M_{transfer,n}(f)$ is the transfer function between wave forces and displacement response, which is given by:

$$\left|M_{transfer,n}(f)\right|^2 = \frac{1}{K_n^2\left\{1+\left(4\zeta_n^2-2\right)\left(\frac{f}{f_n}\right)^2+\left(\frac{f}{f_n}\right)^4\right\}}$$

where:

$$\zeta_n = \frac{C_n}{M_n}$$

- <u>The hotspot stress spectrum</u>

Between displacement and stress range there is the following linear relation:

$$\Delta\sigma(x,t) = -ED\frac{\partial^2 z(x,t)}{\partial x^2} = -ED\sum_{n=1}^{m} Z_n(t)\frac{\partial^2 \psi_n(x)}{\partial x^2}$$

The stress spectrum for a specific point along the beam, is therefore given by:

$$S_{\sigma\sigma}(f,x) = E^2 D^2 \sum_{n=1}^{m}\left(\frac{\partial^2 \psi_n(x)}{\partial x^2}\right)^2 S_{Z_n Z_n}(f)$$

The hotspot stress spectrum is given by:

$$S_{hotspot}(f,x) = (SCF)^2 S_{\sigma\sigma}(f,x)$$

where:

SCF is the stress concentration factor.

The resulting hotspot stress response spectrum will be numerically integrated to obtain the necessary moments m_n that are used for calculating the fatigue damage.

$$m_n = \int_0^\infty f^n S_{hotspot}(f,x)df \qquad \text{for n=0, 1, 2}$$

The zero crossing rate and bandwidth are determined by:

$$T_z = 2\pi\sqrt{\frac{m_0}{m_2}}, \quad \varepsilon = \sqrt{1-\frac{m_2^2}{m_0 m_4}}$$

11.4 Comparisons of Frequency Domain and Time Domain Approaches

A computer program called “FATIGUE” has been developed. Time-domain program consists of two parts:

- part one solves the differential equation of motion,
- part two calculates the fatigue damage

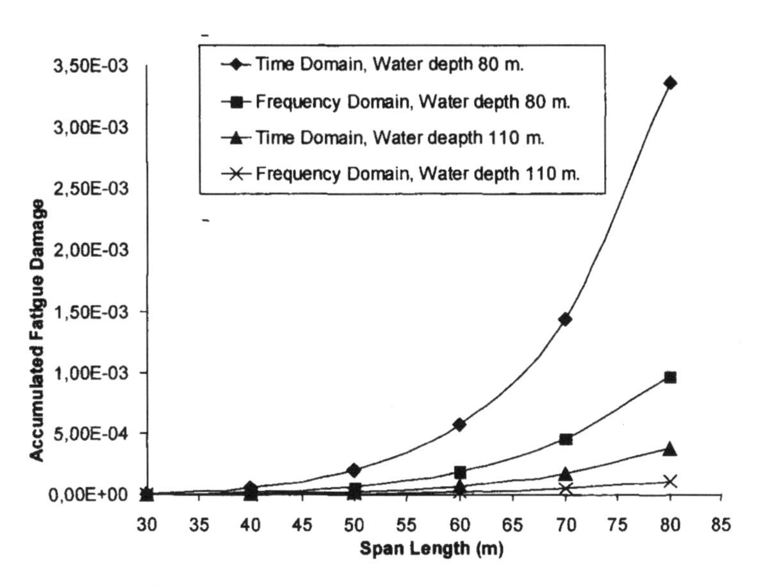

Figure 11.2 Accumulated fatigue damage vs. span length based on time-domain and frequency-domain approaches, for water depths 80m and 110m (42" pipe).

The FATIGUE program has been compared with fatigue calculations by Fyrileiv (1998), see Bai et al. (1998).

It appears that the difference between the results from the time-domain and frequency domain approaches is not small, and further investigation is required. The time domain approach is believed to be more accurate than the frequency domain approach because it accounts for the influence of current velocity and non-linearity's.

A parametric study on fatigue damage assessment is conducted by Xu et al. (1999).

11.5 Conclusions and Recommendations

1. The chapter presented a methodology for analyzing wave-induced fatigue of free spanning pipelines.

2. The analytical equations for the dynamic response analysis of free spans in frequency domain are developed, neglecting current velocity.

3. The equation of motion is solved in time domain for combined regular waves and current velocities with different probability of occurrence.

4. Fatigue damage is calculated by adding contributions from all sea-states and currents using joint probability.

5. The numerical examples illustrate that there is a disturbing difference between the time domain fatigue analysis and frequency domain fatigue analysis. This is due to the non-linear effects of the Morison equation and current velocity and is a subject for further investigation.

6. The fatigue is calculated at 51 points along the pipe span. The computer program predicts fatigue damages reasonably close to those predicted by Fyrileiv (1998).

A general free span may be described in terms of the natural frequencies, modes shapes, damping and modal mass. Beams with spring boundary conditions are considered in this chapter. However, the developed formulation may be easily used to post-process the modal analysis results from in-place finite element models of a pipeline which models the seabed and in-service conditions accurately. The FATIGUE program may then be used to validate the more detailed models, which have been developed as part of the DTA (Design Through Analysis) by JP Kenny A/S (Bai and Damsleth, (1998)).

11.6 References

1. Bai, Y. and Damsleth, P.A., (1998) "Design Through Analysis Applying Limit-state Concepts and Reliability Methods", *Plenary paper for ISOPE'98.*

2. Bai, Y., Lauridsen, B., Xu, T. and Damsleth P.A., (1998) "Force Model and In-line Fatigue of Free-Spanning Pipelines in Waves", *Proc. of OMAE'98.*

3. Blevins, R.D., (1994) "*Flow-Induced Vibration*", Krieger Publishing Company.

4. Det Norske Veritas (1998). Guidelines No. 14, "Free Spanning Pipelines".

5. Fyrileiv, O. et al., (1998) "Fatigue Calculations Using Frequency Domain Approach", Fax dated 30th Jan. 1998.

6. Mørk, K.J., Vitali, L. and Verley, R., (1997) "THE MULTISPAN Project: Design Guideline for Free Spanning Pipelines", *Proceedings of OMAE'97.*

7. Sumer, M. B. and Fredsøe Jørgen. (1997). "Hydrodynamics Around Cylindrical Structures", Advanced Series on Ocean Engineering – Vol 12, Published by World Scientific, Singapore.

8. Tura, F., Bryndum, M.B. and Nielsen, N.J.R., (1994). "Guideline for free Spanning Piplines: Outstanding Items and Technological Innovations", *Proceedings of Conference on Advances in Subsea Pipeline Engineering and Technology*, Aberdeen.

9. Verley, R., (1992) "Gudesp – Hydrodynamic Force Model, in-line", Memo dated 19th June 1992.

10. Xu, T., Lauridsen, B. and Bai, Y. (1999) "Wave-induced Fatigue of Multi-span Pipelines" Journal of Marine Structures, Vol. 12, pp. 83-106.

Part II
Pipeline Design

Chapter 12 Trawl Impact, Pullover and Hooking Loads

12.1 Introduction

The interaction between fishing gear and a pipeline is one of the most severe design cases for an offshore pipeline system. The reason for the severity of the impact, pullover and hooking is not well described by the industry today. The damage to the pipeline (and to the fishing gear & ship) is very dependent on the type of fishing gear and the pipeline conditions, e.g. the weight and velocity of the fishing gear and the wall thickness, coating and flexibility of the pipeline. The most important issue with respect to design of fishing gear resistant pipelines is the ability to make a realistic description of the applied loads and their time history, and pipeline resistance, i.e. the pipeline configuration on the seabed including freespan and the pipe stiffness. The summary of loads, response analysis and acceptance criteria are listed in Table 12.1.

Table 12.1 Summary trawl impact, pullover and hooking.

	Time	Load	Solution	Design acceptance criteria	Design parameters
Impact	mseconds	mass velocity	mass-spring system dynamics	dent damage in pipe <0.035D	energy absorption capacity of pipe coating
Pullover	seconds	time history of horizontal & vertical loads	time domain dynamics	allowable moment. allowable stress/strain	Heights & length of free spans
Hooking	minutes	vertical displacement	static solution	allowable moment. allowable stress/strain	fishing gear frequency free spans

12.2 Trawl Gears

12.2.1 Basic Types of Trawl Gear

Bottom trawling is typically conducted with two types of trawl gear in North Sea: Otter and Beam. Otter trawling occurs down to depths of more than 400 m. Generally beam trawling occurs in water depths down to 100m. The Otter trawl board is a more or less rectangular steel board which holds the trawl bag open, while the beam trawl consist of a long beam which holds the trawl open. The beam has beam shoes on each end and an impact is assumed to be from these beam shoes.

12.2.2 Largest Trawl Gear in Present Use

Table 12.2 indicates presently applicable data for the largest trawl boards in use in the North Sea in 1995 as follows:

Table 12.2 Data for largest trawl boards in use in North Sea (in 1995).

	Consumption		**Industrial**
	Polyvalent	**V-board**	**V-board**
mass (kg)	3500	2300	1525
length x breadth (m)	4.8 x 2.8	3.8 x 2.25	3.7 x 2.4
trawl velocity (m/s)	2.8	2.8	1.8

In the detailed impact and pullover analyses, the Consumption Polyvalent Board is considered as this assumes the most severe case (mass m=3500 (kg) and velocity V=2.8 (m/s)).

As for future developments or changes in equipment, these must be accounted for by investigating possible changes within the lifetime of the pipeline. Trends are going towards improved design in order to optimize trawl board shape and in this way reduce the power needed to drag the trawl, hence minimizing fuel consumption and improving the economy. Although there may be fewer, but larger trawlers in the future, this indicates that there will be a negligible increase in the mass of trawl board and the velocity of trawling.

12.3 Acceptance Criteria

The acceptance criteria corresponding to accidental loads and environmental loads from NPD (1990), is that *no leak should occur*. The acceptance criteria 'no leak' is interpreted below.

12.3.1 Acceptance Criteria for Impact Response Analyses

When the trawl loads are considered as accidental loads, the present study proposes a dent depth acceptance criterion as below.

In past practices, Dent Depth was limited to 2% of OD (Outer Diameter) according to ASME B31.8 (1992). This was a conservative assumption. A rational criterion on dent acceptability can be argued based on residual strength assessment. Up to 5% of OD can be allowed based on the following considerations:

- serviceability limit state: the limit for allowing pigging operation is 5% OD

- burst strength: the pipe corrosion coating is not likely to be penetrated by the impact. It is then assumed that no cracks (gouges) will be given to pipe steel wall due to impact. Therefore, burst strength of the pipeline will not be reduced significantly because the dent depth is 5% OD with no cracks in the dented area.

- fatigue strength: the required fatigue life is that no fatigue failure should occur before the subsequent inspection in which possible dent damage can be detected and repaired. Based on information from an American Gas Association (AGA) study Fowler et al. (1992), it can be documented that a dent depth of 5% OD might be acceptable from the point of view of fatigue due to cyclic internal pressure.

- buckling/collapse: the collapse pressure will be reduced because of dents. The allowable strain is reduced from the viewpoint of Strain-based Design Criteria.

Internal pressure can reduce the dent depth. However, the reduction of dent depth due to internal pressure is neglected.

Strictly, it is necessary to check the local stress and strain to ensure that no leak occurs during the impact process. Since pullover loads are much higher than impact loads, such leak check is to be done only for pullover loads.

12.3.2 Acceptance Criteria for Pullover Response Analyses

In the pullover response analyses, 'no leak' means satisfaction of the strength requirements to local buckling and fracture/plastic as discussed in Chapter 4. Especially, girth weld fracture shall be a governing failure mode because local buckling strain is considered to be large.

According to STATOIL (1996), free-spans are generally permitted in areas where trawling occurs, provided that the above criteria are satisfied.

12.4 Impact Response Analysis

12.4.1 General

The impact analysis is carried out in order to define the impact energy that must be absorbed by the coating and the testing requirements for the coating.

For concrete coated pipelines, the impact energy is generally assumed protected by the coating and no further analysis is required by STATOIL (1996).

12.4.2 Methodology for Impact Response Analysis

The analysis will be carried out following the procedure recommended in the document STATOIL (1996). The finite element model recommended in this design guide is similar to that proposed by Bai and Pedersen (1993).

This kind of detailed analysis is carried out because the traditional impact analysis, assuming the impact energy will be totally absorbed by the steel and insulation coating as deformation energies, is too conservative. Kinetic energies absorbed by the trawl board and the pipe can be large. Only a fraction of the kinetic energy of the trawl board is absorbed by the steel pipe locally.

A Level-2 analysis is shown in Figure. 12.1. The notations used in this figure are defined as below:

Trawl Board

- m_a and m are the added mass and steel mass of the trawl board
- k_b and k_i are the trawl board out-of-plane and in-plane stiffness

Coating and Steel Shell

- k_{c1} represents the coating stiffness
- k_{c2} denotes possible effect the coating has on the steel shell stiffness
- k_s is the local shell stiffness of the steel pipeline

Pipe and Support

- m_p is the effective mass of the pipe including hydrodynamic added mass
- k_{pb} is the effective bending stiffness of the pipe
- k_{ps} is the effective soil stiffness acting on the pipe

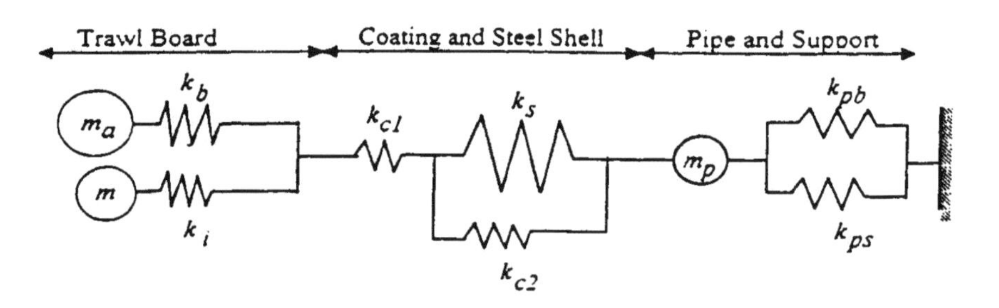

Figure 12.1 Physical model (mass-spring system) for simulation of impact between trawl board and pipeline.

The local indentation curve including both steel pipe and insulation coating can be obtained by finite element analyses using Static Local Shell Model. The steel pipe should be modeled using geometrical and material nonlinear elements. Large deflection should be considered but small strain theory might be applied. The sophistication of the elements for insulation coating shall largely depend on the availability of the material properties from the insulation-coating manufacturer.

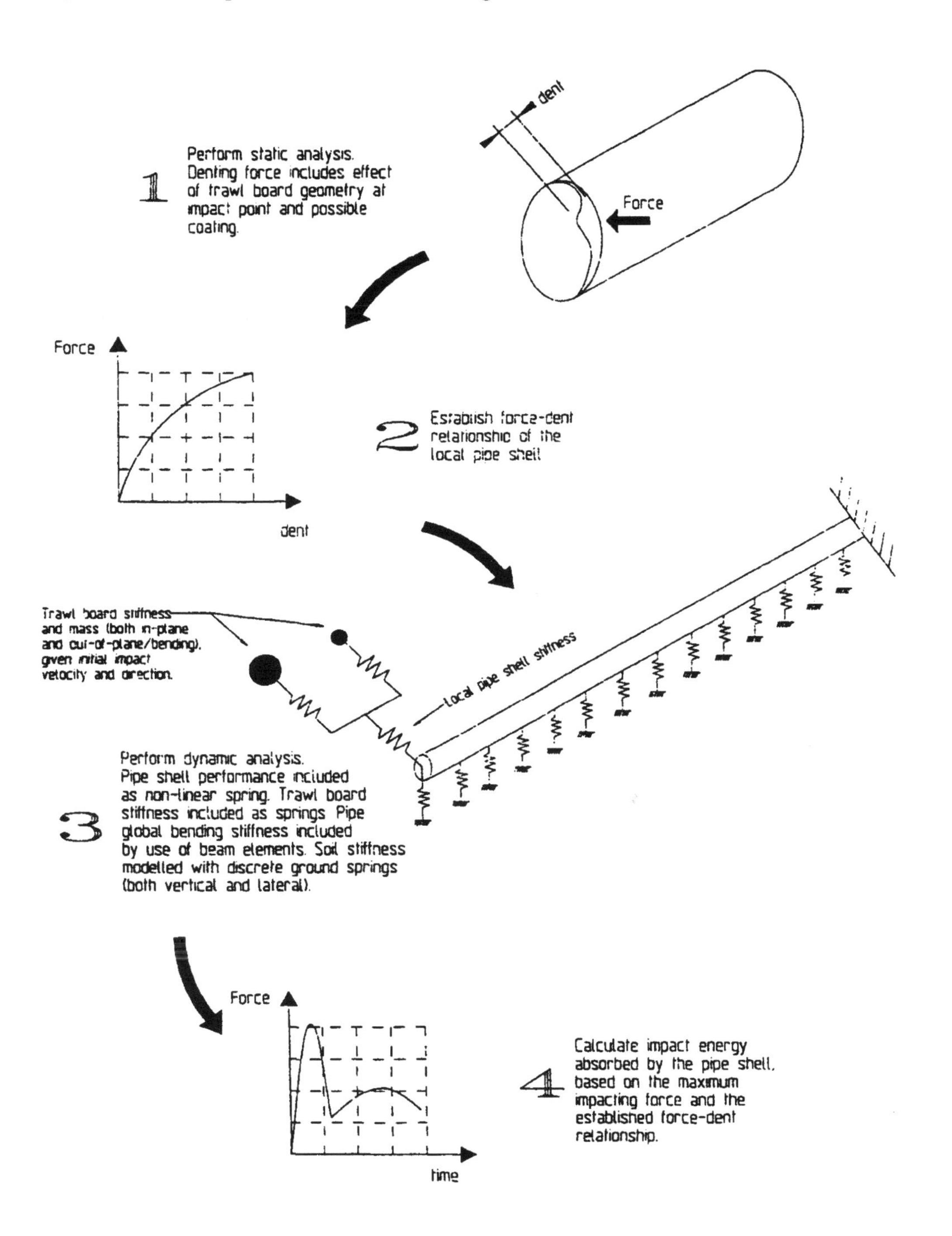

Figure 12.2 Scheme for simulation of impact between trawl board and pipeline (STATOIL, 1996).

The energy absorption process will be simulated applying dynamic global pipe model. Nonlinear beam elements with pipe section can be used for the simulation of pipeline global behavior. The indentation curve for steel pipe and insulation coating are modeled using non-linear spring elements capable of accommodating compression forces only. The dynamic analysis is carried out assuming the steel and added masses of the trawl board with an initial velocity. The finite element modeling will be similar to that described in Chapter 7. The difference is that the pipeline length to be considered can be much shorter for the impact response analyses. Figure 12.2 shows the principle that will be employed.

As a result of the dynamic global pipe model, a dent size will be obtained as the description of the damage to the pipe steel. In addition, analyses results also include time histories of deformation in steel pipe and coating, and impact force between the trawl board and the pipeline.

For a balanced consideration of coating material costs and pipeline safety, the impact energy absorption capability of the coating should be determined based on impact response of pipelines to trawl board loads. An analytical method is developed to determine the initially assumed energy absorption capability of the coating. Detailed impact response analyses of the dynamic system are carried out using non-linear finite element programs to confirm the assumed energy absorption capability.

A coefficient C_h (=0.85) will be applied together with the trawling velocity to arrive at an effective velocity.

12.4.3 Steel Pipe and Coating Stiffness

12.4.3.1 General

The local stiffness of the pipe is represented by the stiffness of,

- the local shell stiffness of the steel pipe, k_s
- the coating stiffness, k_{c1}
- Possible effect the coating has on the steel shell stiffness, k_{c2} - This is because the coating will distribute the impact load to a wider area on the steel shell, and possibly transfer certain forces tangentially.

The deformation energy to be absorbed by the steel pipe and the insulation coating are:

$$E = E_S + E_{C1} + E_{C2} \tag{12.1}$$

where:

E_S : deformation energy absorbed by the steel pipe while coating is not used (bare steel pipe);

E_{C1} : deformation energy absorbed by coating;

E_{C2} : effect of the coating on the energy absorption.

12.4.3.2 Steel Pipe Stiffness, *ks*

The indentation (δ-F) curve recommended by STATOIL (1996) for steel pipe is:

$$\delta = \left(\frac{1}{25\sigma_y^2 t^3}\right) F^2 \tag{12.2}$$

where:

δ : deformation (indentation) of the steel pipe;

F : impact force between trawl board and steel pipe;

σ_y : yield stress of the steel pipe;

t : wall-thickness of the steel pipe.

12.4.3.3 Coating Stiffness, k_{C1}

In general two types of insulation coating are used in the industry: rubber and plastic. Both rubber and plastic coatings will distribute the load to the steel underneath while they absorb part of the impact energy. It is recommended that finite element analysis and/or experimental tests are to be carried out in order to obtained load-indentation curves for insulation coating (k_{c1}) and possible effect the coating has on the steel shell stiffness (k_{c2}). At early design stage, no information is available with respect to k_{c1}, it is proposed to represent the coating indentation curve (k_{c1}) by empirical equations as below:

Case 1:

$$\delta_C = \alpha F^2 \tag{12.3}$$

or Case 2:

$$\delta_C = \beta F \tag{12.4}$$

where α and β are empirical coefficients to be calculated by equating the energy calculated from the above empirical equation with the energy absorption capability of the coating, E_C, obtained from the coating tests conducted by the manufacturer.

$$E_C = \int_0^{t_c} F\, d\delta_C = \frac{2t_c^{1.5}}{3\alpha^{0.5}} \tag{12.5}$$

and

$$E_C = \int_0^{t_c} F\, d\delta_C = \frac{t_c^2}{2\beta} \tag{12.6}$$

where:

t_c : coating thickness;

δ_c : deformation in coating.

Solving the above energy equation, we may get:

$$\alpha = \frac{4t_c^{\ 3}}{9E_C^{\ 2}} \tag{12.7}$$

and

$$\beta = \frac{t_c^{\ 2}}{2E_C} \tag{12.8}$$

and the indentation curve (δ_c - F curve) for the insulation coating:

For Case 1:

$$\delta_C = \left(\frac{4t_c^{\ 3}}{9E_C^{\ 2}}\right) F^2 \tag{12.9}$$

and for Case 2:

$$\delta_C = \left(\frac{t_c^{\ 2}}{2E_C}\right) F \tag{12.10}$$

Through finite element simulation, it will be possible to know the indentation of coating $\delta_{coating}$. In such cases, energy actually absorbed by the coating is:

For Case 1:

$$E_{Coating} = \int_0^{\delta_{coating}} F\, d\delta_C = \frac{2\delta_{coating}^{\ 1.5}}{3\alpha^{0.5}} \tag{12.11}$$

and for Case 2:

$$E_{Coating} = \int_0^{\delta_{coating}} F\, d\delta_C = \frac{\delta_{coating}^{\ 2}}{2\beta} \tag{12.12}$$

The safety factor in the design of coating energy absorption capacity may then be calculated as:

$$\text{Safety Factor} = \frac{E_C}{E_{Coating}} \tag{12.13}$$

We may get for Case 1:

$$\text{Safety Factor} = \left(\frac{t_c}{\delta_{coating}}\right)^{1.5} \tag{12.14}$$

and for Case 2:

$$\text{Safety Factor} = \left(\frac{t_c}{\delta_{coating}} \right)^2 \tag{12.15}$$

12.4.3.4 Coating Effect on Steel Pipe Stiffness, k_{C2}

The coating effect on steel pipe stiffness, k_{C2}, may be established through finite element analysis. However, this requires material stress-strain curves for the coating from the manufacture. No information on the coating material properties is available for the preparation of the present technical note. The coating effect on steel pipe stiffness is *conservatively* neglected.

12.4.4 Trawl Board Stiffness, Mass and Hydrodynamic Added Mass

12.4.4.1 General

There are two masses associated with the trawl board:

- The Mass of the steel, m
- The hydrodynamic added mass, m_a

It is assumed that:

- Steel Mass, m-3500 kg
- Added Mass, m_a-2.14 m (STATOIL (1996)) for P-board

12.4.4.2 Trawl Board In-plane Stiffness Connected with Steel Mass

The mass of the steel, *m*, is connected to a spring which simulates the in-plane stiffness, k_i, of the board.

It is suggested by STATOIL (1996) that:

$$k_i = 500 \ \ \text{(MN/m)} \tag{12.16}$$

12.4.4.3 Trawl Board Out-of-plane Stiffness Connected with Added Mass

The added mass, m_a, is connected to a spring which simulates the bending stiffness, k_b, of the board.

It is suggested by STATOIL (1996) that:

$$k_b = 10 \text{ (MN/m)} \tag{12.17}$$

12.4.4.4 Pipe Stiffness, Mass and Added Mass

The mass of the activated pipe, m_p, is a function of time. The length of the activated pipe will increase during the impact.

The mass of the pipe consists of:

- The mass of the content within the pipe
- The mass of the steel pipe
- The mass of the coating
- The hydrodynamic added mass related to the pipe. According to DNV'81 pipeline rules, the added mass is 2.29 times the mass of the displaced water for pipes resting on the bottom, 1.12 times the mass of the displaced water for pipes with 1m elevation.

Pipe added mass is calculated for every case using the following equation:

$$M_a = C_m \cdot \frac{\pi \cdot OD^2}{4} \cdot \rho_w \tag{12.18}$$

where:

C_m : 2.29, (added mass coefficient);

M_a : added mass;

OD : pipe outside diameter, including all coatings;

ρ_w : water density.

The bending stiffness of the pipe, k_{pb}, is also a function of time and decreases over the time.

The pipe stiffness and mass are simulated using beam elements.

Seabed Soil Stiffness

The soil stiffness, k_{ps}, is a function of time and decreases over time. The soil stiffness will only be of concern for impacts with a downward vertical component, or when the soil forms a support to the pipeline in the opposite direction of the impact. No soil stiffness is assumed for pipeline free spans. If the pipeline is laid freely on seabed, not trenched or buried. No sticking effect is applied for pipe - soil interaction.

The soil is represented by a spring stiffness k_{ps} in the vertical direction and friction m in the horizontal direction.

In this study a constant soil spring stiffness is assumed due to the very small time period of the impact response.

12.4.5 Impact Response

12.4.5.1 General

In this section, detailed impact response analyses will be described. The purpose of this section is to model a trawl board impact on a pipeline resting on the seabed. The main acceptance criteria on assessing pipe behavior under dynamic load are pipe shell dent, coating deflection and forces and stresses in pipe body. The objective is to identify the minimum for coating characteristics to provide sufficient protection of the pipelines against impact load of trawl board - pipeline interaction.

12.4.5.2 Conclusions of the Analyses

The acceptance criteria for trawl board impact response under accidental loads is no leak, STATOIL (1996). This implies that:

- no direct contact between trawl board and steel pipe (no gouge should occur due to impact) - this means deformation of coating should be sufficiently less than the coating thickness.
- the dent depth of the steel pipe should be less than 5% of steel pipe diameter.
- the local equivalent stress in the pipe should be sufficiently small to avoid possible bursting - this failure mode is not considered to govern in the impact analysis.
- the maximum tensile strain should be sufficiently small to avoid possible fracture at girth weld - this failure mode is not considered to govern in the impact analysis.

Detailed modeling of local strain is also necessary in case we evaluate possible fracture at girth welds during impact process. However, bursting due to over-stress and fracture due to over-strain will be more critical during pullover process. These failure modes are therefore not considered in the present impact response analysis and will be evaluated in the pullover response analysis.

12.4.5.3 Finite Element Model

In the finite element calculations, it is assumed that:

- coating density $r_c = 1200\ kg/m^3$
- coating indentation curve is $\delta_C = \alpha\ F^2$

A 60 meters long pipe section is fixed at both ends. Pipe shell and coating are modeled by two nonlinear springs acting in direction of the impact and placed one after the other. The springs do not provide any reaction force on tensile deformation and the springs are unloaded along line parallel to slope at origin of loading curve (plastic behavior).

The trawl board is modeled by two structural masses: board steel & added mass. Both masses are connected in parallel by springs to the end of coating spring. These two springs have stiffness of trawl board's in-plane and out-of-plane stiffness respectively.

The soil to pipe interaction is modeled as Chapter 5.

In fact, as a result of investigations into effect of soil model on dent depth it can be concluded that for different soil - pipe models' results are very close. This can be explained by the very short time of impact. At the moment of time when impact force reaches maximum, pipe displacements are very small, so soil reaction is negligible.

In the first moment (initial conditions) all springs and contact elements are not loaded. Both masses have velocities in a direction of 45° to the "soil" plane.

12.5 Pullover Loads

The maximum horizontal force applied to the pipe model, F_p, is given by (DNV, 1997):

$$F_p = C_F V (mk)^{1/2} \gamma \quad (12.19)$$

where:

m : the trawl board steel mass;

k : the warp line stiffness $= \dfrac{3.5 * 10^7}{3 * d} [N/m]$, (for single 38mm ø wire);

d : water depth;

V : the tow velocity;

γ : load factor = 1.3.

The coefficient C_F is calculated as below:

$$C_F = 6.6\left(1 - e^{-0.8\overline{H}}\right) \text{ for polyvalent and rectangular boards;} \quad (12.20)$$

$$C_F = 4.8\left(1 - e^{-1.1\overline{H}}\right) \text{ for v-shaped boards.} \quad (12.21)$$

where:

$\overline{H}$: dimensionless height, $\overline{H} = \dfrac{H_{sp} + D/2 + 0.2}{B}$ (12.22)

H_{sp} : span height;

D : pipe diameter;

B : half-height of the trawl board.

For trawl boards maximum vertical force acting in the downward direction can be accounted for as,

$$F_Z = F_p(0.2 + 0.8e^{-2.5\overline{H}}) \text{ for polyvalent and rectangular boards} \quad (12.23)$$

$$F_Z = 0.5F_p \text{ for v-shaped boards} \quad (12.24)$$

The total pullover time, T, is given by

$$T_p = C_T C_F (m/k)^{1/2} + \delta_p / V \quad (12.25)$$

where:

$$\delta_p \approx 0.1(C_T C_F (m/k)^{1/2}) \quad (12.26)$$

C_T : coefficient for the pullover duration given as:

$C_T = 2.0$ for trawl boards.

The fall time for the trawl boards may be taken as 0.6 seconds, unless the total pullover time is less than this, in which case the fall time should be equal to the total time. Figure 12.3 shows typical time history of vertical forces and horizontal forces.

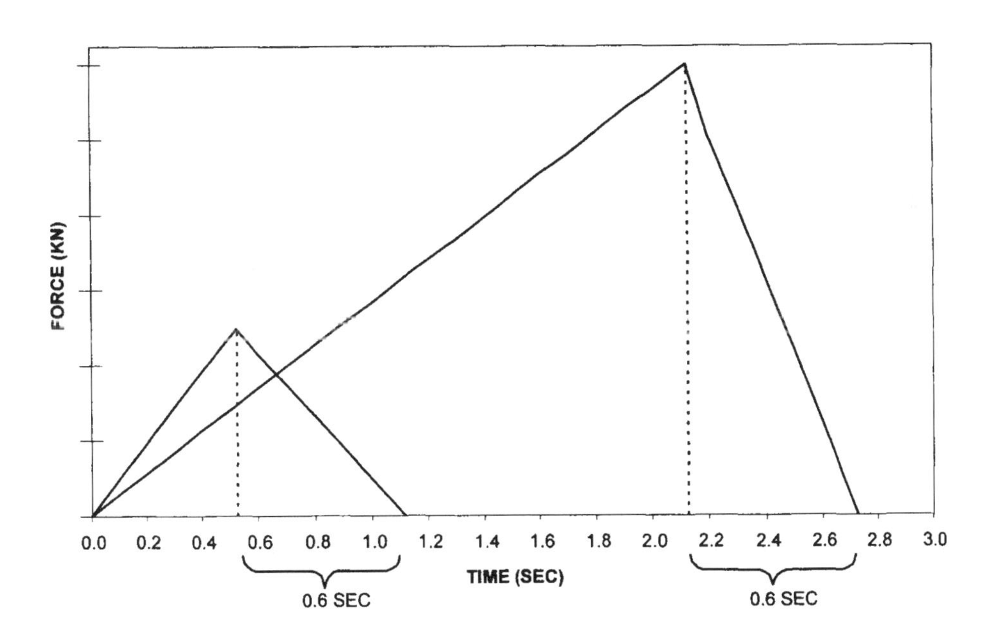

Figure 12.3 Pullover load vs. time history.

12.6 Finite Element Model for Pullover Response Analyses

12.6.1 General

Pullover response analysis procedures are presented in this section. Pullover is the stage of the trawl gear interaction in which the trawl board is held behind the pipeline and as the warp is tensioned by the movement of the vessel, the trawl gear is pulled over the pipeline.

In general, two base cases have been considered:

- pipeline in full contact with a flat seabed, and
- pipeline suspended in a free span on a real seabed.

In the first base case an ideally flat seabed profile has been assumed. In the latter base case a real seabed profile giving free spanning is imported from the most up to date pipeline route profiles and has been implemented to establish a span model. In the first base case, the seabed profile has been expanded to create a quasi 3-D seabed surface while in the second base case a certain width of real seabed has been used to represent the 3-D seabed.

The main objectives of these analyses are:

- to establish whether a pipeline with continuous contact with the seabed can withstand a pullover load;
- to establish whether a pipeline in a free span can withstand a pullover load or whether it requires protection, or if a maximum allowable span length is required to ensure the structural integrity of the span in the event of a pullover load;
- to investigate the influence of different parameters such as vertical soil stiffness, seabed friction etc. through sensitivity analyses.

- In the analyses presented herein, the pipeline is subjected to the most severe pullover loads.

12.6.2 Finite Element Models

The purpose of the analyses presented in this section is to study the pullover response. Non linear transient least-plastic finite element analyses using the model described in Chapter 7 is performed to achieve this purpose.

The length of the model has been selected so that full axial anchoring is achieved at a distance away from the pullover point, i.e. the effective force is unchanged after the time of impact.

12.6.3 Analysis Methodology

A real 3-D seabed surface from in-site survey data, with real spans which have been determined in in-place analyses, has been adopted for analyses of pullover on spans.

- The pullover load is modeled as a dynamic transient analysis.

- The pipeline has in general been assumed to be in its operating condition prior to pullover that is at its full design pressure and ambient temperature, and with operating content. Additional sensitivity analyses on the influence of soil friction, pre-buckled pipe, empty condition, different trawl board weight, low seabed stiffness and a packing condition with different internal temperature has been undertaken for the flat seabed model.

- Pipeline material stress-strain relationship is based on Ramberg-Osgood parametric curves at the design temperature.

- Added mass of the pipeline has been taken into account by attaching point mass elements to the pipeline nodes. An added mass coefficient of 2.29 is assumed in the analysis.

- Due to symmetry, only half of a pipeline section is modeled for the flat seabed cases and thus only half of the total pullover load is applied to the symmetric plane.

- The large-deflection option and material nonlinear option in ANSYS are activated. This means that geometric nonlinearly and material nonlinearly are taken into account, i.e. the change in overall structural stiffness due to geometrical changes of the structure as it responds to loading.

- A Coulomb friction model is assumed.

- No additional damping effect has been included in the analysis model.

The pullover load for the flat seabed cases is applied as a force vs. time history at the model's second end node (symmetry plane). For the real seabed cases the pullover forces are applied at the middle of the span investigated. Two force components are applied at the pullover point: one acting in the horizontal (lateral) plane and one acting vertical downward which tends to punch the pipe into the seabed and thus increasing the lateral restraint. The time-history of the pullover loads applied to the model are presented in Figure 12.3.

As industrial practice, upheaval buckling and lateral buckling have been considered as displacement-controlled situation, and strain-criteria are applied to check load effects.

However, free-spanning pipeline and pullover response have been cheated as load-controlled structures and moment criteria are to be applied to check load effects.

Ideally conditional load factors should have been defined for pullover response of pipelines for different slenderness (as Euler beam). This is an area of future research.

The following main assumptions in the pullover analysis have been made:

- Dents and ovalisation are not accounted for in the pipe elements, i.e. the pipe cross-section is always circular during deformation.
- An equivalent pipe wall density is used to obtain the correct submerged weight, accounting for the effect of concrete, corrosion coating and buoyancy.

12.7 Case Study

12.7.1 General

It has been common industrial practice in the North Sea, to trench or cover all pipelines less than 16" in order to protect them from fishing gear interference. To trench a pipeline is costly and may lead to an additional requirement to cover it with backfill plus rockdumping in order to restrain it from buckling out of the trench.

A 3-D non-linear transient Finite Element model has been developed to investigate the structural response of pipelines subjected to pullover loads. A realistic 3-D model of an uneven seabed is simulated by importing survey data directly into the model, see Tørnes et al. (1998).

Through a case study it will be shown how a 10" High Pressure – High Temperature flowline was found to be able to withstand pullover loads when left exposed on an uneven seabed, see Tørnes et al. (1998) for full documentation.

12.7.2 Trawl Pull-over for Pipelines on an Uneven Seabed

Pipelines installed in areas with uneven seabed will have a number of free spans along the pipeline route. Furthermore, a HP/HT pipeline laid on an uneven seabed may have undergone global buckling prior to being exposed to trawl pullover loads. To assess the structural response of the line under these circumstances, it is necessary to apply the pullover loads on a 3D in-place model for a given load case. In the following examples, pullover simulations have been applied to a small diameter HP/HT flowline that has undergone global buckling prior to being exposed to the trawl load. Intermittent rock berms have been applied to control the thermal buckling behaviour and the model is therefore limited to the section of the flowline between two adjacent rock berms.

The vertical and horizontal configuration of the flowline in its as-laid condition, at its maximum pressure (370 barg) and temperature (135 deg), are shown in Figure 12.4. A large horizontal buckle has formed across the large span at KP 4.250 and a further span has formed at KP3.700. In this particular design case, the approach has been to use intermittent rock dumping as a means of controlling the buckling behaviour. The extent of the model has therefore been limited to approximately 1500 m, i.e. the distance between two rock berms. The effective axial compressive force prior to pullover has been reduced to about 5 tonnes (i.e. significantly less than for the flat seabed examples) due to the release of thermal and pressure strain into the lateral buckle. A friction factor of 0.3 has been applied.

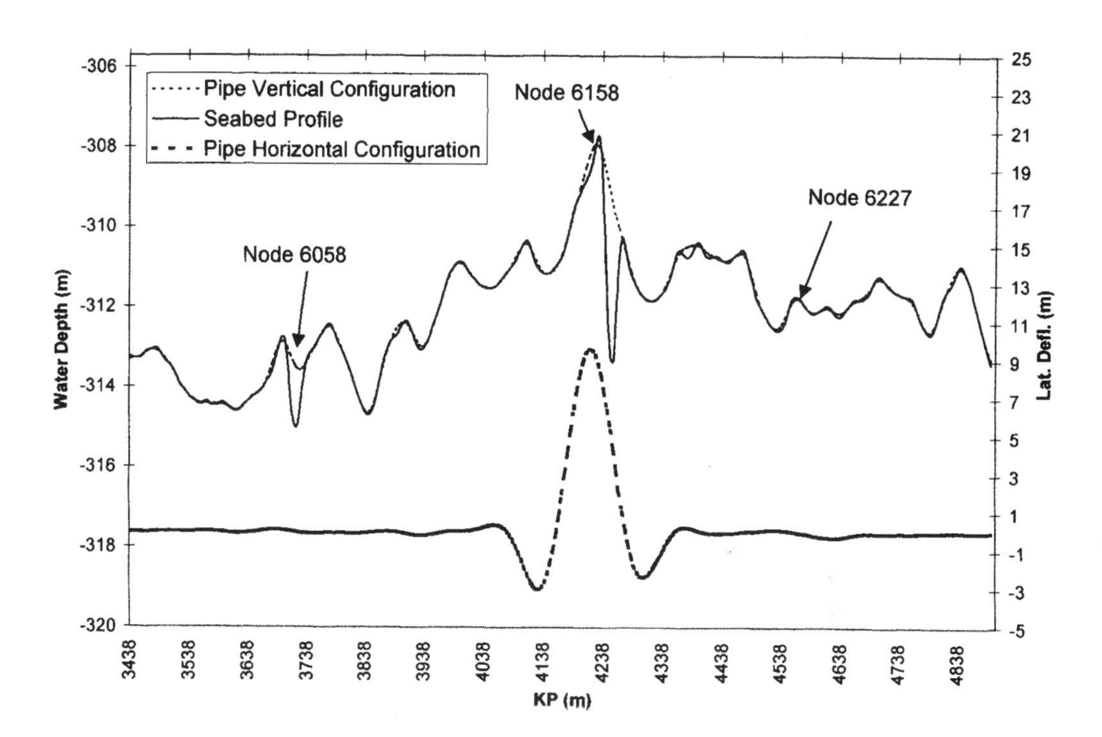

Figure 12.4 Uneven seabed: flowline configuration prior to pullover.

Basic Case: Pullover On Section In Contact With Seabed

The pullover load is applied at a point where the pipeline is in full contact with the seabed. Figure 12.5 shows the resulting pipeline configuration as a function of time. It shows how a new "buckle" has appeared at the point of impact with a permanent amplitude of 2.4 m, which is close to the result for the flat seabed (2.8 m).

In the example with uneven seabed, the effective compressive force was only 4 tonnes, which reduced to approximately 2 tonnes after the passage of the trawl board. In other words global buckling, which is associated with a large release of axial force did not take place. However, the pipeline is also here being "fed" into the "buckled" area, not because thermal strains are released, but because the axial "slack" in pipeline is being recovered. For a HP/HT pipeline, this "slack" is due to existing spans in the adjacent area, as in this case, or in existing buckles adjacent to the point of impact. This is an important observation; When analysing the flowline on a flat seabed, with the same axial effective force (4 tonnes), a considerably smaller lateral deflection occurred.

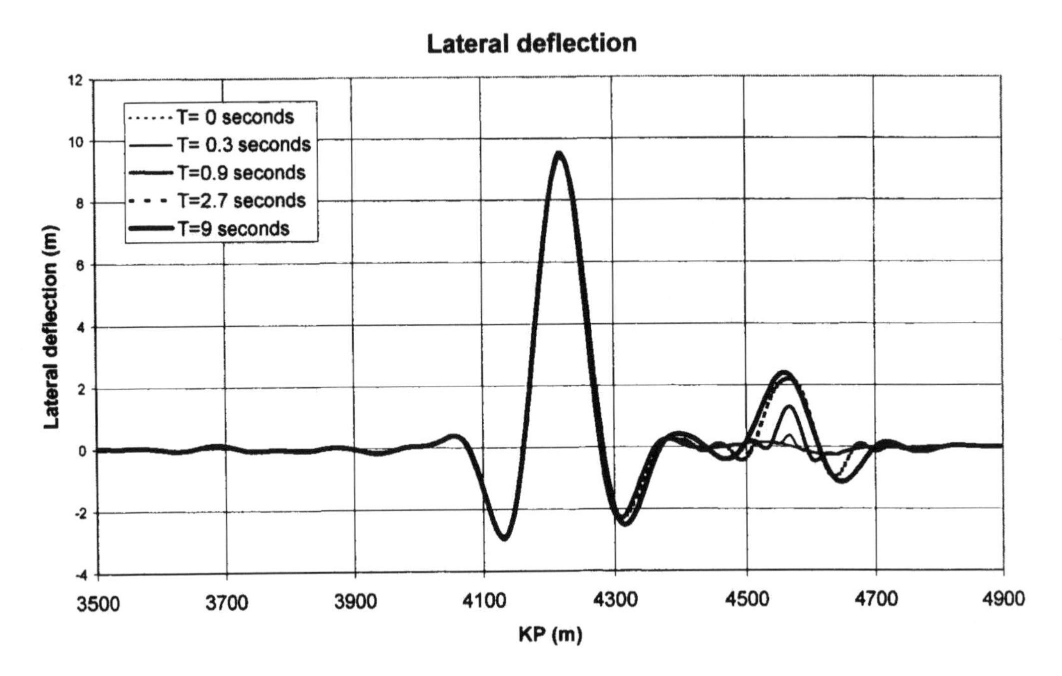

Figure 12.5 Lateral flowline configurations as function of time.

The resulting equivalent stress distribution after t= 0 second, t= 0.3 seconds and t=9 seconds are presented in Figure 12.6. Figure 12.6a shows the stress distribution in the flowline prior to trawl impact. The highest equivalent stress of 385 MPa has occurred in the buckled section and approximately 315 MPa at the hit point. In Figure 12.6b, there is a peak in the equivalent stress of about 395 MPa at the hit point after 0.9 seconds, which reduces to about 320 MPa after the passage of the board.

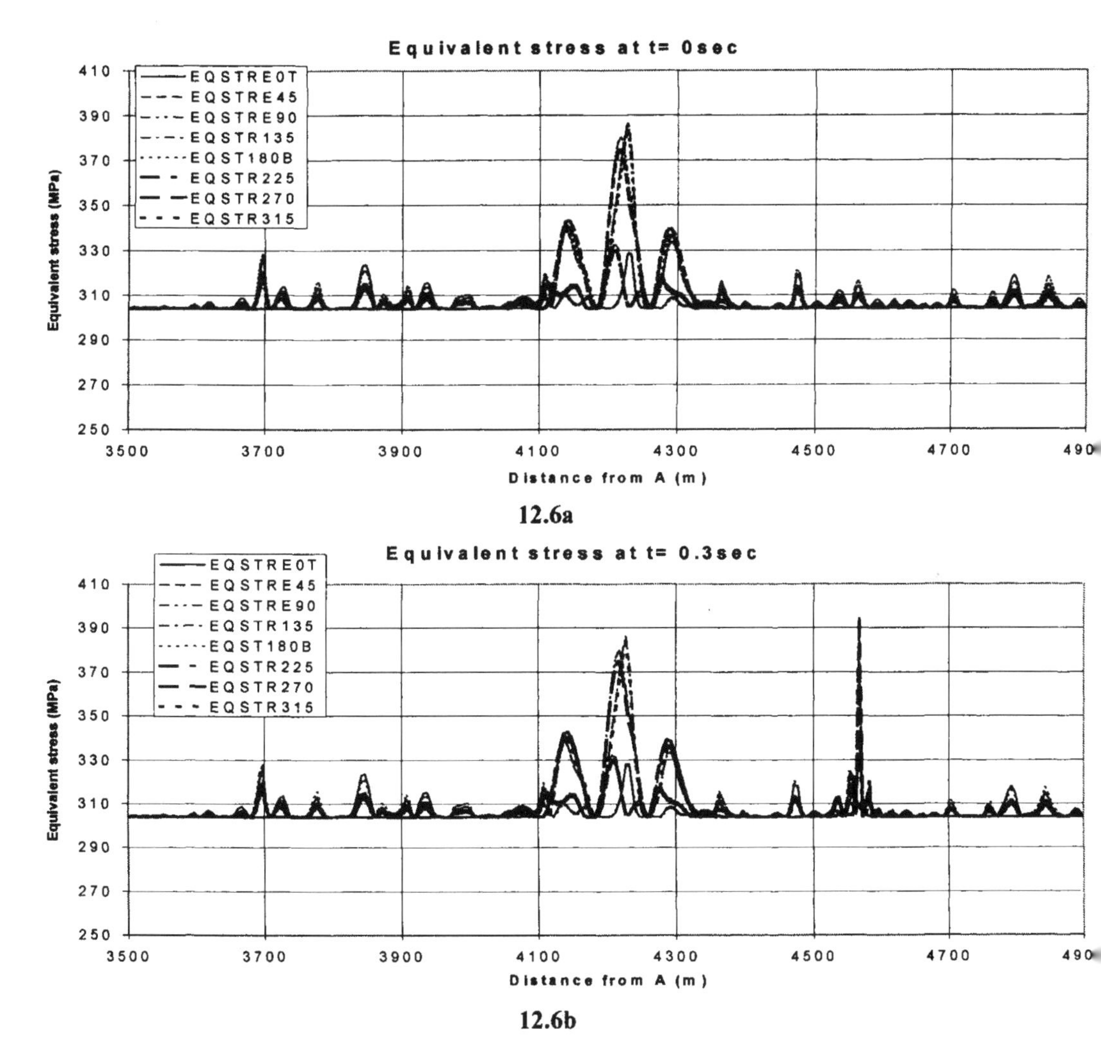

Figure 12.6 Equivalent stress distributions as a function of time.

Span Acceptance Criteria for Pull-over Loads for a 10" Flowline

The effective axial force in the line will vary from load case to load case. The 10-inch flowline are in effective compression during normal operation and pressure test and in tension in the temporary phases. The feed-in of expansion and resulting buckle amplitude during the trawl board pull-over will be larger when the pipe is in compression prior to being pulled over, which affects the stresses in the pipeline.

The location of the hit-point relative to neighboring spans and buckles will also affect the amount of feed-in into the buckled sections of the pipe during pull-over, and in turn the flexibility of the pipe. However, for the 10-inch flowline discussed above, it was demonstrated that span height is the governing parameter in structural response to trawl board pull-overloads.

In order to establish the critical span height with respect to trawl board pullover, a series of finite element analyses were performed for the 10-inch flowline. The analyses considered trawl-board pullover loads applied to the flowline at various free-spans along the route. The spans analyzed had different heights ranging from 0.1 to 1.2 m.

The critical span heights based on equivalent stresses and axial strains criteria were found to be as listed in Table 12.3:

Table 12.3 Critical span heights for trawl pull-over.

Critical Span Height (m)		
Operation	**Temporary**	
	Cool-Down	**Shut-Down**
0.25	0.40	+ 1.00

Note: Cool-down means ambient temperature combined with full design temperature. In the shut-down case, the temperature is ambient with no internal overpressure.

As long as the pullover load used as input to the analysis is a strong function of the gap height, other variables such as span length and axial force in the 10-inch flowline prior to impact, did not significantly affect the response.

The pullover loads and durations, based on Eqs. (12.19) to (12.26) for a 10-inch flowline, are presented in Figure 12.7 below. Among others, the following parameters influence the pullover loads:

- Flexibility, which is governed by pipeline diameter, wall-thickness, span length and supporting condition.
- Geometrical effects, for instance due to the relative position and motion between the trawl gear and the pipeline, front geometrical shape of trawl gear and the location where wire rope is attached. See Figure 12.8.

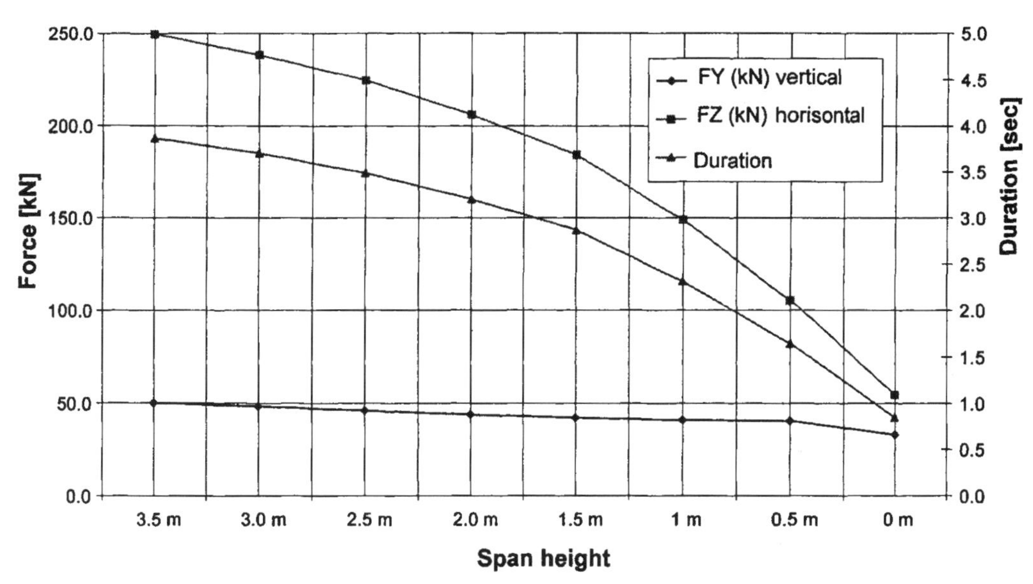

Figure 12.7 Pullover loads and duration for different span heights.

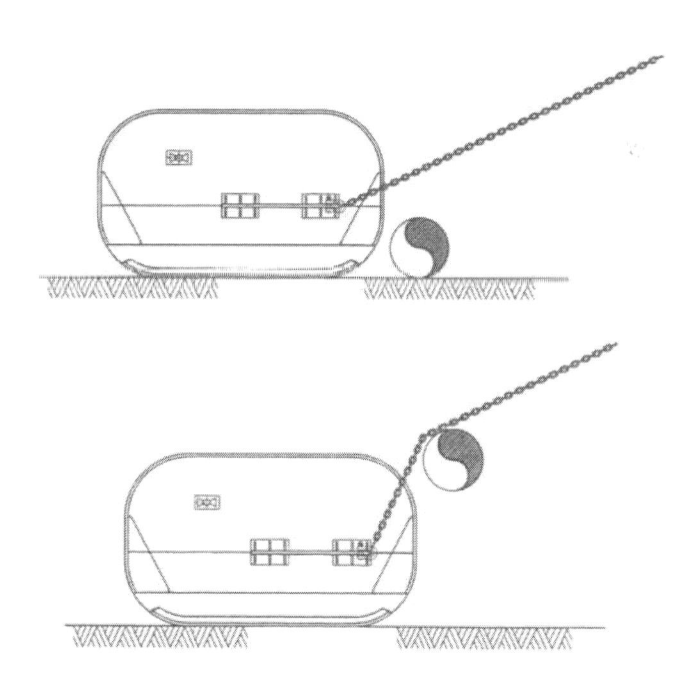

Figure 12.8 Geometrical effects on pull-over loads for pipelines on seabed and on a free-spanning pipeline.

12.8 References

1. ASME B31.8, (1992) "Code for Gas Transmission and Distribution Piping Systems", American Society of Mechanical Engineers (1994 Addendum).
2. Bai, Y. and Pedersen, P.T. (1993) "Impact Response of Offshore Steel Structures," International Journal of Impact Engineering, Vol. 13(1), pp. 99-115.
3. DNV, (1997) "Guideline NO.13 – Interference between Trawl Gear and Pipelines", Det Norske Veritas, Sept.
4. NPD, (1990) "Guidelines to Regulations Relating to Pipeline Systems in the Petroleum Activities".
5. Statoil, (1996) "Design Guidelines for Trawl Loads on Pipelines", Document No. UoD/FLT- 95051.
6. Tørnes, K., Nystrøm, P., Kristiansen, N.Ø., Bai, Y. and Damsleth, P.A., (1998) "Pipeline Structural Response to Fishing Gear Pullover Loads", Proc. of ISOPE'98.

Part II

Pipeline Design

Chapter 13 Pipe-in-pipe and Bundle Systems

13.1 Introduction

The main feature of pipe-in-pipe and pipeline bundle systems is that the pipeline is comprised of concentric inner and outer pipes. The inner pipe or pipes within sleeve pipes carry the production fluids and are insulated, whilst the outer pipe or carrier pipe provides mechanical protection.

The first known pipe-in-pipe system was installed in 1973 by Pertamina Offshore Indonesia. This pipeline was 8 miles long extending from shore to a single point mooring facility. The outer and inner diameters of this pipeline were 40" and 36" respectively.

Up till now nearly 36 pipeline bundles have been installed by controlled depth tow method (CDTM). The first one was installed at the Murchison field in 1980. The longest pipeline bundle is the one being designed, constructed and installed in Norwegian Sector by Rockwater. This bundle is 14 km long with 46" carrier pipe and three production lines.

This Chapter presents the design procedure and strength acceptance criteria for pipe-in-pipe and pipeline bundle system. The design should ensure that adequate structural integrity is maintained against all possible failure modes. All relevant failure modes for pipelines described in Chapter 4 are to be considered in the design of pipe-in-pipe and bundle system.

13.2 Pipe-in-pipe System

13.2.1 General

Many of the newly emerging generation of high pressure high temperature (HP/HT) reservoirs in the North Sea are being exploited using pipe bundles and single pipe-in-pipe configurations as part of subsea tie-backs to existing platforms. Not only are reservoir conditions more harsh but there is a need to insulate the flowlines to prevent wax and hydrate formation as the product cools along the length of the pipeline.

With the use of pipe-in-pipe systems come additional design features that are not present in conventional pipeline design. Challenging engineering problems rang from structural design of spacers and internal bulkheads to the understanding of the structural behavior both globally and locally under a variety of loading regimes. Due to the increased number of components in a pipe-in-pipe system compared with conventional pipelines, the design process is therefore more iterative in nature as the interactions of the components may necessitate design alteration.

A pipe-in-pipe system is essentially made up of an insulated inner pipe and a protective outer pipe. The function of the inner pipe is to convey fluids and therefore is designed for internal pressure containment. The inner pipe is insulated with thermal insulation materials to achieve the required arrival temperature. The outer pipe protects the insulation material from external hydrostatic pressure and other mechanical damage. Concrete weight coating is not normally required due to high submerged weight and usually low ocean current speeds in deepwater areas.

For the exploitation of HP/HT reservoirs, pipe-in-pipe system can provide the necessary thermal insulation and integrity for transporting hydrocarbon at high temperature (above 120 ^{0}C) and high pressure (in excess of 10000 psi). Pipe-in-pipe system comprises a rigid steel flowline inside a rigid sleeve pipe. The two pipes are kept apart by some form of spacer at the ends of each joint, and by bulkheads at the ends of the pipeline. The various proprietary systems in the market differ in the details of the spacers and bulkhead arrangements. The air gap between the inner and outer pipes provides the means of achieving the high thermal insulation. This air gap accommodates the insulation, which typically consists of either granular material poured into the inter-pipe annulus, or of a blanket form, which is wrapped around the inner pipe. In either case, the insulation material needs to be kept dry in order to maintain its insulation properties.

13.2.2 Why Pipe-in-pipe Systems

There are several conditions under which pipe-in-pipe systems (including bundles in this definition) may be considered for a particular flowline application over a conventional or flexible pipeline.

a) Insulation- HP/HT reservoir conditions

HP/HT flowlines require thermal insulation to prevent cool down of the wellstream fluid to avoid wax and hydrate deposition. There are many thermal coatings available that can be applied to conventional steel pipe but they tend not to be particularly robust mechanically and have not been proven at the temperatures now being encountered in HP/HT field, typically 150°C and above. A similar problem exists for flexibles in this respect. An alternative is to place the flowlines(s) inside another larger pipe, often called a carrier or outer sleeve pipe. The annulus between them can then be used to contain the insulating material whether it be granular, foam, gel or inert gas.

b) Multiplicity of flowlines

The bundle concept (pipes-in-pipes) is a well established one and a number of advantages can be achieved by grouping individual flowlines together to form a bundle. For specific projects the complete bundle may be transported to site and installed with a considerable cost saving relative to other methods. The extra steel required for the carrier pipe and spacers can be justified by a combination of the following cost advantages.

- A carrier pipe can contain more than one flowline. Common applications have also contained control lines, hydraulic hoses, power cables, glycol lines etc.
- Insulation of the bundle by the use of gel, foam or inert gas is usually cheaper than individual flowline insulation.
- In most cases there is no trenching or burial requirement due to the carrier pipe's large diameter. Since there are multiple lines within the carrier, seabed congestion within the filed is also minimized.

Bundle installation is commonly carried out through use of the Control Depth Tow Method (CDTM). The main limitation to the CDTM is the permissible length of bundle that can be installed, currently around 7.8 km. This is due to a combination of construction site and in-shore launch area size.

c) Trenching and Rock-dumping

Traditionally, flowlines less than 16-inch in diameter are trenched and/or buried. When contained within a sleeve pipe, which could be anything from 18-inch to 24-inch in diameter for single pipe-in-pipe systems and much larger for bundles, a reasoned argument for non-trenching can be made demonstrating that the line will not pose a risk to human life or the environment, nor will it become a hazard to other users of the sea. The cost associated with needing to trench, backfill and rock dump is often greater than that of the installation cost of the pipeline. By not trenching, buckling of the pipeline will only occur in the lateral direction across the seabed and there are methods to control such an event, e.g. mid-line spools or laying in a 'snaking' configuration. Upheaval buckling through the seabed, which is the more severe situation, can only be controlled through sufficient over burden being placed on the line in the form of rock dumping. These issues are addressed later.

In terms of impact from trawl boards or fishing gear, the external pipe acts as the first line of defense and although it may be breached, the integrity of the flowline will not compromised.

For certain applications, pipe-in-pipe systems offer significant cost saving over conventional pipelines, particularly when the need to trench, backfill and rockdump can be eliminated with additional mechanical and structural benefits as well.

13.2.3 Configuration

Various configurations of pipe-in-pipe can be used. The followings should be considered when determining the configuration.

- Gap thickness between the internal and external pipes: This should be optimized to maintain the heating;
- Thermal stability;
- Overall feasibility.

13.2.4 Structural Design and Analysis

There are four main structural parts to a pip-in-pipe system, these being the flowline, the sleeve pipe, the sleeve pipe connection (field joint) and internal components such as insulation material, spacers and bulkheads. Each component is designed to individual specifications initially and then the combined system must be analyzed to ensure that the local and global response to various loading regimes is satisfactory. In this way the interaction of all the components is checked, which is important as one component's behavior may affect the behavior of others. The design of pipe-in-pipe systems is therefore a more iterative process than the systematic approach used for conventional pipelines.

The carrier or sleeve pipe is line pipe but is sized in accordance with the requirements of the overall system. Diameter is usually dependent upon volume of insulating material required and wall thickness is generally determined on a hydrostatic collapse criteria, i.e. operating water depth. Sleeve pipe dimensions have a direct economic impact in that a larger pipe, whether it be diameter or wall thickness, means more steel and probably longer offshore welding time at each station on the installation barge.

As the sleeve pipe is not a pressure containing structure it is not subject to the same design codes as the flowline. In fact there is no applicable design code, as it is only a structural member, and therefore the design requirement is fitness for purpose. A general basis of 2% strain can be used for limiting design as this is of the order of strain seen by reeled pipe. Obviously, it is not desirable for the sleeve pipe to be at this level of strain for the duration of its lifetime but short excursions to this level can be tolerated, such as during installation in the limiting sea states.

If the flowline is designed to 0.1% plastic strain and this governs the limiting installation sea state (i.e. maximum permissible bend radius), then the sleeve pipe will be at a higher level of strain due to its larger diameter.

Of all the components it is the design of the sleeve pipe that is most flexible for achieving specific system characteristics. Optimization of sleeve pipe size and other advantages that are to be gained from a particular size sleeve pipe are dependent on the global behavior of the system which is addressed later in this section.

In terms of structural behavior, pipe-in-pipe system is categorized as being either compliant or non-compliant, depending on the method of load transfer between the inner and outer pipes. In compliant systems, the load transfer between the inner and outer pipes is continuous along the length of the pipeline, and no relative displacement occurs between the pipes, whereas in non-compliant system force transfer occurs at discrete locations. The structural design of pipe-in-pipe system is more complex than that of a single pipe system. The number of components in the system, comprising inner the outer pipes, spacers and bulkheads, and the increased amount of welding required.

The structural behavior of a pipe-in-pipe system is dependent on both the overall behavior of the system, and the mechanism of load transfer between inner and outer pipes. The overall effective axial force developed in the system is dependent on the operating conditions of temperature and pressure, and if the pipeline is in the end expansion zone, on the friction forces developed between the outer pipe and the soil. The stresses that develop within the pipe-in-pipe assembly are governed by the type of system used, i.e. compliant or non-compliant, and the presence of end bulkheads.

A pipeline lying on the seabed will develop effective axial compressive forces within the system when subjected to operating temperature and pressure. As the pipeline expands under operating conditions, soil friction forces between the outer pipe and the seabed oppose the free thermal expansion of the assembly and results in an overall effective axial compressive force developing within the system. The magnitude of the maximum overall effective axial force depends on whether or not the pipeline develops full axial constraint. If the pipeline is operating in the end expansion zone, then the overall effective axial force is a function of the soil friction and submerged weight and distance from the spool, given as:

$$P_{eff} = \int W_S \cdot \mu_a + R \tag{13.1}$$

where, P_{eff} is the overall effective axial force (compression positive), W_S is the submerged weight, μ_a is the pipe-soil axial friction coefficient and R is the resistance provided by the spool. The overall effective axial force increases from the spool location up until it develops full axial constraints, given by:

$$P_{eff} = -P_{true} + P_i - P_e \tag{13.2}$$

where:

P_{true} : true wall force;

P_i : $p_i.A_i$, force due to internal pressure;

P_e : $p_e.A_e$, force due to external pressure;

p_i : internal pressures;

p_e : external pressures;

A_i : inside areas of the inner pipe;

A_e : outside areas of the outer pipe.

The true wall forces for a pipe-in-pipe system comprises both contribution from the inner and outer pipes, *i.e.*

$$P_{true} = P_{t1} + P_{t2} \tag{13.3}$$

where:

P_{t1} : true wall forces in the inner pipe;

P_{t2} : true wall forces in the outer pipe.

At full constraint, the true wall forces are given by:

$$P_{t1} = EA_{S1}\alpha\Delta T_1 - \frac{\nu p_i D_1 A_{S1}}{2t_1} \tag{13.4}$$

and

$$P_{t2} = EA_{S2}\alpha\Delta T_2 - \frac{\nu p_e D_2 A_{S2}}{2t_2} \tag{13.5}$$

where:

E : Young's modulus;
A_S : cross sectional area of the pipe wall;
α : coefficient of linear expansion;
ΔT : temperature of the pipe wall;
ν : Possion's ratio;
p_i : internal pressure;
p_e : external pressure;
D : pipe diameter;
t : wall thickness of the pipe;
A_i : cross sectional area of the inside of the pipe.

subscripts 1 and 2 refer to inner and outer pipes respectively.

13.2.5 Wall-thickness Design and Material Selection

Compared to conventional pipeline, there are several issues practical considerations associated with the pipe-in-pipe system including insulation methods and insulation capabilities currently available, material and construction costs, ease of repair, and structure integrity issues.

The inner pipe will be designed to resist bursting under internal operating pressure and hydrotest pressure. The inner pipe may also be designed to resist collapse under external hydrostatic pressure and local buckling in case of leakage in the outer pipe.

For the outer pipe, the governing criteria are usually collapse and local buckling under combined loading of hydrostatic pressure and bending. Resistance to busting may also be required so hat fluid containment can be maintained in case of leakage in the inner pipe. This would be a contingency measure and would not be considered as a normal operating condition. For deepwater pipeline, the use of buckle arrestors is more economical to limit the extent of a buckle than having a thick wall to resist buckle propagation. This is particularly true for pipe-in-pipe systems self-weight needs to be kept low to ensure that the pipeline is installable.

The factors to be considered in material selection include adequate material toughness for fracture and fatigue performance and practical weld defect acceptance criteria, and whether or not sour service is required throughout the design life.

In situ stress conditions need to be assessed in wall thickness and material selection. Any stress locked into the inner and outer pipes as a result of installation procedures needs to be accounted for. Also, the tension or towing capacity of the installation vessel needs to be checked for both normal lay and contingency conditions.

13.2.6 Failure Modes

Failure modes described for single pipeline in Chapter 4 are applicable for the pipe-in-pipe system. Additional failure modes need to be considered for the design and assessment of the pipe-in-pipe system. The complexity and the differing load carrying capacity of the system add to the number of failure modes. In addition, the risk and consequence of each particular failure mode, and the impact on the pipe-in-pipe system will differ from a single pipeline.

Bursting

The burst capacity of the pipe-in-pipe system is determined based on the inner pipe subjected to the full internal pressure, and the outer pipe subjected to the full external pressure.

Fatigue and fracture

Pipe-in-pipe systems are subjected to both low cycle and high cycle fatigue due to daily operational fluctuations and start-up/shut-down conditions. One area particularly prone to fatigue is the weld joint. Typically, the weld joint for pipe-in-pipe systems comprise butt weld on the internal pipe, and either split shells or some form of sleeve arrangement for the external pipe connection. Special attention should be given to the fatigue assessment for the inner face of internal pipe since it is subjected to corrosive environment, and the outer face of the external pipe subjecting to seawater environment.

Global buckling

Due to effective axial force and the present of out-of-straightness (vertically and horizontally) in the seabed profile, pipe-in-pipe systems are subjected to global buckling, namely upheaval buckling and lateral buckling. The upheaval buckling should be investigated if the pipe-in-pipe system is intermittent rockdumped. Lateral buckling should be investigated for all the case.

13.2.7 Design Criteria

13.2.7.1 Stress Based Design Criteria

Stress-based design criterion is that the hoop and equivalent stresses are limited to some fraction of the SMYS depending upon the considered design cases.

The hoop stress criterion may be used with due consideration of material derating factor for both internal and external pipes.

The equivalent stress criterion limits von Mises stress to a fraction of SMYS. For D/t ratio larger than 20, the bi-axial form of equivalent stress should be calculated and the criterion reads as:

$$\sigma_e = \sqrt{\sigma_h^2 + \sigma_l^2 - \sigma_h \sigma_l + 3\tau^2} \leq \eta \cdot SMYS \tag{13.6}$$

where:

σ_h : hoop stress;

σ_l : longitudinal stress;

τ : shear stress;

η : usage factor .

For high pressure pipes with D/t ratio less than 20, the tri-axial form of equivalent stress should be calculated and the criterion reads as:

$$\sigma_e = \sqrt{\frac{1}{2}\left[(\sigma_h - \sigma_l)^2 + (\sigma_l - \sigma_R)^2 + (\sigma_h - \sigma_R)^2\right]} \leq \eta \cdot SMYS \tag{13.7}$$

where:

σ_R : radial stress

13.2.7.2 Strain Based Design Criteria

For high temperature (e.g. above 120^{0}C) pipe, the stress based design criteria might severely limit the hoop stress capacity due to internal pressure. In this case, the strain based design criteria can be applied instead of stress based criteria. The maximum equivalent plastic strain shall be calculated by

$$\varepsilon_p = \sqrt{\frac{2}{3}\left(\varepsilon_{pl}^2 + \varepsilon_{ph}^2 + \varepsilon_{pr}^2\right)} \tag{13.8}$$

where:

ε_{pl} : longitudinal plastic strain;

ε_{ph} : hoop plastic strain;

ε_{pr} : radial plastic strain.

The accumulated plastic strain shall satisfy $\varepsilon_p < 0.5\%$. Otherwise, fracture assessment shall be performed.

13.2.7.3 Local Buckling Design Criterion

Local buckling capacity check for both internal and external pipes shall in line with the criterion given in Chapter 3.

13.2.7.4 Global Buckling Analysis

To simplify the global buckling analysis, the pipe-in-pipe system could be modeled as single pipe by use of equivalent section concept. In contrast to upheaval buckling, lateral buckling is accepted if it does not result in unacceptable stresses and strains.

13.2.8 Insulation Considerations

Thermal analysis is fundamental to the design of a pipe-in-pipe system. Detailed descriptions about heat transfer and insulation are given in Chapter 19. The main drivers for an insulated flowline system are:

- To ensure that the product arrives at the topsides with a temperature above the wax appearance temperature.
- To ensure that hydrates do not form anywhere in the system.
- To reduce the rate of cool down in the event of a shutdown in order to allow sufficient time to re-establish flow or inject wax and hydrate inhibiting chemicals before the product reaches the WAT or hydrate formation temperature at any point in the system. The required cool down duration usually ranges from several hours to a few days.

Some of the typical thermal analyses are briefly described in the following:

- Flashing analysis of production fluid to determine hydrate curve. From this data the critical minimum temperature is established.
- Global thermal hydraulic analysis of the flowline system to determine the required overall heat transfer coefficient (OHTC) at each point in the system and length weighed average overall heart transfer coefficient for the system as a whole and hence determine if the insulation is required.
- The required OHTC determine the type and thickness of insulation to be used and hence determines the required cross-section of the pipe-in-pipe system. At this stage a trade off between the cost of insulation and the cost of injecting inhibition chemicals during operation may be feasible.
- Local heat transfer analysis to calculate the heat transfer coefficient (HTC) for each component of the pipe-in-pipe system.
- Based on the calculated HTC's performance a global thermal hydraulic analysis of the insulated flowline system to determine the temperature distribution along pipeline and check if it satisfies the required value.
- Perform local transient heat transfer analysis at strategic points along the system to develop cool down curves and hence determine cool down times to the critical minimum allowable temperature at each location.

13.2.9 Fabrication and Field Joints

Dependent upon the installation method chosen, a pipe-in-pipe flowline system may be the ideal candidate for utilizing onshore fabrication to reduce offshore fabrication time, as any such offshore operation will be fairly time consuming leading to low production rates in comparison to a single wall flowline. On shore fabrication site requirements will depend on system design and local availability of resources.

Prior to offshore installation, for most pipe-in-pipe system, onshore fabrication of the individual pipe-in-pipe joints is required. The inner pipe must be placed within the outer pipe and the annulus filled with the insulating material, or the inner pipe pre-coated with the insulation material must be slide into the outer pipe. The joint fabrication method would depend on the pipe-in-pipe system selected and the installation method and vessel selected. As part of the pipe-in-pipe joint fabrication, the joints could be made up as double, quad or hex joints in single operation to suit the installation method.

Field joint is a critical area for S-lay and J-lay installation. A suitable method that allows the welding of pipe joints in an efficient manner that maintains the integrity of the insulation and mechanical properties is essential. There are two basic methods available. The first, which is more applicable to J-lay is to allow the outer pipe to slide over the inner pipe after the inner pipe field weld has been made. The outer pipe is then welded after the field joint area has been insulated with suitable insulating material .t he technique is required and the integrity the pipe-in-pipe system during the sliding operation needs to be closely examined. This system may not be used for S-lay as the outer pie cannot be slid over the inner pipe in the firing line over multiple weld stations.

13.2.10 Installation

13.2.10.1 Installation Methods

The total submerged weight of the pipe section suspended in the water column increases as a faster rate than the water depth. A pipe-in-pipe system is generally much heavier than its single wall counterparts, therefore the tension capacity of the installation vessel becomes an important design factor given the generally low tension capacities of the existing installation vessels available on the market.

The methods for the installation of deepwater pipelines are S-lay, J-lay, reeling and towing. Detailed accounts on these methods have been made by various authors. A brief summary is given here to capture some the key characteristics of each method.

The S-lay method is tall active with the use of S-lay vessels with dynamic positioning and with stinger capable of very deep departure angles. With its long firing line and many work stations, an S-lay vessel can be reasonably productive. Limitations o the use of the S-lay technique are tension capacity and potential high strains in the overbend region, hence restrictions on combination of large pipe diameter and water depth.

The J-lay method results in a reduction in lay tension requirements. Also, large J-lay vessels have better motion characteristics and hence lower dynamic pipe stress especially at the stinger tip as compared to S-lay. However, productivity can be low due to limited number of work stations and rather confined working space. This shortcoming may be offset somewhat by the use of pre-fabricated quad or even hex joints. J-lay is generally not suitable for shallow water applications.

The reeling method can be very efficient, particularly for relative short length pipelines with the number of reloads can be minimized. The method is suitable for outer pipe diameter up to 16" with no concrete coating. Plastic strains developing reeling and the unreeling, and hence particular attention needs to be given to stress/strain conditions at bulkheads. Up till now, its application has been restrained by the relatively low tension capacity currently available. There are several large reel lay vessels under construction. The vessels will have very high tensions capacity, large drum and near vertical departure angle, make reel lay a strong alternative to J-lay for small to medium diameter, short to medium length pipelines.

Towing methods include several arrangements, mid-depth off-bottom and on –bottom tows. It can be very cost effective for flowline of short lengths. However, it may be restrained by factors such as the maximum pull of the tow vessels, the ocean current conditions, availability for a suitable onshore fabrication and launching site, and the seabed topography and soil condition along the tow route. Towing is particularly suitable for the installation of pipeline bundles where a large outer pipe can be used to house several flowlines and umbilicals.

13.2.10.2 Installation Analysis

In this section, various aspects of installation analysis for a J-lay operation starting with are considered. First of all, layability checks are performed for the worst cases as part of the wall thickness and steel grade selection exercise. Once the wall thickness and steel grade are finalized, detailed pipeline analyses are carried out covering both normal lay and contingency operations.

Static normal lay analysis establishes the optimum lay parameters, i.e., barge tension and J-lay tower inclination angle, for various sections along the entire pipeline route. Dynamic analyses on selected static conditions are then performed to confirm that the resultant stress/strain is acceptable and the vessel tensioner capacity is adequate. The cumulative fatigue damage a weld is estimated accounting for the various stress ranges it experiences as the pipe is lowered towards the seabed or is suspended below the water surface during undesirable weather conditions.

13.3 Bundle System

13.3.1 General

A pipeline bundle system consists of an outer carrier pipe, inner sleeve pipe, several internal flowlines, insulation system and appurtenances such as spacer, valves, chains, supports, etc. The carrier pipe is a continuous tubular structure that contains flowlines, sleeve pipe and is used to provide additional buoyancy to the bundle components during installation, structural strength mechanical protection during operation and corrosion free environment for the flowlines. Sleeve pipe is used to provide dry pressurized compartment for internal flowlines.

The internal flowlines are continuous linepipes without frequent branches used for transportation of fluids within the filed. Spacers are non-stress distribution elements provided to locate and support flowlines within the bundle configuration. Usually, the pipeline bundle is terminated by bulkheads which are stress distribution diaphragms. To facilitate the installation, ballast chains are attached to the bundle for the purpose of submerged weight adjustment and to suit installation by towing. Spools, which are short rigid flowlines, are needed to facilitate tie-in of the bundles/flowlines to structures.

The advantages of this concept include onshore fabrication and pressure test, simplified installation, thermal insulation and mechanical protection for the internal flowlines, etc.

13.3.2 Bundle Configurations

Bundle configurations can be grouped into two kinds namely conventional configuration and innovative configuration.

Conventional configuration is mainly carrier-based system as sketched in the figure below. All the rests of the bundle system are within the carrier pipe which provides buoyancy and acts as mechanical protection.

Innovative configuration is the one without carrier pipe or no inner sleeve pipes. One option is to use external single or multiple buoyancy pipe(s) for all flowlines as sketched in Figure 13.1.

The bundle configuration shall fulfill the weight and buoyancy requirements. In addition, the following principles shall be considered when selecting the bundle configuration:

- The center of gravity of the bundle has to be situated as near as possible to the vertical center line of the carrier and as far as possible below the horizontal center line of the carrier.
- The bundle should be configured at the bulkhead such that sufficient distance exists between the flowlines to allow for access during fabrication.
- The minimum clearance between the flowlines and the sleeve pipe shall be selected to allow for heat transfer.
- The configuration must allow for the design of suitable spacers.

13.3.3 Design Requirements for Bundle System

As a minimum requirement, the bundle system is to be designed against the following potential failure modes:

- Service Limit State
- Ultimate Limit State
- Fatigue Limit State
- Accidental Limit State

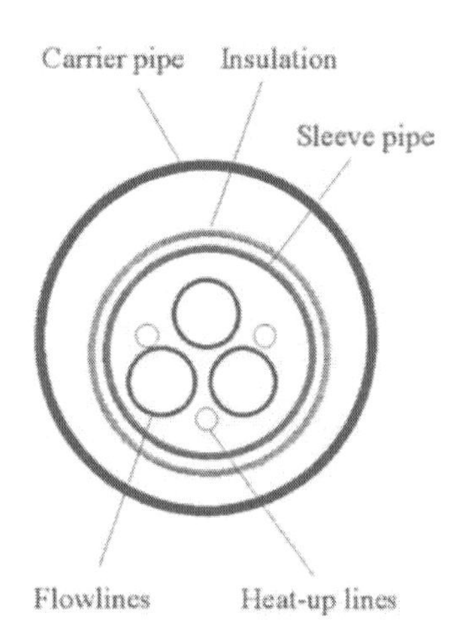

Figure 13.1 Bundle configurations.

13.3.4 Bundle Safety Class Definition

The safety class for flowlines, sleeve and carrier pipes may be tabulated below unless specified by clients.

Table 13.1 Safety class definition for the bundle

Pipes	Launch & Installation	Operation
Carrier and Sleeve	Normal	Low
Flowlines	Low	Normal
Heat-Up Lines	Low	Low

13.3.5 Functional Requirement

Design Pressure

The general design pressures for the bundle system shall be based on processing data. The internal pressure for carrier and sleeve pipes during installation should be at least 1 bar higher then the expected water depth.

Hydrotest Pressure

Hydrotest pressure of the flowlines, heat-up and service lines systems shall be based on 1.2×design pressure.

Design Temperature

Design temperature shall be based on the processing data. Significant temperature drop along the bundle system must be avoided.

Pigging Requirements

If the bundle system is designed for pigging, the geometric requirement shall be fulfilled. The minimum bend radius should be 5 times the nominal internal diameter of the pipe to be pigged. The pipe shall be gauged as a part of onshore and offshore testing of the system.

13.3.6 Insulation and Heat-up System

The following requirements related to insulation and heat-up functional requirement for the bundle system should be satisfied:

- During normal operation the temperature shall be above the hydrate formation temperature for the system.
- Minimum arrival temperature for the production lines shall be above hydrate formation temperature.
- A minimum of certain hours shutdown shall be accepted before the fluid in the production lines has reached hydrate formation temperature.
- To melt wax it shall be possible to bring temperature in the system up to above a certain degree.

The cool down time is the critical factor to determine the bundle insulation requirements. Therefore, the design of insulation thickness shall be based on the minimum cool down time. The insulation combined with active heating shall fulfil the heat-up requirement where applicable.

The following factors shall be considered for the bundle thermal design:

- Maximum / minimum operating temperature
- Cool down time
- Heat-up time

The following conditions shall be analyzed for bundle thermal analysis:

- Steady state

The evaluation of bundle steady state thermal performance includes the calculation of U-value and process fluid properties.

- Bundle cool down

The evaluation of bundle cool down thermal performance includes the transferring initial cool down properties from steady state analysis and calculating the bundle cool down time.

- Bundle heat-up

The evaluation of bundle heat-up thermal performance includes the calculation of the process fluid properties at initial and final heat-up conditions and calculation of the bundle heat-up time.

The following heat-up system operating parameters have been subject of the design:

- Maximum heating medium flowrate
- Heat-up time
- Heat-up system volume

The heat transfer inside the bundle and the bundle heat-up time are dependent on the following factors:

- Bundle configuration and the relative positions of the components
- Bundle length
- Heating medium temperature
- Properties of the fluid contained inside the flowlines

13.3.7 Umbilicals in Bundle

The general functional design requirements for umbilical are as follows:

- The level of redundancy shall be the same as for a system with separate umbilicals.
- The control system shall be protected during fabrication and testing period.
- Testing of the system shall be catered for during/after fabrication and during/after installation on the seabed.
- Seamless tubing, all welded, shall be used.
- The electrical cable shall be in continuous lengths to avoid splices and shall have suitable outer isolation to provide
- The electrical connectors shall be electrically isolated from the cathodic protection system of the bundle to avoid build up of calcareous layer on the metal parts.
- Components shall be located to minimize temperature effects.
- It must be possible to individually replace any pipeline or control jumper as well as any electrical jumper.
- Connection system should be compatible with the flowlines connection system.
- Future extension of the system should be planned for.

13.3.8 Design Loads

Load Definition

The following loads shall be considered in the design of the bundle system:

Weight : Self-weight of pipe, coatings, attachments, and transported contents;

Buoyancy : Displacement of carrier pipe;

Pressure : Internal, external, hydrotest, pressurization of carrier and sleeve pipe;

Expansion : Due to product temperatures and pressures;

Prestressing: Permanent curvature, permanent elongation;

Constraint : Of bulkheads, carrier pipe, towhead assemblies;

Launch Loads: Launch tension, environmental loading;

Tow/Installation: Loads due to tow and installation tension, tow and installation environmental loading and bundle submerged weight;

Accidental : Dropped objects etc.;

Environmental: Waves, current and other environmental phenomena plus loads due to third party operations;

Hydrodynamic: Loads induced due to the relative motion between the bundle and the surrounding seawater.

Tie-in : Loads induced due to pull-in and connection operation as well as running of the connection tool.

Temporary Phase Loads

The temporary phase loading to be addressed as part of the bundle component analyses are summarized below:

- Lifting and support during fabrication;
- Sheathing of the sleeve pipes with applied tension to flowline and sleeve sections;
- Sheathing of the carrier pipe over the sleeve pipe;
- Support of inner bundle on the sleeve pipe;
- Hydrostatic testing of the completed flowlines and subsequent leak testing of the sleeve and carrier pipes;
- Launch after tie-in of the towheads;
- Tow from fabrication site to the field;
- Installation at the designated infield locations;
- Flooding of the bundle flowlines and carrier pipe with subsequent hydrotesting of the flowlines.

Operational Phase Loads

The operational phase loads to be addressed as part of the flowline analyses are summarized below:

Hydrostatic collapse of flowlines considering axial tensions and support conditions

- Expansion loading during operation, considering thermal and pressure induced forces, support conditions within the carrier pipe and sleeve pipe and carrier free spanning.
- Stability of the bundles, considering environmental forces during extreme wave events, operational conditions within the flowlines and residual bundle curvature/displacement present after installation.
- Carrier expansion considering existing operational conditions within the bundle, including thermal/pressure induced loads, residual installation curvature/displacement of the carrier and possible free spans.

Additionally, where a sleeved pipe insulation system is present within the bundle, the effect of thermal/pressure effects upon the integrity of the sleeve and associated insulation bulkheads will be evaluated.

Load Combinations

The most onerous of the following loading conditions shall be applied to the bundle design.

- Functional loads alone.
- Functional loads plus simultaneous environmental loading.
- Accidental loads.

Functional loads are described as all loads arising from the flowline bundle system normal function and include in addition the loads imposed during launch and installation.

Environmental loading is usually direct loading resulting from wave, wind or current but may also include indirect loads, due to the environment, which are transmitted to the bundle system during installation.

Design Procedure and Acceptance Criteria

The design of bundle system shall ensure that the system satisfies the functional requirements and adequate structural integrity is maintained against all the failure modes. In principle, the design procedure and acceptance criteria for conventional single pipeline could be applied for the bundle system. Some special design considerations are needed, which are presented in this section.

Design Procedure for Bundle System

The first requirement of the bundle design is to determine the carrier pipe size. Having fixed the carrier size the bundle is considered with regard to its on-bottom stability, tow stresses, mechanical protection etc. The recommended design procedure is shown in Figure 13.2.

The carrier pipe is generally regarded as an installation aid. After installation the carrier pipe provides the flowlines with protection from impact. Consideration of this policy is required when carrier pipes contain flowing fluids to provide either a cool down or heat up process, as they may be considered as a pipeline.

Flowlines are usually sized according to processing data. The wall thickness of the flowlines depends on the internal pressure containment. However, in high temperature applications especially with CRA material thermal loadings must be considered with regard to flowline sizing and more likely material properties that will apply at elevated temperatures.

The insulation "U" value will be determined from thermal and processing analysis. The weight and volume of the insulation are needed for bundle design. Thick coatings of polymer insulation can result in carrier pipes of large cross section enclosing relatively small weights producing in excess buoyancy. In such cases consideration may have to be given to flooding a flowline to provide additional weight.

Pipe-in-pipe type insulation presents a good balance between volume and weight. Post installation annulus fill insulation does not in itself affect the bundle weight but this type of insulation may require pipelines to accommodate expansion or filling requirements.

The weight of all the bundle component parts must be determined. Generally only carrier pipe displacement is considered in the buoyancy calculation. The displacement of external anodes, clamps and valves is accounted for by using a submerged weight for these items in the weight calculations. The objective of this weight and displacement determination is to arrive at a carrier pipe diameter that will provide a resultant buoyancy of 200 N/m +3% of steel pipe weight. The 3% figure stems form the weight tolerance and the 200 N/m figure is suited to carrier pipe diameters of 32-inch or greater. If diameters are less than 32-inch then 100 N/m should be used. The optimum carrier could be found through a reiterative process to calculate the weights and buoyancy.

The installed submerged weight, expansion analysis and flowline equivalent stresses are considered to ensure that the bundle design will be stable once installed and has no adverse effect on the permanent flowlines.

The submerged weight of the bundle is compared with the minimum submerged weight of the bundle required to satisfy Morison's classical two dimensional theories. The cyclic varying horizontal velocity in the water particles introduced by the design wave is superimposed on the steady bottom current velocity at the height of the bundle. The bundle is considered stable when actual submerged weight is greater than calculated minimum, applying a safety factor of 1.1. Bundle expansion due to the temperature and pressure of the product in the flowlines shall be considered. The geometry of the installed bundle is such tat the flowlines are attached to the carrier pipe at the extreme ends only via a solid bulkhead. Over their length the flowlines are supported by spacers maintaining their positions relative to and parallel with, each other and the carrier pipe. The carrier pipe is supported over its length by the seabed. Expansion of the flowlines exerts loads on the bulkhead which in turn mobilizes the carrier pipe which is itself constrained by seabed friction. The bundle is a system with boundaries at a free end and at the associated anchor point. The bundle expansion analysis includes determining external and internal forces acting on the system, calculating axial strain of system and integrating the axial strain of unanchored bundle to determine the expansion.

The check on flowline equivalent stresses is important especially with high temperature applications not only for the higher thermal loads produced but also the reduction in yield strength with some materials.

Carrier stresses shall be looked at during the carrier selection and should be reviewed in light of the stresses determined during the expansion analysis.

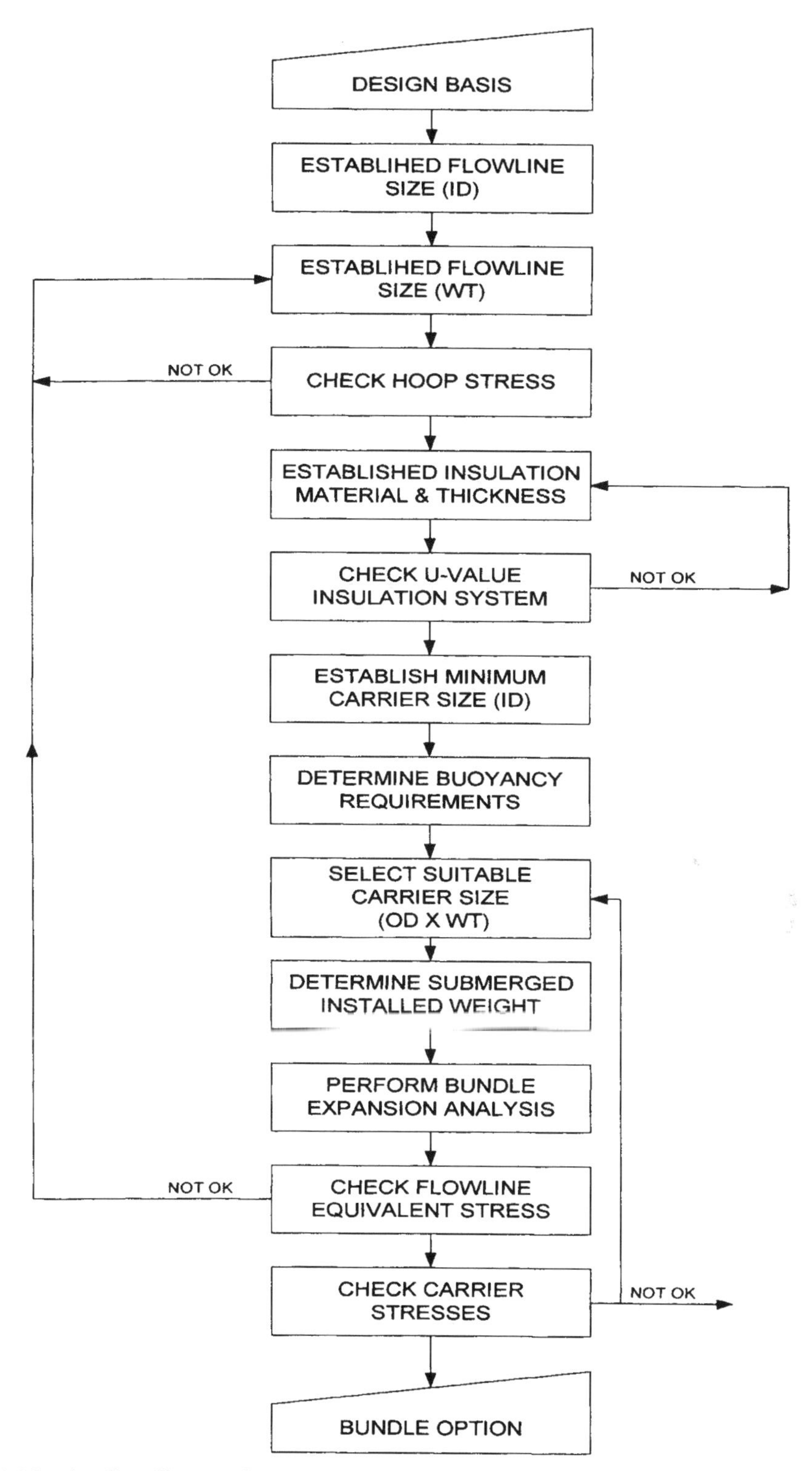

Figure 13.2 Design flow diagram for bundle system.

Design Criteria for Bundle System

The carrier and the sleeve pipes shall be designed in accordance with the criteria given in Chapter 4. The deduction of large D/t ratio of carrier and sleeve pipes on bending moment should be accounted for in the maximum allowable bending moment criterion defined in Chapter 4.

Global equivalent stress in the carrier pipe during installation shall be limited to 72% of yield stress. Local equivalent stress in the carrier pipe during installation shall be limited to 90% of yield stress.

The hydraulic, chemical and temperature monitoring tubes located within the bundles could also be designed according to those Chapters.

The design criteria shall be applicable for the design situations of installation, hydrotest and operating.

Wall Thickness Design Criteria

The wall thickness design of pipes within the bundle system shall take into account the following.

Hoop stress

The hoop stress criterion is in principle applicable for the bundle system with the following special considerations.

The allowable hoop stress for flowlines and heat-up lines in Safety Zone 1 and those inside the bundle carrier in Safety Zone 2 shall be limited to 0.8*SMYS.

The allowable hoop stress for the riser pipe, spool pieces and towhead piping in Safety Zone 2 shall be limited to 0.67*SMYS.

Collapse due to external hydrostatic pressure

Wall thickness shall be designed to avoid pipe collapse due to external hydrostatic pressure. A collapse analysis may be performed in accordance with Chapter 3. The carrier pipe and the sleeve pipe shall be pressurized to a pressure one bar greater than the maximum external pressure at the deepest point in the installation area to prevent collapse.

Local buckling

The flowlines, sleeve pipe and carrier pipe shall be designed to withstand local buckling due to the most unfavorable combination of external pressure, axial force and bending. The design may be carried out in accordance with Chapter 4.

On-bottom stability

The wall thickness design shall be adequate to ensure the on-bottom stability of the bundle without any additional means.

Installation stress

The wall thickness shall be adequate to withstand both static and dynamic loads imposed by installation operations.

Hydrotest and operational stresses

The wall thickness shall be adequate to ensure the integrity of the flowlines, sleeve pipe and carrier pipe under the action of all combinations of functional and environmental loads experienced during hydrotest and operation.

On-Bottom Stability Design

The bundle system shall be stable on the seabed under all environmental conditions encountered during installation, testing and throughout the design life. The bundle shall be designed with sufficient submerged weight to maintain its installed position, or to limit movement such that the integrity of the bundle system is not adversely affected.

On-bottom stability shall be verified in accordance with Chapter 9 for the following cases:

Environmental Load Combination	Flowline Content	Phase
1 year significant wave and 1 year current	Empty	Installation
10 year significant wave and 100 year current (if current dominates)	Product filled	Operation
100 year significant wave and 10 year current (if waves dominates)	Product filled	Operation

Allowable Free Spans Design

Maximum allowable span lengths for the bundle system shall be calculated for both static and dynamic loading conditions. Analyses shall consider all phases of the bundle design life including installation, testing and operation. 1 year wave and current shall be considered for as-installed and hydrotest conditions. Combination of 10 year wave and 100 year current or 100 year wave and 10 year current resulting in the higher environmental load shall apply to operational conditions.

Bundle Expansion Design

The design shall take into account expansion and/or contraction of the bundle as a result of pressure and/or temperature variation. Design pressure and the maximum design temperature shall be used in bundle expansion analysis. The presence of the sleeve pipe shall be taken into account.

Bundle Protection Design

The bundle system shall be designed to against trawl loads outside the trawl free around the installations. The flowlines and umbilicals shall be protected against dropped objects around the installations. Carrier pipe and bundle towheads shall offer sufficient protection against the dropped objects with impact energy of 20 kJ.

Impact loads from dropped objects for the protection structure design shall be treated as a PLS condition.

Corrosion Protection Design

Cathodic protection design shall be performed according to relevant codes.

Cathodic protection together with an appropriate protective coating system shall be considered for protection of the bundle external steel surfaces from the effects of corrosion.

Sleeve pipe will be protected from corrosion by provision of chemical inhibitors within the carrier annulus fluid.

Flowlines within the sleeve pipe will be maintained in a dry environment. Therefore, cathodic protection system is not required.

Bulkheads and Towhead Structure Design

The bulkheads will form an integral part of the towhead assemblies. The towhead structures shall remain stable during all temporary and operational phases. Stability shall be addressed with respect to sliding and overturning with combinations of dead-weight, maximum environmental and accidental loads applied. The design of towhead structure shall be in accordance with relevant structural design code like API RP2A.

Bundle Appurtenances Design

Relevant design codes should be applied for the design of bundle appurtenances based on their functional requirements.

Fabrication and Construction Design

Design check and analysis shall be performed to confirm the adequacy of the selected pipe wall thickness to withstand loads during fabrication and construction phases and to ensure that the pipe stress values remain within the specified limits.

13.3.9 Installation by CDTM

The most feasible and reliable way of bundle installation is by use of Controlled Depth Tow Method (CDTM), which is a subsea pipeline installation system.

The principle of the CDTM involves the transportation of the bundle towed between two lead tugs and one trail tug. By controlling the tow speed in combination with the tension maintained by the trailing tug the trailing towhead, the bundle configuration and its deflections are kept under control during the tow. The essential parameters are continuously monitored during the two, and adjusted if necessary to maintain the desired bundle configuration, well clear of the seabed, nominal position during tow is some 30 m below sea surface.

The complete installation of bundle system includes the following main activities

Launch

Upon completion of fabrication and testing, the bundle will be outfitted for tow and installation with ballast chain, telemetry system and other installation aids. Breakout and pull forces during various stages of the launch shall be calculated and assessed for the bundle system.

Pre-tow preparation

The pre-tow preparation will commence with the activities including towhead inspection, trimming, bundle submerged weight check and tow preparation.

Tow to field

The bundle system will be towed to the field by use of CDTM along a pre-surveyed route. During tow, the drag on the ballast chain creates a 'lift force', and so reduces the bundle submerged weight. This lift force will result in a complete lift off from the seabed into CDTM mode. When the two arrives near the field, the bundle will be lowered to the seabed in the designated parking area situated in front of the bundle installation area.

Infield installation

The infield installation of the bundle system will be carried out by remote intervention, which will be carried out directly by ROV.

The bundle will be towed at a slow speed in off-bottom mode into the installation area. After adding weight to the bundle, the off-bottom tow can commence. During the off-bottom tow the bundle position must be monitored at all times.

The bundle will be pulled in at a straight line. A temporary target box will be determined for the leading towhead.

When the towheads and bundle position have been confirmed, flooding down of the bundle can commence.

CDTM involves the transportation of prefabricated and fully tested flowlines, control lines and umbilicals in a bundle configuration suspended between two tugs. A further vessel accompanies the tow as a patrol/survey vessel. To maintain control during tow, the bundle is designed and constructed within specified tolerances with respect to its submerged weight.

The bundle is designed to have buoyancy, this being achieved by encasing the bundled pipelines, control lines, umbilicals etc. inside a carrier pipe. Ballast chains are attached to the carrier pipe at regular intervals along its length to overcome the buoyancy and provide the desired submerged weight.

The tow speed has a direct lift and straightening effect on the bundle. By controlling the tow speed in combination with the tension exerted by the tugs the bundle tow characteristics and deflections are maintained.

The tow is controlled by adjustment of the tow wire length, tow wire tension, tow speed and the tug's relative positions. In this manner the tow depth, catenary shape, stresses and movement are kept within specified operational limits under given environmental conditions.

During tow the bundle is kept well clear of the seabed to enable a safe and unobstructed passage. The towheads are kept below the surface to minimize the effect of surface waves. The towhead depth is normally about 30m below the surface but this controlled depth can be increased or reduced by adjustment of the tow wire lengths.

On arrival in the field the bundle is gradually lowered by adjustment of the controlling parameters (tow wire length, forward speed and tension) and the bundle settles in a position of equilibrium above the seabed with the lower portion of the chains resting on the seabed. Once in this position the bundle can easily be maneuvered in the off-bottom mode to its final position and the towheads located in the required target areas. The carrier annulus is flooded with inhibited seawater and the bundle settles on the seabed.

13.4 References

1. Carmichael, R., Fang J., and Tam, C., (1999) "Pipe-in-pipe Systems for Deepwater Developments", Proc. of Deepwater Pipeline Technology Conference, New Orleans.
2. Dixon, M. and Patel, M., "Analysis of Methods for Pipe-in-Pipe Systems".
3. McKelvie, M., "Bundles – Design and Construction", Integrated Graduate Development Scheme, Heriot-Watt University.

Part II
Pipeline Design

Chapter 14 Seismic Design

14.1 Introduction

Oil and gas pipeline routes often pass through large geographical areas, from the supply point to the end-user, crossing seismic-active areas. Earthquake damage to oil and gas pipelines can cause significant financial loss, including secondary losses resulting in service interruption, fires, explosions, and environmental contamination. Examples of such catastrophes include the 1964 Alaska Earthquake; the San Fernando Earthquake of 1971; the Guatemala Earthquake in 1976; the 1987 Ecuador Earthquake; the Kobe Earthquake in 1995 and the 2003 Algeria Earthquake. A general conclusion drawn from a review of many earthquake events shows that, for buried steel pipelines, the direct effect of seismic ground wave on the integrity of long and straight pipelines is generally not significant. Where there is permanent ground deformation due to soil failure, there may be a severe influence upon pipeline integrity. For unburied pipelines, both seismic ground wave and permanent ground deformation can cause severe damage to pipelines, depending on the pipeline geometry and connected structures.

Damage to pipeline systems during an earthquake, whether onshore or offshore, can arise from the traveling ground waves and permanent ground deformation due to soil failures. The primary soil failures are:

- Faulting;
- Landslides;
- Liquefaction;
- Differential Settlement;
- Ground-cracks.

Seismic ground waves produce strains in buried pipelines. However, because there are little or no inertia effects from dynamic excitation, the strains tend to be small and often are well within the yield rupture threshold of the pipeline material. The direct effect of seismic waves is, therefore, generally not expected to cause rupture or buckling failure to buried pipelines. Nonetheless, seismic waves can cause damage to unburied pipeline systems, especially in the interfacing area, such as in the pipeline transition section from buried-to-unburied and the pipeline tie-in spool to the subsequent structure. In general, the seismic analyses of the permanent ground deformation for buried pipes and unburied pipes, and seismic ground waves for unburied pipes are required for designing pipeline systems.

Offshore pipelines are normally buried for stability and mechanical protection; otherwise they are laid on the seabed. This Chapter will:

- Address available seismic design codes and standards for offshore pipelines;
- Discuss a general design and analysis methodology for fault crossing and seismic ground wave;
- Present design and analysis examples using a static model for buried pipe, subjected to permanent ground deformations due to the foundation failure and a time history dynamic model for unburied pipelines subjected to seismic ground waves.

14.2 Pipeline Seismic Design Guidelines

The American Society of Civil Engineers, ASCE (1984) collected some published systematical papers in seismic analysis and design as a standard, giving seismic design guidelines for oil and gas pipeline systems. These guidelines provide valuable information on seismic design considerations for pipelines, primarily onshore-buried pipelines, and also force-deformation curves of the pipe-soil interactions for pipelines buried in both clay and sand. ASCE (2001, 2002) has also developed seismic design guidelines for onshore piping systems and buried pipes, but not for petroleum pipelines and offshore pipelines. The American Society of Mechanical Engineers (ASME) states that the limit of calculated stresses due to occasional loads, such as wind or earthquake, shall not exceed 80% of SMYS of the pipe, but this specification does not provide guidance for the design method. Det Norske Veritas (DNV) in the code of "Submarine Pipeline Systems" classifies the earthquake load into accidental or an environmental load depending on the probability of earthquake occurrence. It also does not provide an earthquake design method for offshore pipelines.

The current Design Code and Guidelines for pipeline systems basically specifies the loads for analysis and the acceptable stress/strain levels for the system design. For buried pipelines, the parameters of interest are the displacement, stress and strain under the imposed permanent ground deformation due to foundation failure. Although the mechanism of the seismic foundation failure varies for different types, a pipeline response model can be generated with only minor modifications. For the unburied pipeline, earthquake design motions are typically presented in the form of a seismic time history ground motion or a design response spectrum, which is based upon the estimated ground waves and characteristics of the ground structure.

14.2.1 Seismic Design Methodology

Several seismic analysis approaches for pipeline design were developed to predict the pipeline behavior in response to differential ground movements. Two main structural response models are considered:

1. Static Model for Buried Pipelines, subjected to fault crossing due to soil failure.
2. Dynamic Analysis Model for Unburied Pipelines, subjected to ground wave load.

14.2.1.1 Static Analysis of Fault Crossing

Fault crossing is one of the major hazards to offshore pipelines, whether buried or unburied. Numerous investigations have been carried out for fault crossing with different soil movements. The ability of a pipeline to deform in the plastic range under tension helps prevent rupture at fault crossings. If compression of the pipeline in a fault crossing is unavoidable, the compressive strain should be limited to within the local buckling criteria.

The amount and type of ground surface displacement is the main factor for designing pipelines to resist permanent ground deformation at fault crossings. Bonilla (1982) summarized a simple equation relating the maximum displacement at ground surface to the earthquake surface-wave magnitude as:

$$\log L = -6.35 + 0.93 M_s \tag{14.1}$$

where, L is the maximum surface displacement in meters and M_s is the earthquake surface-wave magnitude. The earthquake magnitude is one of the design criteriae based on the historical seismicity and geological data. Displacement data from the fault of similar earthquakes might be used in selecting a value for designing pipelines because of a big deviation in earthquake surface displacement data, which the equation is based on.

Two typical analytical methods under certain assumptions were suggested for the fault crossing analysis, Newmark-Hall (1975) and Kennedy et al (1977). Kennedy and others extended the ideas of Newmark and Hall and incorporated some improvements in the method for evaluation of the maximum axial strain. They considered the effects of lateral interaction in their analyses. The influence of large axial strains on the pipe's bending stiffness is also considered. O'Rourke and Liu (1999) reported that the Kennedy model for strike slip faulting, which results in axial tension, provides the best match to ABAQUS finite element results, based on an independent comparison of the available analytical approaches. The ASCE Guidelines give a detailed description of both the Newmark-Hall and Kennedy schemes. It must be emphasized that both schemes are only valid for pipe under tension, since this condition may not be guaranteed under other various combined modes of fault movement.

Due to the largely non-linear nature of the problem, a finite element analysis (FEA) is the most general tool for pipeline fault crossing design. Non-linear finite element modeling allows accurate determination of pipeline stress/strain at various locations along the pipeline route with a wide range of parameters. The pipe-soil interaction can be modeled as discrete springs in three dimensions. The pipeline is represented as a sequence of finite straight beam elements supported on the bottom by the bearing springs. The imposed fault movement is then input into the FE model as a static displacement boundary condition. The analysis is performed to determine the equilibrium nodal position of the pipe, bending moment, axial force, strains and stresses. The next section explains a detailed example of finite element analysis for the fault crossing using ABAQUS software.

14.2.1.2 Ground Wave Analysis

Both permanent ground deformation and seismic ground wave can cause severe damages to unburied pipelines and connected equipments. There are three basic methods available for analyzing the responses of a structure subjected to seismic ground wave,

1. Static Analysis;
2. Response Spectra Analysis;
3. Time History Analysis.

In general, a static analysis is sufficient for the long-term response of a structure to applied loads. However, if the duration of the applied load is short, such as in the case of an earthquake event, a time history dynamic analysis is required.

Static Analysis

The pipeline is divided into individual spans or into a series of segments. Static seismic loads are considered to be in direct proportion to the weight of pipe segments. The peak acceleration from the response spectrum is applied as a lateral force distributed along the pipe and bending stresses and support reactions are calculated. The seismic static coefficients are usually obtained from the seismic "zone", which is corresponding to a level of seismic acceleration. Many design software programs can perform static analysis, but these methods are primarily used in building seismic design.

Response Spectra Analysis

In response spectra analysis, the ground motion vs. frequency method is used. The maximum acceleration for a given frequency and damping is determined based on seismic maps and soil characteristics. The higher the damping, the lower its acceleration will be. The responses of displacements (translations and rotations), loads (forces and moments) and stresses at each point for each natural frequency of the system and for each direction are obtained after analysis. The calculated loads, displacements and stresses of the piping system are typically calculated by taking the square root sum of squares of the response in each of the three directions. The response spectra method is approximate, but is often a useful, inexpensive method for preliminary design studies.

Time History Analysis

This analysis method involves the actual solution of the dynamic equation of motion throughout the duration of the applied load and subsequent system vibration, providing a true simulation of the system response at all times. In time history analysis, the seismic time history ground motions (displacement, velocity or acceleration as a function of time) of seismic ground waves in three directions are applied to a finite element model of a system to obtain time history excitations of the system, including stresses, strains and reaction forces. Time history analysis is a more accurate, more computationally intensive method than response spectrum analysis, and is best suited to the transient loadings where the profile is known.

An example of time history analysis with a finite element model for the ground wave movement with ABAQUS software is detailed in the next section. ABAQUS is the selected program to develop finite element models of ground soil, pipelines and subsea manifold connection because of its capability to accurately simulate solid objects, pipes, elbows, material and geometric non-linearities, and interactions between soil and pipelines. ABAQUS also provides analytical models to describe the pipe-soil interaction. These models describe the elastic and perfectly plastic behavior by defining the force exerted on the pipeline and its displacement. These definitions are suitable for use with sands and clays and can be found in detail in the ASCE guidelines for the Seismic Design of Oil and Gas Pipeline Systems.

14.2.2 Seismic Level of Design

Two design levels are normally adopted for the design criteria:

1. Contingency design earthquake (CDE), and
2. Probable design earthquake (PDE).

The CDE represents a higher-level earthquake, established on the basis of a geo-seismic evaluation with a typical return period of 200 to 1000 years for pipelines. The intensity of CDE is taken as the design limits, exceeding causes of pipe failure, or at least sufficient damage to cause an interruption of service. On the other hand, the PDE is a lower level earthquake, which assumes only minor damages to the pipeline system without interrupting the service. These events are likely to occur during the life of the pipeline and are therefore incorporated as part of the design environmental load. PDE is usually taken to have a return period of 50 to 100 years.

14.2.3 Analysis Examples

To explore the seismic responses of offshore pipeline systems, two study examples are presented here:

- Static response of a 42-inch buried pipeline to permanent ground deformations where the pipeline is fully buried under the natural seabed.
- Dynamic response of a 42-inch unburied pipeline system to seismic waves where the pipeline is laid on the seabed and connected to a subsea manifold.

14.2.3.1 Buried Pipeline Responses for a Fault Crossing

A buried steel pipeline with a 42-inch diameter and a 0.875-inch wall thickness, material of API 5L Grade-X65, contains oil at a specific gravity of 0.8. The pipeline is backfilled with a 3-foot sand depth median, with a density of 120 pounds per cubic foot and a friction angle of 35°.

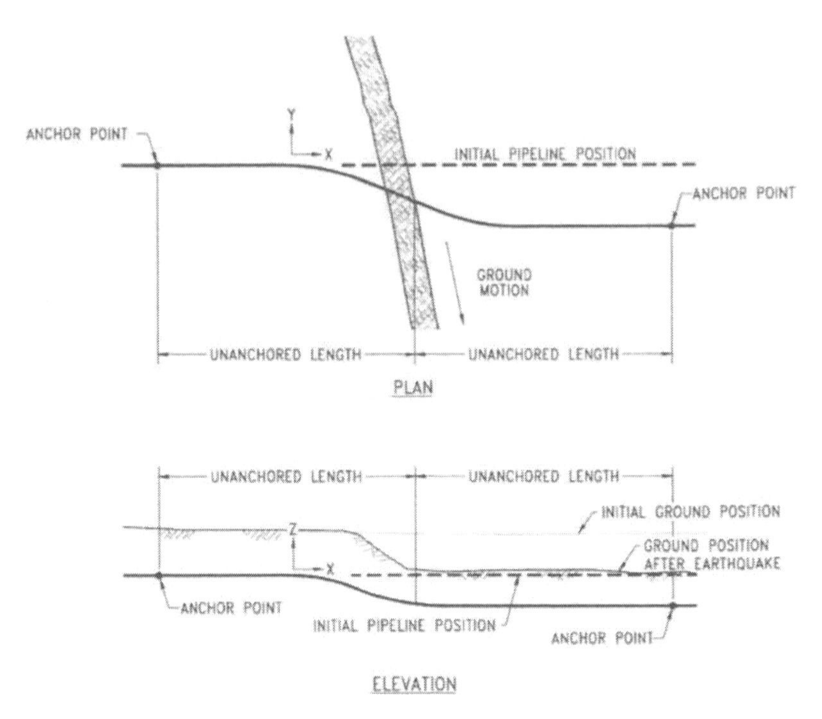

Figure 14.1 Buried pipeline under a fault crossing.

Figure 14.1 shows a sketch of a buried pipeline under a fault crossing due to an earthquake. The fault length in the plan direction is set as 1.2 m, in the vertical direction, with set as 1.0 m. A static analysis of buried pipeline was analyzed by using ABAQUS, the Finite Element software. Here, the unanchored length varies depending on the pipeline size and axial pipe-soil interaction force (friction force). The 1000 m long pipeline, with both ends fixed, is modeled by using pipe elements in the example.

Non-linear pipeline-soil interactions in axial, lateral, and vertical directions are modeled with pipe-soil interaction elements and soil characteristics in f_t-x_t, f_p-y_p and f_q-z_q force-deformation curves. Based on the formulas suggested in the ASCE guidelines, the maximum axial interaction force per unit length at the pipe-soil interface (ft is 36.6 kN/m, and corresponding maximum deformation, x_t) is 0.004 m. The maximum lateral interaction force per unit length (f_p is 175.4 kN/m, and corresponding maximum deformation, y_p) is 0.00 m. The maximum upward interaction force per unit length f_q is 38.0 kN/m and corresponding maximum deformation zq is 0.044 m. The maximum downward interaction force per unit length f_q is 1450 kN/m and corresponding maximum deformation z_q is 0.13 m.

Figure 14.2 shows the displacements of the pipeline in y and z directions under the fault crossing. The corresponding stress distribution at the bottom wall along the pipeline is shown in Figure 14.3. The maximum stress exceeds 80% of SMYS of the pipe, which is within ASME criteria. Therefore, the designed buried pipeline is not suitable for the seismic level which can cause inputted fault distances.

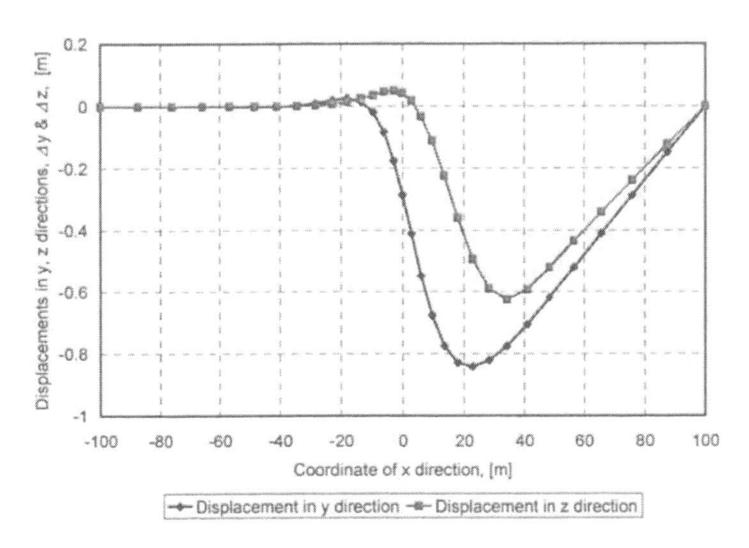

Figure 14.2 Deformations of pipeline in y and z directions.

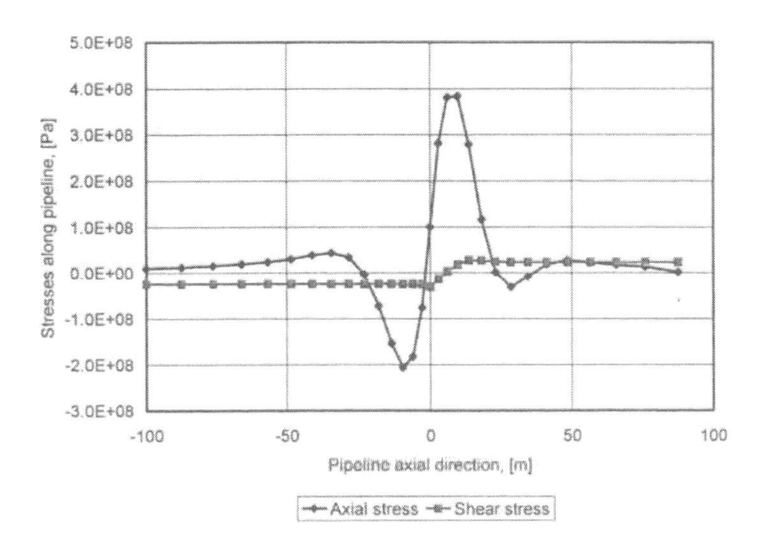

Figure 14.3 Stress distributions at the bottom wall along the pipeline.

Sensitivity calculations of different buried depths of the pipeline also show that the maximum stress and strain of the pipeline are proportional to the buried depth, when other parameters are the same. To decrease the damage of the pipeline, in the possible area of the seismic fault cross, the pipeline should not be buried.

14.2.3.2 Responses of Unburied Pipelines for a Ground Wave

A seismic dynamic analysis was performed, using ABAQUS, for an offshore pipeline system. This analysis consisted of two 42" OD x 0.875" WT (API X65 pipelines) and a 300 metric ton subsea manifold, as shown in Figure 14.5. The pipelines contained oil at a specific gravity of 0.8 with an internal pressure of 600 psi. A settlement of 0.1 m for the subsea manifold due to sand liquefaction in the earthquake, is considered.

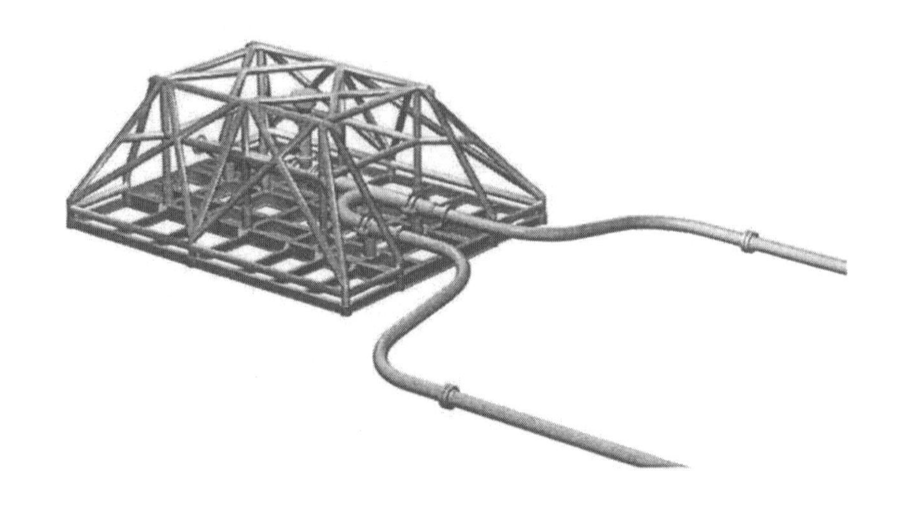

Figure 14.5 Offshore pipeline system, with a subsea manifold.

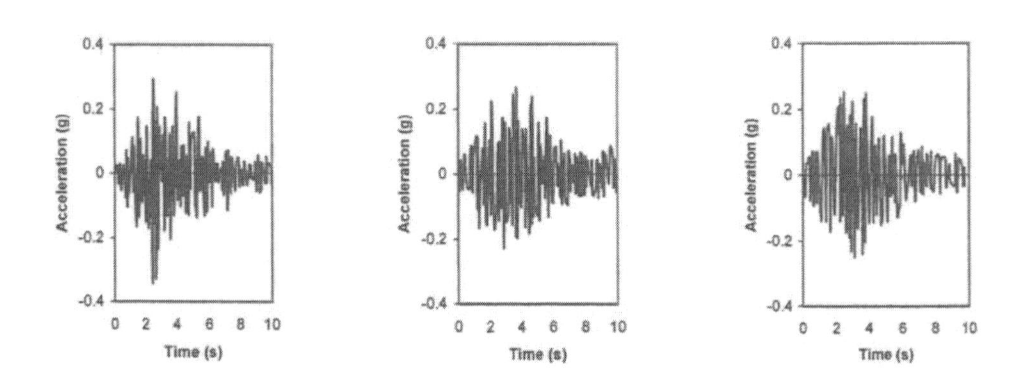

Figure 14.6 Seismic ground motions: E-W, N-S and vertical accelerations.

A 10-second seismic event was used in the dynamic analysis. Figure 14.6 shows the acceleration time history in the E-W, N-S and Vertical directions. The maximum accelerations are 0.34g, 0.26g and 0.25g for E-W, N-S and vertical directions, respectively.

In the ABAQUS model, the subsea manifold was modeled as a solid box. The straight and curved pipeline sections were modeled as 3D beam elements and elbow elements, respectively. The seabed was modeled as a rigid surface with frictions in both longitudinal and lateral directions. The pipeline-soil interaction was modeled by a linear contact pressure relationship. The accelerations in three directions were applied to the seabed. As shown in Figure 14.7, the maximum Von Mises stress of 191.9 MPa (27.8 ksi) occurs at the spools. Figure 14.8 shows the time history of the maximum Von Mises stress in the pipelines.

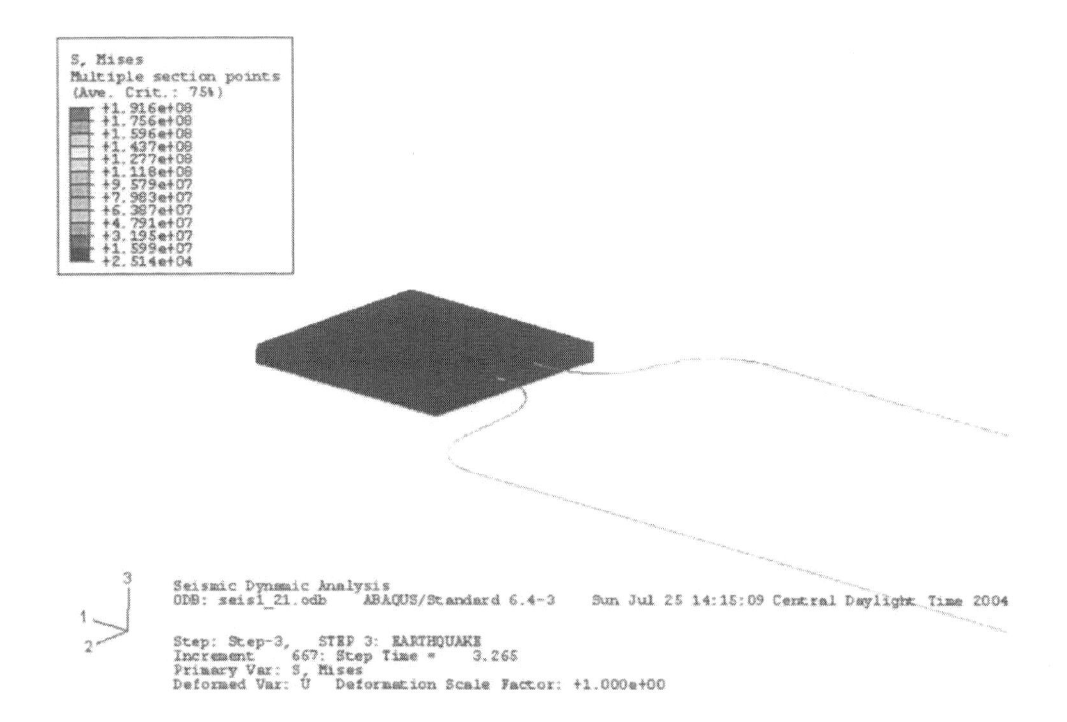

Figure 14.7 Maximum Von Mises stress in the pipelines and tie-in spools.

The maximum Von Mises stresses in the time history always occurs in the spool areas. The difference of natural frequencies and weights for the subsea manifold and pipelines causes the response difference between subsea manifold and pipelines. Therefore, the maximum stress occurs in the spool areas.

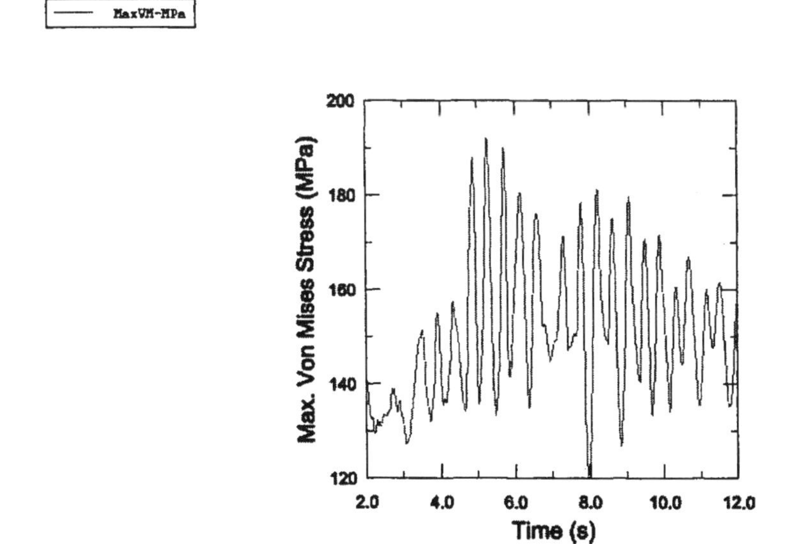

Figure 14.8 Time history of maximum Von Mises stress.

14.3 Conclusions

This seismic design and analysis methodology as presented here was developed for offshore pipeline design. It has been successfully applied in seismic analyses of buried pipelines under fault crossing and unburied pipelines with a subsea manifold by using a static analysis and a dynamic time history analysis. The sensitivity analysis results show that the buried depth of buried pipeline and the soil stiffness in the pipeline-soil interaction are the primary factors affecting pipeline stress in an earthquake. As discussed, the seismic analysis within this technical note is intended for assistance in developing seismic analysis and design guidelines for offshore pipelines.

14.4 References

1. Bai, Q., Zeng, W., and Tao, L. (2004), "Seismic Analysis of Offshore Pipeline Systems", Offshore, Vol. 64, No. 10, 2004, pp.100-104.
2. ASCE, (1984), "Guidelines for the Seismic Design of Oil and Gas Pipeline Systems".
3. ASCE, (2001), "Guideline for the Design of Buried Steel Pipe".
4. ASCE, (2002), "Seismic Design and Retrofit of Piping Systems".
5. ASME B31.4, (1998), "Pipeline Transportation System for Liquid Hydrocarbons and Liquids".
6. Bai, Y., (2003), "Marine Structural Design", Elsevier.
7. Bonilla, M. G., (1982), "Evaluation of Potential Surface Faulting and other Tectonic Deformation", Open File Report 82-732, U.S. Geological Survey.
8. DNV-OS-F101DNV, (2000), "Submarine Pipeline Systems", Det Norske Veritas.
9. Kennedy, R. P., Chow, A. W., and Williamson, R. A., (1977), "Fault Movement Effects on Buried Oil Pipeline", Journal of the Transportation Engineering Division, ASCE, Vo. 103, No. TE5, pp. 617-633.
10. Newmark, N. M. and Hall, W. J., (1975), "Pipeline Design to Resist Large Fault Displacements", Proc. US National Conference on Earthquake Engineering, Ann Arbor, Michigan.
11. O'Rourke, M. J. and Liu, X., (1999), "Response of Buried Pipelines Subject to Earthquake Effects", Monograph No.3, Multidisciplinary Center for Earthquake Engineering Research.

Part II

Pipeline Design

Chapter 15 Corrosion Prevention

15.1 Introduction

Infrastructures such as steel pipelines are susceptible to corrosion. This Chapter deals with coatings and external corrosion protection such as cathodic protection (CP). The preferred technique for mitigating marine corrosion is use of coatings combined with CP. Coatings can provide a barrier against moisture reaching the steel surface therefore defense against external corrosion. However, in the event of the failure of coatings, a secondary CP system is required.

Corrosion is the degradation of a metal by its electro-chemical reaction with the environment. A primary cause of corrosion is due to an effect known as galvanic corrosion. All metals have different natural electrical potentials. When two metals with different potentials are electrically connected to each other in an electrolyte (e.g. sea water), current will flow from the more active metal to the other causing corrosion to occur. The less active metal is called the cathode, and the more active, the anode. In Figure 15.1, the more active metal Zn is anode and the less active metal steel is cathode. When the anode supplies current, it will gradually dissolve into ions in the electrolyte, and at the same time produce electrons, which the cathode will receive through the metallic connection with the anode. The result is that the cathode will be negatively polarized, and hence protected against corrosion.

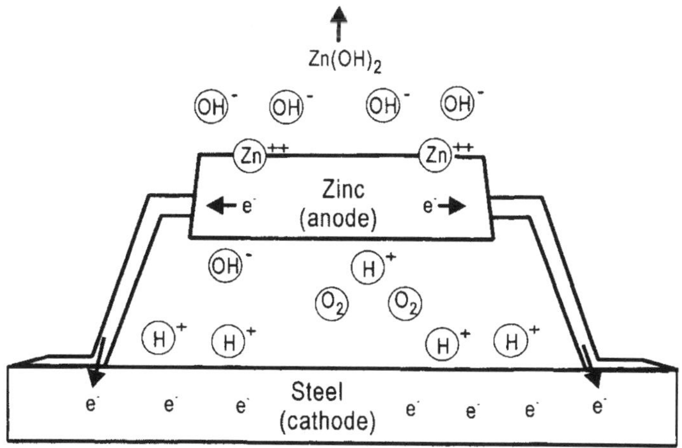

Figure 15.1 Galvanic corrosion.

15.2 Fundamentals of Cathodic Protection

Carbon steel structures exposed to natural waters generally corrode at an unacceptably high rate unless preventative measures are taken. Corrosion can be reduced or prevented by providing a direct current through the electrolyte to the structure. This method is called cathodic protection (CP) as showed in Figure 15.2.

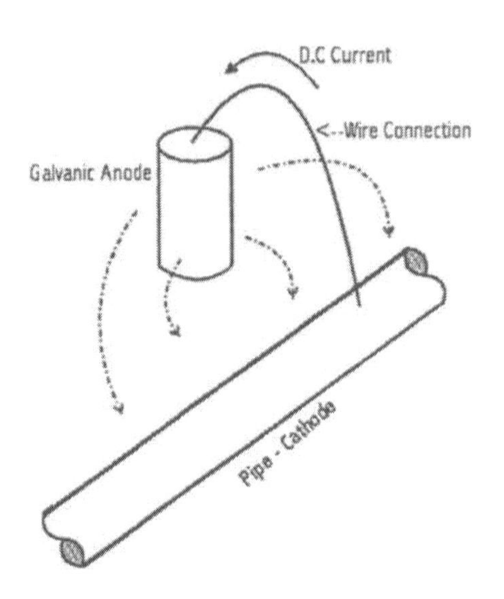

Figure 15.2 Cathodic protection of pipeline.

The basic concept of cathodic protection is that the electrical potential of the subject metal is reduced below its corrosion potential, and that it will then be incapable of going into corroding. Cathodic protection results from cathodic polarization of a corroding metal surface to reduce the corrosion rate. The anodic and cathodic reactions for iron corroding in an aerated near neutral electrolyte are,

$$Fe \rightarrow Fe^{2+} + 2e^- \quad (15.1)$$

and

$$O_2 + 2H_2O + 4e^- \rightarrow 4OH^-, \quad (15.2)$$

respectively. As a consequence of reaction (15.2), pH of the seawater immediate to a metal surface increases. This is beneficial because of the precipitation of solid compounds (calcareous deposits) by the reactions:

$$Ca^{2+} + HCO_3^- + OH^- \rightarrow H_2O + CaCO_3 \quad (15.3)$$

and

$$Mg^{2+} + 2OH^- \rightarrow Mg(OH)_2. \quad (15.4)$$

These deposits decrease the oxygen flux to the steel and hence the current necessary for cathodic protection. As a result, the service life of the entire cathodic protection system is extended.

Offshore pipelines can be protected as a cathode by achieving a potential of -0.80 $V_{Ag/AgCl}$ or more negative, which is accepted as the protective potential (E_c^o) for carbon steel and low alloy steel in aerated water. Normally, it is the best if the potentials negative to -1.05 $V_{Ag/AgCl}$ are avoided because these can cause a second cathodic reaction (Jones, 1992):

$$H_2O+e^- \rightarrow H+OH^- \tag{15.5}$$

which results in 1) wasted resources, 2) possible damage to any coatings, and 3) the possibility of hydrogen embrittlement.

Cathodic protection systems are of two types: impressed current and galvanic anode. The latter has been widely used in the oil and gas industry for offshore platforms and marine pipeline in the last 40 years because of its reliability and relatively low cost of installation and operation. The effectiveness of cathodic protection systems allows carbon steel, which has little natural corrosion resistance, to be used in such corrosive environments as seawater, acid soils, and salt-laden concrete.

15.3 Pipeline Coatings

15.3.1 Internal Coatings

The primary reason of applying internal coatings is to reduce the friction and therefore enhance flow efficiency. Besides, the application of internal coatings can improve corrosion protection, pre-commissioning operations and pigging operations. Increased efficiency is achieved through lowering the internal surface roughness since the pipe friction factor decreases with a decrease in surface roughness. In actual pipeline operation the improved flow efficiency will be observed as a reduction in pressure drop across the pipeline.

The presence of free water in the system is one of the reasons to cause the corrosion of inner pipeline. An effective coating system will provide an effective barrier against corrosion attack. The required frequency of pigging is significantly reduced with a coated pipeline. The wear on pig discs is substantially reduced due to the smoother pipe's surface.

The choice of a coating is dictated by both environmental conditions and the service requirements of the line. The major generic types of coatings used for internal linings include epoxies, urethanes and phenolics. Epoxy based materials are commonly used internal coatings because of their broad range of desirable properties which include sufficient hardness, water resistance, flexibility, chemical resistance and excellent adhesion.

15.3.2 External Coatings

Oil and gas pipelines are protected by the combined use of coatings and cathodic protection. The coating systems are the primary barrier against the corrosion therefore highly efficient at reducing the current demand for cathodic protection. However, they are not feasible to supply sufficient electrical current to protect a bare pipeline. Cathodic protection prevents corrosion at areas of coating breakdown by supplying electrons.

Thick coatings are often applied to offshore pipelines to minimize the holidays and defects and to resist damage by handling during transport and installation. High electrical resistivity retained over long periods is a special requirement, because cathodic protection is universally used in conjunction with coatings for corrosion control. Coatings must have good adhesion to the pipe surface to resist disbondment and degradation by biological organisms, which abound in seawater. Pipe coating should be inspected both visually and by a holiday detector set at the proper voltage before the pipe is lowered into the water. Periodic inspection of the pipeline cathodic protection potential is used to identify the coating breakdown areas.

Coatings are selected based on the design temperature and cost. The principal coatings, in rough order of cost are:

- Tape wrap
- Asphalt
- Coal tar enamel
- Fusion bonded epoxy (FBE)
- Cigarette wrap polyethylene (PE)
- Extruded thermoplastic PE and polypropylene (PP)

The most commonly used external coating for offshore pipeline is Fusion Bonded Epoxy (FBE) coatings. They are thin film coatings, 0.5-0.6 mm thick. They consist of thermosetting powders which are applied to a white metal blast cleaned surface by electrostatic spray. The powder will melt on the pre-heated pipe (around 230 °C), flow and subsequently cure to form thicknesses of between 250 and 650 microns.

15.4 CP Design Parameters

This section specifies parameters to be applied in the design of cathodic protection system based on sacrificial anodes.

15.4.1 Design Life

The design life t_r of the pipeline cathodic protection system is to be specified by the operator and shall cover the period from installation to the end of pipeline operation. It is normal practice to apply the same anode design life as for the offshore structures and submarine pipelines to be protected since maintenance and repair of CP system are very costly.

15.4.2 Current Density

Current density refers to the cathodic protection current per unit of bare metal surface area of the pipeline. The initial and final current densities, i_c (initial) and i_c (final), give a measure of the anticipated cathodic current density demands to achieve cathodic protection of bare metal surface. They are used to calculate the initial and final current demands that determine the number and sizing of anode.

The initial design current density is necessarily higher than the average final current density since the calcareous deposits developed during the initial phase reduces the current demand. In the final phase, the developed marine growth and calcareous layers on the metal surface will reduce the current demand. However, the final design current density shall take into account the additional current demand to re-polarize the structure if such layers are damaged. The final design current density is lower than the initial.

The average (or maintenance) design current density is a measure of the anticipated cathodic current density, once the cathodic protection system has attained its steady-state protection potential. This will simply imply a lower driving voltage and the average design current density is therefore lower than both the initial and final design value. Tables 15.1 gives the recommended design current density used for the cathodic protection system of non-buried offshore pipelines under the various seawater conditions in different standards. For bare steel surfaces fully buried in sediments a design current density of 20 mA/m^2 is recommended irrespective of geographical location or depth.

Table 15.1 Summary of recommended design current densities for bare steel.

Organization	Location	Water Temp. (°C)	Design Current Density (mA/m^2)		
			Initial	Mean	Final
NACE	Gulf of Mexico	22	110	55	75
	U.S. West Coast	15	150	90	100
	N. North Sea	0-12	180	90	120
	S. South Sea	0-12	150	90	100
	Arabian Gulf	30	130	65	90
	Cook Inlet	2	430	380	380
	Buried/Mud Zone	all	10-30	10-30	10-30
DNV	Tropical	>20	150/130	70/60	90/80
	Sub-Tropical	12-20	170/150	80/70	110/90
	Temperate	7-12	200/180	100/80	130/110
	Arctic	<7	250/220	120/100	170/130
	Buried/Mud Zone	all	20	20	20
ISO	Non-Buried	>20	-	70/60	90/80
		12-20	-	80/70	110/90
		7-12	-	100/80	130/110
		<7	-	120/80	170/130
		all	20	20	20
Note: DNV and ISO format: "(depths less than 30 m)/(depth greater than 30 m)"					

15.4.3 Coating Breakdown Factor

The coating breakdown factor describes the extent of current density reduction due to the application of coating. f_c=0 means the coating is 100% electrically insulating. f_c=1 implies that the coating can not provide any protection.

The coating breakdown factor is a function of coating properties, operational parameters and time. The coating breakdown factor f_c can be described as

$$f_c = k_1 + k_2 \cdot t \tag{15.6}$$

where t is the coating life time, k_1 and k_2 are constants that are dependent on the coating properties.

There are four paint coating categories defined for practical use based on the coating properties in DNV (1993):

- Category I: One layer of primer coat, about 50 μm nominal DFT (Dry Film Thickness)
- Category II: One layer of primer coat, plus minimum one layer of intermediate top coat, 150 to 250 μm nominal DFT
- Category III: One layer of primer coat, plus minimum two layers of intermediate/top coats, minimum 300 μm nominal DFT
- Category IV: One layer of primer coat, plus minimum three layers of intermediate top coats, minimum 450 μm nominal DFT

The constants k_1 and k_2 used for calculating the coating break-down factors are given in Table 15.2.

Table 15.2 Constants (k_1 and k_2) for calculation of paint coating breakdown factors (Sunde, 1968)

Depth (m)	Coating Category			
	I	II	III	IV
	k_1=0.1	k_1=0.05	k_1=0.02	k_1=0.02
	k_2	k_2	k_2	k_2
0-30	0.1	0.03	0.015	0.012
>30	0.05	0.02	0.012	0.012

For cathodic protection design purpose the average and final coating breakdown factors are to be calculated by introducing the design life t_r

$$f_c(average) = k_1 + k_2 \cdot t_r / 2 \tag{15.7}$$

$$f_c(final) = k_1 + k_2 \cdot t_r \tag{15.8}$$

15.4.4 Anode Material Performance

The performance of a sacrificial anode material is dependent on its actual chemical composition. The most commonly used anode materials are Al and Zn. Table 15.3 gives the electrochemical efficiency ε of anode materials applied in the determination of required anode mass.

Table 15.3 Design electrochemical efficiency values for Al and Zn based sacrificial anode materials (DNV, 1993)

Anode Material Type	Electrochemical Efficiency (Ah/kg)
Al-base	2000 (max 25 °C)
Zn-base	700 (max 50 °C)

The closed circuit anode potential used to calculate the anode current output shall not exceed the values listed in the Table 15.4.

Table 15.4 Design closed circuit anode potentials for Al and Zn based sacrificial anode materials (DNV, 1993)

Anode Material Type	Environment	Closed Circuit Anode Potential (V rel. Ag/AgCl seawater)
Al-base	seawater	-1.05
	sediments	-0.95
Zn-base	seawater	-1.00
	sediments	-0.95

15.4.5 Resistivity

The salinity and temperature of seawater have influence on its resistivity. In the open sea, the salinity doesn't vary significantly. The temperature becomes the main factor. The resistivities of 0.3 and 1.5 ohm·m are recommended to use to calculate the anode resistance in seawater and marine sediments respectively when the temperature of surface water is between 7 to 12°C (DNV, 1993).

15.4.6 Anode Utilization Factor

The anode utilization factor indicates the fraction of anode material that is assumed to provide cathodic protection current. Performance becomes unpredictable when the anode is consumed beyond a mass indicated by the utilization factor. The utilization factor of an anode is dependent on the detailed anode design, in particular dimensions and location of anode cores. Table 15.3 gives the anode utilization factor for different types of anodes (DNV, 1993).

Table 15.5 Design utilization factors for different types of anodes.

Anode Type	Anode Utilization Factor
Long 1) slender stand-off	0.90
Long 1) flush-mounted	0.85
Short 2) flush-mounted	0.80
Bracelet, half-shell type	0.80
Bracelet, segmented type	0.75

15.5 Galvanic Anodes System Design

15.5.1 Selection of Anodes Type

Pipeline anodes are normally of the half-shell bracelet type (see Figure 15.6). The bracelets are clamped or welded to the pipe joints after application of the corrosion coating. Stranded connector cables are be used for clamped half-shell anodes. For the anodes mounted on the pipeline with concrete, measures shall be taken to avoid the electrical contact between the anode and the concrete reinforcement.

Normally, bracelet anodes are distributed at equal spacing along the pipeline. Adequate design calculations should demonstrate that anodes can provide the necessary current to the pipeline to meet the current density requirement for the entire design life. The potential of pipeline should be polarized to -0.8 V $_{Ag/AgCl}$ or more negative. Figure 15.4 shows the potential profile of a pipeline protected by galvanic bracelet anodes.

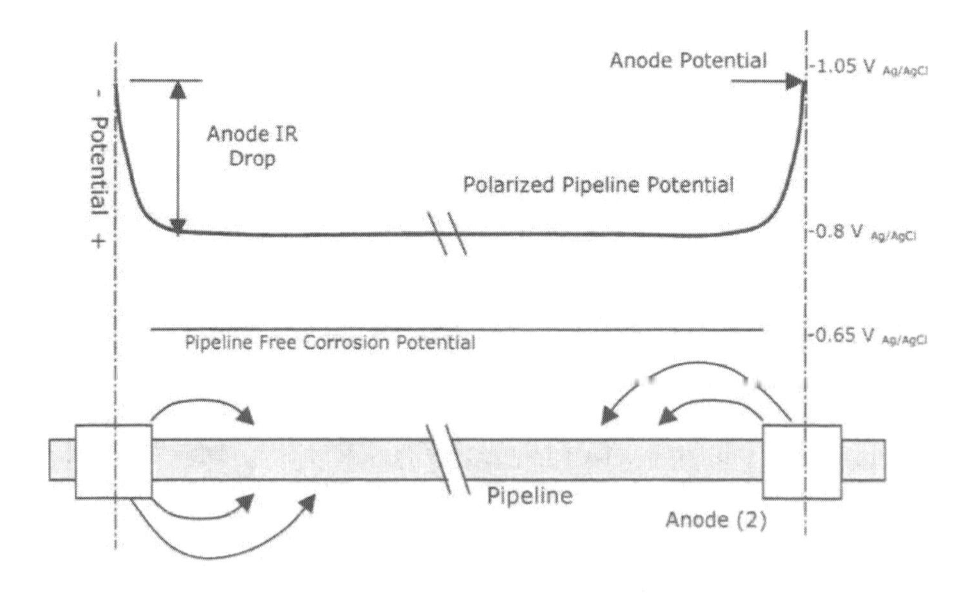

Figure 15.4 Potential profile of a pipeline protected by bracelet anodes.

Since the installation expense is the main part of CP design, larger anode spacing can reduce the overall cost. However, the potential is not evenly distributed along the pipeline. The pipeline close to the anode has a more negative potential. The potential of middle point on the pipeline between two anodes is more positive and must be polarized to -0.80 $V_{Ag/AgCl}$ or more negative in order to achieve the cathodic protection for the whole pipeline. Increased anode spacing brings bigger mass per anode therefore cause more uneven potential distribution. The potential close to the anode could be polarised to negative than -1.05 $V_{Ag/AgCl}$, which should be avoided because of reaction 1.5. Figure 15.5 schematically illustrates the anticipated potential attenuation for situations of large anode spacing (Hartt et al, 2004).

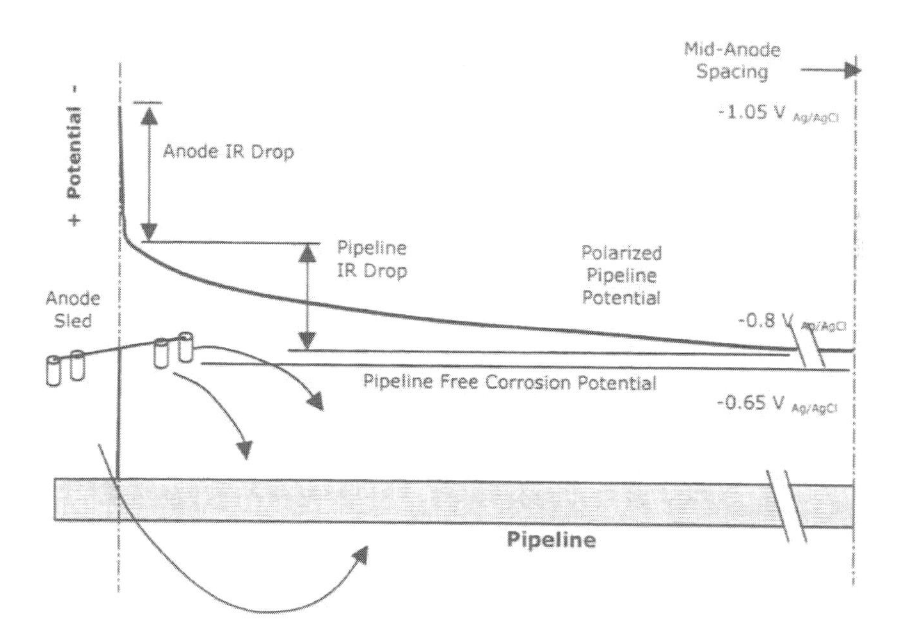

Figure 15.5 Pipeline potential profile for large anode spacing.

15.5.2 CP Design Practice

Offshore pipeline CP design includes the determination of the current demand I_c, required anode mass M and number and current output per anode I_a. The current demand is a function of cathode surface area, A_c, a coating breakdown factor, f_c, and current density, i_c, and can be expressed as (DNV, 1993):

$$I_c = A_c \cdot f_c \cdot i_c \tag{15.9}$$

where i_c depends upon water depth, temperature, sea water versus mud exposure, and whether or not the mean or final life of the CP system is being evaluated. Current density i_c is normally in the range 60-170 mA/m^2 (DNV, 1993). As the initial polarization period proceeding steady-state conditions is normally quite short compared to the design life, the mean (time-averaged) design current density i_m becomes very close to the steady state current density. Therefore, it is used to calculate the minimum mass of anode material necessary to maintain cathodic protection throughout the design life. Correspondingly, *M* can be calculated as:

$$M = \frac{8{,}760 \cdot i_m \cdot T}{u \cdot C} \tag{15.10}$$

where u is a utilization factor, C is anode current capacity and T is design life. The cathode potential is assumed to be spatially constant. Therefore, the current output per anode can be calculated by:

$$I_a = \frac{\phi_c - \phi_a}{R_a} \tag{15.11}$$

where ϕ_c and ϕ_a are the closed circuit potential of the pipe and anode, respectively, and R_a is the anode resistance.

15.5.3 Anode Spacing Determination

Bethune and Hart (2000) have proposed a newly attenuation equation to modify the existing design protocol interrelating the determination of the anode spacing L_{as}. L_{as} can be expressed as:

$$L_{as} = \frac{(\phi_c - \phi_a)}{\phi_{corr} - \phi_c} \cdot \frac{\alpha \cdot \gamma}{2\pi \cdot r_p \cdot R_a} \tag{15.12}$$

where

ϕ_{corr} : the free corrosion potential;

α : the polarization resistance;

γ : the reciprocal of coating breakdown factor *f*;

r_p : the pipe radius.

Assumptions have made in this approach: 1) total circuit resistance equal to anode resistance; 2) all current enter the pipe at holidays in the coating (bare areas); and 3) ϕ_c and ϕ_a be constant with both time and position. The ISO standards recommend the distance between bracelet anodes should not exceed 300 m (ISO 14489, 1993).

15.5.4 Commonly Used Galvanic Anodes

The major types of galvanic anodes for offshore applications are slender stand-off, elongated flush mounted and bracelet (Figure 15.6). The type of anode design to be applied is normally specified by the operator, and should take into account various factors, such as anode utilization factor and current output, costs for manufacturing and installation, weight and drag forces exerted by ocean current. The slender stand-off anode has the highest current output and utilization factor among these commonly used anodes.

15.5.5 Pipeline CP System Retrofit

Cathodic protections systems retrofits become necessary as the pipeline systems age. An important aspect of such retrofitting is determination of when such action should take place. Assessment of cathodic protection systems upon pipelines is normally performed based upon potential measurements. As galvanic anodes waste, their size decreases; and this causes a resistance increase and a corresponding decrease in polarization. Models have been constructed for potential change that occurs for a pipeline protected by galvanic bracelet anodes as these deplete were also developed. Anodes' depletion is time dependent in the model.

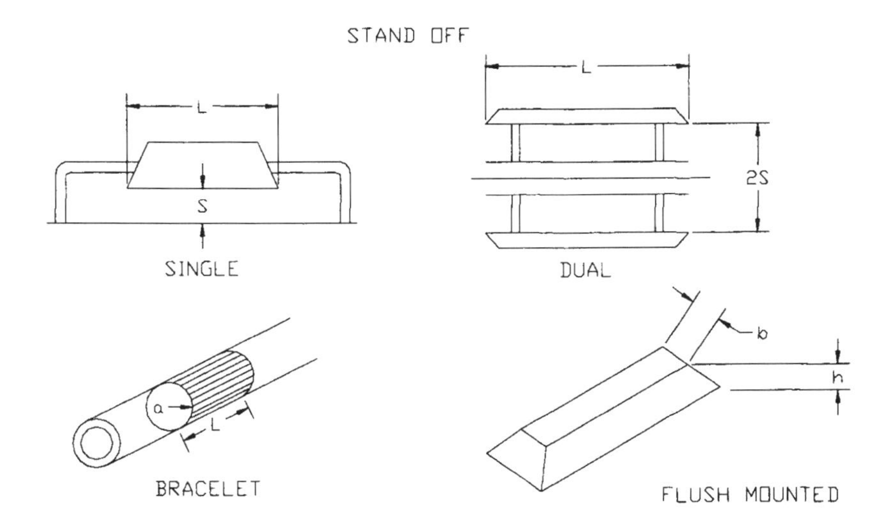

Figure 15.6 Commonly used anodes.

Bracelet anodes have been used for cathodic protection of marine pipelines, especially during the "early period" (roughly 1964-1976), when many oil companies had construction activities in the Gulf of Mexico. According to recent survey data, many of these early anode systems have depleted or are now depleting. Retrofitting of old anode systems on pipelines installed in 1960's and 1970's and even newer ones is required since these are still being used for oil transportation. Anodes can be designed as multiples or grouped together to form an anode array (anode sled) (See Figure 15.3). Anode arrays typically afford a good spread of protection on a marine structure. They are a good solution for retrofitting old cathodic protection systems.

15.5.6 Internal Corrosion Inhibitors

Corrosion inhibitors are chemicals that can effectively reduce the corrosion rate of the metal exposed to the corrosive environment when added in small concentration. They normally work by adsorbing themselves to form a film on the metallic surface (www.corrosion-doctors.org).

Inhibitors are normally distributed from a solution or dispersion. They reduce the corrosion process by either:

- Increasing the anodic or cathodic polarization behavior;

- Reducing the movement or diffusion of ions to the metallic surface;
- Increasing the electrical resistance of the metallic surface.

Inhibitors can be generally classified as follows (www.corrosion-doctors.org):

- Passivating inhibitors;
- Cathodic inhibitors;
- Precipitation inhibitors;
- Organic inhibitors;
- Volatile corrosion inhibitors.

The key to select an inhibitor is to know the system and anticipate the potential problems in the system. The system conditions include water composition (such as salinity, ions and pH), fluid composition (percentage water versus hydrocarbon), flowrates, temperature and pressure. Application of the inhibitors can be accomplished by batch treatments, formation squeezes, continuous injections or a slug between two pigs.

Inhibitor efficiency can be defined as

$$\text{Inhibitor efficiency } (\%) = 100 * (CR_{uninhibited} - CR_{inhibited}) / CR_{uninhibited}$$

Where $CR_{uninhibited}$: the corrosion rate of the uninhibited system;

$CR_{inhibited}$: the corrosion rate of the inhibited system.

Typically the inhibitor efficiency increases with an increase in inhibitor concentration

15.6 References

1. Bethune, K. and Hartt, W.H. (2000), "A Novel Approach to Cathodic Protection Design for Marine Pipelines: Part II-Applicability of the Slope Parameter Method", presented at Corrosion, paper no.00674.
2. DNV Recommended Practice RP B401 (1993), Cathodic Protection Design, Det Norke Veritas Industry AS, Hovik.
3. Hartt, W.H., Zhang, X. And Chu, W. (2004), "Issues Associated with Expiration of Galvanic Anodes on Marine Structures". Presented at Corrosion, paper no. 04093.
4. Jones, D.A. (1992), Principles and Prevention of Corrosion, First Edition, Macmillan Inc., New York, pg. 437-445.
5. NACE Standard RP 0176 (1994), "Corrosion Control of Steel-Fixed Offshore Platforms Associated with Petroleum Production", Houston, TX.
6. ISO/TC 67/SC 2 NP 14489, "Pipeline Cathodic Protection-Part 2: Cathodic Protection of Offshore Pipelines" (1993), International Organization for Standardization, Washington, DC.
7. Sunde, E.D. (1968), Earth Conduction Effects in Transaction Systems, Dover Publishing Inc., New York, NY, pg. 70-73.
8. www.corrosion-doctors.org/Inhibitors/lesson11.htm.

Part II

Pipeline Design

Chapter 16 Åsgard Flowlines Design Examples

16.1 Introduction

This chapter shall focus on the Åsgard Flowlines Project, mainly based on Bai et al (1999). In the last portion of this Chapter, the experience of designing the Åsgard export trunklines (the Åsgard Transport Project) will also be briefly outlined.

The Åsgard Field is a joint effort by Statoil and Saga Petroleum to develop the Smørbukk, Smørbukk Sør and Midgard fields located in the Haltenbanken area of the Norwegian Sea. Floating production and offloading vessels will be linked to 300 km of flowlines in a two-phased field installation. The subject of this presentation refers mainly to the Phase 1 development consisting of 90 km of 10" insulated flowlines operating at 120° to 145° C and 390 Bar (Damsleth and Dretvik, 1998). These were installed in spring 1998 and tied-back to the Åsgard A FPSO for early recovery of oil.

J P Kenny A/S in Stavanger, were contracted in October 1995 by the joint Statoil/Saga development project to perform the conceptual and detailed engineering of the flowlines. The integrated project team benefited from recent developments in subsea equipment, pipeline research and codes to continuously improve the design. The early engineering phases identified several aspects of the design where innovation would have an important impact on the performance of the project and the capital expenditure (CAPEX):

- Selection of 13% Chrome as flowline material;
- Using the Limit-State approach to developing strength criteria;
- Using the Design Through Analysis method to optimize seabed intervention.

The combination of these aspects influenced the course of the flowline design that was focused on minimizing the main CAPEX cost drivers: procurement and marine operations. This presentation outlines the evolutionary design process and highlights the choices made along the way.

16.2 Wall-thickness and Linepipe Material Selection

16.2.1 General

The Smørbukk and Smørbukk Sør well-streams are highly corrosive due to the presence of CO_2 combined with 140°C temperature and 390 bar pressure which places extraordinary demands on material performance. The careful selection of linepipe and coating benefited from recent development work by Statoil's materials research department in co-operation with the suppliers (Andersen, 1997).

16.2.2 Linepipe Material Selection

Four flowline materials were considered for the 10" flowlines: X65 steel, Duplex, Inconel-cladded carbon steel and 13% Chrome. More commonly used for well casing, 13%Cr had not been used previously as a flowline material and there was little experience of how the material would perform when subject to production welding methods and elastic-plastic deformation during installation and operation. While Statoil were engaged in a qualification program for welding 13%Cr linepipe, the conceptual engineering compared several different flowline materials, each of which involved cost trade-offs.

A carbon steel line could be expected to need at least one replacement during the field lifetime due to the high corrosion rate. The significant strength de-rating of Duplex above 100°C meant that a High Integrity Pressure Protection System (HIPPS) was required to limit the pressure in the lines to 170 Bar. Availability of sufficient quantities of the material was also a problem. The cladded pipe was considered to be too costly to weld offshore and unsuitable for the reeling installation method. The material and welding qualification program carried out by Statoil confirmed that 13%Cr was suitable for the Åsgard flowlines due to its resistance to CO_2 corrosion, high strength at elevated temperatures, and its availability.

16.2.3 Wall-thickness Design

The primary requirement of the pipe wall is to sustain stresses from the internal pressure. The tensile hoop stress is due to the difference between internal and external pressure, and is not to exceed the permissible value of 0.8 SMYS. This higher utilization value achieves a 5% reduction of wall-thickness over the previous 0.72 usage factor and represents about 1 mm reduction in wall thickness for the Åsgard flowlines.

The resulting wall thickness for were in the range 11.6 mm to 12.6 mm for the Åsgard flowlines and 15.3 mm for gas injection lines. As these were compatible with the reeling pipelay method, it was also the most cost-effective combination of material and wall-thickness.

16.3 Limit State Strength Criteria

16.3.1 General

This section presents limit-state based strength criteria for typical flowlines under consideration in the Åsgard Flowlines, Phase 1 Detail Engineering Design. Development of the limit-state based strength criteria specific to the Åsgard flowlines was carried out by Bai and Damsleth (1997), before DNV (1996) rules were issued.

The most severe case (thinnest wall-thickness, largest D/t) is chosen to demonstrate that these failure modes will not be governing for the Åsgard Phase 1 flowlines. The most severe design case regarding strength criteria is considered to be the following production flowlines:

- Pipeline Systems: P101, P102, S101, S102
- Area: Smørbukk Sør
- Design Pressure: 370 barg
- Wall-thickness: 11.6 mm
- Outside diameter: 251.8 mm

16.3.2 Bursting Under Combined Loading

For a 10" flowline subject to pull-over loads, finite element analyses revealed that a plastic hinge occurs at the location where a pull-over load is applied (Tørnes et al., 1998). It is therefore concluded that the pull-over response of the 10" flowlines is predominately load-controlled. Investigation of fishing activities documented that there is very little fishing activity in the area where the flowlines are to be installed. Because the flowlines in the Åsgard field are doubled-up for round-trip pigging, the consequences of a structural failure in one of the flowlines is not considered to be as severe as structural failure of an export pipeline. Therefore, an equivalent stress criterion is used with allowable stress of 1.0 SMYS for the bursting failure mode.

16.3.3 Local Buckling/Collapse

An allowable bending moment was defined for load-controlled situations. It was, however, concluded that local buckling for external over-pressure is not a governing failure mode because the D/t values are low and water depth (300 m) is not deep enough for external pressure to be dominant. Calculations also show that propagation buckling will not occur for the Phase 1 flowlines.

Allowable axial compressive strains are estimated for displacement controlled situations. A pipeline is made up of 12 m long pipe joints welded together. The difference in the stiffness of the pipe at the welded joint due to the discontinuity of the external coating and possibly the effect of local heat treatment can be significant. This is accounted for in the displacement-controlled situations by applying a Strain Concentration Factor (SNCF).

16.3.4 Fracture

The stress-strain curves used in converting stress to strain are very conservative because stress-strain curves are usually based on the lowest yield stress and lowest ultimate stress. In addition, PD6493 has been derived for load-controlled situations, and the allowable strain is applied to deformation controlled situations. The flow stress is defined as the average of yield and tensile stress, which is also conservative.

16.3.5 Low-cycle Fatigue

Tests at SINTEF in 1997, based on the author' definition, have established the fatigue performance of the welded pipes for the Åsgard Phase 1 flowlines. The tests simulated loading from reeling installation, vortex induced vibrations, temperature and internal pressure. The results, as shown in Table 16.1, were used to generate fatigue life design curves.

Table 16.1 Low-cycle fatigue test results for pipes with butt weld.

Specimen No	Internal Pressure (bar)	$\Delta\varepsilon_{nom}$ [%]	ε_{max} [%]	ε_{min} [%]	R-ratio	N1[1)]	N_f[2)]	Remarks
1	330	0.29	0.59	0.30	0.51	9600	24768	Surface as welded
2	330	0.47	0.69	0.22	0.32	1900	3600	Surface as welded
3	330	0.30	0.60	0.30	0.50	2021	5620	Simulated undercut in weld
4	330	0.46	0.69	0.23	0.33	1800	3490	Simulated undercut in weld
5	330	0.48	0.70	0.22	0.31	2800	4665	Simulated undercut in weld
6	330	0.17	0.53	0.36	0.68	30175	49045	Simulated undercut in weld
7	330	0.30	0.60	0.30	0.50	3100	7853	Simulated undercut in weld
8	0	2.07	1.80	-0.27	-0.15	91	101	Simulated undercut in weld

Notes:

1) N_1 is first indication of crack by gauges

2) N_f is leakage through crack

3) $\Delta\varepsilon_{nom}$, ε_{max}, ε_{min} are nominal strain range, maximum strain and minimum strain respectively,

4) $R = \varepsilon_{min}/\varepsilon_{max}$

Constant amplitude sinusoidal loads with a frequency range of between 0.5 Hz and 1.0 Hz were used and the specimens were subjected to an internal pressure of 330 bar. All tests were carried out in air. Based on the laboratory tests on the given loading sequence of 8" pipe with weld undercuts, a design curve - with the common 97.5% probability of survival - was established. The design curve was slightly on the conservative side of the AWS-2 $\Delta\varepsilon$-N curve (Marshall, 1992). It is therefore concluded that the AWS-2 $\Delta\varepsilon$-N curve shall be applied to low-cycle fatigue design of Phase 1 lines. The AWS $\Delta\varepsilon$-N curves are expressed as below:

$$\Delta\varepsilon = 0.055N^{-0.4} \qquad \textit{for } \Delta\varepsilon \geq 0.002$$

and

$$\Delta\varepsilon = 0.016N^{-0.25} \qquad \textit{for } \Delta\varepsilon \leq 0.002$$

The Åsgard Phase 1 flowlines are designed to the following cyclic strains:

- 2 strain cycles during reel on and reel off
- 1 strain cycle during bending over stinger and bending in sag bend
- 200 cycles of planned and unplanned shut-downs

The planned cycles of shutdowns are estimated 40-50. However, there will be an unknown number of unplanned production cool-down/re-start cycles. Conservatively, an upper limit of the total shut-down cycles during the entire operation period, s assumed to be 200. It is conservatively assumed that the strain range during shutdowns is a constant.

Based on the AWS curves and the laboratory tests conducted for this project, it is found that the allowable strain range for low cycle fatigue is 0.3%.

16.3.6 Ratcheting

Industrial codes state that ratcheting of high pressure and high temperature flowlines (HP/HT) during start-up and shut-in/shut-down cycles places a stringent limit and allowable equivalent plastic strain is 0.1 %. This allowable limit is conservative because it neglects the effect of strain-hardening and large deflection, see Kristiansen et al (1997) and Bai et al (1999) "Simulation of Ratcheting of HP/HT Flowlines".

A finite element simulation of a flowline subjected to pressure and curvature cycles from installation and repeated shut-down/start-up cycles shows a gradual accumulation of ovalisation and circumferential strain. The accumulation of plastic deformation occurs during the initial stages of the cyclic loading and stabilizes before the end of the series of load cycles demonstrating a characteristic "shake down" behavior.

The results are illustrated for the cyclic load case $P_c/P_o = 0.5$ and $\kappa/\kappa_1 = 0.21$ (See Bai et al (1999)). The changes in minimum and maximum diameters, are recorded in Figure 16.1. The minimum and maximum diameter increase and decrease with number of applied cycles, respectively, and the accumulation of ovalisation stabilises after approximately 30 load cycles.

In Figure 16.2, strain versus number of shut-down cycles is illustrated for combined cyclic bending and internal pressure loading. The circumferential strain increase to approximately 0.60 % in 28 load cycles, and stays constant thereafter. Hence, the limiting strain of 1 % that control brittle fracture of pipes under circumferential tension is not reached.

For the combined curvature and internal pressure case (Figure 16.2) the axial strain increases for the first seven load cycles before stabilizing at approximately 0.3 % strain.

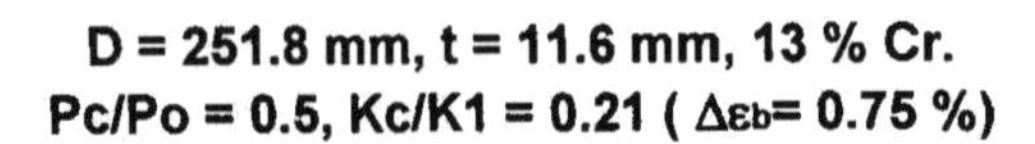

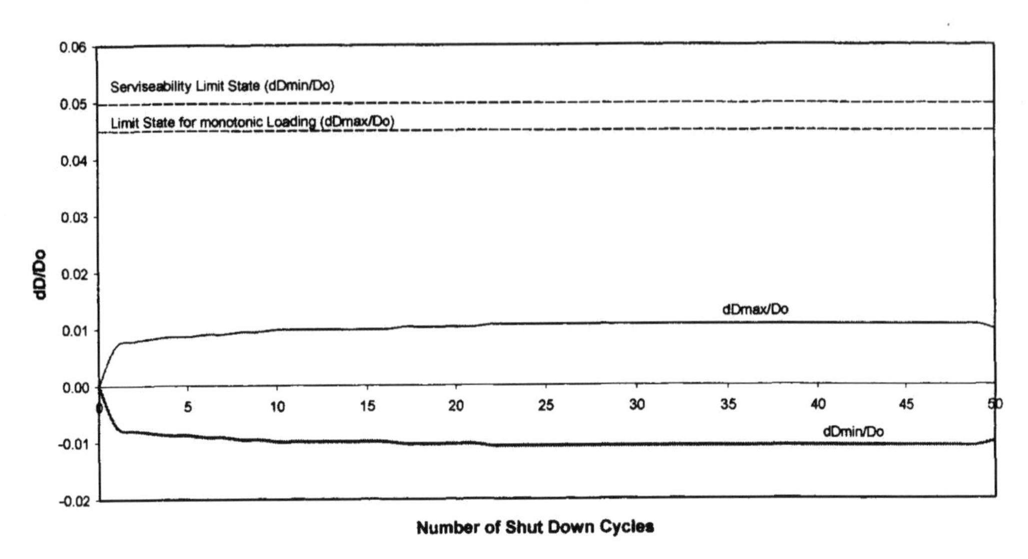

Figure 16.1 Effect of cyclic curvature on ovalisation.

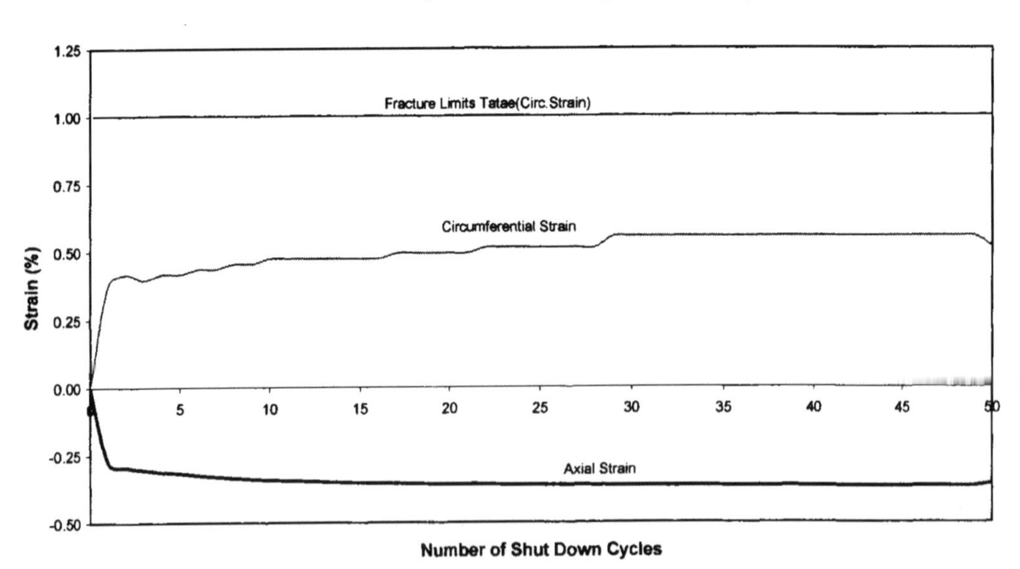

Figure 16.2 Effect of cyclic curvature on axial and circumferential strain ovalization for serviceability.

The fabrication ovality tolerance given by the Statoil specifications "Seamless Linepipe-Austenstic Steel" (1990) and "Welded Linepipe-Ferritic Steel" (1990) is 0.75%. Ovality increases due to the reeling installation process where the pipe is subject to reverse bending and the effect of this on subsequent straining should be considered. Generally, numerical methods (e.g. finite element methods) must be used to predict the reeling-induced ovality, in which bending against a surface, axial load and repeated bending are considered. The fabrication tolerance given by the Statoil specification (0.75 %) is applied as the initial ovalisation prior to reeling.

The most severe case is chosen for the ratcheting analysis. It has been found that the maximum ovalisation of the flowlines at the end of design life is 1.5%, significantly less than the 3% limit. Therefore, it is concluded that the ovalisation for the serviceability limit state will not be violated.

16.4 Installation and On-bottom Stability

16.4.1 Installation Design

The vessel Apache was used to install the flowlines using the reeling method. During the reeling process, the pipeline undergoes plastic deformation on the reel. The pipeline is plastically deformed in the opposite direction during off-reeling as it passes through the straighteners located on the adjustable stern ramp. The installation analysis was carried out by use of the OFFPIPE computer program where the 60 degree ramp angle was considered to be the base case for the Åsgard flowlines.

The design criterion used is that the pipeline strain is to be within the allowable 0.15 % for the static installation analysis, and that there should be no pipeline/support contact (i.e. small gap) on the lower support (lower roller box). The pipeline configuration and residual lay tension in the pipeline is then calculated as an input to the in-place analysis.

The installation analysis shows that the 10" Åsgard Phase 1 flowlines can be installed by the reel method with relatively low residual lay tension as presented in Table 16.2. Some of the lines required different coating density along the route which resulted in tension variations.

The pipeline strains are less than 0.05 % for all cases for the pipeline both on the ramp and in the sagbend. The results show that the pipeline residual lay tensions in general are in the area of 5 to 9 tonnes, depending on the coating characteristics, and up to 12 tonnes zone 2 sections.

As the residual lay tension has a considerable impact on the seabed intervention, possible ways of reducing the residual lay tension were investigated. The flowline departure angle could be increased by bending the pipeline over the layramp roller supports or by adding a ramp extension whereby the residual lay tension could be reduced by 46 % to 58 %. Because the sensitivity analyses showed considerable reductions in the pipeline residual tension were feasible by modifying the standard ramp configuration, the requirement for low lay tension was included in the installation contract in order to reduce the post-installation seabed intervention.

Table16.2 Installation analysis results for phase I flowlines.

Pipeline	Zone	Pipeline Section (Kp.)	Steel data OD/WT (mm/mm) 2)	Coating data WT/density (mm/ kg/m^3)	Pipeline submerged Weight (kg/m)	Residual lay tension (kN)
P101 & P102	2	0 - 0.36	258.8 / 15.6	53 / 1190	46.4 (24.7)	121
	1	0.36 – 1.5	251.8 / 11.6	53 /1190	24.7	83
	1	1.5 - 8.6	251.8 / 11.6	53 / 1000	15.1	53
S101 & S102	2	0 - 0.36	258.8 / 15.6	53 / 1000	37.0 (15.6)	94
	1	0.36 – 7.4	251.8 / 11.6	53 / 1000	15.6	54
	1	7.4 – 11.6	251.8 / 11.6	53 / 1220	26.2	85

16.4.2 On-bottom Stability

The purpose of the analysis is to recommend coating thickness and equivalent coating density to ensure on-bottom stability of the untrenched flowline throughout its design life. The following design conditions are considered:

- Installation phase:1 year wave+ 10 year current or 10 year wave+ 1 year current
- Operational phase: 10 year wave+ 100 year current or 100 year wave + 10 year current

In the analysis, the shear strength of the clay is varied between 2 to 20 kPa, 5 kPa at critical locations, while the corresponding coefficient of friction may vary between 0.15 and 0.3, depending on the pipe penetration. The flowlines are subjected to environmental loads, spring/autumn wave and current and all year wave and current, corresponding to installation and operation phases respectively. Both wave and current flow are assumed to act perpendicular to the flowline.

The long-term extreme sea states are transformed to water particle velocity data at seabed level. Since only the longest period (14 to 25 sec.) surface waves have any significant effect at 300m depths, the two-peaked Torsethaugen wave spectrum is used. The peak combined wave and current velocities form the basis for the long-term hydrodynamic loading.

Two principally different design checks are made for the stability control of the pipeline:

The first design check is a static equilibrium calculation of a pipeline trenched and or buried in the soil, sand or clay. The design check is based on static equilibrium between the hydrodynamic design loads and the soil capacity.

The second check is based on a specified permissible pipeline displacement for a given design load condition (return period) generated through series of pipeline response simulations with PONDUS, a purpose-made FE stability program. For the on-bottom design check on clay, a critical weight is calculated to fulfill the "no breakout criteria". The results in Table 16.3 represent the worst case where directional wave and current flows are applied concurrently. The extreme wave and current data correspond to all-year directional conditions.

Compared to observed pipe penetration into the seabed, the calculated values shown in Table 16.3 are underestimated. The effect of concentrated loads at freespan shoulders is not modeled by PONDUS. An ABAQUS model was later developed to account for the effect of local seabed penetration on stability, including the reduction in lift force where the flowline is in a freespan. It was demonstrated that, while there was local movement, the overall pipeline remained in position and could be considered as stable.

Table16.3 Equivalent insulation coating density for stability.

On-Bottom Stability Results according to RP E305								PONDUS	
Pipeline Section	Flowline	Soil shear strength Su (kPa)	Water depth (m)	Water depth used in analysis (m)	Pipeline Submerged Weight (kg/m)	Coating Thick. (mm)	Min. Coating Density (kg/m^3)	Pipeline Movement Su = 20 kPa (m)	Pipeline Penetration in Clay soil Su = 20 kPa (mm)
KP	Production lines								
0 - 1.5	P101/102	20	306 - 322	306	31.0	53	1170	0.26	3
1.5 - 8.6	"	20	306 - 322	306	22.4	53	1000	0	2
0 - 7.4	S101/102	20	299 - 322	299	22.9	53	1010	0	2
7.4-11.6	"	20	299 – 322	299	33.5	53	1220	0	3

16.5 Design for Global Buckling, Fishing Gear Loads and VIV

16.5.1 General

The seabed intervention design through analysis is conducted as Figure 16.3:

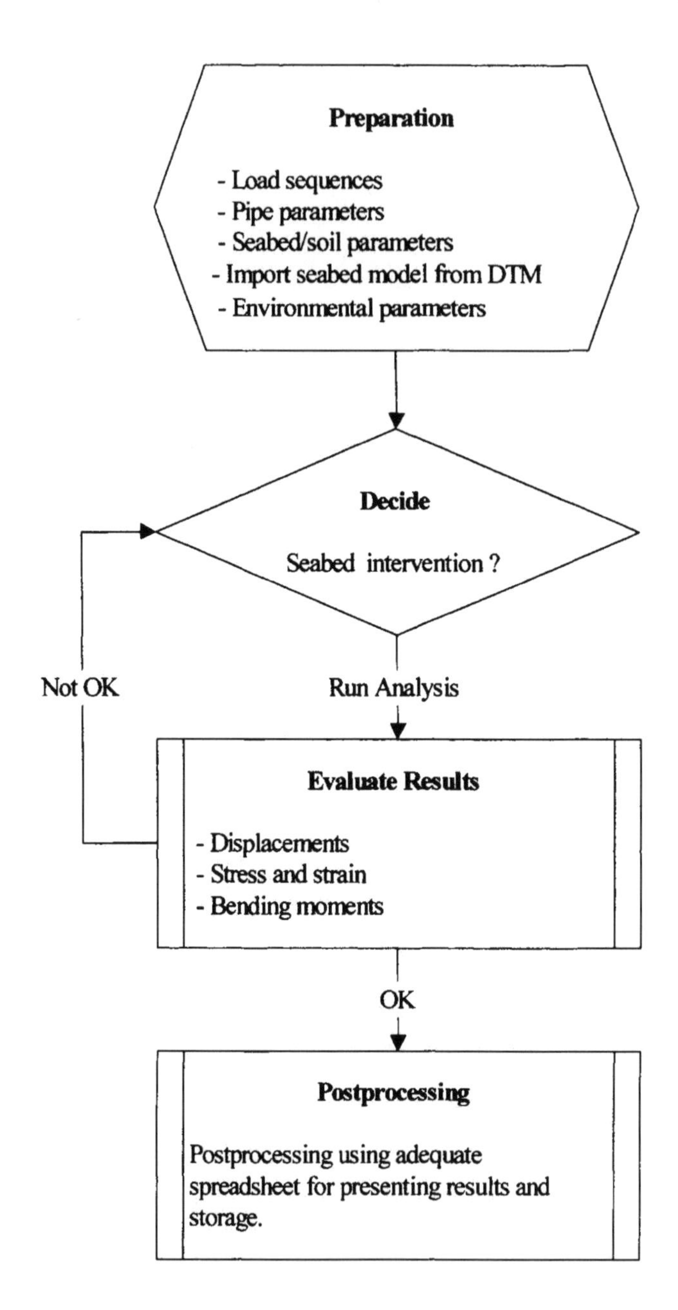

Figure 16.3 Flow chart for seabed intervention design for static strength.

16.5.2 Global Buckling

A pipeline may buckle as seabed friction builds up to resist axial expansion caused by temperature and pressure loads (Hobbs 1984, Pedersen and Jensen 1988). The compressive forces set up by the seabed friction, in addition to any other external forces such as restraint forces at tie-in points, is commonly referred to as the effective axial force.

Other important parameters that govern the buckling behavior are the size and shape of any out-of-straightness (OOS), the structural stiffness and the seabed friction coefficients. For flowlines resting on a very uneven seabed terrain, the vertical seabed imperfections result in more abrupt curvature than the horizontal imperfections created during laying. As a result, the axial load that initially forces the pipe vertically is found to be lower than the corresponding load needed to buckle the pipe laterally. It is observed that the pipe initially moves in the vertical plane as a gradual magnification of the initial imperfection due to its large initial OOS. As the pipe lifts off the seabed, a length of the pipe becomes laterally unrestrained and thus the critical axial force needed for lateral buckling decreases. Eventually, if sufficient pipe has lifted off the seabed, the lateral buckling force becomes less than the force required to lift the pipeline further in the vertical plane. At that moment, an interaction with the horizontal mode occurs and the pipe goes through dynamic 'snap' movement.

On an uneven seabed, it appears that the critical lateral buckling force is little affected by the lateral friction coefficient. The 3D lateral buckling results show significantly lower stresses and strains in the buckles compared to 2D vertical buckling results for the same seabed and loads.

The cyclic shutdown/startup analyses show that the flowline bending tends to 'shake down' to an elastic state after a few cycles. For all cases, the largest stresses/strains are found to occur during the first operation start up. For the case with a higher friction coefficient ($\mu = 1.0$), the lateral displacement converges towards a maximum level of approximately 9.0 m after approximately 10 cycles, as shown in Figure 16.4 (Nystrøm et al 1997).

A buckling force of 220 kN only represents approximately 10 % of the fully restrained axial effective force which indicates that the flowlines are closer to being unrestrained than to being restrained. This important finding indicates that, unless the flowlines are restrained by other means (trenching, rock dump protection), the forces into the templates, riser bases etc. will be relatively low.

In-place analysis of the 10" Production flowlines was conducted for P02 and SSP2. The latter line has many small imperfections ranging in size between 1m and 2m and experiences buckling forces from 200 kN to 260 kN while the P02 line, with larger seabed imperfections from 2m to 4m, experiences buckling forces ranging between 140 kN and 180 kN.

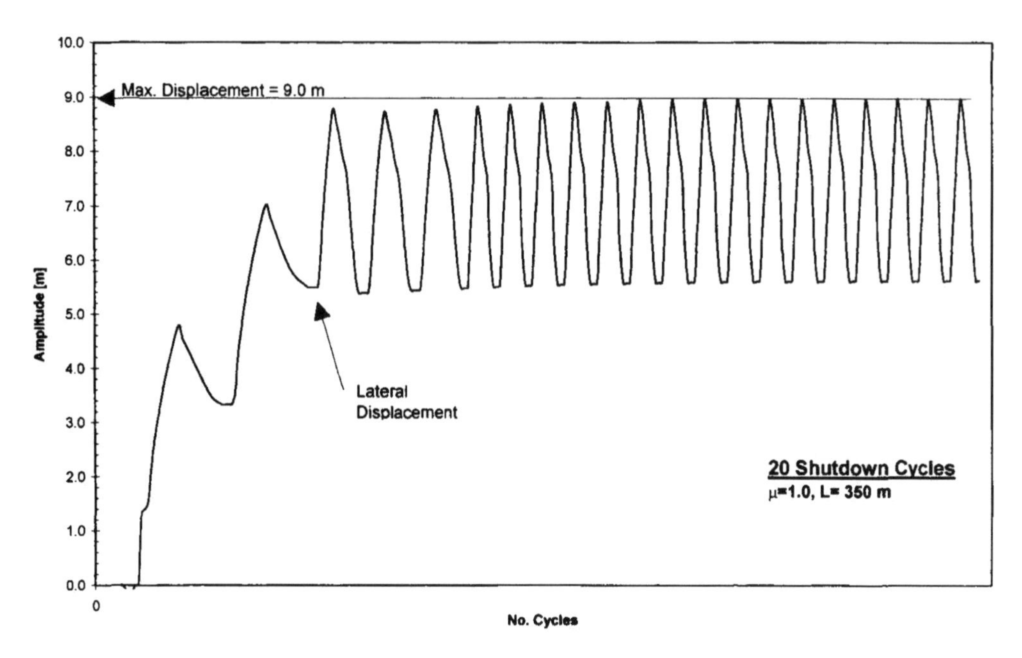

Figure16.4 Displacement vs. no. shutdown cycles.

Should the vertical imperfection be of a size and shape that the vertical buckling force stays lower that the lateral buckling force, even when the pipe has lifted off, the buckle should theoretically remain in the vertical plane. However, interaction with the lateral plane will certainly occur at some stage due to current forces, side slopes, etc. The effect of current and wave forces can be incorporated into the 3D in-place analysis using Morison's equations (Ose et al, 1999).

16.5.3 Trawlboard

In areas where fishing with bottom trawl equipment is likely, the industrial practice in the North Sea has been to protect all pipelines with diameter less than 16-inch from trawl interaction loads by trenching or cover. Larger diameter pipelines are left exposed on the seabed and protected by concrete coating. The 16-inch rule-of-thumb stems from the considerable amount of research conducted in the 70's, Moshagen and Kjeldsen (1980) based on trawl tests and analyses of simple beams subjected to transverse pull-over loads.

To trench a pipeline is costly and may lead to an additional requirement to cover it with backfill plus rock dumping in order to restrain it from buckling out of the trench.

The pull-over load is a result of a dynamic interaction between the fishing gear and the pipeline. An increased flexibility (longer span length, or reduced pipe diameter or wall-thickness) is expected to reduce the pull-over load. However, the geometrical effect is believed to be the governing parameter influencing the pull-over load. Therefore, pull-over loads are expressed as a function of span height, see Verley et al (1992).

A 3-D non-linear transient Finite Element model has been developed to investigate the structural response of pipelines subjected to pull-over loads. The analyses results indicated that this particular flowline would be able to withstand pull-over loads when the impact point on the flowline was in full contact with the seabed prior to the load being applied (Tørnes et al 1998). However, further investigation was required to establish whether any limit should be defined for maximum allowable span length or span height. In order to investigate this, a vertical and horizontal pull-over force of 105 kN and 40 kN respectively were applied at the large span at KP3.720, which corresponds to a span height of 0.5 m. (The span height is actually 1.5 m high, however preliminary analyses had showed that the resultant stresses were far above acceptable levels). Figure 16.5 shows an elevated view of the flowline configuration on the seabed as a function of time. As the pull-over load is applied to the span, the adjacent section of the flowline is set in tension so that a new span appears on the right hand side of the existing span. In other words, the "slack" in the neighbouring section is being pulled in towards the point where the pull-over load is applied, very much in the same way as was found in the previous example where the pipe was in full contact with the seabed. For this span the resulting equivalent stress of 520 MPa was found to exceed the allowable stress of 1*SMYS (=500 MPa).

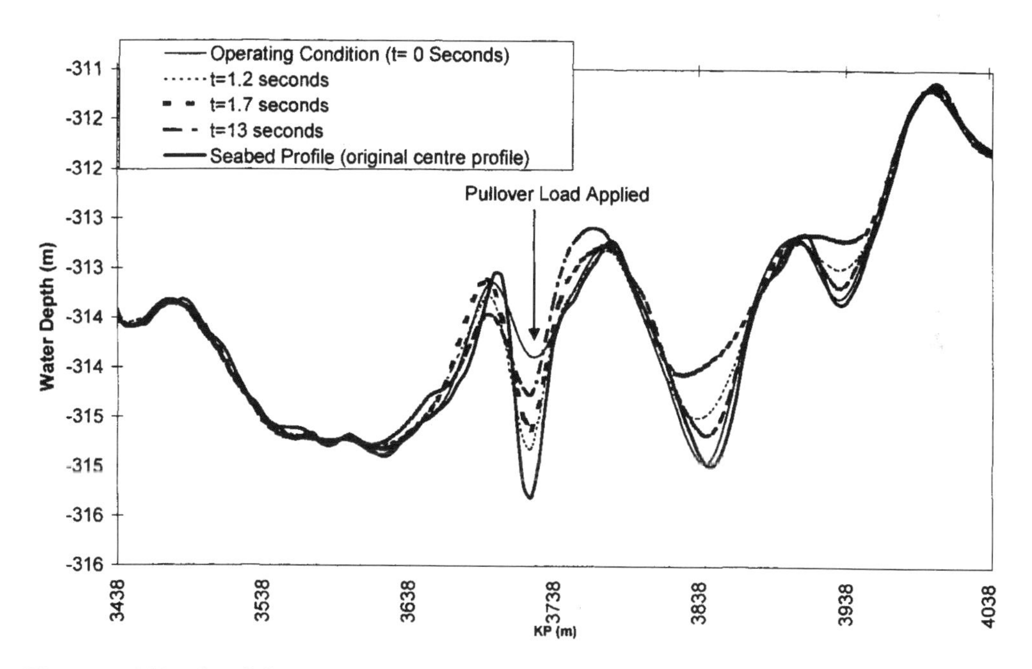

Figure 16.5 Flowline deflected shape (elevated view) as function of time.

A number of further analyses of spans with different heights and lengths revealed that the resulting stresses were not much influenced by span lengths, which agrees with the observations made by Verley (1994). The strong influence of the span gap is due to the way the load is specified by the DNV (1997) Guideline; Both the vertical and horizontal load components increases dramatically with the gap between the seabed and the pipeline.

However, for the 10-inch flowline discussed above, it was demonstrated that span height is the governing parameter in structural response to trawl board pull-overloads.

The main conclusion from a large number of pull-over simulations are:

(1) Span height: For a small diameter pipeline, the span height is a governing parameter in the assessment of pipe structural behavior during trawl board pull-over. For the high strength 10-inch pipeline considered, it was concluded the maximum allowable gap should be limited to 0.25m in the permanent operating conditions. The span length was not found to be an important parameter for the structural pull-over response.

(2) Trawling frequency: Based on detailed information, it has been found that the trawling frequency with in the Åsgard field is less than 1 impact per km *year. Therefore trawling frequency may be classified as "low".

(3) Impact response: Through dynamic non-linear analyses it has been demonstrated that only a fraction of the impact energy is absorbed locally by the steel pipe. The flowlines can accommodate the identified impact loads, provided that the insulation coating has an energy absorption capability of 12 KJ and a minimum thickness of 30 mm. The flowlines will experience denting (plastic deformation), but within allowable limits.

(4) Pull-over response of pipelines on flat seabed: Where the flowlines are in contact with a flat seabed they can accommodate the maximum trawlboard pull-over loads even while they are subject to high compressive effective force.

(5) Pull-over response of pipelines on uneven seabed: While it was confirmed that the pipelines resting on the seabed can resist the highest pull-over loads considered for this area, the effect of span height on pull-over loads and structural response was analysed. For flowlines in operating conditions, it was found that the allowable span height is 0.25 m if both equivalent stress (with usage factor 1.0) and strain criteria are applied; the allowable span height is 0.6 m if only strain criteria are applied.

The length of the finite element model is limited by the length of flowline between two intermittent rock berms. In the seabed intervention design, the rock berm will be designed such that the highest axial load may be resisted. The rock berm may be considered as fixing the ends of the pipelines.

(6) Soil sensitivity study: The effect of varying the axial friction coefficient is small on trawl pull-over response because the axial feed in of pipe to the deflected zone is already limited by the low axial force and intermittent rock dumping. The peak stress and strain values reduce as the lateral friction coefficient increases. For the base case with a axial friction coefficient of 0.3, increasing the lateral friction component from 0.3 to 0.6 results in a reduction in peak equivalent stress and strain of 18% and 6%, respectively.

During trawl board pull-over the pipe is pushed downwards resulting in an increased lateral resistance locally near the hit point. Increasing the lateral friction to 2.0 over a length of 50 m with axial friction coefficients at 0.1 and 0.3, results in a reduction in the peak stress and strain of up to 30 and 40%, respectively. In both cases, the growth in the amplitude of the global buckle induced by trawl pull-over was limited by the higher lateral friction. Therefore, the final bending configuration was less severe than with low friction.

(7) Generalization of trawl pull-over analysis results: It has been demonstrated that span height is the governing parameter in the assessment of pipe behavior during trawl board pull-over. The critical span heights based on equivalent stress and axial strain criteria are listed in Table 16.4.

Table16.4 Critical span heights for trawl board pull-over.

Line	Critical Span Height, (m)		
	Operation	Temporary	
		Cool-Down	Shut-Down
R101	0.30	0.40	+ 0.80
P101	0.25	0.40	+ 1.00

As long as the pull-over load used as input to the analysis is a strong function of the gap height, other variables such as span length and axial force in the line prior to impact, do not significantly affect the response.

16.5.4 Vortex Induced Vibrations (VIV)

The in-place analysis of the flowlines simulates the behavior of the lines on the seabed from the installation phase, including flooding and pressure testing, to operation and repeated shutdown cycles. The 2D FEM multispan modal analysis provides the cross-flow natural response frequencies of free spans while the in-line mode frequencies will be determined from 3D span analyses (Kristiansen et al, 1998). The freespans that are present during the installation and shutdown periods, tend to disappear when in operation because the flowlines expand. Therefore, any proposed seabed intervention measures need to consider the consequences to later stages of operation, including start-up and shutdown cycles.

The criteria applied in design of the Phase 1 flowlines are:

- Onset of in-line VIV may occur during any phase of the design life, provided the accumulated fatigue damage is acceptable.
- Onset of cross-flow VIV will be allowed during any phase of the design life, provided allowable stress and fatigue limits are not exceeded.

Spans that are found to be critical with respect to VIV are usually rectified by placing rock berms below the pipe, in order to shorten the span lengths and thus increase the natural frequency of the spans. In addition to the cost implication of placing a large number of rock berms on the seabed, the main disadvantage of this approach is that feed in of expansion into the spans will be restricted. It was demonstrated through in-place analysis, that allowing the flowline to feed into the spans, reduces the effective force which is the prime factor in the onset of pipeline buckling. It is therefore advantageous with respect to minimizing buckling that the number of rock-berm free-span supports is kept to a minimum.

The span analysis methodology and criteria are based on the design guideline from DNV(1998) for steady current loading and combined wave and current loading. In-house Excel spread sheets and Mathcad validation sheets have been prepared to process the large number of span assessments anticipated for the detail design of the Åsgard Phase 1 flowlines.

Two types of VIV assessment have been performed; a Level 2 assessment and a more advanced Level 3. The Level 2 assessment uses simplified single span FE models and conservative VIV onset criteria to derive critical span lengths. The level 3 assessment uses full multi-span FE analysis to establish natural frequencies and associated mode shapes, and fatigue analysis to calculate the fatigue lives due to in-line and cross-flow vibrations.

When the non-linear effect of sagging is ignored, the effect of axial tension is to stiffen the pipe and increase the natural frequency while axial compression has a tendency to lower it. The effect of varying effective force on the frequency versus span length behaviour is illustrated in Figure 16.6 (Kristiansen et al, 1998). The results are obtained for the P101 line under operational loading. It can be seen that varying the tension/compression does not alter the frequency significantly for short span lengths. However, for span lengths larger that 40 m where non-linear geometry effects come into play, the natural frequency is strongly influenced by the effective force. This is particularly the case for long spans in compression.

It can be concluded that when the axial tension can not be estimated accurately, it may not be conservative to calculate the natural frequency for zero tension. In the case of axial compression, increasing the loading has a lowering effect on the natural frequency for very long spans (>40m). A conservative estimate of natural frequency associated with maximum allowable span length, can be obtained only if the maximum axial compression is adopted in the calculations.

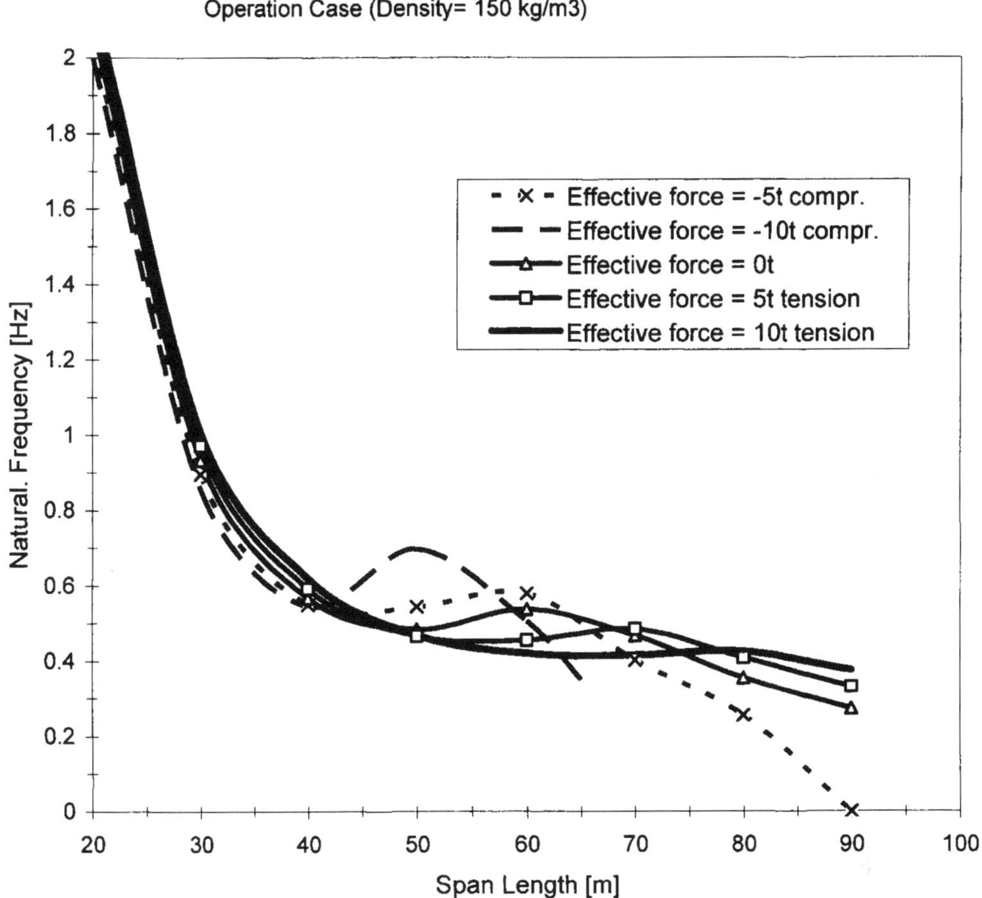

Figure 16.6 First vertical natural frequency as a function of span length and effective force (single span level 2 model).

The effect of the axial stiffness on the frequency is illustrated in Figure 16.7 (Kristiansen et al, 1998), where single span results with pinned-pinned boundary conditions and different axial restraints are illustrated. Curves representing the frequency versus span length for the three stiffnesses; zero, infinite and a stiffness equal to the axial stiffness of the spanning pipe, are illustrated in Figure 16.7. For span lengths longer than 20 m axial restraint is obviously an important variable.

From Figure 16.7, it can be seen that reducing the axial stiffness results in a reduction in the frequency, and hence, a lower allowable span length for a given critical frequency. The axial stiffness is represented by the two components; structural stiffness of the connecting pipe system, and the pipe-soil interaction. The first component that is significantly larger than the second, will be a function of the pipe axial stiffness, shoulder length and span interaction.

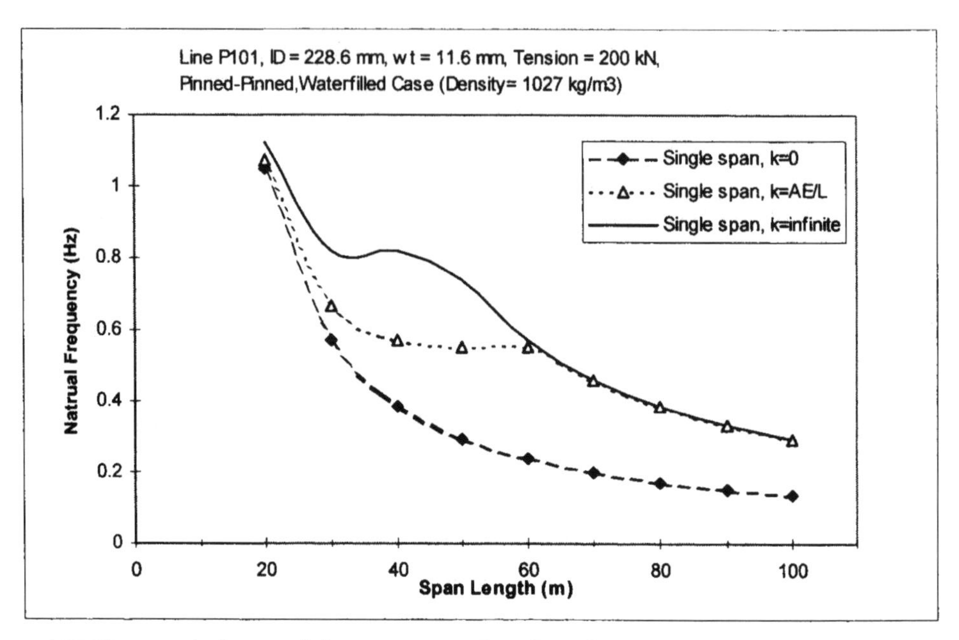

Figure 16.7 First vertical natural frequency as a function of span length and axial restraint (single span level 2 model).

16.6 Åsgard Transport Project

The Åsgard Transport (ÅT) diameter 42-inch export pipeline transports rich gas from the Åsgard field (located 150km from the coast in mid Norway) to Kårstø (located 30km North of Stavanger, Norway). The line traverses a total distance of 700km in water depths varying from 60m to 370m. The design life of ÅT is 50 years. The seabed varied from very soft clays to hard clays, the most notable point being the extremely rough terrain over 200km of the route. The line has 5 pre-installed tees to permit gas from fields along the route to transport gas in ÅT. The line is to be installed in 1998/9 and operated in year 2000.

The cost of the pipeline was estimated at approximately 7 billion NOK ($1 billion US), where approximately 66% is material costs, 29% installation costs, 2% management and design and the remaining 3% insurance and miscellaneous items.

The design approach for ÅT large diameter pipeline is similar to that applied to Åsgard flowlines.

16.7 References

1. ABAQUS Version 5.5, Users Manual, Theory Manual, Verification Examples Manual, Hibbit, Karlsson and Sorensen, Inc., 1995.
2. Andersen, O. (1997) "Overview of Åsgard and Gullfaks Satellite Flowlines Systems", OMAE'97.

3. Bai, Y. and Damsleth, P.A. (1997) "Limit State Design of Flowlines", Proceedings of the International Conference on Offshore Mechanics and Arctic Engineering (OMAE), 1997.
4. Bai, Y. and Damsleth, P.A. (1998) "Design Through Analysis applying Limit State Design Concepts and Reliability Methods", ISOPE'98.
5. Bai, Y., Damsleth, P.A. and Dretvik, S., (1999) "The Åsgard Flowlines Project – Limit-State Design Experience", IBC Conference on Risk-based & Limit-State Design & Operation of Pipelines, Oslo, October 4, 1999
6. Bai et al (1999) "Simulation of Ratcheting of HP/HT Flowlines", ISOPE'99.
7. Damsleth, P. and Dretvik S. (1998) "The Åsgard Flowlines – Phase 1 Design and Installation Design", Proc. of Offshore Pipeline Technology Conference, 1998.
8. Denys R.M. (1992) "A Plastic Collapse-based Procedure for Girth Weld Defect Acceptance", Int. Conf. on Pipeline Reliability, June 2-5, 1992, Calgary.
9. DNV (1988) "RP E305, On-Bottom Stability Design of Submarine Structures".
10. DNV (1996). "Rules for Submarine Pipelines", Dec. 1996.
11. DNV (1997): "Interference between Trawl Gear and Pipelines".
12. DNV (1998): "Guideline no. 14 - Free Spanning Pipelines".
13. Hobbs, R.E, (1984) "In-service Buckling of Heated Pipelines", J. Transp. Eng., Vol.110 (2), 1984.
14. Holme, R., Levold, E., Langford, G. and Slettebø, H. (1999) "Åsgard Transport – The Design Challenges for The Longest Gas Trunkline in Norway", OPT'99.
15. Jee, T., Hawkins, M. and East, A. (1995) "To Trench or Not to Trench", Offshore Pipeline Technology Conference, Oslo 1998.
16. Jiao, G. et al (1996) "The Superb Project: Wall-thickness Design Guideline for Pressure Containment of Offshore Pipelines", Proc.of OMAE'96.
17. Kristiansen, N.Ø., Bai, Y. and Damsleth, P.A. (1997) "Ratcheting of HP/HT Pipelines", OMAE'97.
18. Kristiansen, N.Ø., Tørnes, K., Nystrøm P.R. and Damsleth, D. (1998): "Structural Modeling of Multi-Span Pipe Configurations Subjected to Vortex Induced Vibrations", ISOPE'98.
19. Marshall, P.W. (1992) "Design of Welded Tubular Connections - Basis and Use of AWS Code Provisions", Elsevier, 1992.
20. McKinnon, C. and Ellinas C. (1998)."Reliability Based Limit State Methods Applied to Upheaval Buckling", Risk Based & Limit State Design & Operation of Pipelines, Aberdeen, 20th & 21st Oct.
21. Moshagen, H. and Kjeldsen, S.P. (1980) "Fishing Gear Loads and Effects on Submarine Pipelines", Proc 12th Int. Offshore Technology. Conf., OTC 3782.
22. Murphey and Langner (1986) "Ultimate Pipe Strength Under Bending, Collapse and Fatigue", OMAE'86.
23. Nystrøm, P.R., Tørnes, K., Bai, Y. and Damsleth, P. (1997) "3-D Dynamic Buckling and Cyclic Behaviour of HP/HT Flowlines", ISOPE'97.
24. Nødland, S., Bai, Y. and Damsleth, PA. (1997) "Reliability Approach to Optimize Corrosion Allowance", Proc. of Int. Conf. on Risk based & Limit-state Design & Operation of Pipelines.

25. Ose, B.A., Bai, Y., Nystrøm, P.R., and Damsleth, P. (1999): "A Finite Element Model for In-situ Behavior of Offshore Pipelines and Its Application to On-bottom Stability", ISOPE'99.
26. PD 6493 (1991): "Guidance on Methods for Assessing the Acceptability of Flows in Fusion Welded Structures", 1991.
27. Pedersen, PT and Jensen JJ (1988) "Upheaval Creep of Buried Heated Pipelines with Initial Imperfections", Journal of Marine Structures.
28. SINTEF (1997) "Åsgard flowlines- Fatigue Testing of Cr13 Tube", May 1997.
29. Statoil Technical Specification, R-SF-260, Pipeline Welding Specification, 1991.
30. Statoil Technical Specification, R-SP-230, Welded Linepipe-Ferritic Steel, 1990.
31. Statoil Technical Specification, R-SP-233, Seamless Linepipe-Austenstic Ferritic Stainless Steel, 1990.
32. Stewart G. et al (1994) "An Analytical Model to Predict the Burst Capacity of Pipelines" OMAE'94.
33. Tørnes, K., Nystrøm, P.R., Damsleth, P., and Sortland, L.H. (1997): "The behaviour of High Pressure, High Temperature Flowlines on Very Uneven Seabed", ISOPE'97.
34. Tørnes, K., Nystrøm, P.R., Kristiansen, N. Ø., Bai, Y. and Damsleth, P. (1998) "Pipeline Structural Response to Fishing Gear Pull-Over Loads by 3D Transient FEM Analysis", ISOPE'98.
35. Verley, R.L.P., Moshagen, B.H. and Moholdt, N.C. (1992) "Trawl Forces on Free-Spanning Pipelines", Proc 3rd Int. Offshore and Polar Eng. Conf.
36. Zimmerman et al (1992) "Development of Limit-states Guideline for the Pipeline Industry", 1992.

PART III: Flow Assurance

Part III

Flow Assurance

Chapter 17 Subsea System Engineering

17.1 Introduction

Flow assurance is an engineering analysis process to assure hydrocarbon fluids are transmitted economically from the reservoir to the end user over the life of a project in any environment, in which the knowledge of fluid properties and thermal-hydraulic analysis of the system is utilized to develop strategies for controlling the solids such as hydrates, wax, asphaltenes, and scale from the system.

The term "Flow Assurance" was first used by Petrobras in the early 1990s, it originally only covered the thermal hydraulics and production chemistry issues encountered during oil and gas production. While the term is relatively new, the problems related with flow assurance have been a critical issue in the oil/gas industry from very early days. Hydrates were observed causing blockages in gas pipelines as early as the 1930s and were solved by chemical inhibition using methanol by the pioneering work of Hammerschmidt.

17.1.1 Flow Assurance Challenges

Flow assurance analysis is a recognized critical part in the design and operation of subsea oil/gas systems. Flow assurance challenges mainly focus on the prevention and control of solid deposits which could potentially block the flow path. The solids of concern generally are hydrates, wax and asphaltenes. Sometimes scale and sand are also included. For a given hydrocarbons fluid these solids appear at certain combinations of pressure and temperature and deposit on the walls of the production equipment and flowline. Figure 17.1 shows the hydrate and wax depositions formed in hydrocarbons flowlines, which ultimately may cause plugging and flow stoppage.

The solids control strategies of hydrates, wax and asphaltenes include:

- thermodynamic control -- keeping the pressure and temperature of whole system out of the regions where the solids may form;
- kinetic control -- controlling the conditions of solids formation so that deposits do not form;
- mechanical control -- allowing solids to deposit, but periodically removing them by pigging.

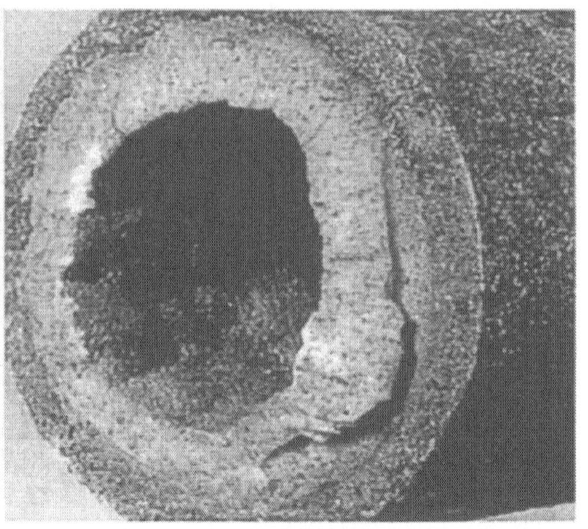

Hydrate Wax

Figure 17.1 Solid depositions formed in hydrocarbons flowlines (Kaczmarski and Lorimer, 2001).

Flow assurance becomes more challenging in the subsea field developments involving long distance tie-backs and deepwater in recent years. The challenges include a combination of low temperature, high hydrostatic pressure for deepwater and economic reasons for long offsets. The solutions of solids deposition problems in subsea systems are different for gas and oil systems.

For gas systems, the main concern of solids usually is hydrates. Continuous inhibition with either methanol or mono-ethylene-glycol (MEG) is a common and robust solution, but low-dosage hydrate inhibitors (LDI's) are finding more applications in gas systems. The systems using methanol for inhibition are generally operated on a once-through basis. The methanol partitions into gas and water phases, and is difficult to recover. Systems using MEG on the other hand normally involve reclaimation of the MEGas an inhibitor are reclaimed. If a hydrate plug forms, remediation may be depressurization.

For oil systems, both hydrates and paraffins are critical issues. In the Gulf of Mexico (GoM), a "blowdown strategy" is commonly used. The strategy relies on insulation coating on the flowline to keep the fluids out of the hydrate and paraffin deposition regions during operation. During start-ups and shutdowns, a combination of inhibition, depressurization and oil displacement is performed to prevent hydrate and paraffin deposition. Wax is removed by pigging. The strategy is effective, but depends on successful execution of relatively complex operational sequences. If a hydrate plug forms, it is necessary to depressurize the line to pressure usually below 200 psi for deep water subsea system and wait for the plug to disassociate which could take a very long time in a well-insulated oil system.

17.1.2 Flow Assurance Concerns

Flow assurance is only successful when the operations generate a reliable, manageable and profitable flow of hydrocarbon fluids from reservoir to sales point. Some flow assurance concerns are:

- System deliverability: pressure drop versus production, pipeline size & pressure boosting, and slugging & emulsion; which will be discussed in detail in Chapter 18, Hydraulics.
- Thermal behavior: temperature distribution, temperature changes due to start-up, shutdown, and insulation options & heating requirements; this will be described in Chapter 19, Heat Transfer.
- Solids and chemistry inhibitors: hydrates, waxes, asphaltenes, and scaling; this will be detailed in Chapters 20 ad 21.

17.2 Typical Flow Assurance Process

17.2.1 General

Flow assurance is an engineering analysis process of developing design and operating guidelines for the control of solids deposition, such as hydrates, wax, and asphaltenes in subsea systems. Depending on the characteristics of the hydrocarbons fluids to be produced, corrosion, scale deposition, and erosion may also be considered in the flow assurance process. The bulk of the flow assurance analysis should be done prior to or in the earlier Front End Engineering and Design (FEED). The requirements for each project are different and therefore project specific strategies are required for flow assurance problems. However, over the last several decades, the flow assurance process itself has become standardized and a typical procedure is shown in Figure 17.2. The main issues associated with the flow assurance process include:

- fluid characterization and flow property assessments;
- steady state hydraulic and thermal performance analyses;
- transient flow hydraulic and thermal performances analyses;
- system design and operating philosophy for flow assurance issues.

Detailed explanations for each issue are listed in the following sections. Some issues may occur in parallel, and there is considerable "looping back" to earlier steps when new information, such as a refined fluids analysis or a revised reservoir performance curve, becomes available.

17.2.2 Fluid Characterization and Property Assessments

The validity of the flow assurance process is dependent on careful analyses of samples from the wellbores. In the absence of samples, an analogous fluid, such as one from a nearby well in production may be used. This always entails significant risks because fluid properties may vary widely, even within the same reservoir. The key fluid analyses for the sampled fluid are PVT properties, such as phase composition, GOR (gas/oil ratio), bubble point, etc.; wax properties, such as cloud point, pour point, or WAT; and asphaltene stability.

Knowledge of the anticipated produced water salinity is also important, but water samples are seldom available and the salinity is typically calculated from resistivity logs. The composition of the brine is an important factor to the hydrate prediction and scaling tendency assessment.

In the case of without brine sample, a prediction of composition can be made based on information in an extensive database of brine composition for deep water.

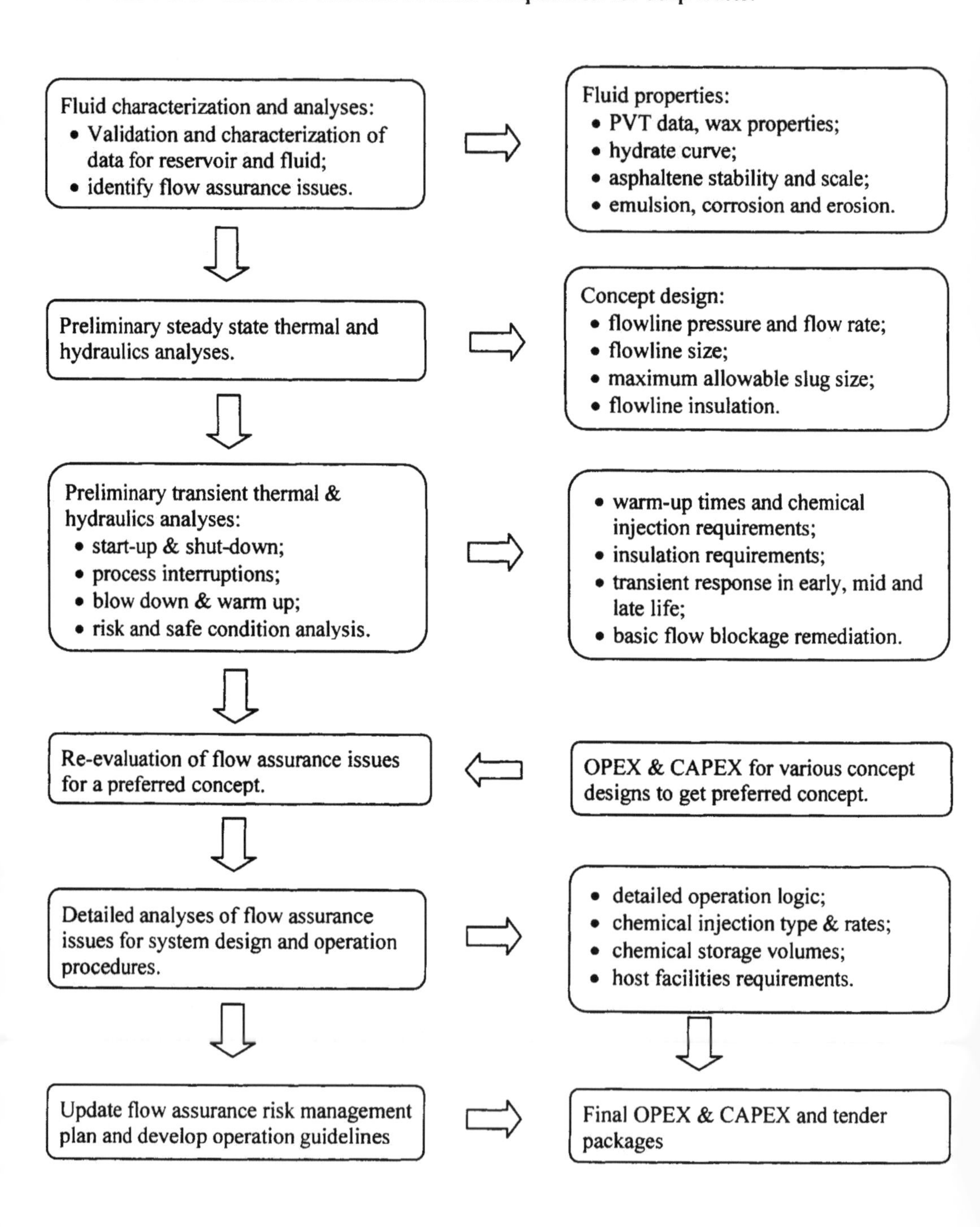

Figure 17.2 Typical flow assurance process.

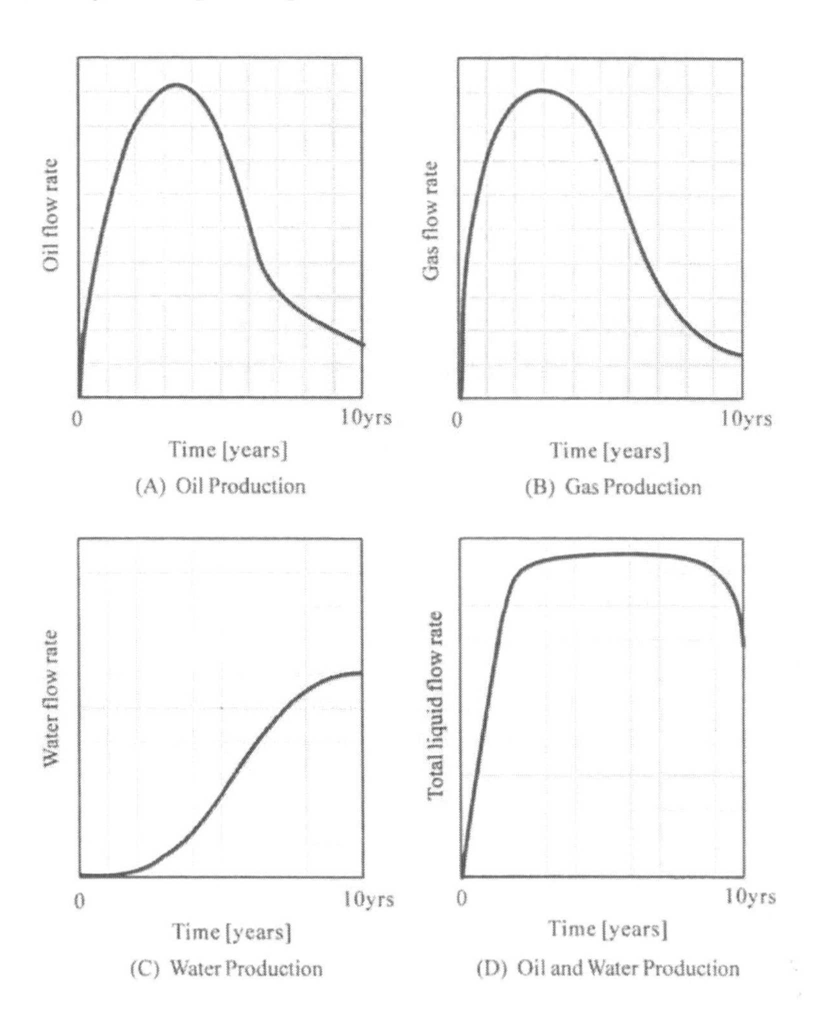

Figure 17.3 Typical oil, water and gas production profiles.

The hydrate stability curves are developed based on PVT data and salinity estimates, and methanol dosing requirements are obtained. A thermal-hydraulic model of the well(s) is developed to generate flowing wellhead temperature and pressures for a range of production conditions. The wellbore temperature transient analyses are carried out. Figure 17.3 demonstrates typical production profiles of oil, water and gas, in which water content increases with time. Water content (water cut) is very important for choosing a flow assurance strategy to prevent hydrate formation.

Hydrate stability curves show the stability of natural gas hydrates as a function of pressure and temperature, which may be calculated based on the hydrocarbon phase and aqueous phase compositions utilizing a thermodynamic package such as Multiflash. The hydrate curves define the temperature and pressure envelope. The dosing calculations of hydrate inhibitor, such as Methanol or MEG, indicate how much inhibitor must be added to the produced water at a given system pressure to ensure that hydrates will not form at the given temperature. Hydrate inhibitor dosing is the way hydrate formation is controlled when system temperatures drop into the range in which hydrates are stable during steady state or transient state of subsea system in the activities of normal operation, shut-down or start-up. The inhibitor dosing requirements are used to determine the requirements for the inhibitor storage, pumping capacities, and number of inhibitor flowlines in order to ensure that the inhibitor can be delivered at the required rates for treating well and subsea system during normal operation, startup and shut-in operations.

17.2.3 Steady State Hydraulic and Thermal Performance Analyses

The steady state flowline model can be generated with the software such as PIPESIM, HYSYS. The steady state modeling has several objectives:

- determine the relationship between flow rate and pressure drop along the flowline; decide flowline size based on the maximum allowable flow rate and the minimum allowable flow rate.
- check temperature and pressure distributions along flowline in steady state condition to ensure that the flowline never enters the hydrate region during steady state operation.
- choose an insulation combination that prevents the temperature at riser base of a tie-back subsea system from falling below the minimum value for cooldown at the maximum range of production rates. Riser base temperature is determined as a function of flow rate and the combined well bore/flowline insulation system.
- determine the maximum flow rate in the system to assure the arrival temperatures do not exceed any upper limits set by the separation and dehydration processes or by the equipment design.

17.2.4 Transient Hydraulic and Thermal Performances Analyses

Transient flowline system models can be constructed with the software packages, such as OLGA and ProFES. The transient flowline analyses generally includes the following scenarios,

- start-up and shut-down;
- emergent interruptions;
- blow down and warm up;
- Ramp up/down
- oil displacement
- pigging / slugging

During these scenarios, fluid temperatures in the system must exceed the hydrate dissociation temperature corresponding to the pressure at every location; otherwise, a combination of insulation of pipeline and injecting chemical inhibitors to the fluid must be simulated in the transient processes to prevent hydrate formation.

Start-up

Hydrate inhibitor should generally be injected downhole and at the tree during start-up. When the start-up rate is high, inhibitor is not required downhole, but the hydrocarbon flow should be treated with inhibitors at the tree. Otherwise, the hydrocarbon flow is required to be treated with inhibitors downhole. Once the tree is outside the hydrate region, hydrate inhibitor can be injected at the tree and the flow rate increased to achieve system warm-up. The Start-up scenario is different for the combination of cold well with cold flowline and hot flowline (commingling production with that from other wells).

Shut-down

Shut-down scenarios include planned shutdown and unplanned shutdown from steady state, and unplanned shutdown during warm up. In general, the planned and unplanned shutdowns from steady state are the same with the exception that for a planned shutdown, hydrate inhibitor can be injected into the system prior to shutdown. Once the system is filled with inhibited produced fluids, no further inhibitor injection or depressurization is needed prior to startup.

After shut-down, the flowline temperature will decrease because of heat transfer from the system to surrounding water. The insulation system of the flowline is designed to keep the temperature of fluid above the hydrate dissociation temperature until the "no-touch time" has passed. The "no touch time" in the minimum cooldown time is the one in which operators can try to correct problems without having to take any action to protect the subsea system from hydrates. Operators always want a longer "no touch time", but it is a cost benefit balance problem and is decided based on the project. Analyses of platform operation experience in West Africa indicate that many typical process and instrumentation interrupts can be analyzed and corrected in 6-8 hours.

A tie-back subsea system in West Africa is used as an example. If the system is shutdown from a steady state, the first step is to see if the system can be restarted within 2 hours. If so, startup should begin. If not, one option for hydrate control is that the riser will be bullheaded with MeOH (if MeOH is chosen as a hydrate inhibitor) to ensure no hydrates are forming in the base of the riser where fluids are collecting. Next the tree piping will be dosed with methanol. After that, the fluid in the flowline will begin to be fully treated with methanol. Once 8 hours has been reached, operations must determine if the system can be started up or not. If it can be started, they will proceed to the startup procedure outlined previously. If it cannot be started up, the flowlines will be depressurized. The intention with depressurization is to reduce the hydrate dissociation temperature to below the ambient sea temperature. Once the flowlines are depressurized, the flowlines, jumpers, trees are in a safe state. If the wells have been shut in for two days without the system starting back up, then the wellbores need to be bullheaded with MeOH to fill the volume of the wellbore down to the SCSSV. Once these steps have been taken, the entire system is safe.

Table 17.1 shows a typical sequence of events during unplanned shutdown of a flowline system for the tie-back subsea system in West Africa. The shut-in event is followed by three hours of no touch time. After these three hours the wellbore and the jumpers are treated with methanol. Simultaneously, the flowline is depressurized. To ensure that this procedure can be finished in three hours the blowdown is carried out with gas lift assisted. Blowdown is followed by dead oil displacement.

Table 17.1 Unplanned shutdown sequence of events.

Time (hrs)	Activity	
0	shut-in	
3	no touch time	
6	blowdown (gas lift assisted)	Jumper/tree, MeOH injection
12	dead oil displacement	

Blowdown

To keep the flowline system out of the hydrate region when shut-in time is longer than the cooldown period, flowline blowdown or depressurization may be an option. The transient simulation of this scenario shows how long blowdown takes and liquid carryover during blowdown, as well as indicating if the target pressure to avoid hydrate formation can be reached. Figure 17.4 shows that shorter blowdown times are accompanied by greater liquid carryover. The blowdown rate may also need to be limited to reduce the amount of Joule-Thompson cooling downstream of the blowdown valve, to prevent the possibility of brittle fracture of the flowline. During blowdown, sufficient gas must be evolved to ensure that the remaining volume of depressurized fluids exerts a hydrostatic pressure that is less than the hydrate dissociation pressure at ambient temperature. Until the blowdown criterion is met, the only way to protect the flowline from hydrate formation is to inject inhibitor.

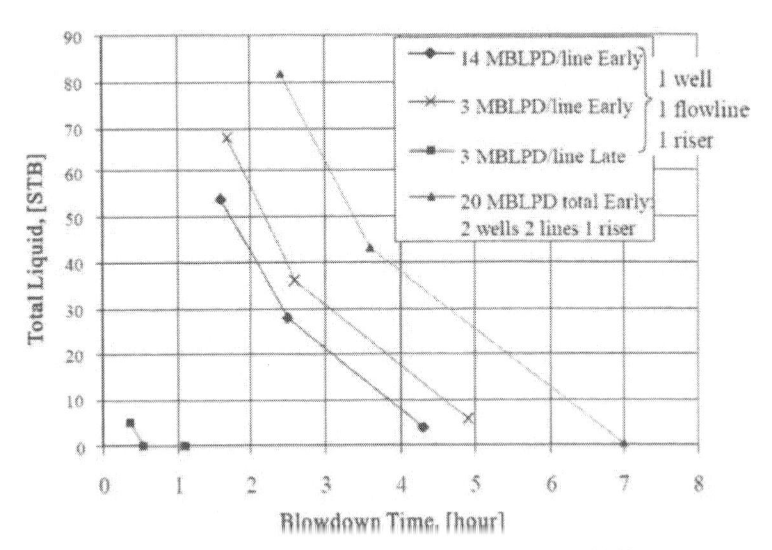

Figure 17.4 Liquid carryover vs. blowdown time (Lorimer and Ellison, 2000).

Warm up

During the warm-up process, hydrate inhibitor must be injected until the flowline temperatures exceed the hydrate dissociation temperature at every location for a given pressure. Figure 17.5 shows the effects of insulation material to the warm-up time. Hot oiling has two beneficial effects; first, it reduces or eliminates the time required to reach the hydrate dissociation temperature in the flowline. Once the minimum void fraction is reached, methanol injection can be safely stopped. Reduction of the methanol injection time is a tremendous advantage for projects with limited available methanol volumes. Hot oiling also warms up the pipeline and surrounding earth, resulting in a much longer cooldown time during the warm-up period than is accomplished by warming with produced fluids. This gives more flexibility at those times when the system must be shut-in before it has reached steady state.

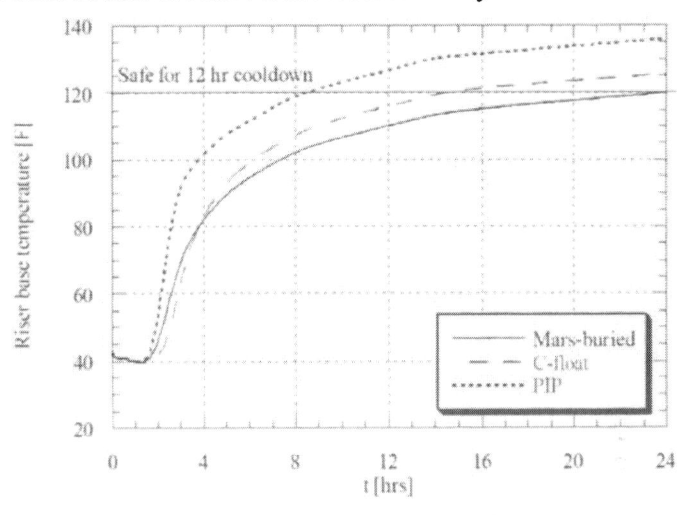

Figure 17.5 Effect of flowline insulation to warm up from cold earth (Lorimer and Ellison, 2000).

Riser Cooldown

The most vulnerable portion of the subsea system, in terms of hydrate formation, is typically the riser base. The steady state temperatures at riser base are near the lowest point in the whole system. The available riser insulation systems are not as effective as the pipe-in-pipe insulation that is used for some flowlines. The riser is subject to more convective heat transfer due to higher current velocity than pipeline, and finally, may be partially or completely gas filled during shut-in conditions, leading to much more rapid cooling.

The desired cooldown time before the temperature at riser base reaches the hydrate temperature, is determined by the following formula,

$$t_{\min} = t_{no\ touch} + t_{treat} + t_{blowdown} \tag{17.1}$$

where, $t_{\min}$: minimum cooldown time, hrs;

$t_{no\ touch}$: no touch time, 2-3 hrs;

t_{treat} : time to treat well bores, trees, jumpers, and manifolds with inhibitors, hrs;

$t_{blowdown}$: time to blowdown flowlines, hrs.

Cooldown times are typically on the order of 12 – 24 hours. Figure 17.6 shows cooldown curves for an 8 inch oil riser with various insulation materials. The increase of desired cooldown time requires better flowline insulation and or higher minimum production rates. The desired cooldown dictates a minimum riser base temperature for a given riser insulation system. This temperature becomes the target to be reached or exceeded during steady state operation of the flowline system.

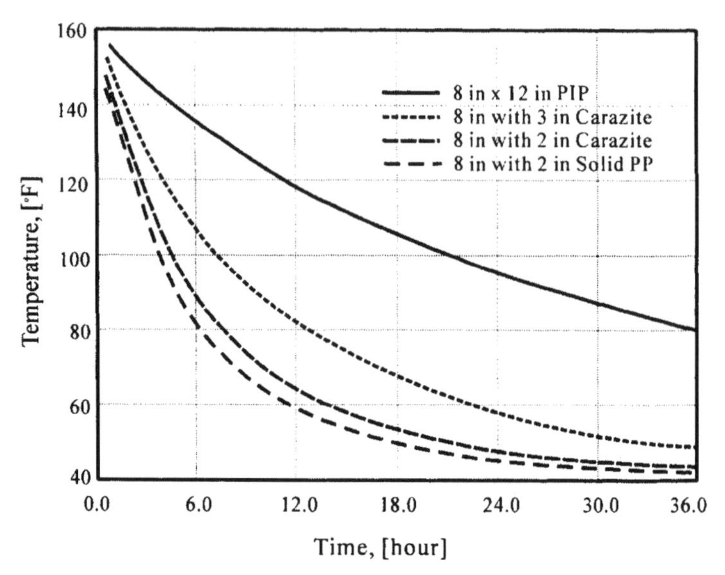

Figure 17.6 8-inch oil riser cooldown curves for different insulation materials (Lorimer and Ellison, 2000).

17.3 System Design and Operability

In a system design, the entire system from the reservoir to the end user (separators) has to be considered to determine applicable operating parameters, flow diameters, flow rates, insulation for tubing, flowline, and manifold, chemical injection requirements, host facilities, operating strategies and procedures, etc. to ensure that the entire system can be built and operated successfully and economically. All production modes including start-up, steady state, flow rate change and shut-down throughout the system life, must be considered.

Operating strategies and procedures for successful system designs are developed with system unknowns and uncertainties in mind and can be readily adapted to work with the existing system, even when that is different from that assumed during design. In a deepwater project, the objective of operating strategies is to avoid the formation of hydrate or wax plugs at any time, especially hydrates in the subsea system including wellbores, trees, well jumpers, manifold, and flowlines during system operation.

Although the operations are time, temperature and pressure dependent, a typical operating procedure can be listed as follow,

- operate the flowlines in an under-packed condition during steady state, e.g., maintain sufficient gas void fraction to allow successful depressurization to below the hydrate dissociation pressure at ambient temperatures.
- upon a platform shutdown, close the boarding valves and tree as close to simultaneously as possible in order to trap the under-packed condition.
- design the insulation system to provide a enough cool-down time to address facilities problems before remedial action is needed and perform the interventions.

- inject hydrate inhibitor into the well, tree, jumpers and manifolds.
- blowdown the flowline to the pressure of fluid below the hydrate formation region.
- flush flowline with hot oil prior to restart from blown down condition.
- start up the wells in stages, while injecting hydrate inhibitor. Continue hydrate inhibitor injection until the warm up of well, tree, jumpers and manifold and enough gas enters the flowline to permit blowdown.

The systems are normally designed to have a 3 hour "no-touch" time during which no hydrate prevention actions are required. Blowdown is carried out only during longer, less frequent shutdowns, and about three times per year. The logic charts of startup and shutdown serve as an outline for the operating guidelines. Figure 17.7 shows a typical logic chart of startup for the cold well start-up of a deep water field. The startup logic begins with the pigging of the flowlines, assuming the flowlines have been blown down. The flowlines must be pigged to remove any residual, uninhibited fluids from the flowlines. The flowlines are then pressured up to system pressure to avoid any problems caused by a severe pressure drop across the subsea choke and to make manifold valve equalization easier when it is time to switch flowlines. The next step is to start up the well that heats up the fastest, remembering to inject hydrate inhibitor at upstream of the choke while this well is ramping up. When the system is heated up, the hydrate inhibitor injection can stop. For this case, once the tree has reached 150°F, MeOH injection can be stopped. The temperature of 150 °F has been determined to provide 8-hours cooldown for the tree.

17.3.1 Well Start-up & Shut-in

Figure 17.8 shows a simplified typical tree schematic. The subsea tree's production wing valve (PWV) is designated as the underwater safety valve (USV). The production master valve (PMV) is manufactured to USV specifications, but will only be designated for use as the USV if necessary. The well has a remotely adjustable subsea choke to control flow. The subsea choke is used to minimize throttling across the subsea valves during shut-in and start-up.

Well Start-up

Below is the start up philosophies for the case that well start-up poses a risk for flowline blockage, particularly with hydrates based on the flow assurance studied in design phase indicates:

- Wells will be started-up at a rate that allows minimum warm-up time, while considering drawdown limitations. There will be a minimum flow rate below which thermal losses across the system will keep the fluids in the hydrate formation region.
- It may not be possible to fully inhibit hydrates at all water-cuts, particularly in the wellbore. High water-cut wells will be brought on-line without being fully inhibited. Procedures will be developed to minimize the risk of blockage if an unexpected shut-in occurs.

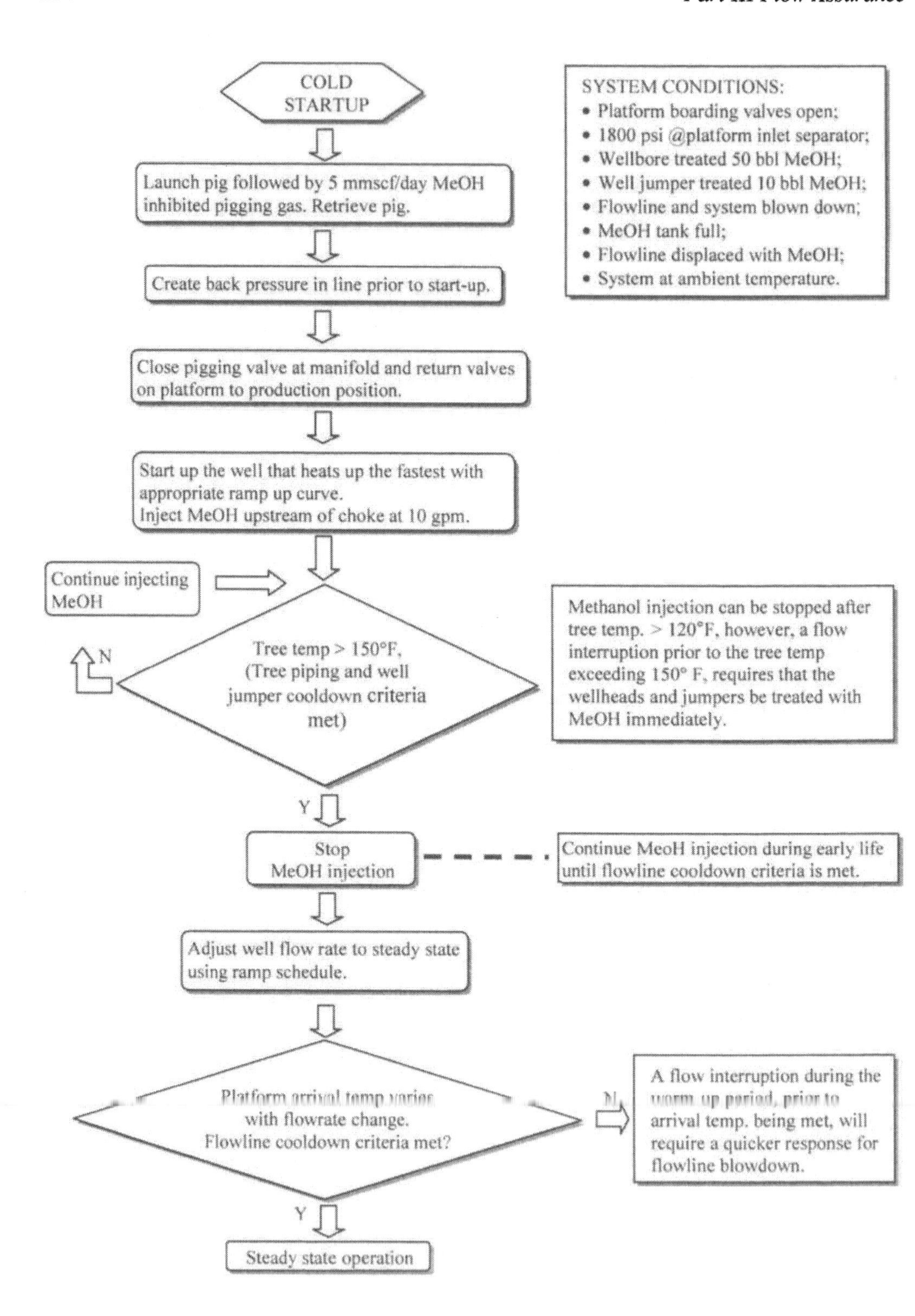

Figure 17.7 Operating logic chart of start up for a cold well and a cold flowline.

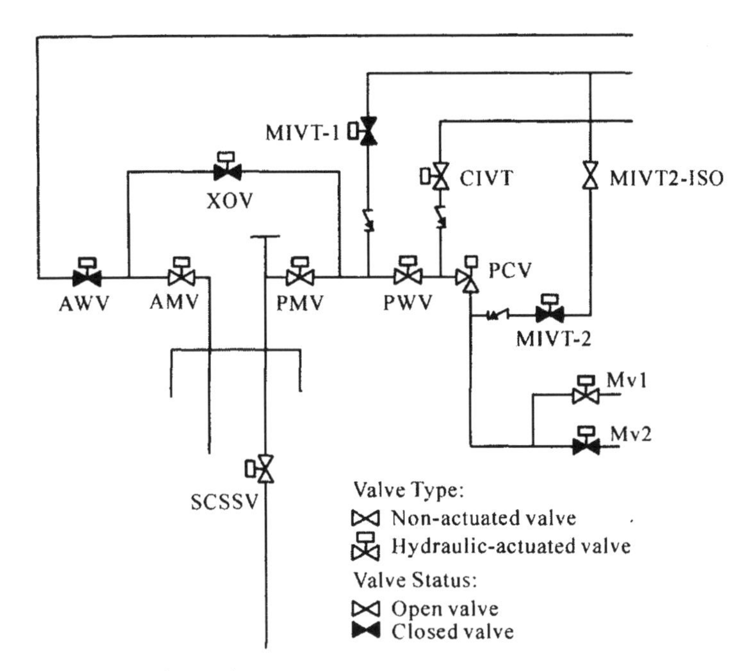

Figure 17.8 Simplified tree schematic.

- The system is designed to inject hydrate inhibitor (typically methanol) at the tree during start-up or shut-in. The methanol injected during initial start-up will inhibit the well jumpers, manifold, and flowline if start-up is interrupted and blowdown is not yet possible.

Well Shut-in

Well shut-in also poses a significant hydrate risk. The following philosophies will be adopted during shut-in operations:

- The subsea methanol injection system is capable of treating or displacing produced fluids with hydrate inhibitor between the manifold and SCSSV following well shut-in to prevent hydrate formation.
- Hydrate prevention in the flowlines is accomplished by blowing the flowline pressure down to less than the hydrate formation pressure at ambient seabed temperatures.
- Most well shut-ins will be due to short-duration host facility shut-ins. The subsea trees, jumpers, and flowlines are insulated to slow the cooling process and allow the wells to be re-started without having to initiate a full shutdown operation.

17.3.2 Flowline Blowdown

17.3.2.1 Why do we blowdown?

The temperature of flowline system is kept from forming hydrates by the heat from the reservoir fluids moving through the subsea flowlines during steady-state operation. When well shut-in occurs, and the pressure in the system is still high but the temperature of system will drop down because of heat is transferred to the ambient environment, hydrates may form in the flowline. Once the blowdown of the fluid pressures to below the hydrate formation pressure corresponding to the environmental temperature has been performed, it is safe from the hydrate formation point of the view to leave the system in this condition indefinitely.

17.3.2.2 When do we blowdown?

The blowdown is carried out based on several factors:

- Flowline pressure;
- Available thermal energy in the system;
- Ability of the insulation to retain heat.

When the flowline is shut-down, the countdown to hydrate formation temperature clock begins. There are several hours of "no touch cooldown" before hydrate inhibitor injection or blow down are required. The base of the risers is the most at risk for hydrate formation in the subsea tie-back system. They have the least amount of insulation, lowest temperature fluid at the sea floor and once the system is shut in, fluids in the vertical section condense and flow downhill to pool at the riser base. Therefore, this section of the flowline must be treated early to prevent hydrate formation. The hydrate inhibitor is injected at the platform and this will mix with the fluids at the riser base and will effectively lower the temperature at which gas/water interfaces form hydrates. This will allow us to avoid blowdown for several more hours.

Once a blowdown operation is begun, the topsides PLC timer will be used to determine the length of time to blow the system down. Following a shut-in, all of the flowlines will need to be completely blown down by the end of 12th hour, which is different depending on the project's requirement. Following an extended shut in that resulted in blowdown of the subsea system, the remaining fluids must be removed from the flowlines before the flowlines can be repressurized. The fluids remaining in the flowlines will have water present and the temperature will be the same as the water temperature. When cold, high pressure gas is introduced to water, hydrates could be formed. One option is pigging the flowline to remove residual fluids, sometimes displacement with hot oil is performed without pigging. Prior to beginning a pigging operation, ensure adequate quantity's of methanol are onboard the platform for startup operations and subsequent shutdown or aborted startup.

17.4 References

1. Kaczmarski, A. A. and Lorimer, S. E. (2001), "Emergence of Flow Assurance as a Technical Discipline Specific to Deepwater: Technical Challenges and Integration into Subsea Systems Engineering", OTC 13123, Offshore Technology Conference, Houston
2. Hammerschmidt, E. G. (1939), "Gas Hydrate Formation in Natural Gas Pipelines", Oil & Gas J. , Vo, 37, No.50,66.
3. Sloan, E. D. (2000), Hydrate Engineering, Edited by Bloys, J. B., SPE.
4. Pattee, F.M., Kopp, F. (2000), "Impact of Electrically-Heated Systems on the Operation of Deep Water Subsea Oil Flowlines", OTC 11894, Offshore Technology Conference, Houston.
5. Lorimer S. E., and Ellison B. T. (2000), "Design Guidelines for subsea Oil Systems", Facilities 2000: Facilities Engineering into the Next Millennium.

Part III

Flow Assurance

Chapter 18 Hydraulics

18.1 Introduction

To ensure system deliverability of hydrocarbon products from one point in the flowline to another, the accurate prediction of the hydraulic behavior in the flowline is essential. From the reservoir to the end user, the hydrocarbon flow is impacted by thermal behavior of heat transfer and phase change of the fluid in the system. The hydraulic analysis method and results are complicated and different for different fluid phases. For a hydrocarbon hydraulic problem with heat transfer and phase change, an adequate knowledge of fluid mechanics, thermodynamics, heat transfer, vapor/liquid equilibrium and fluid physical properties for multi-component hydrocarbon systems is needed. In this Chapter, the composition and phase behavior of hydrocarbon are explained first, then hydraulic analyses of single-phase flow and multi-phase flow, which includes pressure drop versus production flow rate and pipeline sizing are discussed.

18.2 Composition and Properties of Hydrocarbons

18.2.1 Hydrocarbons Composition

The petroleum fluids from reservoirs normally are multi-phase and multi-component mixtures, primarily consisting of hydrocarbons, which can be classified into the following three groups:

- Paraffins;
- Naphthenes;
- Aromatics.

In addition to hydrocarbons, water (H_2O), nitrogen (N_2), carbon dioxide (CO_2), and hydrogen sulfide (H_2S) are often found in petroleum mixtures. Table 18.1 lists some typical physical properties of the main components of petroleum. The temperature of boiling point increases with the increase of the carbon number of the component formula. If the carbon number of a component is less than 5, the component is in gas phase condition at an atmospheric pressure. When the mixture contains larger molecules, it is a liquid at normal temperatures and pressures. A typical petroleum fluid contains thousands of different chemical compounds, and trying to separate it into different chemical is impractical. A composition of gas condensate is shown in Table 18.2. In the usual composition list of hydrocarbons, the last hydrocarbon is marked with a "+", which indicates a pseudo- component, and lumps together all the heavier components.

Table 18.1 Physical properties of main petroleum components (Pedersen, et al., 1989).

Component	Formula	Boiling temperature at 1 atm [°C]	Density at 1 atm and 15°C, [g/cm³]
Paraffins			
Methane	CH_4	-161.5	--
Ethane	C_2H_6	-88.3	--
Propane	C_3H_6	-42.2	--
i-Butone	C_4H_{10}	-10.2	--
n-Butane	C_4H_{10}	-0.6	--
n-Pentane	C_5H_{12}	36.2	0.626
n-Hexane	C_6H_{14}	69.0	0.659
i-Octane	C_8H_{18}	99.3	0.692
n-Decane	$C_{10}H_{22}$	174.0	0.730
Naphthenes			
Cyclopentane	C_2H_6	49.5	0.745
Methyl cyclo-pentane	C_2H_6	71.8	0.754
Cyclohexane	C_2H_6	81.4	0.779
Aromatics			
Benzene	C_6H_6	80.1	0.885
Toluene	C_7H_8	110.6	0.867
o-Xylene	C_8H_{10}	144.4	0.880
Naphthalene	$C_{10}H_8$	217.9	0.971
Others			
Nitrogen	N_2	-195.8	--
Carbon Dioxide	CO_2	-78.4	--
Hydrogen sulfide	H_2S	-60.3	--

Table 18.2 Typical composition of a gas condensate.

Component		Composition (mole %)
Hydrogen Sulfide	H_2S	0.05
Carbon dioxide	CO_2	6.50
Nitrogen	N_2	11.71
Methane	C_1	79.06
Ethane	C_2	1.62
Propane	C_3	0.35
i-Butane	$i\text{-}C_4$	0.08
n-Butane	C_4	0.10
i-pentane	$i\text{-}C_5$	0.04
n-Pentane	C_5	0.04
Hexanes	C_6	0.06
Heptanes plus	C_7+	0.39

The characterization of a pseudo-component including defining molecular weight, density, pseudo-critical properties and other parameters should be calculated ahead of the flow assurance analysis. The characterization should be based on the experimental data and analyses. As is shown in Table 18.1, hydrocarbons with a same formula and carbon atom numbers may have very different normal boiling points and other physical properties.

18.2.2 Equation of State

For oil and gas mixtures, the phase behavior and physical properties such as densities, viscosities, and enthalpies are uniquely determined by the state of the system. The equations of state (EOS) for petroleum mixtures are mathematical relations between volume, pressure, temperature and composition, which are used for describing the system state and transitions between states. Most thermodynamic and transport properties in engineering analyses are derived from the EOS. Since the first EOS for representation of real systems was developed by the van der Waals in 1873, hundreds of different EOS have been proposed, and they are distinguished by Leland (1980) into four families:

1. van der Waals family;
2. Benedict-Webb-Rubin family;
3. Reference-fluid equations;
4. Augmented-rigid-body equations.

The cubic equations of the van der Waals family are widely used in the oil and gas industry for engineering calculations because of their simplicity, and relative accuracy for describing multiphase mixtures. The simple cubic equations of state, Soave-Redlich-Kwong (SRK) and Peng-Robison (PR) are used in the flow assurance software (e.g. PIPESIM). They require limited pure component data and are robust and efficient, and usually give broadly similar results.

- The SRK equation, (Soave, 1972)

$$p = \frac{RT}{v-b} - \frac{a}{v(v+b)} \tag{18.1}$$

where p is the pressure, T the temperature, v the molar volume, R the gas constant and a and b are the EOS constants, which are determined by the critical conditions for a pure component.

$$a = 0.42748\frac{R^2T_c^2}{p_c}\left[1 + m\left(1 - \sqrt{T_r}\right)\right]^2$$

$$b = 0.08664\frac{RT_c}{p_c}$$

and $m = 0.480 + 1.574\omega - 0.176\omega^2$

For a mixture, a and b are found as follows,

$$a = \sum_i \sum_j z_i z_j a_{ij}$$

$$b = \sum_i z_i b_i$$

where, z_i and z_j are the mole fraction of components i and j, respectively and $a_{ij} = \sqrt{a_i a_j}(1 - k_{ij})$, k_{ij} is a binary interaction coefficient, which is usually considered equal to zero for hydrocarbon-hydrocarbon interactions, and different from zero for interactions between a hydrocarbon and a non-hydrocarbon, and between unlike pairs of non-hydrocarbons.

Peng-Robinson, (Peng and Robinson, 1976)

$$p = \frac{RT}{v - b} - \frac{a}{v(v + b) + b(v - b)} \tag{18.2}$$

The EOS constants are given by,

$$a = 0.45724 \frac{R^2 T_c^2}{p_c} \left[1 + m(1 - \sqrt{T_r}\right]^2$$

$$b = 0.07780 \frac{RT_c}{p_c}$$

and $m = 0.3764 + 1.54226\omega - 0.26992\omega^2$

18.2.3 Hydrocarbons Properties

Oil and gas are very complex fluids composed of hydrocarbon compounds that exist in petroleum in a wide variety of combinations. Physical properties change with the component composition, pressure, and temperature. No two oils have the same physical properties. In considering hydrocarbons flow in pipes, the most important physical properties are density and viscosity.

18.2.3.1 Density

Dead oil is defined as oil without gas in solution. Its specific gravity, γ_o is defined as the ratio of the oil density and the water density at the same temperature and pressure.

$$\gamma_o = \frac{\rho_o}{\rho_w} \tag{18.3}$$

API gravity, with a unit of degree (°) is defined as,

$$^oAPI = \frac{141.5}{\gamma_o} - 131.5 \tag{18.4}$$

where γ_o is the specific gravity of oil at 60 °F, and the water density at 60 °F is 62.37 lb/ft^3.

The effect of gas dissolved in the oil should be accounted for in the calculation of the in-situ oil (live oil) density. The oil density can be calculated from,

$$\rho_o = \frac{350.4\gamma_o + 0.0764\gamma_g R_s}{5.615B_o} \tag{18.5}$$

where, ρ_o : oil density, lb/ft^3; γ_o: oil specific gravity; γ_g: gas gravity; R_s: dissolved gas, scf/STB, and B_o: formation volume factor, volume/standard volume.

Gas density ρ_g is defined as,

$$\rho_g = \frac{pM}{ZRT} \tag{18.6}$$

where, M is the molecular weight of gas, R, universal gas constant, p, absolute pressure, T, absolute temperature, and Z is the compressibility factor of gas. The gas specific gravity γ_g is defined as the ratio of the gas density and the air density at the same temperature and pressure.

$$\gamma_g = \frac{\rho_g}{\rho_a} = \frac{M}{29} \tag{18.7}$$

For liquids, the reference material is water, and for gases it is air.

18.2.3.2 Viscosity

Dynamic viscosity

The dynamic viscosity of a fluid is a measure of the resistance to flow exerted by a fluid and for a Newtonian fluid it is defined as,

$$\mu = \left(\frac{\tau}{dv/dn}\right) \tag{18.8}$$

where, τ: shear stress;
v: velocity of the fluid in the shear stress direction;
dv/dn: gradient of v in the direction perpendicular to flow direction.

Kinematic Viscosity:

$$\nu = \mu/\rho \tag{18.9}$$

where ρ: density.

The viscosity of Newtonian fluids is a function of temperature and pressure. If the viscosity of fluid varies with the thermal or time history, the fluid is called non-Newtonian. Most of the hydrocarbon fluids are Newtonian fluid, but in some cases, the fluid in the flowline should be considered as a non-Newtonian. The viscosity of oil decreases with increasing temperature, while the viscosity of gas increases with increasing temperature, and the general relations can be expressed as,

Liquid: $\mu = A\exp(B/T)$ (18.10)

Gas: $\mu = CT^2$ (18.11)

Where, *A*, *B* and *C* are constants which are determined from the experimental measurement. The viscosity increases with increased pressure. The equation of viscosity can be expressed in term of pressure as,

$$\mu = \exp(A + BP) \tag{18.12}$$

The most commonly used unit of the viscosity is poise, which can be expressed as $\text{dyne}\cdot\text{sec/cm}^2$, or $\text{g/cm}\cdot\text{sec}$. The centipoise or cP is equal to 0.01 poise. The SI unit of viscosity is $\text{Pa}\cdot\text{s}$, which is equal to 0.1 [poise]. The viscosity of the pure water at 20 °C is 1.0 cP.

The viscosity of crude oil with dissolved gas is an important parameter in the pressure loss of hydraulic calculation for hydrocarbons flow in pipeline. High viscous pressure drops in the pipeline segment may impact significantly on the production deliverability. The oil viscosity should be determined in the laboratory for the required pressure and temperature ranges. There are many empirical correlations to calculate the oil viscosity based on the system parameters such as temperature, pressure, oil and gas gravities in case of no measured viscosity available.

To minimize pressure drop along the pipeline for viscous crude oils it is beneficial to insulate the pipeline to retain a high temperature. The flow resistance is less because the oil viscosity is lower at higher temperatures. The selection of insulation is dependent upon the costs and the operability issues. For most developed fields insulation may not be important, however, for a heavy oil, high viscous pressure drops in the connecting pipeline between the subsea location and the receiving platform, insulation may have a key role to play. Viscoelastic drag reducing additives of crude oil should be considered if it is possible. These have found application in, for example, the Alyeska 48 inch crude line, the largest oil line in the United States.

18.3 Emulsion

18.3.1 General

A separate water phase in pipeline can result in hydrate formation or oil/water emulsion under certain circumstances in a suitable condition. Emulsions have high viscosities, which may be an order of magnitude higher than the single-phase oil or water. The effect of emulsions on frictional loss may lead to a significantly larger error in calculated pressures loss than expected. In addition to affecting pipeline hydraulics, emulsions may present severe problems to downstream processing plants. Specifically the separation of oil from the produced water can be hindered.

An emulsion is a heterogeneous liquid system consisting of two immiscible liquids with one of the liquids intimately dispersed in the form of droplets in the second liquid. In most emulsions of crude oil and water, the water is finely dispersed in the oil. The spherical form of the water globules is a result of interfacial tension, which compels them to present the smallest possible surface area to the oil. If the oil and water are violently agitated, small drops of water will be dispersed in the continuous oil phase and small drops of oil will be dispersed in the continuous water phase. If left undisturbed, the oil and water will quickly separated into layers of oil and water. If any emulsion is formed, it will be between the oil above and the water below. In offshore engineering most emulsions are the water-in-oil type.

The formation of an emulsion needs following three conditions:

- two immiscible liquids;
- sufficient agitation to disperse one liquid as droplets in the other;
- an emulsifying agent.

The agitation necessary to form an emulsion may result from any one or a combination of several sources:

- bottom-hole pump;
- flow through the tubing, wellhead, manifold, or flowlines;
- surface transfer pump;
- pressure drop through chokes, valves, or other surface equipment.

The greater the amount of agitation, the smaller the droplets of water dispersed in the oil. Water droplets in water-in-oil emulsions are of widely varying sizes, ranging from less than 1 to about 1,000 μm. Emulsions that have smaller droplets of water are usually more stable and difficult to treat than those that have larger droplets. The most emulsified water in light crude oil, i.e., oil above 20°API, is from 5 to 20 vol. % water, while emulsified water in crude oil heavier than 20° API is from 10 to 35%. Generally, crude oils with low API gravity (high density) will form a more stable and higher-percentage volume of emulsion than will oils of high API gravity (low density). Asphaltic-based oils have a tendency to emulsify more readily than paraffin-based oils.

Emulsifying agent determines whether an emulsion will be formed and the stability of that emulsion. If the crude oil and water contain no emulsifying agent, the oil and water may form a dispersion that will separate quickly because of rapid coalescence of the dispersed droplets. Emulsifying agents are surface-active compounds that attach to the water-drop surface. Some emulsifiers are thought to be asphaltic, sand, silt, shale particles, crystallized paraffin, iron, zinc, aluminum sulfate, calcium carbonate, iron sulfide, and similar materials. These substances usually originate in the oil formation but can be formed as the result of an ineffective corrosion-inhibition program.

18.3.2 Effect of Emulsion on Viscosity

Emulsions are always more viscous than the clean oil contained in the emulsion. Figure 18.1 shows the tendency of viscosity to water fraction which was developed by Woelflin, which is widely used in petroleum engineering. However, at higher water cuts (greater than about 40%), it tends to be excessively pessimistic, and may lead to higher pressure loss expectations than are actually likely to occur. Guth and Simha presented a correlation of emulsion viscosity as,

$$\mu_e / \mu_o = 1 + 2.5C_w \tag{18.13}$$

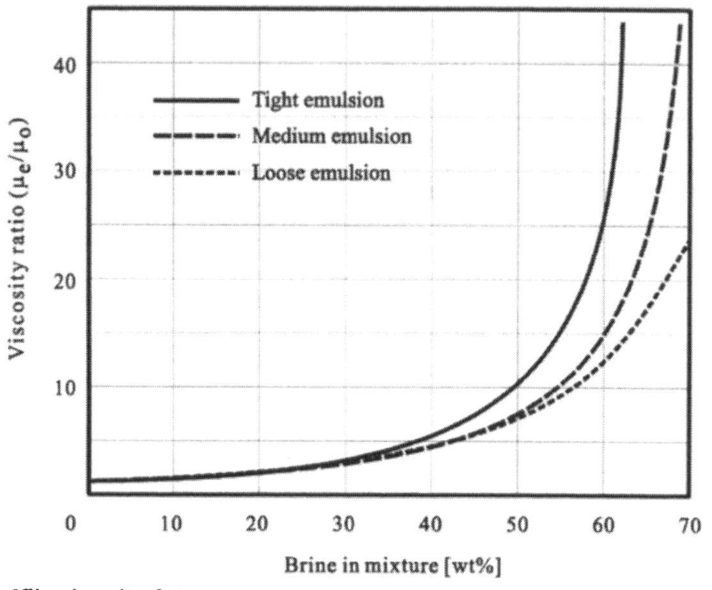

Figure 18.1 Woelflin viscosity data.

where, μ_e : viscosity of emulation [cP or mPa.s],

μ_o : viscosity of clean oil [cP or mPa.s],

C_w : volume fraction of water phase, ratio of volumetric flow rate of oil and liquid.

This correlation is similar to that of Woelflin in that it predicts the emulsion viscosity to be some multiple of the oil viscosity. The factor is determined solely as a function of the water cut. Up to a water cut of about 40%, the two methods will give almost identical results. Above that, the effective viscosity rises much more slowly with the Guth and Simha correlation, and computed pressure losses will thus be lower at higher water cuts.

Laboratory tests can give useful information on emulsion, but the form of the emulsion obtained in practice will be dependent upon the shear history, temperature and composition of the fluids. Limited comparison of calculated emulsion viscosity with experimental data suggests that the Woelflin correlation over estimates the viscosity at higher water fractions. The following equation recommended by Smith and Arnold can be used if no other data are available,

$$\mu_e / \mu_o = 1 + 2.5C_w + 14.1C_w^2 \tag{18.14}$$

18.3.3 Prevention of Emulsion

If all water can be excluded from the oil and/or if all agitation of hydraulic fluid can be prevented, no emulsion will form. Exclusion of water in some wells is difficult or impossible, and the prevention of agitation is almost impossible. Therefore, production of emulsion from many wells must be expected. In some instances, however, emulsification is increased by poor operating practices.

Operating practices that include the production of excess water as a result of poor cementing or reservoir management can increase emulsion-treating problems. In addition, a process design that subjects the oil/water mixture to excess turbulence can result in greater treating problems. Unnecessary turbulence can be caused by over-pumping and poor maintenance of plunger and valves in rod-pumped wells, use of more gas-lift gas than is needed, and pumping the fluid where gravity flow could be used. Some operators use progressive cavity pumps as opposed to reciprocating, gear, or centrifugal pumps to minimize turbulence. Others have found that some centrifugal pumps can actually cause coalescence if they are installed in the process without a downstream throttling valve. Wherever possible, pressure drop through chokes and control valves should be minimized before oil/water separation.

18.4 Phase Behavior

A multi-component mixture exhibits an envelope for liquid/vapor phase change in the pressure /temperature diagram, which contains a bubble-point line and a dew-point line, compared with only a phase change line for a pure component. Figure 18.1 shows a classification of the reservoir types of oil and gas systems based on the phase behavior of hydrocarbon in the reservoir, in which following five types of reservoirs are distinguished:

- Black oils;
- Volatile oils;
- Condensate (Retrograde gases);
- Wet gases;
- Dry gases.

The amount of heavier molecules in the hydrocarbon mixtures varies from large to small from the black oils to the dry gases. The various pressures and temperatures of hydrocarbons along a production flowline from the reservoir to separator are presented in a line A-A2 of Figure 18.1. Mass transfer occurs continuously between the gas and the liquid phases within the two-phase envelope.

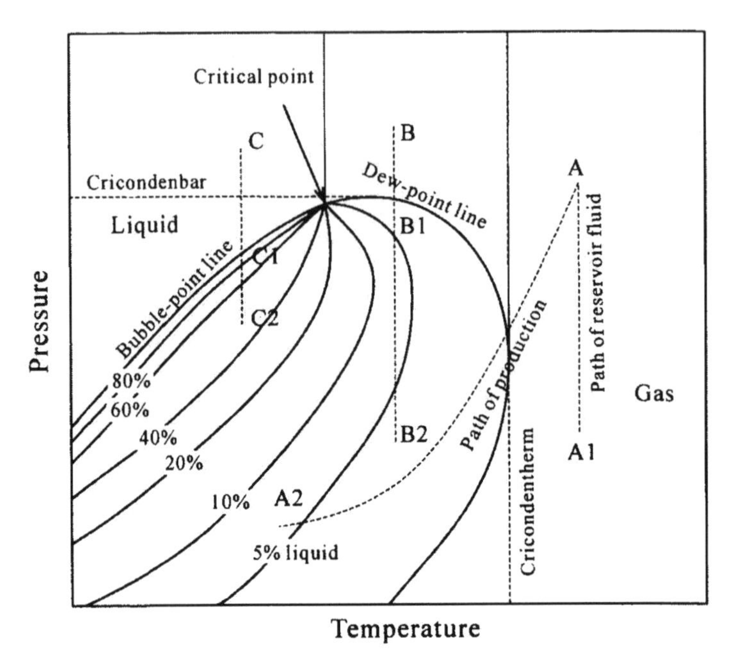

Figure 18.2 Typical phase diagram of hydrocarbons.

18.4.1 Black Oils

Black oil is liquid oil and consists of a wide variety of chemical species including large, heavy and non-volatile molecules. Typical black oil reservoirs have temperatures below the critical temperature of the hydrocarbon mixture. *C* point in Figure 18.2 presents unsaturated black oil; some gases are dissolved in the liquid hydrocarbon mixtures. *C1* point in the same figure shows saturated back oil, which means that oil contains as much dissolved gas as it can take and that a reduction of system pressure will release gas to form gas phase. In transport pipelines, black oils are transported in liquid phase throughout the transport process, while in production flowlines, produced hydrocarbon mixtures are usually in thermo dynamical equilibrium with gas.

18.4.2 Volatile Oils

Volatile oils contain fewer heavy molecules but more ethane through hexane than black oils. Reservoir conditions are very close to the critical temperature and pressure. A small reduction in pressure can cause the release of a large amount of gas.

18.4.3 Condensate

Retrograde gas is the name of a fluid that is gas at reservoir pressure and temperature. However, as pressure and temperature decrease, large quantities of liquids are formed due to retrograde condensation. Retrograde gases are also called retrograde gas condensates, gas condensates or condensates. The temperature of condensate is normally between the critical temperature and the cricondentherm as shown in Figure 18.2. From reservoir to riser top, as shown in *B* to *B2* in the figure, the fluid changes from single gas and partly transfer to condensate liquid and evaporate again as gas, causing a significant pressure drop.

For a pure substance a decrease in pressure causes a change of phase from liquid to gas at the vapor-pressure line, likewise, in the case of a multi-component system a decrease in pressure causes a change of phase from liquid to gas at temperatures below the critical temperature. However, consider the isothermal decrease in pressure illustrated by a line *B-B1-B2* in the figure. As pressure is decreased from point B, the dew point line is crossed and liquid begins to form. At the position indicated by point *B1*, the system is 5% liquid. A decrease in pressure has caused a change from gas to liquid. Table 18.2 lists the composition of a typical condensate gas. Line A-A2 illustrates the variations of pressure and temperature along the flowline from reservoir to the separator on platform.

18.4.4 Wet Gases

A wet gas exists as a pure gas phase in the reservoir, but becomes liquid/gas two-phase in flowline from well tube to the separator at topside of platform. During pressure drop in the flowline, liquid condensate appears in the wet gas.

18.4.5 Dry Gases

Dry gas is primarily methane. The hydrocarbon mixture is solely gas under all conditions in pressure and temperature encountered during the production phases from reservoir conditions to transport and process conditions. In particular no hydrocarbon based liquids are formed from the gas although liquid water can condense. Dry gas reservoirs have temperatures above the cricondentherm.

Table 18.3 Hydrocarbons composition of typical reservoirs (Whitson and Brule, 2000).

Component	Composition (mole %)				
	Black oil	Volatile oil	Condensate	Wet gas	Dry gas
CO_2	0.02	0.93	2.37	1.41	0.10
N_2	0.34	0.21	0.31	0.25	2.07
C_1	34.62	58.77	73.19	92.46	86.12
C_2	4.11	7.57	7.80	3.18	5.91
C_3	1.01	4.09	3.55	1.01	3.58
$i\text{-}C_4$	0.76	0.91	0.71	0.28	1.72
$n\text{-}C_4$	0.49	2.09	1.45	0.24	0.0
$i\text{-}C_5$	0.43	0.77	0.64	0.13	0.50
$n\text{-}C_5$	0.21	1.15	0.68	0.08	0.0
C_6	1.61	1.75	1.09	0.14	0.0
C_{7+}	56.40	21.76	8.21	0.82	0.0
Total	100.0	100.0	100.0	100.0	100.0

Table 18.4 Hydrocarbons properties of typical reservoirs (Whitson and Brule, 2000).

Properties	Black oil	Volatile oil	Condensate	Wet gas	Dry gas
M_{7+}	274	228	184	130	
γ_{7+}	0.92	0.858	0.816	0.763	
GOR, [scf/STB]	300	1490	5450	105,000	∞
OGR, [STB/mmscf]			180	10	0
γ_{API}	24	38	49	57	
γ_g	0.63	0.70	0.70	0.61	
P_{sat} [psi]	2,810	5,420	5,650	3,430	
ρ_{sat} [lb/ft^3]	51.4	38.2	26.7	9.61	

18.4.6 Computer Models

Accurate prediction of physical and thermodynamic properties is a pre-requisite to successful pipeline design. Pressure loss, liquid hold up, heat loss, hydrate formation, and wax deposition all require knowledge of the fluid states.

In flow assurance analyses, the following two approaches have been used to simulate hydrocarbon fluids:

- "black-oil" model: defines the oil as a liquid phase that contains dissolved gas, such as hydrocarbons produced from the oil reservoir. The "black-oil" accounts for the gas that dissolves (condenses) from oil solution with a parameter of R_s that can be measured from the laboratory. This model predicts fluid properties from the specific gravity of the gas, the oil gravity, and the volume of gas produced per volume of liquid. Empirical correlations evaluate the phase split and physical property correlations determine the properties of the separate phases.
- Composition model: for a given mole fraction of a fluid mixture of volatile oils and condensate fluids, a vapor/liquid equilibrium calculation determines the amount of the feed that exists in the vapor and liquid phases and the composition of each phase. It is possible to determine the quality or mass fraction of gas in the mixtures. Once the composition of each phase is known, it is also possible to calculate the interfacial tension, densities, enthalpies, and viscosities of each phase.

The accuracy of the compositional model is dependent upon the accuracy of the compositional data. If good compositional data are available, selection of an appropriate EOS is likely to yield more accurate phase behaviour data than the corresponding "black oil" model. This is particularly so if the hydrocarbon liquid is a light condensate. In this situation complex phase effects such as retrograde condensation are unlikely to be adequately handled by the "black oil" methods. Of prime importance to hydraulic studies is the viscosity of the fluid phases. Both "black oil" and compositional techniques can be inaccurate. Depending on the correlation used, very different calculated pressure losses could result. With the uncertainty associated with viscosity prediction it is prudent to utilise laboratory measured values.

GOR may be defined as the ratio of the measured volumetric flow rates of the gas and oil phases at meter conditions (ambient conditions) or the volume ratio of gas and oil at the standard condition (14.7 psi, 60°F) with unit of scf/STB. When water is also present, the watercut is generally defined as the volume ratio of the water and total liquid at standard conditions. If the water contains salts, the salt concentrations may be contained in the water phase at the standard condition.

18.5 Hydrocarbon Flow

18.5.1 General

The complex mixture of hydrocarbon compounds or components can exist as a single-phase liquid, a single-phase gas, or as a multi-phase mixture, depending on its pressure, temperature, and the composition of the mixture. The fluid flow in flowlines is divided into three categories based on the fluid phase condition,

- Single-phase; black oil or dry gas transport flowline, export flowline, gas or water injection flowline, and chemical inhibitors service flowlines such as methanol, glycol lines and etc.
- Two-phase; oil + released gas flowline, gas + produced oil (condensate) flowline.
- Three-phase; water + oil + gas (typical production flowline).

The flowlines after oil/gas separation equipment generally flow single phase hydrocarbon fluid, such as transport flowlines and export flowlines, while in most cases, the production flowlines from reservoirs have two or three-phase, simultaneously, and the fluid flow is then called multi-phase flow.

In a hydrocarbon flow, the water should be considered as a sole liquid phase or combination with oils or condensates, since these liquids basically are insoluble in each other. If the water amount is small enough that it has little effect on flow performance, it may be acceptable to assume a single liquid phase. At low velocity range, there is considerable slip between the oil and water phase. As a result, the water tends to accumulate in low spots in the system. This leads to high local accumulations of water, and thereby a potential for water slugs in the flowline. It may also cause serious corrosion problems.

Two phase (gas/liquid) models are used for black oil systems even when water is present. The water and hydrocarbon liquid are treated as a combined liquid with average properties. For gas condensate systems with water, three-phase (gas/liquid/aqueous) models are used.

The hydraulic theory underlying single-phase flow is well understood and analytical models may be used with confidence. Multiphase flow is significantly more complex than single-phase flow. However, the technology to predict multiphase-flow behavior has improved dramatically in the past decades. It is now possible to select pipeline size, predict pressure drop, and calculate flow rate in the flowline with an acceptable engineering accuracy.

18.5.2 Single-phase Flow

The basis for calculation of changes in pressure and temperature with pipeline distance is the conservation of mass, momentum, and energy of the fluid flow. In this section, the steady state, pressure–gradient equation for single-phase flow in pipelines is developed. The procedures to determine values of wall shear stress are reviewed and example problems are solved to demonstrate the applicability of the pressure-gradient equation for both compressible and incompressible fluid. A review is presented of Newtonian and non-Newtonian fluid flow behavior in circular pipes. The enthalpy-gradient equation also is developed and solved to obtain approximate equations that predict temperature changes during steady-state fluid flow.

18.5.2.1 Conservation equations

Mass conservation

Mass conservation of flow means that the mass in, $m_{in,}$ minus the mass out, m_{out} of a control volume must equal the mass accumulation in the control volume. For a control volume of a one-dimensional pipe segment, the mass conservation equation can be written as,

$$\frac{\partial \rho}{\partial t}+\frac{\partial(\rho v)}{\partial L}=0 \tag{18.15}$$

where, ρ the fluid density, v the fluid velocity, t the time, and L is the length of pipe segment. For a steady-flow, no mass accumulation occurs, and the equation (18.15) becomes,

$$\rho v = \text{constant} \tag{18.16}$$

Momentum conservation

Based on the Newton's second law applied for the fluid flow in a pipe segment, the rate of momentum change in the control volume is equal to the sum of all forces on the fluid. The linear momentum conservation equation for the pipe segment can be expressed as,

$$\frac{\partial(\rho v)}{\partial t}+\frac{\partial(\rho v^2)}{\partial L}=-\frac{\partial p}{\partial L}-\tau\frac{\pi d}{A}-\rho g\sin\theta \tag{18.17}$$

The terms on the right hand side of Eq. (18.17) are the forces on the control volume. $\partial p/\partial L$ the pressure gradient, $\tau\pi d/A$ the surface forces, and $\rho g \sin\theta$ the body forces. Combining Eqs (18.16) and (18.17), the pressure gradient equation is obtained for a steady-state flow,

$$\frac{\partial p}{\partial L}=-\tau\frac{\pi d}{A}-\rho g\sin\theta-\rho v\frac{\partial v}{\partial L} \tag{18.18}$$

By integrating both sides of the above equation from section 1 to section 2 of the pipe segment shown in Figure 18.3, and adding the mechanical energy by hydraulic machinery such as pump, an energy conservation equation, Bernoulli equation, for a steady one-dimension flow is obtained as,

$$\frac{p_1}{\gamma_1}+\frac{V_1^2}{2g}+z_1=\frac{p_2}{\gamma_2}+\frac{V_2^2}{2g}+z_2+\sum h_f+\sum h_l-h_m \tag{18.19}$$

where, p/γ, is pressure head; $V^2/2g$, velocity head, z, elevation head, $\sum h_f$, sum of fiction head loss between section 1 and 2 caused by frictional force, $\sum h_l$, sum of local head loss, and h_m is mechanical energy per unite weight added by hydraulic machinery. The pipe friction head loss is the head loss due to fluid shear at the pipe wall. The local head losses are caused by local disruptions of the fluid stream, such as valves, pipe bend, and other fittings.

The energy grade line (EGL) is a plot of the sum of the three terms in the left hands side of Eq. (18.19), and is expressed as,

$$EGL=\frac{p}{\gamma}+\frac{V^2}{2g}+z \tag{18.20}$$

The hydraulic grade line (HGL) is the sum of only the pressure and elevation head, and can be expressed as,

$$HGL=\frac{p}{\gamma}+z \tag{18.21}$$

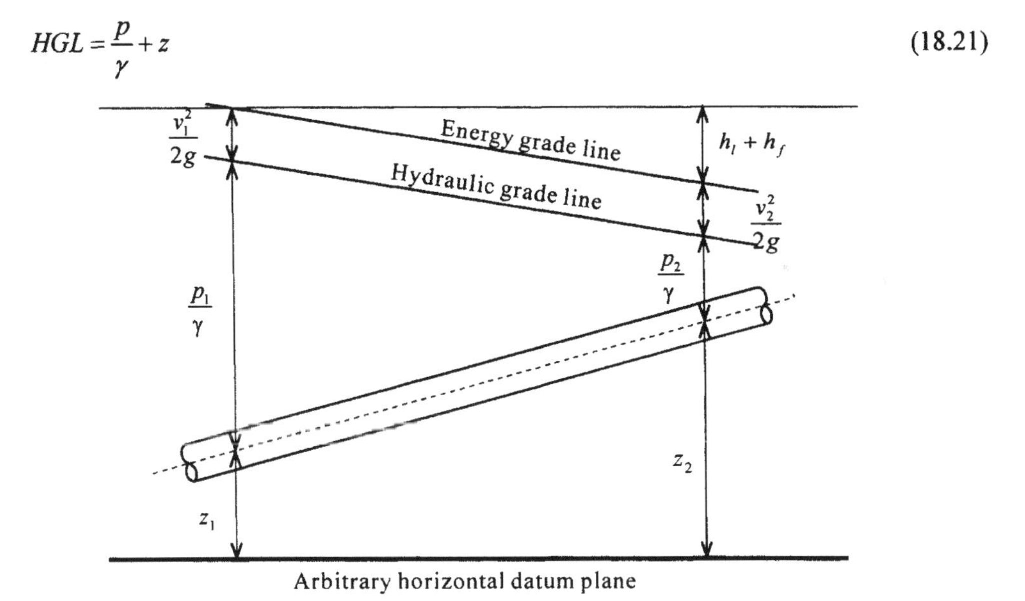

Figure 18.3 Relationship between energy/hydraulic grade lines and individual heads.

Figure 18.3 shows the relation of the individual head terms to the EGL and HGL lines. The head losses are the variation of EGL or HGL between section 1 and 2. In general, in a preliminary design, the local losses may be ignored, but must be considered in the detailed design stage.

As shown in Equation (18.19), the total pressure drop in a particular pipe segment is the sum of the friction head losses of pipe segment, valves and fitting, the local head losses due to valves and fittings, the static pressure drop due to elevation change and the pressure variation due to the acceleration change.

18.5.2.2 Friction factor equation

The Darcy-Weisbach equation is one of the most versatile friction head loss equations for a pipe segment, which is expressed as,

$$h_f = f\frac{L}{D}\frac{V^2}{2g} \tag{18.22}$$

$$f = f(\mathrm{Re}, \varepsilon / D) \tag{18.23}$$

where, f is the friction factor, $\mathrm{Re} = VD\rho / \mu$ Reynolds number, and ε/D is the equivalent sand-grain roughness (relative roughness) of the pipe, which is decided by the pipeline material. Figure 18.4 shows the graphical representation of friction factor at different flow regions, which is also called as Moody diagram.

Table 18.5 lists common values of absolute roughness for several materials. For steel pipe, the absolute roughness is 0.0018" unless alternative specifications are specified. The absolute roughness of 0.0018" is for aged pipe, but accepted in design practice because it is conservative. Flexible pipe is rougher than steel pipe, therefore, requires a larger diameter for the same maximum rate. For a flexible pipe, the roughness shall be given by ε = ID/250.0 unless alternative specification is given. The roughness may increase with use at a rate determined by the material and nature of the fluid. As a general guide, factors of safety of 20 to 30 % on the friction factor will accommodate the change in roughness conditions for steel pipe with average service of 5 to 10 years.

Table 18.5 Pipeline roughness (Larock et al.).

Material	ε, equivalent sand-grain roughness (in)
Concrete	0.012-0.12
Case iron	0.010
Commercial or welded steel	0.0018
PVC, glass	0.00006

As shown in Figure 18.4, the function f (Re, ε/D) is very complex. A laminar region and a turbulent region are defined in the Moody diagram based on the flow characteristics.

Laminar flow

In the laminar flow (Re < 2100), the friction factor function is a straight line, and is not influenced by the relative roughness,

$$f = 64 / \mathrm{Re} \tag{18.24}$$

The friction head loss is shown to be proportional to the average velocity in the laminar flow. Increasing pipe relative roughness will cause an earlier transition to turbulent flow. When Reynolds number is in the range from 2000 to 4000, the flow is in a critical region where the flow can be either laminar or turbulent depending upon several factors. These factors include changes in section or flow direction. The friction factor in the critical region is indeterminate, which is in a range between friction factors for laminar flow and for turbulent flow.

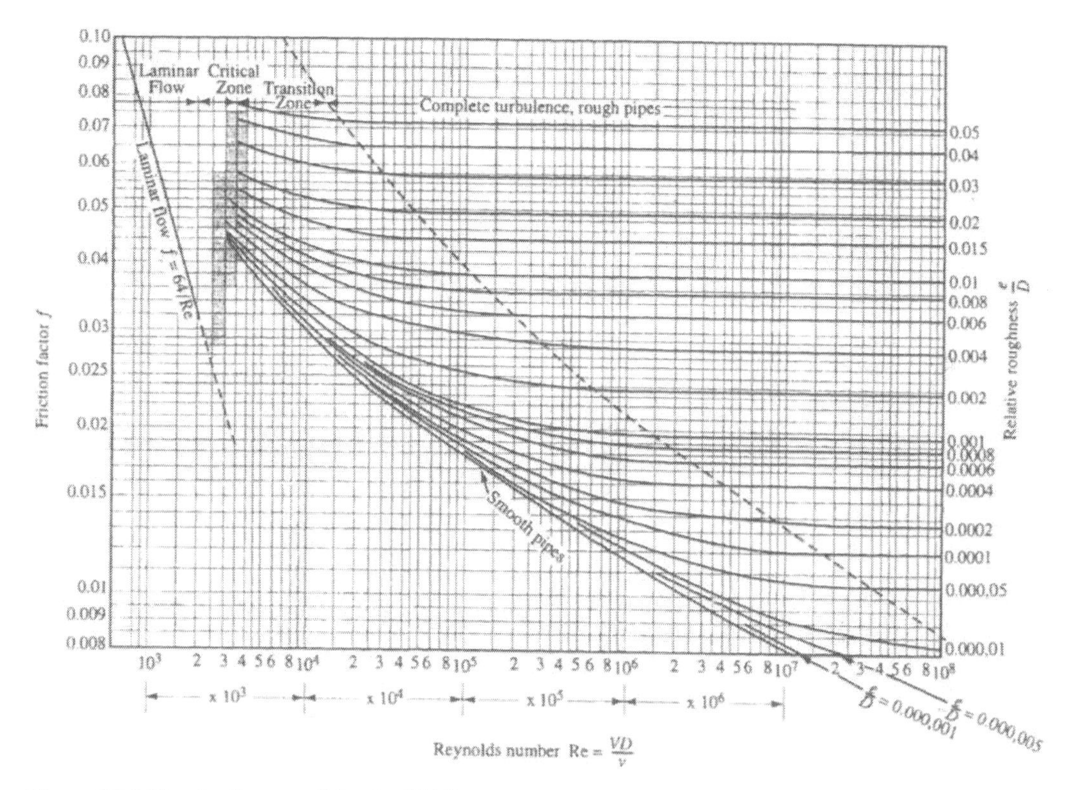

Figure 18.4 Moody diagram (Moody, 1944).

Turbulent flow

When Reynolds number is larger than 4000, the flow inside the pipe is turbulent flow; the fiction factor depends not only upon Reynolds number but also upon the relative roughness, ε/D and others. In the complete turbulence region, the region above a dashed line in the upper right part of Moody diagram, the friction factor f is a function only of roughness ε/D. For a pipe in transition zone, the friction factor decreases rapidly with increasing Reynolds number and decreasing pipe relative roughness. The lowest line in the Figure 18.4 is the line of friction factor for the smoothest pipe, where the roughness ε/D is so small that it has no effect.

Although single-phase flow in pipe has been studied extensively, it still involves an empirically determined friction factor for turbulent flow calculations. The dependence of this friction factor on pipe roughness, which must usually be estimated, makes the calculated pressure gradients subject to considerable error. Table 18.6 summarizes the correlations of

Darcy-Weisbach friction factor for different flow ranges. The correlations for smooth pipe and complete turbulence regions are simplified from the one for transitional region. For smooth pipe, the relative roughness term is ignored, while for the complete turbulence region, Reynolds term is ignored.

18.5.2.3 Local losses

In addition to pressure head losses due to pipe surface friction, the local losses are the pressure head loss occurring at flow appurtenances, such as valves, bends, and other fittings, when the fluid flow through the appurtenances. The local head losses in fittings may include:

- surface fiction;
- direction change of flow path;
- obstructions in flow path;
- sudden or gradual changes in the cross-section and shape of flow path.

Table 18.6 Darcy –Weisbach friction factor correlations (Larock et al.).

Flow Region	Friction factor, f	Reynolds Range
Laminar	$f = 64/\mathrm{Re}$	$\mathrm{Re} < 2100$
Smooth pipe	$1/\sqrt{f} = 2\log_{10}(\mathrm{Re}\sqrt{f}) - 0.8$	$\mathrm{Re} > 4000$ and $\varepsilon/D \rightarrow 0$
Transitional, Colebrook-White correlation	$1/\sqrt{f} = 1.14 - 2\log_{10}(\varepsilon/D + 9.35/(\mathrm{Re}\sqrt{f}))$	$\mathrm{Re} > 4000$ and transitional region
Complete turbulence	$1/\sqrt{f} = 1.14 - 2\log_{10}(\varepsilon/D)$	$\mathrm{Re} > 4000$ and complete turbulence

The local losses are also termed as minor losses. These descriptions are misleading for the process piping system where fitting losses are often much greater than the losses in straight piping sections. It is difficult to quantify theoretically the magnitudes of the local losses, so the representation of these losses mainly is determined using experimental data. Local losses are usually expressed in a form similar to that for the friction loss. Instead of $f\,L/D$ in friction head loss, the loss coefficient K is defined for various fittings. The head loss due to fitting is given,

$$h = K\frac{V^2}{2g} \tag{18.25}$$

where, V is the downstream mean velocity normally. Two methods are used to the determination of K value for different fittings, such as valve, elbow etc. One method is to choose a K value directly from a table that is invariant with respect to size and Reynolds number. Table 18.7 gives K values for several types of fittings. In this method the data scatter can be large and some inaccuracy is to be expected. The other approach is to specify the K for a given fitting in terms of the value of the complete turbulence friction factor f_T for the nominal pipe size. This method implicitly accounts for the pipe size. The Crane Company

Technical Paper 410 detailed the calculation methods of K value for different fittings, which is commonly accepted by the piping industry.

Manufacturers of valves, especially control valves, express valve capacity in terms of a flow coefficient C_v, which gives the flow rate through the valve in gallon/minute of water at 60°F under a pressure drop of 1.0 psi. It is related to K by,

$$C_v = \frac{29.9d^2}{\sqrt{K}} \tag{18.26}$$

where d is the diameter of the valve connections in inches.

Table 18.7 Loss coefficients for fittings (Larock et al.).

Fitting	K
Globe valve, fully open	10.0
Angle valve, fully open	5.0
Butterfly valve, fully open	0.4
Gate valve, full open	0.2
3/4 open	1.0
1/2 open	5.6
1/4 open	17.0
Check valve, swing type, fully open	2.3
Check valve, lift type, fully open	12.0
Check valve, ball type, fully open	70.0
Elbow, 45°	0.4
Long radius elbow, 90°	0.6

18.5.3 Multi-phase Flow

18.5.3.1 General

Multi-phase transportation is currently receiving much attention throughout the oil and gas industry. The combined transport of hydrocarbon liquids and gases, immiscible water and sand can offer significant economic savings over the conventional, local, platform based separation facilities. The possibility of hydrate formation, the increasing water content of the produced fluids, erosion, heat loss and other considerations create challenges to the hydraulic design procedure.

Much research has been carried out on two-phase flow beginning in the 1950's. The flow behavior of two-phase flow is much more complex than that of single-phase flow. Two-phase flow is a process involving the interaction of many variables. The gas and liquid phases normally do not travel at the same velocity in the pipeline because of the differences in density and viscosities. For an upward flow, the gas phase which is less dense and less viscous tends to flow at a higher velocity than the liquid phase. For a downward flow, the liquid often flows faster than the gas because of density differences. Although the analytical solutions of single-phase-flow are available and the accuracy of prediction is acceptable in industry, multiphase flows, even when restricted to a simple pipeline geometry, are in general quite complex.

Calculations of pressure gradients in two-phase flows require values of flow conditions such as velocity, fluid properties such as density, viscosity and surface tension. One of the most important factors to decide the flow characteristic of two-phase flow is **flow pattern**. The flow pattern description is not merely an identification of laminar or turbulent flow for a single flow, the relative quantities of the phases and the topology of the interfaces must be described. The different flow patterns are formed because of relative magnitudes of the forces that act on the fluids, such as buoyancy, turbulence, inertia, and surface tension forces, which vary with flow rates, pipe diameter, inclination angle, and fluid properties of the phases. Another important factor is **liquid holdup**, which is defined as the ratio of the volume of a pipe segment occupied by liquid to the volume of the pipe segment. Liquid holdup is a fraction, which varies from zero for pure gas flow to one for pure liquid flow.

For a two-phase flow, most analyses and simulations solve mass, momentum, and energy balance equations based on one-dimensional behavior for each phase. Such equations, for the most part, are used as a framework in which to interpret experimental data. Reliable prediction of multiphase flow behavior generally requires use of data or experimental correlations. Two-fluid modeling, in which the full three-dimensional partial differential equations of motion are written for each phase, treating each as a continuum, occupying a volume fraction which is a continuous function of position, is a developing technique made possible by improved computational methods.

18.5.3.2 Horizontal flow

In horizontal pipe, flow patterns for fully developed flow have been reported in numerous studies. Transitions between flow patterns are recorded with visual technologies. In some cases, statistical analysis of pressure fluctuations has been used to distinguish flow patterns. Figure 18.5 shows a typical flow pattern map, which gives an approximate prediction of the flow patterns. Commonly used flow pattern maps such as Mandhane and Baker are based upon observed regimes in horizontal, small diameter, air-water systems. Scaling to the conditions encountered in the oil and gas industries is uncertain.

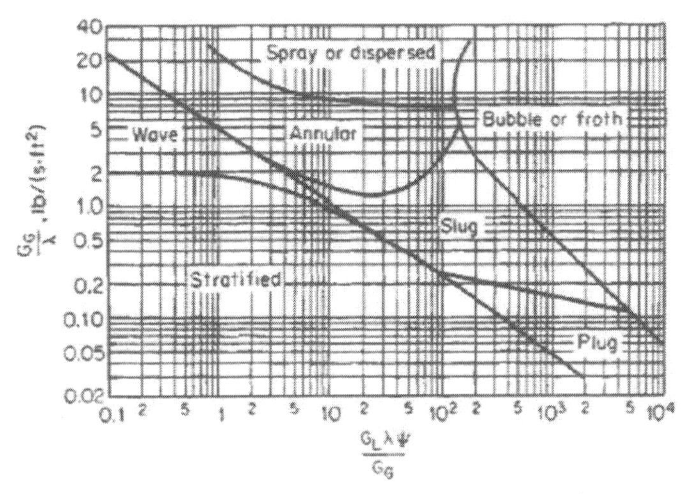

Figure 18.5 Flow pattern regions for two-phase flow through horizontal pipe.

Figure 18.6 shows seven flow patterns for the horizontal gas/liquid two-phase flow.

- **Bubble flow;** gas is dispersed as bubbles that move at a velocity similar to that of liquid and tend to concentrate near the top of the pipe at lower liquid velocities.
- **Plug flow;** alternate plugs of gas and liquid move along the upper part of the pipe.
- **Stratified flow;** liquid flows along the bottom of the pipe and the gas flows over a smooth liquid/gas interface.
- **Wavy flow;** occurs at greater gas velocities and has waves moving in the flow direction. When wave crests are sufficiently high to bridge the pipe, they form frothy slugs that move at much greater than the average liquid velocity.
- **Slug flow;** may cause severe and/or dangerous vibrations in equipment because of impact of the high-velocity slugs against fittings.
- **Annular flow;** liquid flows as a thin film along the pipe wall and gas flows in the core. Some liquid is entrained as droplets in the gas core.
- **Spray;** at very high gas velocities, nearly all the liquid is entrained as small droplets. This pattern is also called **dispersed**, or **mist flow**.

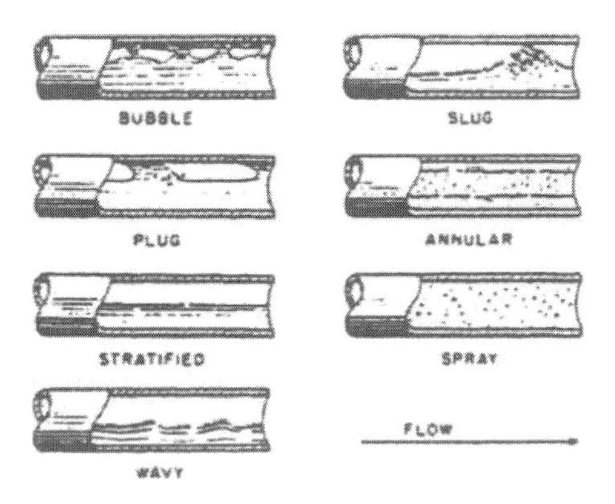

Figure 18.6 Two-phase flow patterns in horizontal pipeline (Beggs and Brill, 1973).

18.5.3.3 Vertical flow

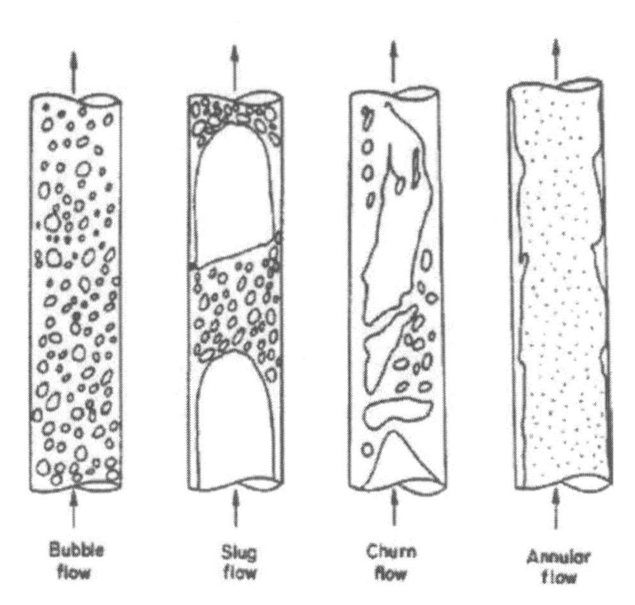

Figure 18.7 Two-phase flow patterns in vertical pipe (Taitel et al., 1980).

Two-phase flow in vertical pipe may be categorized into four different flow patterns, consisting of bubble flow, slug flow, churn flow and annular flow, as shown in Figure 18.7.

- **Bubble flow;** the liquid is continuous with the gas phase existing as bubbles randomly distributed. The gas phase in bubble flow is small and contributes little to the pressure gradient except by its effect on the density.
- **Slug flow;** both the gas and liquid phases significantly contribute to the pressure gradient. The gas phase in slug flow exists as large bubbles and separated by slugs of liquid. The velocity of the gas bubbles is greater than that of the liquid slugs, thereby resulting in a liquid holdup that not only affects well and riser friction losses but also flowing density.
- **Churn flow**, the liquid slugs between the gas bubbles essentially disappear, and at some point the liquid phase becomes discontinuous and the gas phase becomes continuous. The pressure losses are more the result of the gas phase than liquid phase.
- **Annular flow** is characterized by a continuous gas phase with liquid occurring as entrained droplets in the gas stream and as a liquid film wetting the pipe wall.

18.5.4 Comparison of Two-phase Flow Correlations

There are many two-phase flow correlations developed in last several decades, most of them were developed based on water/air systems and every correlation has its limit because of data used. Most software used in flow assurance has several options for models or correlations for two-phase flow. Baker Jardine conducted a major data matching study to identify the most suitable two phase flow correlations for oil wells, condensate wells, oil pipelines and gas/condensate pipelines, and the analysis results are summarized in following sections.

18.5.4.1 Description of correlations

Duns & Ros (D-R)

The Duns & Ros correlation was developed for vertical two-phase flow in wells based on extensive experimental research of oil and air mixtures. Separate correlations were developed for bubble, plug, froth flow, slug flow and mist flow regimes. These regions have low, intermediate and high gas throughputs respectively.

Orkiszewski (OR)

The Orkiszewski correlation was developed for the prediction of two phase pressure drops in vertical pipe. The flow regimes of bubble, slug, annular-slug transition, and annular mist were considered. The method can accurately predict within 10% to the two-phase pressure drops in flowing and gas lifted production wells over a wide range of well conditions based on 148 measured pressure drops.

Hagedorn & Brown (H-B)

The Hagedorn and Brown correlations was developed based on an experimental study of pressure gradients in small diameter vertical conduits. A 1500 ft experimental well was used to study flow through 1", 1.25" and 1.5" nominal size tubing for widely varying liquid flow-rates, gas-liquid ratios and liquid viscosities.

Beggs & Brill Original (B-BO)

The Beggs & Brill correlation was developed following a study of two-phase flow in horizontal and inclined pipe. The correlation is based upon a flow regime map. The model inclines flow both upwards and downwards at angles of up to ±90°.

Beggs & Brill Revised (B-BR)

The following enhancements to the original method are used; (1) an extra flow regime of froth flow is considered which assumes a no-slip holdup, (2) the friction factor is changed from the standard smooth pipe model, to utilise a single phase friction factor based on the average fluid velocity.

Mukherjee & Brill (M-B)

The Mukherjee & Brill correlation was developed following a study of pressure drop in two-phase inclined flow. Results agreed well with the experimental data and correlations were further verified with Prudhoe Bay and North Sea data.

Govier, Aziz & Fogarasi (G-A)

The Govier, Aziz & Fogarasi correlation was developed following a study of pressure drop in wells producing gas and condensate. Actual field pressure drop vs. flowrate data from 102 wells with gas-liquid ratios ranging from 3,900 to 1,170,000 scf/bbl were analysed in detail.

No-Slip (NS)

The No-Slip correlation assumes homogeneous flow with no slippage between the phases. Fluid properties are taken as the average of the gas and liquid phases and friction factors are calculated using the single phase Moody correlation.

OLGAS-89,92 (O-89, O-92)

OLGAS is based on data from the SINTEF two-phase flow laboratory. The test facilities were designed to operate at conditions similar with field conditions. The test loop was 800m long and of 8" diameter. Operating pressures were between 20 and 90 barg. Gas superficial velocities of up to 13 m/s, and liquid superficial velocities of up to 4 m/s were obtained. Pipeline inclination angles between ±1° were studied in addition to flow up or down a hill section ahead of a 50m high vertical riser. Over 10,000 experiments were run on the test loop. OLGAS considers four flow regimes, stratified, annular, slug and dispersed bubble flow.

Ansari (AN)

The Ansari model was developed as part of the Tulsa University Fluid Flow Projects (TUFFP) research program. A comprehensive model was formulated to predict flow patterns and the flow characteristics of the predicted flow patterns for upward two-phase flow. The comprehensive mechanistic model is composed of a model for flow pattern prediction and a set of independent models for predicting holdup and pressure drop in bubble, slug, and annular flows. The model was evaluated by using the TUFFP well databank that is composed of 1775 well cases, with 371 of them from Prudhoe Bay data.

BJA for Condensates (BJ)

Baker Jardine & Associates have developed a correlation for two phase flow in gas-condensate pipelines with a no-slip liquid volume fraction of lower than 0.1. The pressure loss calculation procedure is similar in approach to that proposed by Oliemans, but accounts for the increased interfacial shear resulting from the liquid surface roughness.

AGA & Flanigan (AGA)

The AGA & Flanigan correlation was developed for horizontal and inclined two phase flow of gas-condensate systems. The Taitel Dukler flow regime map is used which considers five flow regimes, stratified smooth, stratified wavy, intermittent, annular dispersed liquid, and dispersed bubble.

Oliemans (OL)

The Oliemans correlation was developed following the study of large diameter condensate pipelines. The model was based on a limited amount of data from a 30", 100 km pipeline operating at pressures of 100 barg or higher.

Gray (GR)

This correlation was developed by H. E. Gray of Shell Oil Company for vertical flow in gas and condensate systems which are predominantly gas phase. Flow is treated as single phase, and water or condensate is assumed to adhere to the pipe wall. It is considered applicable for vertical flow cases where the velocity is below 50 ft/s, the tube size is below 3 ½", the condensate ratio is below 50 bbl/mmscf, and the water ratio is below 5 bbl/mmscf.

Xiao (XI)

The Xiao comprehensive mechanistic model was developed as part of the TUFPP research program. It was developed for gas-liquid two-phase flow in horizontal and near horizontal pipelines. The data bank included large diameter field data culled from the AGA multiphase pipeline data bank, and laboratory data published in literature. Data included both black oil and compositional fluid systems. A new correlation was proposed which predicts the internal friction factor under stratified flow.

18.5.4.2 Analysis Method

The predictions of pressure drop and flow rate of the each correlations built in PIPESIM were compared with available experimental data from well/pipeline tests, reservoir fluid analysis and drilling surveys/pipeline measurements collected by Baker Jardine. The % error for data point is calculated by,

$$\% \text{ error} = (\text{calculated value} - \text{measured values})/\text{measured value} \times 100\%$$

Table 18.8 lists the evaluation method with scores for each correlation. The overall scores for pressure drop and flowrate predictions are averaged for each correlation. The score is calculated with the sum of score for each database divided by 3 × total database number. The evaluation results for each correlation with data base of vertical oil wells, high deviated oil wells, gas/condensate wells, oil pipelines and gas/condensate pipelines are demonstrated in Figure 18.8.

Table 18.8 Evaluation of correlations with score.

Error Range, %	Evaluation	Score
- 5.0 to + 5.0	Very good	3.0
-10.0 to + 10.0	Good	2.0
-20.0 to + 20.0	Moderate	0
Outside above range	Poor	-3.0

The Moody correlation is recommended for single phase flow in vertical oil flow, horizontal oil flow, vertical gas flow and horizontal gas flow.

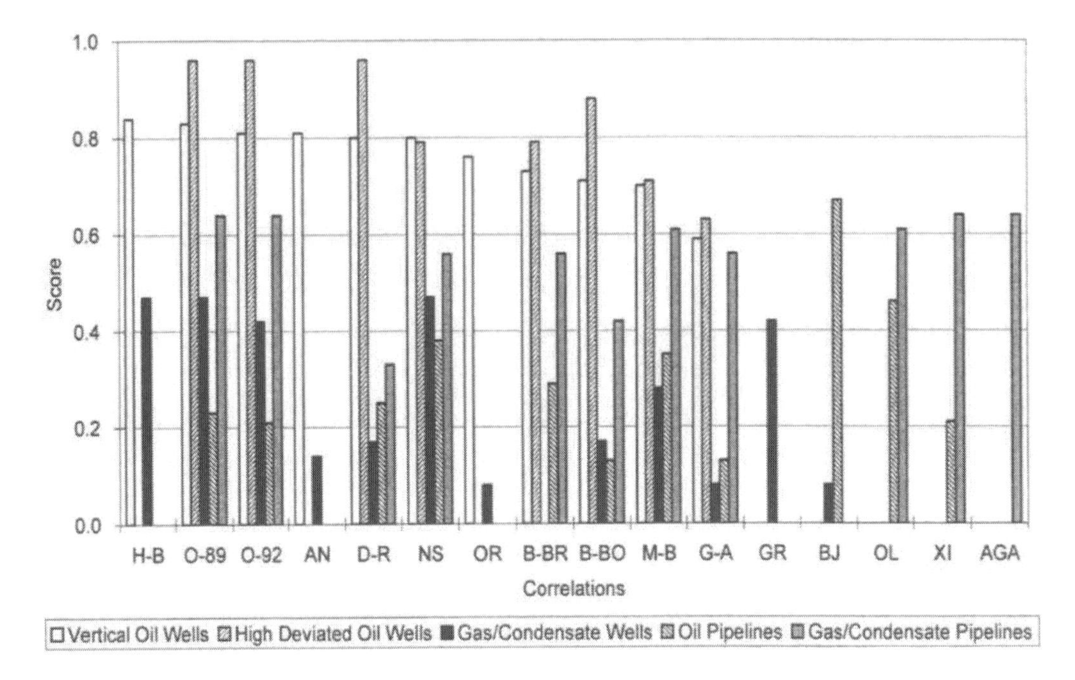

Figure 18.8 Evaluation of two-phase flow correlations used in PIPESIM.

18.6 Slugging and Liquid Handling

18.6.1 General

The occurrence of slug flow in a transportation pipeline can cause many problems in design and operation processes, which include kinetic force on fittings and vessels, pressure cycling, control instability and inadequate phase separation. Slugging affects greatly the design of receiving facilities. In gas condensate systems, larger lines result in more liquid being retained in the pipeline at low rates. When the flow rate is increased, much of the liquid can be swept out, potentially overwhelming the liquid handling capability of the receiving facilities. The facilities can be flooded and damaged if the slugs larger than the slug catcher capacity. Therefore, quantifying the slug size, frequency and velocity is necessary prior to equipment design.

The pressure at the bottom of the riser can vary if the holdup in the riser is not about the same as in the line feeding it. If the riser holdup is too large and the gas velocity is too small to provide continuous liquid lift, too much of the liquid reverses and flows downward. Liquid accumulates at the base, causing an unstable pressure situation. This is relieved by large liquid slugs periodically leaving the riser at a high velocity. The changes in liquid amount and the corresponding pressure changes can be dramatic. A large "Slug catcher" installation can be provided onshore, but it is uneconomic to set on the platform. This is one of the practical reasons why a pipeline section immediately ahead of the riser should be horizontal or have a slightly upward slope of 2 to 5 degrees. The section length probably should be several times the riser height. The upward incline eliminates a possible "sump" effect and serves to decrease pressure/holdup up instabilities. Severe slugging in the riser can be enhanced by a negative pipeline inclination just prior to it. Actually, severe slugging is unlikely if there is a positive inclination. Figure 18.9 shows the effect of mass flow rate of two-phase flow to the flow stability. Higher flow rate helps to decrease slug and increase flow stability in flowline. Higher system pressures also increase the tendency for a stable flow, holding a back pressure (choking) at the top of the riser can be used to minimize severe slugging. The flow in a riser may differ from that in a wellbore which has a relatively long horizontal flowline at the end of it. Holdup and surging from that horizontal flowline is transmitted to the relatively short riser. The riser may have to handle far more liquid than a well because the flowline can feed it liquid surges that far exceed those possible by gas-lift or reservoir mechanisms.

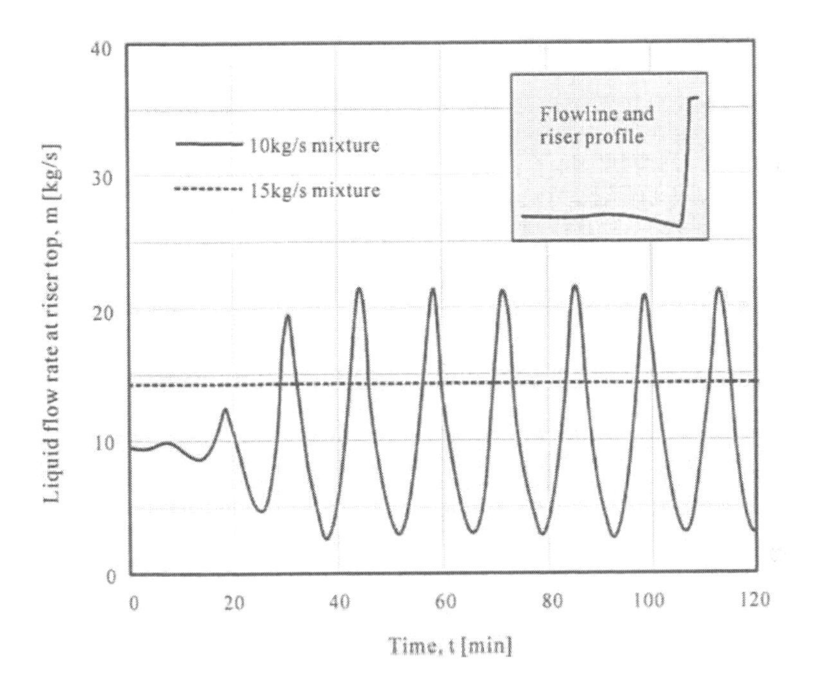

Figure 18.9 Riser flow stability vs. flow rate.

In many oil and gas developments incorporating multiphase flowlines, the possibility of slugs or surges is one of the most important flow assurance concerns due to the excessive demands large changes in oil/gas flow rates place upon the processing facilities. Multiphase surges come in three forms:

1. Hydrodynamic Slugs; formed from the stratified flow regime due to instability of waves at certain flow rates.
2. Terrain Induced Slugs; caused by accumulation and periodic purging of liquid in elevation changes along the flowline, particularly at low flow rates.
3. Operationally Induced Surges: formed in the system during operation transfer between steady state and transient state; for example, during start-up or pigging operations.

18.6.2 Hydrodynamic Slugging

Hydrodynamic slugs are initiated by the instability of waves on the gas-liquid interface in stratified flow under certain flowing conditions. Figure 18.10 shows a formation process of hydrodynamic slugging from a stratified flow. (a) the interface of gas/liquid is lifted to the top of the pipe when the velocity difference between gas phase and liquid phase is high enough. This wave growth is triggered by the Kelvin-Helmholtz instability. (b) once the wave reaches the top of pipe, it forms a slug. The slug is pushed by the gas and so travels at a greater velocity than the liquid film, and more liquid is then swept into the slug. (c) gas entrainment reduces the average liquid holdup in the slug, increasing the turbulence within the slug.

A main part of the frictional pressure drop in multiphase flow is thought to be due to the turbulent region within the slug. Thus the size of the turbulent region can have a significant effect on the frictional pressure losses in a pipeline. Two-phase flow pattern maps indicate hydrodynamic slugging, but slug length correlations are quite uncertain. Tracking of the development of the individual slugs along the pipeline is necessary to estimate the volume of the liquid surges out of the pipeline.

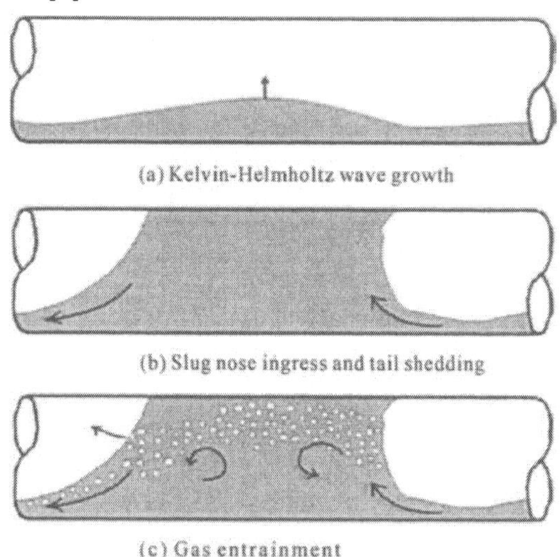

Figure 18.10 Formation of hydrodynamic slugging (Feesa Ltd, 2003).

Slugging simulations need to be performed over the flow rates, watercuts, and GORs for the field life. The effects of any artificial lift should be included in the simulations. In general, simulation results are presented as liquid and gas flow rates at the separator, slug lengths at the base and top of the riser, and pressure at key locations as a function of time. The key locations in a well are upstream of well flow control devices and bottom hole. If processing equipment is included in the model, the separator pressure, separator levels, and outlet gas and liquid rates from the separator as a function of time are presented.

In cases where the predicted slugging causes liquid or gas handling problems, the effects of additional choking upstream of the separator need to be determined. The evolution of slugs is very sensitive to the pipe inclination and changing the inclination by less than a degree can be sufficient to change the balance causing a flow regime transition. Thus, peaks and troughs along the pipeline profile of relatively small elevation change may have a very significant effect.

Figure 18.11 compares the flow patterns of a simple horizontal topography and one with some undulations. The multiphase flow in undulate flowline switches between the stratified and slug flow regimes, implying that not only could the slug sizes differ markedly but the pressure drops could be very different too. Few pipelines have constant inclinations, most undulate following the natural terrain. When modeling multiphase flow in lower flow rates, it is important to represent these undulations as faithfully as possible. At higher flow rates, undulations may not have much impact on predictions.

18.6.3 Terrain Slugging

Figure 18.12 shows a terrain slug formation process in a pipeline-riser system. (a) low spots fills with liquid and flow is blocked; (b) pressure builds up behind the blockage; (c) and (d) when the pressure becomes high enough, gas blows liquid out of the low spot as a slug. Liquid accumulates at the base of the riser, blocking the flow of gas phase. This liquid packet builds up until the gas pressure is sufficient high to overcome the hydrostatic head and blow the liquid slug from the riser. Slug lengths can be 2-3 times the riser height.

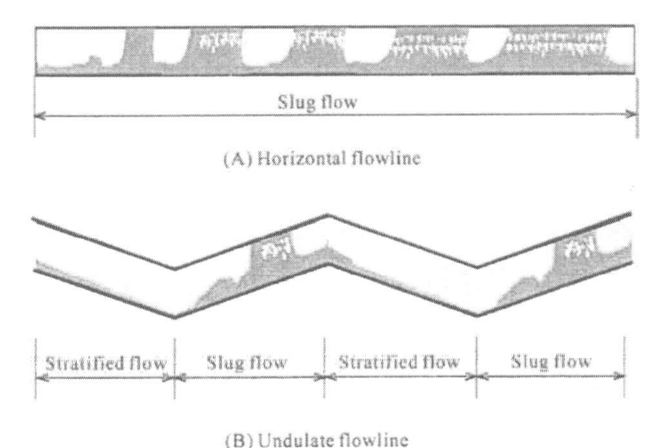

Figure 18.11 Effect of pipeline topography to flow pattern (Feesa Ltd, 2003).

Figure 18.12 Formation of riser slugging.

Terrain slugs can be very severe, causing large pressure variations and liquid surges out of pipeline. Terrain slugging is a transient situation which requires a dynamic model to predict and describe. When the minimum flow rate is defined as the terrain slugging boundary, a region without severe slugging should be determined as a function of watercut and GLR (gas to liquid ratio), including gas lift. In cases where the predicted slugging causes liquid or gas handling problems, the effect of additional choking upstream of the separator should be determined.

It may be difficult to design a slug catcher to cope with the magnitude of terrain slugs. If the transportation system terminates in a vertical riser onto a receiving platform, the passage of the slug from the horizontal pipeline to the vertical riser results in cyclic flow effects. As the slug decelerates into the vertical riser the following gas bubble is compressed, compression continues until sufficient energy is generated to accelerate the slug from the riser. When the wellhead pressure is limited, the vertical riser from the seabed into the platform may form a backpressure due to slugs, and it will limit well production. In such circumstances the vertical pressure which loss can be reduced if a slug catcher is located on the seabed, and the gas and liquid phases are separated; the liquid is pumped to the surface with the gas free flowing to the platform through a separate riser. An alternative is the inject gas at the base of the riser, which will lighten the fluid column and minimizes vertical pressure losses.

18.6.4 Start-up Slugging

Slugs forms in the start-up operation process because of transformation from a stead-state to a transient process. The start-up simulations should be performed starting from shut-in conditions at representative operating conditions throughout the field life. A range of start-up rates, consistent with reservoir management constraints, should be evaluated. If necessary, artificial lift to mitigate start-up slugs should be evaluated. For gas lift, the required gas lift rate should be determined. If lift gas is unavailable until it is obtained from the production, this operating constraint should be included in the simulations.

18.6.5 Pigging

Figure 18.13 shows the flowrate variations of liquid and gas at a riser base due to pigging. When the slug is big enough, the full riser base will be filled with liquid phase. The simulations of the pigging operation should be performed where pigging is required. The time for the pigging operation and the pig velocity as a function of time should be reported. For a round trip pigging, the inlet and outlet temperature, pressure and flow rate as a function of time should be analyzed. For a round trip pigging with liquid, the backpressure necessary to maintain single-phase flow behind the pig should be analyzed.

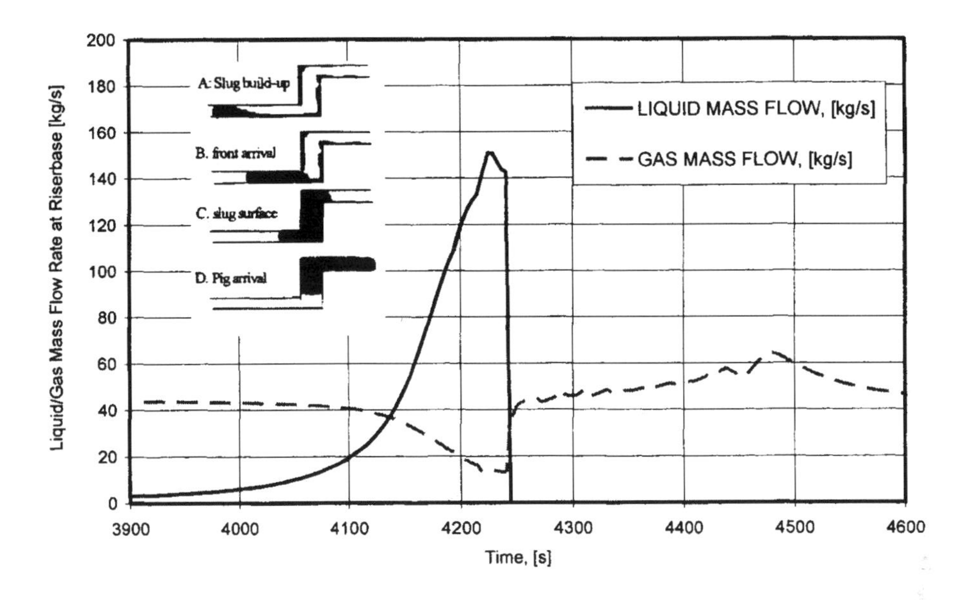

Figure 18.13 Flowrate variations due to pigging.

18.6.6 Slugging Prediction

The slugging prediction can be carried out using PIPESIM (Schlumberger), OLGA (Scandpower Petroleum Technology), ProFES (Aspentech) and TACITE (Simulation Science). Steady state multiphase simulation software such as PIPESIM can predict hydrodynamic slug distributions along the flowline and riser slugs. However, only OLGA 2000 (standard version), OLGA 2000 slug tracking module and ProFES are good at simulating transient multi-phase flow and predicting the liquid holdup variation along the flowline, terrain slugs and start-up/shutdown transient slugs.

Ramp-up slugs are of primary importance for gas condensate systems where the increased flow rate can sweep out large volumes of liquid. Simulations of flow rate ramps should be performed from turndown rates to intermediate and full production rates over the life of the field. OLGA 2000 slug tracking is generally not required for ramp-ups in gas condensate systems except in cases with hilly terrain or high liquid loadings (>50 bbl/mmscf). Results may be presented as outlet liquid and gas rates as a function of time.

The production system should be modeled starting at the reservoir using inflow performance relationships provided by reservoir modeling. At the outlet, inclusion of relevant portions of the separation and level, pressure control is recommended. If this is not possible, a sensitivity study should be performed to ensure that random fluctuations in the outlet pressure do not significantly alter the slugging. Artificial boundary condition of a constant pressure tends to dampen slugging and should be avoided whenever possible for slugging simulations. True boundary conditions on the flowline/riser can be obtained using a transient pipeline model coupled to a dynamic process model. Integrated, dynamic pipeline and process simulation is rarely necessary for the design of wells, flowlines and risers. However, in certain instances, integrated modeling is advisable for the design and control of the process facilities.

18.6.7 Slug Detection and Control Systems

Although slug control is very important to avoid facility damage and upsets, control options are limited. Facility design must be considered for slug handling and control. The potential impact of slugging on the topsides system operation must be addressed and then analyses of the subsea system carried out to assess the effects. Usually there is a trade off between the design of slug catchers and the optimization of the flowline to reduce slugging.

Slugs have been successfully detected using gamma densitometers located on the riser, acoustical measurements, and measurements of pressure at the base of the riser. Slug detection systems should be considered when predicted slugging is expected to give operational difficulties and/or when an advanced control system is to be used for slug mitigation. In this case, the slugging simulations should include the advanced control to test the control algorithm. Flow assurance and process disciplines should demonstrate advanced controls critically dampen predicted slug volumes and frequency.

18.6.8 Slug Catcher Sizing

Slug catchers should be sized to dampen surges to a level that can be handled by downstream processing equipment. Before dynamic models of the topsides facilities are available, the level of acceptable surging is unknown and designers are often forced to make assumptions vis-a-vis surge volumes, such as designing for the 'one in a thousand' slug.

Surge volume for gas condensate requirements are determined from the outlet liquid rates predicted in the ramp-up, startup, and pigging cases. The required slug catcher size is dependent on liquid handling rate, pigging frequency, and ramp-up rates. An iterative process may be required to identify optimum slug catcher size, pigging frequency, liquid handling rate, and acceptable ramp-up rates. For this optimization, the results of the simulations should be presented as surge volume requirements as a function of liquid handling rate for representative ramp-up rates and pigging frequencies.

Separator volumes for black oil systems are typically set by separation requirements rather than liquid slug handling capacity. Consequently, the ability of the separator to accommodate the slugs from all operations should be confirmed based on the results of the slugging simulations.

18.7 Pressure Surge

18.7.1 Fundamentals of Pressure Surge

An important consideration in the design of single liquid phase pipelines is pressure surge, also known as water hammer. Typical surge events in a pipeline or piping system are generally caused by a pump shutdown or a valve closure. The kinetic energy of flow is converted to pressure energy. The velocity of the pressure wave propagation is determined by fluid and pipeline characteristics. Typical propagation velocities range from 1100 ft/sec for propane/butane pipelines to 3300 ft/sec for crude oil pipelines and up to 4200 ft/sec for heavy-wall steel water pipelines.

A rough estimate of the total transient pressure in a pipeline/piping system following a surge event can be obtained from the equations below:

$$v_s = \sqrt{\frac{144Kg/\rho}{1+[KdC/Et]}} \tag{18.27}$$

where, v_s : speed of sound in the fluid, propagation speed of the pressure wave, ft/s;

K : fluid bulk modulus,10 psi;

g : gravitational constant, 32.2 ft/s^2;

ρ : density of liquid, lbf/ft^3;

d : inside diameter of pipe, in;

E : pipe material modulus of elasticity, psi;

t : pipe wall thickness, in;

C : constant of pipe fixity; 0.91 for an axially restrained line; 0.95 for unrestrained.

The surge pressure wave travels upstream and is reflected downstream, oscillating back and forth until its energy is dissipated in pipe wall friction. The amplitude of the surge wave, or the magnitude of the pressure surge P_{Surge}, is a function of the change in velocity and the steepness of the wave front and is the inverse of the time it took to generate the wave:

$$P_{surge} = \rho v_s \Delta v/(144g) \qquad \text{for} \quad T < 2L/v_s \tag{18.28}$$

where, Δv : total change in velocity, ft/s (m/s);

$2L/v_s$: propagation time, s;

T : valve closing time, s.

This is only a rough estimation of the surge pressure magnitude for cases limited by the stated time of closure criteria. An accurate analysis of the maximum surge pressure, location of critical points in the system may be obtained from a dynamic simulation by using software (e.g. Pipe Simulator). The surges are attenuated by friction, and the surge arriving at any point on the line is less than at the origin of the surge wave. Nevertheless, when the flow velocity is high and stoppage is complete, or when a pump station is bypassed suddenly, the surge energy generated can produce pressures high enough to burst pipe. Based on the ASME codes, the level of pressure rise due to surges and other variations from normal operations shall not exceed the internal design pressure at any point in the piping system and equipment by more than 10%.

18.7.2 When Is Pressure Surge Analysis Required?

Surge analysis should be performed during a project's early design and planning phases. This analysis will help to ensure the achievement of an integrated and economical design. Surge analysis provides assurance that the selected pumps/compressors, drivers, control valves, sensors, and piping can function as an integrated system in accordance with a suitable control system philosophy. A surge analysis becomes mandatory when one repairs a ruptured pipeline/piping system to determine the source of the problem.

All long pipelines/piping systems designed for high flow velocities should be checked for possible surge pressures that could exceed the maximum allowable surge pressure (MASP) of the system piping or components. A long pipeline/piping system is defined as one that can experience significant changes in flow velocity within the critical period. By this definition, a long pipeline/piping system may be 1500 feet long or 800 miles long.

A steady-state design cannot properly reflect system operation during a surge event. On one project in which a loading hose at a tanker loading system in the North Sea normally operated at 25-50 psig, a motor-operated valve at the tanker manifold malfunctioned and closed during tanker loading. The loading hose, which had a pressure rating of 225 psig, ruptured. Surge pressure simulation showed that the hydraulic transient pressure exceeded 550 psig. Severe surge problems can be mitigated through the use of quick-acting relief valves, tanks, and gas-filled surge bottles, but these facilities are expensive single-purpose devices.

18.8 Line Sizing

18.8.1 Hydraulic Calculation

Hydraulic calculation should be carried out from wells, pipelines and risers to the surface facilities in line sizing. The control factors in the calculation are fluids (oil, gas or condensate fluids), flow line size, flow pattern, and application region. Unlike single-phase pipelines, multiphase pipelines are sized accounting for the limitations imposed by production rates, erosion, slugging and ramp-up speed. Artificial lift is also considered during line sizing to improve the operational range of the system.

Design conditions:

- Flow rate and pressure drop allowable established, determine pipe size for a fixed length.
- Flow rate and length known, determine pressure drop and line size.

Usually either of these conditions requires a trial and error approach based upon assumed pipe sizes to meet the stated conditions. Some design problems may require determination of maximum flow for a given line size and length, but this is a reverse of the conditions above.

During the FEED stage of a project, the capability of a given line size to deliver the production rates throughout the life of the field is determined by steady-state simulation from the reservoir to the surface facilities throughout the field life. Sensitivities to the important variables such as GOR, water cut, viscosity, and separator pressure should be examined.

The technical criteria for line sizing of pipelines are stated in next section; however, optimum economic line size is seldom realized in the engineering design. Unknown factors such as future flow rate allowances, actual pressure drops through certain process equipments etc., can easily over balance that the design predicted on selecting the optimum. In sizing a line pipe one is always faced with a compromise between two factors. For a given flow rate of a given fluid, piping cost increases with diameter. But pressure loss decreases, which reduces potential pumping or compression costs. There is an economic balance between material costs and pumping costs in downstream flowlines. The optimum pipe size is found by calculating the smallest capitalization/operating cost or using the entire pressure drop available; or increasing velocity to highest allowable.

18.8.2 Criteria

The line sizing of the pipeline is governed by the following technical criteria:

- allowable pressure drop;
- maximum velocity (allowable erosional velocity) and minimum velocity;
- system deliverability;
- and slug consideration if applicable.

Other criteria considered in the selection of the optimum line size include:

- standard versus custom line sizes; generally, standard pipe is less expensive and more readily available. For long pipelines or multiple pipelines for the same size, custom line sizes may be cost effective;
- ability of install; particularly in deep water, the technical installation feasibility of larger line sizes may constrain the maximum pipe size;
- future production; consideration should be given to future production that may utilize the lines;
- number of flowlines and risers; if construction or flow assurance constraints require more than two production flowlines per manifold, optional alternatives including subsea metering processing, bundles etc should be explored;
- low temperature limits; subsea equipment including trees, jumpers, manifolds and flowlines have minimum temperature specifications. For systems with Joule-Thompson cooling in low ambient temperature environments, operating philosophies and possibly metallurgical section should be adjusted;
- high temperature limits; flexible pipe has a maximum temperature limit that depends on the materials of construction, watercut, water composition, and water pH. For systems with flexible pipe, the flow rate may be limited, requiring a smaller pipe;
- roughness; flexible pipe is rougher than steel pipe, and therefore, requires a larger diameter for the same maximum rate. To smooth flexible pipe, internal coating may be applied.

18.8.3 Maximum Operating Velocities

Liquid velocity is usually limited because of erosion effects at fittings. Erosion damage can occur in flowlines with multiphase flow because of the continuous impingement of high velocity liquid droplets. The damage is almost always confined to the place where the flow direction is changed, such as elbows, tees, manifolds, valves, risers and so on. The erosional velocity is defined as the bulk fluid velocity which will result in the removal of corrosion product scales, corrosion inhibitors, and/or other protective scales presented on the inter-surface of pipeline.

The pipeline fluids velocity should be constrained as follows:

- the fluid velocity in single-phase liquid lines varies from 0.9 m/sec to 4.5 m/sec (3 ft/sec to 15 ft/sec);
- gas/liquid two-phase lines do not exceed the erosional velocity as determined from following equation, which is recommended in API RP 14E.

$$V_{max} = C / \rho_m^{0.5} \tag{18.29}$$

where, V_{max} : Maximum allowable mixture erosional velocity, m/sec (ft/sec);

ρ : Density of the gas/liquid mixture, kg/m^3 (lb/ft^3);

C : Empirical constant, which determined in Table 18.9.

The mixture density, ρ_m is defined as,

$$\rho_m = C_L \rho_L + (1 - C_L)\rho_g \tag{18.30}$$

where, ρ_L : liquid density;

ρ_g : gas density;

C_L : flowing liquid volume fraction, $C_L = Q_L / (Q_L + Q_g)$

Table 18.9 Empirical constant in the equation.

Service Type	Operational Frequency	
	Continuous	Intermittent
Two phase flow without sand	100	125

If possible, the minimum velocity in two-phase lines should be greater than 3 m/sec (10ft/sec) to minimize slugging.

Assuming on erosion rate of 10 mils per year, the following maximum allowable velocity is recommended by Salama and Venkatesh, when sand appears in oil/gas mixture flow:

$$V_M = \frac{4d}{\sqrt{W_s}} \tag{18.31}$$

where, V_M : Maximum allowable mixture velocity, ft/s;

d : pipeline inside diameter, in;

W_s : rate of sand production (bbl/ month).

Erosion limits or maximum velocities for the flexible pipes should be specified by the manufacturer. API erosion velocity limit is normally very conservative and the practice of this limit could vary depend on different operators.

18.8.4 Minimum Operation Velocities

The following items may effectively impose minimum velocity constraints:

- Slugging; slugging severity typically increases with decreasing flow rate. The minimum allowable velocity constraint should be imposed to control the slugging in multi-phase flow for assuring the production deliverability of the system.
- Liquid handling; in gas condensate systems, the ramp up rates may be limited by liquid handling facilities and constrain the maximum line size.
- Pressure drop; for viscous oils, there is a minimum flow rate necessary to maintain fluid temperature such that the viscosities are acceptable. Below this minimum, production may eventually shut itself in.
- Liquid loading; a minimum velocity is required to lift the liquids, and prevent wells and risers from loading up with liquid and shutting in. The minimum stable rate is determined by transient simulation at successively lower flow rates. The minimum rate for the system is also a function of GLR.
- Sand bedding; the minimum velocity is required to avoid sand bedding.

18.8.5 Wells

During FEED analysis, the production system should be modeled starting at the reservoir using inflow performance relationships. OLGA2000 may be used to evaluate the hydraulic stability of the entire system including the wells, flowlines and risers subject to artificial lift to get minimum stable rates as a function of the operating requirements and artificial lift.

In the hydraulic and thermal analyses of subsea flowline system, the wellhead pressure and temperature are generally used as the inlet pressure and temperature of the system. The wellhead pressure and temperature are the functions of reservoir pressure, temperature, productivity index, and production rate and can be obtained form steady and transient hydraulic and thermal analyses of the well bore. The hydraulic and thermal models may be simulated using the commercial software PIPESIM or OLGA2000.

The hydraulic and thermal process of a well bore is simulated for the following purposes;

- steady-steady wellhead pressure and temperature for flowline hydraulics and thermal analyses;

- determining the minimum production rates for preventing hydrate formation in the wellbore;
- determining the minimum production rate for preventing wax deposition in the well bore and tree. The wellhead temperature is required to be higher than the critical wax deposition temperature in the steady state flowing condition;
- determining cooldown time and warm-up time from transient well bore analysis to prevent hydrate formation in well bore.

The transient analyses determine how hydrates can be controlled during startup and shut-in process. As, vacuum insulated tubing (VIT) is one method to help wellbore out of the hydrate region. Use of 3000 – 4000 ft of VIT results in warm-up of the wellbore to temperatures above the hydrate temperature in typically 2 hours or less. These warm-up rates are rapid enough to ensure that little or no hydrates form in the wellbore. This makes it possible to eliminate the downhole injection of methanol above the subsurface safety valve during warm-up. Instead, methanol is injected upstream of the choke to prevent any hydrates that may be formed during warm-up from plugging the choke. One drawback of VIT is that it cools more quickly than bare tubing since the insulation prevents the tubing to gain heat from the earth which also warms up during production. As a result of this quick cooling, it is necessary to quickly treat the wellbore with methanol in the case that a production shut-in occurs during the warm-up period. For a higher reservoir temperature, VIT may not be used in the wellbore and the cooldown speed of wellbore is very slow upon a process interruption from steady state production.

18.8.6 Gas Lift

Artificial lift can be used to increase production rates and/or reduce line sizes. Line sizing may include the affects of gas lift.

For systems employing gas lift, models should extend from the reservoir to the surface facilities incorporating the gas lift gas in addition to the produced fluids. The composition of the gas lift gas is provided by the topside process but should meet the dew point control requirement. The preferred location (downhole, flowline inlet, or riser base) is determined based on effectiveness and cost.

The affect of gas lift on the severe slugging boundary maybe simulated with OLGA 2000 for the situations, when it is required) there is needed at riser base. The simulation may decide the minimum flow rate without severe slugging as a function of gas lift rates and watercuts. When the gas lift injection point does not include a restriction or valve to choke the gas flows, OLGA 2000 transient simulations may model the transient flow in the gas lift line by including the gas lift line as a separate branch to determine maximum variation in mass rate and pressure and still maintain system stability. Maximum production rates throughout the field life are determined as a function of gas lift rate. In FEED design target gas lift rate and pressure requirements should be recommended, including a measurement and control strategy for the gas lift system.

18.9 References

1. Ansari, A., Sylvester N, D., Shoham, O., and Brill, J. P., (1990), "A Comprehensive Mechanistic Model for Upward Two-Phase Flow in Wellbores", SPE 20630. SPE Annual Technical Conference.
2. API RP 14E, (1984), "Recommended Practice for Design and Installation of Offshore Platform Piping System", 4th Edition, Apr. 1984.
3. Aziz, KL, Govier, G, W. and Fogarasi, M., (1972), "Pressure Drop in Wells Producing Oil and Gas," J. Cdn., Pet. Tech. 38-48.
4. Baker, A. C., Nielsen, K., and Gabb, A., (1988), "Pressure loss, liquid-holdup calculations developed". Oil & Gas Journal, Mar 14, 1988.
5. Beggs H . D., and Brill, J. P.,(1973), "A Study of Two Phase Flow in Inclined Pipes," J. Pet Tech., 607-617, Trans., AIME, 240.
6. Beniksen, K. H., Malnes, D., Moe, R. and Nuland, S., (1990), "The Dynamic Two-Fluid Model OLGA: Theory and Application". SPE 19451.
7. Brill, J. P., and Mukherjee, H., (1999), *Multiphase Flow in Wells*, Monograph Vol. 17, Henry, L. Doherty Series, SPE.
8. Crane Company, *Flow of Fluids through Valves, Fittings and Pipe*, Technical Paper No. 410, 25th printing, 1991.
9. Dukler, E. A., et al, (1969), "Gas-Liquid Flow in Pipelines, I. Research Results," AGA-API Project NX-28.
10. Duns, H., and Ros, N. C, J., (1963), "Vertical Flow of Gas and Liquid Mixtures in Wells," 6th World Pet Congress, 452.
11. Fayed, S. F., and Otten, L., (1983), "Comparing Measured with Calculated Multi-phase Flow Pressure Drop", OGJ, Aug 22, 1983, p136.
12. Feesa Ltd, (2003), "Hydrodynamic Slug Size in Multiphase Flowlines", http://www. feesa. net/flowassurance.
13. Flanigan, O., (1958), "Effect of Uphill Flow on Pressure Drop in Design of Two-Phase Gathering Systems," Oil and Gas J., 56,132.
14. Gray, W. G., (1978), "Vertical flow correlation - gas wells" API Manual 14BM.
15. Gregory, G. A., (1991), "Erosional Velocity Limitations for Oil and Gas Wells", *Technical Note* No. 5, Neotechnology Consultants Ltd.
16. Gregory, G. A., (1985), "Viscosity of Heavy Oil/Condensate Blends", *Technical Note* No. 6, Neotechnology Consultants Ltd.
17. Gregory, G. A., (1990), "Pipeline Calculations for Foaming Crude Oils and Crude Oil-Water Emulsions", *Technical Note* No. 11, Neotechnology Consultants Ltd.
18. Guth, E., and Simha, R., (1936), Kolloid-Zeitschrift, Vol. 74, p.266.
19. Hagedom, A, R. and Brown, K. E., (1965), "Experimental Study of Pressure Gradients Occurring During Continuous Two-Phase Flow in Small-Diameter Vertical Conduits," J. Pet Tech., 475-484.
20. Larock, B. E., Jeppson, R. W., and Watters, G. Z., *Hydraulics of Pipeline Systems*, CRC Press.

21. Leland, T. W., (1980), *Phase Equilibria and Fluid Properties in the Chemical Industry*, Frankfurt/Main: DECHEMA, 1980, pp. 283-333.
22. Mohinder, L., N. (Editor in Chief), *Piping handbook*, sixth Edition, Mcgraw-Hill.
23. Moody, L. F., (1944), "Friction Factors for Pipe Flow", *Trans. ASME*, 66, 1944, pp. 671-678.
24. Mukherjce, H. and Brill, J. P.,(1983), "Liquid Holdup Correlations for Inclined Two-Phase Flow", Journal of Petroleum Technology,1003-1008.
25. Oliemana, R, V. A., (1976), "Two-Phase Flow in Gas-Transmission Pipeline", ASME paper 76-Pet-25, presented at Pet. Div, ASME meeting Mexico City.
26. Qrkifizewski, J., (1967), "Predicting Two-Phase Pressure Drops in Vertical Pipes," J, Pet Tech. , 829-838.
27. Pedersen, K. S., Fredenslund, A., and Thomassen, P., (1989), Properties of Oils and Natural Gases, Gulf Publishing Company.
28. Peng, D. Y., and Robinson, D. B., (1976), "A New Two-constant Equation of State", *Ind. Eng. Chem. Fundam.*, 15, 1976, pp.59-64.
29. PIPESIM Course, (1997), "Information on Flow Correlations used within PIPESIM".
30. Reid, R. C., Prausnitz, J. M., and Sherwood, T. K., (1977), *The Properties of Gases and Liquids*, 3rd ed., New York, McGraw-Hill, 1977.
31. Salama, M. M., and Venkatesh, E. S., (1983), "Evaluation of API RP 14E Erosional Velocity Limitations for Offshore Gas Wells", OTC 4485, Houston.
32. Scandpower, "OLGA 2000", OLGA School, Level I, II.
33. Smith, H. V., and Arnold, K. E., (1987), "Crude Oil Emulsions", *Petroleum Engineering Handbook*, Editor, Bradley, H.B.
34. Soave, G., (1972), "Equilibrium Constants from a Modified Redilich-Kwong Equation of State", *Chem. Eng. Sci*, 27, 1972, pp.1197-1203.
35. Taitel, Y. M., Barnea, D., and Dukler, A. E., (1980), "Modeling Flow Patter Transitions for Steady Upward Gas-Liquid Flow in Vertical Tubes", AIChE J, Vol 26, 245.
36. Whitson, C. H., and Brule, M. R., (2000), *Phase Behavior*, Monograph Volume 20, Henry, L. Doherty Series, SPE.
37. Woelflin, W., "The Viscosity of Crude Oil Emulsions", Drill and Prod. Prac., API, p148.
38. Wu, J. C., (2001), "Benefits of Dynamic Simulation of Piping and Pipelines", Technotes of Paragon.
39. Xiao, J. J., Shoham, O., and Brill, J. P., (1990), "A Comprehensive Mechanistic Model for Two-Phase Flow in Pipelines", SPE 20631, SPE Annual Technical Conference,

Part III

Flow Assurance

Chapter 19 Heat Transfer and Thermal Insulation

19.1 Introduction

The thermal performance of subsea production system is controlled by the hydraulic behavior of fluid in the flowline; conversely, it also impacts the hydraulic design indirectly through the influence of temperature on fluid properties such as GOR, density, and viscosity. Thermal design, which predicts the temperature profile along the flowline, is one of the most important parts in the flowline design; and this information is required for pipeline analyses including expansion analysis, upheaval or lateral buckling, corrosion protection, hydrate prediction and wax deposition analysis. In most cases, the solids managements (hydrate, wax, asphaltenes, and scales) determine the requirements of hydraulic and thermal designs. In order to maintain a minimum temperature of fluid to prevent hydrate and wax deposition in the flowline, insulation layers may be added to the flowline.

Thermal design includes both steady state and transient heat transfer analyses. In steady state operation, the production fluid temperature decreases as it flows along the flowline due to the heat transfer through pipe wall to the surrounding environment. The temperature profile in the whole pipeline system should be higher than the requirements for prevention of hydrate and wax formation during normal operation and is determined from steady-state flow and heat transfer calculations. If the steady flow conditions are interrupted due to a shut-in or restarted again during operation, the transient heat transfer analysis for the system is required to make sure the temperature of fluid is out of the solid formation range within the required time. It is necessary to consider both steady state and transient analyses in order to ensure that the performance of the insulation coatings will be adequate in all operational scenarios.

Thermal management strategy for flowlines can be divided into passive control and active heating. Passive control includes flowlines insulated by external insulation layers, pipe-in-pipe (PIP), bundle and burial; and active heating includes electrical heating and hot fluid heating.

In addition, if the production fluid contains gas, the fluid can have a temperature drop due to the Joule-Thompson (JT) effect. The JT effect is primarily caused by pressure-head changes, which predominantly occur in the flowline riser and may cause the flowline temperature to fall below ambient temperatures. JT cooling cannot be prevented by insulation. Thus, the JT effect will not be explicitly discussed other than its effects being implicitly included in some of the numerical results.

The purpose of this Chapter is to discuss thermal behavior and thermal management of a flowline system, which includes,

- heat transfer fundamentals;
- overall heat transfer coefficient calculation;
- steady state analysis of thermal design;
- transient heat transfer analyses of thermal design;
- insulation or heating management.

19.2 Heat Transfer Fundamentals

Figure 19.1 shows the three heat transfer modes occurring in nature; conduction, convection and radiation. Heat is transferred by any one or combination of these three modes. When a temperature gradient exists in a stationary medium, which may be gas, liquid and solid, the conduction will occur across the medium; if a surface and a moving fluid have a temperature difference, the convection will occur between the fluid and surface; all solid surfaces with a temperature will emit energy in the form of electromagnetic waves, which is called radiation. Although these three heat transfer modes occur at all subsea systems, for typical flowlines, heat transfer from radiation is relatively insignificant compared with the heat transfer from conduction and convection because the system temperature is below 200 °C in generally. Therefore, conduction and convection will be solely considered here.

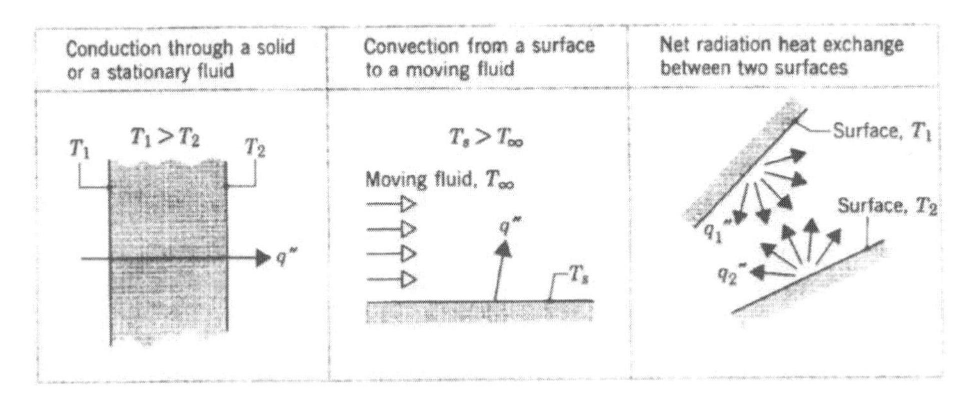

Figure 19.1 Conduction, convection, and radiation modes (Incropera and DeWitt, 1996).

19.2.1 Heat Conduction

For a one-dimensional plane with a temperature distribution $T(x)$, the heat conduction is quantified by Fourier equation,

$$q'' = -k \cdot \frac{dT(x)}{dx} \tag{19.1}$$

where, q'' : heat flux, Btu/(hr·ft^2) or W/m^2, heat transfer rate in the x direction per unit area;

k : thermal conductivity of material, Btu/(ft-hr-°F) or W/(m-K);

dT/dx : temperature gradient in x direction, °F/ft or °C/m.

When the thermal conductivity of material is constant along the wall thickness, the temperature distribution is linear and the heat flux becomes,

$$q'' = -k \cdot \frac{T_2 - T_1}{x_2 - x_1} \tag{19.2}$$

Once the temperature distribution is known, the heat flux at any point in the medium may be calculated from Fourier equation. By applying an energy balance to a 3D differential control volume and temperature boundary condition, the temperature distribution may be acquired from the heat diffusion equation as below,

$$\frac{\partial}{\partial x}\left(k\frac{\partial T}{\partial x}\right) + \frac{\partial}{\partial y}\left(k\frac{\partial T}{\partial y}\right) + \frac{\partial}{\partial z}\left(k\frac{\partial T}{\partial z}\right) + \dot{q} = \rho c_p \frac{\partial T}{\partial t} \tag{19.3}$$

where, $\dot{q}$: heat generation rate per unit volume of the medium, Btu/(hr·ft^3) or W/m^3;

ρ : density of the medium, lb/ft^3 or kg/m^3;

c_p: specific heat capacity, BTU/(lb ·°F) or kJ/(kg·K);

x,y,z : coordinates, ft or m;

t : time, second.

For a cylindrical coordinates, the heat diffusion equation may be rewritten as,

$$\frac{1}{r}\frac{\partial}{\partial r}\left(kr\frac{\partial T}{\partial r}\right) + \frac{1}{r^2}\frac{\partial}{\partial \phi}\left(k\frac{\partial T}{\partial \phi}\right) + \frac{\partial}{\partial z}\left(k\frac{\partial T}{\partial z}\right) + \dot{q} = \rho c_p \frac{\partial T}{\partial t} \tag{19.4}$$

where, r,z : radius and axial directions of cylindrical coordinates;

ϕ : angle in radius direction.

For most thermal analyses of flowline systems, the heat transfer along the axial and circumferential directions may be ignored and therefore, the transient heat conduction without heat source will occur in the radial direction of cylindrical coordinates, the above equation is simplified as,

$$\frac{1}{r}\frac{\partial}{\partial r}\left(kr\frac{\partial T}{\partial r}\right) = \rho c_p \frac{\partial T}{\partial t} \tag{19.5}$$

The temperature change rate ($\partial T / \partial t$) depends not only on the thermal conductivity of material k, but also the density, ρ and specific heat capacity, c_p. This equation can be solved numerically. For a steady heat transfer, the right side of equation is equal to zero; the total heat flow per unit length of cylinder is calculated by following equation,

$$q_r = -2\pi k \frac{T_2 - T_1}{\ln(r_2 / r_1)} \tag{19.6}$$

where, r_1, r_2 : inner and outer radii of the cylinder medium, ft or m;

T_1, T_2 : temperatures at corresponding points of r_1, r_2 , °F or °C;

q_r : heat flow rate per unit length of cylinder, Btu/(hr·ft) or W/m.

19.2.2 Convection

Both internal and external surfaces of subsea flowline contact with fluids, convection heat transfer will occur when there is a temperature difference between pipe surface and fluid. The convection coefficient is also called as film heat transfer coefficients in flow assurance field because convection occurs at a film layer of fluid adjacent to the pipe surface.

Internal convection

Internal convection heat transfer occurs between internal flowing fluid and pipe internal surface; it depends on the fluid properties, the flow velocity, and the pipe diameter. For the internal convection of pipelines, Dittus and Boelter (1930) proposed the following dimensionless correlation for fully turbulent flow of single-phase fluids,

$$\mathrm{Nu}_i = 0.0255 \cdot \mathrm{Re}_i^{0.8} \cdot \mathrm{Pr}_i^{n} \tag{19.7}$$

where, Nu_i: Nusselt number, $\mathrm{Nu}_i = \dfrac{h_i D_i}{k_f}$;

Re_i : Reynolds number, $\mathrm{Re}_i = \dfrac{D_i V_f \rho_f}{\mu_f}$;

Pr_i : Prandtl number, $\mathrm{Pr}_i = \dfrac{C_{pf} \mu_f}{k_f}$;

n : 0.4 if the fluid is being heated, and 0.3 if the fluid is being cooled;

h_i : internal convection coefficient, BTU/(ft^2-hr-°F) or W/(m^2-K);

D_i : pipeline inside diameter, ft or m;

k_f : thermal conductivity of the flowing liquid, BTU/(ft-hr-°F) or W/(m-K);

V_f : velocity of the fluid, ft/s or m/s;

ρ_f : density of the fluid, lb/ft^3 or kg/m^3;

μ_f : viscosity of the fluid, lb/(ft-s) or Pa.s;

C_{pf} : specific heat capacity of the fluid, BTU/(lb-°F) or J/(kg-K).

All fluid properties are assumed to be evaluated at the average fluid temperature. This correlation gives satisfactory results for flows with Reynold number greater than 10,000, Prandtl number of 0.7 to 160 and a pipeline length greater than 10D.

If the flow is laminar (*i.e.* $\mathrm{Re}_i < 2100$), h_i may be calculated using Hausen equation (1943) as follows,

$$\mathrm{Nu}_i = 3.66 + \frac{0.0668\left(\dfrac{D_i}{L_o}\right)\mathrm{Re}_i\,\mathrm{Pr}_i}{1 + 0.4\left[\left(\dfrac{D_i}{L_o}\right)\mathrm{Re}_i\,\mathrm{Pr}_i\right]^{2/3}} \tag{19.8}$$

where, L_o : distance from the pipe inlet to the point of interest. In most pipelines case, $D_i/L_o \approx 0$, therefore Eq. (19.8) becomes,

$$\mathrm{Nu}_i = 3.66 \tag{19.9}$$

For the transition region ($2100 < \mathrm{Re}_i < 10^4$), the heat transfer behavior in this region is always uncertain because of the unstable nature of the flow, especially for the multiphase flow in pipeline systems. A correlation proposed by Gnielinski may be used to calculate h_i in this region,

$$\mathrm{Nu}_i = \frac{(f/8)(\mathrm{Re}_i - 1000)\mathrm{Pr}_i}{1 + 12.7(f/8)^{1/2}\left(\mathrm{Pr}_i^{\,2/3} - 1\right)} \tag{19.10}$$

where, the fiction factor f may be obtained from the Moody diagram, which is discussed in the chapter 18, or for smooth tubes,

$$f = [0.79 \ln(\mathrm{Re}_i) - 1.64]^{-2} \tag{19.11}$$

This correlation is valid for $0.5 < \mathrm{Pr}_i < 2000$ and $3000 < \mathrm{Re}_i < 5\times10^6$.

Table 19.1 shows typical ranges of internal convection coefficients for turbulent flow. For most pipelines with multiphase flow, an approximate value of internal convection coefficient based on the table generally is adequate.

Table 19.1 Typical internal convection coefficients for turbulent flow (Gregory, 1991).

Fluid	Internal convection coefficient, hi	
	BTU/(ft²·hr·°F)	W/(m²·K)
Water	300 - 2000	1700 - 11350
Gases	3 - 50	17 - 285
Oils	10 - 120	55 - 680

Tables 19.2 to 19.5 show thermal conductivities and specific heat capacities for a variety of typical oils and gases, which can be used for most purposes with a sufficient accuracy if these parameters are not known.

Table 19.2 Typical thermal conductivities for crude oil/hydrocarbon liquids (Gregory, 1991).

Temperature	0 °F (-18 °C)		200 °F (93 °C)	
API Gravity	BTU/(ft·hr·°F)	W/(m·K)	BTU/(ft·hr·°F)	W/(m·K)
10	0.068	0.118	0.064	0.111
20	0.074	0.128	0.069	0.119
30	0.078	0.135	0.074	0.128
40	0.083	0.144	0.078	0.135
50	0.088	0.152	0.083	0.144
60	0.093	0.161	0.088	0.152
80	0.103	0.178	0.097	0.168
100	0.111	0.192	0.105	0.182

Table 19.3 Typical thermal conductivities for hydrocarbon gases (Gregory, 1991).

Temperature	50 °F (10 °C)		100 °F (38 °C)		200 °F (93 °C)	
Gas Gravity	BTU/(ft·hr·°F)	W/(m·K)	BTU/(ft·hr·°F)	W/(m ·K)	BTU/(ft·hr·°F)	W/(m·K)
0.7	0.016	0.028	0.018	0.031	0.023	0.039
0.8	0.014	0.024	0.016	0.028	0.021	0.036
0.9	0.013	0.022	0.015	0.026	0.019	0.033
1.0	0.012	0.021	0.014	0.024	0.018	0.031
1.2	0.011	0.019	0.013	0.022	0.017	0.029

Table 19.4 Typical specific heat capacities for hydrocarbon liquids (Gregory, 1991).

Temperature	0 °F (-18 °C)		100 °F (38 °C)		200 °F (93 °C)	
API Gravity	BTU/(lb·°F)	kJ/(kg·K)	BTU/(lb·°F)	kJ/(kg·K)	BTU/(lb·°F)	kJ/(kg·K)
10	0.320	1.340	0.355	1.486	0.400	1.675
30	0.325	1.361	0.365	1.528	0.415	1.738
50	0.330	1.382	0.370	1.549	0.420	1.758
70	0.335	1.403	0.375	1.570	0.430	1.800

Table 19.5 Typical specific heat capacities for hydrocarbon gases (Gregory, 1991).

Temperature	0 °F (-18 °C)		100 °F (38 °C)		200 °F (93 °C)	
Gas Gravity	BTU/(lb·°F)	kJ/(kg·K)	BTU/(lb·°F)	kJ/(kg·K)	BTU/(lb·°F)	kJ/(kg·K)
0.7	0.47	1.97	0.51	2.14	0.55	2.30
0.8	0.44	1.84	0.48	2.01	0.53	2.22
0.9	0.41	1.72	0.46	1.93	0.51	2.14
1.0	0.39	1.63	0.44	1.84	0.48	2.01

External convection

The correlation of average external convection coefficient suggested by Hilpert (1933) is widely used in industry,

$$\mathrm{Nu}_o = C\,\mathrm{Re}_o^{\,m}\,\mathrm{Pr}_o^{\,1/3} \tag{19.12}$$

where, Nu_o : Nusselt number, $\mathrm{Nu}_o = \dfrac{h_o D_o}{k}$;

Re_o : Reynolds number, $\mathrm{Re}_o = \dfrac{D_o V_o \rho_o}{\mu_o}$;

Pr_{sf} : Prandtl number, $\mathrm{Pr}_o = \dfrac{C_{p,o}\mu_o}{k_o}$;

h_o : external convection coefficient, BTU/(ft^2·hr·°F) or W/(m^2·K);

D_o : pipeline outer diameter, ft or m;

k_o : thermal conductivity of the surrounding fluid, BTU/(ft·hr·°F) or W/(m·K);

V_o : velocity of the surrounding fluid, ft/s or m/s;

ρ_o : density of the surrounding fluid, lb/ft^3 or kg/m^3;

μ_o : viscosity of the surrounding fluid, lb/(ft·s) or Pa·s;

$C_{p,o}$: specific heat capacity of surrounding fluid, BTU/(lb·°F) or J/(kg·K);

C, m : constants, dependent on the Re number range, and listed in Table 19.6.

All the properties used in the above correlation are evaluated at the temperature of film between external surface and surrounding fluid.

Table 19.6 Constants of correlation, Equation (12).

Re_o	C	m
$4\times10^{-1} - 4\times10^{0}$	0.989	0.330
$4\times10^{0} - 4\times10^{1}$	0.911	0.385
$4\times10^{1} - 4\times10^{3}$	0.683	0.466
$4\times10^{3} - 4\times10^{4}$	0.193	0.618
$4\times10^{4} - 4\times10^{5}$	0.027	0.805

When the velocity of surrounding fluid is less than approximately 0.05 m/s in water and 0.5 m/s in air, natural convection will have the dominating influence and the following values may be used:

$$h_o = \begin{cases} 4\ \text{W/(m}^2\text{K), Natural convection in air;} \\ 200\ \text{W/(m}^2\text{K), Natural convection in water.} \end{cases} \tag{19.13}$$

Note:

- many of the parameters used in the correlation are themselves dependent upon temperature. Because the temperature drop along most flowlines is relatively small, average values for physical properties may be used.
- the analysis of heat transfer through wall to surrounding fluid does not address the cooling effect due to Joule-Thompson expansion of the gas. For a long gas flowline or flowline with two-phase flow, an estimation of this cooling effect should be made.

19.2.3 Buried Pipeline Heat Transfer

Pipeline burial in subsea engineering occurs due to,

- placement of rock, grit or seabed material on the pipe for stability and protection requirements;
- gradual infill due to sediment of a trenched pipeline;
- general embedment of the pipeline into the seabed because of the seabed mobility or the pipeline movement.

While seabed soil can be a good insulator, porous burial media, such as rock dump, may provide little insulation because water can flow through the spaces between the rocks and transfer heat to the surroundings by convection.

Fully buried pipeline

For a buried pipeline, the heat transfer is not symmetrical. To simulate a buried pipeline, a pseudo-thickness of the soil is used to account for the asymmetries of the system in pipeline heat transfer simulation software, e.g., OLGA, and PIPESIM.

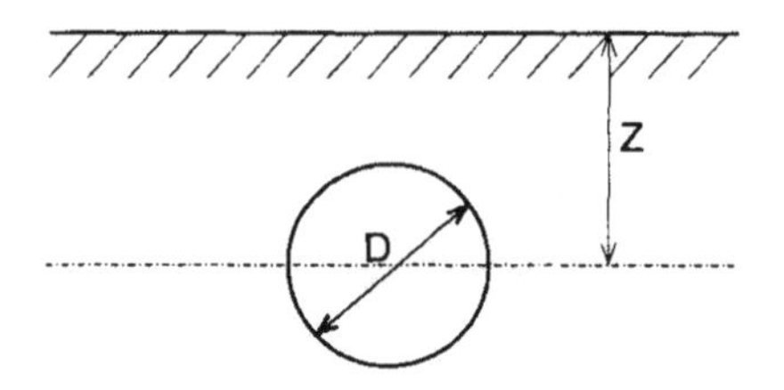

Figure 19.2 Cross section of a buried pipeline.

By using a conduction shape factor for a horizontal cylinder buried in a semi-infinite medium shown in Figure 19.2, the heat transfer coefficient for a buried pipeline can be expressed as,

$$h_{soil} = \frac{k_{soil}}{\left(\frac{D}{2}\right)\cosh^{-1}\left(\frac{2Z}{D}\right)} \tag{19.14}$$

Where: h_{soil} : heat transfer coefficient of soil, BTU/(ft^2·hr·°F) or W/(m^2·K);
k_{soil} : thermal conductivity of soil, BTU/(ft·hr·°F) or W/(m·K);
D : outside diameter of buried pipe, ft or m;
Z : distance between top of soil and center of pipe, ft or m.

For the case of Z > D/2, $\cosh^{-1}(x)$ can be simplified as $\ln[x + (x^2 - 1)^{0.5}]$, therefore,

$$h_{soil} = \frac{2k_{soil}}{D\ln\left[\frac{2Z + \sqrt{4Z^2 - D^2}}{D}\right]} \tag{19.15}$$

Partially Buried Pipeline

The increase of insulation effect for a partially buried pipeline is not large compared with a fully buried pipeline. Heat flows circumferentially through the steel to the section of exposure. Even exposure of just the crown of the pipeline results in an efficient heat transfer to the surroundings due to the high thermal conductivity of the steel pipe. A trenched pipeline (partially buried pipeline) experiences less heat loss than an exposed pipeline but more than a buried pipeline. Engineering judgment must be used for the analysis of trenched pipelines. Seabed currents may be modified to account for the reduced heat transfer, or the heat transfer may be calculated using a weighted average of the fully buried pipe and exposed pipe as,

$$h_o = (1 - f)h_{o,buried} + f\,h_{o,exposed} \tag{19.16}$$

where, f : fraction of outside surface of pipe exposed to the surrounding fluid, [-];

h_o : external heat transfer coefficient, [BTU/(ft^2·hr·°F) or W/(m^2·K)].

For exposed pipeline the theoretical approach assumes that water flows over the top and bottom of the pipeline. For concrete coated pipelines, where the external film coefficient has little effect on the overall heat transfer coefficient, the effect of resting on the seabed is negligible. However for pipelines without concrete coating these effects must be considered.

A pipeline resting on the seabed is normally assumed to be fully exposed. A lower heat loss can give rise to upheaval buckling, increased corrosion and overheating of the coating. It is necessary to take into account changes in burial levels over field life, and hence changes in insulation values.

19.2.4 Soil Thermal Conductivity

Soil thermal conductivity has been found to be a function of dry density, saturation, moisture content, mineralogy, temperature, particle size/shape/arrangement and the volumetric proportions of solid, liquid and air phases. A number of empirical relationships have been developed to estimate thermal conductivity based on these parameters, e.g. Kersten (1949). For a typical unfrozen silt-clay soil, the Kersten correlation is expressed:

$$\kappa_{soil} = [0.9 \cdot \log(\omega) - 0.2] \times 10^{0.01 \times \rho} \qquad (19.17)$$

where, κ_{soil} : soil thermal conductivity, [BTU- in/(ft^2-hr-°F)];

ω : moisture content in percent of dry soil weight;

ρ : dry density, [pcf].

The correlation for the silt-clay was based on five soils and is valid for moisture contents of seven percent or higher.

Table 19.7 Thermal conductivities of typical soil surrounding pipeline (Gregory, 1991).

Material	Thermal conductivity, k_{soil}	
	BTU/(ft-hr-°F)	W/(m-K)
Peat (dry)	0.10	0.17
Peat (wet)	0.31	0.54
Peat (icy)	1.09	1.89
Sand soil (dry)	0.25 – 0.40	0.43 – 0.69
Sandy soil (moist)	0.50 – 0.60	0.87 – 1.04
Sandy soil (soaked)	1.10 –1.40	1.90 – 2.42
Clay soil (dry)	0.20 – 0.30	0.35 – 0.52
Clay soil (moist)	0.40 – 0.50	0.69 – 0.87
Clay soil (wet)	0.60 – 0.90	1.04 – 1.56
Clay soil (frozen)	1.45	2.51
Gravel	0.55 – 0.72	0.9 – 1.25
Gravel (sandy)	1.45	2.51
Limestone	0.75	1.30
Sandstone	0.94 –1.20	1.63 – 2.08

Table 19.7 lists the thermal conductivities of typical soils surrounding pipeline. Whilst the thermal conductivity of onshore soils has been extensively investigated, until recently there has been little published thermal conductivity data for deep-water soil (e.g. Power et al., 1994; von Herzen and Maxwell, 1959). Many deepwater offshore sediments are formed with predominantly silt and clay sized particles, since sand sized particles are rarely transported this far from shore. Hence convective heat loss is limited in these soils and the majority of heat transfer is due to conduction. Recent measurements of thermal conductivity for deepwater soils from the Gulf of Mexico (MARSCO, 1999) have shown values in the range of 0.7 to 1.3 W/(m·K), which is lower than that previously published for general soils and is approaching that of still seawater, 0.65 W/(m·K). This is a reflection of the very high moisture contents of many offshore soils, where liquidity indices well in excess of unity can exist, which are rarely found onshore. Although site-specific data is needed for the detailed design most deepwater clay is fairly consistent. Figure 19.3 shows the thermal conductivity test results of subsea soil worldwide. The sample results prove that most deepwater soils have similar characteristics.

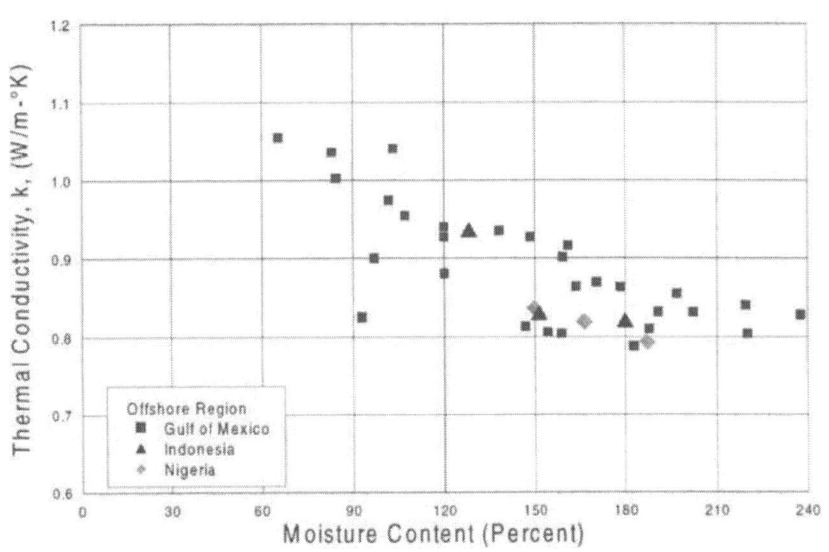

Figure 19.3 Soil thermal conductivity test results for worldwide offshore (Young *et al*, 2001).

19.3 U-value

19.3.1 Overall Heat Transfer Coefficient

Figure 19.4 shows the temperature distribution of a cross-section for a composite subsea pipeline with two insulation layers. Radiation between internal fluid and pipe wall; flowline outer surface and environment is ignored because of the relatively low temperature of subsea system. The following convection and conduction occur:

- convection from the internal fluid to the flowline pipe wall;
- conduction through the pipe wall and exterior coatings, and/or to the surrounding soil for buried pipelines.
- convection from flowline outer surface to the external fluid.

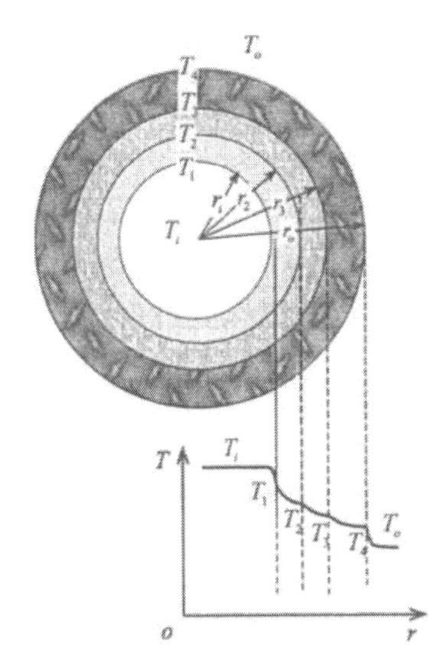

Figure 19.4 Cross-section of insulated pipe and temperature distribution.

For internal convection at the flowline inner-surface, the heat transfer rate across the surface boundary is given by Newton equation,

$$Q_i = A_i \cdot h_i \cdot \Delta T = 2\pi\, r_i\, L\, h_i\, (T_i - T_1) \tag{19.18}$$

where, Q_i : convection heat transfer rate at internal surface, Btu/hr or W;

h_i : internal convection coefficient, BTU/(ft^2·hr·^{o}F) or W/(m^2·K);

r_i : internal radius of flowline, ft or m;

L : flowline length, ft or m;

A_i : internal area normal to the heat transfer direction, ft^2 or m^2;

T_i : internal fluid temperature, ^{o}F or ^{o}C;

T_1 : temperature of flowline internal surface, ^{o}F or ^{o}C.

For external convection at flowline outer-surface, the heat transfer rate across the surface boundary to the environment is,

$$Q_o = A_o \cdot h_o \cdot \Delta T_o = 2\pi\, r_o\, L\, h_o\, (T_4 - T_o) \tag{19.19}$$

where, Q_o : convection heat transfer rate at outer surface, Btu/hr or W;

h_o : outer convection coefficient, Btu/(ft^2·hr·^{o}F) or W/(m^2·K);

r_o : outer radius of flowline, ft or m;

A_o : outer area normal to the heat transfer direction, ft^2 or m^2;

T_o : environment temperature, ^{o}F or ^{o}C;

T_4 : outer surface temperature of flowline, ^{o}F or ^{o}C.

Conduction in the radial direction of a cylinder can be described by Fourier's equation in radial coordinates as,

$$Q_r = -2\pi\, r\, L\, k \frac{\partial T}{\partial r} \tag{19.20}$$

where, Q_r : conduction heat transfer rate in radial direction, Btu/hr or W;

r : radius of cylinder, ft or m;

k : thermal conductivity of cylinder, BTU/(ft-hr-°F) or W/(m-K);

$\partial T / \partial r$: temperature gradient, °F/ft or °C/m.

Integration of Equation (19.20) gives,

$$Q_r = \frac{2\pi\, L\, k(T_1 - T_2)}{\ln(r_2 / r_1)} \tag{19.21}$$

Temperature distribution in radial direction can be calculated for a steady state heat transfer between the internal fluid and pipe surroundings where heat transfer rates of internal convection, external convection and conduction heat transfer are same and following heat transfer rate equation is obtained,

$$Q_r = \frac{T_i - T_o}{\dfrac{1}{2\pi r_i L h_i} + \dfrac{\ln(r_1/r_i)}{2\pi k_1 L} + \dfrac{\ln(r_2/r_1)}{2\pi k_2 L} + \dfrac{\ln(r_o/r_2)}{2\pi k_3 L} + \dfrac{1}{2\pi r_o L h_o}} \tag{19.22}$$

Heat transfer rate through a pipe section with length of L, due to a steady state heat transfer between the internal fluid and the pipe surroundings, is also expressed as,

$$Q_r = UA(T_i - T_o) \tag{19.23}$$

where, U : overall heat transfer coefficient (OHTC), based on the surface area A, BTU/(ft^2·hr·°F) or W/(m^2·K);

A : area of heat transfer surface, A_i or A_o, ft^2 or m^2;

T_o : ambient temperature of the pipe surroundings, °F or °C;

T_i : average temperature of the flowing fluid in the pipe section, °F or °C.

Therefore, the OHTC based on the flowline internal surface area A_i is,

$$U_i = \frac{1}{\dfrac{1}{h_i} + \dfrac{r_i \ln(r_1/r_i)}{k_1} + \dfrac{r_i \ln(r_2/r_1)}{k_2} + \dfrac{r_i \ln(r_o/r_2)}{k_3} + \dfrac{r_i}{r_o h_o}} \tag{19.24}$$

and the OHTC based on the flowline outer surface area A_o,

$$U_o = \frac{1}{\dfrac{r_o}{r_i h_i} + \dfrac{r_o \ln(r_1/r_i)}{k_1} + \dfrac{r_o \ln(r_2/r_1)}{k_2} + \dfrac{r_o \ln(r_o/r_2)}{k_3} + \dfrac{1}{h_o}} \tag{19.25}$$

The OHTC of the flowline is also called the U-value of the flowline in subsea engineering. It is a function of many factors, including the fluid properties and fluid flow rates, the convection nature of the surroundings, and the thickness and properties of the pipe coatings and insulation. Insulation manufacturers typically use a U-value based on the OD of pipeline, while pipeline designers use a U-value based on the ID. The relationship of these two U-values is,

$$U_o \times \mathrm{OD} = U_i \times \mathrm{ID} \tag{19.26}$$

The *U*-value for a multi-layer insulation coating system is easily obtained from an electrical-resistance analogy between heat transfer and direct current. The steady-state heat transfer rate is determined by,

$$Q_r = UA(T_i - T_o) = (T_i - T_o)/\sum R_i \tag{19.27}$$

where, *UA* is correspondent with the reverse of the cross-section's thermal resistivity that comprises three primary resistances: internal film, external film and radial material conductance. The relationship is written as follows:

$$\frac{1}{UA} = \sum R_i = R_{film,in} + R_{pipe} + \sum R_{coating} + R_{film,ext} \tag{19.28}$$

The terms on the right hand side of the above equation represent the heat transfer resistance due to internal convection, conduction through steel well of pipe, conduction through insulation layers and convection at the external surface. They can be expressed as follows.

$$R_{film,in} = \frac{1}{h_i A_i} \tag{19.29}$$

$$R_{pipe} = \frac{\ln(r_1/r_i)}{2\pi L k_{pipe}} \tag{19.30}$$

$$\sum R_{coating} = \frac{\ln(r_{no}/r_{ni})}{2\pi L k_n} \tag{19.31}$$

$$R_{film,ext} = \frac{1}{h_o A_o} \tag{19.32}$$

where, r_{no} and r_{ni} represent the outer radius and inner radius of the coating layer *n*, respectively. k_n is the thermal conductivity of coating layer *n*, BTU/(ft-hr-°F) or W/(m-K); The use of U-values is appropriate for steady-state simulation. However, U-values cannot be used to evaluate transient thermal simulations; the thermal diffusion coefficient and other material properties of the wall and insulation must be included.

19.3.2 Achievable U-values

Following U-values may be used as the lowest possible value for initial design purposes,

- conventional insulation: 0.5 BTU/ft²·hr·°F (2.8 W/m²·K);
- polyurethane Pipe-in-Pipe: 0.2 BTU/ft²·hr·°F (1.1 W/m²·K);
- ceramic insulated Pipe-in-Pipe: 0.09 BTU/ft²·hr·°F (0.5 W/m²·K);
- insulated bundle: 0.2 BTU/ft²·hr·°F (1.1 W/m²·K).

Insulation materials exposed to ambient pressure must be qualified for the water depth where they are to be used.

19.3.3 U-value for Buried Pipe

For insulated flowlines, thermal insulation coating and/or burial provide an order of magnitude more thermal resistance than both internal and external film coefficients. Therefore, the effects of internal and external film coefficients to U-value of the flowlines can be ignored and the U-value of the flowlines can be descried as,

$$U_i = \frac{1}{\sum \frac{r_i \ln(r_{no}/r_{ni})}{k_n} + \frac{r_i \cosh^{-1}(2Z/D_o)}{k_{soil}}} \tag{19.33}$$

where the first term in the denominator represents the thermal resistance of the radial layers of steel and insulation coatings, and the second term in the denominator represents the thermal resistivity due to the soil and is valid for $H > D_o/2$. If the internal and external film coefficients are to be included, the external film coefficient is usually only included for non-buried flowlines. For buried bare flowlines, where the soil provides essentially all the flowline's thermal insulation, there is a near linear relationship between U-value and k_{soil}. For buried and insulated flowlines, the effect of k_{soil} on the U-value is less than for buried bare flowlines.

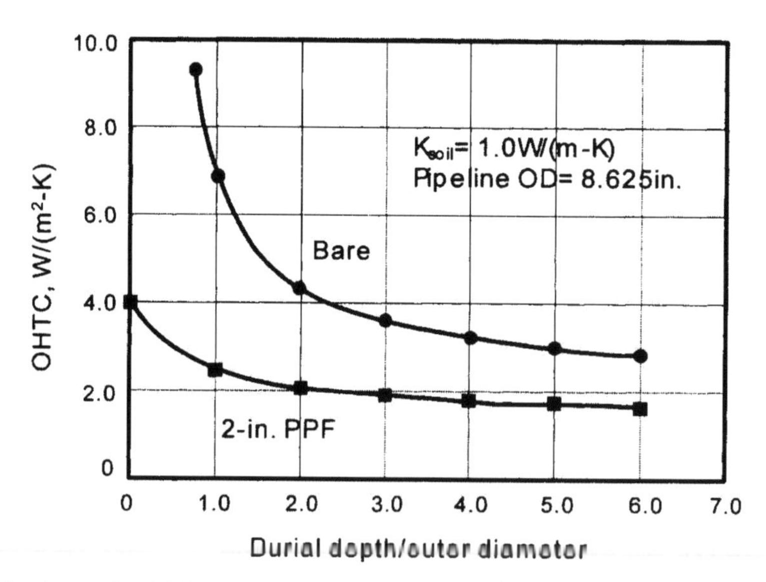

Figure 19.5 U-value vs. burial depth for bare pipe and 2 inch PPF coated pipe (Loch, 2000).

Figure 19.5 presents the relationship between U-value and burial depth for both a bare pipe and a 2-inch PPF coated pipe. The U-value decreases very slowly with the burial depth, when the ratio of burial depth/outer diameter is greater than 4.0. Therefore, it is not necessary to bury the flowline to a great depth to get a significant thermal benefit. However, there are practical minimum and maximum burial depth limitations for using modern flowline burial techniques. In addition, issues such as potential seafloor scouring and flowline upheaval buckling need to be considered in designing the appropriate flowline burial depth.

19.4 Steady State Heat Transfer

Steady state thermal design is aimed at ensuring a given maximum temperature drop over a length of insulated pipe. In the steady state condition, the flow rate of flowing fluid and heat loss through pipe wall from the flowing fluid to the surroundings is assumed to be a constant at any time.

19.4.1 Temperature Prediction along Pipeline

The accurate predictions of temperature distribution along a pipeline can be calculated by coupling velocity, pressure and enthalpy given by mass, momentum, and energy conservation. The complexity of these coupled equations prevents an analytical solution, but they can be solved by a numerical procedure.

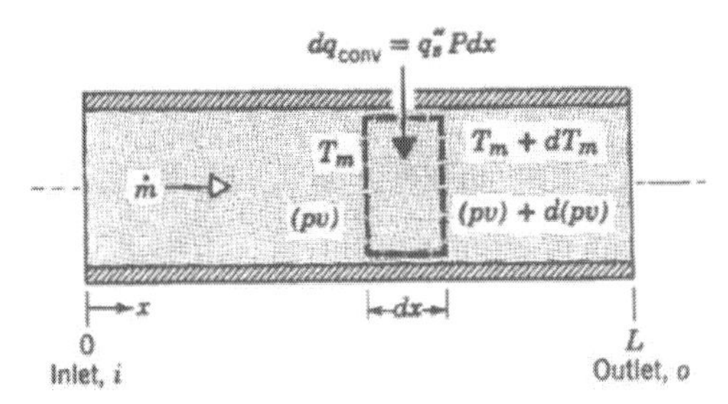

Figure 19.6 Control volume for internal flow in pipe (Incropera and DeWitt, 1996).

Figure 19.6 shows a control volume of internal flow for a steady-state heat transfer analysis. In a fully developed thermal boundary layer region of flowline, the velocity and temperature radial profiles of a single-phase fluid are assumed to be approximately constant along the pipeline. A mean temperature can be obtained by integrating the temperatures over the cross section, and the mean temperature profile of fluid along the pipeline is obtained by the energy conservation between the heat transfer rate through pipe wall and the fluid thermal energy change as the fluid is cooled by conduction of heat through the pipe to the sea environment, which is expressed as,

$$UdA(T_m - T_o) = -\dot{m}c_p dT_m \tag{19.34}$$

where, dA : surface area of pipeline, $dA = Pdx = \pi\, Ddx$, ft^2 or m^2;

$\dot{m}$: mass flow rate of internal fluid, lb/s or kg/s;

c_p : specific heat capacity of internal fluid, Btu/(lb ·°F) or J/(kg·K);

U : U-value based on pipe diameter D, Btu/ft^2·hr·°F, W/m^2·K;

D : ID or OD, ft or m;

T_m : mean temperature of fluid, °F or °C.

Separating variables and integrating from pipeline inlet to the point with pipe length of x from inlet,

$$\int_{T_{in}}^{T(x)} \frac{dT_m}{T_m - T_o} = \int_0^x -\frac{U\pi Ddx}{\dot{m}c_p} \tag{19.35}$$

where, T_{in} : inlet temperature of fluid, °F or °C;

$T(x)$: temperature of fluid at distance of x from inlet, °F or °C.

From integration of Equation (19.35), the following temperature profile equation is obtained,

$$T(x) = T_o + (T_{in} - T_o)\exp(\frac{-U\pi Dx}{\dot{m}c_p}) \quad (19.36)$$

Here, the thermal decay constant, β is defined as,

$$\beta = \frac{U\pi DL}{\dot{m}c_p} \quad (19.37)$$

where, L : total pipeline length, ft or m. However, if two temperatures and their corresponding distance along the route are known, fluid properties in the thermal decay constant are not required. The inlet temperature and another known temperature are used to predict the decay constant and then develop the temperature profile for the whole line based on the exponential temperature profile that is given by the following equation,

$$T(x) = T_o + (T_{in} - T_o)\exp\left(\frac{-x\beta}{L}\right) \quad (19.38)$$

If the temperature drop due to Joule-Thompson cooling is taken into account and a single-phase viscous flow is assumed, the temperature along the flowline may be expressed as:

$$T(x) = T_o + (T_{in} - T_o)\exp\left(\frac{-x\beta}{L}\right) + \frac{[p(x) - p_{in}]/\rho - g[z(x) - z_{in}]}{c_p} \quad (19.39)$$

where, p_{in} : fluid pressure at pipe inlet, Pa;

$p(x)$: fluid pressure at point with distance x from inlet, Pa;

z_{in} : flowline elevation at pipe inlet, m;

$z(x)$: flowline elevation at point with distance x from inlet, m;

g : acceleration due to gravity, m/s^2;

ρ : density of fluid, kg/m^3.

19.4.2 Steady State Insulation Performance

The following insulation methods of subsea flowlines are used in the field,

- external coatings;
- flowline burial;
- pipe-in-pipe (PIP);
- electrical heating;
- hot water annulus.

The most popular methods used in subsea flowlines are external coatings, pipe-in-pipe and flowline burial.

Table 19.8 presents the U-values for various pipe size and insulation system combinations. Bare pipe has little resistance to heat loss, but buried bare pipe is equivalent to adding approximately 60 mm of external coating of polypropylene foam (PPF). Combining burial and coating (B & C) achieves U-values near those of PIP. If lower soil thermal conductivity and/or thicker external coating are used, the B & C system can achieve the same level U-value as typical PIP. Considering realistic fabrication and installation limitations, PIPs are chosen and designed in most cases to have a lower U-value than B & C. However, the fully installed costs of equivalent PIP systems are estimated to be more than double those of B & C systems.

Table 19.8 U-values for different subsea insulation systems (Loch, 2000).

Flowline OD [inch]	U-value [W/(m2 ·K)]				
	Bare Pipe	Coated Pipe t =50 mm PPF k=0.17 W/(m·K)	Buried Pipe H =1.3 m k=1.0 W/(m·K)	Buried & Coated	Carrier OD of PIP 8.625 inch* 12.75 inch** 18.0 inch***
4.5*	871.0	4.65	4.56	2.52	1.56
8.625**	544.0	4.04	2.87	1.81	1.44
12.75***	362.0	3.81	2.22	1.5	1.02

19.5 Transient Heat Transfer

Transient heat transfer occurs in subsea flowline during shut-down (cooldown) and start-up scenarios. In shut-down scenarios, the energy, kept in the system at the moment the fluid flow stops, goes to surrounding environment through pipe wall. This is no longer a steady state system and the rate at which the temperature drops with time becomes important to hydrate control in flowline.

Flowline systems are required to be designed for hydrate control in the cooldown time, which is defined as the period before the flowline temperature reaches the hydrate temperature at the flowline operating pressure. This period provides the operators with a decision time in which to commence hydrate inhibition or flowline depressurization. In the case of an emergency shut down or cool down, it also allows for sufficient time to carry out whatever remedial action required before the temperature reaches the hydrate formation temperature. Therefore it is of interest to be able to predict how long the fluid will take to cool down to any hydrate formation temperature with a reasonable accuracy. When a pipeline is shutdown for an extended period of time, generally it is flushed (blowdown) or vented to remove the hydrocarbon fluid, since the temperature of the system will eventually come to equilibrium with the surroundings.

The cooldown process is a complex transient heat transfer problem, especially for multiphase fluid systems. The temperature profiles over time may be obtained by considering the heat energy resident in the fluid and the walls of the system, $\sum \dot{m} C_p \Delta T$, the temperature gradient over the walls of the system, and the thermal resistances hindering the heat flow to the surroundings. The software OLGA is widely used for numerical simulation of this process. However, OLGA software generally takes several hours to do these simulations; in many preliminary design cases, an analytic transient heat transfer analysis for the flowline, e.g. lumped capacitance method is fast and with a reasonable accuracy.

19.5.1 Cool Down

The strategies for solving the transient heat transfer problem of subsea flowline system are analytical method (lumped capacitance method), finite difference method (FDM) and finite element method (FEM); those methods are chosen depending on the complexity of problem. The analytical methods are used in many simple cool down or transient heat conduction problems. Finite difference method is by far a relatively fast and gives a reasonable accuracy. The finite element method is more versatile and better for complex geometries, however, it is also more demanding to implement. A pipe has a very simple geometry, and hence obtaining a solution to the cool-down rate question does not require the versatility of the FEA method. For such systems, the FDA approach is convenient and has adequate accuracy. Valves, bulkheads, Xmas trees etc have more complex geometries and require the greater versatility and 3-D capability of FEA models.

Lumped capacitance method

A commonly used mathematical model for the prediction of system cooldown time is the "lumped capacitance model", which may be expressed as follows:

$$T - T_o = (T_i - T_o)\, e^{-U\pi DLt / \sum mC_p} \tag{19.40}$$

where, T : temperature of the body at time t, °F or °C;

t : time, s;

T_o : ambient temperature, °F or °C;

T_i : initial temperature, °F or °C;

D : inner diameter or outer diameter of the flowline, ft or m;

L : Length of the flowline, ft or m;

U : U-value of the flowline based on diameter D, Btu/ft^2·hr·°F or W/m^2·K;

m : mass of internal fluid or coating layers, lb or kg;

c_p : specific heat capacity of fluid or coating layers, Btu/(lb·°F) or J/(kg·K).

However, this model assumes a uniform temperature distribution throughout the object at any moment during the cooldown process, and that there is no temperature gradient inside of the object. This assumption means that the surface convection resistance is much larger than the internal conduction resistance, or the temperature gradient in the system is negligible. This model is only valid for Biot (Bi) numbers less than 0.1. The Bi number is defined as the ratio of internal heat transfer resistance to external heat transfer resistance and may be written as:

$$Bi = \frac{L_c U}{k} \tag{19.41}$$

where, L_c : characteristic length, ft or m;

k : thermal conductivity of the system, Btu/ft·hr·°F or W/m·K .

Since most risers and flowlines are subject to large temperature gradients across their walls and are also subject to wave and current loading, this leads to a Bi number outside the applicable range of the "lumped capacitance" model. A mathematical model is therefore required which takes into account the effects of external convection on the transient thermal response of the system during a shutdown event.

Finite difference method:

Considering a pipe segment shown in Figure 19.4 which has a length L, it is assumed that the average fluid temperature in the segment is $T_{f,o}$ at a steady state flowing condition. The temperature of the surroundings is assumed to be T_a, and constant.

Under the steady state condition the total mass of the fluid in the pipe segment is given by,

$$W_f = \frac{\pi}{4} D_i^2 L \rho_f \tag{19.42}$$

where, W_f : fluid mass in the pipe segment, lb or kg;

D_i : pipe inside diameter, in or m;

L : length of the pipe segment, in or m;

ρ_f : average density of the fluid in the pipe segment when the temperature is $T_{f,o}$, lb/in^3 or kg/m^3.

W_f is assumed to be constant in all time after the shutdown. At any given time, the heat content of the fluid in the pipe segment is given by,

$$Q_f = W_f C_{p,f} \left(T_f - T_{ref}\right) \tag{19.43}$$

where, $C_{p,f}$: specific heat of the fluid, BTU/(lb·°F) or kJ/(kg·°C);

T_{ref} : a constant reference temperature, °F or °C;

T_f : average temperature of fluid in the pipe segment, °F or °C.

The heat content of pipe wall and insulation layers can be expressed in the same way,

$$Q_{pipe} = W_{pipe} C_{p,pipe} \left(T_{pipe} - T_{ref}\right)$$

$$Q_{layer} = W_{layer} C_{p,layer} \left(T_{layer} - T_{ref}\right) \tag{19.44}$$

An energy balance is applied to the control volume, and is used to find the temperature at its centre. It is written in the two forms as below, the first without Fourier's law of heat conduction. Notice from the equation that during a cool-down the quantity of heat leaving is larger than the heat entering.

The heat transfer along the axial and circumferential directions of pipeline system can be ignored, the transient heat transfer without heat source occurs only in the radial direction. The 1st. law of thermodynamics states that the change of energy inside a control volume is equal to the heat going out minus the heat coming in. Therefore, the heat transfer balance for all parts (including contained fluid, steel pipe, insulation materials and coating layers) of the pipeline can be explicitly expressed in following equation,

$$Q_{i,1} - Q_{i,0} = -(q_{i+1/2} - q_{i-1/2})\Delta t \tag{19.45}$$

where, $Q_{i,0}$: heat content of the part i at one time instant, BTU or J;

$Q_{i,1}$: heat content of the part i after one time step, BTU or J;

$q_{i+1/2}$: heat transfer rate from part i to part i+1 at the "0" time instant, BTU/hr or W;

$q_{i-1/2}$: heat transfer rate from part i-1 to part i at the "0" time instant, BTU/hr or W;

Δt : time step, second or hour.

If the heat transfer rate through all parts in pipeline system in the radial direction is expressed with the thermal resistant concept, then

$$q_{i+1/2} = \frac{A_o (T_{i+1,0} - T_{i,0})}{R_i/2 + R_{i+1}/2} \tag{19.46}$$

$$q_{i-1/2} = \frac{A_o (T_{i,0} - T_{i-1,0})}{R_i/2 + R_{i-1}/2} \tag{19.47}$$

Therefore, the average temperatures for a pipeline with two insulation layers after one time step are rewritten as,

$$T_{f,1} = T_{f,0} - \frac{A_o \Delta t}{W_f C_{p,f}} \left[\frac{T_{f,0} - T_{p,0}}{R_{\text{int}} + R_p/2} \right] \tag{19.48}$$

$$T_{p,1} = T_{p,0} - \frac{A_o \Delta t}{W_p C_{p,p}} \left[\frac{T_{p,0} - T_{f,0}}{R_{\text{int}} + R_p/2} + \frac{T_{p,0} - T_{l1,0}}{R_{l1}/2 + R_p/2} \right] \tag{19.49}$$

$$T_{l1,1} = T_{l1,0} - \frac{A_o \Delta t}{W_{l1} C_{p,l1}} \left[\frac{T_{l1,0} - T_{l2,0}}{R_{l2}/2 + R_{l1}/2} + \frac{T_{l1,0} - T_{p,0}}{R_{l1}/2 + R_p/2} \right] \tag{19.50}$$

$$T_{l2,1} = T_{l2,0} - \frac{A_o \Delta t}{W_{l2} C_{p,l2}} \left[\frac{T_{l2,0} - T_{l1,0}}{R_{l2}/2 + R_{l1}/2} + \frac{T_{l2,0} - T_a}{R_{l2}/2 + R_{surr}} \right] \tag{19.51}$$

The solution procedure of finite difference method is based on a classical implicit procedure using matrix inversion. The main principal of solving is to find the temperature at each node using the energy equation above. Once the whole row of temperatures are found at the same time step by matrix inversion, the procedure jumps to the next time step. Finally, the temperature at all the time steps is found and the cool down profile is established. Detailed calculation method and definitions are described in a Mathcad worksheet at the Appendix of this Chapter. To get an accurate transient temperature profile, the fluid and each insulation layer may be divided into several layers and the above calculation method applied. By setting the exponential temperature profile along the pipeline from steady-state heat transfer analysis as an initial temperature distribution for a cooldown transient analysis, the temperature profiles along the pipeline at different times can be calculated. This methodology is easily programmed as an Excel macro or with C++ or other language, but the more layers the insulation is divided into, shorter time step required to obtain a convergent result, because explicit difference is used in the time domain. The maximum time step may be calculated using following equation,

$$\Delta t_{max} = \min\left(\frac{W_i C_{p,i} R_i}{A_0}\right) \quad \text{for all parts or layers } i \tag{19.52}$$

If the layer is very thin, R_i will be very small, which controls the maximum time step. In this case, the thermal energy saved in this layer is very small and may be ignored. The model implemented here conservatively assumes that the content of the pipe has thermal mass only, but no thermal resistance. The thermal energy of the contents is immediately available to the steel. The effect of variations of C_p and k for all parts in the system is of great importance to the accuracy of the cool-down simulation. For this reason, these effects are either entered in terms of fitted expressions or fixed measured quantities. Figure 19.7 shows the variation of measured specific heat capacity for polypropylene solid with material temperature.

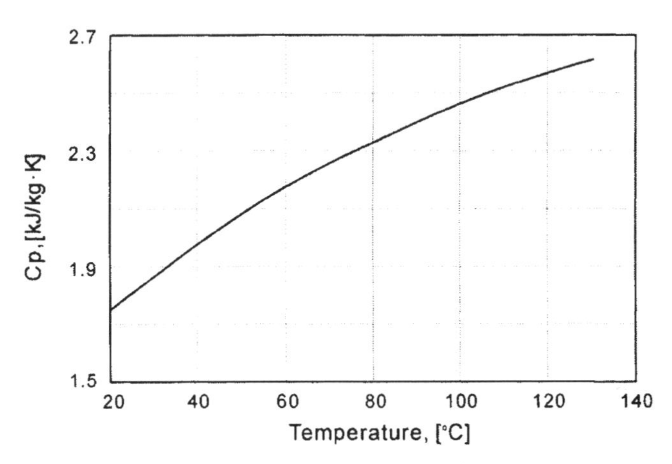

Figure 19.7 Specific heat capacity measurements for generic polypropylene solid.

Guo *et. al* (2004) developed a simple analysis model for predicting heat loss and temperature profiles in insulated thermal injection lines and wellbores. The concept of this model is similar with the finite difference method which is described in this section. The model can accurately predict steady and transient heat transfer of well tube and pipeline.

19.5.2 Transient Insulation Performance

While steady state performance is generally the primary metric for thermal insulation design, transient cool-down is also important, especially when hydrate formation is possible. Figure 19.10 presents the transient cool down behavior at the end of a 27 km flowline. For the case presented in the figure, the PIP flowline had a slightly lower U-vale than that of the B&C flowline. Therefore, the steady state temperature at the base of the riser is slightly higher for the PIP than the B&C flowline. These steady state temperatures are the initial temperatures at the start of a transient shutdown. But the temperature of B&C system quickly surpasses the PIP system and provides much longer cool down times in the region of typical hydrate formation. For the case considered, B&C flowline had approximately three times the cool down time of PIP before hydrate formation. The extra time available before hydrate formation has significant financial implications for offshore operations, such as reduced chemical injection, reduced line flushing and reduced topside equipments.

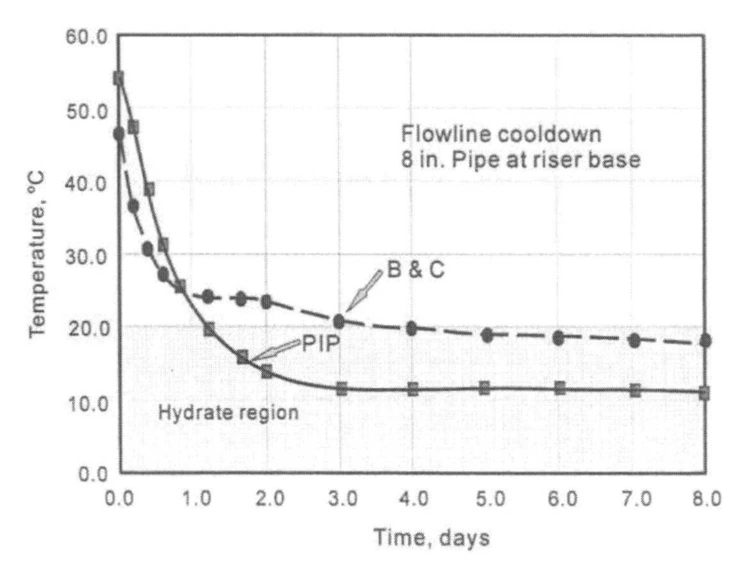

Figure 19.10 Transient cool down behaviour between burial pipe and PIP (Loch, 2000).

The temperature of PIP at the initial cool down condition is higher than that of B & C flowline, but the cooldown curves cross as shown in Figure 19.10. In addition to the U-value, the fluid temperature variation with time is dependent on the fluid's specific heat, C_p, its mass flow rate and the ambient temperature.

19.6 Thermal Management Strategy and Insulation

As oil/gas production fields move into deeper water there is a growing need for thermal management to prevent the build up of hydrate and wax formations in subsea systems. The thermal management strategy is chosen depending on the required U-value, cool-down time, temperature range and water depth. Table 19.9 summarizes the advantages, disadvantages and U-value range of the thermal management strategies used in subsea systems.

Figure 19.11 summarizes subsea pipeline systems with respect to length and U-value, which concludes:

- U-values of single pipe insulation systems are higher than 2.5 W/m^2·K, and mostly installed using reeled pipe installation methods;
- About 6 different operation PIP systems with length less than 30 km;
- Reeled PIP systems have dominated the 5-20 km market with U-values of 1-2 W/m^2·K;
- PIP system in development with length of 66 km offers U value of 0.5-1.0 W/m^2·K;
- About 3 bundles used on tie-backs of over 10 km. With Britannia also incorporating a direct heating system.

Table 19.9 Summary of thermal management for subsea systems (Grealish and Roddy, 2002).

Thermal management method	Advantages	Disadvantages	U-value Range (W/m^2K)
Integral external insulation (including multilayer insulation)	• suitable for complex geometries • high temperature resistance • multi-layer systems can tailor configuration to achieve desired properties. • solid systems have almost unlimited water depth	• difficult to remove • limits on level of insulation that can be achieved • solid systems have almost no buoyancy • need to allow time to set • limits on thickness	1.7-5.0
Insulation modules /bundle systems	• accommodation of auxiliary lines • can remove during service • lower cost than PIP systems • simple implementation to flowlines • water depth limits lower than PIP or solid integral external insulation (except for bundle systems protected by outer pipe)	• gaps between modules can lead to convection currents • large amounts of material required which can be prohibitive • may be a complex geometry to insulate	2.0-8.0
PIP insulation	• best insulation properties (except for vacuum system) • water depth limits defined by capacity of installation vessels	• expensive to install and fabricate • some PIP systems cannot be reeled	0.3-1.5
Vacuum Systems	• ultimate insulation properties • only heat transfer by radiation	• if outer pipe is breached then all insulation properties lost for pipe section. • effort required in preserving vacuum • vacuum level in annulus must be continuously monitored.	<0.01
Burial/ trenching	• surrounding soil provides some of the insulation • provides on-bottom stability for flowlines	• may not be practical to dig trenches for burial • only applicable where large amounts of pipe are lying on the seabed • thermal conductivity of soil varies significantly depending on location • thermal performance uncertain due to variance in soil properties and difficulty in measuring or predicting them.	insulation layers + soil insulating
Electrical heating	• active system; can control cooldown time during shutdown	• power requirements may be unfeasible for project • may need to electrically insulate flowline	depends on power supplied
Hot water/ oil systems	• active system; can control cooldown time during shutdown	• with greater depth longer tie-backs will lose more heat thus reducing thermal efficiency • separate return line required	depends on power supplied
Thin–film/ multi-layer (Horn and Lively, 2001)	• conjunct with Vacuum Insulated Tubing (VIT), get a high thermal resistance. The total thickness of the layers is less than 3 mm	• with a very low thermal storage, causing a fast cool down, suitable for trees, jumpers, manifolds, flowlines, risers, and wellbore tubing.	used in BP Marlin, Shell Oregano and Serrano.

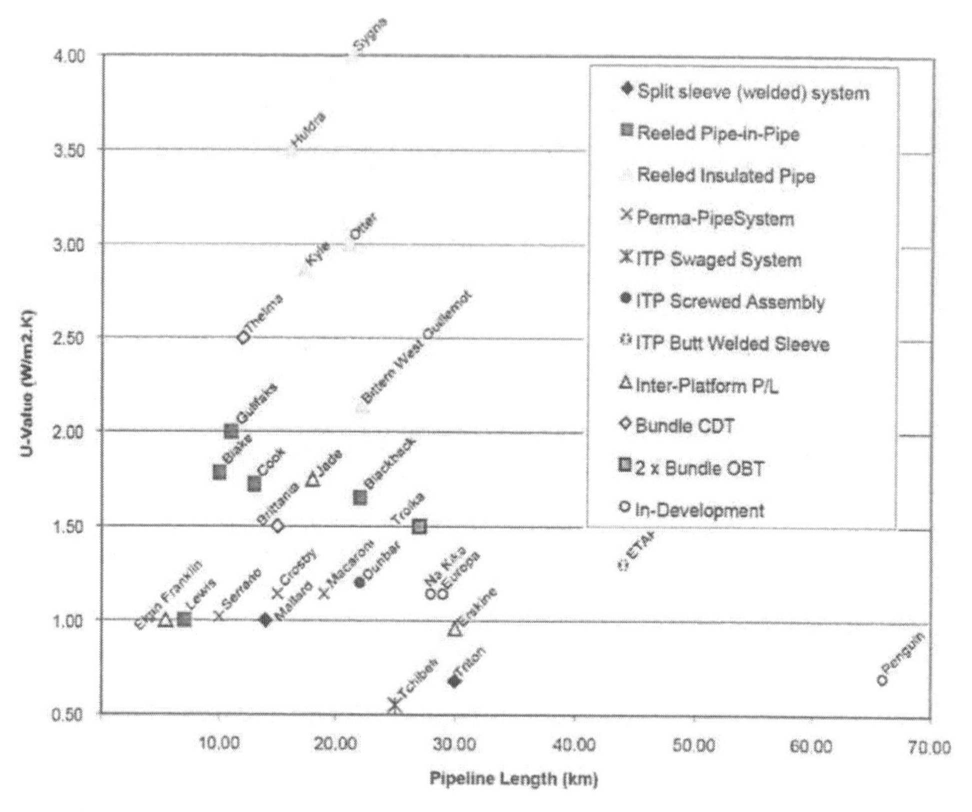

Figure 19.11 Field development vs. pipeline length and U-value (McKechnie and Hayes, 2001).

19.6.1 External Insulation Coating System

Figure 19.12 shows a typical multi-layer coating system which combines the foams with good thermal insulating properties and PP shield with creep resistance. These coatings vary in thickness from 25mm up to 100mm or more. Typically thicknesses over about 65mm are applied in multiple layers.

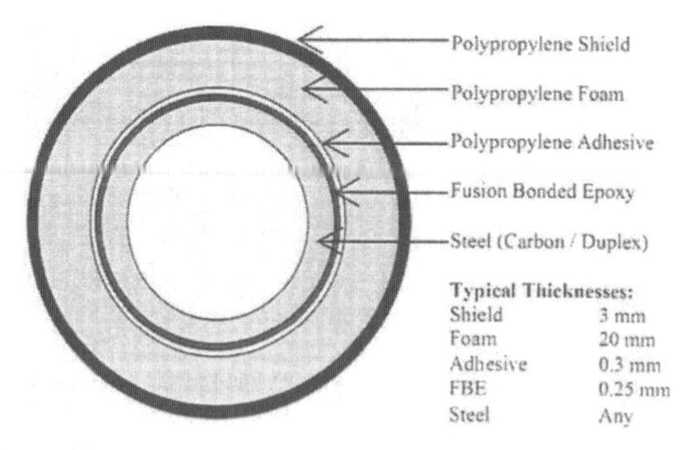

Figure 19.12 Typical multi-layers insulation system (McKechnie and Hayes, 2001).

Coating systems are usually limited by a combination of operating conditions including: temperature, water depth and water absorption. Combinations of temperature and hydrostatic pressure can cause creep and water absorption, with resultant compression of the coating and a continuing reduction in insulating properties throughout the design life. These issues need to be accounted for during design.

Table 19.10 Possible coating systems for thermal insulation of flowlines (Melve, 2002).

Flowline	Temp. Range [°C]	Insulation System
Low temperature	-10 – 70	Polyurethane (PU), Polypropylene (PP), Filled rubbers, Syntactic foams based on epoxy and polyurethanes.
Medium temperature	70 – 120	PP, Rubbers, Syntactic epoxy foams.
High temperature	120 – 200	PP systems, Phenolic foams, PIP with poly-urethane foam or inorganic insulating materials.

Flowlines may be classified according to the flowline temperature. Table 19.10 shows the temperature range and corresponding insulation materials. To find suitable thermal insulation for high temperature subsea flowlines is an extra challenge compared to lower temperature applications.

19.6.1.1 Insulation material

Table 19.11 summarizes the insulation types, limitations and characteristics of the most commonly applied and new systems being offered by suppliers. The systems have progressed significantly in the past 10 years on issues such as: thickness limits, water depth limitations, impact resistance, operating temperature limits, creep with resultant loss of properties, suitability for reel lay installation with high strain loading, life expectancy and resultant U-values and submerged weight.

Brief descriptions of external insulation systems for subsea systems are given below:

- Polypropylene (PP); polyolefin system with relatively low thermal efficiency. Applied in 3-4 layers, can be used in conjunction with direct heating systems. Specific heat capacity 2000 J/(kg·K);
- Polypropylene-Syntactic PPF (SPP); controlled formation of gas bubbles are used to reduce the thermal conductivity, however as ocean depth and temperature increase, rates of compression and material creep increase. Generally not used on its own;
- Polypropylene-Reinforced Foam Combination (RPPF); a combination of the above two systems incorporating a FBE layer, PPF and an outer layer of PP to minimize creep and water absorption;
- Polyurethane (PU); a polyolefin system with relatively low thermal efficiency; has a relatively high water absorption as the temperature increases over 50 °C. Commonly applied with modified stiffness properties to field joints;
- Polyurethane-Syntactic (SPU); one of the most commonly applied systems over recent years; offers good insulation properties at water depths less than 100 m. Specific heat capacity, 1500 J/(kg·K);

- Polyurethane-Glass Syntactic (GSPU); similar to SPU but incorporates glass providing greater creep resistance; specific heat capacity 1700 J/(kg·K).
- Phenolic Syntactic (PhS), Epoxy Syntactic (SEP) & Epoxy Syntactic with Mini-Spheres (MSEP); based on epoxy and phenolic materials which offer improved performance at higher temperatures and pressures. These materials are generally applied by trowel, pre-cast or poured into moulds. For example SEP was used on the 6-mile King development by pouring the material under a vacuum into polyethylene sleeves. Specific heat capacity 1240 (J/kg·K).

PP and PU are the main components of insulation coating systems. The syntactic insulations, incorporating spheres are used to improve insulation strength for hydrostatic pressure.

19.6.1.2 Structural issues

Installation and Operation Loads:

Coating systems for installation through reeling need careful selection and testing, as some systems have experienced cracking particularly at field joints where there is a natural discontinuity in the coating. This can lead to strain localization and pipe buckling. Laying and operation of sub-sea pipelines requires the load transfer through the coating to the steel pipe and from the steel pipe to the coating. The coating should have a sufficient shear load capacity to hold the steel pipe during the laying process. Thermal fluctuations in the operation process lead to expansion and contraction of the pipe. The thermal insulation coating is required to follow such changes without being detached or cracked. These requirements have resulted in most thermal insulation systems being based on bonded geometries, comprised of several layers or cast-in shells.

Hydrostatic Loads:

Deepwater flowlines are subject to significant hydrostatic loading due to the high water depths. The thermal insulation system must be designed to withstand the large hydrostatic loads. The layer of foamed thermal insulation is the weakest member of the insulation system and therefore the structural response of the insulation system will depend largely on this layer. Elastic deformation of the system leads to a reduction in volume of the insulation system, and an increase in density and thermal conductivity of the foam layer, which leads to an increase in the U-value of the system. In order to reduce the elastic deformation of the system stiffer materials are required. The stiffness of the foams increases with increasing density. Therefore to reduce elastic deformation of the system, higher density foams are required.

Creep:

Thermal insulation operating in large water depths for long periods (e.g. 20/30 years) can be subject to significant creep loading. Creep of thermal insulation can lead to some damaging effects, such as structural instability leading to collapse, and loss of thermal performance. Creep of polymer foams leads to a reduction in volume or densification of the foam. The increased density of the material leads to an increased thermal conductivity and therefore, the effectiveness of the insulation system is reduced. To design effective thermal insulation systems for deepwater applications the creep response of the insulating foam must be known.

Table 19.11 External insulation systems (McKechnie and Hayes, 2001).

Name of Coating System	Abbreviation	Reelable (yes/no)	Max. Temp. [°C]	Water Absorption	Conductivity increase at end of life [%]	Compression at end of life [%]	Max. Thickness if <100mm	Density [kg/m^3]	Max. Depth [m]	Conductivity [W/m^{2}K]
Polypropylene – Solid	PP	Y	145	<0.5%	<1%	<2%	60	900	3000	0.22
Polypropylene – Syntactic PPF (e.g. Carizite)	SPP	Y	115	<5.0%	<1%	<5%	-	600 -800	-	0.13 –0.22
Polypropylene – Reinforced Foam Combination	RPPF	Y	115-140	<0.5%	<1%	<0.5%	multi-layer [Note 1]	600 -800	600 - 3000	0.16 – 0.18
Polyurethane – Soild	PU	Y	115	<5.0%	<10%	<2%		1150		0.19- 0.20
Polyurethane – Syntactic (plastic beads)	SPU	Y	70 -115	<5.0%	<10%	<2%	multi-layer	750 -780	100@ 115C 300 @ 90C	0.12 – 0.15
Polyurethane – Glass Syntactic	GSPU	Y	55 - 90	<2.2%	Variable on thickness	<5%	multi-layer	610 -830	2000 -3000	0.12 – 0.17
Polyurethane – Reinforced Foam Combination	RPUF	N.A	n.a.	n.a.	n.a.	n.a.	multi-layer	448	-	0.080
Phenolic Syntactic	PhS	Y	200		<5%	<1%		500	-	0.080
Epoxy Syntactic	SEP	N	75-100	<5.0%	<10%	<2%	75	590-720	2000 -3000	0.10 -0.135
Epoxy Syntactic with Mini-Spheres	MSEP	N	75		<10%	<4%		540		0.12

Note 1: For the multi-layer coatings the properties quoted are for the insulation layer;

Note 2: Value measured at mean product temperature of 50 °C;

Note 3: Values are approximate only. All values used in design must be provided by the manufacturer.

The factors affecting the creep response of polymer foams include temperature, density and stress level. The combination of high temperatures and high stresses can accelerate the creep process. High density foams are desirable for systems subject to large hydrostatic loads for long time periods.

The design of thermal insulation systems for deepwater applications is complex and involves a number of thermal and structural issues. These thermal and structural issues can sometimes conflict. In order to design effective thermal insulation systems, the thermal and structural issues involved must be carefully considered to achieve a balance. For long distance tie-backs the need for a substantial thickness of insulation will obviously have an impact on the installation method due to the increase in pipe outside diameter and the pipe field-jointing process. As most insulation systems are buoyant, the submerged weight of the pipe will decrease. It may be necessary to increase pipe wall thickness to achieve a submerged weight suitable for installation and on-bottom stability.

19.6.2 Pipe-in-pipe System

A method of achieving U-values of 1 $W/m^2{\cdot}K$ or less requires pipe-in-pipe (PIP) insulation systems, in which the inner pipe carrying the fluid is encased within a larger outer pipe. The outer pipe seals the annulus between the two pipes and the annulus can be filled with a wide range of insulating materials that do not have to withstand hydrostatic pressure.

The key features for application of PIP system are,

- Jointing method;
- Protection of the insulation from water ingress, water stops/bulkheads;
- Field joint insulation method;
- Response to applied load;
- Offshore fabrication process and lay rate of installation;
- Submerged weight;
- Limitations, cost and properties of insulation materials.

Typical PIP field joints are summarized below:

- Sliding sleeve with fillet welds or fillet weld-butt weld combination;
- Tulip assemblies involving screwed or welded components;
- Fully butt welded systems involving butt welded half-shells;
- Sliding outer pipe over the insulation with a butt weld on the outer pipe, which has recently been proposed for J-lay installation systems;
- Mechanical connector systems employing a mechanical connector on the inner pipe, outer pipe or both pipes to minimize installation welding;

For all these systems, except reel-lay, the lay vessel field-joint assembly time is critical to installation costs and the use of quadruple joints has been used to speed this up.

Table 19.13 PIP insulation materials (McKechnie and Hayes, 2001).

Insulation material	Density [kg/m³]	Conductivity [W/mK]	Thickness [mm]	Annulus Gap(if any)	Max. temp [°C]	U-value [W/m²·K] 16" inner pipe	Comment
Mineral wool	140	0.037	100	clearance	700	1.6 (40 mm)	Rockwool or Glava, usually in combination with Mylar reflective film.
Alumina silicate microspheres	390-420	0.1	no limit	none	1000+	3.9 (100 mm)	Commonly referred to as Fly-ash, injected to fill the annulus.
Thermal cement	900-1200	0.26	100	none	200	-	Currently being investigated under a JIP to provide collapse resistance with reduced carrier pipe wall thickness.
LD PU foam	60	0.027	125	none	147	0.76 (100 mm)	Pre-assembled as single or double jointed system, used on the Erskine Replacement.
HD PU foam	150	0.035	125	none	147	1.2 (100 mm)	
Microporous silica blanket	200-400	0.022	24	clearance	900	0.4 (100 mm)	Cotton blanket, calcium based powder, glass and titanium fibers.
Vacuum insulation panels	60-145	0.006-0.008	10	clearance	160	0.26 (100 mm)	Foam shells formed under vacuum with aluminium foil and uses gas absorbing 'getter' pills to absorb any free gas thereafter.

Table 19.13 summarizes some insulation materials used in PIP, their properties and U-values for a given thickness. For most insulation systems, thermal performance is mainly based on the conductive resistance. However heat convection and radiation will transfer heat across a gas voids in annulus if there is one. The convection and radiation in the gas at large void spaces result in a less effective thermal system than the one that is completely filled with the insulation material. For this reason PIP systems often include a combination of insulation materials. An inert gas such as nitrogen or argon can be filled in the gap or annulus between the insulation and the outer pipe to reduce convection. A near vacuum in the annulus of PIP greatly improves insulation performance by minimizing convection in the annulus and also significantly reducing conduction of heat through the insulation system and the annulus. The difficulty is to maintain a vacuum over a long time.

19.6.3 Bundling

Bundles are used to install a combination of flowlines inside an outer jacket or carrier pipe. Bundling offers attractive solutions to a wide range of flow assurance issues by providing cost-effective thermal insulation and capability of circulating heating medium. Other advantages of bundles include:

- install multiple flowlines in a single installation;
- outer jacket helps to resist the external hydrostatic pressure in deep water;
- possibility of including heating pipes and monitoring systems in the bundle.

The disadvantages are the bundle length limit and the need for a suitable fabrication and launch site. The convection in the gap between the insulation and pipes needs to be specifically treated.

Bundle modeling

- FEM model;
- OLGA 2000, FEMTherm module.

FEM method is easily to simulate a wide range of bundle geometries and burial configurations, FEM software maybe used to simulate the thermal interaction among the production flowlines, heating lines and other lines that are enclosed in a bundle to determine U-values and cooldown times. The OLGA2000 FEMTherm module can model bundles with a fluid medium separating the individual production and heating lines, but it is not applicable to cases in which the interstitial space is filled with solid insulation. Once OLGA2000 FEMTherm model has been done, the model can be used for steady state and transient simulations for these components. A comparison of the U-values and cooldown times for the finite element and bundle models is helpful to get a confident result.

19.6.4 Burial

Trenching and backfilling can be an effective method of increasing the insulation, since the heat capacity of soil is significant and acts as a natural heat store. The effect of burial can typically decreases the U-value and significantly increase the pipeline cool down period as shown in the Section 19.5.2. Attention to the material ageing characteristics in burial conditions needs to be considered.

The cool down time for a buried insulated flowline can be greater than a PIP system due to the heat capacity of the system. The cost of installing an insulated and buried line is approximately 35-50% that of a PIP system. In order to confidently use the insulating properties of the soil, reliable soil data is required. The lack of this can lead to over-conservative pipeline systems.

Table 19.12 Comparison of impact of flowline insulation between PIP and B&C system.

PIP Systems	B&C Systems
• Maintains steady state flow out of hydrate and wax regions for greatest portion of production life. • "Threshold" for wax deposition < 5000 bpd; • hydrate formation < 3000 bpd. • Insulation hinders hydrate remediation (estimates are of the order of 20 - 22 weeks); • Wax deposition occurs mainly in the riser at low flowrates.	• Flowline cools down more slowly than PIP system but starting temperature is lower (lower steady state temperatures); • Lower insulation characteristic and heat capacity of earth facilitates hydrate remediation (estimates are of the order 8 -10 weeks); • Wax deposition is more evenly distributed along the entire flowline at low flowrates; • Flow rate "threshold" for wax & hydrate deposition occurs earlier in field life; • wax deposition < 12000 bpd; • hydrate formation < 5000 bpd.

19.6.5 Direct Heating

Actively heated systems generally use hot fluid or electricity as a heating medium. The main attraction of active heating is its flexibility. It can be used to extend the cool down time by continuously maintaining a uniform flowline temperature above the critical levels of wax or hydrate formation. It is also capable of warming up a line from seawater temperature to a target operating level and avoids the requirements for complex and risky start-up procedures.

Direct heating by applying an electrical current in the pipe has been used since the 1970's in shallow waters and could mitigate hydrate and wax deposition issues along the entire length of the pipeline, and has found wide application in recent years. Electrically heated systems have also been recognized as an effective method in removing hydrate plugs within estimated times of 3 days, while depressurization methods employed in deepwater developments can take up to several months. The most efficient electrical heating systems can provide the heat input as close to the flowline bore as possible and have minimum heat losses to the environment. Trace heating is believed to provide the highest level of heating efficiency and can be applied to bundles or PIP system. The length capability of an electrical system depends on the linear heat input required and the admissible voltage. Trace heating can be applied for very long tie-backs (10's of kilometres) by either using higher voltage or by introducing intermediate power feeding locations, fed by power umbilicals via step down transformers. Table 19.13 summarizes the projects that are currently using EH or have EH as part of their design. It has been reported that there are potential CAPEX savings of 30% for single electrically heated pipelines over a traditional PIP system for lengths of up to 24 km. Beyond these distance the CAPEX can surpass PIP systems, however operational shutdown considerations could negate these cost differences. OPEX costs may be reduced when offset against reduced chemical use, reduced shutdown and start-up times, no requirement for line depressurization, no pigging requirements and enhanced cool-down periods.

Table 19.13 Projects currently using electrical heating systems (Cochran, 2003).

Operator & Project	Line Diameter & Length	Water Depth (m)	Year of Installation	Electrical Heating Method	Mode of Operation	Status
Shell Nakika	10-in. x 16-in. PIP	1900	2003	PIP Direct	Remediation	Engineering & qualification testing
Shell Serrano / Oregano	6-in. x 10-in. PIP two lines 10 and 12 km	1000	2001	PIP Direct	Temperature maintenance for shutdown	In operation
Statoil Huldra	8-in. single pipe 15 km	300 to 400	2001	Earthed Direct	Temperature maintenance of 25°C during shutdown	In operation
Statoil Asgard	8-in. single pipe 6 lines 43 km total	300 to 400	2000	Earthed Direct	Temperature maintenance of 27°C during shutdown	In operation
Statoil Sleipner	20-in. single pipe 12.6 km	-	1996	Induction	Temperature maintenance	In operation

The disadvantages of the systems include the requirement for transformers and power cables over long distances, the cost and availability of the power and the difficulties associated with maintenance. Typical requirements are 20-40 W/m for an insulated system of $U=1.0W/m^2 \cdot K$. The method can not be used in isolation and requires insulation applied to the pipe. There are three basic approaches for electrically heated systems, which consist of the closed and open return systems for wet insulated pipes using either cable or surrounding seawater to complete the electrical circuit, and dry PIP systems.

19.6.5.1 Closed return systems

This system supplies DC electrical energy directly to the pipe wall between two isolation joints. The circuit is made complete by a return cable running along the pipeline. Research has shown that the power required for this system is approximately 30% less than that required for an open system. The required power is proportional to the insulation U value.

19.6.5.2 Open return systems (earthed-current)

Currently implemented on the "Åsgard" and "Huldra" fields in the North Sea, this system uses a piggybacked electrical cable to supply current to the pipeline. The return circuit comprises of a combination of the pipe wall and the surrounding seawater to allow the flow of electrons. Pipeline anodes are used to limit the proliferation of stray current effects to nearby structures.

19.6.5.3 PIP System

Use a combination of passive insulation methods and heat trace systems in which direct or alternating current circulate around the inner pipe. Development of optical fiber sensors running the along the inner pipe can provide the operator with a real time temperature profile.

AC systems being offered have been recommended for distances of 20 km. The installation would comprise of five 20 km sections joined together by T-boxes and transformers.

19.6.6 Hot Fluid Heating (Indirect Heating)

The use of waste energy from the reception facility to heat and circulate hot water (similar to a shell and tube heat exchanger) has been applied on integrated bundles, e.g. Britania. In this instance the length of the bundle was limited to less than 10 km with the heat medium only applied during pipeline start-up or shutdowns. The system would normally have the flexibility to deliver the hot medium to either end of the pipeline first.

In order to be sufficiently efficient, it is necessary to inject large flow of fluid at a relatively high temperature. This involves storage facilities and significant energy to heat the fluid and to maintain the flow.

Active heating by circulation of hot fluid is generally more suited to a bundle configuration because large pipeline cross sections are required. There is a length limitation to hot fluid heating because it generally involves a fluid circulation loop along which the temperature of the heating medium decreases.

19.7 References

1. Cochran, S., (2003), "Hydrate Control and Remediation Best Practices in Deepwater Oil Developments", OTC 15255, 2003 Offshore Technology Conference, Houston, Texas.
2. Dittus, F. W., and Boelter, L. M. K., (1930), University of California, Berkeley, Publications on Engineering, Vo. 2, p.443.
3. Guo B., Duan S., and Ghalambor A., (2004), "A Simple Model for Predicting Heat Loss and Temperature Profiles in Thermal Injection Lines and Wellbores with Insulations", SPE 86983, SPE international Thermal Operations and Heavy Oil Symposium and Western Regional Meeting, Bakerfield, California, March 2004.
4. Grealish, F. and Roddy, I., (2002), "State of the Art on Deep Water Thermal Insulation Systems", OMAE2002-28464, June 23-28, 2002, Oslo, Norway.
5. Gregory, G. A., (1991), "Estimation of Overall Heat Transfer Coefficient for the Calculation of Pipeline Heat Loss/Gain", Technical Note No. 3, Neotechnology Consultants Ltd.
6. Hausen, H., (1943), "Darstellung des Warmeuberganges in Rohren durch verallgemeinerte Potezbeziehungen", Z. VDI Beih. Verfahrenstechnik, No. 4, p. 91.
7. Hilpert, R., (1933), "Warmeabgabe von geheizen Drahten und Rohren", Forsch. Gebiete Ingenieurw., Vol. 4, p. 220.
8. Horn, C., and Lively, G., (2001), "A New Insulation Technology: Prediction vs. Results from the First Field Installation", OTC13136, 2001 Offshore Technology Conference, Houston, Texas.
9. Incropera, F. P., and DeWitt D. P., (1996), Introduction to Heat Transfer, third Edition, John Wiley & Sons.

10. Kersten, M.S., (1949). "Thermal properties of soils". Univ. of Minnesota Eng. Exp. Station Bull. No. 28, Univ. of Minnesota. Inst.

11. Loch, K., (2000), "Flowline Burial: An Economic Alternative to Pipe-in Pipe", OTC 12034, 2000 Offshore Technology Conference, Houston, Texas.

12. MARSCO, (1999), "Thermal soil studies for pipeline burial insulation - laboratory and sampling program Gulf of Mexico", internal report completed for Stolt Offshore.

13. Melve, B., (2002), "Design Requirements for High Temperature Flowline Coatings", OMAE2002-28569, June 23-28, 2002, Oslo, Norway.

14. McKechnie, J. G., and Hayes, D. T., (2001), "Pipeline Insulation Performance for Long Distance Subsea Tie-Backs", Long Distance Subsea Tiebacks, Nov 26-28, 2001, Amsterdam.

15. Newson, T.A., Brunning, P., and Stewart, G., (2002), "Thermal Conductivity of Offshore Clayey Backfill", OMAE2002-28020, June 23-28, 2002, Oslo, Norway.

16. Parker, J. D, Boggs, J. H., and Blick, E. F., (1969), Introduction to Fluid Mechanics and Heat Transfer, Addison-Wesley Pub.

17. Pattee, F. M. and Kopp, F., (2000), "Impact of Electrically Heated System on the Operation of Deep Water Subsea Oil Flowline", OTC 11894, 2000 Offshore Technology Conference, Houston, Texas.

18. Power, P.T., Hawkins, R.A., Christophersen, H.P. and McKenzie, I. (1994), "ROV assisted geotechnical investigation of trench backfill material aids design of the Tordis to Gullfaks flowlines". ASPECT 94, 2nd Int. Conf. on Advances in Subsea Pipeline Engineering and Technology, Aberdeen.

19. Von Herzen, R. and Maxwell, A.E. (1959), "The measurement of thermal conductivity of deep sea sediments by a needle probe method", Journal of Geophysical Research, Vol. 64, No. 10, p. 1557-1563.

20. Wang, X., Chin, Doreen Y., Perera, R. M., Prescott, C. N. and Raborn, G.O., (2000), "Convection Heat Losses through Installation Gaps between Pipe and Insulation and between Insulation Half Shells", OTC12033, 2000 Offshore Technology Conference, Houston, Texas.

21. Young, A. G., Osborne, R. S., and Frazer, I., (2001), "Utilizing Thermal Properties of Seabed Soils as Cost-Effective Insulation for Subsea Flowlines", OTC 13137, 2001 Offshore Technology Conference, Houston, Texas.

22. Zhang, J. J, Chen, M. J., Wang X., and Chacko, J., (2002), "Thermal Analysis and Design of Hot Water Heated Production Flowline Bundles", OTC 14015, 2002 Offshore Technology Conference, Houston, Texas.

19.8 Appendix: U-value and Cooldown Time Calculation Sheet.

U-VALUE AND COOLDOWN TIME CALCULATION SHEET

JOB NUMBER: CLIENT:
PROJECT NAME:
FLOWLINE:
DATE: 11/19/2004
VERSION: 1.0 DEVELOPER: QIANG BAI
USER: YB CHECKER: QB

Units:

$°C := K$

$°F := R$

$$F(T) := \left[32.0 + \frac{9.0}{5.0}\left(\frac{T}{C}\right)\right]°F$$

Input Information:

Pipeline Geometry

Pipe outer dia meter	$D_o := 0.219 \cdot m$	$D_o = 8.62\,in$
Wall thickness	$t := 14.3mm$	$t = 0.56\,in$

Insulation layer Thickness

Insulation layer 1	$t_1 := 0.3 \cdot mm$	$t_1 = 0.01\,in$
Insulation layer 2	$t_2 := 0.3 \cdot mm$	$t_2 = 0.01\,in$
Insulation layer 3	$t_3 := 76.2 \cdot mm$	$t_3 = 3\,in$
Insulation layer 4	$t_4 := 46 \cdot mm$	$t_4 = 1.81\,in$
Insulation layer 5	$t_5 := 5 \cdot mm$	$t_5 = 0.2\,in$
Length of pipe	$L_p := 1 \cdot m$	$L_p = 3.28\,ft$

Thermal Conductivity

Fluid	$k_f := 0.35 \cdot W m^{-1} K^{-1}$	$k_f = 0.2\,BTU \cdot ft^{-1} \cdot hr^{-1} \cdot °F^{-1}$
Pipe wall	$k_{pipe} := 45 \cdot W m^{-1} K^{-1}$	$k_{pipe} = 26\,BTU \cdot ft^{-1} \cdot hr^{-1} \cdot °F^{-1}$
Insulation layer 1	$k_{I1} := 0.3 \cdot W m^{-1} K^{-1}$	$k_{I1} = 0.17\,BTU \cdot ft^{-1} \cdot hr^{-1} \cdot °F^{-1}$
Insulation layer 2	$k_{I2} := 0.22 \cdot W m^{-1} K^{-1}$	$k_{I2} = 0.13\,BTU \cdot ft^{-1} \cdot hr^{-1} \cdot °F^{-1}$
Insulation layer 3	$k_{I3} := 0.215 \cdot W m^{-1} K^{-1}$	$k_{I3} = 0.12\,BTU \cdot ft^{-1} \cdot hr^{-1} \cdot °F^{-1}$
Insulation layer 4	$k_{I4} := 0.175 \cdot W m^{-1} K^{-1}$	$k_{I4} = 0.1\,BTU \cdot ft^{-1} \cdot hr^{-1} \cdot °F^{-1}$
Insulation layer 5	$k_{I5} := 0.215 \cdot W m^{-1} K^{-1}$	$k_{I5} = 0.12\,BTU \cdot ft^{-1} \cdot hr^{-1} \cdot °F^{-1}$

Specific Heat Capacity

Fluid	$Cp_{fluid} := 3.054 \cdot 10^3 \cdot J kg^{-1} \cdot K^{-1}$	$Cp_{fluid} = 0.73\, BTU \cdot lb^{-1} \cdot °F^{-1}$
Carbon steel	$Cp_{steel} := 2.29 \cdot 10^3 \cdot J kg^{-1} \cdot K^{-1}$	$Cp_{steel} = 0.55\, BTU \cdot lb^{-1} \cdot °F^{-1}$
Insulation layer 1	$C_{pl1} := 1.6 \cdot 10^3 \cdot J kg^{-1} \cdot K^{-1}$	$C_{pl1} = 0.38\, BTU \cdot lb^{-1} \cdot °F^{-1}$
Insulation layer 2	$C_{pl2} := 1.95 \cdot 10^3 \cdot J kg^{-1} \cdot K^{-1}$	$C_{pl2} = 0.47\, BTU \cdot lb^{-1} \cdot °F^{-1}$
Insulation layer 3	$C_{pl3} := 1.95 \cdot 10^3 \cdot J kg^{-1} \cdot K^{-1}$	$C_{pl3} = 0.47\, BTU \cdot lb^{-1} \cdot °F^{-1}$
Insulation layer 4	$C_{pl4} := 1.95 \cdot 10^3 \cdot J kg^{-1} \cdot K^{-1}$	$C_{pl4} = 0.47\, BTU \cdot lb^{-1} \cdot °F^{-1}$
Insulation layer 5	$C_{pl5} := 2.0 \cdot 10^3 \cdot J kg^{-1} \cdot K^{-1}$	$C_{pl5} = 0.48\, BTU \cdot lb^{-1} \cdot °F^{-1}$

Density

Fluid	$\rho_f := 795.8 \cdot kg \cdot m^{-3}$	$\rho_f = 49.68\, lb \cdot ft^{-3}$
Carbon steel	$\rho_p := 7850 \cdot kg \cdot m^{-3}$	$\rho_p = 490.06\, lb \cdot ft^{-3}$
Insulation layer 1	$\rho_{l1} := 1450 \cdot kg \cdot m^{-3}$	$\rho_{l1} = 90.52\, lb \cdot ft^{-3}$
Insulation layer 2	$\rho_{l2} := 900 \cdot kg \cdot m^{-3}$	$\rho_{l2} = 56.19\, lb \cdot ft^{-3}$
Insulation layer 3	$\rho_{l3} := 900 \cdot kg \cdot m^{-3}$	$\rho_{l3} = 56.19\, lb \cdot ft^{-3}$
Insulation layer 4	$\rho_{l4} := 700 \cdot kg \cdot m^{-3}$	$\rho_{l4} = 43.7\, lb \cdot ft^{-3}$
Insulation layer 5	$\rho_{l5} := 890 \cdot kg \cdot m^{-3}$	$\rho_{l5} = 55.56\, lb \cdot ft^{-3}$

PROPERTIES OF AMBIENT SURROUNDING

BURIED PIPELINE

Depth of cover to center line	$Z := 0.0 \cdot m$
Thermal conductivity of soil/clay	$k_s := 1.22 \cdot W m^{-1} K^{-1}$

SURFACE LAID PIPELINE

If the depth of cover, Z, is set to 0, the ambient condition is sea water.

Density of sea-water	$\rho_{h2o} := 1026 \cdot kg \cdot m^{-3}$
Viscosity of sea-water	$\mu_{h2o} := 0.00182 \cdot Pa \cdot s$
Thermal conductivity of seawater	$k_{h2o} := 0.56 \cdot W m^{-1} K^{-1}$
Velocity of sea-water	$v_{h2o} := 1.6 \cdot ft \cdot s^{-1}$
Specific heat capacity	$Cp_{h2o} := 4300 \cdot 10^3 \cdot J \cdot kg^{-1} \cdot K^{-1}$
Ambient temperature	$T_a := 4.28 \cdot °C$

Internal Fluid Flow Parameters

Reynold number of internal fluid: $Re_f := 1.35 \cdot 10^6$

Prandle number of internal fluid: $Pr_f := 3.5$

Initial fluid temperature $Tf_0 := 87.0 \cdot °C$

Time step $\Delta t := 0.01 hr$

Safty operation temperature $T_c := 64 \cdot °C$

Calculations:

Pipe internal diameter $D_i := D_o - 2 \cdot t$ $\quad D_i = 0.19 m$

Inside film heat transfer coefficient $h_i := 0.023 \cdot Re_f^{0.8} \cdot Pr_f^{0.3} \cdot \frac{k_f}{D_i}$ $\quad h_i = 4.94 \times 10^3 \frac{W}{m^2 \cdot K}$

Inside film resistance $R_i := \frac{D_o}{D_i \cdot h_i}$ $\quad R_i = 2.33 \times 10^{-4} \frac{m^2 \cdot K}{W}$

Pipewall resistance $R_{pipe} := \frac{D_o \cdot \ln\left(\frac{D_o}{D_i}\right)}{2 \cdot k_{pipe}}$ $\quad R_{pipe} = 3.41 \times 10^{-4} \frac{m^2 \cdot K}{W}$

Insulation 1 resistance $R_{I1} := \frac{D_o \cdot \ln\left(\frac{D_o + 2 \cdot t_1}{D_o}\right)}{2 \cdot k_{I1}}$ $\quad R_{I1} = 9.99 \times 10^{-4} \frac{m^2 \cdot K}{W}$

Insulation 2 resistance $R_{I2} := \frac{D_o \cdot \ln\left[\frac{D_o + 2 \cdot (t_1 + t_2)}{D_o + 2 \cdot t_1}\right]}{2 \cdot k_{I2}}$ $\quad R_{I2} = 1.36 \times 10^{-3} \frac{m^2 \cdot K}{W}$

Insulation 3 resistance $R_{I3} := \frac{D_o \cdot \ln\left[\frac{D_o + 2 \cdot (t_1 + t_2 + t_3)}{D_o + 2 \cdot (t_1 + t_2)}\right]}{2 \cdot k_{I3}}$ $\quad R_{I3} = 0.27 \frac{m^2 \cdot K}{W}$

Insulation 4 resistance $R_{I4} := \frac{D_o \cdot \ln\left[\frac{D_o + 2 \cdot (t_1 + t_2 + t_3 + t_4)}{D_o + 2 \cdot (t_1 + t_2 + t_3)}\right]}{2 \cdot k_{I4}}$ $\quad R_{I4} = 0.14 \frac{m^2 \cdot K}{W}$

Insulation 5 resistance $R_{I5} := \frac{D_o \cdot \ln\left[\frac{D_o + 2 \cdot (t_1 + t_2 + t_3 + t_4 + t_5)}{D_o + 2 \cdot (t_1 + t_2 + t_3 + t_4)}\right]}{2 \cdot k_{I5}}$ $\quad R_{I5} = 0.01 \frac{m^2 \cdot K}{W}$

Surrounding resistance

$$f(Z) := \begin{cases} \left(\frac{D_o}{2 \cdot k_s} \cdot \ln\left(\frac{2 \cdot Z}{0.5 \cdot D_o}\right)\right) & \text{if } Z > 0 \cdot m \\ \frac{D_o}{\left[0.0266 \cdot k_{h2o} \cdot \left(\frac{D_i \cdot h_i}{k_f}\right)^{0.8} \cdot \left(\frac{\rho_{h2o} \cdot v_{h2o} \cdot D_o}{\mu_{h2o}}\right)^{0.33}\right]} & \text{otherwise} \end{cases}$$

$$R_{surr} := f(Z) \qquad R_{surr} = 7.03 \times 10^{-4} \frac{m^2 \cdot K}{W}$$

Area for heat transfer

$$A_o := \pi D_o \cdot L_p$$

Weight per unit length

Fluid

$$W_f := \frac{\pi \cdot D_i^2 \cdot \rho_f \cdot L_p}{4}$$

Pipe

$$W_p := \frac{\pi \cdot \left(D_o^2 - D_i^2\right) \cdot \rho_p \cdot L_p}{4}$$

Insulation layer1

$$W_{I1} := \frac{\pi \cdot \left[(D_o + 2 \cdot t_1)^2 - (D_o)^2\right] \cdot \rho_{I1} \cdot L_p}{4}$$

Insulation layer 2

$$W_{I2} := \frac{\pi \cdot \left[(D_o + 2 \cdot t_1 + 2 \cdot t_2)^2 - (D_o + 2 \cdot t_1)^2\right] \cdot \rho_{I2} \cdot L_p}{4}$$

Insulation layer 3

$$W_{I3} := \frac{\pi \cdot \left[(D_o + 2 \cdot t_1 + 2 \cdot t_2 + 2 \cdot t_3)^2 - (D_o + 2 \cdot t_1 + 2 \cdot t_2)^2\right] \cdot \rho_{I3} \cdot L_p}{4}$$

Insulation layer 4

$$W_{I4} := \frac{\pi \cdot \left[(D_o + 2 \cdot t_1 + 2 \cdot t_2 + 2 \cdot t_3 + 2 \cdot t_4)^2 - (D_o + 2 \cdot t_1 + 2 \cdot t_2 + 2 \cdot t_3)^2\right] \cdot \rho_{I4} \cdot L_p}{4}$$

Insulation layer 5

$$W_{I5} := \frac{\pi \cdot \left[(D_o + 2 \cdot t_1 + 2 \cdot t_2 + 2 \cdot t_3 + 2 \cdot t_4 + 2 \cdot t_5)^2 - (D_o + 2 \cdot t_1 + 2 \cdot t_2 + 2 \cdot t_3 + 2 \cdot t_4)^2\right] \cdot \rho_{I5} \cdot L_p}{4}$$

Results:

- **Overall heat transfer c oefficient**

U value based on pipeline OD

$$U_o := \frac{1}{R_i + R_{pipe} + R_{I1} + R_{I2} + R_{I3} + R_{I4} + R_{I5} + R_{surr}}$$

$$U_o = 2.38 \frac{W}{m^2 \cdot K}$$

U value based on pipeline ID

$$U_i := U_o \cdot \frac{D_o}{D_i} \qquad U_i = 2.74 \frac{W}{m^2 K}$$

U value based on pipeline ID $U_i := U_o \cdot \frac{D_o}{D_i}$ $U_i = 2.74 \frac{W}{m^2 K}$

- **Cooldown calculation**

Initial temperatures of system for cooldown

Wall $Tp_0 := Tf_0 - U_o \cdot \left(R_i + \frac{R_{pipe}}{2} \right) \cdot (Tf_0 - T_a)$

Insulatioin layer 3 $Tl3_0 := Tf_0 - U_o \cdot \left(R_i + R_{pipe} + R_{l1} + R_{l2} + \frac{R_{l3}}{2} \right) \cdot (Tf_0 - T_a)$

Insulatioin layer 4 $Tl4_0 := Tf_0 - U_o \cdot \left(R_i + R_{pipe} + R_{l1} + R_{l2} + R_{l3} + \frac{R_{l4}}{2} \right) \cdot (Tf_0 - T_a)$

Insulatioin layer 5 $Tl5_0 := Tf_0 - U_o \cdot \left(R_i + R_{pipe} + R_{l1} + R_{l2} + R_{l3} + R_{l4} + \frac{R_{l5}}{2} \right) \cdot (Tf_0 - T_a)$

Inside fluid film heat transfer coefficient for laminar flow
(Velocity of internal fluid is extremely low during cooldown)

$h_l := 3.66 \cdot \frac{k_f}{D_i}$ $h_l = 6.73 \frac{W}{m^2 \cdot K}$

Inside laminar film resistance

$R_l := \frac{D_o}{D_i \cdot h_l}$ $R_l = 0.17 \, m^2 \frac{K}{W}$

Cooldown temperature profiles of fluid, pipe and insulation layer 3,4 and 5

$i := 0..\ 10000$

$$\begin{pmatrix} Tf_{i+1} \\ Tp_{i+1} \\ Tl3_{i+1} \\ Tl4_{i+1} \\ Tl5_{i+1} \end{pmatrix} := \begin{bmatrix} Tf_i - \frac{A_o \cdot \Delta t}{W_f \cdot Cp_{fluid}} \cdot \left(\frac{Tf_i - Tp_i}{R_l + \frac{R_{pipe}}{2}} \right) \\ Tp_i - \frac{A_o \cdot \Delta t}{W_p \cdot Cp_{steel}} \cdot \left(\frac{Tp_i - Tf_i}{R_l + \frac{R_{pipe}}{2}} + \frac{Tp_i - Tl3_i}{R_{l1} + R_{l2} + \frac{R_{l3}}{2} + \frac{R_{pipe}}{2}} \right) \\ Tl3_i - \frac{A_o \cdot \Delta t}{W_{l3} \cdot C_{pl3}} \cdot \left(\frac{Tl3_i - Tl4_i}{\frac{R_{l3}}{2} + \frac{R_{l4}}{2}} + \frac{Tl3_i - Tp_i}{R_{l1} + R_{l2} + \frac{R_{l3}}{2} + \frac{R_{pipe}}{2}} \right) \\ Tl4_i - \frac{A_o \cdot \Delta t}{W_{l4} \cdot C_{pl4}} \cdot \left(\frac{Tl4_i - Tl3_i}{\frac{R_{l3}}{2} + \frac{R_{l4}}{2}} + \frac{Tl4_i - Tl5_i}{\frac{R_{l4}}{2} + \frac{R_{l5}}{2}} \right) \\ Tl5_i - \frac{A_o \cdot \Delta t}{W_{l5} \cdot C_{pl5}} \cdot \left(\frac{Tl5_i - Tl4_i}{\frac{R_{l5}}{2} + \frac{R_{l4}}{2}} + \frac{Tl5_i - T_a}{R_{surr} + \frac{R_{l5}}{2}} \right) \end{bmatrix}$$

$$f(Tf, T_c) := \left| \begin{array}{l} i \leftarrow 0 \\ \text{break} \quad \text{if} \quad T_a = Tf_i \\ \text{while} \quad Tf_i \geq T_c \\ \quad i \leftarrow i + 1 \\ i \end{array} \right.$$

This conditional statement is employed to identify whether the critical condition (i.e. WAT or HFT) is breached.

$$\text{Critical\_Time} := f(Tf, T_c) \cdot \Delta t$$

Critical time is the required time for fluid coolling down to the critical temperature.

Cooldown Time :

$$\text{Critical\_Time} = 19.55\,\text{hr}$$

$$\text{Time}_i := i \cdot \frac{\Delta t}{\text{hr}}$$

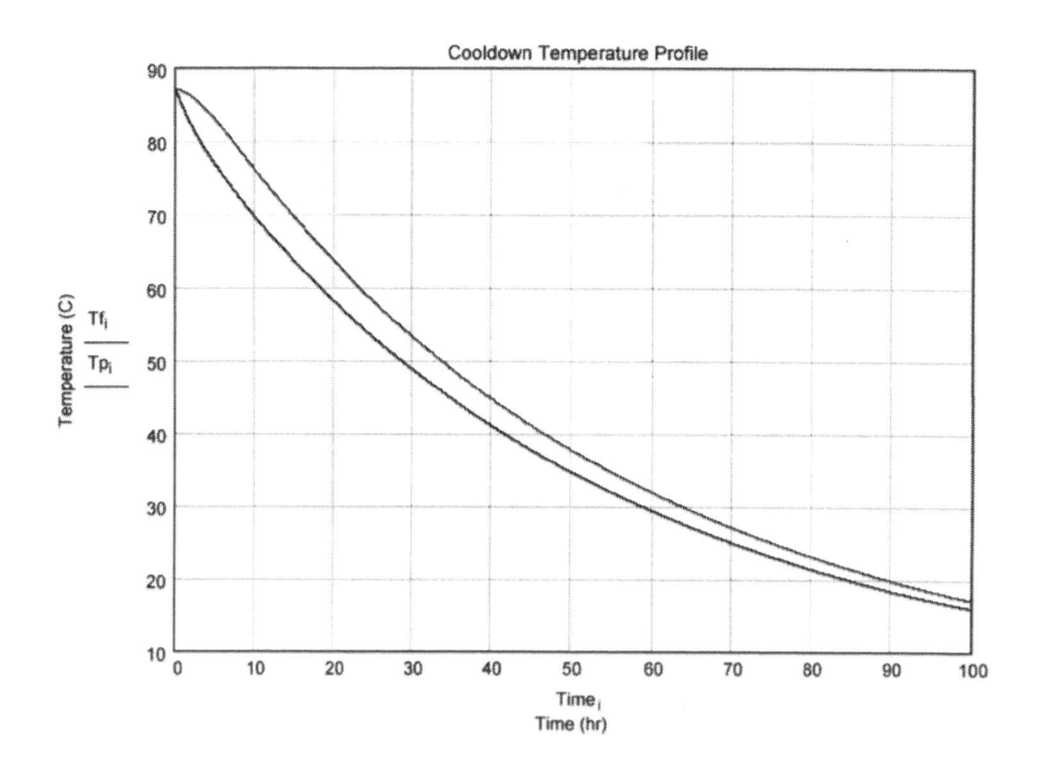

Part III

Flow Assurance

Chapter 20 Hydrates

20.1 Introduction

Natural gas hydrates are crystalline compounds formed by the physical combination of water molecules and certain small molecules in hydrocarbons fluid such as methane, ethane, propane, nitrogen, carbon dioxide and hydrogen sulphide. Hydrates are easily formed when the hydrocarbon gas contains water at high pressure and relatively low temperature.

As development activities in deeper waters are increasing, gas hydrates have become one of the top issues in the offshore industry. Deep water operations are being carried all over the world in locations such as the Gulf of Mexico (GoM), West Africa (WA), and the North Sea (NS). Gas hydrate problems manifest themselves most commonly during drilling and production processes. Hydrates may appear anywhere and anytime in an offshore system when there is natural gas, water and suitable temperature and pressure. Figure 20.1 shows a sketch of areas where hydrate blockages may occur in a simplified offshore deepwater system from the well to the platform export flowline. Hydrate blockages in a subsea flowline system are most likely to be found in direction change areas in well, pipeline and riser parts of the system,

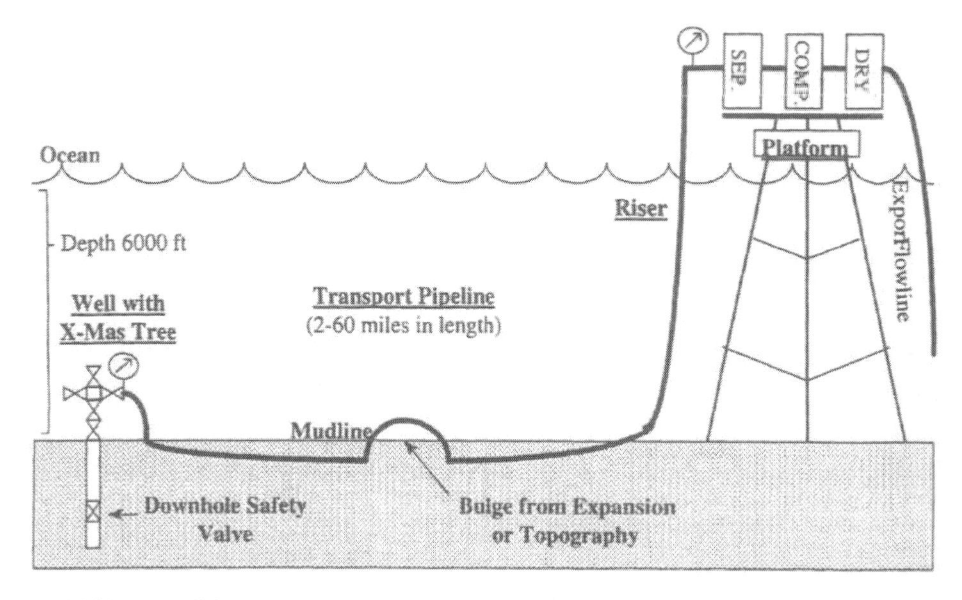

Figure 20.1 Typical offshore well, production pipeline and platform (Sloan, 1998).

and are rarely found in the tubing below the downhole safety valve and the flowlines after platform. In general, the hydrocarbon fluids are at higher pressure and higher temperature before the downhole safety valve, such that the fluid temperature is higher than the hydrate formation temperature corresponding to the local pressure. On the platform, the processes of separation, drying and compression are usually carried out. First the multi-phase hydrocarbons fluids are separated into gas, oil and water phases; then, the gas is dried and dehydrated; lastly, the gas is compressed for export. After gas and water are separated on the platform, hydrates will not form in gas export flowline without the presence of water.

Figure 20.2 shows a typical temperature variation with water depth in the GoM. At the water surface, the temperature deviation in all seasons is about 15 °F. When water depth is deeper than 3000 ft, the temperature deviations are very small and water temperature at the seabed becomes approximately constant at 40 °F. In the wellhead and X-tree, there are many valves to control the production operation. Hydrates may possibly occur during the shut-in/start-up operation in the well and X-tree because the ambient temperatures are typically around 40 °F in deepwater, even though they may not form in the normal operation steady condition in which the flow rate and temperature of hydrocarbon fluid are higher. In most deepwater production, the ambient water temperature is the main factor causing hydrate formation in the subsea system, but numerous examples also exist of hydrates forming due to Joule-Thomson cooling (JT cooling) of gas, where the gas expands across a valve, both in subsea and on the platform. Without insulation or other heat control system for the flowline, the fluids inside a subsea pipeline will cool to the ambient temperature within a few miles from the well. The cooling rate per length depends on the fluid composition, flow rate, ambient temperature, pipe diameter and other heat transfer factors.

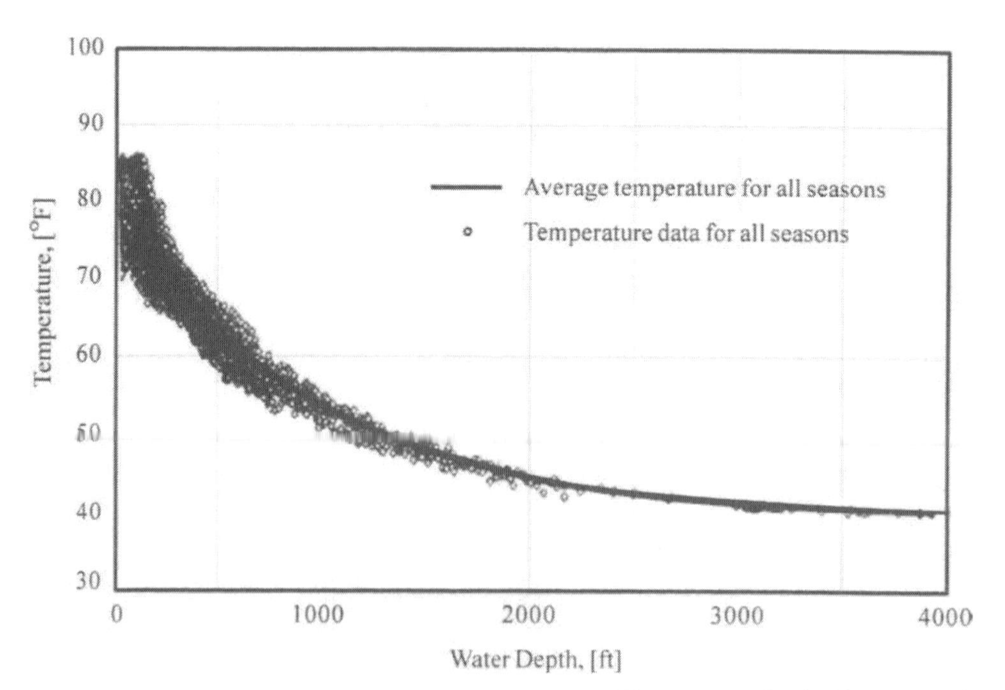

Figure 20.2 Seawater temperatures of GoM for all seasons at different water depths (Lorimer and Ellison, 2000).

Hydrates should be avoided in the offshore engineering since they can plug flowline, valves and other subsea devices. Hydrates are of importantance in deepwater gas developments because ambient temperatures are low enough to be in the hydrate formation region at operating pressures. The presence of a certain amount of water in the hydrocarbon systems can be troublesome due to the formation of hydrates. When temperature and pressure are in the hydrate formation region, hydrates grow so long as water and small molecules hydrocarbons are present. Hydrate crystals can develop into flow blockages, which can be time-consuming to clear in subsea equipment or flowlines and cause safety problems. Lost or delayed revenue and costs associated with hydrates blockages can be significant due to vessel intervention costs and delayed production. Thus, hydrate prevention and remediation are important design factors for deepwater developments.

The following topics about hydrates are discussed in this Chapter:

- Fundamental knowledge of hydrates.
- Hydrate formation process and consequences.
- Hydrate prevention techniques.
- Hydrate remediation.
- Hydrate control design philosophies
- Thermal inhibitor recovery.

20.2 Physics and Phase Behavior

20.2.1 General

Natural gas hydrates are crystalline water structures with low molecular weight guest molecules. They are often referred to as clathrate hydrates. The presence of the gas molecules leads to stability of the crystalline structure, allowing hydrates to exist at much higher temperatures than ice. Natural gas hydrates typically form one of three crystal structures, depending primarily on the sizes of the guest molecules. They are metastable minerals whose formation, stability and decomposition depend upon pressure, temperature, composition and other properties of the gas and water. Hydrate formers include nitrogen, carbon dioxide, hydrogen sulfide, methane, ethane, propane, iso-butane, n-butane, and some branched or cyclic C5-C8 hydrocarbons. Figure 20.3 shows one of the typical hydrate crystal structures found in oil and gas production systems.

Natural gas hydrates are composed of approximately 85 mol% water; therefore they have many physical properties similar to those of ice. For instance, the appearance and mechanical properties of hydrates are comparable to those of ice. The densities of hydrates vary somewhat due to the nature of the guest molecule(s) and the formation conditions, but are generally comparable to that of ice. Thus, hydrates typically will float at the water/hydrocarbon interface. However, in some instances, hydrates have been observed to settle on the bottom of the water phase. If a hydrate plug breaks from the pipe walls, it can be pushed down along the flowline by the flowing of hydrocarbon fluid like an ice bullet, potentially rupturing the flowline at a restriction or bend.

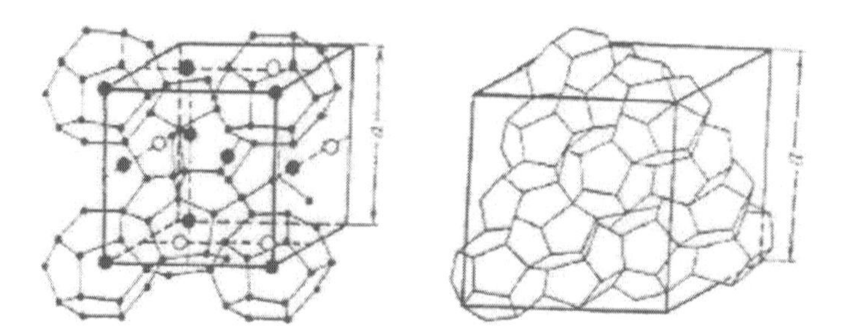

Figure 20.3 Hydrate crystal structures in oil and gas production systems (Lorimer and Ellison, 2000).

Four components are required to form gas hydrates: water, light hydrocarbon gases, low temperature and high pressure. If any one of these components is absent then gas hydrates will not form. Hydrate problems can appear during normal production, but transient operations are often more vulnerable. For instance, during a shut-in, the temperature of the subsea line drops to that of the surrounding environment. Given sufficient time under these high pressures and low temperatures, hydrates will form.

The extent to which the gas, oil and water partition during shut-in somewhat limits the growth of hydrates; though direct contact between the gas phase and the water phase is not needed for hydrate formation, an intervening oil layer slows transport of the hydrate forming molecules. Additionally, hydrates typically form in a thin layer at the water/oil interface, which impedes further contact between the water and gas molecules. Even if the flowlines do not plug during shut-in, when the well is restarted, the agitation breaks the hydrate layer and allows good mixing of the subcooled water and gas. Rapid hydrate formation often leads to a blockage of flowing at low spots where water tends to accumulate. Plugging tendency increases as the water cut increases, since there is a higher likelihood that sufficient hydrate particles will contact each other and stick together. Other typical locations include flow restrictions and flow transitions, such as elbows and riser bases.

Hydrate plugs may also occur in black oil subsea systems. Most deepwater black oil systems are not producing significant volumes of water. As water cuts rise, the incidence of hydrate plugs in black oil lines will certainly increase. Some black oils have a tendency not to plug, even when hydrates are formed. The hydrates remain small particles dispersed in the liquid phase and are readily transported through the flowline. These back oils will eventually plug with hydrates if the water cut gets high enough.

20.2.2 Hydrate Formation and Dissociation

Hydrate formation and dissociation curves are used to define pressure/temperature relationships in which hydrates form and dissociate. These curves may be generated by a series of laboratory experiments, or more commonly, are predicted using thermodynamic software such Multiflash or PVTSIM based on the composition of the hydrocarbon and aqueous phases in the system. The hydrate formation curve defines the temperature and pressure envelope in which the entire subsea hydrocarbons system must operate in at steady state and transient conditions in order to avoid the possibility of hydrate formation.

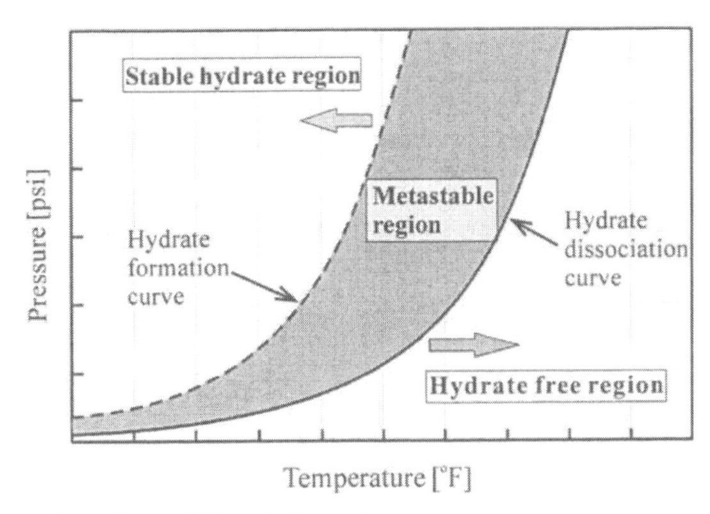

Figure 20.4 Hydrate formation and dissociation regions.

Figure 20.4 shows an example of these curves, which shows the stability of natural gas hydrates as a function of pressure and temperature. To the right of dissociation curve is the region in which hydrates do not form; operating in this region is safe from hydrate blockages. To the left of hydrate formation curve is the region where hydrates are thermodynamically stable and have the potential to form. This does not mean that hydrates will necessarily form or formed hydrates will cause operational difficulties. The stability of hydrates increases with increasing pressure and decreasing temperature. There is often a delay time or the temperature must be lowered somewhat below the hydrate stability temperature in order for hydrates to form. The subcooling of a system is often used when discussing gas hydrates, which is defined as the difference between hydrate stability temperature and the actual operating temperature at the same pressure. If the system is operating at 40 °F and 3000 psi, the hydrate dissociation temperature is 70 °F, the system is experiencing 30 °F of subcooling.

The subcooling of a system without hydrate formation leads to an area between the hydrate formation temperature and the hydrate dissociation temperature, called the metastable region where hydrate is not stable. Some software packages attempt to predict this metastable region. Regularly operating within this metastable region is risky. While such differences in hydrate formation and dissociation temperatures are readily observed in the laboratory, the quantitative magnitude of this hysteresis is apparatus and technique dependent.

Hydrates may not form for hours, days, or even at all even if a hydrocarbon system containing water is at a temperature and pressure condition close to the hydrate dissociation curve. A certain amount of "subcooling" is required for hydrate formation to occur at rates sufficient to have a practical impact on the system. Figure 20.5 shows the variation of hydrate formation time with the subcooling of hydrate formation temperature. When subcooling increases, the hydrate formation time decreases exponentially. In general subcooling higher than 5 °F will cause hydrate formation to occur at the hydrocarbon-water interface in flowlines.

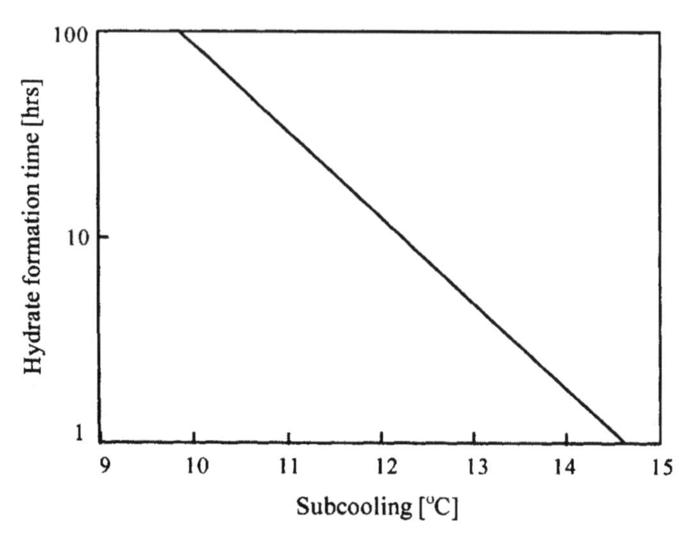

Figure 20.5 Variation of hydrate formation time with subcooling (Ellision and Gallagher, 2001).

The thermodynamic understanding of hydrates indicates the conditions of temperature, pressure, and composition that a hydrate may form. However, it does not indicate where or when a hydrate plug will form in the system. Hydrate plugs can form in just a few minutes, or take several days to block production. There are two mechanisms of plug formation, one in which hydrates slowly build up on the bottom of the pipe, gradually restricting flow, and the other in which hydrates agglomerate in the bulk fluid, forming masses of slush which bridge and eventually block the flow. Both mechanisms have been observed in the field, though the latter is believed to be more prevalent. The mechanics of plug formation are not yet well understood, although it is known that certain geometries, such as flow restrictions at chokes, are prone to hydrate plug formation.

Control of hydrates relies upon keeping the system conditions out of the region in which hydrates are stable. During oil production operation, temperatures are usually above the hydrate formation temperature, even with the high system pressures at the wellhead (on the order of 5000-10000 psia). However, during a system shut-in, even well insulated systems will fall to the ambient temperatures eventually, which in the deep GoM is approximately 38-40 °F. There are many methods available for hydrate formation prediction. Most of them are based upon light gas hydrocarbon systems and vary in the complexity of the factors utilized within the computational procedures. The Peng-Robinson method is one typical equations of state (EOS) method, which is currently extensively utilized to predict hydrate boundaries.

Knowledge about hydrates has significantly improved in the last ten years. Hydrate disassociation can be predicted within one to three degrees with the exception of brines that have high salt concentration. The hydrate disassociation curves typically provide conservative limits for hydrate management design. The effects of thermodynamic hydrate inhibitors, Methanol and ethylene glycols can be predicted with acceptable accuracy.

When the temperature and pressure are in the hydrate region, hydrates grow so long as water and light hydrocarbons are available and can eventually develop blockages. Clearing hydrate blockages in subsea equipment or flowlines poses safety concerns and can be time consuming and costly. Hydrate formation is typically prevented by several methods including controlling temperature, controlling pressure, removing water, shifting thermodynamic equilibrium with chemical inhibitors such as methanol or mono-ethylene glycol, low-dosage hydrate inhibitors and etc.

20.2.3 Effects of Salt, MeOH, Gas Composition

The hydrate dissociation curve may be shifted towards lower temperatures by adding a hydrate inhibitor. Methanol, ethanol, glycols, sodium chloride, and calcium chloride are common thermodynamic inhibitors. Hammerschmidt (1969) suggested a simple formula to roughly estimate the temperature shift of the hydrate formation curve.

$$\Delta T = \frac{KW}{M(100 - W)} \tag{20.1}$$

where, ΔT : temperature shift, hydrate depression [°C]

K : constant [-], which is defined in Table 20.1.

W : concentration of the inhibitor in weight percent in the aqueous phase.

M : molecular weight of the inhibitor divided by the molecular weight of water.

Table 20.1 Constant of Equation (20.1) for various inhibitors.

Inhibitor	K value
Methanol	2335
Ethanol	2335
Ethylene glycol (MEG)	2700
Diethylene glycol (DEG)	4000
Triethylene glycol (TEG)	5400

The Hammerschmidt equation was generated based upon more than 100 natural gas hydrate measurements with inhibitor concentrations of 5 to 25 wt % in water. The accuracy of the equation is 5 % average error compared with 75 data points. The hydrate inhibition abilities are less for larger molecular weight of alcohol, for example, the ability of methanol is higher than that of ethanol and glycols. With the same weight percent, methanol has a higher temperature shift than that of glycols, but MEG has a lower volatility than methanol and MEG may be recovered and recycled more easily than methanol on platforms. Figure 20.6 shows the effect of typical thermodynamic inhibitors on hydrate formation at 20 wt % fraction. Salt, methanol and glycols act as thermodynamic hydrate inhibitors that shift the hydrate stability curve to the left. Salt has the most dramatic impact on the hydrate stability temperature. On a weight basis, salt is the most effective hydrate inhibitor and so accounting correctly for the produced brine salinity is important in designing a hydrate treatment plan. In offshore fields, MEG found more application than DEG and TEG because MEG has a lower viscosity and has more effect per weight.

Figure 20.7 shows the effect of salts in fluid to hydrate formation. Increasing salt content in the produced brine shifts the hydrate curve to lower temperatures at the same pressure. The solubility of salt to water has a limit based on the temperature.

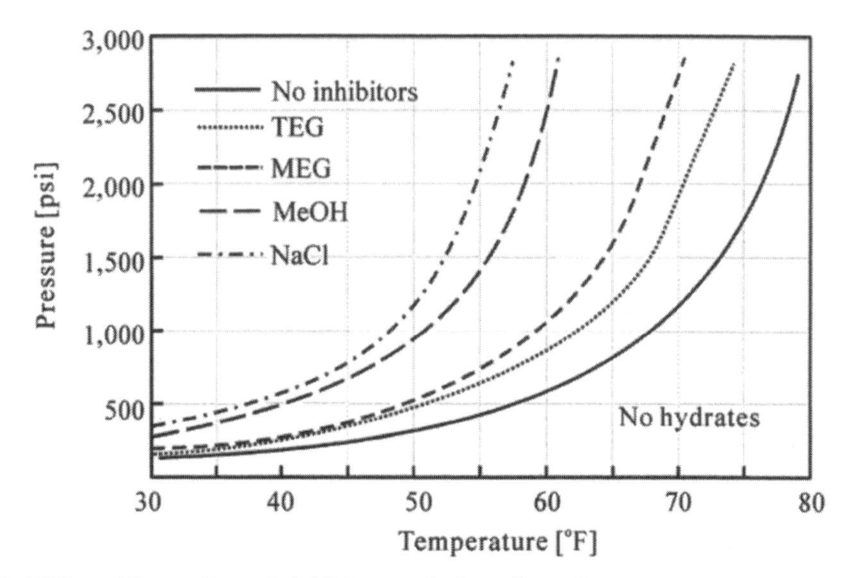

Figure 20.6 Effect of thermodynamic inhibitors on hydrate formation.

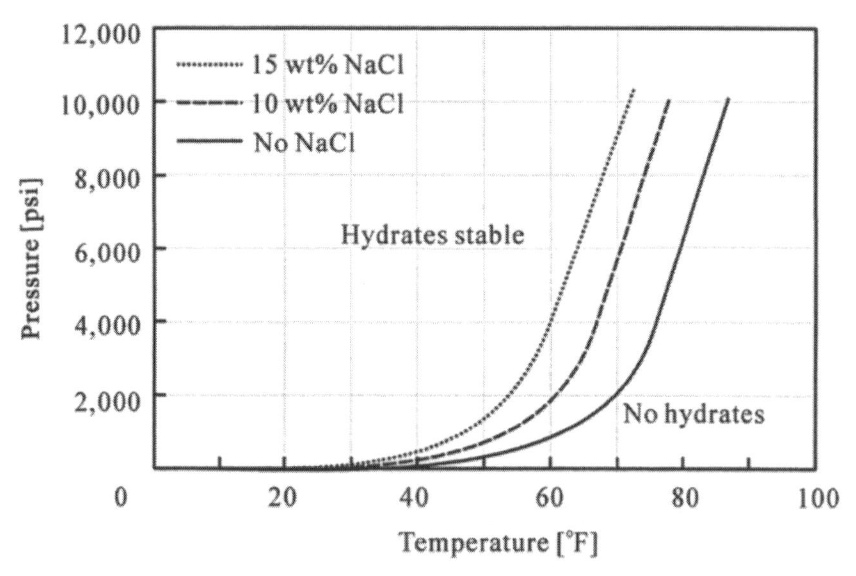

Figure 20.7 Effect of salts on hydrate formation.

Figure 20.8 shows the effect of gas composition on the hydrate formation curve. More small molecular components results in a lower hydrate formation at the same pressure. Figure 20.9 shows the effect of weight percentage of methanol on the hydrate formation curve. More weight percentage of methanol leads to a greater temperature shift of the hydrate formation curve.

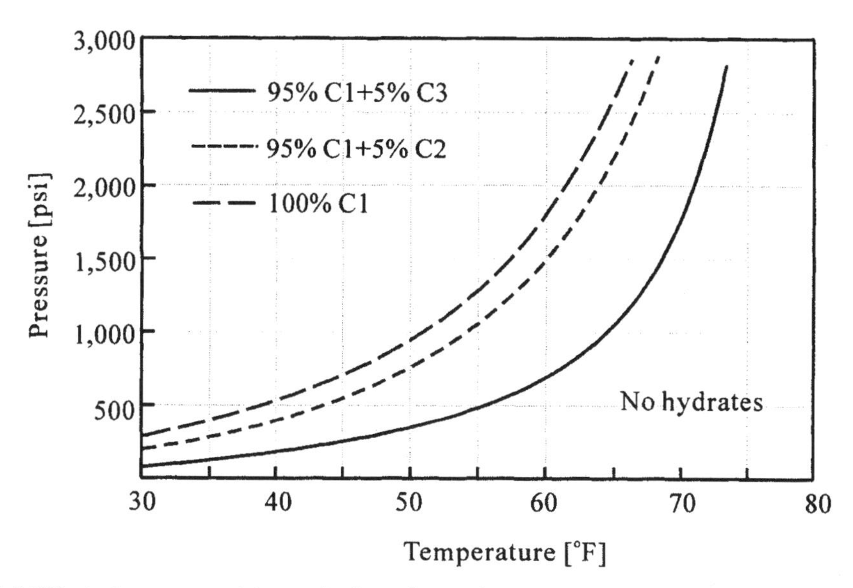

Figure 20.8 Effect of gas composition on hydrate formation.

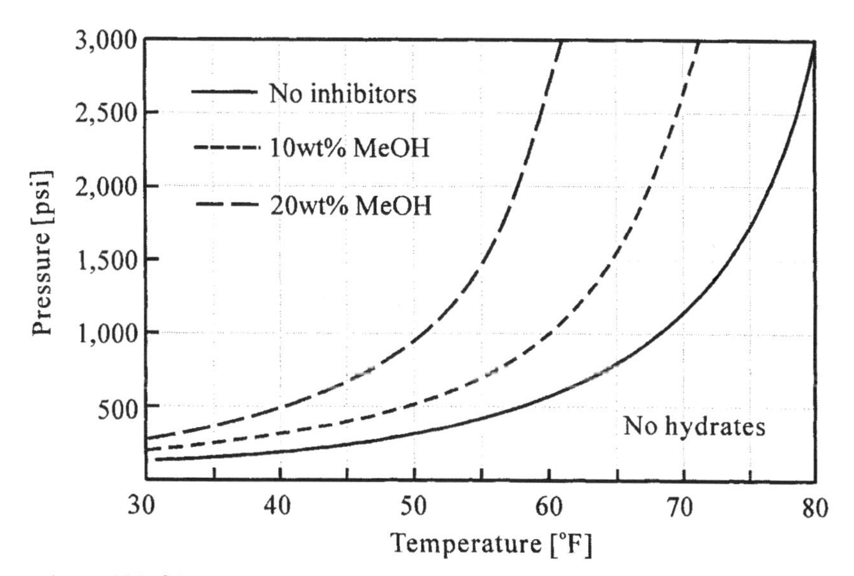

Figure 20.9 Effect of MeOH on hydrate formation.

20.2.4 Mechanism of Hydrate Inhibition

There are two types of hydrate inhibitors used in subsea engineering: thermodynamic inhibitors (THIs) and low-dosage hydrate inhibitors (LDHIs).

The most common thermodynamic inhibitors are methanol and MEG, even though ethanol, other glycols (DEG, TEG), and salts can be effectively used. They inhibit hydrate formation by reducing the temperature at which hydrates form. This effect is the same as adding anti-freeze to water to lower the freezing point. Methanol and MEG are the most commonly used inhibitors,

LDHIs include anti-agglomerants and kinetic inhibitors. LDHIs have found many applications in subsea systems in recent years. LDHIs prevent hydrate blockages at significantly lower concentrations, e.g., less than 1 weight percent, than thermodynamic inhibitors such as methanol and glycols. Unlike thermodynamic inhibitors, LDHIs do not change the hydrate formation temperature. They either interfere with formation of hydrate crystals or agglomeration of crystals into blockages. Anti-agglomerates can provide protection at higher subcooling than kinetic hydrate inhibitors. However, low dosage hydrate inhibitors are not recoverable and they are expensive. The difference in hydrate inhibition mechanism between LDHIs and THIs is shown in Figure 20.10.

LDHIs are preferred to be used for regular operations because they reduce volumes and can work out cheaper. For transient events, the volumes required are not usually that large, so there is no much benefit in LDHIs, and methanol is the preferred inhibitor.

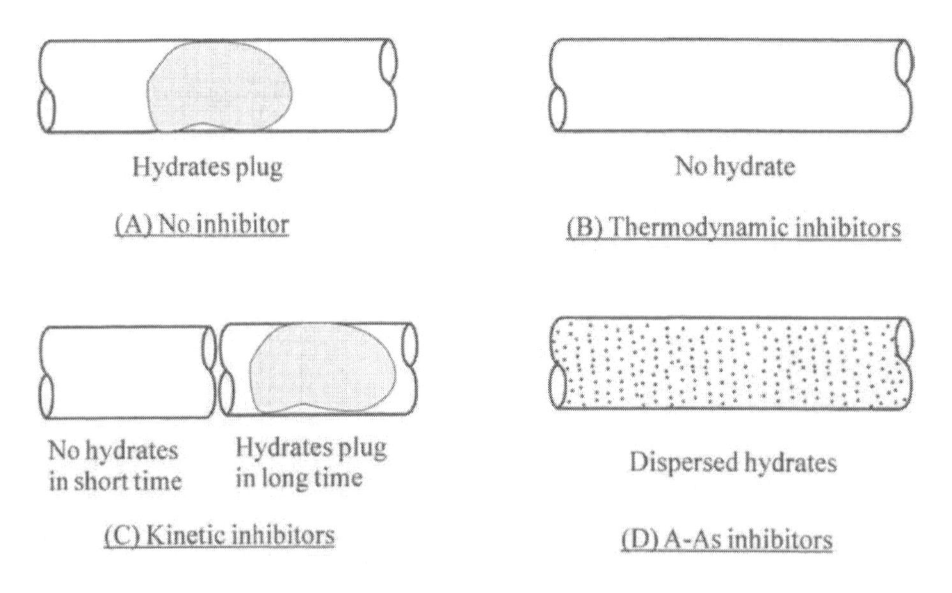

Figure 20.10 Mechanism of hydrate inhibition.

20.2.4.1 Thermodynamic Inhibitors

THIs inhibit hydrate formation by reducing the temperature at which hydrate form by changing the chemical potential of water. This effect is the same as adding anti-freeze to water to lower the freezing point. THIs include methanol, glycols and others. In general, methanol is vaporized into the gas phase of a pipeline, and then dissolves in any free water accumulation to prevent hydrate formation. Hydrate inhibition occurs in the aqueous liquid, rather than in the vapor or oil/condensate. Although most of the methanol dissolves in the water phase, a large amount of methanol remains in the vapor or oil/condensate phase, therefore, the proportions of methanol dissolved in the vapor or oil/condensate liquid phases are usually counted as an economic loss.

20.2.4.2 Low-dosage Hydrate Inhibitors

Kinetic inhibitors (KIs) are low molecular weight water-soluble polymers or copolymers that prevent hydrate blockages by bonding to the hydrate surface and delaying hydrate crystal nucleation and/or growth. They are dissolved in a carrier solvent and injected into the water phase in pipelines. These inhibitors work independently of watercut, but are limited to relatively low subcooling (less than 20 °F), which may not be sufficient for deep-water applications. For a greater subcooling, KI must be blended with a thermodynamic inhibitor. Additionally, KIs inhibition effect is time-limited, and thus their benefit for shut-in is limited. KIs have been applied in the North Sea and the Gulf of Mexica, Long-term shutdowns will require depressurization, which complicates the restart process, and methanol without KIs will be required for restarts. KIs are generally environmentally friendly.

Anti-agglomerants (AAs) are surfactants, which cause the water phase to be suspended as small droplets. When the suspended water droplets convert to hydrates, the flow characteristics are maintained without blockage. They allow hydrate crystals to form but keep the particles small and well dispersed in the hydrocarbon liquid. They inhibit hydrate plugging rather than hydrate formation. AAs can provide relatively high subcooling up to 40 °F, which is sufficient for deepwater applications and have completed successful field trials in deepwater Gulf of Mexico production system. AA effectiveness can be affected by type of oil or condensate, water salinity and watercut. For deepwater gas developments AAs can only be applied where there is sufficient condensate, such that the in-situ watercut is less than 50%. Methanol may still be required for shutdown and restart. AAs have toxicity issues and may transport micro-crystals of hydrate into and remain in the condensed/oil phase.

20.3 Hydrate Prevention

Gas subsea systems typically contain small quantities of water, which allows them to be continuously treated with methanol or glycol to prevent hydrate formation. These inhibitors prevent the formation of hydrates by shifting the hydrate stability curve to lower temperatures for a given pressure. If the systems produce too much water, it is difficult to economically treat with methanol. As a result, the system designs have to incorporate insulation of almost all components of system and develop complex operating strategies to control hydrate formation during transient activities such as system start-up and shut-in.

Hydrate prevention techniques for subsea system include:

- Thermodynamic inhibitors
- Low-dosage hydrate inhibitors (LDHIs)
- Low-pressure operation
- Water removal
- Insulation, and
- Active heating.

20.3.1 Thermodynamic Inhibitors

The most common method of hydrate prevention in deepwater developments is injection of thermodynamic inhibitors, which include methanol, glycols, and others. Methanol and MEG are the most commonly used inhibitors, though ethanol, other glycols, and salts can be effectively used.

Inhibitor injection rates, whether methanol or MEG, are a function of water production and inhibitor dosage. Inhibitor dosage is a function of design temperature and pressure, and produced fluid composition. Water production multiplied by the methanol dosage is the inhibitor injection rate, which will change throughout field operating life due to typically decreasing operating pressures and increasing water production.

Selection of the hydrate inhibitor is an important decision to be made in the early FEED design and can involve a number of criteria:

- Capital costs of topsides process equipment, especially for regeneration;
- Capital costs of subsea equipment;
- Topsides weight/area limitations;
- Environmental limits on overboard discharge;
- Contamination of the hydrocarbon fluid and impacts on downstream transport /processing;
- Safety considerations;
- System operability;
- Local availability of inhibitor.

Technical advantages and disadvantages of methanol and MEG are compared in Table 20.2.

Table 20.2 Comparison of methanol and MEG (Cochran and Gudimetla, 2004).

	Advantages	Disadvantages
Methanol	• move hydrate formation temperature more than MEG in a mass basis; • less viscous; • less likely to cause salt precipitation; • relative cost of regeneration system is less than for MEG, • approximate GoM cost of 1.0 $/gal;	• losses of methanol to gas and condensate phases can be significant, leading to a lower recovery (<80%) ; • impact of methanol contamination in downstream processing; • low flash point; • environmental limitation on overboard discharge.
MEG	• easy to recover with recovery of 99%. • low gas & condensate solubility • approximate GOM cost of 2.5$/gal	• high viscosity, impacts umbilical and pump requirements; • less applicable for restarts, stays with aqueous phase at bottom of pipe; • more likely to cause salt precipitation.

Methanol and MEG are both effective inhibitors if sufficient quantities are injected; for deep water, inhibitor dosages of 0.7 to 1 bbl of inhibitor per bbl water are generally used. Methanol can provide higher hydrate temperature depression but this effect is typically countered by high losses to the hydrocarbon liquid and gas phases. The selection of inhibitor is often based on economics, downstream process specifications, environmental issues and/or operator preferences.

Costs for inhibition systems are driven by up-front capital costs, which are dominated by the regeneration system, and also by makeup costs for inhibitor loss. Methanol is cheaper per unit volume but has greater makeup requirements. Additionally, a methanol regeneration system may be as much as 50% less expensive than a MEG regeneration system. The methanol system starts out cheaper but, with increasing field life, becomes more expensive due to methanol makeup costs.

The risks in using thermodynamic inhibitors include:

- Underdosing, particularly due to not knowing water production rates;
- Inhibitor not going where intended (operator error or equipment failure);
- Environmental concerns, particularly with methanol discharge limits;
- Ensuring remote location supply;
- Ensuring chemical/material compatibility; and
- Safety considerations in handling methanol topsides.

20.3.2 Low-dosage Hydrate Inhibitors

While development of AAs and KIs continues, cost per unit volume of LDHIs is still relatively high, but is expected to decrease as their use increases. A potentially important advantage is that they may extend field life when water production increases.

20.3.3 Low Pressure

Low pressure operation refers to maintaining the system pressure to be lower than the pressure corresponding to the ambient temperature based on the hydrate dissociation curve. For deep water with an ambient temperature of 39°F (4°C), the pressure may need to be 300 psia (20 bara) or less. Operation at such a low pressure in the wellbore is not practical because pressure losses in a deepwater riser or long distance tieback would be significant.

By using subsea choking and keeping the production flowline at a lower pressure, the difference between hydrate dissociation and operating temperatures (*i.e.* subcooling) is reduced. This lower subcooling will decrease the driving force for hydrate formation and can minimize inhibitor dosage.

20.3.4 Water Removal

If enough water can be removed from the produced fluids, hydrate formation will not occur. Dehydration is the common hydrate prevention technique applied to export pipelines. For subsea production systems, subsea separation systems can reduce water flow in subsea flowlines. The advantage for applying subsea separation is not only hydrate control, but also increasing recovery of reserves and/or accelerating recovery by making the produced fluid stream lighter and easier to lift. Another benefit is reduced topsides water handling, treating and disposal.

As a new technology, subsea water separation/disposal systems are designed to separate bulk water from the production stream close to subsea trees on the seafloor. Basic components of such a system include separator, pump to re-inject water, and water injection well. Additional components include instrumentation, equipment associated with controlling the pump and separator, power transmission/distribution, and chemical injection. Watercut leaving the separator may be as high as 10%. Operating experience on the Troll Pilot has shown watercuts of 0.5 to 3%. Because these systems do not remove all free water, and water may condense farther downstream, subsea bulk water removal does not provide complete hydrate protection. These systems need to be combined with another hydrate prevention technique, for example, continuous injection of a thermodynamic inhibitor or LDHI. The main risk associated with the subsea water separation systems is reliability.

20.3.5 Thermal Insulation

Insulation provides hydrate control by maintaining temperatures above hydrate formation conditions. Insulation also extends the cooldown time before reaching hydrate formation temperatures. The cooldown time gives operators time either to recover from the shutdown and restart a warm system or prepare the system for a long-term shutdown.

Insulation is generally not applied to gas production systems, because the production fluid has low thermal mass and also will experience JT cooling. For gas systems, insulation is only applicable for high reservoir temperatures and/or short tieback lengths. One advantage of an insulated production system is that it can allow higher water production, which would not be economical with continuous inhibitor injection. However, shutdown and restart operations would be more complicated. For example, long-term shutdowns will probably require depressurization. An overview of insulation requirements is described in the Chapter 19.

20.3.6 Active Heating

Active heating include electrical heating and hot fluid circulation heating in a bundle. In flowlines and risers, active heating must be applied with thermal insulation to minimize power requirements.

20.3.6.1 Electrical heating (EH)

EH is a very fast developing technology and has found applications in the offshore fields including Nakika, Serrano, Oregano and Habanero in the GOM, and Asgard, Huldra and Sliepner in the North Sea. Advantages of electrical heating include eliminating flowline depressurization, simplifying restart operations, and providing ability to quickly remediate hydrate blockages.

Electrical heating techniques include:

- Direct heating, using the flowline as an electrical conductor for resistance heating.
- Indirect heating, using an electrical heating element installed on the outer surface of flowline.

20.3.6.2 Hot fluid circulation in pipe bundle

Hot fluid heating has many advantages same as electrical heating. Instead of using electricity for supplying heat, hot fluid, typically inhibited water, circulates in the bundles to provide heat to the production fluids. Examples of such bundles include statoil Asgard and Gullfaks South, Conoco Britannia and BP King. These bundles can be complex in design, with thermal and mechanical design, fabrication, installation, life cycle and risk issues that need to be addressed.

Active heating techniques provide a good level of protection. With active heating, hydrate control is simply a matter of power, insulation and time. Active heating can increase operating flexibility of a subsea production system, such that concerns including watercut, startup and operating flowrate and depressurization times are of lesser importance.

Electrically heated flowlines and low dosage hydrate inhibitors are two developing technologies in hydrate prevention for reducing the complexity of the design and operation of subsea systems. Electrically heated flowline technology reduces hydrate concerns in subsea systems. Instead of relying on the lengthy process of blowdown for hydrate remediation, electrical heating provides a much faster way to heat the flowline and remove the plug. The other potential advantages of electrical heating are covered in the Chapter 19. Low dosage hydrate inhibitors reduce the volume of chemicals that must be transported and injected into the subsea system. Methanol treatment rates, for hydrate control, are on the order of one barrel of methanol for each barrel of produced brine. The low dosage hydrate inhibitors may be able to accomplish the same task at dosage rates of or less than 1%. This leads to a reduction in umbilical size and complexity. However, it must be noted that the hydrate inhibitors must be injected continuously to prevent hydrate formation.

20.4 Hydrate Remediation

Like the kinetics of hydrate formation, hydrate dissociation is a poorly understood subject and applying laboratory observations to field predictions has proven difficult. Part of the reason is the complicated interplay of flow, heat transfer, and phase equilibria. The dissociation behaviour of hydrate depends on the hydrate size, porosity, permeability, volume of occluded water, "age" of the deposit, and local conditions such as temperature, pressure, fluids in contact with the plug, and insulation layers over the pipeline.

Two factors combine to make hydrate plugs exceedingly difficult to remove: it takes a large amount of energy to dissociate the hydrate and heat transfer through the hydrate phase is slow. Hydrates also concentrate natural gas. One ft^3 of hydrates can contain up to 182 ft^3 of gas. This has significant implications for safety in depressurizing hydrate plugs. Hydrate dissociation is highly endothermic. If heat transfer through pipeline insulation layer from the surroundings is low, the temperature near a dissociating hydrate can drop rapidly. In addition, as gas is evolved during hydrate dissociation, JT cooling of the expanding gas is alos possible. By either of these mechanisms, additional hydrates and/or ice can form during the dissociation process. For a more complete discussion of gas hydrate structures and properties, the reader is referred to the book by Sloan.

While the design is intended to prevent hydrate blockages, industry must include design and operational provisions for remediation of hydrate blockages. A "Hydrate Blockage Remediation Plan" should be developed for a subsea system where hydrate formation is an issue. This tells operators how to spot when a blockage might be occurring and what to do about it. The state of the art would be to have an "on-line" or "real-time" system using a calculation engine such as OLGA to continuously predict temperatures and pressures in the pipeline and raise an alarm if hydrate formation conditions are detected. Such as system may also be able to pinpoint the most likely blockage location.

Hydrate remediation techniques are similar to hydrate prevention techniques, which include,

- Depressurization from two sides or one side, by reducing pressure below hydrate pressure at ambient temperature, the hydrate will become thermodynamically unstable.
- Thermodynamic inhibitors; the inhibitors can essentially melt blockages with direct hydrate contact.
- Active heating; by increasing temperature to above the hydrate dissociation temperature and providing significant heat flow to relatively quickly dissociate a blockage.
- Mechanical methods; drilling, pigging or scraping have been attempted, but are generally not recommended. thruster or pig inserted from surface vessel with coiled tubing through a work-over riser at launchers. Melting by jetting with MEG.
- Pipeline segment replacement.

20.4.1 Depressurization

Depressurization is the most common technique used to remediate hydrate blockages in production systems. Rapid depressurization should be avoided because it can result in JT cooling, which can worsen the hydrate problem and form ice. From both safety and technical standpoints, the preferred method to dissociate hydrates is to depressurize from both sides of the blockage. If only one side of a blockage is depressurized, then a large pressure differential will result across the plug, which can potentially create a high speed projectile.

When pressure surrounding a hydrate is reduced below dissociation pressure, hydrate surface temperature will cool below seabed temperature, and heat influx from the surrounding ocean will slowly melt the hydrate at the pipe boundary. Lowering pressure also drops hydrate formation temperature and helps prevent more hydrates from forming in the rest of the line. Because most gas flowlines are not insulated, hydrate dissociation can be relatively fast due to higher heat flux from pipeline surface, as compared to an insulated or buried flowline.

The depressurization of the flowlines, known as blowdown, creates many operational headaches. Not only does the host facility have to handle large quantities of gas and liquid exiting the flowlines; it must also be prepared to patiently wait until the plug to dissociates. Since multiple plugs are common, the process can be extremely long and much revenue is lost. Some subsea system configurations, such as flowlines with a number of low spots can be extremely difficult to blow down. The best policy is to operate, if at all possible, in a manner to prevent hydrates from forming in the first place. Depressurization may not be effective due to production system geometry; sufficiently high liquid head in the riser or flowline may prevent depressurization below hydrate conditions. In this case, some methods may be needed to reduce the liquid head. If additional equipment is needed to perform depressurization or remediation, equipment mobilization needs to be factored in the total downtime. System designers need to evaluate cost/benefit of including equipment in the design for more efficient remediation vs. having higher remediation times.

20.4.2 Thermodynamic Inhibitors

Thermodynamic inhibitors can be used to melt hydrate blockages. The difficulty of applying inhibitors lies in getting the inhibitor in contact with the blockage. If the injection point is located relatively close to the blockage, as may be the case in a tree or manifold, then simply injecting the inhibitor can be effective. Injecting inhibitor may not always help with dissociating a hydrate blockage, but it may prevent other hydrate blockages from occurring during remediation and restart.

If the blockage can be accessed with coiled tubing, then methanol can be pumped down the coiled tubing to the blockage. In field applications, coiled tubing has reached as far as 14800 ft in remediation operations, and industry is currently targeting lengths of 10 miles.

20.4.3 Active Heating

Active heating can remediate hydrate plugs by increasing temperature and heat flow to the blockage; however, safety concerns arise when applying heat to a hydrate blockage. During the dissociation process, gas will be released from the plug. If the gas is trapped within the plug, then the pressure can build and potentially rupture the flowline. Heating evenly applied to a flowline can provide a safe, effective remediation.

Active heating can remediate a block age within hours, whereas depressurization can take days or weeks. The ability to quickly remediate hydrate blockages can enable less conservative designs for hydrate prevention.

20.4.4 Mechanical Methods

Pigging is not recommended for removing a hydrate plug because they can compress the plug, which will compound the problem. If the blockage is complete, it will not be possible to drive a pig. For a partial blockage, pigging may create a more severe blockage.

Coiled tubing is another option for mechanical hydrate removal. Drilling a plug is not recommended because it can cause large releases of gas from the blockage. Coiled tubing can be inserted through a lubricator. Coiled tubing access; either at the host or somewhere in the subsea system; should be decided early in the design phase.

20.4.5 Safety Considerations

Knowledge of the location and length of a hydrate blockage is very important in determining the best approach to remediation, although the methodology is not well defined, This information facilitates both safety considerations in terms of distance from the platform and time necessary to dissociate the blockage.

When dissociating a hydrate blockage, operators should assume that multiple plugs may exist both from safety and technical standpoints. The following two important safety issues should be kept in mind:

- Single sided depressurization can potentially launch a plug like a high-speed projectile and result in ruptured flowlines, damaged equipment, release of hydrocarbons to the environment, and/or risk to personnel.
- Actively heating a hydrate blockage needs to be done such that any gas released from the hydrate is not trapped.

20.5 Hydrate Control Design Philosophies

Steady state temperature calculations from the flow assurance process are used to indicate the flow rates and insulation systems that are needed to keep the system above the hydrate formation temperature during normal operation. Transient temperature calculations are used to examine the conditions of transient action such as startup and shut-in process. It is essential that each part of the system has adequate cooldown time. Dosing of hydrate inhibitor is the way hydrate formation is controlled when system temperatures drop into the range in which hydrates are stable during the transient actions. Dosing calculations of thermodynamic hydrate inhibitor indicate how much inhibitor must be added to the produced water at a given system pressure. The inhibitor dosing requirements are used to determine the requirements for inhibitor storage, pumping capacities, and number of umbilical lines in order to ensure that the inhibitor can be delivered at the required rates for treating each subsea device during startup and shut-in operations.

20.5.1 Selection of Hydrate Control

Injection rates of thermodynamic inhibitors are a function of water production and inhibitor dosage. Inhibitor dosage is a function of design temperature and pressure, and produced fluid composition. Water production multiplied by the methanol dosage is the inhibitor injection rate, which will change throughout field operating life due to typically decreasing operating pressures and increasing water production. Hydrate control strategies for gas and oil systems are different. Gas systems are designed for continuous injection of a hydrate inhibitor. Water production is small, typically only the water of condensation. Inhibitor requirements are thus relatively small, of the order of 1-2 bbl MeOH/mmscf. Oil systems produce free water. Continuous inhibition is generally too expensive so some alternative control system is adopted. Insulation is often used to control hydrates during normal production and some combination of blow down and methanol injection used during start-up and shut-in. But insulation is costly, and the more serious drawback may be lengthening the hydrate plug remediation time, because insulation limits the heat transfer required to melt the hydrate solid, in case hydrates have formed. Table 20.3 provides a summary of the applications, benefits and limitations of the three classes of chemical inhibitors.

Table 20.3 Summary of applications, benefits & limitations of chemical Inhibitors (Pickering et al.).

Thermodynamic Hydrate Inhibitors	Kinetic Hydrate Inhibitors	Anti-Agglomerant Inhibitors
	Applications	
1. Multiphase 2. Gas & Condensate 3. Crude Oil	1. Multiphase 2. Gas & Condensate 3. Crude Oil?	1. Multiphase 2. Condensate 3. Crude Oil
	Benefits	
1. Robust & effective 2. Well understood 3. Predictable 4. Proven track-record	1. Lower OPEX/CAPEX 2. Low volumes (< 1wt%) 3. Environmentally friendly 4. Non-toxic 5. Tested in gas systems	1. Lower OPEX/CAPEX 2. Low volumes (< 1wt%) 3. Environmentally friendly 4. Non-toxic 5. Wide range of subcooling
	Limitations	
1. Higher OPEX/CAPEX 2. High volumes (10-60 wt%) 3. Toxic / hazardous 4. Environmentally harmful 5. Volatile – losses to vapour 6. 'Salting out'	1. Limited subcoolings (<10°C) 2. Time dependency 3. Shutdowns 4. System specific – testing 5. Compatibility 6. Precipitation at higher temps 7. Limited exp. in oil systems 8. No predictive models	1. Time dependency? 2. Shutdowns? 3. Restricted to lower watercuts 4. System specific – testing 5. Compatibility 6. Limited experience 7. No predictive models

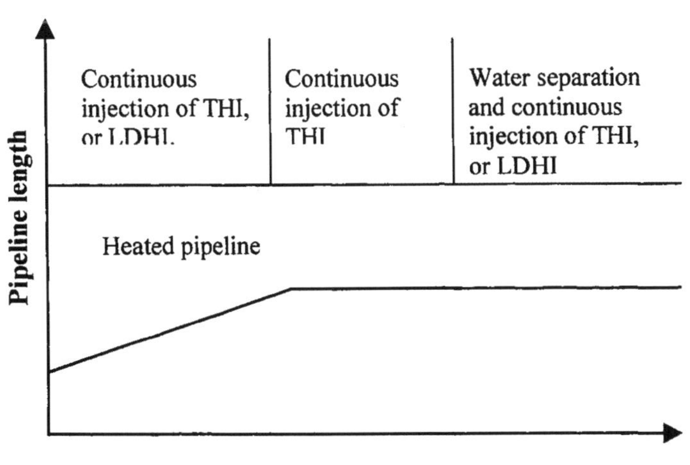

Figure 20.11 Hydrate control method for different water cut and pipeline length.

Figure 20.11 shows the relationship between hydrate control options for different water cut and pipeline length. LDHIs offer the ability to treat production with higher water rates because they are injected at lower quantities. Subsea separation of bulk water in combination with either thermodynamic or low-dosage inhibitors also can enable developments with potentially high water production. Basically, THIs, LDHIs, low-pressure operation, subsea processing (water separation) and thermal management are used in fields for hydrate prevention in subsea systems.

The selection of hydrate mitigation and remediation strategies is based on technical and economic considerations and the decision is not always clear-cut. While continuous injection of thermodynamic inhibitors are expected to remain the most economic and technically feasible approach to hydrate control, LDHIs or subsea processing, will offer advantages for some developments. Whatever the hydrate control strategy, these decisions are critical in the early design stage because of the many impacts on both the subsea and topsides equipment selection and design.

The main benefits of the thermodynamic hydrate inhibitors are their effectiveness, reliability (provided sufficient quantities are injected) and proven track-records. However, these benefits are outweighed by significant limitations, including the high volumes, high associated costs (both CAPEX and OPEX), toxicity and flammability. In addition, they are harmful to the environment and significant disposal into the environment is prohibited.

Kinetic hydrate inhibitors are injected in much smaller quantities compared to thermodynamic inhibitors and therefore offer significant potential costs savings, depending on the pricing policies of major chemical suppliers. They are also typically non-toxic and environmentally friendly. Moreover, considerable field experience is now available following a number of successful trials. However, they have some important limitations, including restrictions on the degree of subcooling (typically only guaranteed for less than 10°C) and problems associated with residence times (implications for shutdowns). In addition, the effectiveness of KHIs appears to be system specific meaning that testing programs are required prior to implementation. Unfortunately adequate testing can require appreciable quantities of production fluids which may not be available, particularly for new field developments. Furthermore, KHIs can interact with other chemical inhibitors (e.g. corrosion inhibitors) and testing programs need to account for this too. Finally, there are no established models for predicting the effectiveness of KHIs which presents difficulties for field developers considering the application of these chemicals.

The benefits and limitations of Anti-Agglomerants are largely similar to those for KHIs, although AAs do not have the same subcooling limitations. However, there is uncertainty about the effectiveness of AAs under shutdown or low flow rate conditions and it is postulated that agglomeration may still proceed. In addition, the one major limitation of AAs compared to KHIs or THIs is that they are limited to lower watercuts due the requirement for a continuous hydrocarbon liquid phase. Finally, compared to both THIs and KHIs, field experience with Anti-Agglomerants appears to be lacking which is reflected by the relatively small number of publications available in the open literature.

Thermal management can assist with maintaining some room for response by assisting with adequate temperatures for both hydrate and paraffin control. Passive thermal management is by insulation or burial. Pipe-in-pipe and bundled systems can extend the production cool down time before continuous hydrate inhibition is required. However, there may not be sufficient thermal capacity to provide the necessary cool down time for the shut-in greater than four to eight hours. Phase change material insulation systems that provide heat "storage" are beginning to appear as commercial systems. Burial can provide extended cool down time due to the thermal mass of the soil. Warm up time and cool down time can be optimized with insulation and thermal mass.

Both electrical and heating media systems belong to the active thermal management. Heating media systems require pipe-in-pipe or bundle flow line designs to provide the flow area required for the circulation of the heating media. When design, installation, and corrosion management issues are successful, the heating media systems can be reliable. However, the heat transported is still limited by the carrier insulation and heat input at the platform.

20.5.2 Cold Flow Technology

Cold flow technology is a completely new concept to solve hydrate blockage problem in steady-state operation condition, which is recently developed and owned by SINTEF. It is different from the chemical-based technology (THIs, LDHIs) and insulation/heating technology for hydrate protection, and concerns cost-effective flow of oil, gas and water mixtures in deepwater production pipelines, from wellhead to processing facility without using chemicals to prevent hydrate and/or wax deposition. In cold flow technology the hydrate formation occurs under controlled conditions in specialized equipment. The formed hydrate particles flow easily with the buck fluid mixture as slurry and not form wall deposits or pipeline blockage. Cold flow technology is environmentally friendly because of envisaged reduction in the use of bulk and specialty chemicals. This technology provides the most exciting challenge and could result in significant economic savings.

Figure 20.12 shows the flow diagram of cold flow hydrate technology process. The flow enters from a wellbore (WELL), and leaves through a cold flow pipeline (CFP). This system has several main process units: a wellhead unit, WHU; a separator unit, SU; a heat exchanger unit, HXU, and a reactor unit, RU. G/L means gas/liquid. Depending on the watercut and gas/oil ratio (GOR) and other fluid properties, more than one set of SU+HXU+RU may be required.

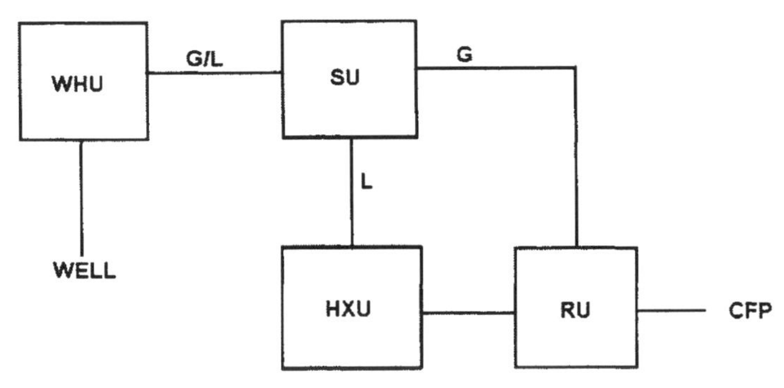

Figure 20.12 Block diagram of cold flow hydrate technology process (Gudmundsson, 2002).

Figure 20.13 demonstrates a schematic draw of the SINTEF-BP Cold Flow project, which develops a very robust process for long-distance transportation of unprocessed well stream containing water by converting the water to a very stable and transportable gas hydrate. This process is successfully proven in a one inch flow loop facility, operated with a variety of field fluids. This technology can reduce CAPEX by 15-30% for subsea tie-backs.

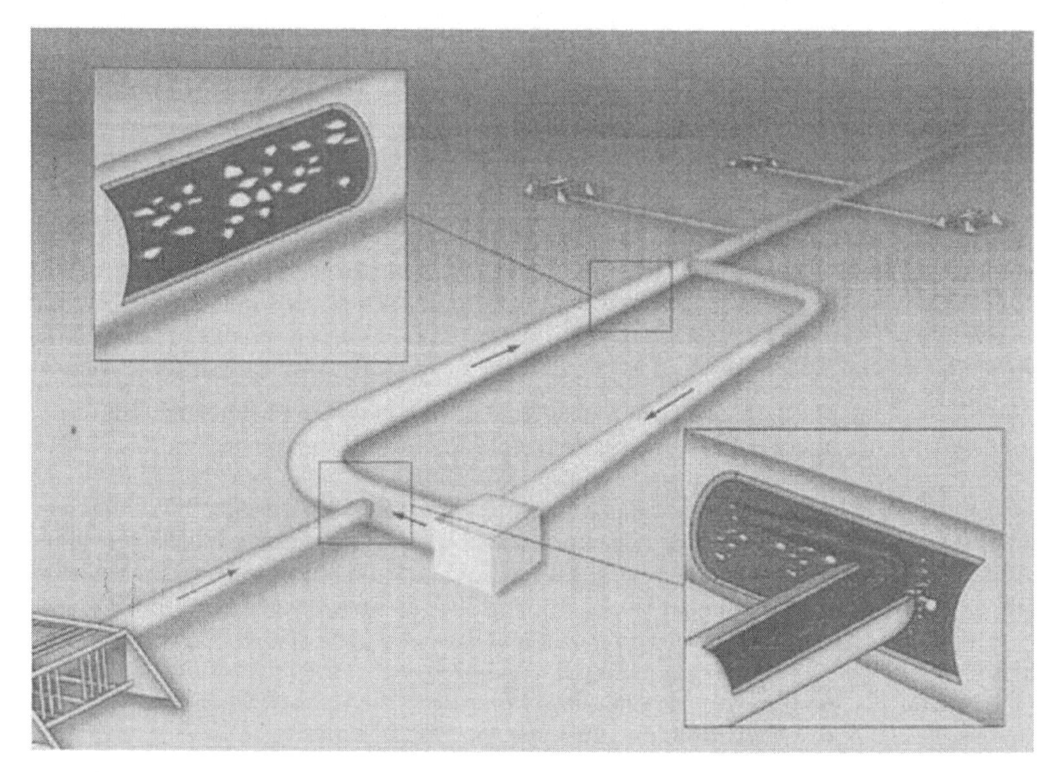

Figure 20.13 A schematic draw of cold flow project (Lysne, SINTEF, 2005).

20.5.3 Hydrates Control Design Process

Hydrate control design process in subsea hydrocarbons system is summarized as follows:

- determine operating conditions: pressure, temperature, GOR, water-cut;
- obtain good representative samples of oil, gas and water;
- measure chemical composition and phase behaviour;
- analyze reservoir fluids to determine hydrate formation conditions;
- perform hydrate prediction calculations;
- estimate the effects of insulation and thermodynamic inhibitors;
- determine thermodynamic hydrate inhibitor dosing and sizes of umbilical and inhibitor storage; consider the use of LDHIs.

20.5.4 Hydrates Control Design and Operation Guideline

The Guidelines for hydrate control utilized in subsea hydrocarbons system design and operations are summarized as:

- Keep the entire production system out of the hydrate formation envelope during all operations. Current knowledge is not sufficient to design a system to operate in the hydrate region without hydrate or blockage formation.
- Inject thermodynamic inhibitors at the subsea tree to prevent the formation of hydrates in the choke and downstream during transient operations.
- Use LDHIs inhibitors only for transient startup/shutdown operations and not for continuous operation.
- Insulate flowlines and risers from heat loss during normal operation and to provide cooldown time during shutdown. Insulating subsea equipment (trees, jumpers, and manifolds) is also included.
- Consider wellbore insulation to provide fast warm-up during restart operations and to increase operating temperatures during low rate operation.
- Determine minimum production rates and flowing wellhead temperatures and check consistency with technical and economic criteria.
- Establish well and flowline startup rates to minimize inhibitor injection while assuring that the system warms in an acceptable amount of time.
- Ramp-up well production rates sufficiently fast to outrun hydrate blockage formation in wellbores.
- Provide system design and operating strategies to ensure the system can be safely shut down.
- Monitor water production from individual wells.
- Locate SCSSV at a depth where the geothermal temperature is higher than hydrate temperature at shut-in pressure.
- Remediate hydrate blockages via depressurization or heating.

20.6 Recover of Thermodynamic Hydrate Inhibitors

Thermodynamic hydrate inhibitors (methanol, glycols) are widely used as a primary hydrate inhibitor in subsea systems. For projects producing oil there is a methanol limit for sales oil quality. The issue of an oxygenated solvent limit for glycols in oil is still under discussion. Even through the recovery units of methanol and glycol are improving, the units require appreciable heat to recover the THIs, and scaling in the methanol and glycol stills can generate operations challenges, especially when the produced water chemistry is not available from the reservoir appraisal during the design phase. Glycol recovery units can be designed to remove the salts that have traditionally limited the glycol quality. To reduce the methanol and glycol to the requirements of oil quality, crude washing requires large volumes of water that must be treated to the seawater injection quality. The recovery units and wash units have a significant weight, and operability impact on the project and operations design. These units are large and heavy.

Figure 20.14 shows a typical flow chart of an offshore gas production facility with methanol as the hydrate inhibitor. As shown in the figure, the fluids from the subsea flowlines enter the separation vessels and then distribute in three separate phases as following:

- Gas export stream to onshore gas plant in vapor phase;
- Condensate or oil export stream to shore in liquid hydrocarbon phase;
- Produced water stream in aqueous phase.

The methanol in the hydrocarbon vapour phase is recovered by adsorption, and the methanol in the hydrocarbon liquid phase is recovered by water wash, mechanical separation, or the combination of them. Most of the injected methanol is in the aqueous phase. A methanol tower is the necessary device to recover the methanol from the aqueous phase. More than 96% of the methanol injected can be recovered, if good engineering judgment and experience are applied,

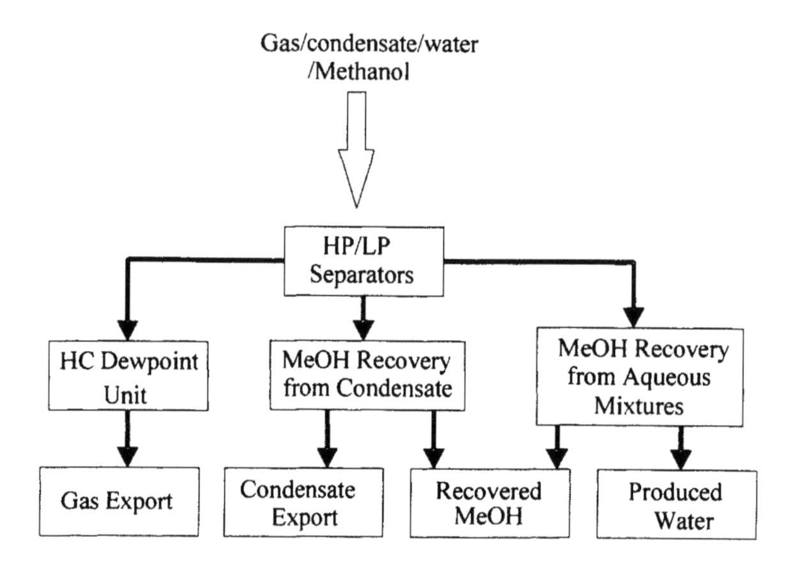

Figure 20.14 MeOH Recovery flowchart of offshore gas facility.

The aqueous phase mixture contains most of the methanol injected to the system. Methanol is a highly polar liquid and is fully miscible with water; therefore the recovery of methanol from water is achieved by distillation rather than phase separation.

Methanol is fully miscible with water, while the solubility of methanol in hydrocarbons is very small. Therefore, water (the solvent) can be used to extract methanol (the solute) from the hydrocarbon condensate (the feed) efficiently. Since water is the solvent, this extraction process is called water wash.

Methanol losses in the hydrocarbon liquid phase are difficult to predict. Solubility of methanol is a strong function of both the water phase and hydrocarbon phase compositions. The recovery of the dispersed and dissolved methanol from the feed to the condensate stabilizer can be achieved by mechanical separation like coalescence, or liquid-liquid extraction, or the combination of the two depending on the feed characteristics, weight, space, and cost of the entire subsea topside operation. Mechanical separation can separate the dispersed methanol from the hydrocarbon liquid phase to a certain extent. Liquid-liquid extraction can recover the smaller droplets of dispersed methanol and the dissolved methanol from the hydrocarbon liquid phase. However, liquid-liquid extraction is more costly than the mechanical separation. The solvent needs to be regenerated and re-used by the extraction tower. In the case, it increases the flowrate, the heating and cooling duties of the methanol distillation tower. For subsea operation, the treating facility space requirement, weight, and cost have to be considered together to determine which system to use to recover the methanol from the liquid hydrocarbon phase.

Methanol in the hydrocarbon vapor phase can be recovered by adsorption. If the hydrocarbon vapor passes through cryogenic system, the methanol in the hydrocarbon vapor phase condenses into liquid phase and can be recovered.

20.7 References

1. Cochran, S, (2003) "Hydrate Control and Remediation Best Practices in Deepwater Oil Developments", OTC 15255.
2. Cochran, S. and Gudimetla, R., (2004) "Hydrate management: Its importance to deepwater gas development success", Word Oil, Vol. 225, pp.55-61.
3. Edmonds, B., Moorwood, R.A.S, Szczepanski, R., Zhang, X., (1999), "Latest Developments in Integrated Prediction Modeling Hydrates, Waxes and Asphaltenes", Focus on Controlling Hydrates, Waxes and Asphaltenes, IBC, Aberdeen, October, 1999.
4. Ellision, B. and Gallagher, (2001), "Baker Petrolite Flow Assurance Course", Houston.
5. Ellison, B. T., Gallagher, C.T., and Lorimer, S. E., (2000), "The Physical Chemistry of Wax, Hydrates, and Asphaltene", OTC 11960.
6. Gudmundsson, J. S., 2002, "Cold Flow Hydrate Technology", 4th International Conference on Gas Hydrates, May 2002, Yokohama, Japan.
7. Hammerschmidt, E. G., (1969) "Possible Technical Control of Hydrate Formation in Natural Gas Pipelines", Brennstoff-Chemie, 50, pp. 117-123.
8. Lysne, D., (2005), "Ultra Long Tie-backs in Arctic Environments with the SINTEF-BP Cold Flow Concept", Oil and Gas Developments in Arctic and Cold Regions, US-Norway Oil & Gas Industry Summit, March , 2005, Houston.
9. Lorimer, S. E., and Ellison, B. T.,(2000), "Design Guidelines for Subsea Oil Systems", Facilities 2000: Facilities Engineering into the Next Millennium.
10. Mehta, A. P.,Hebert, P. B., Cadena, E. R., and Weatherman, J. P., (2002), "Fulfilling the Promise of Low Dosage Hydrate Inhibitors: Journey from Academic Curiosity to Successful Field Implementation", OTC 14057.
11. Pattee, F.M. and Kopp, F., (2000), "Impact of Electrically-Heated Systems on The Operation of Deep Water Subsea Oil Flowlines", OTC11894, 2000 Offshore Technology Conference, Houston, Texas, 1-4 May 2000.

12. Pedersen, K. S., Fredenslund, A., and Thomassen, P., (1989), Properties of Oils and Natural Gases, Gulf Publishing Company.
13. Pickering, P.F., Edmonds, B., Moorwood, R.A.S., Szczepanski, R. and Watson, M.J, (2003), "Evaluating New Chemicals and Alternatives for Mitigating Hydrates in Oil & Gas Production".
14. Sloan, E., (2000),"Offshore Hydrate Engineering Handbook", SPE Monograph, Vol. 21, 2000.

Part III

Flow Assurance

Chapter 21 Wax and Asphaltenes

21.1 Introduction

Waxes or paraffins are typically long-chain, normal alkane compounds that are naturally present in crude oil. When the temperature drops, these compounds can come out of the oil and form waxy and elongated crystals. If the control of waxes deposition is not effective, the waxy deposits can build up significantly with time and cause disruption of production, reduction of throughput, and even complete blockage of the flowlines. Subsea production facilities and pipelines are very susceptible to wax deposits and asphaltene precipitates induced by the lower temperature and decreasing pressure environment.

Asphaltenes are a component of the bitumen in petroleum and are usually black, brittle coal-like materials. They also exist as a thick black sludge. Asphaltenes can flocculate and form deposits under high shear, high velocity flow conditions. Asphaltenes are insoluble in non-polar solvents, but are soluble in toluene or other aromatics based solvents. Asphaltene deposits are very difficulty to remove once they occur. Unlike wax deposits and gas hydrates, asphaltene formation is not reversible. Frequently asphaltene deposition occurs with wax deposition, and makes the combined deposits very hard and sticky, and difficult to remove.

21.2 Wax

21.2.1 General

Wax varies in consistency from that of petroleum jelly to hard wax with melting points from near room temperature to over 100°C. Wax has a density of around 0.8 g/cm^3 and a heat capacity of around 0.140 W/(m·K). Paraffinic hydrocarbon fluids can cause a variety of problems in a production system ranging from solids stabilized emulsions to a gelled flowline. Although the pipeline is thermal insulated, the pipeline will ultimately cool to the ambient temperature given sufficient time in a shutdown condition. Problems caused by wax occur when the fluid cools from reservoir conditions and wax crystals begin to form. Wax is a naturally occurring substance in most crude oils, and depending upon its characteristics it may cause problems. Wax deposition on the pipeline walls increases as the fluid temperature in the pipeline decreases. The deposits accumulate on the pipe walls and over time may result in a drastic reduction of pipeline efficiency. The temperature at which crystals first begin to form is called cloud point or WAT (wax appearance temperature). At temperatures below the cloud point, crystals begin to form and grow. Crystals may form either in the bulk fluid, forming particles that are transported along with the fluid, or deposit on a cold surface where the crystals will build-up and foul the surface. Pour point temperature is the temperature at which the oil stops flowing and gels.

Gel formation and deposition are the main problems that wax may cause in a production system. A crude oil gel forms when wax precipitates from the oil and forms a three dimensional structure spanning the pipe. This does not occur while the oil is flowing because the intermolecular structure is destroyed by shear forces. However, when the oil stops flowing, wax particles will interact, join together and form a network resulting in a gel structure if enough wax is out of solution.

Wax deposition results in flow restrictions or possibly a blockage of a pipeline. Complete blockage of flow due to deposition is rare. Most pipeline blockages occur when a pig is run through a pipeline after deposition has occurred and a significant deposit has built up. In this situation the pig will continue to scrape wax from the pipe wall and build up a viscous slug or candle in front of the pig. However, if the candle becomes too large there will be insufficient pressure for the pig to move. When this occurs the pig becomes stuck and mechanical intervention to remove the candle will be necessary before the pig can be moved.

In subsea system, following issues caused by waxes should be addressed by flow assurance analysis:

- Deposition in flowlines, gradual with time but can block pipelines;
- Gelation of crude oil during shut-in;
- High start up pressures and high pumping pressures due to higher viscosity;
- Insulation for pipeline increases capital expense;
- Wax inhibitors increases operational expense;
- Pigging operation in offshore environment,
- Wax handling in surface facilities - needs higher separator temperature.

21.2.2 Wax Formation

The wax deposit is complex in nature, comprised of a range of normal paraffins of different lengths, some branched paraffins and incorporated oil. The build up of wax over time can eventually reach a point where flow rates are restricted. Wax deposits in the oil production flowlines are primarily comprised of paraffins of $C7^+$ in length. The deposition of paraffins is controlled by temperature. As the temperature in a system drops, paraffins that are in the liquid phase begin to come out of solution as solids. Wax deposits form at the wall of the pipe where the temperature gradient is at its highest.

In addition to wax deposition, the formation of sufficient wax solids can cause oil to "gel" at sufficiently low temperature during a shut-in of the system. Once this occurs, it is difficult, or even impossible to restart flow in the system due to the very high viscosity of the gelled oil.

The wax properties of oils have been characterized by cloud point and pour point measurements. The cloud point essentially measures the point at which wax first visibly comes out of solution as oil is cooled at atmospheric pressure. The pour point is the temperature at which the oil ceases to flow when the vessel is inverted for 5 seconds. These measurements give a general understanding of the temperatures at which wax deposition will become a problem and when crude oil gelling will become a problem.

The key to wax deposition prediction is a precise analysis of the concentration of the normal paraffins in an oil sample, which is carried out using the high temperature gas chromatography technique. The paraffin composition data is used to construct a thermodynamic model for prediction of wax deposition rates in the flowline as well as for predicting the cloud point and pour point dependence on pressure. The thermodynamic model may be combined with the model of flowline using software such as HYSYS, PIPESIM or OLGA, to predict where wax deposits will occur, how fast wax will accumulate, and the frequency at which the line must be pigged.

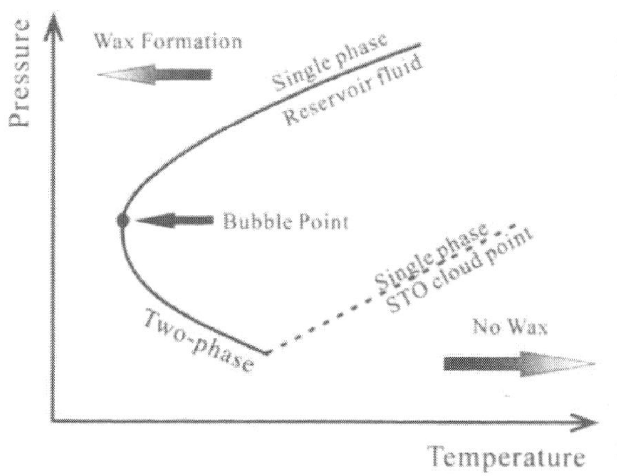

Figure 21.1 Temperature/pressure relationship in formation of wax.

In contrast to hydrates, wax deposits slowly and can be controlled by controlling system temperature and temperature differential at the pipe wall. Cloud points for deepwater GoM crude oils generally fall in the range 80 -100 °F. If the system is operated at a temperature approximately 10 - 20 °F above the cloud point, wax will not deposit. A "rule of thumb" for the deposition temperature which has been frequently used is Cloud Point + 15°F. Although this can usually be achieved for the wellbore and subsea tree, it is often not possible to operate at this high temperature in the flowline. The host arrival temperature may be limited due to processing concerns or by the temperature rating of equipment. During late life, the reservoir temperature may have dropped to the point that host arrival temperatures are substantially below the cloud point leading to significant deposition. Figure 21.1 shows the relationship of temperature and pressure for wax formation. The left hand of the curve shows the area where waxes will deposit, while in the right hand of the curve the waxes will not form.

Wax formed in wellbores can only be removed by a process such as coiled tubing, which is prohibitively expensive for subsea wells. As a result, it is important to control the temperature of the well tubing, tree, and other components that cannot be pigged (such as jumpers) above the point at which wax deposition occurs. This critical wax temperature may be chosen as Cloud Point +15°F. The flowing wellhead temperatures and pressures from flow assurance analysis are used to check that the tubing and tree remain above this critical wax deposition temperature. The need to prevent wax deposition in the wellbore, tree, and jumpers may set a minimum late life production rate, based on the temperature predictions from flow assurance.

High pour point oils present a potentially serious problem somewhat similar to the formation of hydrates. Wax formation during shut-in can be sufficient to make line restart difficult, or even impossible. Even though temperatures may be well above the pour point during steady state operation, it is impossible to know when a shut-in may occur.

Wax deposition will occur once the oil has fallen below the cloud point if there is a negative temperature gradient between the bulk oil and a surface. However, Deepstar has shown that wax deposition also can occur above the dead oil cloud point in some systems. Therefore, to prevent wax deposition, the system temperature should remain greater than 9°C above the dead oil cloud point. Experimentally the tendency of oil to deposit and the rate of deposition can be measured by placing a cold surface in contact with warm oil. Experimental systems include cold fingers, co-axial shearing cells, and pipe loops. The cold finger consists of a test tube-shaped metal finger cooled by flowing chilled fluid through the finger, and a heated stirred container for an oil sample. The co-axial shearing cell is similar to a cold finger but the finger rotates to create uniform shear on the surface. A pipe loop is a pipe in pipe heat exchanger where the cold fluid is pumped through the shell side and the oil is heated and pumped through the tube side. It might be expected that the pipe loop would be the preferred method due to its geometric similarity to a pipeline. However, if an actual field sample is to be used, none of the methods properly simulate the system since dynamic similitude is impossible to achieve without building a system of the same size. However each method can be used to measure the wax flux to the surface. With careful analysis the data from any laboratory method can be used to make predictions about field deposition.

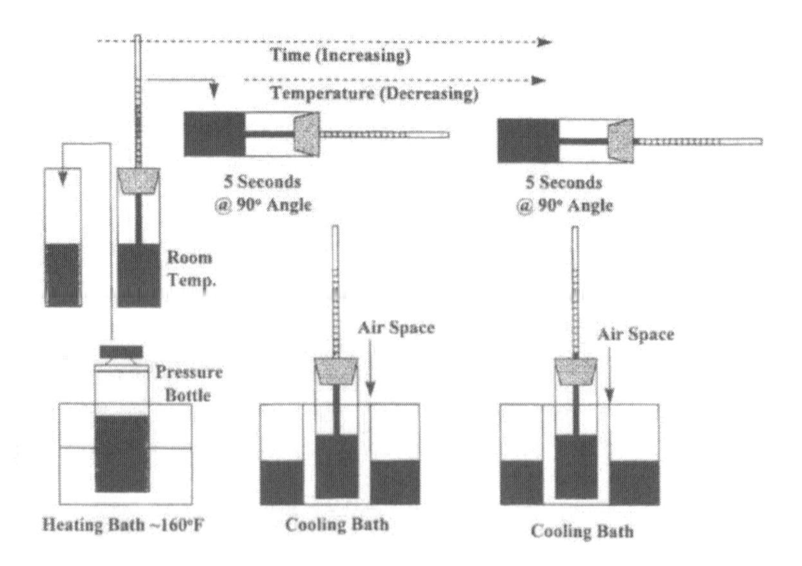

Figure 21.2 ASTM D-5853 pour-point testing (Becker, 1997).

As oil cools far below the cloud point it may begin to gel. While the amount of wax out of solution needed to form a gel structure varies considerably, 2% paraffin is used as a useful rule of thumb. The typical method of measuring a crude oil pour point is ASTM D-5853-95, which specifies the conditioning required for obtaining reproducible values. Because the cooling rate is not controlled in this method, results may vary from lab to lab, but in general the results are accurate to within ±5°F.

21.2.3 Viscosity of Waxy Oil

21.2.4 Gel Strength

The pour point is used to determine whether or not a crude oil will form a gel, the gel strength is defined as the shear stress at which gel breaks. The gel strengths measured in a rheometer are typically of the order of 50 dyne/cm^2. However, the sample in a rheometer will yield uniformly due to the small gap. In a pipe the sample will not yield uniformly. A pressure front will propagate through the oil and the oil near the inlet will yield and begin to flow when a flowline is restarted. Since this problem is not tractable due to the compressibility and non-Newtonian behaviour of a crude oil gel, predictions continue to be made from rheometer measurements or short pipe sections.

Start up pressure predictions can be made using following equation,

$$\Delta P = 4\tau_y\Big|_{wall}\frac{L}{D} \tag{21.1}$$

where ΔP is the pressure drop, τ_y is the yield stress, L is the length of the pipe, and D is the diameter of the pipe. Equation (21.1) assumes that the whole flowline is gelled, will yield simultaneously, and that the yield at the wall is the appropriate parameter to consider. This method typically over-predicts the actual restart pressures.

For export pipelines laboratory pour point methods are very good indicators of potential problems since the export pipelines and blown-down flowlines are full of dead crude. If a flowline were not blown down, then the gel situation would improve due to an increase in light components in the oil.

21.2.5 Wax Deposition

The deposition tendency and rate can also be predicted adequately by calculating the rate of molecular diffusion of wax to the wall by following equation:

$$\frac{dm}{dt} = -\rho D_m A \frac{dC}{dr} \tag{21.2}$$

where, m : mass of deposit, [kg];

ρ : density of wax, [kg/m^3];

D_m : molecular diffusion constant, [m^2/s];

A : deposition area, [m^2];

C : concentration of wax, [%];

r : radial position, [m].

The radial concentration gradient can easily be calculated if broken into two components by applying the chain rule as shown in following equation;

$$\frac{dm}{dt} = -\rho D_m A \frac{\partial C}{\partial T}\frac{dT}{dr} \tag{21.3}$$

where T is temperature. The concentration gradient may be calculated from the wax concentrations predicted by a thermodynamic model for a range of temperatures.

21.2.6 Wax Deposition Prediction

Waxes are more difficult to understand than pure solids because they are complex mixtures of solid hydrocarbons that freeze out of crude oils if the temperature is low enough. Waxes are mainly formed from normal paraffins but isoparaffins and naphthenes are also present and some waxes have an appreciable aromatic content. The prediction of wax deposition potentials comprises two parts, the cloud point temperature or wax appearance temperature, and the rate of deposition on the pipe wall. The cloud point is the temperature below which wax crystals will form in oil, and the rate of deposition determines the wax build-up rate and pigging frequency requirement. The cloud point temperature can be satisfactorily predicted using thermodynamic models. These models require detailed hydrocarbon compositions. Figure 21.3 shows the comparison between experimental data and model simulation. The tendency of pressure effect is accurately predicted.

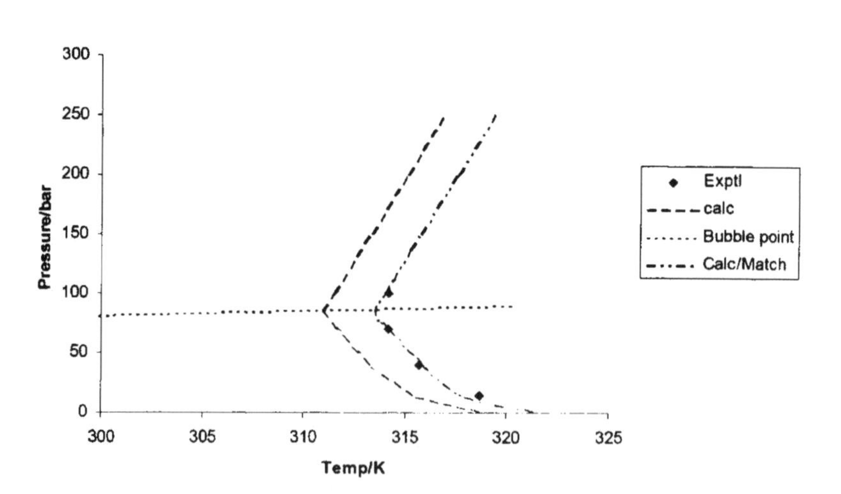

Figure 21.3 Comparison of WAT between experimental data and calculation results at different pressure (Edmonds *et al.*, 1999).

The prediction of deposition rates depends on a thermodynamic model for the amount of wax in oil, and a diffusion rate model. The accuracy of the predictions is not satisfactory at present, but some industrial JIPs (e.g., University of Tulsa) are investigating to improve the accuracy. General used wax models include:

- Tulsa University model;

- Wax module in Profes;
- Olga wax module;
- and BPC model.

21.3 Wax Management

21.3.1 General

Several methods of wax control and management are practiced by production operations, but the transportation of crude oil over a long distance in subsea system demands significant planning and forethought. The wax management strategy generally is based on one or more of the following methods,

- Flowline pigging,
- Thermal insulation and pipeline heating,
- Inhibitor injection,
- Coiled-tubing technology.

The most common method of wax control is flowline pigging if the wax has formed. The solid deposit is removed with regularly removing the wax layer by the scouring action of the pig. Chemical inhibitors can also help control the wax deposition, although these chemicals are not always effective and tend to be expensive. In case that the flowline pigging option for the production lines is not practical, particularly for subsea completions, wax deposition is controlled by maintaining fluid temperatures above the cloud point for the whole flowline.

Coiled-tubing technology has become an important means of conducting well clean-up procedures. This technology involves the redirection of well production to fluid collection facilities or flaring operations while the coiled tubing is in the well. Heavy coiled tubing reels are placed at the wellhead by large trucks, the well fluids are diverted, and high-pressure nozzles on the end of the coiled tubing are placed in the well. Tanker trucks filled with solvent provide the high-pressure pumps with fluids that are used to clean the well tubing as the coiled tubing is lowered into the well. The value of this method is apparent in many areas of the world, since certain integrated production companies maintain a fleet of coiled tubing trucks that remain busy a large percentage of the time.

21.3.2 Thermal Insulation

A good thermal insulation can keep the fluid above the cloud point for the whole flowline and thus eliminate waxes deposition totally. Although line heaters can be successfully employed from the wellhead to other facilities, the physical nature of the crystallizing waxes has not been changed. This can be a problem once the fluids are sent to storage, where the temperature and fluid movement conditions favour the formation of wax crystals and lead to gels and sludge.

21.3.3 Pigging

The rate of deposition can be reduced by flowline insulation and by the injection of wax dispersant chemicals, which can reduce deposition rates by up to five times. However, it must be emphasized that these chemicals do not completely stop the deposition of wax. Therefore, it is still necessary to physically remove the wax by pigging the flowline. To facilitate pigging, a dual flowline system with a design that permits pigging must be built. Pigging must be carried out frequently to avoid the buildup of large quantities of wax. If the wax deposit becomes too thick, there will be insufficient pressure to push the pig through the line as the wax accumulates in front of it. Pigging also requires that the subsea oil system be shutdown, stabilized by methanol injection and blow down, and finally, restarted after the pigging has been completed. This entire process may result in the loss of 1-3 days of production. The deposition models created based on the fluids analysis work and the flow assurance calculations are the key to establishing pigging intervals that are neither too frequent to be uneconomical or too infrequent to run the risk of sticking the pig in the flowline. OLGA models the "slug" of wax pushed ahead of the pig.

21.3.4 Inhibitor Injection

Chemical inhibition is generally more expensive than mechanical pigging, although the cost comparison depends on pigging frequency requirements, chemical inhibition effectiveness, and many other factors. Chemical inhibitors can reduce deposition rates but rarely can eliminate deposition altogether. Therefore pigging capabilities still have to be provided as a backup when chemical inhibition is used. The chemicals must match the chemistry of the oil, at the operating conditions, to be effective. Testing of inhibitor effectiveness is absolutely necessary for each application. The tests should be carried out at likely operating conditions. The chemical inhibitors for wax prevention include:

- Thermodynamic wax inhibitor(TWI): cloud point suppression, viscosity and pour point reduction, requires high volume;
- Pour point depressants: modify wax crystal structure, reduce viscosity and yield stress, but do not reduce rate of wax deposition;
- Dispersants/surfactants: coat wax crystals to prevent wax growth; alter wetting characteristics to minimize wax adhesion to pipe wall or other crystals.
- Crystal modifiers: co-crystallize with wax, reduce deposition rate, but do not prevent formation, modify wax crystals to weaken adhesion and prevent wax from forming on pipe wall, inhibit agglomeration, suitable for steady state and shut-ins, reduce viscosity/pour point, no universal chemical - performance is case specific, high cost, pigging still required, inject above cloud point.

21.4 Wax Remediation

Wax remedial treatments often involve the use of solvents, hot water, the combination of hot water and surfactants, or hot oil treatments to revitalize production. Following methods are used for removal of wax, paraffin and asphaltenes:

- Heating by hot fluid circulation or electric heating;
- Mechanical means (scraping);

- NGS, nitrogen generating system, thermo-chemical cleaning;
- Solvent treatments;
- Dispersants;
- Crystal modifiers;

21.4.1 Wax Remediation Methods

Heating: Removal wax by means of a hot fluid or electric heating works well for downhole and for short flowlines. The hydrocarbon deposit is heated above the pour point by the hot oil, hot water or steam circulated in the system. It is important that the hydrocarbons are removed from the wellbore to prevent re-deposition. This practice, however, has a drawback. The use of hot oil treatments in wax-restricted wells can aggravate the problem in the long run, even though the immediate results appear fine.

Mechanical means: This method is only suitable for cleaning a flowline that is not completely plugged. The wax is cleaned by mechanically scraping the inside of the flowline by pigging. The effectiveness of the pigging operation can vary widely depending on the design of the pigs and other pigging parameters. The pigging strategies in subsea system, and the pigging requirement for subsea equipment, flowline, platforms and FPSO designed were discussed by Gomes et al. Coiled tubing is another effective mechanical means used in wax remediation

NGS: Nitrogen Generating System; introduced by Petrobras in 1992, is a thermo-chemical cleaning method. The NGS process combines thermal, chemical and mechanical effects by controlling nitrogen gas generation to comprise the reversible fluidity of wax/paraffin deposits. Such an exothermal chemical reaction causes the deposits to melt.

Solvent treatments: solvent treatments of wax and asphaltene deposition are often the most successful remediation methods, but are also the most costly. Therefore, solvent remediation methods are usually reserved for applications where hot oil or hot water methods have shown little success. When solvents contact the wax, the deposits are dissolved until the solvents are saturated. If they are not removed after saturation is reached, there is a strong possibility that the waxes will precipitate, resulting in a situation more severe that that prior to treatment.

Dispersants: dispersants do not dissolve wax but disperse it in the oil or water through surfactant action. They may also be used with modifiers for removal of wax deposits. The dispersants divide the modifier polymer into smaller fractions that can mix more readily with the crude oil under low shear conditions.

Crystal modifiers: wax crystal modifiers are those chemically functionalized substances that range from poly-acrylate esters of fatty alcohol to copolymers of ethylene and vinyl acetate. Crystal modifiers attack the nucleating agents of the hydrocarbon deposit and break it down and prevent the agglomeration of paraffin crystals by keeping the nucleating agents in solution.

There are chemicals available that can be tailored to work with a particular crude oil composition but tests should be carried out on samples of the crude to be sure that the chemical additives would prevent the wax deposition. The combined hot water and surfactant method allows the suspension of solids by the surfactant's bipolar interaction at the interface between the water and wax. An advantage of this method is that water has a higher specific heat than oil, and therefore usually arrives at the site of deposition with a higher temperature.

21.4.2 Assessment of Wax problem

The process of assessment for a wax problem can be summarized in follows:

- Obtain a good sample;
- Cloud point or WAT based on solid-liquid-equilibria;
- Rheology: Viscosity, pour point, gel strength;
- Crude oil composition: standard oil composition, HTGC;
- Wax deposition rates: cold finger or flow loop;
- Wax melting point;
- Consider the use of wax inhibitors.

21.4.3 Wax Control Design Philosophies

Wax control guidelines for the subsea devices and flowlines can be summarized as follows:

- Design the subsea system to operate above the WAT by thermal insulation;
- Operate the well at sufficiently high production rates to avoid deposition in the wellbore and tree.
- Remove wax from flowlines by pigging and pig frequently enough to ensure that the pig does not stick.
- Utilize insulation and chemicals to reduce pigging frequency.
- Identify and treat high pour point oils continuously.

21.5 Asphaltenes

21.5.1 General

Asphaltenes are a class of compounds in crude oil that is black colour and not soluble in n-heptane. Aromatic solvents such as toluene, on the other hand, are good solvents for asphaltenes. From an organic chemistry standpoint, they are large molecules comprised of polyaromatic and heterocyclic aromatic rings, with side branching. Asphaltenes originate with the complex molecules found in living plants and animals, which have only been partially broken down by the action of temperature and pressure over geologic time. Asphaltenes carry the bulk of the inorganic component of crude oil, including sulfur and nitrogen, and metals such as nickel and vanadium. All oils contain a certain amount of asphaltene. Asphaltenes only become a problem during production when they are unstable. Asphaltene stability is a function of the ratio of asphaltenes to stabilizing factors in the crude such as aromatics and resins. The factor having the biggest impact on asphaltene stability is pressure. Asphaltenes may also be destabilized by the addition of acid or certain types of completion fluids and by the high temperatures seen in the crude oil refining process.

In general asphaltenes cause few operational problems since the majority of asphaltic crude oils have stable asphaltenes. Typically problems only occur downstream due to blending or high heat. Crude oils with unstable asphaltenes suffer from some severe operational problems, most of which are fouling related and affect valves, chokes, filters, and tubing. Asphaltenes become unstable as the pressure of the well decreases and the volume fraction of aliphatic components increases. If the aliphatic fraction of the oil reaches a threshold limit then asphaltenes begin to flocculate and precipitate. This pressure is called the flocculation point. Figure 21.4 shows the effect of pressure to the asphaltenes stability. In the left hand of the curve, asphaltenes are unstable, while to the right of the curve, asphaltenes are stable.

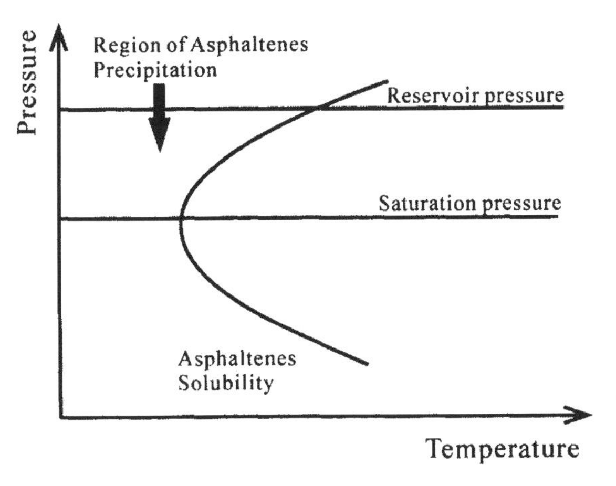

Figure 21.4 Asphaltene stability vs. pressure.

There are currently no standard design and operating guidelines for the control of asphaltenes in subsea systems. Some experience has been gained from asphaltene control programs used for onshore wells. Approaches have varied from allowing the wellbore to completely plug with asphaltenes, then drilling the material out to utilizing periodic solvent washes with coiled tubing to remove material. Relatively few operators have chosen to control asphaltene deposition with dispersants, possibly due to the treating expense and variable results.

21.5.2 Assessment of Asphaltene Problem

One method of characterizing oil is with a SARA analysis. This method breaks the oil down into four pseudo components or solubility classes and reports each as a percentage of the total. The four pseudo components are saturates, aromatics, resins, and asphaltenes. The asphaltene fraction is the most polar fraction and is defined as aromatic soluble and n-alkane insoluble. Asphaltenes are condensed polyaromatic hydrocarbons that are very polar. Whereas wax has a hydrogen to carbon ratio of about 2, asphaltenes have a hydrogen to carbon mole ratio around

1.15. The low hydrogen content is illustrated in Figure 21.5, which is a drawing of a hypothetical asphaltene molecule.

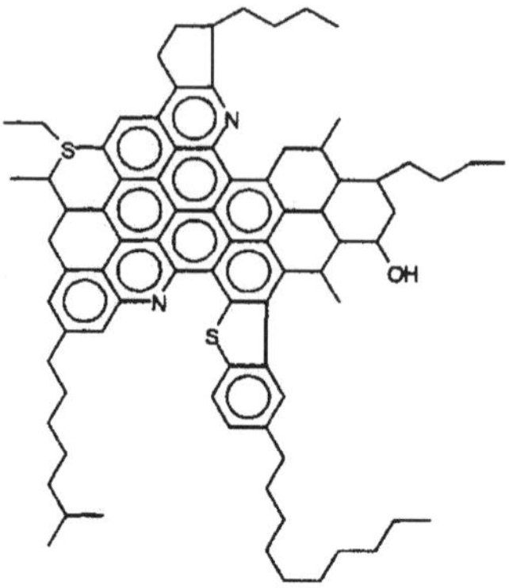

Figure 21.5 Hypothetical asphaltene molecule (Ellison et al., 2000).

Asphaltenes have a density of approximately 1.3 g/cm^3. In oil production systems the asphaltenes are often found mixed with wax. There are two mechanisms for fouling that occur in the formation. The first involves acid; the second is adsorption to formation material. Acidizing is one of the most common well treatments and can cause severe damage to a well with asphaltic crude oil. The acid causes the asphaltenes to precipitate, sludge and form rigid film emulsions which severely affects permeability, often cutting production by over 50%. Formation material, particularly clays, contain metals that may interact with the asphaltenes and cause the chemisorption of the asphaltenes to the clay in the reservoir. A SARA screen, aliphatic hydrocarbon titration, or depressurization of a bottom hole sample are used to determine if asphaltenes are unstable in a given crude. One SARA screen is the colloidal instability index (CII)4. The CII is the ratio of the unfavorable components to the favorable components of the oil as shown in following equation:

$$CII = \frac{S + As}{R + Ar} \tag{21.4}$$

where S is the percentage of saturates, As is the percentage of asphaltenes, R is the percentage of resins, and Ar is the percentage of aromatics in the oil. If the CII is greater than one then amount of unfavorable components exceeds the amount of favorable components in the system then the asphaltenes are likely to be unstable.

There are several aliphatic hydrocarbon titrations that can be used to assess the stability of asphaltenes in a dead crude oil. One method in particular involves the continuous addition of an aliphatic titrant to oil and the measurement of the optical density of the solution. In this method the precipitation point of asphaltenes is detected by monitoring changes in transmission of an infrared laser. The instrument is referred to as the asphaltene precipitation detection unit (APDU). As the hydrocarbon is added to the oil the optical density decreases and the laser transmittance through the sample increases. At the point when enough titrant has been added to the sample that asphaltenes become unstable and precipitate the optical density drops dramatically. This point is called the APDU number, which is defined as the ratio of the volume of titrant to the initial mass of crude oil.

Depressurization of a live bottom hole sample provides the most direct measure of asphaltene stability for production systems. During depressurization, the live oil flocculation point or the pressure at which asphaltenes begin to precipitate in the system is determined by monitoring the transmittance of an infrared laser which passes through the sample. Onset of flocculation will produce a noticeable reduction of light transmittance. If an oil has a flocculation point then the asphaltenes are unstable at pressures between the flocculation point to just below the bubble point. Many oils are unstable only near the bubble point which has led many engineers to believe that problems only occur at the bubble point. However, there are oils that have an instability window of several thousand pounds per square inch. In figure 21.6 an asphaltene solubility curve is depicted. The onset or flocculation point, saturation pressure, asphaltene saturation point and reservoir fluid asphaltene concentration are indicted with dashed lines while the unstable region is indicated by a shaded area between the asphaltene saturation point and the solubility curve. As shown on the curve the region of instability is from 2600 psi to 3800 psi.

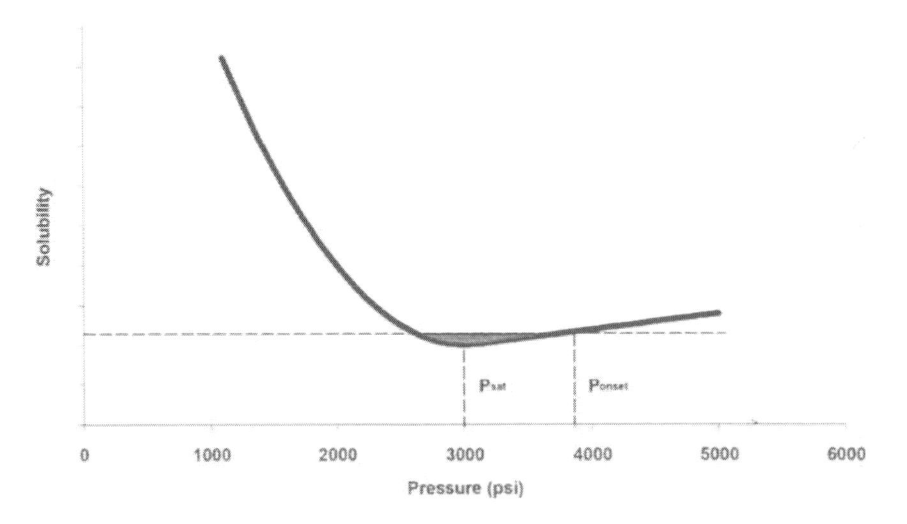

Figure 21.6 Asphaltene solubility as a function pressure (Ellison et al., 2000).

21.5.3 Asphaltene Formation

Asphaltene solids are typically black coal like substances. They tend to be sticky, making them difficult to remove from surfaces. In addition, asphaltenes solids tend to stabilize water-oil emulsions, complicating oil separation and water treatment at the host. It is felt that asphaltenes will be a problem for a relatively small fraction of deepwater projects. Current research efforts have focused on improving screening tests for asphaltene and finding how screening test results relate to field problems. Three screening tools are used. The first, a test known as the P-value test, which involves the titration of the crude oil with cetane, the normal paraffin with a carbon chain length of 16. Additions of normal paraffins tend to destabilize asphaltenes in the crude oil. The stability of the oil increases with the amount of cetane that can be added before visible amounts of asphaltenes come out of solution. The SARA screen test examines the stability of asphaltenes by determining the concentration of the primary components of crude oil, saturated hydrocarbons, aromatics, resins, and asphaltenes. The ratio of saturates to aromatics and asphaltenes to resins are computed and used to determine the stability of the oil. The PVT screen utilizes two values available from a PVT analysis, the in-situ density and the degree of under-saturation (difference between reservoir pressure and bubble point) to make a general assessment of asphaltene stability. In general, increasing under saturation and decreasing in-situ density are associated with decreasing asphaltene stability.

21.5.4 Asphaltene Deposition

After the bubble point has been reached the mass of precipitated asphaltenes and the mass of asphaltenes deposited in the cell is measured. These two measurements provide a means to assess the likely hood of problems due to deposition. We have observed that only a small percentage of total asphaltenes will adhere to a surface during an experiment. Based on these observations we would expect only about 5% of the total asphaltenes to play a role in asphaltene deposition. The mass of deposited asphaltenes can be used to predict the mass of deposition in a production system. While this number is likely to be an overestimate it is a good number for design purposes and for contingency planning.

Asphaltene problems occur infrequently offshore but can have serious consequences on project economics. Since asphaltene deposition is most likely as the produced fluid passes through the bubble point, the deposition often occurs in the tubing. Subsea systems designed to mitigate asphaltene problems generally rely on bottom hole injection of an inhibitor with provision to solvent treat the wellbore when required. In the absence of inhibitors, monthly tubing cleanouts are not uncommon onshore. This frequency would be intolerable in a subsea system making effective inhibitors essential to the development.

21.6 Asphaltenes Control Design Philosophies

In subsea wells, direct intervention with coiled tubing is very expensive and is not a viable means of control. Therefore, the strategy that has been proposed for control of subsea wells utilizes a combination of techniques to minimize deposition. This strategy is as follows:

- Inject an asphaltene dispersant continuously into the wellbore (injection must be at the packer to be effective).
- Install equipment to facilitate periodic injection of an aromatic solvent into the wellbore for a solvent soak.
- Be financially and logistically prepared to intervene with coiled tubing in the well bore to remove deposits.
- Control deposition in the flowline with periodic pigging with solvents

This strategy requires the installation of additional umbilical lines for delivery of asphaltene dispersant and large volumes of solvent, as well as a downhole line for injection of dispersant immediately above the packer. These hardware requirements add considerable project cost. Currently, there are no models for asphaltene deposition as a function of system pressure or other parameters. The flow assurance modelling is helpful in understanding the pressure profile in the subsea system, especially where the bubble point is reached. Since the bubble point is typically the pressure where asphaltenes are least stable, deposition problems would be expected to be the worst at this location.

If steady state flow cannot be reached in a reasonable time due to the high viscosity of the fluids in the pipeline or difficulty in passing the gel segments through the outlet choke, consider displacing the pipeline contents with diluent hydrocarbon such as diesel or perhaps water. Another similar approach is to chemically treat the produced fluids being introduced at the pipeline inlet. Once the pipeline is filled with diesel, water, or treated fluids, flow can be accelerated rapidly due to the low viscosity and no concern about gel segments. If the gel does not break some options to consider are:

- Use coiled tubing system to push a tool into the pipeline and flush out the gel. Commercial coiled tubing equipment is available with extended reach up to several miles.
- Generate pressure pulses in the gel. Since the gels will compress somewhat, a pressure pulse will move the exposed end of the gel and break a portion of it. With successive pressure pulses a small segment of the gel can be broken with each pulse up to a limit.
- An easy first approach is to apply pressure to the gel and wait. There have been field reports of the gel breaking after several hours of exposure to pressure.

In order to plan, design, and operate a subsea pipeline to transport high paraffinic crudes, the following activities are recommended:

- Measure key properties of the crude including cloud point, pour point, gel strength, effectiveness of pour point depressant chemicals, and viscosity as a function of temperature, shear rate, and dissolved gas.
- Assess the thermal conditions to be experienced by the crude for steady state flow and transient flow during startup, shut-in, and low production rates.
- Based on crude properties and thermal data, develop operational plans for shut-in, start-up, and low flow situations. A key decision for start-up is whether to break the gel with pressure or to employ more costly means to prevent gel formation.
- If gel breaking or warm up are required for restart, the necessary flow control equipment and possibly a static mixer at the pipeline outlet will be required.

Asphaltenes design and control guidelines are still in the early stages of development. There is no predictive model for asphaltene deposition, as there is for wax and hydrates. Asphaltene design and control are usually considered with wax design and the following steps and analyses are carried out.

1. Define samples to be taken and analyses to be performed;
2. Perform wax modeling to provide,
   - WAT of live fluids;
   - Location of deposition;
   - Rates of deposition;

- Amount of wax deposited;
- Pipeline design parameters.

3. Assess asphaltene stability under producing conditions;
4. Provide chemical and thermal options;
5. Determine pigging frequency;
6. Design cost effective solutions for prevention and remediation of wax and asphaltenes.

21.7 References

1. Becker J. R., (1997), Crude Oil, Waxes, Emulsions and Asphaltenes, Pennwell Publishing Company.
2. Chin, Y. and Bomba, J., (2000), "Review of the State of Art of Pipeline Blockage Prevention ad Remediation Methods", the Proceedings of 3rd Annual Deepwater Pipeline & Riser Technology Conference & Exhibition.
3. Chin, Y., (2001), "Maintaining Plug-free Flow and Remediating Plugged Pipelines", Offshore, March 2001.
4. Cochran, S. and Gudimetla, R., (2004) "Hydrate management: Its importance to deepwater gas development success", Word Oil, Vol. 225, pp.55-61.
5. Edmonds, B., Moorwood, R.A.S., Szczepanski, R., Zhang, X., (1999), "Latest Developments in Integrated Prediction Modelling Hydrates, Waxes and Asphaltenes", Focus on Controlling Hydrates, Waxes and Asphaltenes, IBC, Aberdeen, 1999.
6. Ellison, B. and Gallagher, "Baker Petrolite Flow Assurance Course", Houston.
7. Ellison, B. T., Gallagher, C.T., and Lorimer, S. E., (2000), "The Physical Chemistry of Wax, Hydrates, and Asphaltene", OTC 11963.
8. Gomes, M. G. F. M., Pereira, F. B., and Lino, A. C. F., (1996), "Solutions and Procedures to Assure the Flow in Deepwater Conditions", OTC 8229, Houston, TX.
9. James, H., and Karl, I., (1999), "Paraffin, Asphaltenes Control Practices Surveyed", Oil & Gas Journal, July 12, 1999, pp. 61-63.
10. Lorimer, S. E., and Ellison, B. T.,(2000), "Design Guidelines for Subsea Oil Systems", Facilities 2000: Facilities Engineering into the Next Millennium.
11. Pereira, A. Z. I., and Khlil, C. N., (1996), "Advances in Multiphase Operations Offshore – New Process for the chemical De-Waxing of Pipelines", Advances in Multiphase Technology – An International Two-day Conference, June 24-25, 1996, Houston TX.

PART IV: Riser Engineering

Part IV

Riser Engineering

Chapter 22 Design of Deepwater Risers

22.1 Description of a Riser System

22.1.1 General

A riser system is essentially conductor pipes connecting floaters on the surface and the wellheads at the seabed. There are essentially two kinds of risers, namely rigid risers and flexible risers. A hybrid riser is the combination of these two.

The riser system must be arranged so that the external loading is kept within acceptable limits with regard to:

- Stress and sectional forces
- VIV and suppression
- Wave fatigue
- Interference

The riser should be as short as possible in order to reduce material and installation costs, but it must have sufficient flexibility to allow for large excursions of the floater.

22.1.2 System Descriptions

The riser system of a production unit is to perform multiple functions, both in the drilling and production phases. The functions performed by a riser system include:

- Production/injection
- Export/import or circulate fluids
- Drilling
- Completion & workover

A typical riser system is mainly composed of:

- Conduit (riser body)
- Interface with floater and wellhead
- Components
- Auxiliary

22.1.3 Flexible Riser Global Configuration

Flexible risers can be installed in a number of different configurations. Riser configuration design shall be performed according to the production requirement and site-specific environmental conditions. Static analysis shall be carried out to determine the configuration. The following basis shall be taken into account while determining the riser configuration:

- Global behavior and geometry
- Structural integrity, rigidity and continuity
- Cross sectional properties
- Means of support
- Material
- Costs

The six main configurations for flexible risers are shown in Figure 22.1. Configuration design drivers include a number of factors such as water depth, host vessel access / hang-off location, field layout such as number and type of risers and mooring layout, and in particular environmental data and the host vessel motion characteristics.

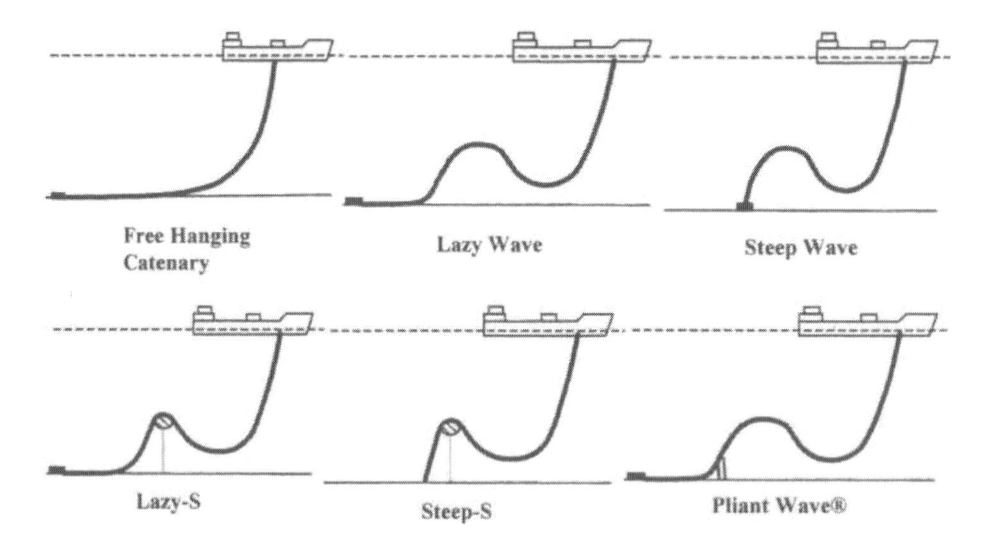

Figure 22.1 Flexible riser configurations

– ***Free Hanging Catenary***

This is the simplest configuration for a flexible riser. It is also the cheapest to install because it requires minimal subsea infrastructure, and ease of installation. However a free hanging catenary is exposed to severe loading due to vessel motions. The riser is simply lifted off or lowered down on the seabed. A free hanging catenary under high vessel motions is likely to suffer from compression buckling at the riser touch down point and tensile armor wire 'birdcaging'. In deeper water the top tension is large due to the long riser length supported.

– ***Lazy wave and steep wave***

In the wave type, buoyancy and weight are added along a longer length of the riser, to decouple the vessel motions from the touch down point of the riser. Lazy waves are preferred to steep waves because they require minimal subsea infrastructure. However lazy waves are prone to configuration alterations if the internal pipe fluid density changes during the riser lifetime. On the other hand, steep wave risers require a subsea base and subsea bend stiffener, and yet are able to maintain their configuration even if the riser fluid density changes.

Buoyancy modules are made of syntactic foam which has the desirable property of low water absorption. The buoyancy modules need to be clamped tightly to the riser to avoid any slippage which could alter the riser configuration and induce high stress in the armor wires. On the other hand the clamping arrangement should not cause any significant damage to the external sheath of the riser as this might cause water ingress into the annulus. Buoyancy modules tend to lose buoyancy over time, and wave configurations are inherently designed to accommodate up to a 10% loss of buoyancy.

– ***Lazy S and steep S***

In the lazy S and steep S riser configuration there is a subsea buoy, either a fixed buoy, which is fixed to a structure at the seabed or a buoyant buoy, which is positioned by e.g. chains. The addition of the buoy removes the problem with the TDP, as described above. The subsea buoy absorbs the tension variation induced by the floater and the TDP has only small variation in tension if any.

'S' configurations are considered only if catenary and wave configurations are not suitable for a particular field. This is primarily due to the complex installation required. A lazy-S configuration requires a mid-water arch, tether and tether base, while a steep-S requires a buoy and subsea bend stiffener. The riser response is driven by the buoy hydrodynamics and complex modeling is required due to the large inertial forces in action. In case of large vessel motions a lazy-S might still result in compression problems at the riser touchdown, leaving a steep-S as a possible alternative.

– ***Pliant wave***

The pliant wave configuration is almost like the steep wave configuration where a subsea anchor controls the TDP, i.e. the tension in the riser is transferred to the anchor and not to the TDP. The pliant wave has the additional benefit that it is tied back to the well located beneath the floater. This makes well intervention possible without an additional vessel.

This configuration is able to accommodate a wide range of bore fluid densities and vessel motions without causing any significant change in configuration and inducing high stress in the pipe structure. Due to the complex subsea installation that is required, it would be required only if a simple catenary, lazy wave or steep wave configurations are not viable.

22.1.4 Component Descriptions

The components of a riser system must be strong enough to withstand high tension and bending moments, and have enough flexibility to resist fatigue, yet be as light as practicable to minimize tensioning and floatation requirements. A short description of the most common riser components and auxiliary components (such as end fittings and bend stiffemers) is given below.

- **Riser joints**

A riser joint is constructed of seamless pipe with mechanical connectors welded on the ends. For drilling risers, choke and kill lines are attached to the riser by extended flanges of the connector. The riser can be run in a manner similar to drill pipes by stabbing one stalk at a time into the string and tightening the connector.

- **Buoyancy modules**

Buoyancy modules can be attached to the riser to decrease the tension required at the surface. These modules may be thin-walled air cans or fabricated syntactic foam modules that are strapped to the riser. These buoyancy modules require careful design and the material for their construction needs to be selected appropriately so as to ensure that they have a long-term resistance to water absorption.

- **Bend Stiffeners and Bellmouths**

One of the critical areas of a flexible riser is the top part of the riser just before the hang-off arrangement. This area is prone to over-bending and hence an ancillary device is incorporated into the design to increase the stiffness of the riser and prevent over-bending of the riser beyond its allowable bend radius. The two devices used for this application are bend stiffeners and bellmouths. Figure 22.2 illustrates a schematic of both devices. Flexible pipe manufacturers tend to have a preference for one or the other device, yet bend stiffeners are known to provide a better performance in applications with high motion vessels. Bend stiffeners also provide a moment transition between the riser and its rigid end connection. The ancillary devices are designed separately from the pipe cross section analysis, and specialized software is used for this purpose. Global loads from the flexible riser analysis are used as input to the ancillary device design.

Bend stiffeners are normally made of polyurethane material and their shape is designed to provide a gradual stiffening to the riser as it enters the hang-off location. The bend stiffener polyurethane material is itself anchored in a steel collar for load transfer. Bend stiffeners are sometimes utilized subsea, such as in steep-S or steep-wave applications to provide support to the riser at its subsea end connection, and to prevent over-bending at this location. Design issues for bend stiffeners include polyurethane fatigue and creep characteristics. Figure 22.3 shows an example of a bend stiffener. It is to be noted that bend stiffeners longer than 20 ft have been manufactured and are in operation in offshore applications.

Bellmouths are steel components that provide the same function as bend stiffeners, i.e. to prevent over-bending of the riser at its end termination topsides. The curved surface of a bellmouth is fabricated under strict tolerances to prevent any kinks on the surface that might cause stress concentrations, and damage to the pipe external sheath.

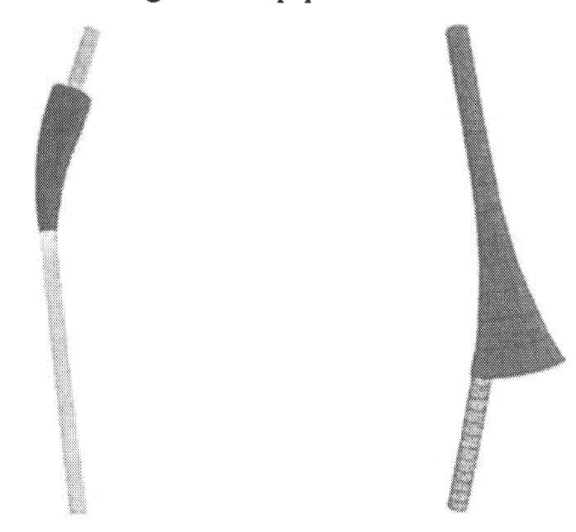

Figure 22.2 Schematic of bend stiffener (left) and bellmouth (right).

Figure 22.3 Example of a bend stiffener.

- Bending Restricter

This is normally located at the bottom and top connections. The purpose is to provide additional resistance to over-bending of the riser at critical points (such as the ends of the riser, where the stiffness is increased to infinity).

Bend restrictors are designed to limit bending on static pipelines. They are made of a hard plastic material and typically used at wellhead tie-ins and at riser bases to restrain the riser tension, bending and shear loads. Bend restrictors provide mechanical locking to prevent over-bending. Figure 22.4 illustrates a bend restrictor used at the end termination of a flexible pipe subsea.

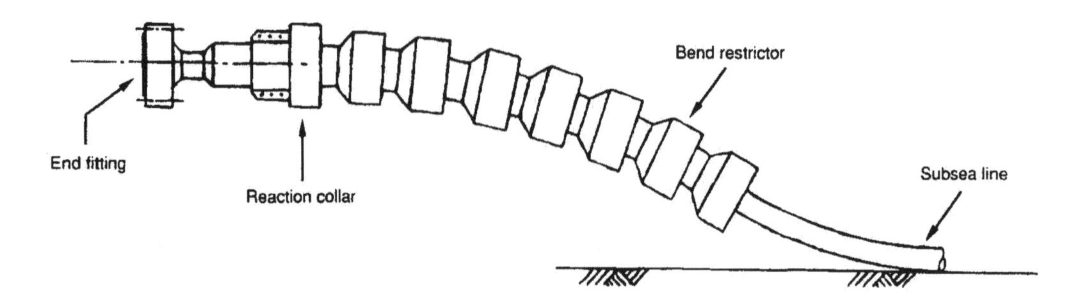

Figure 23.4 Bend restrictor.

22.1.5 Catenary and Top Tensioned Risers

In shallow water it has been practice to use top tensioned risers, but as design for larger water depth is accounted the need for new design practice has increased, see Figure 22.5. The ordinary Top Tensioned riser is very sensitive to the heave movements due to wave and current loads because the rotation at the top and bottom connections is limited. The heave movement also requires top tension equipment to compensate for the lack of tension. If the top tension is reduced it will cause larger bending moment along the riser especially if the riser is located an environment with strong current. If the effective tension becomes negative (i.e. compression) then Euler buckling will occur.

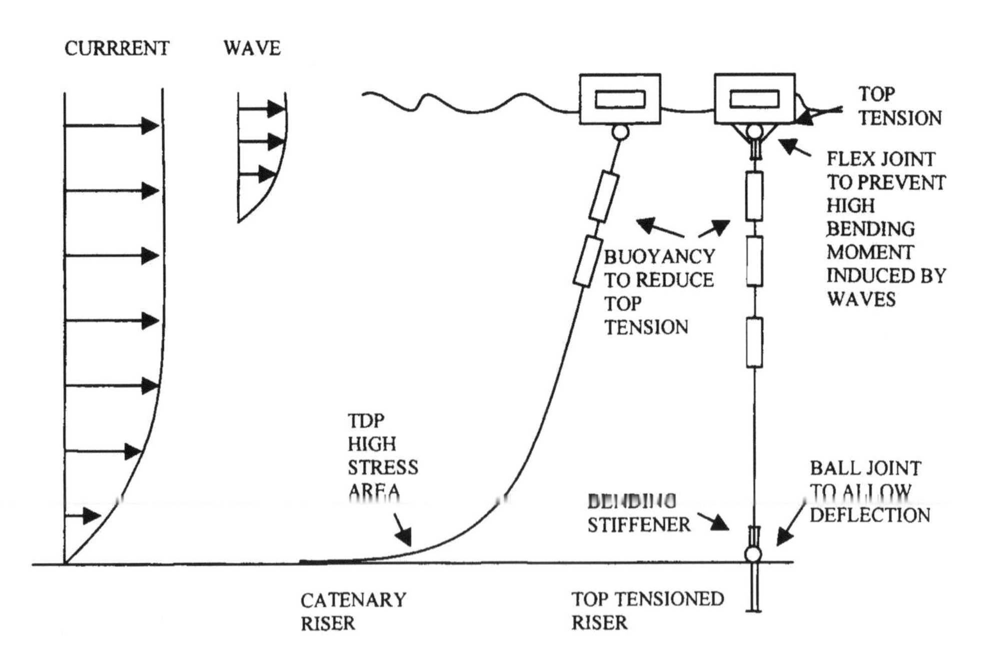

Figure 22.5 Top Tensions Risers and Steel Catenary Risers and Their Components.

The catenary riser is self compensated for the heave movement, i.e. the riser is lifted of or lowered on the seabed. The catenary riser still need a ball joint to allow for rotation induced by waves, current and vessel motion, at the upper end connection.

The catenary riser is sensitive to environmental loads, i.e. wave and current due to the normally low effective tension in the riser. The fatigue damage induced by Vortex Induced Vibration (VIV) can be fatal to the riser. Use of the VIV suppression devices such as helical strakes and fairing can reduce the vibrations to a reasonable level.

22.2 Riser Analysis Tools

Various analysis tools are available for riser design, examples of these are:

- General purpose finite element programs: ABAQUS, ANSYS, etc;
- Riser Analysis Tools: Flexcom, Orcaflex, Riflex, etc;
- Riser VIV Analysis Tools: Shear7, VIVA, VIVANA, CFD based programs;
- Coupled motion analysis programs: HARP, etc;
- Riser Installation Analysis Tools: OFFPIPE, Orcaflex, Pipelay, etc.

Riser analysis tools are special purpose programs for analyses of flexible risers, catenary risers, top tensioned risers and other slender structures, such as mooring lines and pipelines. The most important features for the finite element modeling are listed below:

- Beam or bar element based on small strain theory.
- Description of non-linear material properties.
- Unlimited rotation and translation in 3D space.
- Stiffness contribution from material properties as well as geometric stiffness.
- Allowing varying cross-sectional properties.

- Riser analyses

  Typical analyses are for instance:
  - Strength analysis;
  - Fatigue Analysis;
  - VIV Analysis;
  - Interference Analysis .

  The results from the finite element analysis are listed below:
  - Nodal point co-ordinates;
  - Curvature at nodal points;
  - Axial forces, bending moment, shear forces and torsion.

- Time domain analysis and frequency domain analysis

The purpose of the analysis is to determine the influence of support vessel motion and direct wave induced loads on the system. The results from the frequency domain analysis are the systems eigenfrequencies and eigenvectors. The results from the time-domain analysis are time series of a selected limited response parameters, such as stress, strain and bending moment.

The results from the above analyses are stored in separate files for subsequent post processing, such as plots or calculation. Some of the more interesting output is listed below:

- Plots
  - System geometry;
  - Force variation along lines;
  - Pipe wall forces;
  - Geometry during variation of parameters;
  - Response time series;
  - Vessel motion transfer function;
  - Animation of the dynamic behavior of the complete system including support vessel and exciting waves.

- Tables
  - Support forces;
  - Pipe wall forces;
  - Velocities and accelerations from wave and vessel motion time series;

Statistical time series analysis, estimation of spectral densities

22.3 Steel Catenary Riser for Deepwater Environments

22.3.1 Design Codes

Riser maximum equivalent stresses during extreme storm conditions are limited to 80% yield stress. 100% yield stress is acceptable during abnormal conditions such as a mooring line or tether failure. This approach has been adopted on other (vertically tensioned) riser systems and is in line with API RP 2RD and the ASME Boiler Code.

Higher stress allowables are particular interest at the Touch Down Point (TDP) where stresses are largely displacement controlled. Whilst this offers some scope to the designer to address extreme storm response, caution must be exercised. Designing with higher utilization may lead to an unacceptable fatigue life and the validity of assuming that TDP response is displacement controlled is not always correct. This is particularly true where low-tension levels are observed near the TDP.

22.3.2 Analysis Parameters

Hydrodynamic Loads

There are uncertainties related to vortex-induced vibration (VIV). If the stresses are above the endurance limit of the material then fatigue may take place. In addition, VIV may result in drag amplification that may result in increased stresses. Finally the hydrodynamic interaction between risers may result in riser crashing loads which must be considered. VIV suppression has been used on most SCR's.

Material Properties

The steel material to be used in deepwater SCR's offshore is likely to be steel of API grade X65 or above. The main uncertainty lies in the effect of welding combined with plastic strain (reeling and laying). Until validated S-N curves (Stress range versus Number of cycles to failure curves) are available, SCR design has to be based on conservative assumptions which may limit the use and complicate installation.

Soil Interaction

In most deepwater fields, relatively loose clay is found on the seabed. The pipe will sink into this clay and might be buried over time. The exact behavior of the soil is not known. The soil uplift and sideways resistances are hence important aspects. It is important to properly model riser-soil interaction effects.

Extreme Storm

The primary objective of the extreme storm analysis is to define basic geometry and assess acceptability of response. A large number of analyses need to be conducted when optimizing a steel catenary riser. The approach is highly iterative in order to ensure that the response is optimized for all combinations of load and vessel offset.

22.3.3 Soil-Riser Interaction

When a pipe is placed on soil and subjected to oscillatory motion, there is complex interaction between pipe movements, penetration into the soil and soil resistance. At the touch down point (TDP) region of the riser, transverse (out-of-plane) motions will occur as a consequence of oscillatory forces caused by transverse wave acting on the free hanging part of the riser.

A proper description of the pipe-soil interaction is therefore important for the accuracy in calculation of riser fatigue damage. Depending upon the stiffness and friction of the seafloor, out-of-plane bending stresses will be more or less concentrated in the TDP region when the riser is subjected to oscillatory motion.

In riser response analysis tools, the pipe-soil interaction is commonly modeled by use of friction coefficients (sliding resistance) and linear springs (elastic soil stiffness). However, these parameters must be selected carefully in order to properly represent the complex pipe-soil interaction.

During small and moderate wave loading (the seastates contributing most to the fatigue damage) the riser TDP response in the lateral direction is very small (in the order of 0.2 pipe diameters). This will cause the riser to dig into the top sand soil layer and create its own trench. This effect will gradually decrease as the riser gets closer to the underlying stiff clay soil, where very limited penetration is expected. The width of this trench will typically be 2-3 pipe diameters, which leaves space within the trench for the pipe to move without hitting the trench edges. During a storm build-up, the trench will gradually disappear as a result of larger riser motions in addition to natural back fill. For the extreme strength analysis, the pipe-soil interaction is found to be of minor importance even if higher lateral soil resistance is mobilized.

22.3.4 Pipe Buckling Collapse under Extreme Conditions

Within the industry, there are considerable differences between recommended methods for sizing riser pipe for resistance to collapse and propagation buckling in deepwater particularly for low D/t ratios. Existing formulations are based on empirical data, which attempt to account for variations in material properties and pipe imperfections. Application of these codes to deepwater applications provides scatter of results. Additionally, the effects of tension and bending (dynamic and static) are uncertain, depending on the nature of the loading condition.

22.3.5 Vortex Induced Vibration Analysis

The VIV analyses of the SCR could follow two different approaches: using SHEAR7 (MIT, 1995 and 1996). As the VIV- and wave-induced fatigue damage is established independently, results from both calculations must be combined to get the total distribution of fatigue damage for the SCR's. The areas where significant wave-induced fatigue damage occurs are very distinct. The VIV-induced fatigue damage occurs more evenly distributed (according to the larger variations in mode shapes and their superposition). The total fatigue damage is then obtained by a simple sum of the two contributions. The fact that VIV- and wave-induced response will be more or less perpendicular to each other is conservatively not accounted for ("hot-spots" are assumed to coincide).

22.4 Stresses and Service Life of Flexible Pipes

Calculation of ultimate capacity may be performed with good accuracy by tools estimating the average layer stress. All the available flexible pipe analysis tools, including the manufacturers design programs calculate the average stresses in each layer.

Service life prediction on the other hand requires detail knowledge of the mechanism leading to failure. The manufacturers have established estimation methods based on theory and test results. These analysis methods must be calibrated for each manufacturer, each wire geometry and type of pipe (i.e. additional hoop spirals). The advantage with such empirical methods is that residual stresses from manufacturing, actual tolerance on wire geometry, etc are present in the tests and hence incorporated in the analysis. The problem is that design optimization is hardly possible and independent verification is impossible.

Løtveit and Bjærum (1995) has found that by combining detailed knowledge of flexible pipes with state of the art non-linear FEM programs it is possible to develop an analysis tool that can predict the stresses sufficiently accurately to provide input to service life prediction. SeaFlex has recently developed a second-generation analysis tool, PREFLEX, for analysis of flexible pipes. PREFLEX is based on the general non-linear FEM program MARC. PREFLEX can model each wire with a mesh sufficiently detailed to calculate local hot spot stresses.

Examples of attractive features of PREFLEX are:

- Virtually no modeling limitations. End fitting areas, damaged pipe etc., can be modeled.
- Service life predictions based on a minimum of test results. PREFLEX can accurately calculate the stresses and small-scale tests of the wires may hence be used to define the capacity. The previous analyses tools required results from full-scale test for service life prediction.
- Analyses have shown that the use of simplified analysis tools based on average stresses in the layer may recommend the use of hoop spirals where local stresses are very high. One example is use of a rectangular back-up spiral as an additional hoop strength layer.

22.5 Drilling and Workover Risers

Deepwater drilling and workover is presently performed with jointed steel risers. The vessels and equipment have been upgraded to work in a water depth down to more than 1700 m. In deepwater and harsh environment the challenges related to operation are large due to use of buoyancy, fairings etc.

The drilling contractors are presently building new vessels and upgrading existing vessel to meet the deepwater requirement. Smedvig and Navion have contracted a new drillship MST ODIN to be rented by Statoil. The vessel is fully equipped for drilling in 2500m water-depth. Drilling in even deeper water is planned. The technology status is, however, presently limited to approximately 2500 m.

Two of the critical items for deepwater drilling are riser weight and riser control. In order to reduce the riser weight, alternative materials are considered. SeaFlex and Raufoss have completed a JIP project related to composite risers. At the Heidrun TLP a titanium drilling-riser has been installed. One composite drilling joint has been qualification tested and is ready for offshore trial in the Gulf of Mexico.

22.6 References

1. API RP 2RD, (1998) "Recommended Practice for Design of Risers for Floating Production Systems and TLP's", First Edition, 1998.
2. DNV (1998) "VISFLOW Users Manual", Det Norske Veritas 1998.
3. Hatton, S.A., and Willis, N., (1998) "Steel Catenary Riser for Deepwater Environments-STRIDE", Offshore Technology Conference 1998.
4. Hibbitt, Karlsson & Sorensen (1998), "ABAQUS, Ver. 5.8".
5. Langner, C.G., and Bharat C.S., (1997) "Code Conflicts for High Pressure Flowlines and Steel Catenary Risers", OTC'97.

6. Løtveit, S.A. and Bjærum, R., (1995) "Second Generation Analysis Tool for Flexible Pipes", MarinFlex 95.
7. Lund, K.M., Jensen, P., Karunakaran, D. and Halse, K.H., (1998) "A Steel Catenary Riser Concept for Statfjord C", OMAE'98.
8. Marine Computational Services (MCS), (1994) "FLEXCOM-3D, Version 3.1.1".
9. MIT, (1995) "SHEAR7 Program Theoretical Manual", Department of Ocean Engineering, Massachusetts Institute of Technology.
10. MIT, (1996) "User Guide for SHEAR7, Version 2.0", Department of Ocean Engineering, Massachusetts Institute of Technology.
11. SINTEF (1998) "RIFLEX- Flexible Riser System Analysis Program- User Manual", Marintek and SINTEF Division of structures and concrete report-STF70 F95218.

Part IV

Riser Engineering

Chapter 23 Design Codes for Risers and Subsea Systems

23.1 Introduction

Authorities and classification societies have developed riser design codes such as ISO, API, NPD, HSE, NS, BS, CSA, DNV and ABS. In particular codes relevant for riser design are e.g.

- API RP16Q for drilling risers;
- API 2RD for risers attached to floating systems;
- API 17B and 17J for flexible pipes;
- ISO13628-5 for steel tube umbilicals.

Two design formats have been applied:

- Working Stress Design (WSD)-API
- Limit State Design (LSD)-DNV, ISO

The key issues in structural design are:

- Loads
- Resistance
- Acceptance Criteria

The resistance consists of characteristic resistance and material factors of typically 1.15. The characteristic resistance is defined for the following failure modes:

- yielding- SMYS
- brittle fracture- material toughness
- fatigue- SN curves and SCF

The loads are classified to 3 categories:

- functional loads
- environmental loads
- accidental loads

Loads factors are applied based on the category of the loads. Load combinations include a couple of combinations of different categories of loads with assigned load factors. The load effect is then calculated using FE models.

For example, environmental conditions are:

- Wind
- Wave
- Current
- Snow and Ice
- Earthquake

Wave loads are the most important and complex environmental load, and may be predicted using theory of regular/irregular waves accounting for long-term variation of waves.

For deepwater application, there are technical challenges such as:

- New dominant load combinations
- New concepts
- New materials
- New critical failure modes

A large amount of project activities are on-going to improve e.g. the following aspects:

- Analytical tools
- Coupled analysis
- Materials qualifications
- Failure mode identifications
- Convert from WSD to LSD

23.2 Design Criteria for Deepwater Metallic Risers

23.2.1 Design Philosophy and Considerations

The design philosophy adopted in this chapter is to apply proved technical advances in order to conduct safe and cost-effective design of marine risers. The design of a marine riser system will require consideration of a number of factors in relation to its functional suitability and long term integrity. Considerations should be given to:

- Consistence with laws, acts and regulations;
- Riser integrity: reliability, safety and risk;
- Riser functional requirements;
- Riser operational requirements;

- Riser structural design criteria;
- Materials;
- Installation requirements;
- Fabrication requirements;
- Inspection and maintenance;
- Costs.

23.2.2 Currently Used Design Criteria

The API RP 2RD is a stress-based code using the von-Mises yield criterion for yield strength checks. API RP 2RD provides also design criteria for external overpressure (collapse) and its usage factor is given in Table 23.1.

Table 23.1 Usage factors in API RP 2RD.

Load Combination	Normal Operating	Extreme (100 year)	Survival (1000/10 000 year)
Functional plus environmental	0.67	0.8	1.0

23.2.3 Ultimate Limit State Design Checks

Based on the above discussion on pipe failure, the design of the pipe will be considered fit to resist ultimate limit states when it has been checked that the following conditions will not occur:

- Ductile bursting due to internal overpressure only, excluding any externally applied loads;
- Excessive axial yielding of cross-section (plastic mechanism) under the combined effect of bending, axial force, and pressure differential;
- Collapse due to external overpressure only, excluding any externally applied loads;
- Local buckling due to the combined bending, axial force, and external overpressure.

Such design checks are performed with the design equations, which are based on the capacity formulations that have been rewritten in terms of characteristic values and safety factors. All relevant load cases that can be expected during the lifetime of the riser, including installation conditions, have to be examined to achieve an acceptable design.

23.3 Limit State Design Criteria

23.3.1 Failure Modes and Limit States

The limit states are classified as below:

1. Serviceability limit state (SLS)

Serviceability requirements may also be imposed to avoid permanent deformation of the tubes, which would hamper pigging or other operation of equipment in the pipes.

2. Ultimate limit state (ULS)

ULS are defined for the failures due to yielding, buckling, bursting, collapse and loss of equilibrium of the pipe cross-section. Yielding failure mode for risers is treated as ULS even though it will not result in immediate failure. Because risers are usually made of high strength materials which do not have a considerable strain hardening effect, the yield strength is close to the ultimate tensile strength.

3. Fatigue limit state (FLS)

FLS is defined for the fatigue failure due to dynamic cycle loading effects. Three major issues causing fatigue damage of risers include:

- 1st order wave loading and associated floater motion;
- 2nd order floater motion;
- Vortex induced vibrations (VIV) due to current.

23.3.2 Acceptance Criteria

With reference to the loads and limit states, different acceptance criteria need to be considered for different design conditions. The following acceptance criteria shall be included in riser design:

- Hoop strength acceptance criteria;
- Yielding strength acceptance criteria;
- Bursting strength acceptance criteria;
- Buckling strength acceptance criteria;
- Fracture strength acceptance criteria;
- Fatigue strength acceptance criteria.

23.4 Loads, Load Effects and Load Cases

23.4.1 Loads and Load Effects

1. Functional loads:

The follows shall be considered as functional loads:

- Weight of riser, contents, and corrosion coating;
- Pressures due to internal contents and external hydrostatics;
- Buoyancy;
- Thermal effects;
- Nominal top tension.

The follows should be considered as appropriate:

- Weight of marine growth, attachments, tubing contents;
- Loads due to internal contents flow, surges, slugs, or pigs;
- Loads due to installation;

- Loads due to floater restraints.

2. Environmental loads:

The follows shall be considered as environmental loads:

- Wave loads;
- Current loads;

The follows should be considered as appropriate:

- Wind loads;
- Seismic loads;
- Ice loads.

3. Accidental loads:

The follows shall be considered as accidental loads:

- Dropped objects;
- Partial lost of station-keeping capability;
- Vessel impact;

The follows should be considered as appropriate:

- Tensioner failure;
- Riser interference;
- Explosion and fire;
- Heat flux;
- Operational malfunction.

23.4.2 Definition of Load Cases

Load cases that shall be checked are defined by combining functional, environmental and accidental loads. Depending on riser orientation and directional variation in wave and current loading, extreme combinations of wave and current loading shall be selected. The selection of load cases is also related to the design stage. Critical load cases are needed to be considered for preliminary and conceptual design. All the load cases are needed to be included for detailed design.

23.4.3 Load Factors

Appropriate load factors shall be applied for riser design. For specified design case, those factors can be calibrated based on, for instance, detailed FEM analysis and reliability methodology. To back this calibration, the details should be documented for verification.

23.5 Improving Design Codes and Guidelines

23.5.1 General

ISO standards are presently being prepared for subsea equipment and risers in deepwater. As part of this work, integration of API and ISO standards and preparation of adequate design codes for deepwater applications are ongoing. A new International Standard for flexible pipe, API spec 17J (ISO 13628-2) was issued. Furthermore, the API 2RD code is being revised.

23.5.2 Flexible Pipes

1. Flexible Pipe Guidelines

Flexible pipes have been used for decades. The early pipes and hoses were of the bonded type (vulcanized rubber and armoring). The designs were primarily governed by the ratio burst to design pressure.

From the early seventies, large resources were put into the development of reliable non-bonded flexible pipes. As a result of the product development work, the confidence in flexible pipes increased, and flexible pipes are considered attractive for many applications. The use of flexible pipes was, however, still limited partially because no general industry standard was available. In the middle eighties, Veritec (1987) developed a general design standard for flexible pipes, based on a JIP. These guidelines were based on the design methods used by the manufacturers and the offshore design codes. These design codes represented the state of the art of flexible pipe design. With the exception of Brazil, the use of flexible pipes was still moderate during this period. There was, however, a continuous growth in demand and requirements (temperature, pressure and diameter) to flexible pipes. Many oil companies developed their own specifications for flexible pipes and the industry faced the following problems:

- Many operators had their own design standards;
- The manufacturers used their in-house standards for design. To prepare additional documentation conforming with the operators' standards was often cumbersome and expensive;
- The general design standards were not updated and were considered to be increasingly inadequate.

The design requirements were divided into two categories:

- Mandatory requirements that are auditable should be included in the specification (API Spec 17J).
- Recommendations with respect to how to satisfy the mandatory requirements, as well as guidance for the design of flexible riser systems, are included in a separate Recommended Practice (RP) such as API RP 17B. The RP will include methodology for the design of risers outside the experience range. Deepwater will be one such area.

2. API Specification 17K

The design of flexible pipe is according to API Specification 17J for unbonded pipes, and API Specification 17K for bonded pipes. These documents provide a check-list of all the essential parameters and guidelines that pipe operators need to verify when ordering flexible pipes from pipe manufacturers. The rest of this section shall deal with API Spec 17J for unbonded pipes, but API Spec 17K contains essentially identical information and specifications for bonded pipes.

3. API Specification 17J

API 17J describes the parameters that need to be determined before carrying out a flexible pipe design. These essential parameters (besides the external environmental conditions) are the internal bore characteristics such as pressure, temperature and fluid composition. These parameters determine much of the pipe design such as material selection, and layer thickness.

API 17J lists the flexible pipe system requirements such as inspection and condition monitoring, gas venting, and installation requirements. Another section deals with the allowable loads that can be imposed on the pipe during its lifetime. Once a pipe cross-section is established during design work, calculations are carried out to ensure that these allowable loads are not exceeded throughout the pipe design life. Any anomalies in this work would require a new pipe cross-section to be established. During normal operational conditions the tensile armor layer maximum load cannot exceed 0.67 of the ultimate tensile strength (UTS) of the armor material. The pressure armor layer is allowed a maximum load of 0.55 of UTS. During abnormal and installation conditions the allowable load may be increased to 0.85 of UTS, and during factory acceptance tests the load may be increased up to 0.91 of UTS.

Further sections of API 17J describe other conditions and limitations that need to be considered during design work. One of the most important of these requirements is the minimum bend radius that the pipe is able to withstand without unlocking the pressure armor layer. An important aspect of design work is calculations to ensure that the pipe does not exceed its minimum bend radius under extreme load conditions. The two areas of a pipe most susceptible to over-bending are the touch-down zone and the upper region just before the hang-off location. Once the minimum bend radius is known, ancillary devices such as bend stiffeners or bend restrictors can be designed to ensure that the pipe does not exceed this minimum bend radius under all possible extreme loading conditions.

API 17J contains useful information for the design of the various layers that make up the flexible pipe. Details are also available on the design of the end-fitting arrangement, bend stiffeners, and bend restrictors. Besides the local cross-section design, the flexible pipe also needs to be verified under a global static and dynamic analysis. Since unbonded flexible pipes have a large damping factor (due to the presence of a number of unbonded layers), they do not suffer from fatigue damage induced by vortex-induced vibration (VIV). Hence flexible pipes do not need to be installed with strakes or fairings to limit VIV. This means that fatigue damage is primarily due to wave motions and installation damage. A detailed fatigue life analysis is required and the pipe manufacturer needs to prove that the pipe fatigue life is ten times the pipe required service life.

Procedures are required for pipe installation, since incorrect installation induces a greater risk of exceeding the tensile limits of the armor layer material, over-bending, and causing impact damage to the flexible pipes. There are documented cases of flexible pipe damage during installation, for example the piercing of the pipe outer sheathing that required expensive mitigation measures to be undertaken in order to prevent the replacement of the whole pipe before commencing operations.

API 17J also includes guidelines for the manufacture of the flexible pipe, and the qualification testing that is required before the pipe is issued to the operator.

4. API RP 17B

Another useful document for flexible pipe operators is API Recommended Practice 17B. This document is not a specification, hence it is not binding on any party. However many of the recommendations in API 17B are enforced in practice since they provide additional measures to maintain the integrity of the flexible pipes, and ensure a more efficient and safe operation. API 17B contains useful information on integrity management procedures, and inspection and monitoring measures that can be undertaken to manage any risks of damage or failure modes of the flexible pipes. This document also contains useful information on the design and analysis methods that can be used to verify the pipe design and service life. It discusses various methods for carrying out these design calculations and is a useful tool for pipe operators and manufacturers to ensure an efficient and cost-effective solution for any flexible pipe application.

5. Safety Against Collapse

API Spec 17J is based on working stress design. Present standards have been based on a permissible utilization of 67% of the pipe capacity for external pressure. In practice this means that the stresses in the carcass must be less than 67% of the stresses required to collapse the carcass.

API Spec 17J uses the following formulae, see Table 23.2.

Table 23.2 API Spec. 17J.

Water Depth (D)	Permissible Utilization
$D \leq 300m$	0.67
$300m < D < 900m$	(D-300)/600*0.18+0.67
$D > 900m$	0.85

For water depth less than 300 m the permissible utilization is as before. Due to the negligible uncertainty related to hydrostatic pressure in deepwater, the permissible utilization is gradually increased with water depth. The maximum value of 0.85 is reached at 900 m water depth.

23.5.3 Metallic Risers

API issued a new code on design of metallic risers, API RP 2RD (API, 1998). There are a number of technical challenges that have to be addressed by the new guidelines such as:

- Uncertainties in loads, vortex shedding (e.g. vortex-induced vibrations (VIV) and soil interaction);
- Strength criteria;
- Fatigue life of girth welds;
- Reeling with plastic strain of a dynamic riser influence on fatigue;
- Fatigue qualification tests;
- ECA assessment for determination of weld defect inspection frequency and defect acceptance criteria.

23.6 Regulations and Standards for Subsea Production Systems

One of the first steps in designing a subsea production system should be to determine the regulatory rules applying to the specific situation. Every oil-producing nation has a set of regulations governing the exploration and extraction of its natural resources.

The origin of standardization in the petroleum industry has its roots in the United States, whereby the industry have been dominated by American solutions. This has reflected on the developed standards and specifications, where many of the fundamental standards have been developed by the American Petroleum Institute (API), an organization consisting of American companies involved in petroleum exploration, production, transportation, and refining. As American companies started exploration in other parts of the world they brought the American solutions and standards, due to the lack of similar standards in the applicable countries. The American standards were, as time progressed, adapted to fit the individual countries whereby new national and industry standards were developed.

The choice of standards, which a subsea development is to be designed according to, is the choice of the operator assigned to develop and operate the field. The operator is obliged to follow the applicable country's regulations, and can choose which (if any) standards shall be the basis for design and operation. In practice a field development is always developed according to a number of standards or recommendations, which can be any kind of national, international, industrial, or company specific standards. In order to ensure the security of their investments, shareholders in a development will usually demand the designs and constructions of the development classified by an independent third party classification society, as will the operator with any sub-contractors. The classification society will review and verify the design and construction to ensure compliance with the predefined regulations, standards, rules, and guidelines.

As the use and demand of subsea production systems increased so did the need for separate standards, the first being the API 17 series covering the following areas:

- RP 17A, Design and Operation of Subsea Production Systems;
- RP 17B, Flexible Pipe;
- RP 17C, TFL (Through Flowline) Systems;
- Spec 17D, Subsea Wellhead and Christmas Tree Equipment;
- Spec 17E, Subsea Production Control Umbilicals;
- RP 17G, Design and Operation of Completion/Workover Riser Systems;
- RP 17I, Installation of Subsea Umbilicals;
- Spec 17J, Unbonded Flexible Pipe;
- Spec 17K, Bonded Flexible Pipe.

In 1999 the international organization for standardization (ISO) published the fist standard in the ISO 13628 series, Design and operation of subsea production systems, was published. To date the following ISO 13628 standards have been published:

- ISO 13628-1: General requirements and recommendations;
- ISO 13628-2: Flexible pipe systems for subsea and marine applications;
- ISO 13628-4: Subsea wellhead and tree equipment;
- ISO 13628-3: Through flowline (TFL) systems;
- ISO 13628-5: Subsea control umbilicals;
- ISO 13628-6: Subsea production control systems;
- ISO 13628-7: Workover/completion riser systems;
- ISO 13628-8: Remotely Operated Vehicle (ROV) interfaces on subsea production systems;
- ISO 13628-9: Remotely Operated Tool (ROT) intervention systems.

23.7 References

1. API RP 2RD, (1998) "Recommended Practice for Design of Risers for Floating Production Systems and TLP's", First Edition, 1998.
2. API, Spec. 17J, (1996) "Specification for Unbonded Flexible Pipe", First edition December 1996.
3. ISO 13628-7, (1999) "Petroleum and natural gas industries – Design and operation of subsea production systems", Part 7: "Completion/workover riser systems", International Standardisation Organisation.
4. NPD, (1990) "Regulations Relating to Pipeline Systems in the Petroleum Activities", 1990.
5. Veritec, (1987) "JIP guidelines for flexible pipes", 1987.

Part IV

Riser Engineering

Chapter 24 VIV and Wave Fatigue of Risers

24.1 Introduction

For risers attached to a floater, the fatigue damage is mainly due to:

- 1st order and 2nd order wave loading and associated floater motion
- Vortex induced vibrations (VIV) due to current along the water column
- Vortex-induced hull motions due to loop current
- Installation

Fatigue damage shall be estimated using Miner's rule summation as given below:

$$D = \sum_{i=1}^{k} \frac{n_i}{N_i} \leq \eta \tag{24.1}$$

where, D is the accumulative fatigue damage ratio, n_i is the number of cycles in stress range block i, N_i is the number of cycles to failure corresponding to the stress range, η is the allowable fatigue damage ratio. The value of allowable fatigue damage ratio is typically 0.1 for production and export risers, and 0.3 for drilling risers.

Fatigue assessment is typically conducted using S-N curves. Fracture mechanics may be applied to estimate the crack growth using Paris' law, and may be applied to determine the weld inspection frequency and allowable defect size.

24.2 Fatigue Causes

24.2.1 Wave Fatigue

1. Wave Scatter Diagrams

For fatigue analysis, the wave environmental condition is described using an Hs-Tp (or Hs-Tz) wave scatter diagram. The parameters, which define the seastate spectra, should be provided based on observed data. This may take the form of Pierson-Moskovitz or JONSWAP single peak spectra or a bi-modal spectrum. Further definition of wave loading conditions should consist of a spreading parameter, which gives the directional distribution of wave loading about the predominant direction. This is a cosine function, the power of which varies according to environmental location.

The directional probability of wave loading should be specified for each of at least 8 compass points. These probabilities are used to avoid undue conservatism in estimation of riser fatigue damage that may result from assuming loading from one or two directions.

2. 1st Order Wave Loading and Floater Motion Induced Fatigue

The wave scatter diagram may be divided into several windows. A single seastate from each window is analyzed to determine the response transfer function, or stress RAO, along the riser length. The transfer function is then used to determine riser response in the other seastates of the window, assuming the transfer function is constant across the window. Fatigue damage from each seastate can then be determined based on an assumed statistical distribution of response and the total fatigue damage across the scatter diagram summed.

Careful consideration must be given to the method of scatter diagram windowing, accounting for variation in wave height and period and the effects of selecting different parameters. The mean slow drift offsets of individual seastates must also be considered as these can have a significant influence on TDP fatigue damage distribution. Hence, the linearisation analysis must use wave height, period and offset representative of the window. A preliminary assessment of the scatter diagram can be conducted to identify seastates, which provide the greatest contribution to total fatigue damage.

3. 2nd Order Floater Motion Induced Fatigue

Mean floater drift motions can have a significant influence on riser TDP fatigue damage and must be accounted for in linearisation analyses. In addition, the slowly varying component of drift motions provides a further contribution to total riser fatigue damage.

Analysis of slow drift fatigue damage is based on static analysis of floater motions with no current or wave applied. The scatter diagram is first split into several windows, the seastates in each having similar drift characteristics. For each window, linearisation analyses are conducted in pairs, using a representative mean drift offset and mean plus RMS low frequency drift motion, each applied statically to the riser. The difference in stress between the two static analyses, at each pint along the riser, is assumed to represent the RMS stress amplitude due to drift motions. The fatigue damage from each linearisation seastate is calculated assuming the drift motions are Rayleigh distributed. The total fatigue damage from each window is then calculated assuming the same drift motions apply to each seastate in the window. For each scatter diagram window, the mean and RMS drift offset are conservatively selected based on the extreme values of any of the seastates in the window.

4. Frequency-Domain vs. Time-Domain Analysis

Two methods may be used to calculate wave fatigue:

(1) A spectral fatigue analysis approach can be applied for first order fatigue analysis.

The fatigue seastates are split into a number of windows and one seastate selected from each window for the purpose of response linearisation. Mean second order motions are accounted for as part of first order fatigue analysis by use of appropriate floater offsets when conducting linearisation analyses.

(2) A time domain random sea analyses may account for the non-linearity of the system, particularly around the critical touch down point.

Time-domain random sea analysis is carried out to derive timetraces of tension and bending variations in the riser. The fatigue damage resulting from each seastate is then determined using a rainflow counting method that transfers the stress range series into a stress histogram. The stress histogram describes the number of occurrence for each stress range defined. A Miner's law is applied to add the fatigue damage induced by each stress range.

24.2.2 VIV Induced Fatigue

Vortex-induced vibration (VIV) is probably the single most important design issue for steel catenary risers, particularly for high current locations. High frequency vibration of the riser pipe due to vortex shedding leads to high frequency cyclic stresses, which can result in high rates of fatigue damage.

Vortex-induced vibration occurs anytime when a sufficiently bluff body is exposed to a fluid flow that produces vortex shedding at, or near, a structural natural frequency of the body (see Figure 24.1). Deepwater risers are especially susceptible to VIV because:

1. currents are typically higher in deepwater areas than in shallower areas;
2. the increased length of the riser lowers its natural frequency thereby lowering the magnitude of current required to excite VIV; and
3. deepwater platforms are usually floating platforms so that there are no structures adjacent to the riser to which it could be clamped.

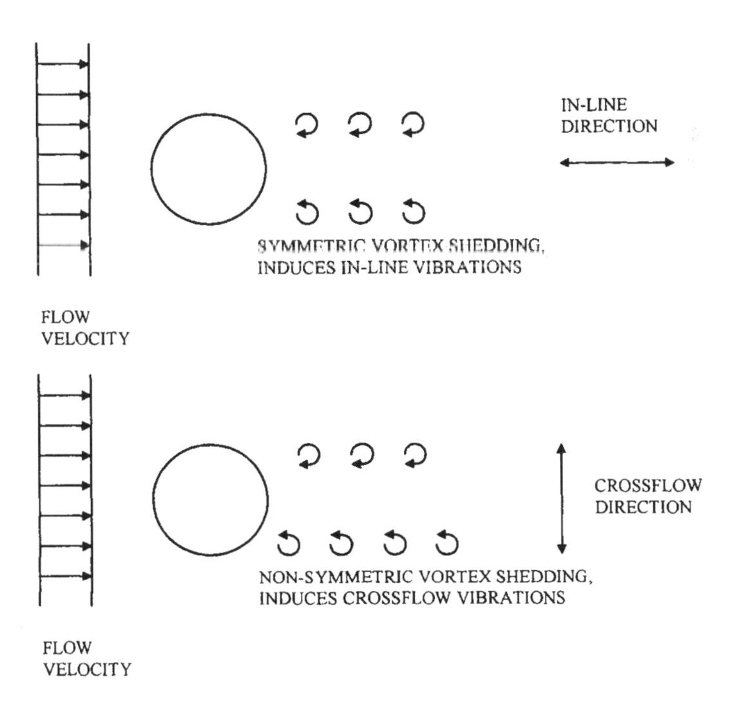

Figure 24.1 Typical flow behind a cylinder.

Deepwater risers are usually so long that significant currents will excite a natural bending mode that is much higher than the fundamental bending mode. Since deepwater currents usually change in magnitude (and direction) with depth, it is possible that multiple modes of the riser can be excited into VIV. This makes deepwater riser VIV prediction much more complex than that for short riser spans typical of fixed platforms in shallow water.

The VIV response of deepwater risers is further complicated by the presence of adjacent tubulars such as risers and tendons. When all, or part, of a riser is in the wake of an upstream tubular, the VIV of the riser can be substantially altered and often worsened. Furthermore, the presence of adjacent tubulars can cause changes in the drag forces acting on a riser, resulting in the possibility of damaging collisions between tubulars.

24.3 Riser VIV Analysis and Suppression

24.3.1 VIV Predictions

1. General

Accurate estimation of the fatigue life of a deepwater riser experiencing vortex-induced vibration depends critically upon accurate estimation of the response amplitude and frequencies (or mode numbers). Accurate estimations of the response amplitude and frequencies are, in turn, dependent upon several "basic" parameters that include:

- The current profile (both magnitude and shape variation with depth);
- The frequency and magnitude of the lift force imparted to the riser by the vortex shedding;
- The excitation and correlation lengths of the lift forces and vortex shedding;
- The hydrodynamic damping;
- The structural properties of the riser including damping, mass, tension, bending stiffness, and the cross sectional geometry (including surface roughness);
- VIV is perhaps more sensitive to the current profile than to any other parameter. For short riser spans the current magnitude determines whether or not VIV will occur. The cross-flow response is more significant than the inline response, see Figure 24.2.

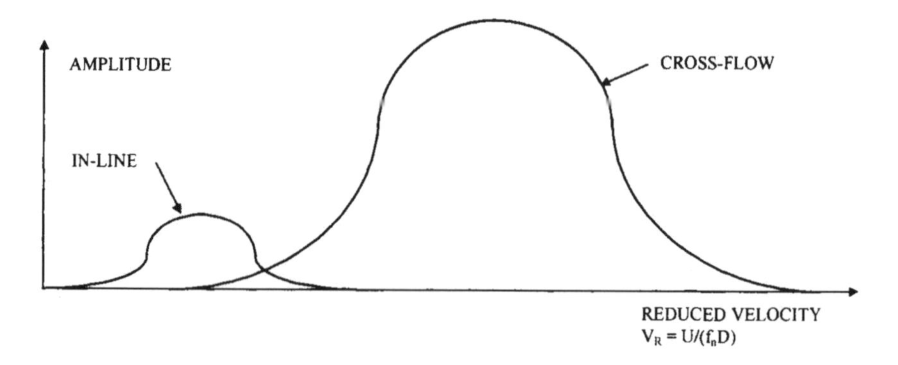

Figure 24.2 Typical amplitude response as function of reduced velocity.

24.3.2 Theoretical Background

The VIV analysis is typically performed using the industry VIV software package Shear7 developed Vandiver and Li (1998) and Vandiver (1998). The theoretical background of the VIV analysis is briefly described as follows. The governing equation for a taut string is given by

$$m_t \ddot{y} + R\dot{y} - Ty'' = P(x,t) \tag{24.2}$$

in which m_t is the mass per unit length, $\ddot{y}$ is the acceleration of the structure, R is the damping per unit length, $\dot{y}$ is the velocity of the structure, T is the tension, y'' is the second derivative of the displacement of the structure with respect to the spatial variable, and $P(x,t)$ is the excitation force per unit length.

The system displacement response can be written as the superposition of modal response:

$$y(x,t) = \sum_r Y_r(x) q_r(t) \tag{24.3}$$

where $Y_r(x)$ is the rth mode shape. Substituting Equation (24.3) into Equation (24.2) leads to

$$M_r \ddot{q}_r + R_r \dot{q}_r(t) + K_r q_r(t) = P_r(t) \tag{24.4}$$

in which M_r is the modal mass, R_r is the modal damping, K_r is the modal stiffness, and P_r is the modal mass.

For pined-pined beam with varying tension, the nth natural frequency is given by:

$$\int_0^L \sqrt{-\frac{1}{2}\frac{T(s)}{EI(s)} + \frac{1}{2}\sqrt{\left[\frac{T(s)}{EI(s)}\right] + 4\frac{m_t(s)\omega_n^2}{EI(s)}}}\, ds = n\pi, \qquad n = 1,2,3 \tag{24.5}$$

where $T(s)$ is the tension, $EI(s)$ is the bending stiffness, $m_t(s)$ is the mass per unit length, and ω_n is the nth natural frequency of the structure.
The nth mode shape is given by:

$$Y_n(x) = \sin\left[\int_0^x \sqrt{-\frac{1}{2}\frac{T(s)}{EI(s)} + \frac{1}{2}\sqrt{\left[\frac{T(s)}{EI(s)}\right]^2 + 4\frac{m_t(s)\omega_n^2}{EI(s)}}}\, ds\right] \tag{24.6}$$

where x is the spatial location. The origin is at the minimum tension end. The nth curvature is given by:

$$Y_n''(x) = \frac{1}{2}\left[\frac{T(x)}{EI(x)} - \sqrt{\left(\frac{T(x)}{EI(x)}\right)^2 + \frac{4\omega_n^2 m_t(x)}{EI(x)}}\right] \sin\left[\int_0^x \sqrt{-\frac{1}{2}\frac{T(s)}{EI(s)} + \frac{1}{2}\sqrt{\left[\frac{T(s)}{EI(s)}\right]^2 + 4\frac{m_t(s)\omega_n^2}{EI(s)}}}\, ds\right] \tag{24.7}$$

The shedding frequency is,

$$f_s = S_t \frac{V}{D} \tag{24.8}$$

where S_t is the Strouhal number, which is a function of the Reynolds number and roughness of the structure, and V denotes the current velocity.

The fatigue damage at location x due to all modes is given by the summation of the individual modal damage:

$$D(x) = \sum_r D_r(x) \tag{24.9}$$

The fatigue damage $D_r(x)$ due to excitation frequency, ω_r, is given by

$$D_r(x) = \frac{\omega_r T}{2\pi K}\left[\sqrt{2}S_{r,rms}(x)\right]^m \Gamma\left(\frac{m+2}{2}\right) \tag{24.10}$$

where, $S_{r,rms}(x)$ is the RMS stress due to r^{th} mode, m and K are constants and defined by the S-N curve $N = KS^{-m}$, where S is the stress range.

24.3.3 Riser VIV Analysis Software

The preliminary design of fatigue resistant risers requires relatively easy-to-use structural dynamic models, which have the capability to estimate dynamic stress levels in the riser as a function of the properties of the structure and imposed velocity profiles (Allen, 1998). The programs must lend themselves to easy parametric variations of current profiles, tension and structural properties. The user must understand the assumptions and program limitations. The most widely used program at the present time is the MIT program SHEAR7 (MIT, 1995 and 1996), see Vandiver and Li (1998) and Vandiver (1998).

SHEAR7 is based on mode-superposition and therefore has a practical limit of about one hundred participating modes. The program was initially written to model straight risers with constant diameter with spatially varying tension. It has been extended to model structures such as catenaries, by hybrid techniques in conjunction with finite element models. As with all existing VIV design programs for risers, SHEAR7 requires calibration with measured data.

The relative lack of data at super-critical Reynolds numbers limits the absolute accuracy of all programs currently available. In many straight riser scenarios in sheared currents, common to the industry today, the likely error in the response amplitude prediction may be as high as a factor of two. Much of the reason for this lack of accuracy is to be found in our poor ability to model the hydrodynamics and in the lack of calibration data at high Reynolds numbers. The hydrodynamics issues were mentioned in the previous section and the calibration issue is addressed in a later section on field data.

As mentioned before, SHEAR7 is based on the mode superposition method, which has practical limitations when the number of excited modes becomes large. Many deepwater production risers will require modeling of dynamic properties that may be best described as typical of structures that behave as if infinite in length. For example, vortex shedding in high velocity surface currents may produce traveling waves at the top of the riser that are damped out before reaching the bottom end. Mode superposition models are poorly suited for such scenarios.

SHEAR7 predicts the cross flow VIV response. The input to SHEAR7 can be calculated in a FEM program. The inputs needed are natural frequencies, mode shapes and modal curvatures from the riser. The results from SHEAR7 include for every node, the RMS values of the displacement and stress, fatigue damage, local drag coefficient, tension and current velocity.

24.3.4 Vortex-induced Vibration Suppression Devices

Often, a deepwater riser will fail to meet the fatigue design criteria due to VIV. The designer may choose, (Howells and Lim, 1999):

- Redesign the riser either by changing the mass (e.g. subtracting buoyancy), increasing the tension, or radically changing the riser design (e.g. using a top tensioned riser instead of a catenary riser); or
- Add VIV suppression devices to reduce the vibration.

The natural frequencies for the modes of interest are, usually dominated by the tension, and not the structural bending stiffness of the risers.

VIV in steel riser systems is typically mitigated using strakes, fairings or a combination of both to the steel risers. In high current regions such as the Gulf of Mexico, which is susceptible to high velocity loop, submerged and bottom currents, many deepwater steel risers utilize vortex suppression devices such as strakes or fairings.

Riser fairings are also used for VIV suppression, streamlining the flow of water around the riser and minimizing vortex shedding, see Figure 24.3. Fairings are designed to rotate, so as to align with the current, efficiently minimizing vortex shedding and drag loading. Fairings operate most efficiently while the riser is in a near vertical configuration. For this reason, they are typically deployed on vertical or near-vertical risers, such as drilling risers, top-tensioned risers and the upper catenary section of SCRs. Shell has successfully employed fairings on steel catenary risers, drilling risers and top tensioned risers, including Auger, Mars, Ursa, Brutus and NaKika developments.

Figure 24.3 Riser fairing (typical, picture courtesy of Shell Global Solutions).

Helical strakes act to disrupt the flow pattern around the riser that creating shorter and weaker vortices. Unlike fairings, strakes tend to increase the drag on risers, which can increase the potential for interference and may increase stresses near hang-off location or flex-joint angles under severe currents. Conversely, however, by increasing drag, they can also reduce wave-induced fatigue by damping the dynamic response caused by vessel motions. 5D pitch strakes are an alternative to the conventional 16D to 17.5D strakes, potentially offering reduced-drag coefficients.

24.3.5 VIV Analysis Example

The analysis presented in this sub-section is based on Qiu et al (2003). A 5-in gas export SCR was selected as a case to be studied. The wall thickness of the SCR is 3/4-in. The hang-off angle is 8.0 degree. The model includes the pull-tube to ensure that the effect of its bending stiffness is accounted for in the modal solution. A minimum fatigue life of 257 years is observed giving a margin of 28.5% above the required minimum of 200 years. The maximum damage is observed in the TDP area, with negligible damage observed away from the TDP.

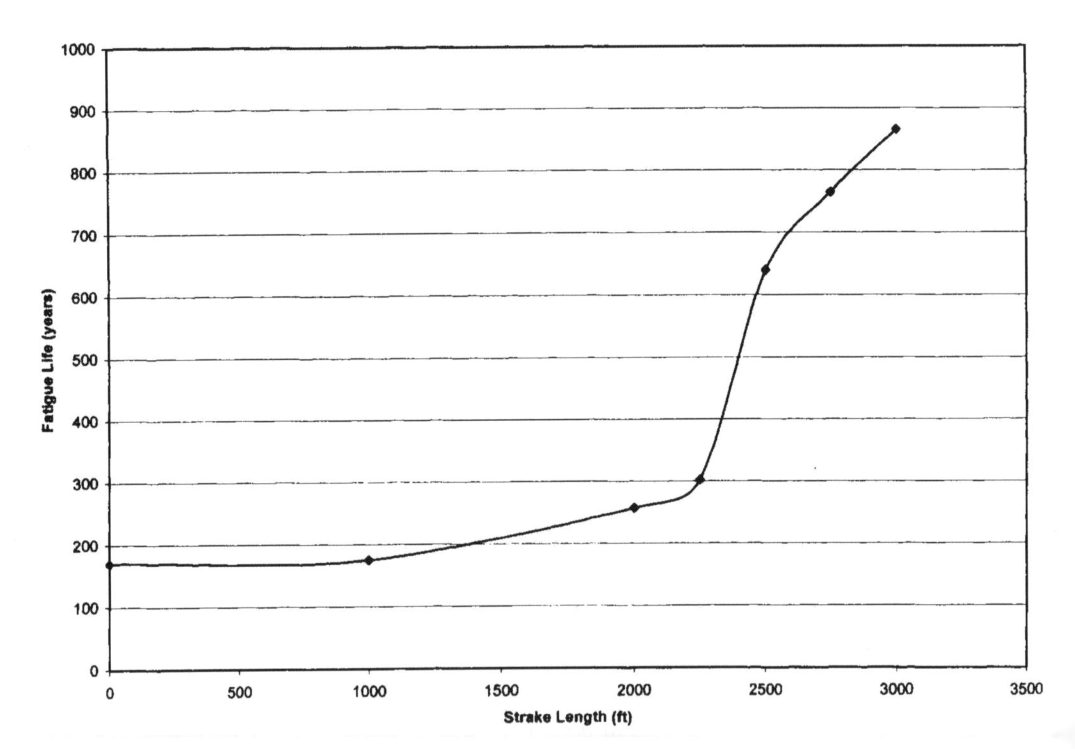

Figure 24.4 Fatigue life versus strake length (Qiu et al., 2003).

Figure 24.4 illustrates the effect of strake length on the fatigue damage alone the riser arc length. It is observed that for no-strake case the fatigue life of the riser is only 170 years, indicating the need for VIV suppression. As the length of strakes applied on the riser reaches 2,250 ft, the fatigue life of the SCR increases dramatically. The results also indicate that 2,000 ft of strakes is necessary to sufficiently reduce the magnitude of the VIV response of the riser.

Table 24.1 shows the effect of single-mode or a multi-mode response on the SCR fatigue life. The results clearly indicate that the single-mode vibration tends to produce higher fatigue damage or lower fatigue life (196 years). The single-mode analysis is relevant to the lower modes of vibrations and yields a much more conservative estimation of the fatigue life, while the multi-mode analysis is relevant to the higher modes vibration and the potential multi-mode excitation.

Table 24.1- Effect of single mode and multi-mode responses (Qiu et al., 2003).

Parameters	Fatigue life (years)
Single-Mode	196
Multi-Mode	257

Table 24.2 presents the effect of current profiles on the SCR fatigue life. Two sets of long-term current profiles have been applied. Current Set 1 is finely defined by 176 current bins and directions, while Current Set 2 is roughly defined by only 35 surface current speeds in 6 directions. It is observed the SCR fatigue life of 1,138 years induced by Current Set 2 is almost 5 times of the fatigue life of 257 years induced by Current Set 1. Therefore, it is very critical to generate current profiles for the VIV fatigue analysis.

Table 24.2- Effect of current sets on predicted fatigue life (Qiu et al., 2003).

Parameters	Fatigue life (years)
Current Set 1	257
Current Set 2	1,138

24.4 Riser Fatigue due to Vortex-induced Hull Motions (VIM)

24.4.1 General

VIM of Spars originates when fluid passing a bluff body causes low-pressure eddies (or vortices) to form down stream of the body (Huang et al, 2003). These vortices are shed periodically at frequencies that are fluid velocity dependent. Primarily, vortex shedding induces loading on the body normal to the direction of current flow. If the frequency of excitation of the vortices is close to the natural frequency of the Spar, resonance will occur. Consequently, large and damaging amplitudes of oscillation may be induced when interaction between the flow (usually current) and structure motion causes *lock-in.* Spar motions induce cyclic ranges of tensions and curvatures in the riser causing fatigue damages. Hence it is important to investigate the effects of VIM on riser fatigue. In the absence of preventive measures, fatigue failure of the risers may occur.

The following features are to be noted in a VIM analysis:

1. There exists no reliable analytical tool for the prediction of Spar VIM. Presently, VIM design data must be obtained from model testing and field measurement.
2. Spar VIM is not only affected by current velocity but also by current profile and turbulence, which are more difficult to measure. The duration for peak current/VIM

can be much longer than peak storm duration. Because of the above issues, it is important to address VIM conservatively in the design stage.

This sub-section is based on Bai et al. (2004).

24.4.2 VIM Amplitudes

Environmental Conditions

It is important to consider all the long-term and short-term current events. In the Gulf of Mexico region (GoM), loop current events have high velocity measured up to 6 knots. Furthermore, these high velocity profiles are not restricted to the surface but extend almost 1200 feet below the mean water level. As a consequence, short-term current events are important in predicting the fatigue damage along the SCR.

Although Spar VIM was detected in GoM under the eddy and hurricane current environments, other types of currents can also induce Spar VIM. Fatigue analyses for Spar VIM are normally conducted for the following conditions:

1. long-term currents;
2. short-term currents: 100 year Loop Currents and Submerged Currents.

VIM fatigue damage is better understood if short-term current events are also taken into consideration along with long-term current events as in many cases in the GoM region short-term currents drive fatigue design.

In-line and Transverse Motions of Spar

The primary motion of a pipe under current is perpendicular or transverse to the current direction, while the secondary motion is in-line with the current. Together, these trace a lemniscate of Bernoulli ("number 8") in the horizontal plane. In general, for a SCR, the locations of primary interest are the hang-off, and touch-down regions as shown in Figure 24.5.

The relation between VIM response amplitude and the system natural period is often given in terms of the Reduced Velocity, which is defined by

$$V_r = V_c \times T_n / D \tag{24.11}$$

Where:

V_c : current velocity, typically highest velocity in the current profile;

T_n : Spar natural period.

D : diameter of the Spar hull.

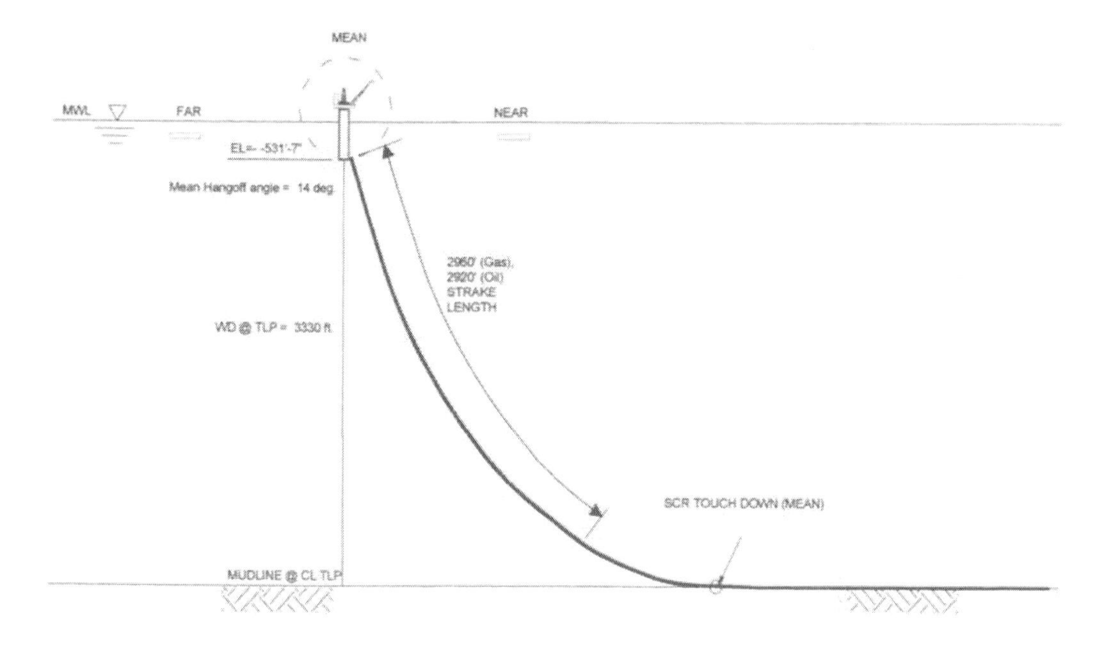

Figure 24.5 Configuration of SCR attached to a Spar, Bai et al (2004).

24.4.3 Riser Fatigue due to VIM

Fatigue Analysis Methodology

The first step in evaluating the VIM response is to determine the relationship between stress in the SCR and VIM amplitude. For each direction (in-plane and out-of-plane), the SCR is offset to the corresponding maximum VIM amplitude. Fatigue damages may be calculated for all nodes in the FE model, or for the regions of primary interest such as the touch -down and hang-off regions. The Influence coefficient (IC) is termed as stress range per unit displacement. It is computed in transverse and in-line directions of the currents, respectively.

$$IC = \text{Stress Range/offset} \tag{24.12}$$

The Stress Ranges are calculated using a non-linear finite element analysis. After determining the influence coefficients, the following steps are applied to calculate the fatigue damage along the riser.

a. The reduced velocity for each current is calculated based on the maximum velocity observed in the current profile.
b. Stress ranges corresponding to each VIM amplitude are calculated using the influence coefficients at the 8 points along the circumference of the SCR.
c. The number of cycles to failure for each stress range is calculated using the API X' SN curve with no endurance limit (API RP 2A) and a stress concentration factor (SCF) of 1.2.
d. For each current profile, the number of cycles occurring per year is calculated based on the probability of each current profile and the Spar's natural period.

The VIM methodology is schematically presented in Figure 24.6.

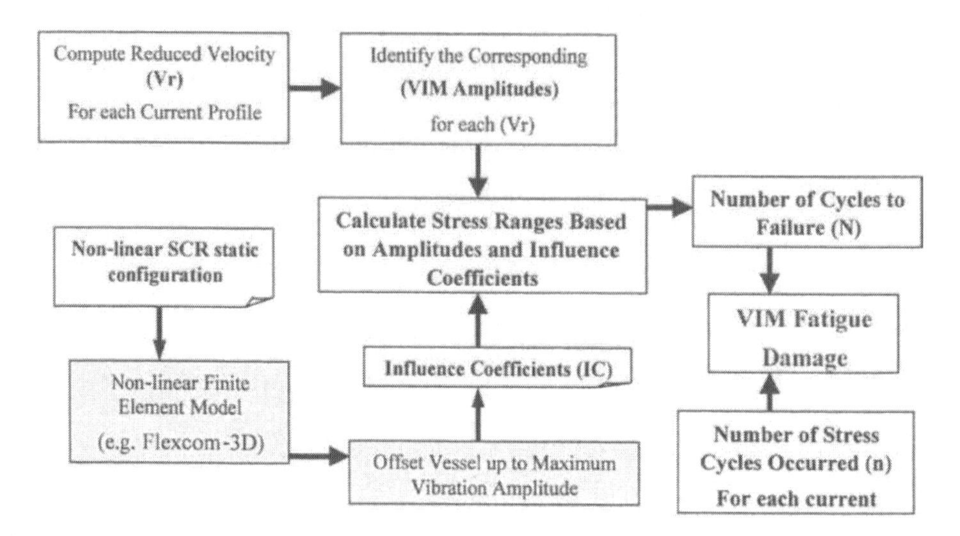

Figure 24.6 VIV analysis methodology (Bai et al., 2004).

The number of stress cycles is obtained as:

$$n = \frac{365 \times 24 \times 3600 \times \text{Probability of Current Profile}}{T_n} \tag{24.13}$$

where: T_n = Spar natural period.

Combined Fatigue Damage

Fatigue damage generated from VIM fatigue analysis has to be appropriately combined with other damage such as riser VIV, installation and wave induced fatigue to get the total fatigue damage of the system. When combining total damage, care has to be taken to ensure that the damage from various sources is accumulated at the same point of the riser.

Fatigue Life at any given point is given by:

$$\text{Fatigue Life} = \frac{1}{\text{FatigueDamage}(\text{VIV} + \text{VIM} + \text{WaveFatigue} + \text{Installation})}$$

24.4.4 VIM Stress Histograms

An Engineering Criticality Assessment (ECA) plays an important role in the determination of the allowable defect sizes at critical locations for risers. The key inputs that can have impact on the results of an ECA are stress histograms at regions where the riser experiences high fatigue damages. To perform ECA, stress ranges and number of cycles is derived from VIM analysis. At a later stage these stress ranges will be combined with those corresponding to VIV, installation and wave induced fatigue.

24.4.5 Sensitivity Analysis

VIM fatigue analysis was carried out for 12" oil export SCR, for different currents. Figure 24.9 presents the distribution of fatigue life along the 12" Oil Export SCR.

24.5 Challenges and Solutions for Fatigue Analysis

The operator's ability to reliably predict deepwater currents is a key requirement for robust riser design. Extensive tank testing has been conducted to collate comprehensive strake/fairing characteristics at high Reynolds Numbers. The VIV analysis software packages have been validated against in-situ VIV fatigue damage measurements, allowing for refining of analysis methodology. For reliable VIV design, the importance of reliable metocean current predictions has been brought into more focus, and several industry initiatives, current monitoring and VIV fatigue monitoring programs are in progress to advance the technology of VIV prediction.

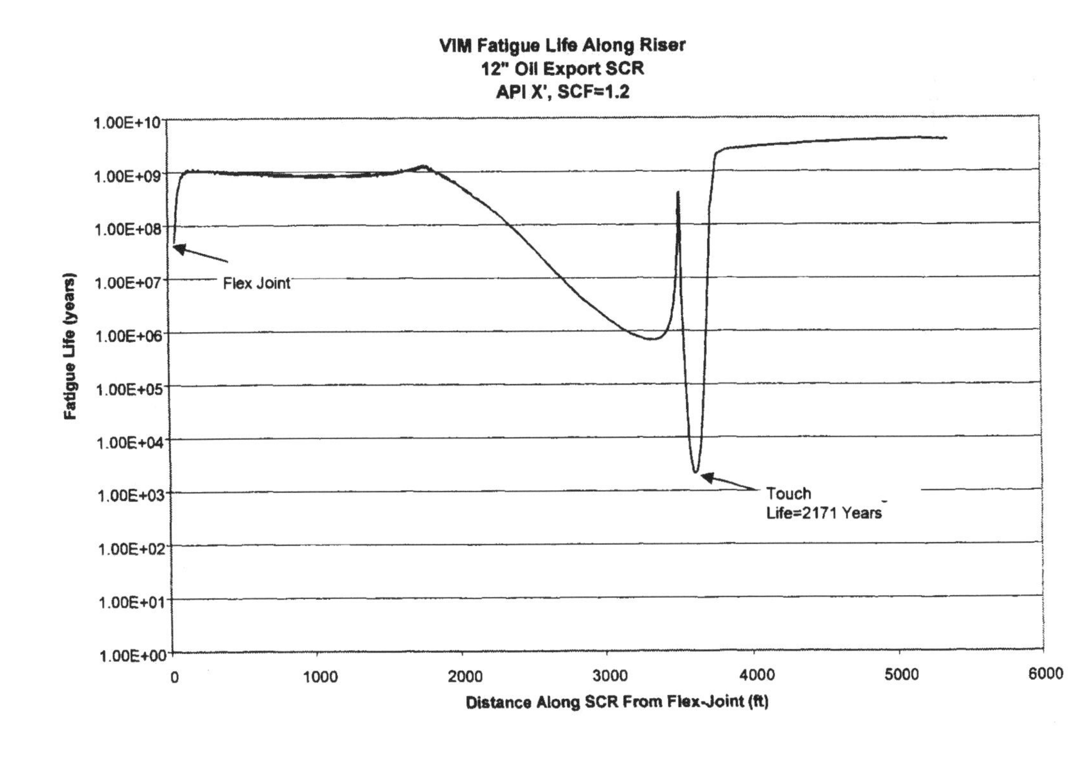

Figure 24.7 VIM fatigue damage along a 12" oil export SCR (Bai et al., 2004).

24.6 Conclusions

Some conclusions from this Chapter are:

- Fatigue is a significant design challenge for SCRs in ultra-deepwater. However, mitigation measures can be identified, including use of the strakes (or fairings) to reduce SCR VIV response.
- The prediction of this VIV carries with it considerable uncertainty, both with regard to current data and VIV amplitude prediction. Robust measurements of the current conditions and a solid understanding (prediction modeling ability) of the different VIV phenomenon are required.
- Host vessel (spar or semi-submersible) vortex-induced motions (VIM) under design near-surface current conditions constitute a potentially significant contributor to SCR fatigue damage when water depth is approximately 3000 ft.

24.7 References

1. Bai, Y., Tang, A., O'Sullivan, E., Uppu, K.C. and Ramakrishnan, S. (2004), "Steel Catenary Riser Fatigue due to Vortex-Induced Spar Motions", OTC16629.
2. Howells, H. & Lim, F. (1999) "Deepwater VIV riser monitoring". Advances in riser technologies, Aberdeen, May 1999.
3. Huang, K., Chen, X. and Kwan, C.T. (2003),"The Impact of Vortex-Induced Motions on Mooring System Design for Spar-Based Installations". OTC 15245.
4. MCS (2004), Flexcom-3D, Version 6.1, User Manual
5. Qiu, W., Bai, Y. and Kavanagh, K. (2003), "Vortex-Induced Vibration Analysis of Deepwater Steel Catenary Risers", Petroleum Science & Technology Forum held at The Petroleum University of China.
6. Vandiver, J.K. & Li, L., (1998) "User Guide for SHEAR7, Version 2.1 & 2.2, For Vortex-Induced Vibration Response Prediction of Beams or Cables With Slowly Varying Tension In Sheared or Uniform Flow", MIT.
7. Vandiver, J.K., (1998) "Research Challenges in the Vortex- induced Vibration Prediction of Marine Risers", Proc. of OTC'98.

Part IV

Riser Engineering

Chapter 25 Steel Catenary Risers

25.1 Introduction

Steel Catenary Risers (SCRs) are a preferred solution for deepwater wet-tree production, water/gas injection and oil/gas export. Statistics, illustrated in Figure 25.1, demonstrates that the number of SCRs currently either under design or planned exceeds the current worldwide inventory of operating SCRs.

The design, welding and installation challenges associated with SCR in ultra-deepwater floating production are primarily related to the higher hang-off tensions caused by the integration of its weight over the water depth, in combination with additional challenges from high pressure, high temperature and sour service.

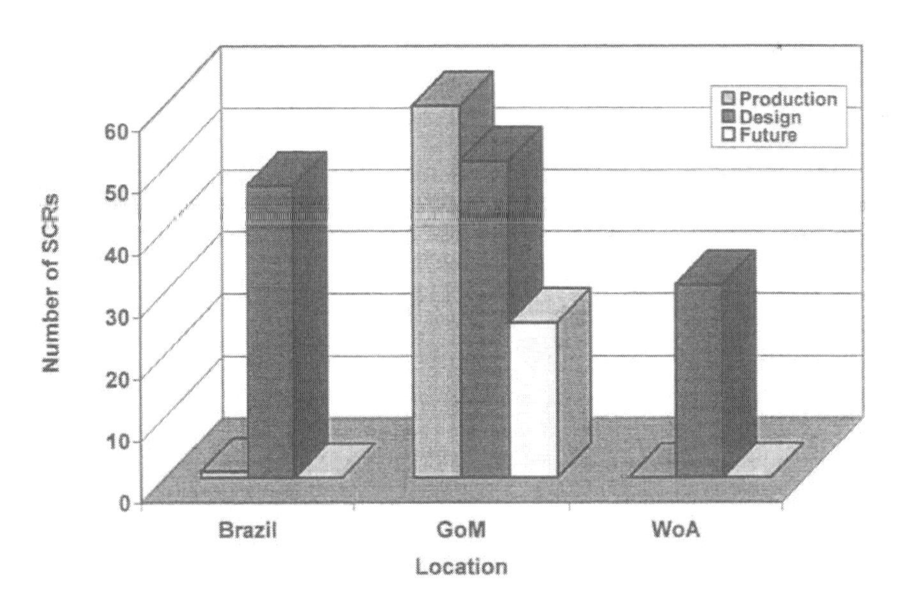

Figure 25.1 Number of SCRs installed/proposed worldwide (215 SCRs) (Bai et al., 2004).

25.2 SCR Technology Development History

To date, SCRs have been designed and installed on TLP, Spar and semi-submersible facilities in the Gulf of Mexico, semi-submersibles offshore Brazil and TLPs offshore Indonesia. SCRs are planned for FPSO production service offshore West Africa for the Bonga, Erha and Dalia projects.

This section presents a brief review of some of the key SCR projects (Bai et al., 2004):

1993 First SCRs were installed on Shell's Auger TLP in the Gulf of Mexico for oil and gas export (Phifer et al., 1994). The SCRs were installed in a water depth of 2860 ft using a flexjoint connected to the TLP pontoon, 70 ft below the mean water level. Their diameter and wall-thickness were 12" and 0.688" respectively. A J-lay method was used to install the riser pipe. For VIV suppression, helical strakes 500 ft long were used at the top portion of the risers. Since then similar SCRs have been installed on Shell's TLP such as Ram-Powell, Mars, Ursa and Brutus.

1997 The first SCR to a semi-submersible was installed in the Marlim Field, offshore Brazil (Serta et al., 1996). The water depth was 1985 ft. The SCR was part of a 10" gas import line with a wall-thickness of 0.812". The material grade was X60 of API 5L. Because the dynamic motions were more severe, it was necessary to employ higher tension in the SCR, in order to reduce bending moments in the sag bend.

1998 Morpeth export/import SCRs were installed on a TLP in a water depth of 1670 ft (Chaudhury and Kennefick, 1999). The J-lay method was used to install the straked portion of the riser and the rest of the riser was installed using the S-lay method.

2001 The Prince SCRs were installed on a TLP located in water depth of approximately 1500 ft (Gore and Mekha, 2002). The SCRs were part of two 12" oil and gas export lines. The SCRs have a departure angle of 24 degrees and a flexjoint angle variation of +/- 20 degrees to accommodate the riser motions in such relatively shallow water.

2001 The two 6" King Kong SCRs were installed in 2001 in 3300 ft of water, and suspended from Allegheny TLP (Korth et al., 2002). 50,000-pounds of buoyancy per riser were required to reduce riser loads acting on the TLP. The SCR pipe wall-thickness is 0.791" and its material grade is API 5L X-65. Titanium stress joints of 27 ft were employed at the vessel interface. The external coating was 12-14 mils FBE, except 1000 ft of touchdown zone where a three-layer polyethylene coating was used. 520 ft of strakes (type 5D) were installed below the stressjoints.

2002 Typhoon 18" Gas Export Riser has the lowest water depth to diameter ratio of any installed SCR to date, resulting in an extremely challenging fatigue design at the seabed touchdown point (Thompson et al., 2003). The Typhoon TLP was located in approximately 2,100-ft water depth. Flexible risers were used for the production wells. A 10" oil export SCR and an 18" gas export SCR were used for product export.

2003 8" and 10" Matterhorn export SCRs were the first reeled SCRs installed on a TLP, located in 2850ft water depth (Kavanagh et al., 2004). The design of these SCRs was one of the first to consider the Cold-Core Eddy Current (or submerged current) in the Gulf of Mexico. Reeled installation of the SCRs also affected the design, fabrication and fatigue testing process, particularly the pipe welding and matching process and the fatigue test and fracture mechanics analysis strategy.

2003 The Na Kika SCRs were the first SCRs being designed and installed as pipe-in-pipe systems (Kopp et al., 2004). The SCRs were installed on a semi-submersible in water depth of approximately 6300 ft. The production SCRs were pipe-in-pipe systems while export SCRs were single walled.

2003 The Marco Polo 12-inch oil and 18-inch gas export SCRs were attached to a TLP located in 4300 ft water (Mekha et al., 2004). During installation, one of the risers was shifted by about 820 ft toward the TLP. It was decided that four extra joints were to be welded to the SCRs, resulting in SCR departure angle change from 12 to 10 degrees.

2004 The Bonga SCRs were the first to be installed on an FPSO vessel offshore West of Africa.

A gas expert SCR and an oil export SCR were installed on Front Runner spar in a water depth of approximately 3000 ft. A VIM fatigue analysis of the SCRs has been presented in an OTC paper by Bai et al. (2004).

Future In addition to the SCRs aforementioned, SCRs have also been designed for use on BP Atlantis and Thunder Horse semi-submersibles and Tahiti spar in the Gulf of Mexico and the Erha and AKPO FPSO in the West of Africa.

25.3 Material Selection, Wall-thickness Sizing, Source Services and Clap Pipe

Commonly used riser pipe material grades are typically API 5L X60, X65 and X-70. The line pipe specification shall consider the following:

- Weldability
- Reservoir properties (HP/HT) and corrosive fluids (H2S, CO2)
- Installation methods and Cost vs. riser fatigue performance
- Topside weight budget

The corrosion coating materials include FBE, 3LPE and 3LPP. The anti-abrasion coatings are HDPE and HDPP. The thermal insulation coatings include: GSPU, multi-layer PP and PUF.

25.3.1 Wall-Thickness Sizing

The wall-thickness for SCR is typically size to pressure containment (hoop and burst strength) requirement such as API RP 1111 and CFR. Collapse analysis is then carried out using API RP 2RD and API RP 1111 particularly for export SCR. Some installation methods such as reeling may have a special requirement on diameter to thickness (D/t) ratio. In certain situations, wall-thickness may have to be increased to meet the requirement of on-bottom axial and lateral stability.

Finally per API 2RD combined (von Mises) stress check should be conducted where the allowable stress is:

- Operating condition: $0.67\sigma_Y$
- Extreme condition or temporary condition: $0.80\sigma_Y$

- Hydro-test: $0.90\sigma_Y$
- Accidental loads: $1.00\sigma_Y$

The maximum wall-thickness that can be welded offshore to date is approximately 1.57" for X-65 welded steel pipe (with or without inconel cladding for sour service). This wall-thickness limit is set by challenges in terms of pipe fabrication, weldability, cladding application technology and successful AUT to achieve the required fatigue performance.

25.3.2 Sour Services and Clad Pipe

Sour service means that the hydrocarbon media inside the riser pipe has a high H2S and CO2 content with a low pH. The acid environment, formed by the combination of H2S and CO2 with water, promotes corrosion and also the H2S enhances the absorption of hydrogen into steel.

The hydrogen and corrosion severity increases with decreasing pH, and increasing H2S and temperature. Effective dehydration, persistence of oil phase and chemical inhibition may reduce corrosion effects and affect the severity of hydrogen charging and therefore the severity of Hydrogen Induced Cracking (HIC) and Sulfide Stress Cracking (SSC).

In order to prevent corrosion defects, over the past ten years there have been several manufacturing methods developed for producing Corrosion Resistant Alloy (CRA) clad/lined pipes. API (1993) has developed a specification for the CRA clad and lined pipe. CRA clad or lined pipes may be classified into two basic categories (Urband, 2001):

- Metallurgically Bonded Pipes: They are those in which there is a metallurgical bond between the structural outer pipe and the corrosion resistant inner pipe.
- Mechanically Lined Pipes: The simplest and most economical method of producing clad pipe is to internally line carbon steel pipes.

The CRA pipe may be joined using welding or threaded & coupled connectors. On-shore welding can produce higher quality welds in prefabricated lengths because it takes place in a more controlled environment. During onshore welding, more time may be used than an offshore welding because there is no cost penalties associated with the use of offshore installation barges. Offshore welding of clad pipe may be a technical challenge and require a longer period of time.

25.4 SCR Design Analysis

25.4.1 Initial Design

In the Pre-FEED phase, an initial design is carried out to define:

- Riser host layout (for interface with other disciplines)
- Riser hang-off system (flex joint, stress joint and pull-tube etc)

- Riser hang-off location, spacing and azimuth angle (hull layout, subsea layout, total risers and interference consideration)
- Riser hang-off angle for each riser
- Riser location elevation at hull (hull type, installation and fatigue consideration)
- Global static configuration

A static configuration may be determined based on catenary theory accounting for hang-off angle, water depth and riser unit weight). The SCR design shall meet basic functional requirements such as SCR internal (and/or external) diameters, submerged tension on host vessel, design pressure/temperature, fluid contents.

Due considerations shall be given to future tie-back porches to accommodate the variations for hang-off system, riser diameter, azimuth angles and the required extreme response and fatigue characteristics.

25.4.2 Strength and Fatigue Analysis

For the preparation of FEED documentation, preliminary analyses are carried out to confirm:

- Extreme response meet the stress criteria per API 2RD and extreme rotation for flexjoints
- VIV fatigue life and the required length of strakes (or fairing)
- Wave fatigue life
- Interference between risers and with floater hull

The preliminary design and analyses will be confirmed in the detail design phase and documented in technical reports.

In the detail design phase, installation analysis and special analyses such as VIM induced fatigue analysis and semi-submersible heave VIV induced fatigue analysis and coupled system analysis.

The host vessel motions are defined by the global performance analysis accounting for wave, wind and current loads using either time-domain analysis or frequency domain analysis. The motion data are expressed as time traces of vessel motions or RAOs defined at center of gravity (CoG) for the floater for pre-defined loading conditions. The motions at the riser hang-off location will be transferred from the CoG via rigid body assumptions. The riser system is considered as a cable under current loads with a boundary condition that is defined as the motions at the hang-off location.

25.5 Welding Technology, S-N Curves and SCF for Welded Connections

25.5.1 Welding Technology

High fatigue demands and new materials place additional requirements on Non-Destructive Evaluation (NDE) processes. It is also necessary to investigate issues such as the required level of overmatching of the weld metal related to the base material, high toughness and high

Yield stress to Tensile stress (Y/T) ratios. The level of mismatching has a great impact on fracture behavior of the welded connections.

25.5.2 S-N Curves and SCF for Welded Connections

Based on the results of tests performed at the University of Texas, Shell has used API X' curve; see Phifer et al. (1994). However, it should be pointed out that these tests were conducted at a considerably higher stress ranges than most of the fatigue causing stress ranges any SCRs are ever going to experience. Based on Department of Energy Offshore Installation code (issued in 1990, 4th Edition) and DNV Classification notes 30.2 (issued in 1984), the most suitable S-N curve for SCR welded connections is the E curve.

Figure 25.2 compares API X' curve, DNV E curve, and E/20 curve. The E-curve will be more conservative than API X' curve for most of the stress ranges an SCR will be subjected to, e.g. stress ranges less than 10 ksi. The E/20 curve is produced by dividing the number of cycles to failure from the E-curve by 20. The S-N Curves for risers in sour services have been discussed by Buitrage and Weir (2002), who indicated that the number of cycles to failure may be reduced by a factor of 20 for risers in sour service.

The SCF value may be calculated by use of the formula derived by Connelly and Zettlemoyer (1993). In practice, SCF values as low as 1.1 have been achieved for large-diameter export risers in offshore service, though significant efforts (e.g. pipe matching, ID boring, tight pipe spec, tight fabrication QC and alignment procedures for welding) are typically required to achieve such low SCFs.

25.6 UT Inspections and ECA Criteria

Ultrasonic (UT) is usually integrated into an Engineering Critical Assessment (ECA) approach. Figure 25.3 shows that the ECA is a tool that links flaw geometry with material properties and the stress ranges. The allowable flaw size calculated from the ECA analysis is used as acceptance criteria for UT inspections, instead of workmanship criteria. (Kopp et al., 2003).

The ECA work is typically performed based on guidance provided in BS 7910 code (formerly PD6493). Maximum critical sizes, which will not initiate a brittle fracture under extreme design loads, are determined using different levels (1 to 3) of assessment as required. A check is made to ensure that a crack starting from an initial defect will not reach the critical size prior to the design life and that the crack growth time will be greater than five times the design life of the SCRs.

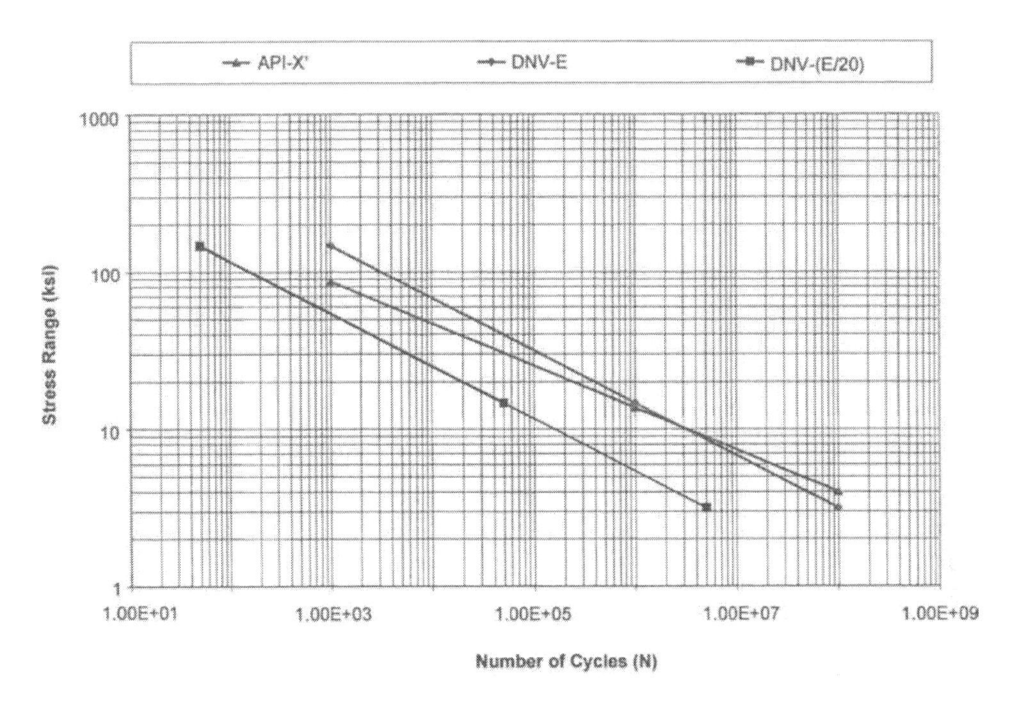

Figure 25.2 Comparison of the S-N curves used in SCR design (Bai et al., 2004).

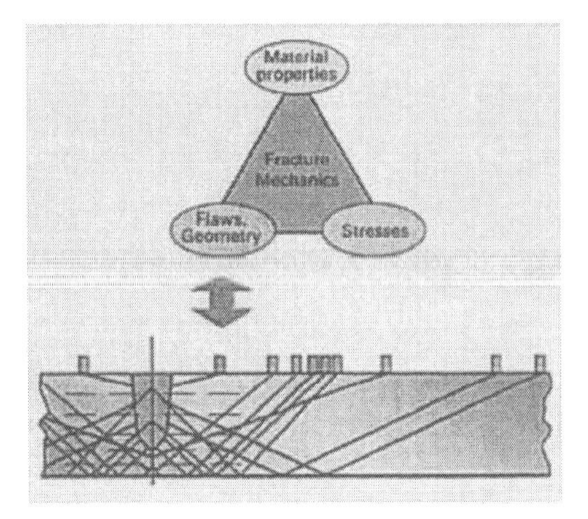

Figure 25.3 Schematic of UT transducer arrangement & ECA as a tool to link allowable flaw geometry with the material properties and stresses (Courtesy of Shaw Pipeline Services Ltd).

ECA can potentially reduce the cost of fabricating risers by:

- reducing the number of unnecessary weld repairs, through the use of optimised acceptance criteria based on fitness-for-purpose principles.
- reducing the cost of qualification for UT inspection procedures, partly by reducing the requirements for individual inspection qualification.

25.7 Flexjoints, Stressjoints and Pulltubes

25.7.1 Flexjoints

A flexjoint allows the riser system to rotate with minimum bending moment. The flexjoint normally exhibits strong nonlinear behavior at small rotation angle and hence should ideally model as a nonlinear rotation spring or a short beam with nonlinear rotational stiffness.

A flexjoint bellows system is illustrated in Figure 25.4 (Hogan, 2002). The bellows protect the elastomeric flex element from explosive decompression caused by large internal pressure fluctuations in a gas-saturated environment. The cavity between the body/flex element and the bellows is sealed and filled with a water-propylene glycol based corrosion inhibiting fluid.

For high pressure service, a large number of thin layers (e.g. 26) is typically required to ensure that the strain in the rubber will be acceptable. The flexjoint stiffness may be customized by changing the percentage of the carbon block, and changing the height of the rubber (number of layers).

For sour service, the inclusion of an inconel bellows improves performance of the flexjoint and shields the elastomer from the temperature of the bore fluid by approximately 15°F.

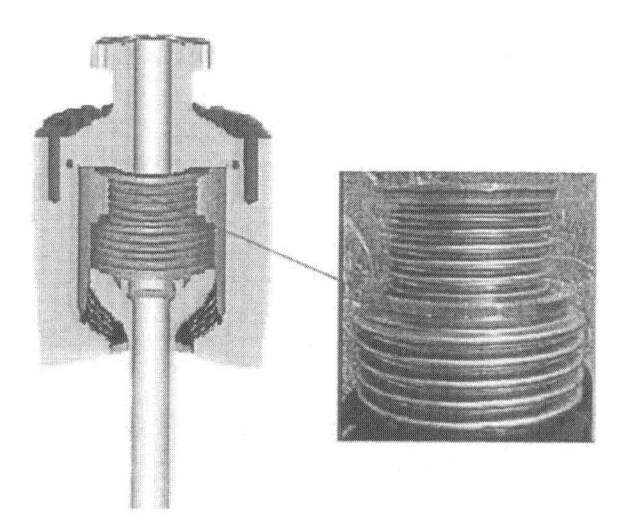

Figure 25.4 Illustration of a flexjoint bellows system for sour services (Hogan, 2002).

The design of flex-joints has generally only accounted for observed rotations (maximum and (fatigue) ranges). However, for ultra deepwater applications, the flex-joint designer should consider the effect of the large hang-off tensions and similarly the tension ranges for fatigue design. The ability to retrieve the SCR with the flex-joint should be considered as a mechanism for the SCR designer to enable detailed inspection. The implementation of a risk based integrity management plan could address the periodic inspection/monitoring of the flex-joint to minimize the risk of failure during the field life.

25.7.2 Stressjoints

For conventional SCRs at lower pressure, a stressjoint's flexibility is typically stiffer than the alternative flexjoint. The rotational stiffness for stressjoints is $\sqrt{EI \times T}$ where EI is pipe bending stiffness and T is axial tension applied. Stressjoints are modeled using beam elements of small size accounting for actual material properties (steel or titanium).

The low modulus, high strength, and excellent fatigue properties of titanium make it an ideal material for stressjoints. The joints should be sufficiently long so that the steel will be out of the region of significant fatigue loading. They are, therefore, particularly applicable where sour service conditions are encountered. A SPO flange may be required from CP design point of view. A steel stressjoint is similar except that the SPO flange connection is not required and the stressjoint is welded directly to the riser steel pipe.

25.7.3 Pulltubes

Pulltubes are used on recent spar SCR projects. The pulltubes are a kind of stressjoints. In the pulltube analysis, finite element analysis is conducted to model contacts between the riser and pulltube at the exit and contact between the guide and pulltube at the last support.

25.8 Strength Design Challenges and Solutions

25.8.1 Strength Design Issues

The SCR is designed to withstand storm events, such as a 100 year hurricane in the Gulf of Mexico. The predicted extreme peak stresses and tensions are generally calculated based on SCR response time histories from a number of realizations of these design storms and/or frequency domain analysis.

Ultra-deepwater SCRs can pose design challenges not normally experienced by riser engineers. The main issues are:

- High SCR hang-off tensions – high von Mises stress near hang-off location;
- SCR touchdown zone effective compression – buckling in beam-column mode ;
- SCR touchdown zone stress – yielding and low-cycle fatigue issues.

25.8.2 SCR Hang-off Tensions

For ultra deepwater SCRs, the water depth alone will play a large role in determining the hang-off tension. In an effort to reduce hang-off tensions, designers have used buoyancy modules attached in the near sea surface section e.g. Allegheny King Kong SCRs.

The large hang-off loads at the facility require more supporting structural steel at the riser porch, which requires larger vessel displacement. In addition to this, the large hang-off tensions and the corresponding large axial stress utilize more of the allowable stress, thus leaving a smaller allowable bending stress.

By using higher-grade steels (e.g.X70), the hang-off tensions may be reduced and the allowable bending stress may be increased.

25.8.3 SCR Touchdown Zone Effective Compression

The SCR touchdown zone motion response is coupled to the hang-off motion response induced by the hull motions. During large seastate events (hurricanes), the vessel heave motions can cause effective bottom compression in the SCR touchdown zone. This may cause upheaval/lateral buckling of the SCRs on the seabed, risking the integrity of the pipe.

25.8.4 SCR Touchdown Zone Stress

Traditionally von Mises stress in the touchdown zone is checked against 80% of material yield stress, per API 2RD. Per new design codes such as the DNV design code (DNV, 2001) and ABS pipeline and riser guide (ABS, 2001), bending moment, axial force and internal/external pressure may be checked, as discussed in the pipeline portion of this book.

25.8.5 Strength Design Solutions

There are a number of solutions available to reduce bottom compression of an SCR, these include:

- Locate the SCR hang-off point as close as practically possible to the center of the host vessel motion.
- Increase the wall thickness in the touchdown region to: i) increase the capacity of the region, and ii) increase the tension (by increasing riser weight).

25.9 Fatigue Design Challenges and Solutions

25.9.1 Fatigue Issues

The total fatigue damage on an SCR is determined from the sum of a number of sources, including:

- Waves and hull motions,
- SCR vortex induced vibrations (VIV),
- Hull VIV induced motions (VIM),
- Hull heave motions, and
- Installation fatigue.

Some of the key factors affecting SCR fatigue design are discussed in the following sub-sections.

25.9.2 VIV Design Challenges

Fatigue due to vortex-induced vibrations (VIV) becomes a critical part of the design for deepwater SCRs. VIV may be caused by long-term current profiles (operational current) and short-term current profiles (100 Year Loop Current and 100 Year Submerged Current). The objective of a VIV analysis is to estimate the length of strakes/fairings required to suppress VIV sufficiently to achieve the required design life.

- The prediction of these VIV carries with it considerable uncertainty, both in the current conditions and the VIV response of the riser system.
- On the current data side, VIV response is critically dependent on design metocean conditions, especially short-term and long-term current conditions.
- On the VIV response side, VIV is predicted by semi-empirical software whose accuracy requires experimental validation, ideally at Reynolds numbers representative of actual conditions. A significant source of that uncertainty is the ability to predict multi-mode as distinct from single-moded VIV response.

SCRs designed for operation in the deepwater (2,000 ft – 5,000 ft) in the Gulf of Mexico, are significantly influenced by the extreme submerged (Cold Core Eddy) currents. However, for ultra deepwater developments, the SCR configurations are expected to be less sensitive to the extreme submerged current, because SCR hang-off tensions increase. In some ultra deepwater developments, strong near bottom slab currents need to be considered as part of the design basis. Substantial SCR VIV suppression coverage (up to full coverage of the catenary portion) may be required to mitigate the fatigue damage caused by bottom currents.

25.9.3 Fatigue Due to Hull Heave Motions and VIM

Methods for Spar VIM induced fatigue are discussed in Chapter 24. If the host facility is a deep draft semi-submersible, a pontoon may be 100 ft below the mean sea level. The Vortex-Induced Motions (VIM) phenomenon is observed when the current direction is such that the total projected area of the columns in the flow is large. In some scenarios, VIM induced fatigue damage is comparable to damage due to VIV fatigue, wave induced fatigue and installation fatigue.

25.9.4 Effect of Wall-thickness Tolerance on Submerged Weight and Fatigue

Significant differences in submerged weight of a deepwater SCR can be caused by decisions made in the specification of wall-thickness tolerance. A wall-thickness tolerance measure, heavily skewed towards positive tolerance, may lead to a confidence that most of the pipe joints are above the required wall-thickness. However this comes at the expense of an increase in mean wall-thickness which can lead to a substantially higher submerged weight, higher hang-off loads for a deepwater risers and a detrimental effect on the fatigue performance of the SCR, especially for gas service, (Thompson et al, 2002). It is important that this issue is not ignored in the interface between cross-section design, line pipe specification and global design of a deepwater SCR riser system.

25.9.5 Effect of Vessel Selection, Hang-off Angle, Riser Orientation

Understanding the effect of hang-off location, SCR hang-off angle, riser orientation and vessel motions on fatigue response at touchdown zone are key issues to maximizing the potential for achieving a feasible SCR design.

Semi-submersibles have significant heave motions, in addition to significant roll and pitch. Predicted SCR fatigue performance is sensitive to the modeling of low-frequency host motions.

25.9.6 Combined Frequency and Time Domain Analysis

The computational efficiencies and short run times of frequency domain (FD) analysis often result in a riser designer adopting this approach for wave-induced fatigue analysis of SCRs. This has proved an effective method for SCR design for Spars & TLPs, where conservative SCR designs can be achieved, without being prohibitive to the riser designer or operator.

However, for SCRs attached to semi-submersibles, it is often difficult to achieve a feasible wave-induced fatigue design using only the frequency domain, due to the non-linear dynamic touchdown response of the SCRs. The linearization process inherent in the frequency domain, can result in an over estimation of SCR fatigue damage in the touchdown zone. Generally, the vast majority of SCR wave-induced fatigue damage is associated with 15 to 20 seastates from the total scatter diagram comprising up to 200 seastates. To expedite the design process, it is feasible to perform full time-domain analysis of the SCR response for the critical seastates and then substitute this damage for that calculated by the frequency domain approach.

25.9.7 Touchdown Soil Effect

Deepwater SCR feasibility is often a combination of strength and fatigue challenges, each of which is affected by TDP soil stiffness. Several approaches are available in riser analysis software for TDP modeling, representing trenching, suction and lateral friction. The riser TDP locations move when the host vessel changes its draft or moves between near and far positions. Riser fatigue damage is actually spread over a length of the SCR. Hence it is important to account for TDP movement in the calculation of SCR fatigue at the TDP.

In general, soil stiffness is determined by soil shear strength. The higher the soil stiffness is, the more fatigue damage may occur in the riser pipe, since vibration frequency and stress ranges will be higher. Secant soil stiffness is generally higher than tangent soil stiffness. It is hence more conservative to use secant soil stiffness. However, the actual tangent soil stiffness may yield better accuracy for fatigue prediction because it better represents soil deformation characteristics for fatigue loading.

25.9.8 Fatigue Design Solutions

For ultra-deepwater SCR projects, several options exist for the mitigation of the uncertainties associated with vessel motion, or VIV-induced riser fatigue in ultra-deepwater, among these are:

- Optimization of the hull shape to minimize the motions due to wave action. This is a route being actively pursued by a number of hull designers with semi-submersible hull shapes evolving with variations in column geometry e.g. reducing the water plane area to reduce heave motions; varying the water plane along the column to reduce the heave and pitch motions.

- Plan to move the vessel location a number of times during the field life, hence changing the touchdown point and spreading out the fatigue damage over a larger length of the SCR touchdown zone.
- Development of metocean prediction models which improve confidence in deepwater current prediction and/or site specific measurements of the deepwater current events.
- The implementation of a risk-related integrity management plan which includes inspection, monitoring, testing or analysis to validate design assumptions or direct measurement of VIV response in-service. VIV monitoring is of particular importance because there is a lack of field experience using SCRs on semi-submersibles and FPSOs. In addition, uncertainties with current data and VIV prediction tools require monitoring of VIV induced fatigue damage, (Franciss, 2001, Grealish et al., 2002).
- Use of a higher safety factor for VIV design to account for the uncertainties in the current data and VIV response predictions.

25.10 Installation and Sensitivity Considerations

25.10.1 Installation Considerations

The actual SCR installation philosophy may not be decided at the start of a project. Therefore the line pipe specifications may have to be drafted with contingency for all installation methods (S-lay, J-lay & Reel-lay). In the case of J-lay, ID boring was likely to be performed; therefore 1 mm was added to the pipe wall thickness to allow for this, resulting in a specification for a heavy pipe.

For smaller diameter SCRs, installation with the reel-lay procedure may be possible. Substantial engineering effort would be needed to qualify these SCRs for reeling:

- Full scale bending trials and testing of centralizers to maintain a constant annulus for the PIP.
- Full scale fatigue testing of the reeling riser welds. It is important to accurately calculate the bending strain in the welds during the reeling operation, to assess crack propagation.
- The allowable bending strain is carefully monitored.

25.10.2 Sensitivity Analysis Considerations

Key sensitivities to be assessed include: currents, waves, length of strakes, seabed soil stiffness, flex joint stiffness and vessel RAOs. Some of the design input (metocean conditions, flex-joint stiffness, vessel motions, linepipe weight, etc.) assumptions inevitability change throughout the project. Therefore, sensitivity analyses to understand the implication of changes to these key inputs is a valuable and worthwhile exercise.

25.11 Integrity Monitoring and Management Systems

25.11.1 Monitoring Systems

The monitoring systems may provide operational assurance on flexjoint angle, riser top tension force and riser stress levels at touchdown point for strength and fatigue assessment.

The measurements include:

- The current profiles along the water column are measured;
  - using ADCP mounted at vessel lower pontoon to monitor the current profile in the top 3000 ft of water column.
  - using ADCP located on the seabed to monitor the current profile in the bottom 2000 ft of water column.
  - using mid-water ADCP to capture the current profile in the mid-water column.
- The flexjoint rotational angles are measured using inclinometers;
- The riser top tension is measured using conventional strain gauges;
- The riser stress levels in the touch down point are measured using strain gauges and accelerometers.

25.11.2 Integrity Management Using Monitored Data

The monitored data may be used to manage structural integrity:

- To define acceptable vessel offset excursion
- To assist inspection planning.
- To calibrate the design tools for accuracy improvement.
- To detect any abnormalities in the riser system due to damages and unforcasted environmental conditions.
- Conduct re-assessment of the structural integrity.

25.12 References

1. ABS (2001), "Guide for Building and Classing Subsea Pipeline Systems and Risers".
2. API (1993), "API 5LD, Specification for CRA Clad or Lined Steel Pipe".
3. API (1998), "RP 1111 - Design, Construction, Operation and Maintenance of Odffshore Hydrocarbon Pipelines (Limit State Design)".
4. API (1998), "RP 2RD-1998, Design of Risers for Floating Production Systems (FPSs) and Tension-Leg Platforms (TLPs)"
5. Y. Bai, A. Tang, E. O'Sullivan, E. Uppu and S. Ramakrishnan (2004), "Steel Catenary Riser Fatigue Due to Vortex Induced Spar Motions", OTC Paper 16629.
6. Y. Bai, E. O'Sullivan, C. Galvin and K. Kavanagh (2004), "Ultra-Deepwater SCRs Design Challenges and Solutions for Semi-Submersibles", Deep Offshore Technology.
7. J. Buitrago and M. S. Weir (2002) "Experimental Fatigue Evaluation of Deepwater Risers in Mild Sour Service", Deep Offshore Technology Conference, New Orleans.

8. G. Chaudhury and J. Kennefick (1999), "Design, Testing, and Installation of Steel Catenary Risers", OTC Paper 10980.
9. L.M. Connelly and N. Zettlemoyer (1993) "Stress Concentration at Girth Welds of Pipeswith Axial Wall Misalignment", Proceedings of the 5th Intl. Symposium on Tubular Structures, Nottingham (UK).
10. DNV (2001), Offshore Standard F201, "Dynamic Risers", 2001.
11. R. Franciss (2001), "Vortex Induced Vibration Monitoring System in the Steel Catenary Riser of P-18 Semi-Submersible Platform", OMAE2001/OFT-1164.
12. C.T. Gore and B.B. Mekha (2002), "Common Sense Requirements (CSRs) for Steel Catenary Risers (SCR), OTC Paper 14153.
13. F. Grealish, D. Lang and P. Stevenson (2002), "Integrated Riser Instrumentation System – IRIS 3D", IBC Conference on Deepwater Risers, Moorings & Anchorings.
14. M. Hogan (2002), "Flex Joints", ASME ETCE SCR Workshop, Houston, Feb. 2002.
15. D.R. Korth, B.S.J. Chou and G.D. McCullough (2002), "Design and Implementation of the First Buoyed Steel Catenary Risers", OTC Paper 14152.
16. F. Kopp, G. Perkins, G. Prentice and D. Stevens (2003), "Production and Inspection Issues for Steel Catenary Riser Welds", OTC Paper 15144.
17. F. Kopp, B.D. Light, T.S. Preli, V.S. Rao and K.H. Stingl (2004), "Design and Installation of the Na Kika Export Pipelines, Flowlines and Risers", OTC Paper 16703.
18. B.B. Mekha, E. O'Sullivan and A. Nogueira (2004), "Design Flexibility Saves Marco Polo Oil SCR during Its Installation", OMAE.
19. E.H. Phifer, F. Kopp, R.C. Swanson, D.W. Allen and C.G. Langner (1994) "Design and Installation of Auger Steel Catenary Risers", OTC Paper 7620.
20. O.B. Serta, M.M. Mourelle, F.W. Grealish, S.J. Harbert and L.F.A. Souza (1996), "Steel Catenary Riser for the Marlim Field FPS P-XVIII", OTC Paper 8069.
21. H.M. Thompson, F.W. Grealish, R.D. Young and H.K. Wang (2002), "Typhoon Steel Catenary Risers: As-Built Design and Verification", OTC Paper 14126.
22. Urband, B. "CRA Clad Downhole Tubing – An Economical Enabling Technology", AADE 01-NC-HO-46, AADE National Drilling Technical Conference.

Part IV

Riser Engineering

Chapter 26 Top Tensioned Risers

26.1 Introduction

Top tensioned risers (TTRs) are used as the conduits between dynamic floating production units (FPU) and subsea systems on the sea floor, for dry tree production facilities such as Spars and tension leg platforms (TLPs), see Figure 26.1.

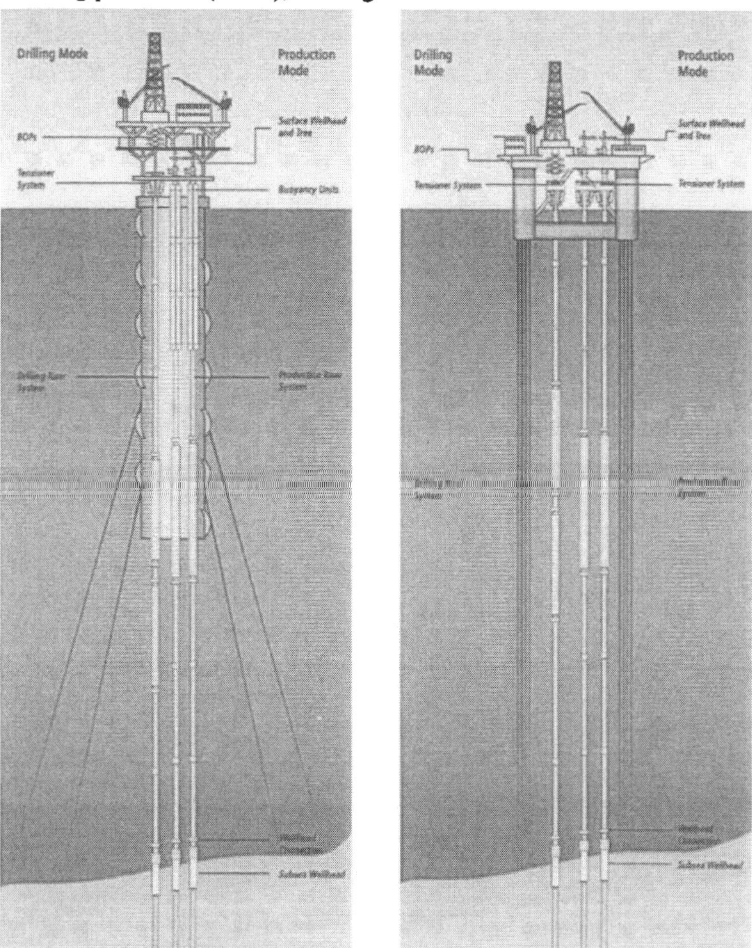

Figure 26.1 Top tensioned risers (TTRs) used on Spar and TLP (The Composite Catalog ® of Oilfield Equipment Services, 2002/03, 45th Edition).

With such systems, the workover operations are possible, without the need to have a separate dedicated drilling rig. In general, TTRs may perform the functions of production and injection, drilling and export. This Chapter is devoted to components and design/analysis for TTRs.

26.2 Top Tension Risers Systems

26.2.1 Configuration

Figure 26.2 presents a typical TTR set-up for use with a TLP. The TTR configuration depends on the riser function and the number of barriers selected (single or dual). In general, the riser configuration comprises the following components:

- Main body is made up of rigid segments known as joints. These joints may be made of steel, titanium, aluminum or composites, though steel is predominantly used.
- Successive joints are linked by connectors such as: threaded, flanged, dogged, clip type, box and pin.

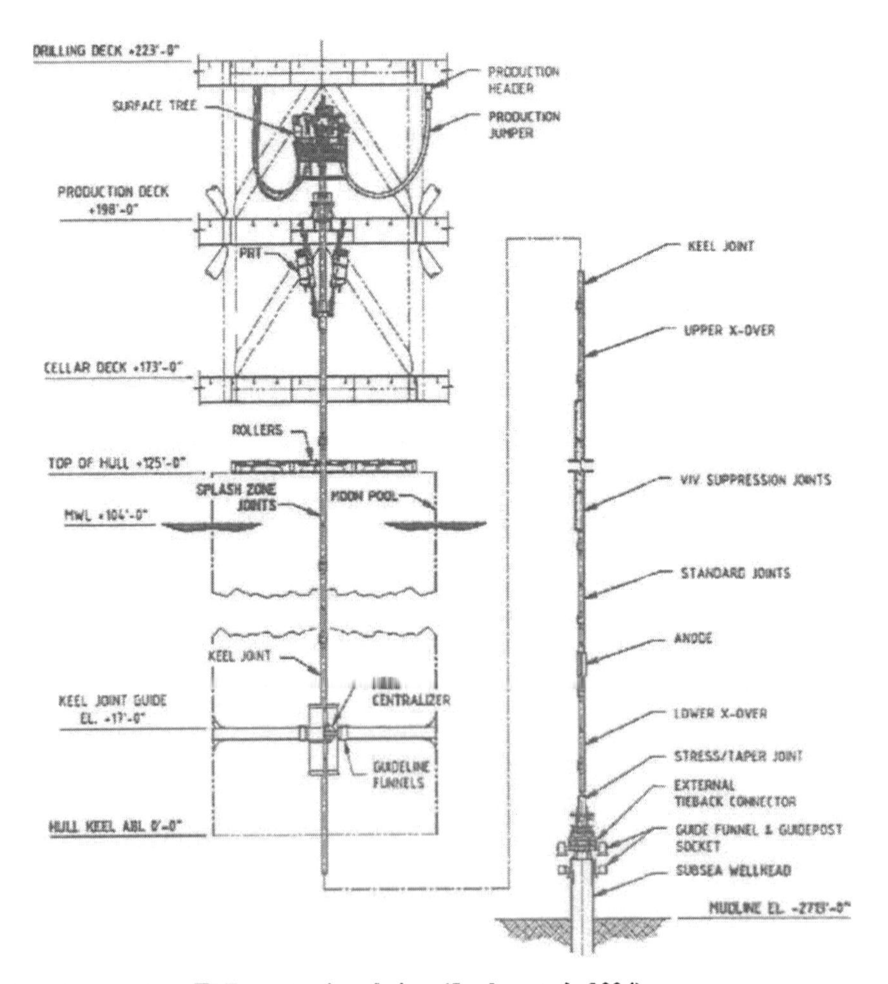

Figure 26.2 Stack up for a TLP top tensioned riser (Jordan et al., 2004).

- The riser is supported by a tensioning system, such as: traditional hydraulic tensioners, air cans, RAM tensioners, tensioner decks and counterweights.

Figure 26.3 presents a buoyancy based TTR for use with a Spar. The buoyancy can system has been used in the following Spars (Yu et al., 2004):

- 1996 Oryx closed end buoyancy cans;
- 1999 Genesis non-integral buoyancy cans utilizing rig for installation;
- 2000 Hoover integral buoyancy cans using heavy lift vessel and rig for installation;
- 2002 Horn Mountain integral buoyancy cans with complaint guides and "U" tube.

The buoyancy cans decouple the vertical riser movement from the vessel, and can be built by the fabrication yard. However, it requires either heavy lift vessel or a specially designed rig for offshore installation. It has large relative vertical motion in storm condition and may generate lateral loads between the buoyancy cans and the spar center wall.

The Holstein top tension risers are vertically supported by the Spar hull by means of individual production riser tensioners (PRTs). The PRT support module is supported by the Spar's hard tank through a riser support structure over the center well and near the top of the Spar hull. Figure 26.4 shows Holstein TTR riser and the tensioning system stack up (Yu et al, 2004).

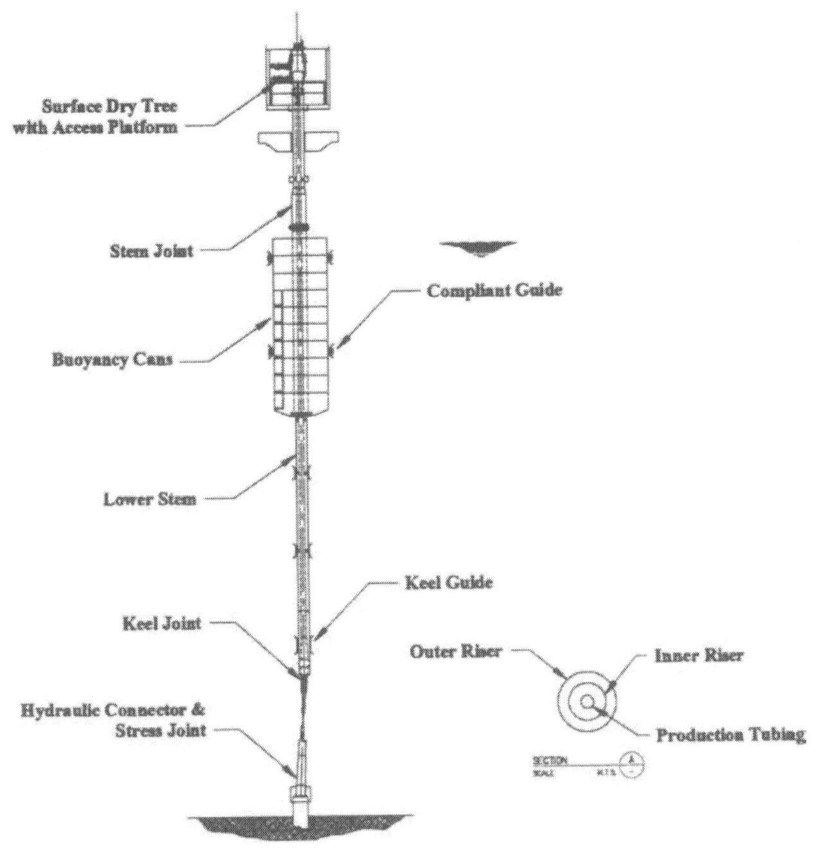

Figure 26.3 Schematic layout of a buoyancy can based Spar top tensioned riser (Yu et al., 2004).

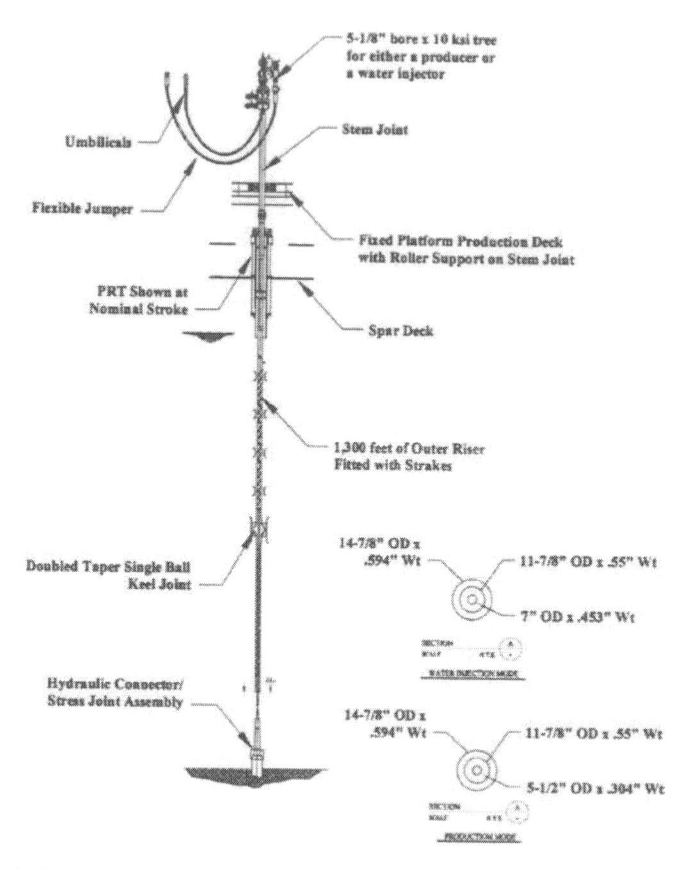

Figure 26.4 Holstein dry tree riser and tensioning system (Yu et al., 2004).

Figure 26.5 shows the sea floor spacing of 35 ft nominal for the outer circle and 28 ft for the inner circle for the future wells. Riser clashing analyses are performed to determine the seafloor well spacing.

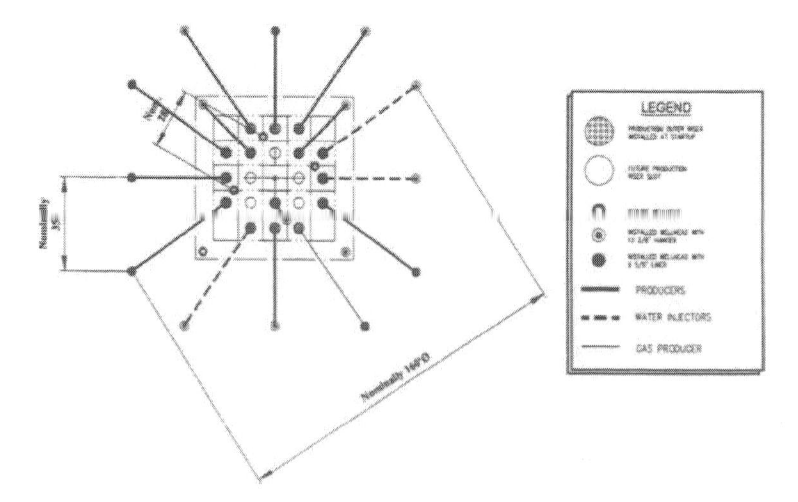

Figure 26.5 Holstein sea floor well patter (Yu et al., 2004).

26.2.2 General Design Considerations

The major design drivers of TTRs are for instance:

- Allowable floater motions;
- Allowable stroke of tensioning system;
- Maximum riser top tension;
- Size of stress joints and flex joints;
- Keel joints;
- Increasing length of riser joints;
- Design criteria for the safety philosophy of liquid barriers, valve and seals;
- Current, interface with array and vortex induced vibration;
- Impact between buoyancy cans and hull guides.

Some of these design drivers will be further discussed below.

The horizontal motions of the floater induce stresses in the riser base and at the riser top end, i.e. near the stress/flex/keel joints. As the water depth increases, the influence of vessel motions will have less influence on the riser sections near the sea floor. However, under certain local flow conditions, deepwater (bottom) currents may induce significant amplification of the deflections and stresses in the riser bottom portion.

Current induced fatigue damage and clashing between adjacent risers may become a driving design parameter for ultra-deep water risers. It is therefore important that current data in the form of scatter diagrams, etc., are available.

The riser top tension requirements become significant at depths beyond 1500-2000m, where composite riser systems may have an advantage compared to metallic ones. At such depths the TTR architecture must be optimized in order to reduce the riser top tension, e.g. by using single casing risers instead of dual casing.

The required tension of the riser increases with the water depths as it must support the riser weight, prevent bottom compression, and limit VIV damage and clashing between adjacent risers. This increased riser tension will affect the size of the tensioning system, the buoyancy requirements, as well as the size of the flex-joints or stress joints. The current trend is to increase the joint lengths form 12m to 24m for TTRs at large water depth in order to reduce the installation time, though, this will impact the handling and manufacturing cost, which must be accounted for.

An essential issue in the cost optimization is the ability to control the implicit conservatism in the design via rational *design criteria and analysis procedures*. The state of practice for riser design is reflected by recommended practices such as API RP 2RD.

26.2.3 Drilling Risers

The main design drivers for drilling risers are as follows:

- Weight;
- Top tension;
- Cost;
- Running time;
- VIV

Syntactic foam is used to provide buoyancy and reduce top tension, although it is costly in terms of material procurement and running time. The syntactic foam may increase drag diameter and accelerate VIV fatigue. Fatigue concerns make it desirable to rely on relatively frequent inspection each time the riser is pulled. In most cases the drilling program will only last a few years. This also needs to be considered.

26.3 TTR Riser Components

26.3.1 General

Specialist Components for a Top Tensioned Riser System will be discussed in the following sub-sections.

26.3.2 Dry Tree Riser Tensioner System

Five tensioner options are considered suitable for deep water applications, as follows:

- Conventional TLP type Hydro-Pneumatic Tensioners;
- Hydro-Pneumatic Tensioners with remote accumulators/APVs;
- Ram style Pneumatic Tensioners system;
- Buoyancy cans;
- Wire-rope tensioners.

Each of the tensioner system may be evaluated under the following criteria:

- System description and drawings;
- Tensions capacity, system redundancy and stroke;
- Motions of FPU;
- Safety, integrity, inspection, access and replaceability;
- Distinguishing features and advantages;
- Limitations

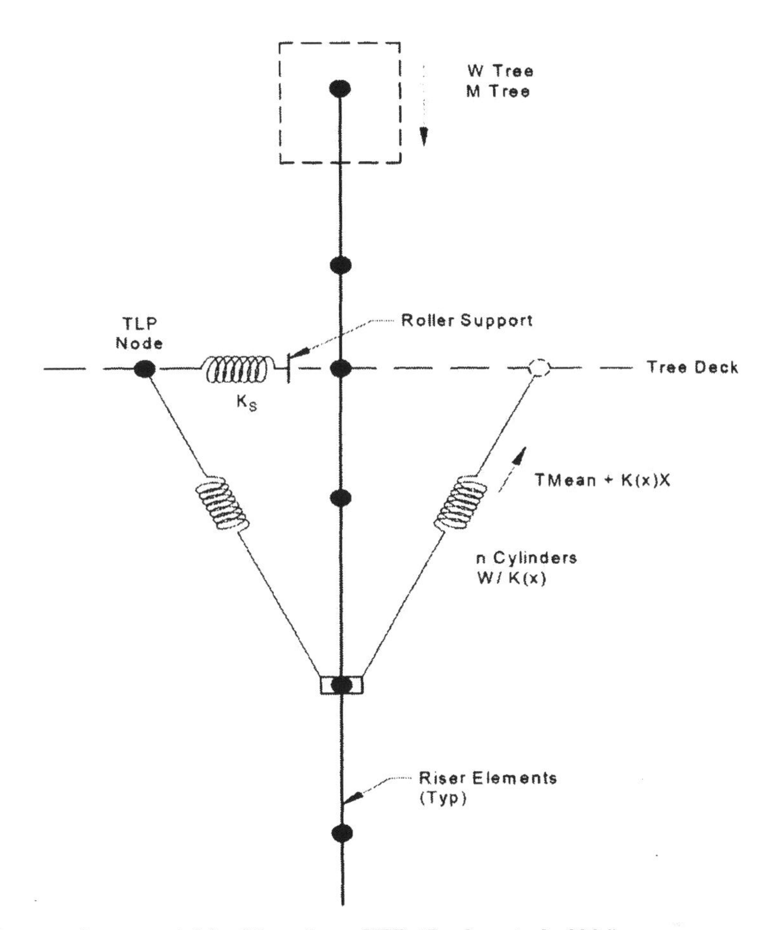

Figure 26.6 Riser tensioner model for Materhorn TTR (Jordan et al., 2004).

26.3.3 Tie-back Connector

A tie-back connector and tapered stress joint (TSJ) are used as the seafloor connection between production risers and completed wells. The tie-back connector and TSJ can be fabricated as an integral component to form a complete end-termination connection assembly.

The tie-back connector must be designed with sufficient pressure, bending and tension load capacity. The pressure integrity and fatigue life must be sufficient to withstand the dynamic forces exerted by the riser over the service life of the development.

In the design process, the riser designer provides interface loads to the perspective suppliers, who then finalize their design using local FE analysis. The interface loads include extreme tensions, bending moments and shear loads.

26.3.4 Keel Joint

Figure 26.7 illustrates the keel centralizer and guide used for the Materhorn TTR system. The keel joint is located at the point at which the riser enters the bottom of a deep draft vessel. The keel joint is used to protect the riser against large bending stresses due to vessel motions by providing the necessary additional wall thickness and load distribution.

The critical design parameters for the design of the keel joint are stroke, riser bending response at the entrance to the vessel, and wear and shear of the joint and guide structure. A typical guideline could be that the keel joint length be 50% greater then the required vessel stroke. The cross-section design of the keel joint and tapering sections is driven by requirement to withstand bending moments generated under a range of design environmental/operational conditions.

Worst case loading conditions are those which will result in the largest riser angular rotations and associated tensions at the intersection of the riser and vessel hull. Maximum rotations and tensions will typically occur in load cases in which severe vessel offsets and motions are applied at surface level, causing the riser string to bend in the region of the hull interface.

The fatigue analysis of the keel joint could initially be carried out using the DOE D S-N curve in combination with an SCF of 1.5. For detailed engineering this would need to be assessed, considering all aspects of the design of the keel joint.

The wear potential of the keel support in the seawater environment can be significant and can even drive the design of the keel joint hull. As a result, the intermittent contact and wear between the keel joint and the hull guide surface should be considered in the detailed design of the riser system.

Figure 26.7 Keel centralizer in Hull Keel Guide (Jordan et al., 2004).

26.3.5 Tapered Stress Joint

A tapered stress joint (TSJ) is a specialized riser joint with a tapered cross section used to distribute bending load over a controlled length so that the bending stresses are acceptable. Typical locations of TSJs on dry tree production riser systems are at wellhead connection, above and below the keel joint in deep draft vessels.

The thickness and length of the tapered stress joint are driven by the requirement to withstand excessive bending under all potential loading combinations expected throughout the design life. Design length can also be limited by handling and fabrication considerations with a reasonable upper limit being around 18m (60ft), though if required the TSJ can be fabricated in more than one section to achieve longer lengths.

The overall design of the TSJ can be performed through global analysis using analytical tools. Key issues include extreme and fatigue response characteristics. Local finite element analysis will be required to verify the TSJ design and to determine stress concentration factors to be used in global fatigue analysis. A critical area occurs at the interface between the TSJ and the connection.

With regard to a TSJ located at wellhead interface, worst case loading conditions are those which will result in the largest angular rotations and associated tensions at this location. Maximum rotations and tensions will typically occur in load cases in which the riser is blown away from the wellhead under extreme environmental conditions with severe vessel offsets and motions applied at surface level. Large vessel offsets/drilling requirements result in an increase in the length of the TSJ section.

The critical region is at the interface between the TSJ and the first standard riser joint. It is possible to use a thicker riser section at this interface in order to optimize response and limit the length of the TSJ section.

26.3.6 Riser Joint Connectors

Individual riser joints are coupled together by mechanical connectors located on the ends of each pipe section. The connector provides a means of connecting/disconnecting successive riser joints, transferring load between joints and providing leak tightness. The most commonly used connector designs are as follows:

- Threaded or grooved connectors assembled by torque or radial interference;
- Flanged connectors (e.g. Standard ANSI flanges);
- Compact flanges (e.g. as supplied by SPO);
- Dogged connectors using radial wedges between pin and box parts;
- Clamped connectors assembled by bolts;
- Connectors that rely on impinging grip or friction, by means of collets or sliding dogs to affect the load transfer (e.g. tie-back connectors).

Figure 26.8 High fatigue joints used on West Seno drilling and production riser system (Utt et al., 2004).

The design of flange connectors is considered to be standard. Thread-on connectors utilize threaded couplings to connect successive riser joints. Each pipe end is threaded to allow connection via the coupling and eliminate the need for any welding. To date the use of thread-on connectors has only been proven viable (i.e. economically) in selected deepwater and ultra-deepwater applications. Such technology is currently employed with TLP and SPAR platforms in the Gulf of Mexico.

To date, a wide variety of proprietary connectors have been used in top tensioned riser systems. Key issues to be considered in selecting the connector type include:

- Production or drilling application (single make-up or multiple);
- Internal or external riser (for dual casing);
- Exposure to water;
- Available space (e.g. for internal riser);
- Riser joint material (welding issues with high strength steels);
- Splash zone requirements;
- Previous experience/history of connectors;

- Loads (extreme axial, pressure and bending);
- Fatigue characteristics (SCF and S-N curve);
- Sealing mechanisms and durability;
- Riser installation/running requirements (load shoulders).

The current standard practice is to use the following:

- Internal casing of dual casing production risers – typically grade 110 steel joints with upset ends and integral threaded connectors (no welding). In some cases the connectors are machined on to non-upset plain pipe.
- External casing (single or dual casing production risers) – typically grade X80 steel joints with welded on proprietary connectors.
- Drilling risers – typically grade X80 steel with standard flange or proprietary flange connectors.

The key issue with drilling risers is the selection of connectors that can be made up and broke-out on numerous occasions. To date flange type connectors are the preferred choice, though threaded connectors have been used for internal casings in dual casing drilling risers.

26.3.7 Tension Joint & Ring

The tension joint should be designed to transfer the required tensioning force to the top of the production riser under all foreseeable design conditions.

The ring is typically threaded to allow for adjustments to be made during installation. The tension ring is purpose-built to suit the rig tensioner system and number of tensioning lines required. An adequate number of pad-eyes should be provided to allow sufficient alignment with rig tensioners in order to minimize torque in the riser system. Alternatively a swivel ring may be used to eliminate this effect. Each pad-eye should be adequately sized to support the required tensioning loads and associated connectors.

The joint itself should provide continuity of production and annulus bores. The upper end of the joint should be capable of connecting to the Christmas tree or riser spool, the bottom should be capable of connecting to a standard riser joint. The connection between the upper end of the tension joint and Christmas tree is usually made by a spool joint. The length of this spool joint is selected to achieve the appropriate total stack-up height.

The joint should be designed to accommodate the asymmetrical pull caused by failure of at least one tensioner. The global design of the tensioner point can be performed using riser analysis tools such as **Flexcom**. The design of the tensioner ring and tensioner joint interface will need to be performed using general purpose analysis tools such as **ANSYS** or **ABAQUS**.

26.3.8 Riser Joint in Splash Zone

Splash-zone riser joints are special joints which incorporate riser sections of increased wall thickness and/or increased length. These heavy duty joints are required when a large corrosion allowance is required in the aggressive environment of the splash zone. The higher internal forces in the riser close to the tensioner due to effective tension and bending moment can also lead to increases in the thickness of the splash-zone riser joints.

Typically, no joint couplings are located in the splash zone of the riser. Hence, the length and position of the joint must be carefully planned and allowed for in the design and construction of the system make-up. If more than one joint is required in the splash zone, it is recommended to use joints with flange connections.

Corrosion protection can be applied to riser joints using a neoprene coating on the surface of the pipe. In addition, sacrificial aluminum anodes are typically clamped to each joint to provide protection for the riser. A copper-nickel sheeting can be applied on top of the neoprene coating to provide anti-fouling protection in regions where heavy marine growth is expected. A further corrosion protection option is thermal sprayed aluminum (TSA).

26.3.9 Flexible Jumper between Surface Tree and Deck-based Manifold

Flexible jumpers have a limited temperature capacity (e.g. approximately 120°C for unbonded flexibles). For higher temperature operations, a system consists of short sections of rigid steel pipe connected by articulation joints, is applied to give some flexibility to the system.

The bonded pipe is a rubber based pipe with all layers bonded in a composite construction, whereas the unbonded pipe is a polymer based pipe with slippage allowed between the individual layers. Bonded pipes are typically significantly cheaper. The key concerns with bonded hoses are degradation of the rubber, gas permeations and rapid decompression damage to the material in a production environment.

Once a suitable pipe structure is selected, the design configuration is developed through detailed analysis using the design range of riser motions specified from the global analysis of riser system under extreme conditions.

26.3.10 Tubing/Casing Hanger

The hanger systems are used to control the load distribution between different riser casings and to allow load transfer from the tensioning device into the riser system components, see Figure 26.9. Design considerations for the casing/tubing hanger are similar to that for standard riser joints/connections.

A key issue with the design of the hanger is to be able to generate the required preload on the tubing and/or internal casing, accounting for thermal and pressure induced effects. In many cases the design philosophy was to consider all axial loads as being taken by the outer casing on a dual casing riser.

In many cases the design philosophy was to consider all *axial loads* as being taken by the outer casing on a dual casing riser. In more recent projects there is tendency to more accurately consider the load distribution across the three tubulars in a dual casing riser. This requires a careful identification of all possible load scenarios, in particular to identify the various thermal states across the riser cross-section, as thermal stresses have a major impact on the load distribution. For multiple casing riser it is not required that all tubular have positive tension. However, keeping the individual lines in tension does help to:

- Enhance fatigue life of the threaded connector;
- Minimizes the centralizer requirement near the mudline;
- Simplifies and increases the confidence in the computation of bending moments in the multiply strings.

The hanger should be designed and detailed in such a manner that it is simple and safe to make-up. All seals should be adequately designed to ensure the overall integrity and pressure/leak tightness of the connection.

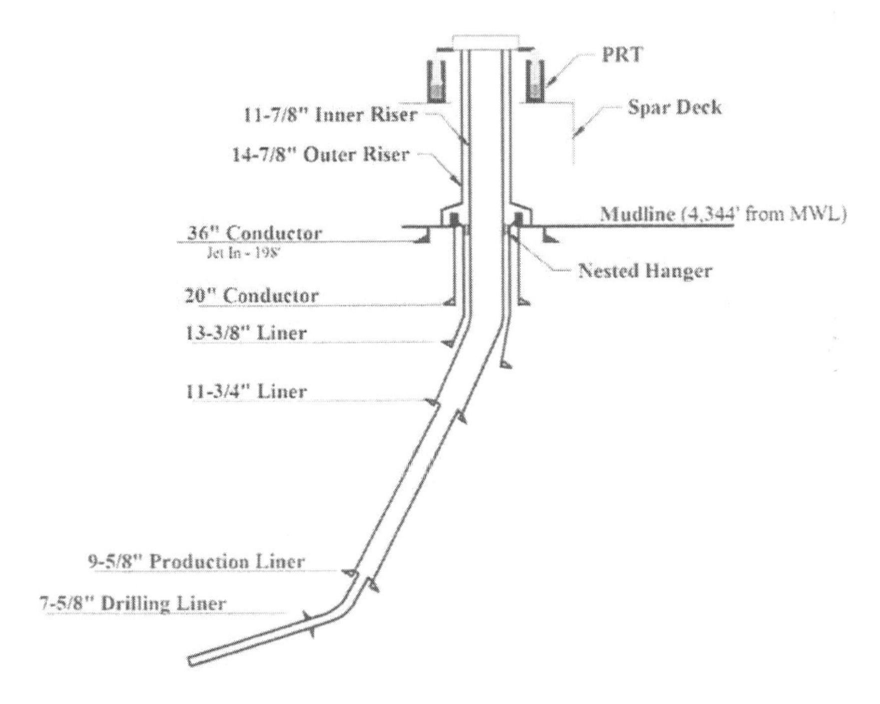

Figure 26.9 Typical casing program for a horizontal well (Yu et al., 2004).

26.3.11 Air Cans

Air cans are the central component of buoyancy based tensioning system and are typically used to provide top tension to risers in deep draft vessel applications. The air can tensioning system is in principle very simple and has the advantage of being self-tensioning and requiring relatively low supervision in comparison to mechanical tensioning systems.

Typical air can assemblies consist of the air cans, keel stem and upper stem. The keel stem and upper stem are typically connected to the air can by mechanical connectors. The can and stem assembly can be either "integral" or "non-integral". In an integral system the outer wall of the riser forms the inner diameter of the air can while in a non-integral system, both air can and associated stem are independent of the riser outer casing. More recent designs have the air cans jointed together through a stem joint. In this case the full tension is transmitted to the riser at a single point, i.e. at the surface well head.

A major design issue associated with the integral type system is the stepped increase in bending stiffness experienced by the riser as a result of the intermittent presence of the air cans. Coupled with relative lateral motion between air cans and guides the effective increases in bending stiffness affects the modal response of the riser in this region and promotes vibrations or a resonant type response of the riser within the hull. This in turn can lead to stress concentrations and increased fatigue damage of the riser within the hull. This problem can be avoided when a non-integral system is employed.

In general, the efficiency of non-integral air cans is less than that of integral cans. Therefore for a given number of cans on a riser, a larger air can diameter is required for non-integral cans. This is an important consideration in hull design as the diameter of the cans and the number of adjacent risers in the wellbay will normally dictate the diameter of a hull.

26.3.12 Distributed Buoyancy Foam

Top distributed buoyancy foam is used to provide lift and to reduce the submerged weight of riser joints in deepwater applications. A typical distributed buoyancy system incorporates buoyancy modules, Kevlar straps and thrust collars. The modules are usually supplied in half shells and strapped around the riser joint with the thrust collars fitted just below and above the upper and lower couplings of the riser joint.

In the design and specification of the buoyancy system the surface area to volume ratio should always be minimized in order to keep buoyancy loss due to water absorption to a minimum. Also, the provision of a compatible external protective coating to the modules should be considered. This can help with the minimization of water absorption and provide resistance to impact and abrasion encountered in normal handling operations prior to and during installation. The buoyancy material should be selected accounting for the water depth to which it is exposed, exposure to chemicals and extreme temperatures and the extent of loads encountered during service life of the system.

The thrust collars should be designed to transfer the buoyancy force from the system to the riser joint and be able to support the full weight in air of the system. The internal profile of each shell is fabricated specially for each line to allow for the presence of auxiliary lines and riser clamps.

Finally, the buoyancy modules should be designed so that they do not present any major difficulties in storage, handling, transportation or installation process.

26.4 Modelling and Analysis of Top Tensioned Risers

26.4.1 General

In the preliminary design phase, the key issues are (Burke, 2004):

- Top tension factor determination;
- Pipe sizing;
- Stroke analysis;
- Tensioning system sizing;
- Riser Component Sizing;
- Preliminary VIV assessment;
- Interference Analysis;
- Strength Analysis;
- Fatigue Analysis.

The detailed design phase includes more detailed strength and fatigue analysis, analysis of centralizer spacing, and riser running and installation analysis. By its nature, the design of the riser system is an iterative process, as summarized below:

1. Review all aspects of the system operation and identify all conditions to be considered in the design. Also, select whether single or dual casing is to be used. Select riser stack-up arrangement.
2. Select initial wall thickness sizes, material and other relevant design factors (e.g. corrosion allowance, dimensional tolerances).
3. Perform static analyses to confirm stack-up and define tension requirements.
4. Perform extreme, VIV and fatigue response analyses.
5. Perform interference assessment.
6. If the results from points 3, 4, 5, or 6 generate the requirement for design changes, update the system design and redo all relevant analyses.
7. Finalize design of specialist components, such as TSJs or keel joints. Typically initial design of the specialist components is incorporated into the global model at the start of the design process, with these initial designs selected based on previous experience. These designs can then be updated in the various design stages, e.g. fatigue analysis results can be used to update TSJ profiles. In parallel it may also be necessary to undertake local FE analysis to verify the component design and to generate stress concentration factors (SCFs).
8. Perform installation analysis.
9. Complete all design reports.

In all stages of the design, consideration should be given to the requirement to perform sensitivity studies on key parameters.

26.4.2 Stack-up Model and Tension Requirement

The methodology used for the stack-up model selection and tension requirement is outlined in the following steps:

1. Considering all the components that contribute to the riser stack-up, calculate the apparent weight of the riser for each of the internal fluid conditions.
2. Calculate the minimum top tension required in the riser, by considering the range of loading conditions. The tensioning requirement for the riser can be referred to in terms of 'Tension Factor', described as follows:
3. Tension Factor = Tension Provided by Tensioning System / Apparent weight of riser from Seabed TSJ to Xmas Tree
4. If the design strategy is such that a base case Tension Factor of 1.05 for the riser in its 'heaviest' condition is applied while assuming 'one tensioning component down' scenario, then the resulting capacity of the tensioning system will be adequate under all other loading combinations. In other words, under the most onerous riser apparent weight condition, the tensioning system will still provide adequate tension even if one aircan (or tensioner) is not in operation. Tension factors for all other internal fluid conditions and tension redundancy conditions can be extrapolated based on this requirement.
5. In establishing the heaviest condition of the riser, e.g. workover or wellkilled conditions, the weight of tubing, control/injection lines, centralizers, and surface tree (BOP) are considered. Note that in some systems supplemental tensioning can be used to support the BOP and this need to be accounted for in the calculations.
6. From the minimum top tension required for the heavy riser condition, establish the resulting riser top tension factor for all other operating conditions, assuming both full capacity and redundant tensioning system conditions.
7. Develop a finite element model of the production riser stack up.
8. Perform static finite element analyses at vessel design offsets to determine the stress along the riser, tapered stress joint and keel joint under static offset conditions defined in the riser analysis basis. If necessary, alter the design of the riser or TSJ (length and wall thickness) to ensure stresses remain within allowable limits and re-iterate on required top tension calculations.
9. Verify that for all load cases, the axial stress in the pipe is acceptable, and that there is sufficient margin available in the pipe cross-section areas to accommodate extreme offset, current, and dynamic loading effects.
10. Evaluate the stroke range of the riser.

Initial analyses are performed to ensure the riser tensioning system is capable of providing adequate tension to the riser for a range of riser operating conditions including normal production, workover, completion, wellkilled and shut in. Based on the possible permutations of internal fluid pressure, density and functional loading of the riser assembly that can occur in the riser for such operating conditions, a matrix comprising of operating conditions considered most critical is identified. These cases form the basis for the evaluation of the tension factor design and tension distributions within the riser. Note, the tension factor is considered to be the ratio of the applied top tension to apparent weight of the entire riser assembly.

Following these preliminary calculations, a finite element model of the riser should be established.

26.4.3 Composite Riser Section

The finite element model typically used to represent a cross section of this type comprises a single beam element with properties which represent the equivalent composite structure. Equivalent properties are therefore firstly calculated. The calculation of these equivalent properties is discussed below.

Bending and Torsional Stiffness. Equivalent values of bending and torsional stiffness are calculated to account for the stiffness characteristics of the production tubing and any gas lift or services lines in the overall stiffness of the model. The equivalent stiffness are calculated as follows (assuming Young's Modulus E and Shear Modulus G is the same for all lines):

$$EI_{model} = EI_{casings} + EI_{tubing} + EI_{otherlines} \tag{26.1}$$

$$GJ_{model} = GJ_{casings} + GJ_{tubing} + GJ_{otherlines} \tag{26.2}$$

Axial Stiffness. The axial stiffness value to be used in the finite element model depends on the make up on the pre-tensioning sequence used during installation of the riser. It is generally reasonable to assume that both the outer and inner casings will withstand the entire dynamic tension experienced by the riser, in which case the axial stiffness of the equivalent model is expressed as follows, where A is the cross-sectional area of the riser:

$$EA_{model} = EA_{outercasing} + EA_{innercasing} \tag{26.3}$$

The contribution from the tubing and other lines can be considered not to influence the axial stiffness of the riser.

Mass. An equivalent mass is also calculated accounting for the mass of the production tubing and gas lift or service lines in the annulus:

$$Mass_{equiv} = Mass_{casings} + Mass_{tubing} + Mass_{otherlines} \tag{26.4}$$

Internal Fluid. An equivalent internal fluid density must be calculated to account for the different internal fluid densities in the production tubing and the annulus of the riser. The equivalent density is calculated using a ratio of the fluid areas:

$$\rho_{annulus} A_{equiv} = \rho_{annulus} A_{annulus} + \rho_{tubing} A_{tubing} + \rho_{gaslift} A_{gaslift} - Mass_{additional} \tag{26.5}$$

where

$$Mass_{equiv} = Mass_{casings} + Mass_{tubing} + Mass_{otherlines} + Mass_{additional} \tag{26.6}$$

Buoyancy Forces. The equivalent buoyancy diameter to be used in the finite element model is that of the outer casing:

$$D_{buoyancy\ model} = D_{buoyancy\ outercasing} \tag{26.7}$$

Drag Diameter. As only the outer casing is exposed to relative hydrodynamic loading the equivalent drag diameter is that of the outer casing:

$$D_{drag\,model} = D_{dragoutercasing} \tag{26.8}$$

Finite Element Discretisation

While no hard and fast rule applies to the correct element lengths and aspect ratios to be used for a given set of conditions, there are some guidelines available on recommended lengths for beam elements, as follows:

$Element\ Length \le (EI/T)^{1/2}$ Close to a boundary condition

$Element\ Length \le \pi/\omega(T/m)^{1/2}$ Away from a boundary condition

where

ω : highest lateral frequency included in the analysis;

T : applied tension;

m : distributed mass.

Also the aspect ratio for elements should not exceed 1:1.5. Sensitivity analyses on mesh density should be performed to demonstrate the appropriateness of the discretisations being applied.

26.4.4 Vessel Boundary Conditions

Vessel boundary conditions can be applied in the lateral vessel plane to a node on the keel elevation of the riser. These boundary conditions cause the riser to displace laterally with vessel offsets while still allowing the riser to stroke. However, as the vessel boundary conditions are attached to a specific riser node the elevation of the boundary conditions move with the riser as it strokes.

Vessel boundary conditions can also be modelled as guide surfaces. The location, orientation, length and distance of the guide surfaces from the riser are specified. These guide surfaces are not attached to a specific node and remain at the same elevation during riser stroke. Longitudinal and transverse friction coefficients can be applied to these surfaces to model the frictional interaction between the riser and the surfaces. The modelling of the guide surfaces can include a gap between the surface and the riser, which allows small motions of the riser to occur before a boundary condition is applied.

26.4.5 Soil Conditions

For a template structure the riser support at the seabed is generally considered to be fixed, as the template will ensure little or no motions of the subsea wellhead. For individual wellhead layouts it will be necessary to include the wellhead and a reasonable length of conductor (e.g. 80m) in the riser model, so that the effect of the riser-soil interaction on the overall system loads can be determined. In detailed engineering it will also be necessary to evaluate the effect of installation tolerances on the riser response (e.g. effect of misalignment, wellhead angle from vertical, etc).

The behaviour of conductor casing embedded in the seabed is an important consideration in the design of the conductor casing and in determining moments at the base of a riser. If the soil immediately below the seabed consists of weak clay, as is often the case in deep water, reduced resistance is provided against lateral deflection in this region and significant movement can occur.

Generally, under lateral loading, clay soils behave as a plastic material which makes it necessary to relate conductor casing lateral deformation to soil resistance. This is achieved by lateral soil resistance-deflection (p-y) curves, with the ordinate of these curves being soil resistance per unit length, p and the abscissa being lateral deflection, y.

The stiffness of the spring is taken as the instantaneous slope of the F-y curve for a particular deflection. Therefore, in order to use a p-y curve as input to a nonlinear spring element the p-y curve must be converted into an F-y curve. This is done using the following simple relationship, for a particular deflection:

$$F = p(y).L \tag{26.9}$$

where

F	: axial force in the spring, (N);
p	: soil resistance, (N/m);
y	: deflection, (m);
L	: length of casing over which force acts.

26.4.6 Modelling of Riser Components

1. Aircan

Integral aircans are attached along the riser at intervals. They lie parallel to the riser and are directly connected to the riser at the top of each aircan. These aircans can be modelled as single vertical point force on the riser at the aircan connection point, or the buoyancy and weight of the aircan can be incorporated into the riser model over the length of the aircan. Riser joints with integral aircans attached are modelled with equivalent geometric properties to correctly model the combined stiffness of the riser joints and aircans. The external diameter of the aircans is used as the drag diameter of the section. The point force will account for both gravity and buoyancy load of the aircans. The equivalent bending stiffness of the model will be the sum of that from the cans, casings and tubings as follows:

$$EI_{model} = EI_{casings} + EI_{tubing} + EI_{aircan} \tag{26.10}$$

Non-integral aircans are modelled in a similar manner to that described for the keel. The aircan is represented by two lines of co-linear elements, one representing the riser, the other representing the aircan. The aircan elements are joined rigidly to the riser elements at the point where the tension from the aircan is transferred to the riser. The buoyancy can is joined to the riser at the bottom by means of a centralizing element and articulation, this centralizes the stem and aircan on the riser without transmitting bending or axial loads. The aircan is specified in the model in terms of weight, outside diameter and wall thickness. The bending stiffness of the riser remains unchanged, and therefore the following applies:

$$EI_{model} = EI_{casings} + EI_{tubing} \tag{26.11}$$

The net buoyancy provided by the aircans must be modified for vessel offsets to account for the loss of buoyancy in aircans with riser stroke-down.

The above modelling guidelines apply to aircans that are open at the base. As an alternative the aircan may also be sealed, in which the buoyancy generated by the can is constant. The enclosed aircan is modelled as an integral component in the riser, i.e. aircan and riser are incorporated into a composite model, with increase in mass, stiffness, buoyancy and drag characteristics.

2. Tensioner System

Hydro-pneumatic tensioner systems can be modelled in different ways depending on the level of detail required for the design. All the tension can be applied by a single vertical spring element or a detailed tensioner assembly can be modelled using individual spring elements to model each of the tensioners explicitly.

Single Spring Model:

One end of the non-linear spring element is attached to the riser at the elevation of the tensioner ring via an articulation element and the other end is fixed vertically above the articulation element with vessel boundary conditions. The articulation element connecting the non-linear spring element to the riser model has negligible rotational stiffness and therefore loads will only be applied in the translational degrees of freedom at this point and no moments will be induced at the connection point. Thus the loading will be similar to what the tensioners will apply in practice.

The vessel boundary conditions ensure the tension is applied in the vertical axis of the vessel, as is the case with hydro-pneumatic tensioner systems.

The tension value required for the riser system is provided by the spring element in its mean position. The slope of the curve for the non-linear spring element corresponds with the stiffness of the tensioner system and is specified for both extension and compression of the element. This accurately models the increase in tension with riser stroke-down and decrease in tension with riser stroke-up.

Explicit Tensioner Model:

If explicit modelling of the tensioners is required the same method for modelling the single spring system is used, but with an individual spring element used to model each one of the tensioners. The spring elements are orientated in the same angle as the hydro-pneumatic tensioners.

3. Keel Joint Interface

The Lower Keel Transition Joint (LKTJ) and Upper Keel Transition Joint (UKTJ) are modelled as linear tapered sections, with the keel joint section in between.

In the finite element model, the keel joint is represented by two lines of co-linear elements, one representing the riser, the other representing the keel sleeve. The keel sleeve is joined with articulation elements with a rotational stiffness at the top and bottom of the riser joint. The axially elastic element has a nominal axial stiffness, but a bending and torsional stiffness corresponding to that of the keel sleeve. This arrangement effectively centralizes the sleeve on the joint and ensures that there is no moment transfer mechanism between the keel sleeve and the keel joint.

4. Flexjoint

Flexjoints, where applicable, are modelled by articulation elements with a specified rotational stiffness equivalent to the flexjoint stiffness. The stiffness of the flexjoint should be selected to reflect the expected loading scenario, i.e. the stiffness of the flexjoint can be much higher for low rotations than for large rotations. For a fatigue analysis, accurate modelling of the flexjoint nonlinear stiffness characteristics is critical.

5. Wellhead Taper Stress Joint and Wellhead Connector

Tapered stress joints are simply modelled as a series of elements with linearly varying external diameters. The element mesh is refined in the region of the tapered stress joint to ensure an accurate representation of the varying diameter of the stress joint is obtained.

In order to evaluate loads at the subsea wellhead, the wellhead and tieback connector are modelled with equivalent geometric properties, weight and stiffness. Consideration must be given to subsea wellhead installation tolerances to correctly model the riser at the mudline. For preliminary feasibility/conceptual design studies the wellhead connector need not be included in the riser model. However for detailed design of the riser the inclusion of the wellhead connector in the model will depend on the level of detail required.

6. Marine Growth

Marine growth can be considered to be either soft or hard. Soft is identified as a grassy type material whereas hard marine growth is generally comprised of layers of shells. The impact of marine growth is incorporated in a finite element model by augmenting the mass, buoyancy and drag characteristics of the riser model in the area affected by marine growth.

7. Strakes and Fairings

Various VIV suppression devices are currently commercially available. The most widely used types are strakes and fairings. They prevent VIV by disrupting the pattern of fluid flow around the pipe, thus preventing the formation of eddies or vortices. The impact of suppression devices is incorporated in a finite element model by augmenting the mass, buoyancy, drag, and inertia characteristics of the riser model in the portion of riser to which they are attached.

The above drag and inertia coefficients are typical for a straked riser. However, consideration should be given to the height and pitch of the strakes in the selecting of the final hydrodynamic coefficients. Note also that fairing parameters can be calculated using the strake method.

In ***SHEAR7***, VIV suppression sections are modelled using the strake base diameter and applying a lift reduction factor. The lift reduction factor is somewhat a VIV suppression efficiency parameter. A typical lift coefficient reduction factor for a strake or fairing is 0.2 (assuming 80% strake efficiency).

8. Hydrodynamic Coefficients

The following are sample hydrodynamic coefficients used to model a bare riser joint in the ambient wave field:

- Normal Drag: 1.0 (0.6 –0.7 post-critical, > 1.2 sub critical flow);
- Tangential Drag: 0.0
- Normal Inertia: 2.0
- Tangential Added Mass: 0.0

For most projects, a low normal drag coefficient (e.g. 0.7) is used for fatigue analysis (where high drag would damp out the fatigue loads), and use a high coefficient (e.g. 1.2) for extreme response analysis (where high drag would increase the extreme loads).

26.4.7 Installation Analysis

In a multi-tube analysis there are many design considerations, including:

1. Casing and tubing elongation;
2. Internal fluid and seawater effects;
3. Riser temperature distribution under installation and operating conditions;
4. Equivalent geometric properties for the inner casing and tubing;
5. 3D riser, wellbay geometry and wellhead elevation.

A finite element approach is used to predict the hook load, buoyancy can elevation and buoyancy (if aircan tensioners were used), riser casing tensions, and tubing tension during installation and surface tree fit-up, to ensure the appropriate in-service distribution of tension between outer and inner (if present) riser casing and production tubing.

26.5 Integrated Marine Monitoring System

26.5.1 General

This section is based on an integrated marine monitoring system (IMMS) developed on the Horn Mountain Spar project by Edwards et al (2003). The IMMS provides real-time operational decision support and archived data on a common time bases for forensic analysis and engineering studies.

26.5.2 IMMS System

The IMMS is composed of the following measurement sub-systems (Edwards et al, 2003):

1. Top Tensioned Riser Monitoring Sub-system
   (a) Direct top tension and bending moment measurement on ten risers with three compression load cells per riser located in the load path between the riser top connector and the buoyancy can stem up.
   (b) Buoyancy Chamber and Stem Pressure via (9) pressure sensors per riser air can
   (c) Riser Stroke-measured on 4 risers with string potentiometers
2. Platform Position Monitoring Sub-system
   (a) Dual Redundant GPS Units with Combination UHF/Satellite antenna.
3. Precision Static Inclination and 6 Degree of Freedom Motion Measurement Sub-system
   (a) Three precision angular rate sensors
   (b) Three precision linear accelerometers
   (c) Software to provide, in real time, heel and trim and pitch and roll
4. Ballast, Draft and Void Leakage
5. Meteorological Monitoring
   (a) Wind Speed and Direction (dual redundant)
   (b) Air Temperature and Barometric Pressure
   (c) Wave Height and Period (2 radar based Ait Gap Sensors)
6. Mooring Line Tension and Payout
7. Current Profile Measured with a 75 KHz Acoustic Doppler Current Profile (ADCP)
8. Compliant Riser Guide Monitoring System (CRGMS) with 16 subsea LVDT's monitoring the lateral deflection of the compliant riser guides on the "B" and "I" slots.

The IMMS computer processes the data from these sub-systems. It has the following functionality (Edwards et al, 2003):

(a) Acquire all sensor data synchronized in time stamped files
(b) Identify and alarm malfunctioning sensors and other components in the IMMS
(c) Provide "user friendly" and timely displays of operationally important information
(d) Archive the measured data on a common time base.

26.5.3 Use of the Monitored Data

Examples of the use of real time data on Horn Mountain are (Edwards et al, 2003):

1. Production riser tension – insure that the risers are properly tensioned;
2. Production Riser Stroke – warn the operator of a situation where the up and down stops are approached;
3. Buoyancy Can Chamber Pressure – In case loss of tension is observed, identify the source of the problem- identify the leaking chamber;
4. Platform Position, Mooring Line Tension and Payout – Positioning of the Spar over the subsea well template;
5. Platform Draft, Trim and Heel, and Ballast Tank Status – guidance on weight and ballast control to maintain platform attitude. Cratain drilling and riser running operations are intolerant of excessive trim and heel of the platform;
6. Wind Velocity and Direction, Wave Height and Barometric Pressure – guidance for helicopter, crane and boat operations;
7. Dynamic Tilt (Pitch and Roll) – guidance for BOP handling operations;
8. Current Profile – guidance for ROV operation or riser mating to the subsea well heads.

Edwards et al (2003) also presented design validations by comparing design and field measurements.

26.6 References

1. API (1998), "RP 2RD-1998, Design of Risers for Floating Production Systems (FPSs) and Tension-Leg Platforms (TLPs)", American Petroleum Institute.
2. Burke, R. (2004), "TTR Design and Analysis Methods", Deepwater Riser Engineering Course, Clarion Technical Conferences.
3. Edwards, R., Shilling, R., Thethi, R. and Karakaya, M. (2003), "BP Horn Mountain Spar – Results of Comprehensive Monitoring of Platform and Riser Response", DOT.
4. Jordan, R., Otten, J., Trent, D. and Cao, P. (2004): "Matterhorn TLP Dry-Tree Production Risers", OTC 16608.
5. Utt, M., Emerson, E and Yu, J. (2004): "West Seno Drilling and Production Riser Systems", OTC 16526.
6. World Oil (2002/03): "The Composite Catalog ® of Oilfield Equipment & Services", the 45th Edition.
7. Yu, A., Allen, T. and Leung, M. (2004): "An Alternative Dry Tree System for Deepwater Spar Applications", Deep Oil Technology Conference, New Orleans.

Part IV

Riser Engineering

Chapter 27 Steel Tube Umbilical & Control Systems

27.1 Introduction

27.1.1 General

The plan for umbilical delivery typically includes an overall schedule plan for the following:

- Feasibility study;
- Umbilical specifications & request for quotation;
- (Fatigue and other) qualification tests – specifications & execution;
- Long-lead item procurement;
- Bid evaluation;
- Supplier selection;
- Project sanction & umbilical procurement;
- Detailed umbilical design and analysis by the supplier;
- 3rd party design verification by an analysis specialist;
- Prototype qualification tests;
- Umbilical manufacturing (typically requires a period of one year);
- System integration test;
- Umbilical delivery to host vessel;
- Commissioning;
- System start-up;
- Project management, QA/QC.

This Chapter provides an overview of steel tube umbilical design. The first task in an umbilical design is to size the umbilical cross-section. Some design checks and calculations that are carried out for initial steel tube sizing and cross-section design are described in Sections 27.3 and 27.4. This part of the design process is followed by an extreme wave analysis, which is described in Section 27.5. Following the feasibility study, a more detailed design of the umbilical is carried out. This involves a detailed fatigue analysis, aspects of which are discussed in Sections 27.6 – 27.8. The fatigue analysis shall cover all sources of fatigue damage such as manufacturing, in-place wave-induced stresses, installation, vortex-induced vibrations (VIV), and vortex-induced motions (VIM).

One of the first papers dealing exclusively with steel tube umbilical design was 'Metal Tube Umbilicals – Deepwater and Dynamic Considerations' presented in OTC in Swanson et al (1995). Another useful publication for further information is ISO 13628-5 which is used as the standard for umbilical design and operation.

The most recognized subsea umbilical manufacturers are Nexans, DUCO, Oceaneering Multiflex, Kvaerner Oil Products. Figure 27.1 shows an umbilical cross-section, courtesy of Nexans (Bjornstad, 2004).

Figure 27.1 Umbilical cross-section (Bjornstad, 2004).

27.1.2 Feasibility Study

The project activities in a feasibility study phase include:

- Cross-section design & sizing;
- Analysis of extreme response, bend stiffener sizing and riser interference;
- Determination of umbilical azimuth angle, departure angle and lay route;
- Installation sequence and methods;
- Specification as part of request for quotation (RFQ);
- Umbilical delivery plan.

Umbilical design commences with a feasibility study, which involves sizing the steel tubes that make up the umbilical. Tension-angle plots from the extreme wave analysis are used to design an adequate bend stiffener or bellmouth for the umbilical. A final aspect of feasibility study is an interference analysis in order to ascertain that the umbilical does not collide with adjacent risers, mooring lines or structures when under the influence of extreme waves or currents. The main deliverables from the feasibility study are the umbilical cross-section design, the required drag to diameter ratio and azimuth angle derived from an interference analysis, as well as the departure angle from extreme response analysis.

The umbilical procurement specification that goes with the RFQ document includes:

- General requirements;
- Met-ocean data and vessel motion data;
- Umbilical requirements;
- Umbilical configuration;
- Umbilical components;
- Testing requirements;
- Failure analysis and rework;
- Quality control, testing, inspection, & QA/QC surveillance;
- Project management; Documentation;
- Equipment packaging, shipping, and storage;
- Safety.

This procurement specification shall govern the design, materials, fabrication, inspection, qualification of the umbilical. Umbilical suppliers shall bid according to the RFQ specifications. The bids are then subject to technical evaluation, and price comparisons. The objective of the technical evaluation is to identify any show stoppers (feasibility issues) and potentials for contract deviations that may require extra cost and time period. Ideally the technical quality for individual bids should also be ranked. When comparing prices, the technical quality and service quality as well as the product reliability (simplicity) and durability should be given consideration.

27.1.3 Detailed Design and Installation

The most critical design challenge for an umbilical is usually fatigue. In some designs it is necessary to iteratively design bend stiffeners such that the fatigue design requirements are met. Another important fatigue aspect to consider is fatigue damage due to vortex-induced vibrations (VIV) and vortex-induced vessel motions (VIM). The procedure for analyzing VIV and VIM damage for an umbilical is similar to the methods used for a SCR.

A detailed design shall be conducted by the umbilical supplier. Some oil companies also require a third-party verification of the detailed design. The design and design verification include a full set of dynamic analysis such as: analyses of extreme response, interference, fatigue due to VIV, VIM , wave loading and installations, etc.

We should be aware that the umbilical installation requires a large installation vessel, and in certain scenarios, the availability for such a large vessel may be an issue. This is particularly true for deep and ultra-deep water umbilicals. The installation of flying leads and other components may be conducted using smaller vessels.

27.1.4 Qualification Tests

The metal tubes, electrical cables, fibre optic conductors, and umbilical section shall be subjected to tests as defined in the umbilical procurement specifications. All component designs shall be qualified for their appropriate requirements prior to manufacture.

In the past years, tube fatigue tests have been conducted in order to qualify the S-N curves used for umbilical fatigue analysis. However the industry has now acquired enough confidence in the typical metals used to construct the umbilical tubes. Hence umbilical design now does not need to include the documentation of the S-N curves used to qualify the metal, unless special/new metals are used in the design.

27.2 Control Systems

27.2.1 General

The wells are controlled by hydraulic and/or electrical signals through an umbilical, which also provides for the injection of chemicals.

The control umbilical is a bundled cable, connecting subsea equipment with the topside facilities, containing all the required components to operate and serve the subsea equipment. Depending on the type of subsea control system being used, the control umbilical may contain required fluid conduits (hoses), electrical power/signal conductors, chemical injection lines, vent lines, and required spare lines. If the umbilical is subjected to dynamic loads from wave action and currents, it may be equipped with steel armoring. The composition of the control umbilical varies considerable from one control system to the next, and it is determined primarily by the requirements of the particular system being used.

27.2.2 Control Systems

The subsea production control system facilitates the operation of valves and chokes on subsea completions and manifolds. Typically five types of control systems are used in the operation of subsea production systems, those being:

- Direct hydraulic;
- Piloted hydraulic;
- Sequential hydraulic;
- Electro hydraulic;
- Multiplex electro hydraulic

The *direct hydraulic* control system is the least complex of the five alternatives and has the fewest number of subsea components. Each subsea function requires a hydraulic flow path from the surface. Actuation of a valve on the surface control panel results in pressurized fluid being routed through a dedicated flow path to the selected subsea tree valve actuator. This system with one line per subsea function is best suited for applications where the control distance is relative short (actuation of a valve 4000m away from the surface facility can take as long as three hours), and a limited number of subsea functions to be operated. As the number of subsea functions increases, so does the outer diameter of the control umbilical and its cost.

Piloted hydraulic systems improve the response time by storing the hydraulic pressure energy at the site. Actuators are then activated by sending a hydraulic signal to a pilot valve, which opens and allows fluid from the accumulated storage to activate the actuator.

The *sequential hydraulic* control system has, as with the piloted system, an accumulator and control valves placed subsea. Control is then achieved by sending a pressure signal to a sequence valve that is pre set to operate or shift at specific pressure levels. At this signal pressure level the sequence valve shifts and hydraulic fluid from the accumulators is routed to a group of pre-selected gate valve operators. As signal pressure is increased in a series of discrete steps, the sequence valve shifts and operates the next group of pre-selected tree valves. This system is suitable for operating at long distances and is cost effective at these distances due to the small diameter umbilical. This system is normally configured to operate several gate valves at a time placing the subsea tree in a particular operational mode. The system can accommodate up to six modes. The disadvantage of this type of system is the complex hydraulic circuitry and the inability to operate each tree valve individually.

The *electro hydraulic* system is capable of controlling a limited number of subsea valves very quickly at large distances. The system works, as with the piloted and sequential system, by supplying hydraulic fluid at pressure through a large supply line within the umbilical, which is accumulated in a pressure vessel at the tree. A switch on the surface control panel directs electrical current to a solenoid pilot valve. This valve shifts and directs hydraulic fluid from the pressure vessel to the associated actuator. One electrical signal conductor is required for each subsea valve or function to be operated. As the number of valves to be operated increase so does the size, cost, and complexity of the control umbilical.

The *multiplex electro hydraulic* system is capable of controlling large numbers of subsea valves very quickly at large distances using a relative small diameter control umbilical coupled to a multiplexer that normally operates a large number of solenoid pilot valves. Hydraulic fluid at pressure is made available through a supply line within the umbilical and stored in pressure vessels or accumulators. When a signal is sent to the multiplexer it energizes the selected solenoid valve, thereby directing the hydraulic fluid from the supply umbilical and accumulators to the associated actuator. The multiplexer makes the transmission of data such as pressure, temperature, and valve position possible by means of electrical signals. The multiplex system is very flexible but also very complex and incorporates a large number of subsea electrical components.

27.2.3 Elements of Control System

The main elements of a subsea production control system typically include the following:

A *hydraulic power unit (HPU)* provides a stable and clean supply of hydraulic fluid to the remotely operated subsea valves. The fluid is supplied via the umbilical to the subsea hydraulic distribution system, and to the SCMs to operate subsea valve actuators.

The *Master control station* (MCS) is the central control "node" containing application software required to control and monitor the subsea production system and associated topside equipment such as the HPU and EPU.

The *electrical power unit* (EPU) supplies electrical power at the desired voltage and frequency to the subsea users. Power transmission is performed via the electrical umbilical and the subsea electrical distribution system.

The *Modem unit* modulates communication signals for transmission to and from the applicable subsea users.

The *uninterruptible power supply (UPS)* is typically provided to ensure safe and reliable electrical power to the subsea production control system.

In a piloted-hydraulic or electro hydraulic control system, the *Subsea control module (SCM)* is the unit, which upon command from the MCS directs hydraulic fluid to operate subsea valves. In an electro hydraulic system the SCM also gathers information from subsea located sensors and transmits the sensor values to the topside facility.

Subsea distribution systems covers distribution systems that distribute electrical, hydraulic and chemical supply from the umbilical termination(s) to the subsea trees, manifold valves, injection points, and the control modules of the subsea production control system.

Subsea located sensors are sensors located in the SCM, or on subsea trees or manifolds, providing data to help monitor operation of the subsea production system.

Control fluids are typically oil-based or water-based liquids that are used to convey control and/or hydraulic power from the surface HPU to the SCM and subsea valve actuators.

27.2.4 Umbilical Technological Challenges and Solutions

Some of the technological challenges are discussed below:

(1) Deepwater

The deepest umbilical installation to date is in 2,316 m of water, at Shell's Na Kika project. Some other deepwater umbilicals are the Thunder Horse umbilical in 1,880 m water depth, and the Atlantis umbilical in 2,134 m (Terdre, 2004). A challenge in design is that steel tubes are under high external pressure as well as high tensile loads. At the same time, the increased weight may also cause installation problems. This is particularly true for copper cables as yield strength for copper is low. In ultra-deep water, a heavy dynamic umbilical may present a problem to installation and operation as its hang-off load is high.

For design and analysis of ultra-deep water umbilical, it is important to correctly model the effect of stress and strain on an umbilical and the friction effect. Sometimes, bottom compression may be observed for umbilical under 100-year hurricanes. In this scenario, the design solution may be to use lazy-wave buoyancy module or to use fibre carbon rods. The use of carbon fibre rods allows umbilicals in a simple catenary configuration, without the need for expensive, inspection/maintenance demanding buoyancy modules. The carbon fibre rods enhance axial stiffness as they have a Young's modulus close to the value of steel but with only a fraction of the weight.

One of the concerns for use of the carbon fibre rods is their capacity for compressive loads. It is hence beneficial to conduct some tests that document the minimum bending radius and compressive strength of the umbilical.

If the currents in severe for ultra-deepwater umbilical, it might be necessary to use strakes for VIV protection, although the use of strakes has so far not been required yet. The strakes may for instance be a 16D triple start helix with a strake height of 0.25D.

(2) Long-Distance

The length for Na Kika, Thunder Horse and Atlantis umbilicals is 130 km, 65 km and 45 km, respectively. The longest yet developed is 165 km in a single length, for Statoil's Snohvit development off northern Norway. One of the constraints on umbilical length is the capacity of the installation equipment. The Nexans-operated installation vessel, Bourbon Skagerrak, can carry up to 6,500 tons of cable, that equals to a length of 260 km, assuming umbilical unit weight is 25 kg/m.

(3) High Voltage Power Cables

The design constraints are the low yield strength of copper, which requires an increasing amount of protection as depths increase, and the weight of steel armoring employed to provide that protection as depths increase. Fatigue of copper cables in dynamic umbilical is another technical challenge.

(4) Integrated Production Umbilical (IPU®)

Heggadal (2004) presented an integrated production umbilical (IPU®) where the flowline and the umbilical are combined in one single line, see Figure 27.2. The IPU cross-section consists of the following elements:

- A 10 ¾" flowline with a 3 layer PP coating (its thickness is 4 mm and 14 mm for static portion and dynamic portion, respectively).

- Around the flowline, there is an annular shaped PVC matrix that keeps in place the spirally wound umbilical tubes and cables and provides thermal insulation to the flowline.
- Embedded in the PVC matrix, but sliding freely with it, the various metallic tubes for heating, hydraulic and service fluids, the electrical/fibre optic cables for power and signal, and the high voltage cables for powering the subsea injection pump.
- An outer protective sheath of polyethylene 12 mm thick.

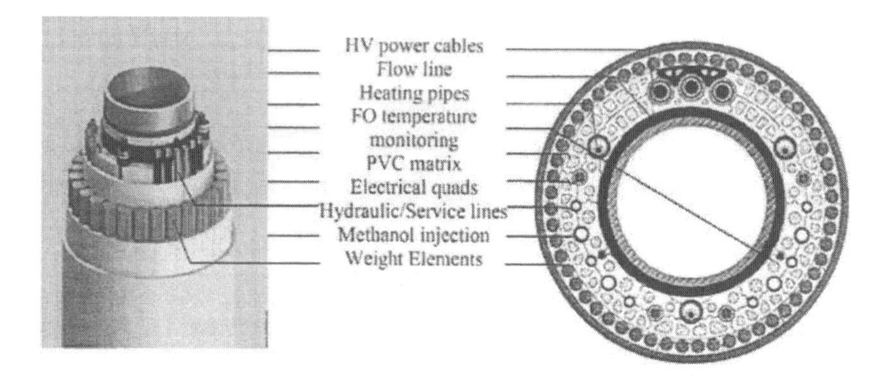

Figure 27.2 IPU dynamic cross-section, super duplex flowline (Heggadal, 2004).

To qualify a new design concept like this, a series of analysis and qualification tests were conducted as below (Heggadal, 2004):

(1) Analysis

- Global riser analysis & fatigue analysis;
- Corrosion and hydrogen induced cracking assessment;
- Thermal analysis;
- Structural analysis (prod. pipe, topside and subsea termination);
- Reeling analysis;
- Electrical analysis;
- Reeling, trawl interaction and on-bottom studies.

(2) Basic Tests

- Mechanical material tests, fatigue, corrosion etc.

(3) Fabrication Tests

- Fabrication & closing test;
- STS injection test;
- QC tests and FAT;
- Pre/post installation tests.

(4) Prototype Tests

- External hydostatic test;
- Impact test;
- Model tensioner test;
- Reeling and straightening trials;
- Stinger roller trial;
- Repair trial;
- Vessel trial;
- System test;
- Dynamic riser full scale testing.

27.3 Cross-sectional Design of the Umbilical

One of the initial stages in the design of an umbilical is the placing of the components of the umbilical in the cross-section design. The cross-section of an umbilical could include various items such as steel tubes for transporting hydraulic and other fluids, electrical cables, fiber optic cables, steel rods or wires for strength capacity, polymer layers for insulation and protection, and polymer fillers to fill in the spaces between the components and keep them in place.

Various issues need to be taken into consideration during this stage of the design. A number of these issues are listed here:

- Tension and bending forces in the umbilical set up stress in the steel tubes of the umbilical. This stress needs to be limited to an allowable value (defined in standards such as ISO 13628-5). If the stress exceeds the allowable value, the design needs to be changed. One method to reduce the stress in the steel tubes is to add steel rods or wires in the cross-section so that the stress is distributed amongst them as well as the steel tubes.
- The maximum yield is the limitation to determine the umbilical breaking loading. The maximum breaking loading calculation based on the steel tube and steel rods material grade. In some cases, steel rods are excluded in the calculation of the maximum breaking loading. The maximum working loading may be taken conservatively as 20% of the maximum breaking loading. The maximum installation loading may be taken as 55% of the maximum breaking loading. However, all the loadings need to be confirmed by the manufacture when the detail design is taken place.
- The umbilical drag to weight ratio is carefully controlled at this stage of the design as this has a significant effect on the dynamic behavior of the umbilical. If this ratio is unsatisfactory and is expected to cause problems in the dynamic behavior of the umbilical (such as clashing with adjacent risers or structures), the drag to weight ratio can be modified by changing the diameter of the umbilical or the weight. It is much easier to increase the weight or the diameter, rather than decrease them. The diameter

can be increased by spacing the components and adding more fillers, or adding more components such as steel rods or steel tubes. The addition of steel rods is the most straightforward method of increasing the umbilical weight.

- Manufacturing constraints could have an impact on the cross-section design. For example some umbilical manufacturers are only able to place 12 functional components in one layer of the umbilical.
- Components might need to be placed towards the outside of the umbilical for easy access for repair and maintenance. This is especially true for electrical and fiber optic cables.
- Electrical cables might need to be placed in a certain order for them not to cause electrical interference with each other. The use of insulation might also be required for electrical cables. This is especially true for high voltage cables. The use of insulation could also have a detrimental effect by causing an increase in temperature in the umbilical due to heat generated by the current flowing through the electrical cables.
- Heavy components in the umbilical might need to be placed in a certain order to prevent torsional unbalance of the umbilical.

Once the cross-section has been designed, an initial static analysis may be performed for the umbilical in the mean vessel position. The length of umbilical from the hang-off point to the touchdown point and its projected horizontal length (which is the horizontal distance from the hang-off point to the touchdown point) may be estimated using simple catenary theory. The static catenary analysis may also predict the effective tension at the hang off point, and at the TDP region, respectively. The minimum bend radius at the TDP region may be calculated using the catenary theory, and be compared with an acceptance criterion.

27.4 Steel Tube Design Capacity Verification

The functional design capacities of the steel tubes of an umbilical need to be evaluated during the design process. Specifically, calculations should be performed to ascertain that the steel tubes are adequately designed in terms of pressure containment, collapse, static loading, buckling and ovality. ISO 13628-5 is a useful reference for carrying out these calculations and for determining the utilization ratios applicable to the umbilical. This section describes some of the equations and methods that are used to verify the umbilical steel tube design.

27.4.1 Pressure Containment

The hoop stress in the steel tubes should be calculated in accordance with the methods described in Sections 7.10.2.2.2 and 7.10.2.3.2 of ISO 13628-5. Most likely, the wall-thickness of the steel tubes is determined by the pressure containment requirements. However, buckling and collapse as well as von Mises stress criteria need also to be checked for completeness.

27.4.2 Allowable Bending Radius

One of the processes during the design of a steel tube umbilical is the determination of the minimum allowable bending radius of the umbilical. This bending radius is calculated by determining the stress that is experienced by the umbilical during different phases of its life and also under various operating conditions. The stress prevalent in the steel tubes depends on three components: the pressure in the tubes, the axial stress due to the tension in the tubes, and the bending stress.

(1) The pressure in the tubes generally remains constant at the standard operating or design pressure. It is only likely to change during extreme environmental or operational conditions, or emergency situations.

(2) The tension and hence the axial stress in the tubes changes depending on environmental conditions.

(3) Thus this stage of the design process usually involves calculating the minimum allowable bending radius of the umbilical for a variety of environmental conditions and top tensions, including the installation phase. These allowable bending radii are not the same as the umbilical storage minimum bending radius (MBR). The storage MBR is defined by the manufacturer and refers to the smallest radius that the umbilical can be bent to when stored on reel. This storage MBR is the minimum radius that does not cause any damage to the steel tubes of the umbilical when stored on a reel, and under no effect from internal pressure and tension in the umbilical.

Knapp (1988) describes a method for calculating the allowable bending stress in the umbilical, based on the pressure and axial stress experienced by the steel tubes. The paper gives a theoretical background to determining compound stress in helical wires of a cable bent through a planar circular arc. The following assumptions have been made:

(1) Only round wires are considered;
(2) The geometry of deformation of a helical wire can be described by its centurial axis;
(3) The helical wire lay angle and pitch radius remain constant;
(4) For the frictionless case, the transverse section of the cable is allowed to warp as wires slip freely;
(5) For the full friction model, plane sections of the cable remain planar after bending;
(6) Inter-wire contact stresses are neglected;
(7) The helical wires are homogeneous, isotropic and linearly elastic.

The paper shows both frictionless (Lower bounding) and full friction (Upper bounding) models to evaluate the bend radius. It also indicates that the frictionless model is more likely to be realistic, based on the experiments conducted. The allowable minimum bending radius of the umbilical can be easily determined once the allowable bending stress is known.

27.5 Extreme Wave Analysis

An important aspect of the umbilical design process is an analysis of extreme environmental conditions. A finite element model of the umbilical is analyzed with vessel offsets, currents and wave data expected to be prevalent at the site where the umbilical is to be installed. For example, in the Gulf of Mexico, this would include an analysis for a 100 year hurricane, 100 year loop current, and submerged current. The current and wave directions are applied in a far, near and cross condition. This analysis is used to determine the top tension and angles that the hang-off location of the umbilical is likely to experience. These values are then used to design an adequate bend stiffener that will limit the umbilical movements and provide adequate fatigue life for the umbilical. Design analysis resulted from extreme analysis are:

(1) The touchdown zone of the umbilical is analyzed to ensure an adequate bending radius that is larger than the minimum allowable bending radius. It is also important to check that the umbilical does not suffer compression and buckling at the touchdown zone.

(2) A polyurethane bend stiffener has been designed to have a base diameter of xxx inch, and cone length of yyy ft. This design is based on the maximum angle and its associated tension, and maximum tension and its associated angle from dynamic analysis results using the pinned FE model.

(3) The maximum analyzed tension in the umbilical was found to occur at the hang-off point for the 1000yr hurricane wind load case when the vessel is in far position.

(4) The minimum tension in the umbilical may be found to occur in the TDP region for the 1000yr hurricane wind load case when vessel is in the near position.

(5) The Minimum Bend Radius (MBR) is estimated over the entire umbilical, over the TDP region and the bend stiffener region, respectively. They are to be larger than the allowable dynamic MBR.

(6) The minimum required umbilical on-seabed length is estimated assuming that when it is subject to the maximum value of the extreme bottom tensions.

27.6 Manufacturing Fatigue Analysis

A certain amount of fatigue damage is experienced by a steel tube umbilical during manufacturing, and this needs to be evaluated during fatigue analysis. The two main aspects of umbilical manufacturing fatigue analysis that require attentions are accumulated plastic strain and low cycle fatigue. These are explained below.

27.6.1 Accumulated Plastic Strain

Accumulated plastic strain is defined as 'the sum of plastic strain increments, irrespective of sign and direction' (DNV, 2000). Accumulated plastic strain can occur in the steel tubes of an umbilical during fabrication and installation. The accumulated plastic strain needs to be maintained within certain limits to avoid unstable fracture or plastic collapse for a given tube material and weld procedure. Accumulated plastic strain is the general criteria used by umbilical suppliers to determine whether the amount of plastic loading on the steel tubes is acceptable. An allowable accumulated plastic strain level of 2% is recommended for umbilical design.

Figure 27.3 shows a schematic of deformations that are likely to take place during the fabrication and installation of a steel tube umbilical. All the processes shown in this diagram are likely to induce plastic strain in the umbilical.

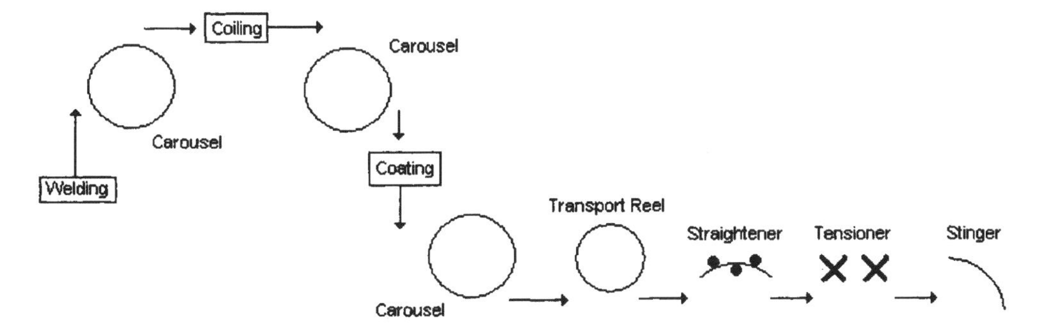

Figure 27.3 Diagram of deformations during fabrication and installation.

27.6.2 Low Cycle Fatigue

The umbilical steel tubes are subject to large stress / strain reversals during fabrication and installation. Fatigue damage in this low cycle regime is calculated using a strain-based approach.

For each stage of fabrication and installation the fatigue damage is calculated by considering the contributions from both the elastic and plastic strain cycles. The damage calculated from low frequency fatigue is added to that from in-service wave and VIV conditions to evaluate the total fatigue life of each tube of the umbilical.

27.7 In-place Fatigue Analysis

The methodology used to assess wave induced in-place fatigue damage of the umbilical tubes can be summarized as follows:

- Selection of Seastate Data from Wave Scatter Diagram;
- Analysis of Finite Element Static Model;
- Umbilical Fatigue Analysis Calculations;
- Simplified or Enhanced Approach;
- Generation of Combined Stress History;
- Rainflow Cycle Counting Procedure or Spectral Fatigue Damage;
- Incorporation of Mean Stress Effects in Histogram.

These aspects of fatigue analysis are described below. The main difference between fatigue analysis for an umbilical and a SCR is the effect of friction when the tubes in the umbilical slide against their conduits and each other due to bending of the umbilical. The methodology discussed here for umbilical in-place fatigue analysis is based on two OTC Papers: Paper 13203 by DUCO (Hoffman, 2001), and Paper 16631 by MCS (Kavanagh et al., 2004). In-place fatigue analysis is required to prove that the fatigue life of the umbilical is ten times the design life.

27.7.1 Selection of Seastate Data from Wave Scatter Diagram

The wave scatter diagram describes the seastate environment for the umbilical in service. It is not practical to run a fatigue analysis with all the seastates described in a wave scatter diagram. Hence the usual methodology is to group a number of seastates together and represent these 'joint seastates' with one significant wave height and wave period. The values of the wave height and wave period are chosen to be conservative.

This methodology results in the reduction of the wave scatter diagram to a 'manageable' number of seastates (say about 20 – 50). This enables the analysis to be carried out in a reasonable amount of time. It is also very important to accurately consider the percentage of time that the umbilical is expected to be affected by these different seastates.

27.7.2 Analysis of Finite Element Static Model

A finite element static analysis is carried out for a model representing the steel tube umbilical. The static solution is used as a starting point for a time domain or frequency domain dynamic finite element analysis.

27.7.3 Umbilical Fatigue Analysis Calculations

Fatigue damage in an umbilical is the product of three types of stress. These are axial (σ_A), bending (σ_B) and friction stress (σ_F). The equations defining these stress terms are as follows:

$$\sigma_A = 2\sqrt{2}SD_T / A \quad (27.1)$$

$$\sigma_B = 2\sqrt{2}E\,R\,SD_k \quad (27.2)$$

where, SD_T: standard deviation of tension;

A : steel cross-sectional area of the umbilical;

E : Young's Modulus;

R : outer radius of the critical steel tube, and

SD_k : standard deviation of curvature.

The critical steel tube is the tube in the umbilical that experiences the greatest stress. This is usually the tube with the largest cross-sectional area and furthest from the centerline of the umbilical.

The friction stress experienced by the critical tube is the minimum of the sliding friction stress (σ_{FS}) and the bending friction stress (σ_{FB}). This is based on the theory that the tube experiences bending friction stress until a point is reached when the tube slips in relation to its conduit. At this point the tube experiences sliding friction stress. This is represented by Figure 27.4.

Therefore, according to Kavanagh et al. (2004)

$$\sigma_F = \min(\sigma_{FS}, \sigma_{FB}) \quad (27.3)$$

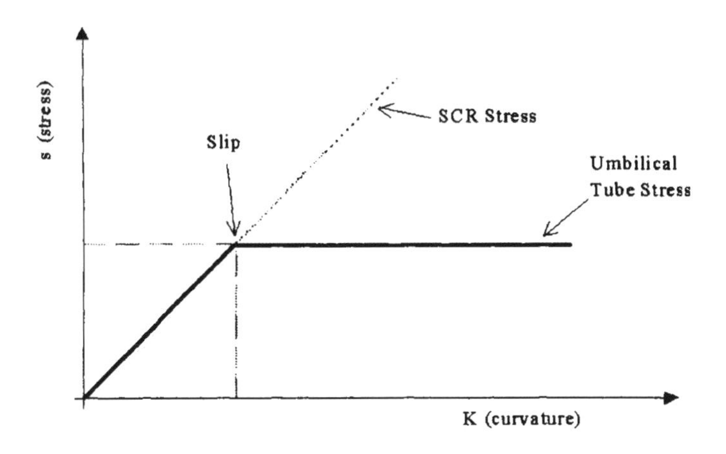

Figure 27.4 Representation of umbilical friction stress (Kavanagh et al., 2004).

$$\sigma_{FS} = \frac{\mu F_C}{A_t} \tag{27.4}$$

$$\sigma_{FB} = E\,R_L\,\sqrt{2}\,SD_k \tag{27.5}$$

where, μ : friction coefficient,

F_C : contact force between the helical steel tube (defined below),

A_t : cross-sectional area of the critical steel tube within the umbilical, and

R_L : layer radius of the tube (this being the distance from the center of the umbilical to the center of the critical steel tube).

$$F_C = \left[\frac{T\sin^2\phi}{R_L} + \frac{EI_{tube}\sin^4\phi}{R_L^3}\right]L_P \tag{27.6}$$

where, T : mean tension,

φ : tube lay angle (this being the angle at which the tubes lie relative to the umbilical neutral axis),

EI_{tube} : individual tube bending stiffness, and

L_P : tube pitch length/2.

27.7.4 Simplified or Enhanced Approach

A finite element dynamic analysis is performed for each of the selected seastates using an appropriate steel tube umbilical model. The dynamic analysis needs to incorporate wave loading and first order vessel motions. It should also include low frequency vessel motions. The dynamic analysis can be carried out in two ways: either a simplified approach, or an enhanced approach.

In the simplified approach, the finite element dynamic analysis is run with regular wave seastates. This approach results in a phase inconsistent combination of axial, bending and friction stress. Thus this conservatively assumes all stress to be in phase. The tube friction stress is also calculated as the minimum of the sliding friction or bending friction stress. The tube is therefore assumed to be either always in slip or stick condition. The total stress damage is then calculated based on an approximate cycle counting method.

In the enhanced approach, the finite element dynamic analysis is run with irregular wave seastates. The stress is combined in a phase-consistent approach. A time or frequency domain analysis can be used. The time domain approach provides more accurate results, but requires larger computer processing time. The friction stress time history is the minimum of sliding friction or bending friction. The tube alternates between the two, and both sliding friction and bending friction is calculated for each time step of the analysis in order to determine the minimum of the two. Figure 27.5 illustrates a typical friction-time history and shows the alternation between bending friction and sliding friction.

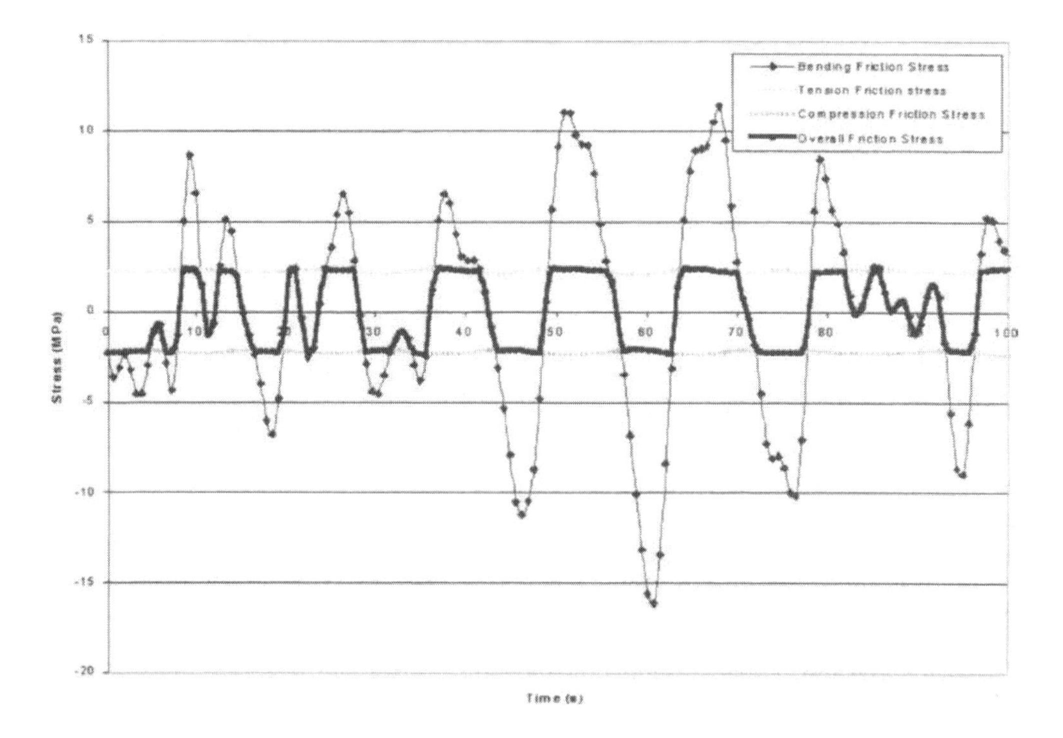

Figure 27.5 Friction-time history using the enhanced approach (Kavanagh et al., 2004).

The enhanced approach requires a longer processing time, but results in a more conservative result.

27.7.5 Generation of Combined Stress History

Whether a simplified or enhanced approach is adopted, the aim of this part of the analysis is to produce a combined stress history. This is calculated as the sum of the axial, bending and friction stress (using the equations defined in Section 27.7.3).

27.7.6 Rainflow Cycle Counting Procedure or Spectral Fatigue Analysis

The stress history for each seastate is compiled to produce a combined stress history using the percentage occurrence for each seastate. The stress ranges experienced by the critical tube in the umbilical are then defined using a rainflow cycle counting procedure. The stress ranges obtained using this procedure are used to calculate the fatigue life of the steel tube using the Palmgren-Miner Rule.

The fatigue damage for a time domain analysis is calculated using the well-known formula:

$$D = \frac{n(\Delta\sigma)^m}{K} \tag{27.7}$$

For a spectral fatigue analysis, this formula is modified as follows:

$$D = \frac{n(8m_0)^{m/2}\Gamma(1+m/2)}{K} \tag{27.8}$$

where, m_0 : 0th moment of the stress spectrum. (The stress standard deviation is defined as the square root of this 0th moment),

Γ : factor used to account for assumption that the spectral fatigue life follows a Rayleigh distribution. Γ is tabulated in mathematical handbooks. See Almar-Naess (1985), for further details.

27.7.7 Incorporation of Mean Stress Effects in Histogram

S-N curves that are used for welded joint fatigue calculations by the offshore industry are defined using pure reverse bending fatigue testing methods. However the actual umbilicals that are commissioned have associated mean stress distributions due to weight, internal fluid density and pressure, vessel offset and the local current flow regime.

A useful method to take account of mean stress effects is through the use of constant-life diagrams. The three best known constant-life models are due to Gerber, Goodman and Soderberg, and are shown in Figure 27.6, see Suresh (1991). In these models, different combinations of the stress amplitude and mean stress are plotted to provide a constant fatigue life.

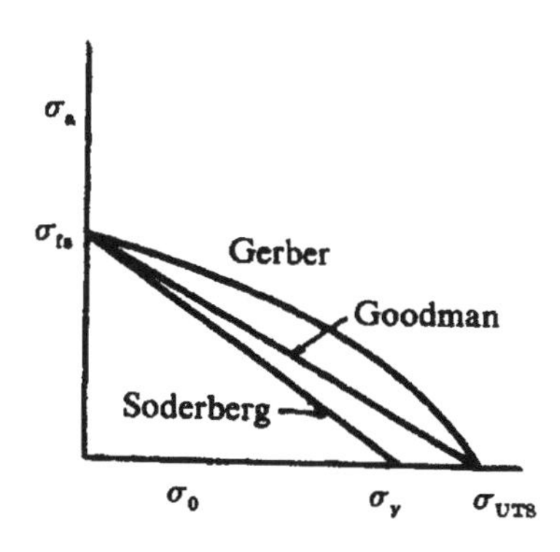

Figure 27.6 Constant life diagrams for fatigue loading with a non-zero mean stress.

The life plots represented in Figure 27.6 are described by the following equations:

Soderberg relation:
$$\sigma_a = \sigma_{fs}\left\{1 - \frac{\sigma_0}{\sigma_y}\right\} \tag{27.9}$$

Goodman relation:
$$\sigma_a = \sigma_{fs}\left\{1 - \frac{\sigma_0}{\sigma_{UTS}}\right\} \tag{27.10}$$

Gerber relation:
$$\sigma_a = \sigma_{fs}\left\{1 - \left(\frac{\sigma_0}{\sigma_{UTS}}\right)^2\right\} \tag{27.11}$$

Where, σ_a : stress amplitude denoting the fatigue strength for a non-zero mean stress;

σ_{fs} : fatigue strength for fully reversed loading;

σ_0 : mean stress;

σ_y : yield strength; and

σ_{UTS} : ultimate tensile strength.

As a general rule-of-thumb the following observations can be made on these constant-life models: the Soderberg relation provides a conservative estimate of fatigue life for most engineering alloys; the Goodman relation matches experimental observations closely for brittle metals, but is very conservative for ductile alloys; the Gerber relation is generally good for ductile alloys.

The Goodman equation is generally used for umbilical applications. However the Goodman relation results in an overly-conservative fatigue life when the mean stress is high. In this case the Gerber relation is preferred.

27.8 Installation Analysis

The issues that need to be considered when dealing with fatigue damage during installation of steel tube umbilicals are as follows:

- The contribution to accumulated plastic strain during reeling and potential retrieval.
- Low Cycle Fatigue during reeling and potential retrieval.
- Dynamic wave frequency fatigue contributions during the critical stages of installation, *i.e.* midlay and handover / pull-in.

The methodology for accounting for accumulated plastic strain and low cycle fatigue has already been considered in Section 27.5. The calculations for accumulated plastic strain and low cycle fatigue are carried out for both fabrication and installation together.

The methodology for the calculation of wave induced fatigue damage during the critical stages of installation is similar to the in-place fatigue assessment described in Section 27.6. However there are some aspects of installation fatigue analysis that do not apply to in-place fatigue analysis. Primarily,

- Since the umbilical changes configuration and is subject to different loads during various stages of installation, different umbilical models are needed to model the various stages of installation that require analysis.

- For installation fatigue analysis it is appropriate to use a time-domain approach. A frequency domain analysis would not adequately predict the fatigue damage suffered during installation due to the highly irregular loading that the umbilical experiences during this stage of its life.

27.9 Required On-seabed Length for Stability

From the on-seabed stability point of view, before a curve may be taken in the field layout, the length of the umbilical on the seabed should be long enough so that the friction force provided by the on-seabed portion of the umbilical is larger than the extreme bottom tension from the dynamic analysis.

Assuming an on-seabed length of L ft, the friction force is calculated as following:

Friction Force = Friction coefficient * Submerged weight * L

Therefore,

L = Friction Force/(Friction coefficient * Submerged weight)

Assuming that the maximum value of the bottom tension is 7.28 kips, the submerged weight is 9.71 (lbs/ft), the required on-seabed length for a friction coefficient of 0.5 is

L = 7.28 *1000 (lbs)/[0.5 * 9.71 (lbs/ft)]

= 1450 ft

Since the umbilical changes configuration and is subject to different loads during various stages of installation, different umbilical models are needed to model the various stages of installation that require analysis.

27.10 References

1. Almar-Naess, A. (1985), "Fatigue Handbook: Offshore Steel Structures", Tapir.
2. API 17E: "Specification for Subsea Umbilical", The American Petroleum Institute.
3. API Recommended Practice 2RD, (1998), "Design of Risers for Floating Production Systems (FPSs) and Tension Leg Platforms", 1st Edition.
4. Bjornstad, B. (2000), "Umbilical Stretches Subsea Performance", E & P Magazine, August 2004.
5. DNV Offshore Standard OS-F101 (2000), "Submarine Pipeline Systems".

6. DNV Recommended Practice, RP-C203 (2000), "Fatigue Strength Analysis of Offshore Steel Structures".
7. GE Research & Development Center (2000), "Influence of Mean Stress on Low Cycle Fatigue in High Temperature Water", Doc. No. 2000CRD025, May 2000.
8. Heggdal, O. (2004): "Integrated Production Umbilical (IPU ®) for the Fram Ost (20 km Tie-Back) Qualification and Testing, Deep Oil Technology (DOT), Dec. 2004, New Orleans.
9. Hoffman, J., Dupont W., Reynolds B. (2001), "A Fatigue-Life Prediction Model for Metallic Tube Umbilicals", OTC Paper 13203.
10. ISO 13628-5, 2000, "Petroleum and Natural Gas Industries – Design and Operation of Subsea Production Systems", Part 5: Subsea Umbilicals, 2002.
11. Kavanagh, W. K., Doynov, K., Gallagher, D., Bai, Y. (2004), "The Effect of Tube Friction on the Fatigue Life of Steel Tube Umbilical Risers – New Approaches to Evaluating Fatigue Life using Enhanced Nonlinear Time Domain Methods", OTC Paper 16631.
12. Knapp, RH, (1988): "Helical Wire Stresses in Bent Cables", Journal of Offshore Mechanics and Arctic Engineering, No. 110, pages 55 – 61, February 1988.
13. MCS (2004), Flexcom, Version 7, User's Manual.
14. Stephens R, Fuchs HO (2001), "Metals in Fatigue Engineering", 2nd Edition, Wiley and Sons Inc., 2001.
15. Suresh, S (1991), "Fatigue of Materials", Cambridge Solid State Science Series.
16. Swanson, RC, Rao, VS, Langner, CG, Venkataraman, G. (1995), "Metal Tube Umbilicals –Deepwater and Dynamic Considerations", OTC Paper 7713 (1995).
17. Terdre N (2004), "Nexans Looking beyond Na Kika to Next Generation of Ultra-Deep Umbilical", Offshore Magazine, March 2004.

Part IV

Riser Engineering

Chapter 28 Flexible Risers and Flowlines

28.1 Introduction

Flexible risers and flowlines (referred to as flexible pipes in the remainder of this chapter) trace their origins to pioneering work carried out in the late 1970s. Initially flexible pipes were used in relatively benign weather environments such as offshore Brazil, the Far East and the Mediterranean. However since then flexible pipe technology has advanced rapidly and today flexible pipes are used in various fields in the North Sea and are also gaining popularity in the Gulf of Mexico. Flexible pipe applications include water depths down to 8,000 ft, high pressure up to 10,000 psi, and high temperatures above 150°F, as well as the ability to withstand large vessel motions in adverse weather conditions.

The main characteristic of a flexible pipe is its low relative bending to axial stiffness. This characteristic is achieved through the use of a number of layers of different material in the pipe wall fabrication. These layers are able to slip past each other when under the influence of external and internal loads, and hence this characteristic gives a flexible pipe its property of a low bending stiffness. The flexible pipe composite structure combines steel armor layers with high stiffness to provide strength, and polymer sealing layers with low stiffness to provide fluid integrity. This construction gives flexible pipes a number of advantages over other types of pipelines and risers such as steel catenary risers. These advantages include prefabrication, storage in long lengths on reels, reduced transport and installation costs, and suitability for use with compliant structures.

28.2 Flexible Pipe Cross Section

There are two types of flexible pipes in use: bonded and unbonded flexible pipes. In bonded pipes, different layers of fabric, elastomer and steel are bonded together through a vulcanization process. Bonded pipes are only used in short sections such as jumpers.

Figure 28.1 shows a bonded flexible pipe cross-section.

Figure 28.2 shows a picture from the Heidrun TLP where 5 ½ " production and 2" gas lift hoses are connected to production tree.

Figure 28.1 Bonded flexible pipe (Antal et al., 2003).

Figure 28.2 Production and gas lift hoses on the Heidrun TLP (Antal et al., 2003).

On the other hand, unbonded flexible pipes can be manufactured for dynamic applications in lengths of several hundred meters. Unless otherwise stated, the rest of this Chapter shall deal with unbonded flexible pipes. Figure 28.3 shows a typical cross-section of an unbonded flexible pipe. This figure clearly identifies the five main components of the flexible pipe cross-section. The space between the internal polymer sheath and the external polymer sheath is known as the pipe annulus. The five main components of the flexible pipe wall shall be dealt with in the following sections.

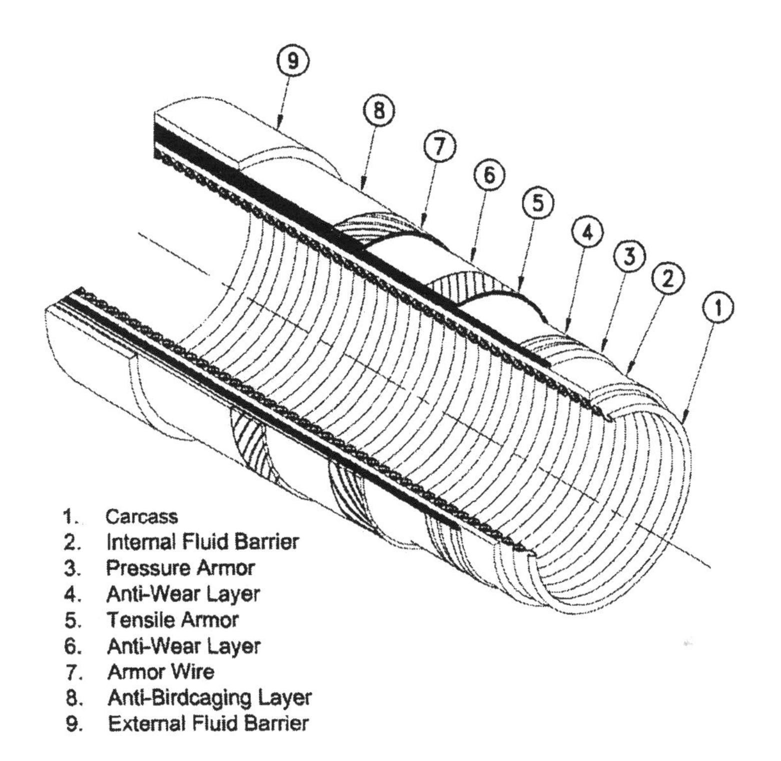

Figure 28.3 Typical cross-section of an unbonded flexible pipe (Zhang et al., 2003).

28.2.1 Carcass

The carcass forms the innermost layer of the flexible pipe cross-section. It is commonly made of a stainless steel flat strip that is formed into an interlocking profile as seen in Figure 28.2. Different steel grades can be used to form the carcass, and the choice of material usually depends on the internal fluid characteristics. The most common grades used to manufacture the carcass are AISI grades 304 and 316, and Duplex. The inner bore fluid is free to flow through the carcass profile and therefore the carcass material needs to be corrosion - resistant to the bore fluid.

The main function of the carcass is to prevent pipe collapse due to hydrostatic pressure or build-up of gases in the annulus. The build-up of gases in the annulus could be a potential failure mode for the pipe, and occurs in hydrocarbon-carrying pipes when gases from the inner pipe bore diffuse through the internal polymer sheath into the annulus. In the case of a well shut-down and subsequent depressurization and evacuation of the inner bore, the annulus gas pressure could cause pipe collapse. Thus the steel carcass is designed to withstand this collapse pressure. Pipes that do not carry any hydrocarbon fluid (e.g. water injection pipes) can be constructed without a carcass if there is no potential for gas build-up in the annulus to cause pipe collapse.

28.2.2 Internal Polymer Sheath

The internal polymer sheath provides a barrier to maintain the bore fluid integrity. Exposure concentrations and fluid temperature are key design drivers for the internal sheath. Common materials used for the internal sheath include Polyamide-11 (commercially known as Rilsan®), high density polyethylene (HDPE), cross-linked polyethylene (XLPE), and PVDF. Polyamide-11 and HDPE are the two materials most commonly used. They both can withstand temperatures up to about 149°F (65°C), and have an allowable strain of 7%. PVDF can be used for applications that require a higher temperature tolerance. This material can withstand a temperature of 266°F (130°C), but its allowable strain is only 3.5%. The polymer sheath layer thickness is a function of various parameters such as inner bore fluid temperature, composition and inner bore pressure. The average size of this sheath is about 5 – 8 mm, but pipes with up to 13 mm of internal polymer sheath have also been manufactured.

28.2.3 Pressure Armor

The role of the pressure armor is to withstand the hoop stress in the pipe wall that is caused by the inner bore fluid pressure. The pressure armor is wound round the internal polymer sheath and is made of interlocking wires. These wire profiles (as seen in Figure 28.4) allow bending flexibility and control the gap between the armor wires to prevent internal sheath extrusion through the armor layer. In order to best resist the hoop stress in the pipe wall, the pressure armor is wound at an angle of about 89° to the pipe longitudinal axis.

The material used for the pressure armor wire is typically high strength carbon steel. The choice of wire typically depends on whether the pipe is qualified for 'sweet' or 'sour' service. ('Sour' service is defined by NACE MR 01-75.) The highest strength steel wires used in flexible pipe applications have an ultimate tensile strength (UTS) of 1,400 MPa (200 ksi). However these high-strength wires are prone to hydrogen induced cracking (HIC) and sulfide stress cracking (SSC). Hence it might not be possible to use such high strength steel wires for 'sour' pipe applications. The alternative would be to use additional steel layers with a UTS as low as 750MPa (105 ksi).

28.2.4 Tensile Armor

The tensile armor layers are always cross-wound in pairs. As their name implies, these armor layers are used to resist the tensile load on the flexible pipe. The tensile armor layers are typically made of flat rectangular wires and laid at about 30° - 55° to the longitudinal axis. A lay angle of 55° results in a torsionally-balanced pipe, and this angle is used in pipe designs where the hoop stress is also resisted by the tensile armor layers, and no pressure armor layer exists.

The tensile armor layers are used to support the weight of all the pipe layers, and to transfer the load through the end fitting to the vessel structure. High tension in a deepwater riser may require the use of four tensile armor layers, rather than just two. The tensile armor wires are made of high strength carbon steel like the pressure armor wires. 'Sweet' or 'sour' service conditions are a determining factor on the wire strength that can be used, because high strength wire is more prone to HIC and SSC.

28.2.5 External Polymer Sheath

The external polymer sheath can be made of the same materials as the internal polymer sheath. The main function of the external sheath is as a barrier against seawater. It also provides a level of protection for the armor wires against clashing with other objects and during installation.

28.2.6 Other Layers and Configurations

Besides these five main layers of the flexible pipe, there are other minor layers which make up the pipe cross-section. These layers include anti-friction tapes wound round the armor layers and whose purpose it is to reduce friction and hence wear of the wire layers when they rub past each other as the pipe flexes due to external loads. Anti-wear tapes can also be used to make sure that the armor layers maintain their wound shape. These tapes ensure that the wires do not twist out of their pre-set configuration, a phenomenon called 'birdcaging' which results from hydrostatic pressure causing axial compression in the pipe.

In some flexible pipe applications, because of high tensile loads it is required to use high tensile wires for the tensile armor layers, and yet the presence of a 'sour' environment means that these wires would suffer an unacceptable rate of HIC / SSC. A solution to this situation is to fabricate a pipe cross-section with two distinct annuli rather than one. The inner annulus could contain the pressure armor layer which need not be made of very high tensile steel, and which would therefore not suffer serious corrosion problems due to a high concentration of H_2S. An extra polymer sheath could then be laid between the pressure armor layer and the tensile armor layers. This polymer sheath would prevent a high H_2S concentration in the outer annulus. A certain amount of H_2S would still be able to diffuse through this polymer sheath from the inner to the outer annulus. However the concentration of H_2S in the outer annulus could now be low enough to permit the use of high tensile wires for the tensile armor layers.

28.3 End Fitting and Annulus Venting Design

28.3.1 End Fitting Design and Top Stiffener (or Bellmouth)

The end fitting design is a critical component of the global flexible pipe design process. The main functions of the end fitting are to transfer the load sustained by the flexible pipe armor layers onto the vessel structure, and to complement the sealing of the polymer fluid barrier layers.

A number of critical issues need to be considered during the design and manufacture of end fitting arrangements. Tight manufacturing tolerances are essential for the pressure sheath and sealing ring dimensions, pressure armor termination, and bolt torquing to ensure the adequate transfer of load from the steel layers of the pipe onto the vessel structure. Epoxy filling should be carried out using the appropriate techniques to ensure no air gaps are produced. The correct positioning and functioning of the annulus vent ports are also important to ensure no build-up of gases in the annulus (discussed further in Section 28.3.2).

The most severe location for fatigue damage in the risers is usually in the top hang off region. The riser is protected from over bending in this area by either a bend stiffener or a bellmouth. The detailed local analysis for the curvature or bellmouth is carried out using 2D finite element model. Figure 28.4 shows the basic arrangement for the top stiffener and bellmouth.

The bending stiffener is modeled by a 2D tapered unsymmetric beam element and the pipe is simulated by a 2D beam element. The interface between the pipe and the bend stiffener is represented by a 2D general contact element. Both to non-linear stress strain curve of the bend stiffener and the non-linear bending curvature hysteresis loop are considered in the analysis. If a bellmouth is used, a steel 2D solid element is adopted.

Knowing the top tensions and angles for each load case from the global dynamic analysis, the curvature distribution along the flexible riser can be found by applying these tensions and angles at the bottom of the models as shown in Figure 28.4.

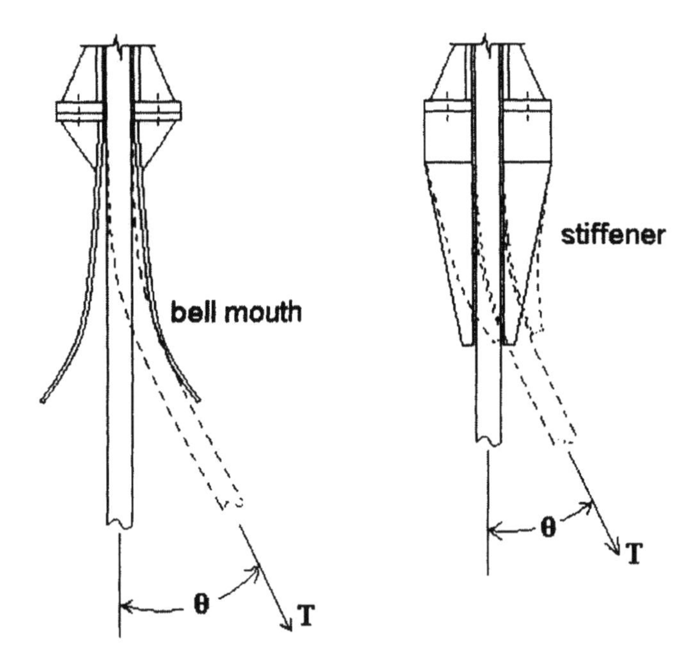

Figure 28.4 General arrangement for local curvature analysis at the bellmouth or stiffener (Zhang et al., 2003).

28.3.2 Annulus Venting System

Over time the fluids that are transported in the pipe bore will diffuse through the internal polymer sheath into the pipe annulus. These diffused gases include water, CO_2 and H_2S, and their presence in the annulus could have a deleterious effect on the steel layers. Water and CO_2 have the tendency to cause general or pitting corrosion in the pressure and tensile armor layers. The presence of water could also have a negative effect on the fatigue life of the steel layers. The presence of H_2S causes HIC and SSC in steel and its concentration is carefully assessed during the design stage because a pipe qualified for 'sweet' service could make use of high strength tensile steel (that would otherwise suffer unacceptable corrosion in a 'sour' environment).

Besides the negative corrosion and fatigue effects that the presence of these diffused gases could have on the steel layers of the pipe, the build-up of pressure in the annulus due to the presence of the gases could also cause collapse of the internal polymer sheath of the pipe. In case of a sudden pressure loss in the inner bore of the pipe (say through an emergency shutdown of the system), the pressure in the annulus due to these diffused gases might be greater than the pressure in the inner bore. This could lead to collapse of the internal polymer sheath, loss of fluid integrity and failure of the pipe. The steel carcass is designed to withstand this collapse pressure due to gas build-up in the annulus. However pipe designs without a steel carcass also exist, especially for non-hydrocarbon carrying pipes, and this failure mode has been documented to occur in flexible pipes.

In order to prevent the build-up of gases in the annulus due to diffusion, a venting system is incorporated into the pipe structure to enable the annulus gases to be vented out to atmosphere. Three vent valves are incorporated into both end fitting arrangements of a pipe. The vent valves are directly connected to the annulus and are designed to operate at a pre-set pressure of about 30 – 45 psi. The vent valves in the end fitting arrangement located subsea are sealed to prevent any ingress of seawater into the annulus.

28.4 Flexible Riser Design

28.4.1 Design Analysis

The essential tasks for design and analysis of flexible risers are similar to those described for other types of risers, see below.

1. ***Design Basis Document***: The document should as minimum include the following:
    - host layout and subsea layout;
    - wind, wave and current data and vessel motion that are applicable for riser analysis;
    - applicable design codes and company specifications;
    - applicable design criteria;
    - porch and I-tube design data;
    - load case matrices for static strength, fatigue and interference analysis;
    - applicable analysis methodology.

2. ***FE Modeling and Static Analysis***: a finite element model is built and a nonlinear static analysis is carried out assuming the vessel is in NEAR, FAR and CROSS positions.

3. ***Global Dynamic Analysis***: A global regular wave dynamic analysis is carried out assuming the vessel is in NEAR, FAR and CROSS positions. A sensitivity study is performed on critical parameters such as wave periods, effect of marine growth and hydrodynamic coefficients.

4. ***Interference Analysis***: A dynamic regular wave analysis is carried out to check the minimum clearance between the risers and with the vessel system along the water column, for various predefined load cases. The interference analysis shall confirm that selection of riser hang-off angles and departure angles etc.

5. ***Cross-sectional Model***: a detailed cross-section model is built to calculate key cross-sectional properties such as bending stiffness, axial stiffness etc., FAT pressure etc.

6. ***Extreme and Fatigue Analysis***: The wire and tube stresses are calculated at design pressure. An extreme response analysis is carried out using regular wave theory to estimate tensions and cyclic angles etc. The cross-sectional model is then used to perform to fatigue analysis.

7. ***Design Review***: This includes check of global configuration, bell mouth design, interference and fatigue design etc. In some special situations, upheaval buckling and on-bottom stability of flexible flowlines are also checked, following pipeline design practice.

Typically, a detailed design of flexible pipe is carried out by the supplier for the flexible pipe materials. A 3rd party, normally a riser engineering company, is engaged to carried out a verification of the design, as aforementioned.

28.4.2 Riser System Interface Design

Figure 28.5 shows a flowchart for integrated mooring and riser system design. Due to the complexity of a field development like Barracuda/Caratinga project, the design of most topside and subsea equipment directly affected and was affected by the riser mooring system design. The interfaces between the mooring and risers and other components of the field development facilities are an integral part of the overall mooring and riser system design. For example, estimates of maximum expected riser loads are required fore the riser supporting structure design. This impacts the modifications required to the hull structure to support the external riser porches.

Figure 28.6 shows riser system interface design for flexible risers. The overall riser configuration impacts the design of the bell-mouths at the base of the individual riser I-tubes, and also the design of the bend-stiffeners required to provide bending resistance at the vessel interface. Additional riser global configuration considerations include the on-bottom stability of the risers and flowlines in order to mitigate against possible interference between the large number of individual lines during both installation and in-service conditions.

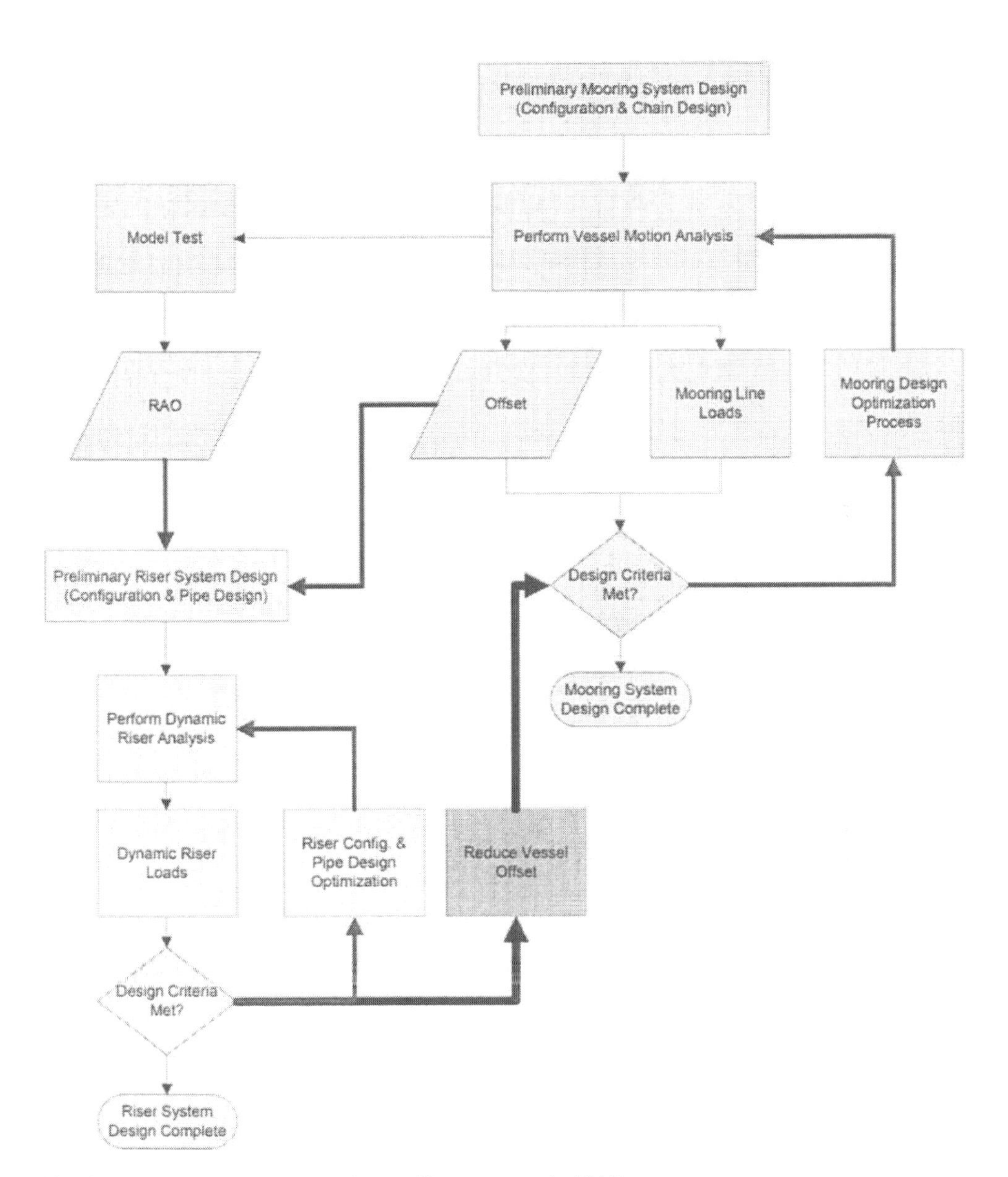

Figure 28.5 Mooring and riser system design (Seymour et al., 2003).

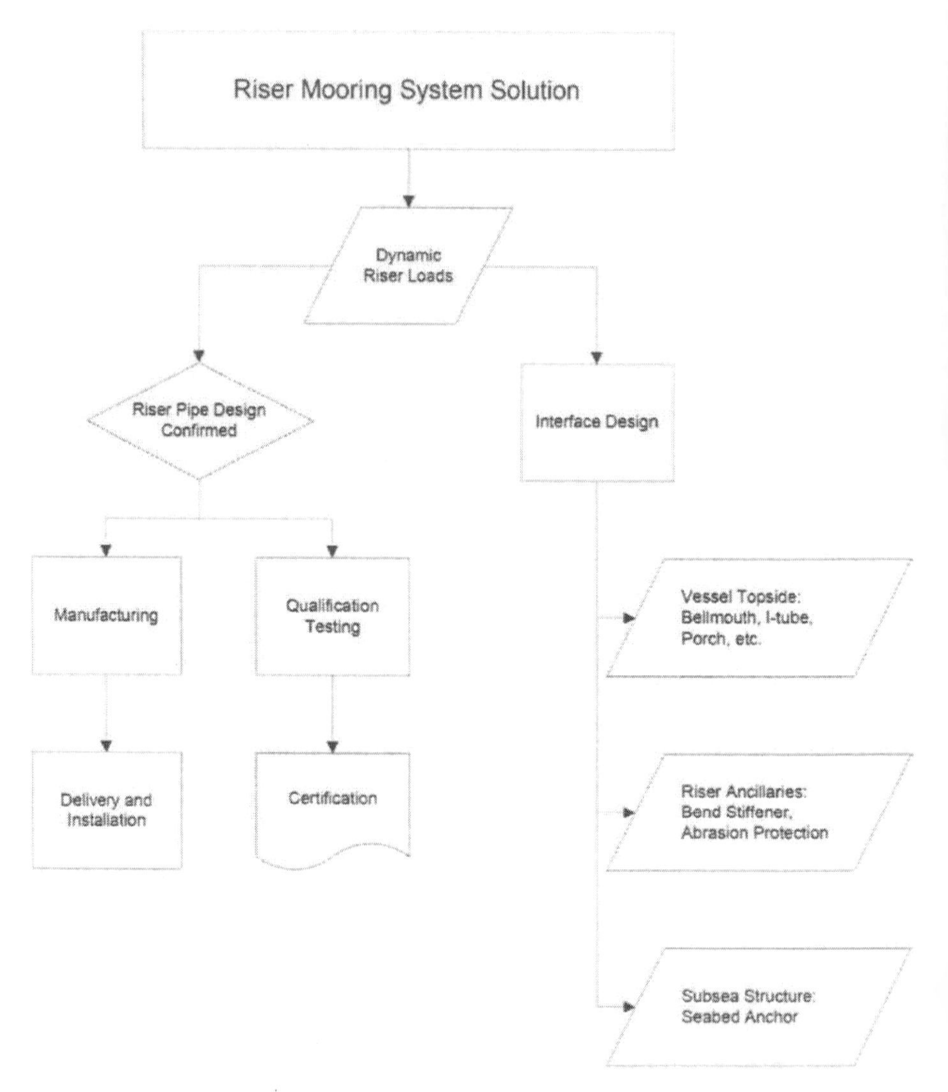

Figure 28.6 Overview of riser system interface design (Seymour et al., 2003).

28.4.3 Current Design Limitations

Figure 28.7 gives an indication of the current design limitations for flexible pipes. The existing water injection lines and the newly qualified pipe are collected by Remery et al. (2004).

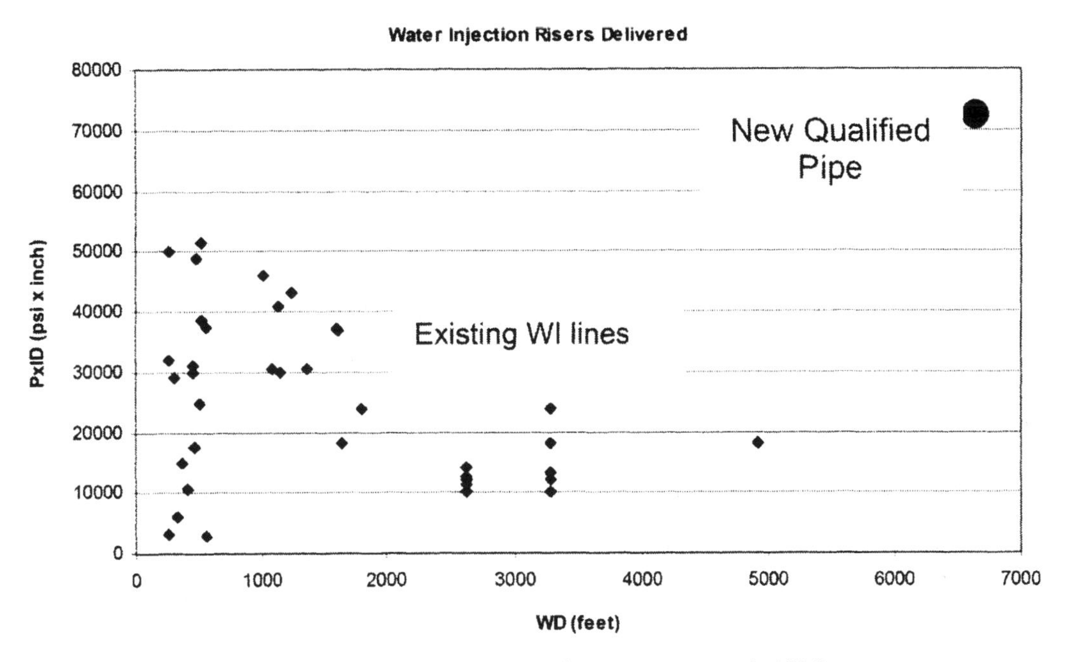

Figure 28.7 Water injection flexible pipe technology limits (Remery et al., 2004).

28.5 References

1. Antal, S., Nagy, T., Boros, A. (2003), "Improvement of Bonded Flexible Pipe according to New API Stnadrad 17K", OTC 15167.
2. API Specification 17J (1999), "Specification for Unbonded Flexible Pipe".
3. API Specification 17K (2001), "Specification for Bonded Flexible Pipe".
4. API Recommended Practice 17B (2002), "Recommended Practice for Flexible Pipe".
5. NACE MR 01-75, "Sulfide Stress Cracking Resistant Metallic Materials for Oilfield Equipment".
6. Remery, J., Gallard R. and Balague, B. (2004), "Design and Qualification Testing of a Flexible Riser for 10,000 psi and 6300 ft WD for the Gulf of Mexico", Deep Oil Technology Conference, New Orleans.
7. Seymour, B., Zhang, H. and Wibner, C. "Integrated Riser and Morring Design for the P-43and P-48 FPSOs", OTC 15140.
8. Zhang, Y., Chen, B., Qiu, L., Hill, T. and Case, M. "State of the Art Analytical Tools Improve Optimization of Unbonded Flexible Pipes for Deepwater Environments", OTC 15169.

Part IV

Riser Engineering

Chapter 29 Hybrid Risers

29.1 Introduction

In this Chapter, we shall focus on two most popular types of the Hybrid Riser (HR): Single Hybrid Risers (SHR) and Bundled Hybrid Risers (BHR).

The 1st use of hybrid risers is a deepwater production riser installed in 1988 in the Green Canyon Block 29 (GC29) by Cameron (Fisher and Berner, 1988). The system was re-used in 1995 for deployment in Garden Banks 388 (GB388) in the Gulf of Mexico by Enserch Exploration Inc and Cooper Cameron Co. (Fisher et al, 1995). This first generation hybrid riser was installed through the moonpool of a drilling vessel.

In recent years, HR are popular for use in the west of Africa offshore, e.g. for the development of Total's Girassol field as shown in Figure 29.1. This second-generation hybrid riser was bundled, and it was fabricated at an onshore site with installation by tow out and upending.

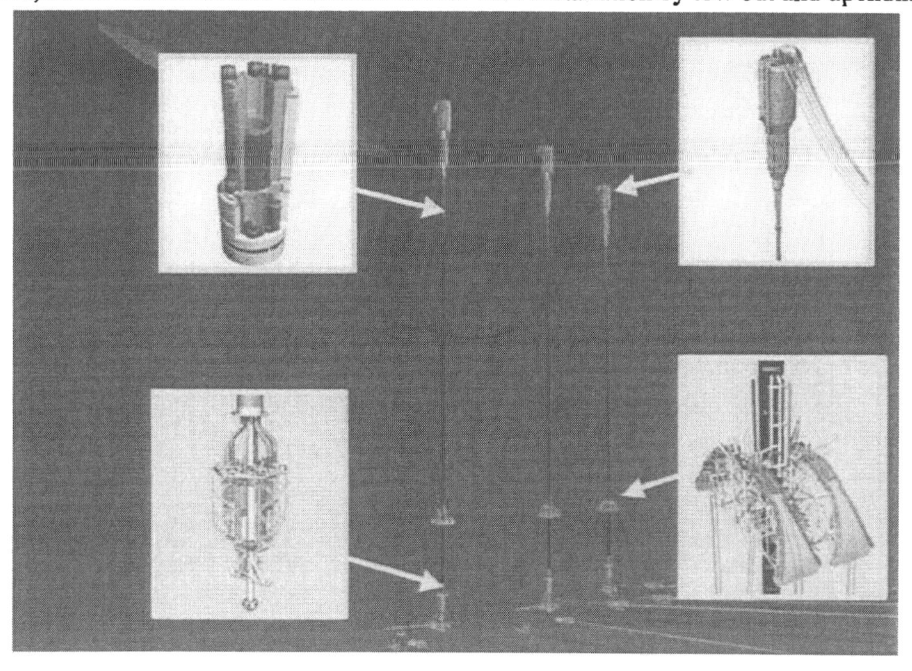

Figure 29.1 Hybrid riser towers at the Girassol Field (Alliot et al., 2004).

The 3rd generation hybrid riser was a single hybrid riser, and it was installed in ExxonMobil's Kizomba A and Kizomba B fields (des Deserts, 2000 and Hatton et al, 2002). The hybrid riser may be installed from a drilling vessel.

One of the advantage for hybrid risers is it offers a clear and well-organized subsesa layout, see Figure 29.2.

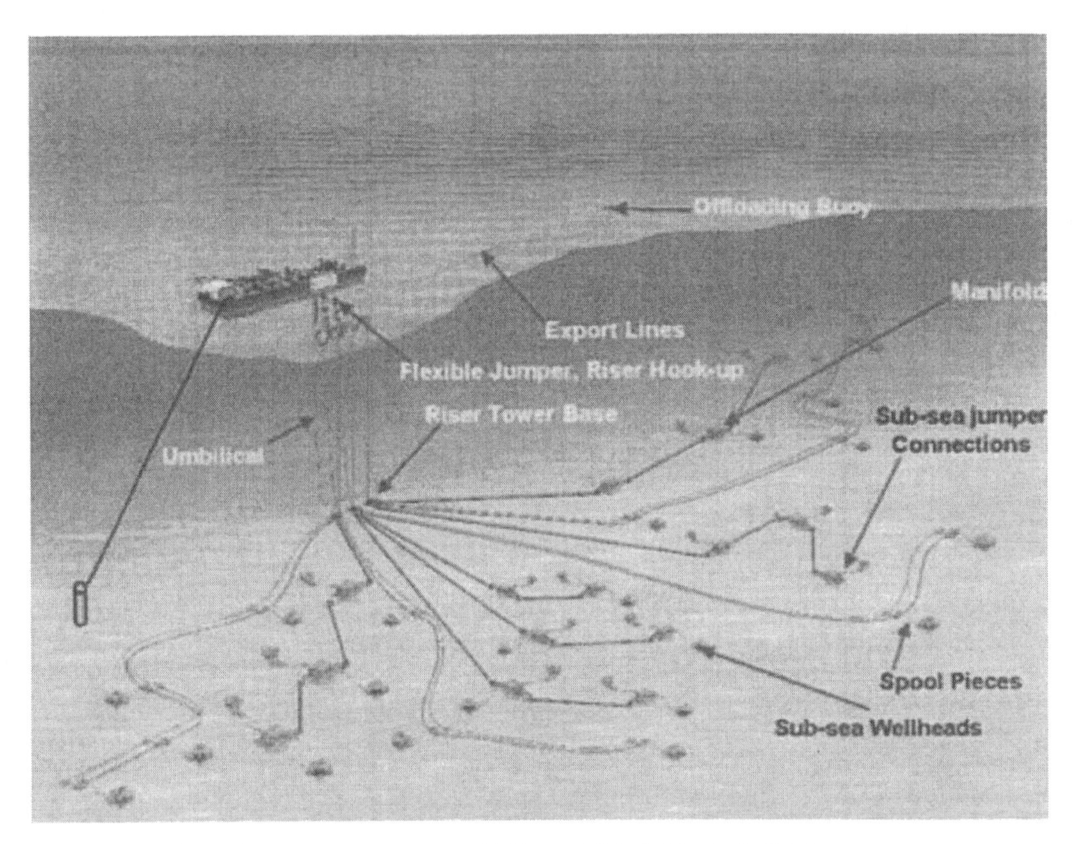

Figure 29.2 Hybrid riser towers and subsea field layout (D'Aloisio et al., 2004).

The objective of the HR riser design is:

- To define the main dimension, material, MTO of the HR;
- To define the main dimension, material, configuration, weight or tension of the buoyancy tank, riser base anchor, flexible and rigid jumpers;
- To provide inputs for design drawings.

In this Chapter, we shall discuss the basic riser components and their sizing as well as strength and fatigue analyses.

29.2 General Description of Hybrid Risers

Fisher and Berner (1988) presented the systems details, engineering design, testing and installation of the 1st hybrid risers. Some of the key system details are:

- *Collect connector holds riser to base.* The connector is designed to provide a 100-year-storm bending capacity of 6.4-million ft-lb. The connector's upper end is a bolted flange connection to the stress joint that was made before the connector was shipped to the rig. The connector is operated through a hose umbilical attached to the riser as the riser is run and is separate from the two three-control bundles installed later.
- *Titanium stress joints handles bending moments.* The stress joint provides required flexibility and stress reduction between the riser connector and lowermost riser joint.
- *Riser joints support flowlines.* Riser joints provide the structural support for production, annulus, and sales lines. The structural member is a 50-ft x 42"-diameter ¾-inch wall tubular with 24-bolt flanges welded to each end.
- *Upper end "free" but held in place.* The upper riser connector package acts as the interface point between the rigid riser and flexible-flowline jumpers to the semi-submersible.
- *Riser management system "holds station".* The riser management system keeps the riser and vessel within a 160-ft watch circle during adverse weather: the riser is current driven and the vessel wind driven, and the two could be moved apart.

In the past years, there have been some developments in HR system and component technology. In the Girassol field development, a bundled hybrid riser concept was developed and applied. In the Kizomba-A and Kizomba-B field developments, single hybrid riser concept was applied.

In this sub-section, we shall present the recent Hybrid Riser concept including the following main components:

- The anchor base foundation,
- The bottom section including rigid jumper spools,
- The free-standing vertical sections of the HR,
- The sub-surface buoyancy tank,

Flexible jumper between the top of the riser and the hang-off point in the Floating Production System (FPS).

29.2.1 Riser Foundation

The foundation structure for the HR consists of a steel suction anchor (and/or a gravity base) which will support the HR installation aids (such as pulling winches) and the flex element (Roto-latch connector is proposed).

Suction Anchor (and/or a gravity base)

A suction anchor is a large diameter cylinder with an open bottom and a closed top end. It is penetrated into the sea floor utilizing hydrostatic forces. When lowered under water, the bottom of the suction anchor will initially penetrate the sea floor by its own weight. Its interior will be sufficiently sealed to develop an under pressure when extracting entrapped water from the suction anchor with an underwater ROV operated pump. By doing so, a differential pressure, also called suction pressure, is created between interior and exterior. The differential pressure causes the suction anchor to penetrate the sea floor.

Alternatively, a gravity base or a combination of the gravity base and suction anchor (or driven anchor) could be used as the base foundation for HR, if required.

Roto-latch / Flex connection

A "Roto-latch" style connector is proposed at the base of the tower mounted on the foundation structure. Derived from TLP tether technology, the connector allows rotation of the tower in a manner similar to a conventional flex joint.

Figure 29.3 Lifting of suction pile from a barge (D' Aloisio et al., 2004).

29.2.2 Riser Base Spools

At the bottom of the HR, a forged Y-shaped tee allows:

- vertical section to be connected to the flex element,

- lateral branch to be ended by a CVC to the rigid jumper spool.

The spool is designed to allow for relative displacements between the HR base and the flowline that may be caused by the thermal expansion of the flowline or/and angular movements (due to environmental loading) at the HR base.

29.2.3 Top and Bottom Transition Forging

Transition forging on the bottom of the riser connects the standard riser to flowline spool.

A short thickened-wall section of pipe (adapter joint), typically 3m long, provides a transition between the standard riser pipe and the large stiffness of riser top assembly. A flanged connection mates with the flow spool that connects the flexible jumper and the standard riser.

29.2.4 Riser Cross-section

Wall-thickness of Riser Pipe

The wall-thickness may be sized according to the bursting and collapse criteria of API RP 1111. The main design loads for wall-thickness sizing are:

- combined internal pressure and axial tension for operating conditions at riser tops and
- combined external pressure and axial force for shut-down conditions near sea-bottom. In addition to internal (or external) pressure, tension from the buoyancy tank, gravity and thermal loads shall also be accounted.

Corrosion Coatings and Anodes

Primary protection against external riser corrosion will be via coating. Fusion Bonded Epoxy (FBE), of minimum thickness 0.45 mm, is selected for the corrosion protection of riser steel pipes and spools.

Secondary protection will be provided via cathodic protection, using bracelet type Galvalum III, aluminum sacrificial anodes. In the detail design phase, the quantities (dimensions and spacing) of anode material for all risers may be determined in accordance with DNV RP-B401.

29.2.5 Buoyancy Tank

A steel structured buoyancy tank is designed based on the following functional requirements:

- To thrust the HR dead weight.
- To provide sufficient pulling tension for the dynamic equilibrium of the HR.
- To limit maximum angular deflection of the HR with regards to its static position.

The buoyancy tank tether will be made of a section of mooring line with suitable bending stiffeners at both ends. The tether isolates the buoyancy tank response from the riser tower, thus reducing fatigue.

29.2.6 Flexible Jumpers

Flexible jumper is used to connect the riser pipe with the FPS hull. It will be designed in accordance with API SPEC 17J and API RP 17B. While behavior of the vertical sections of the HRs is quite static, riser jumper behavior is dynamic, mostly due to the wave loads and FPS motions and offsets. The design pressure and temperature for flexible jumper are the same as those for the riser.

The end fitting provides the link between the flexible riser structure and the end termination (I-tube) at FPS. It fulfils the following main functions:

- Seal the external and the intermediate plastic layers of the flexible pipe structure,
- Anchor the tensile armour to ensure the transmission of the axial loads to the support structure,
- Allow the fitting of a pulling/laydown head for the initiation and laydown purpose.

29.3 Sizing of Hybrid Risers

The key design issues for arrangement/sizing of the HR include:

1. Overall arrangement - footprints at bases; hang-off locations/spacing.
2. General arrangement of HR including:
    - Riser thickness determination;
    - Riser Top: Jumper w/end fittings characteristics and geometry, geometry for the buoyancy tank;
    - Riser Base: forging and foundation.
3. Sizing of buoyancy tank and suction anchor;
4. Sizing of rigid jumper spool (at HR base) connected to the gas export pipeline via collet connectors

29.3.1 Riser Cross-section

This section describes the basic sizing of the HR cross-section.

The wall-thickness is first sized according to the bursting and collapse criteria of API RP 1111. The linepipe material purchase, fabrication and manufacturing may be based on API 5L specification. The following basis data are used in the design:

- The water depth;
- The design pressure. The hydro-test (internal) pressure is 1.25 times of the corresponding design (internal) pressure;
- Seawater density at sea surface & at sea bed;
- Riser pipe containment density at the design pressure;
- Field design life;

- The selected steel material grade and weight density;
- Corrosion allowance and insulation requirement;
- The maximum allowable offset of production vessel as percentage of the water depth for intact condition and for the one line damaged condition;
- For foam buoys used in buoyancy tank, water ingress after 20 years is assumed to correspond to x % of the initial uplift.

Once the maximum tension at riser top has been determined, the equivalent stress criteria as per API 2RD will be checked against the allowable stress, for riser top in operating condition and riser bottom in shut-down condition. The API 5L standard sizing for diameter, wall-thickness, nearest to (and higher than) that required by API RP 1111 and API 2RD criteria, will then be selected.

Unless the maximum design temperature is higher than 121°C, no temperature derating factor needs to be included in the wall-thickness calculation because the API RP 1111 code does not require consideration of temperature derating.

The following API RP 1111 criteria have been applied to calculate the required steel pipe wall-thickness:

- Burst strength criteria for pipe under hydro-test pressure;
- Collapse strength criteria for pipe in operation;
- Collapse strength criteria for pipe installation.

Seamless pipe is used as the riser pipe. Even DSAW pipe offers more competitive pricing opportunities, it provides less collapse factor for API 1111 collapse stress check. Cost of DSAW pipe in this case is about the same as seamless pipe. Thus, seamless pipe is used in the HR design.

The steel pipe is subject to combined radial stress, axial tension and external (or internal) pressure. Based on the stress utilization factors obtained, it is confirmed that the equivalent stress criteria of API 2RD is satisfied for the gas export riser. The stress utilization factors are checked for riser pipe near sea-bottom (in shut-down condition) and at riser top (in maximum operating condition).

29.3.2 Buoyancy Tanks

The section describes the basic sizing of the buoyancy tank for the hybrid risers. Two design cases for the buoyancy tank are considered: the base case (subsurface modular foam buoys) and an alternative case (nitrogen pressurized steel buoyancy tank).

For the base case, rectangular modular buoys will be used to provide the necessary buoyancy to the riser tower. These modular buoys linked together using either chain or beam structure.

Cost are expected to be lower than the alternative case as the buoys are in light steel structure and standardized. Application of the modular buoys also enables easy storage transportation.

For the alternative case, the steel plate material for buoyancy tank and suction anchor is to be in accordance with API Spec. 2B and 2H, respectively concerning rolled steel plate and material for structural pipe specification.

The design issues for buoyancy tanks are:

- Size optimization to minimize drag and maximize buoyancy;
- Size and number of chambers;
- Redundancy for loss of one or more chambers;
- Ballast control system;
- Tension monitoring in services;
- Method of connecting to riser.

The buoyancy tank is designed to provide sufficient uplift force through the design life. The global uplift force is the sum of the minimum tension at bottom and the submerged weight of the pipe/coatings and equipment. 5% contingency has been applied to riser steel weight to account for the weight of corrosion coating and riser joints etc. Based on previous projects experience, the submerged weight of BT equipment (ballasting and pressurizing systems, valves…) may be assumed to be 30% of the steel tank weight (structural steel BT option).

29.3.3 Riser Foundation

In this Section, the base case for riser foundation is to use suction anchor as the anchoring foundation. Alternatively, gravity foundation or a combination of the above solutions may be considered if the long-term settlement of foundation is acceptable.

The foundation design is governed by the permanent uplift created by the net buoyancy of the HR tank in operation. The soil characteristics, design loads and associated safety factors as well as differential under pressure at installation will all influence the foundation design.

The suction anchor is sized according to API RP 2T and API RP 2A. The suction anchor design is typically cylindrical with a diameter of 6.5 meters and wall-thickness of 35 mm. Space is available at the top of the anchor for cathodic protection and installation aids such as pull-down winches or for the pump skid used for the installation of the anchor itself.

The soil contribution to the vertical holding capacity includes the internal and external skin friction, submerged weight of suction pile. The calculation procedure for the pile design follows the equation below, derived from API RP 2T and API RP 2A.

$$W_b + \alpha \int_{H_0}^{H} c * dA_S > T * \gamma_f \tag{29.1}$$

where,

W_b : submerged weight of the foundation;

α : a coefficient accounting for the remoulding of clay (taken as 0.4);

H : total length of anchor;

H_0 : length of anchor with no contribution to soil resistance. (taken as 1 m to account for pile without penetration to the soil);

c : undrained shear strength;

A_s : embedded side surface area of anchor considering both internal and external skin friction;

T : maximum vertical uplift;

γ_f : safety factor. (taken as 2).

29.3.4 Flexible Jumpers

Figure 29.4 shows a flexible jumper being installed. Flowline jumpers are used to connect the riser base with flowline base. The flowline jumpers are conventional subsea technology, requiring 5D bends for pigging.

Figure 29.4 Lifting of suction pile from a barge (D' Aloisio et al., 2004).

This section specifies the design requirements for the flexible jumper, and describes the calculations of the loads acting on the Floating Production System (FPS) hull and on the forging beneath the buoyancy tank. The flexible jumper is modeled as a single catenary line laid between these two hang-off locations.

The maximum tension calculated from the catenary analysis needs to be less than the maximum pull provided by the flexible. The calculation takes into account the contribution of both the current and the FPS offset, which are the major components in the evaluation of the max. jumper loads. In the study, the horizontal distance between the HR base and the nominal location of the corresponding hang-off point is assumed to be 200 m.

The tension acting on the FPS Hull and HR tank from riser jumper and the azimuthal angles of the jumper have been evaluated for two load cases:

Case 1: FPS is under damage condition (one mooring line broken) and its maximum offset is 10% of water depth. The HR and flexible jumper are in operation are filled with exported gas.

Case 2: FPS is in operation and its maximum offset is 8% of water depth. In other words, the HR and flexible jumper are considered to be filled with seawater (hydrotesting or accidentally flooded).

For each load case, the following HR positions are considered.

- neutral HR position, no FPS offset case,
- far HR position, max FPS offset case,
- near HR position, min FPS offset case.

The jumper tension results are very preliminary since no dynamic is considered. Further study needs to be employed using finite element programs (Flexcom, Orcaflex, Deeplines) in the next design phase.

29.4 Preliminary Analysis

A single string approach may be taken to model the hybrid riser system:

- Mass, bending stiffness and axial stiffness are the sum of all members;
- Buoyancy diameter is taken as the outside diameter of the buoyancy modules;
- Air can is model using pipe elements with appropriate buoyancy and drag properties;
- Flexible jumpers are modelled individually or in group.

The key objective of the preliminary analysis is to get indication of the requirements for (Wald, 2004):

- Buoyancy and tension distribution along the riser;
- VIV suppression devices.

Low submerged weight and low tension can make the riser susceptible to VIV. However, it is recommended to avoid over specification of strakes, which can increase drag, and may have negative effect on buoyancy requirements and stresses at riser base.

29.5 Strength Analysis

In the engineering design of the first hybrid riser system, Fisher and Berner (1988) conducted engineering and analysis work on operating stress levels and system fatigue life. An overview of results shows that stress levels were well under conservative design values throughout the system. Specific design requirements for the rigid riser system include:

- Simultaneous operation with a drilling or workover riser under 1-year storm conditions;
- Survival of 100-year storm conditions with a broken mooring line;
- Survival with selected buoyancy loss at discrete points along the riser or at its top;
- Disconnect capability at the top to facilitate maintenance.

The strength analysis is conducted to optimize the following (Wald, 2004):

- Distributed buoyancy requirements;
- Top tension;
- Tethering tension (if any);
- Foundation loading.

The optimization is carried out by performing a sensitivity analysis on the aforementioned parameters to investigate their effect on riser response.

The strength analysis may be performed for a range of possible loading conditions, including: Distributed buoyancy requirements;

- Extreme wave;
- Extreme current;
- Extreme vessel drift;
- Winds and currents from opposing directions.

The selection of the optimum arrangement is usually based on component availability, cost and installation considerations.

The design details that need to be analyzed included:

- Heat loss analysis of peripheral lines;

- Analysis of riser base piping and transition to peripheral lines;
- Design of goosenecks.

29.6 Fatigue Analysis

In the fatigue analysis of the first hybrid riser system, Fisher and Berner (1988) estimated that the titanium flex joint has a fatigue life of over 500 years, and the rest of the riser has a minimum 60-year fatigue life. Riser joints exhibited about 70 years fatigue life without vortex-shedding strakes and 75 years with strakes.

Like any other riser analysis, a fatigue analysis of the hybrid riser should include vessel drift motions, first order wave action, VIV and installation induced fatigue.

Hydrodynamic loading on the upper part of the riser is the main sources of the first order fatigue damage, and may be reduced by lowering the top of the riser. Fatigue due to jumper hoses may be reduced by making the hoses longer.

Fatigue due to vessel drift motions may be significant at the connection between flexible jumper hoses and riser and near the base of stress joint.

A VIV suppression system was developed for GC29 hybrid riser based on the results of analysis and basin testing (Fisher et al., 1995). This configuration was exposed to a severe loop current event on GC29 which lasted several days. Loop currents greater that 3.5 knots were measured and no evidence of VIV was observed. Since high tension force applied to the riser pipe may significantly increase VIV fatigue life, one of the most powerful VIV suppression techniques is to increase the riser tension in a severe loop current event.

Installation fatigue may be significant, if the hybrid riser is towed using a controlled depth towing method. This is because towing may induce oscillations at riser natural frequencies. An analysis should be conducted for a range of wave heights, periods, and directions to determine the conditions that may cause riser fatigue damage. Model testing and full scale measurement may be warranted to verify the analysis. This was done on the Girassol project.

29.7 Structural and Environmental Monitoring System

Thrall and Poklandnik (1995) presented a Structural and Environmental Monitoring System for the Garden Banks 388 (GB388) Deepwater Production Riser. The riser was instrumented with bonded resistance strain gauge so that the remaining operating fatigue life can be quantified. It was also instrumented with strain gauges, inclinometers and accelometers at five locations along its length so that tension, bending, orientation, and motion of the riser can be monitored at these locations, including bottom joint. Correlation of the riser response data with the excitation or environmental data, including wave motion, current velocities, wind velocities, and vessel mooring tensions and positions is enhanced by acquiring and archiving all data on a single common system having multiple redundant elements for reliability.

The structural and environmental monitoring system developed by Thrall and Poklandnik (1995) for GB388 Deepwater Production Riser is still valid as a general approach for riser monitoring. However, it didn't address VIV monitoring. Recent development on VIV monitoring may be found from Kaasen et al. (2000) and Franciss (2001).

29.7.1 Riser Fatigue Monitoring Approach

The primary data monitored is for a cumulative fatigue damage for the titanium stress joint. Since the primary contribution to riser fatigue is bending stress, and riser bending is primarily concentrated in the lowermost joint, connected to the template on the seafloor, the most important parameter to monitor in order to track fatigue damage is the maximum or outer fiber bending stress in this joint.

Outer fiber bending strain can be monitored directly through the use of bonded resistance strain gauges applied to the outer diameter of the joint. The stress can be calculated from the strain values using the material elastic modulus. Fatigue damage is predicted using stress time history and a time-domain fatigue analysis methodology. The time-domain fatigue analysis methodology, such as a Rainflow counting method, may convert stress time-history into stress histograms on which a Miner's law may be applied to calculate cumulative fatigue damage.

29.7.2 Structural Monitoring System

The riser structural monitoring system comprises a surface computer in the production control room and two fully redundant distributed data acquisition systems, each consisting of four subsea computers and associated sensors and connected cables.

The riser system sensors include the following (Thrall and Poklandnik, 1995):

1. **Strain Gages**

Bonded resistance strain gauges are installed on the outer surface of the riser at four equally-spaced locations around each instrumented cross-section. The strain gages are aligned with the longitudinal axis of the riser and wired to be sensitive to longitudinal tensions or compressions.

The strain data are used to calculate fatigue damage, riser tension, bending stress, and bending azimuth for each of the ten instrumented cross-sections.

2. **Accelerometers and Inclinometers**

Two accelerometers are mounted to sense horizontal acceleration and two inclinometers to monitor the tilt of the production riser.

The information monitored – riser strain, tension, bending stress and azimuth, riser acceleration and tilt, are displayed on screens in the production control room and the ballast control room as an aid to the riser operators, and are archived as desired in the event of a storm, loop current, or other unusual event.

29.7.3 Environmental Monitoring System

The riser environmental monitoring system includes the following (Thrall and Poklandnik, 1995):

1. Current Profile

The current profile data together with the riser response data will be used both to confirm the riser analysis models and to provide information to operations, primarily of the drilling riser. An acoustic Doppler current profiler (ADCP) unit capable of monitoring the current velocities at a number of depths up to maximum depth of 1800 ft is suspended from the floating production facility. The current data is displayed and archived with the riser structural data. The current profiles along with the one-year storm current profile, and the drilling riser limiting current profile, may provide a decision tool for the drilling personnel when a loop current is or may be in the vicinity of the vessel.

2. Surface Current

Surface current and wind both contribute to vessel offset, which affects the riser bending stress by displacement of the upper end of the riser from a position directly above its base. An electromagnetic current meter is mounted on the vessel to monitor current above the ADCP, which is located below the keels of the vessel.

3. Wave

Wave action perturbs the vessel and the upper end of the riser and so contributes to riser fatigue. A radar wave-monitoring system is mounted on the vessel, well above the waterline where access and maintenance are easy.

A radar sensor monitors the distance to the sea surface, while an accelerometer signal is doubled-integrated to provide vessel heave information. The signals are combined to give a real-time wave-height signal.

Significant wave height, wave period, vessel heave and sensor-to-sea distance are all displayed in the computer screens.

4. Wind

An anemometer system is mounted to top of the derrick. It provides wind speed and direction data to the distributed digital data acquisition system.

29.7.4 Vessel Mooring and Position

1. **Vessel Mooring System**

Four mooring stations, one in each corner of the semi-submersible, each anchor and control three mooring lines. Each mooring line uses a strain gauge instrumented transducer to provide a local tension readout for the mooring control operator. The twelve mooring line tensions, resultant mooring line force and direction, wind speed and direction, and surface current speed and direction are all provided on one screen to facilitate moving the vessel from one location to another over the template and maintaining the vessel in the desired location.

2. **Vessel Position**

An acoustic transponder system is used to provide relative locations of the vessel, the upper end of the production riser, the upper riser connector package, the riser base, and the lower marine riser package, as well as the ROV when it is in use.

29.8 References

1. Alliot, V., Legras, J.L. and Perinet, D. (2004), "A Comparison between Steel Catenary Riser and Hybrid Riser Towers for Deepwater Field Developments", Deep Oil Technology Conference.
2. D'Aloisio, G., Fosoli, P., Sykes, C. and Vila, J. (2004), "Single Hybrid Risers: Development and Installation of a Novel Deepwater Riser Concept", Deep Oil Technology Conference.
3. API RP 1111 (1999), "Design, Construction, Operation and Maintenance of Offshore Hydrocarbon Pipelines (Limit State Design)", 3rd Edition, July 1999.
4. API RP 2RD (1998), "Design of Risers for Floating Production Systems (FPSs) and Tension-Leg Platforms (TLPs)", 1st Edition, June 1998.
5. API 5L, "Linepipe", 2000.
6. API RP 2T (1997), "Recommend Practice for Planning, Designing, and Constructing Tension Leg Platforms", 2nd edition, August 1997.
7. API RP 2A (1993), "Recommend Practice for Planning, Designing and Constructing Fixed offshore platforms – Load and Resistance Factor Design", 1st edition, July 1993.
8. DNV RP-B401 (1993), "Recommended Practice RP B401 Cathodic Protection Design.", Det Norske Veritas.
9. des Deserts, L (2000), "Hybrid Riser for Deepwater Offshore Africa", OTC 11875, Proceedings of the 33rd Offshore Technology Conference.
10. D' Aloisio, G., Foscoli, P., Sykes, C. and Vila, J. (2004), "Singlke Hybrid Risers: development and Installation of a Novel Deepwater Riser Concept", Deep Oil Technology Conference.
11. Fisher, EA, Berner, PC (1988), "Non-Integral Production Riser for Green Canyon Bllock 29 Development", OTC 5846, Proceedings of the 20th Offshore Technology Conference.

12. Fisher, EA, Holley, P and Brashier, S. (1995), "Development and Deployment of a Freestanding Production Riser in the Gulf of Mexico", OTC7770, Proceedings of the 27th Offshore Technology Conference.
13. Franciss R. (2001), "Vortex Induced Vibration Monitoring System in Steel Catenary Riser of P-18 Semi-Submersible Platform", OMAE2001/OFT-1164, Proceedings of OMAE'01.
14. Hatton, S, McGrial, J and Walters, D (2002), "Recent Developments in Free Standing Riser Technology", 3rd Workshop on Subsea Pipelines, Dec. 3-4, 2002, Rio de Janeiro, Brazil.
15. Kaasen, KE, Lie, H, Solaas, F and Vandiver JK (2000), "Norwegian Deepwater Program: Analysis of Vortex-Induced Vibrations of Marine Risers Based on Full-Scale Measurements", OTC 11997, Proceedings of the 33rd Offshore Technology Conference.
16. Thrall, DE, Poklandnik, RL (1995), "Garden Banks 388 Deepwater Production Riser Structural and Environmental Monitoring System", OTC 7751, Proceedings of the 27th Offshore Technology Conference.
17. Wald, G. (2004), "Hybrid Riser Systems", Deepwater Riser Engineering Course, Clarion Technical Conferences.

Part IV

Riser Engineering

Chapter 30 Drilling Risers

30.1 Introduction

Floating drilling risers are used on drilling semi-submersibles and drilling ships. As the water depth increases, integrity of drilling risers is a critical issue. The design and analysis of drilling risers are particularly important for dual operation, DP based semi-submersible rigs. For the integrity assurance purpose, a series of dynamic analysis needs to be carried out. The objective of the dynamic analysis is to determine vessel excursion limits and limits for running/retrieval and deployment. In recent years, qualification tests are also required to demonstrate fitness for purpose for welded joints, riser coupling and sealing systems. For risers installed in the Gulf of Mexico, vortex-induced vibrations are a critical issue. Some oil companies encourage use of monitoring systems to measure real-time vessel motions and riser fatigue damage. The monitoring results may also be used to verify the VIV analysis tools that are being applied in the design and analysis.

Figure 30.1 Drilling semi-submersibles, drilling ships and drilling risers ("The Composite Catalog ® of Oilfield Equipment & Services", 2002/03, 45th Edition).

In this Chapter, after a brief outline of the floating drilling equipments and subsea systems, the riser components and vessel data are outlined. Various methods of riser analysis are presented.

30.2 Floating Drilling Equipments

30.2.1 Completion and Workover (C/WO) Risers

There are two different types of risers for installation and intervention in and on the well: the completion riser and the workover riser.

A completion riser is generally used to run the tubing hanger and tubing through the drilling riser and BOP into the wellbore. The completion riser may also be used to run the subsea tree. The completion riser is exposed to external loading such as curvature of the drilling riser, especially at the upper and lower joints.

A workover riser is typically used in place of a drilling riser to re-enter the well through the subsea tree, and may also be used to install the subsea tree. The workover riser is exposed to ocean environmental loads such as hydrodynamic loads from waves and current in addition to vessel motions.

Figure 30.2 is a typical figure for C/WO riser, taken from ISO code for Completion and Work Riser Systems. The completion and workover riser can be a common system with items added or removed to suit the task being performed. Either type of risers provides communication between the wellbore and the surface equipment. Both resist external loads and pressure loads and accommodates wireline tools for necessary operations.

Riser connectors are one of the most important riser components. Figure 30.3 shows a bolt-based riser connectors. As drilling depths have increased, riser connectors have evolved to address issues concerning high internal pressure and external pressure, increasing applied bending moment and tension loads and extreme operating conditions such as sweet and sour services. For the connector design, the material selection and fabrication of bolts are critical issues.

Figure 30.4 shows key components in a typical drilling riser system.

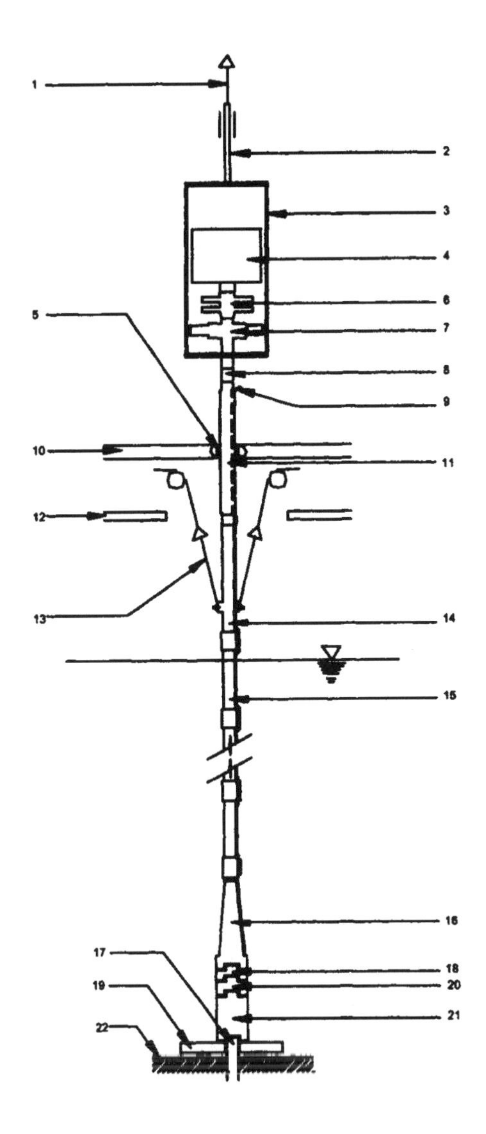

Key

1	Top drive	9	Umbilical (to hose reel)	17	Wellhead
2	Drill sub	10	Drill floor	18	Emergency disconnect package
3	Surface tree tension frame	11	Slick (cased wear) joint	19	Guide base
4	Coiled tubing injector	12	Moon pool area	20	Lower riser package
5	Roller bushing	13	Riser tension wires	21	Subsea tree
6	Surface BOP	14	Tension joint	22	Seabed
7	Surface tree	15	Standard riser joints		
8	Surface tree adapter joint	16	Stress joint		

Figure 30.2 Stack-up model for a completion and workover riser, ISO13628-7.

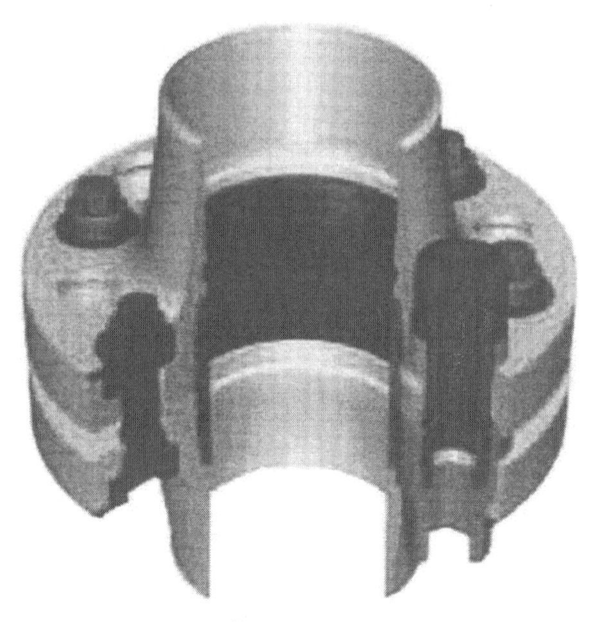

Figure 30.3 Riser connector (The Composite Catalog ® of Oilfield Equipment & Services, 2002/03, 45th Edition).

The *spider* is a device with retractable jaws or dogs used to hold and support the riser on the uppermost connector support shoulder during running of the riser. The spider usually sits in the rotary table on the drill floor.

The *gimbal* is installed between the spider and the rotary table. It is used in reducing shock and to evenly distribute loads, caused by rig roll/pitch motions, on the spider as well as the riser sections.

A *slick joint,* also known as telescope joint, consists of two concentric pipes that telescope together. It is a special riser joint designed to prevent damage to the riser and control umbilical where they pass through the rotary table. Furthermore it protects the riser from damage due to rig heave.

Riser joints are the main members that make up the riser. The joints consist of a tubular midsection with riser connectors in the ends. Riser joints are typically provided in 9.14 m to 15.24 m (30 ft to 50 ft) lengths. For the sake of operating efficiency, riser joints may be of 75 ft length. Shorter joints, pup joints, may also be provided to ensure proper space-out while running the subsea tree, tubing hanger, or during workover operations.

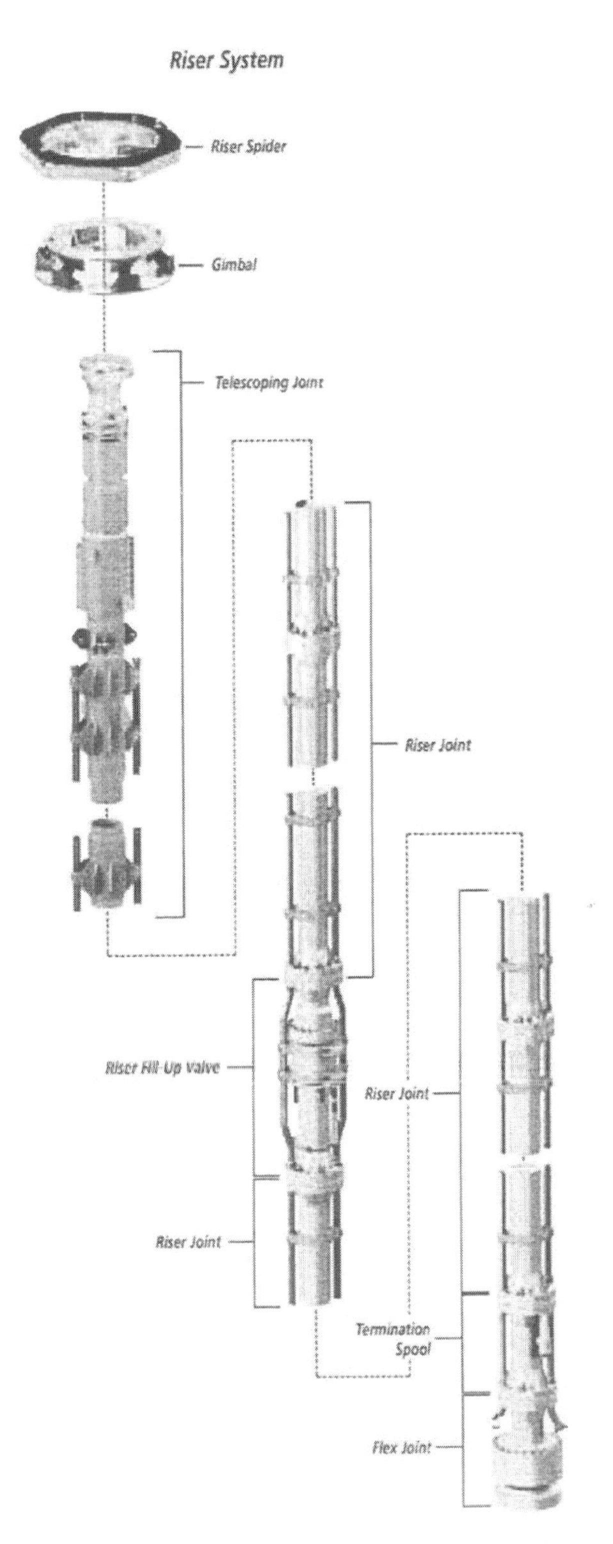

Figure 30.4 Key components in a drilling riser system (The Composite Catalog ® of Oilfield Equipment & Services, 2002/03, 45th Edition).

Depending on configuration and design, a drilling riser system also consists of the following components:

- The *BOP adapter joint* is a specialized C/WO riser joint used when the C/WO riser is deployed inside a drilling riser and subsea BOP to install and retrieve a subsea tubing hanger.

- The *lower workover riser package (LWRP)* is the lowermost equipment package in the riser string when configured for subsea tree installation/workover. It includes any equipment between the riser stress joint and the subsea tree. The LWRP permits well control and ensures a safe operating status whilst performing coiled tubing/wireline and well servicing operations.

- An *emergency disconnect package (EDP)* is an equipment package that typically forms part of the LWRP and provides a disconnection point between the riser and subsea equipment. The EDP is used when it is required to disconnect the riser from the well. It is typically used in case of a rig drift-off or other emergency that could move the rig from the well location.

- The *stress joint* is the lowermost riser joint in the riser string when the riser is configured for workover. The joint is a specialized riser joint designed with a tapered cross section, in order to control curvature and reduce local bending stresses.

- The *tension joint* is a special riser joint, which provides means for tensioning the C/WO riser with the floating vessels tensioning system during open sea workover mode. The tension joint is often integrated in the lower slick joint.

- The *surface tree adaptor joint* is a crossover joint from the standard riser joint connector to the connection at the bottom of the surface tree;

- The *surface tree* provides flow control of the production and/or annulus bores during both tubing hanger installation and subsea tree installation/workover operations.

30.2.2 Diverter and Motion Compensating Equipment

Diverter is similar to a low-pressure BOP. When either gas or other fluids from shallow gas zones enter the hole under pressure, the diverter is closed around the drill pipe or kelly and the flow is diverted away from the rig.

All floating drilling units have motion compensating equipment (as shown in Figure 30.5) installed to compensate for the heave of the rig. Compensators function as the flexible link between the force of the ocean and the rig. The equipment consists of the drill string compensator, riser tensioners, and guideline and podline tensioners. The drilling string compensator, located between the traveling block and swivel and kelly, permits the driller to maintain constant weight on the bit as the rig heaves. Riser tensioners are connected to the outer barrel of the slip joint with wire ropes. These tensioners support the riser, and the mud within it, with a constant tension as the rig heaves. The guideline and podline maintain constant tension on guideline wire ropes, and wire ropes that support the BOP control podlines as the rig heaves.

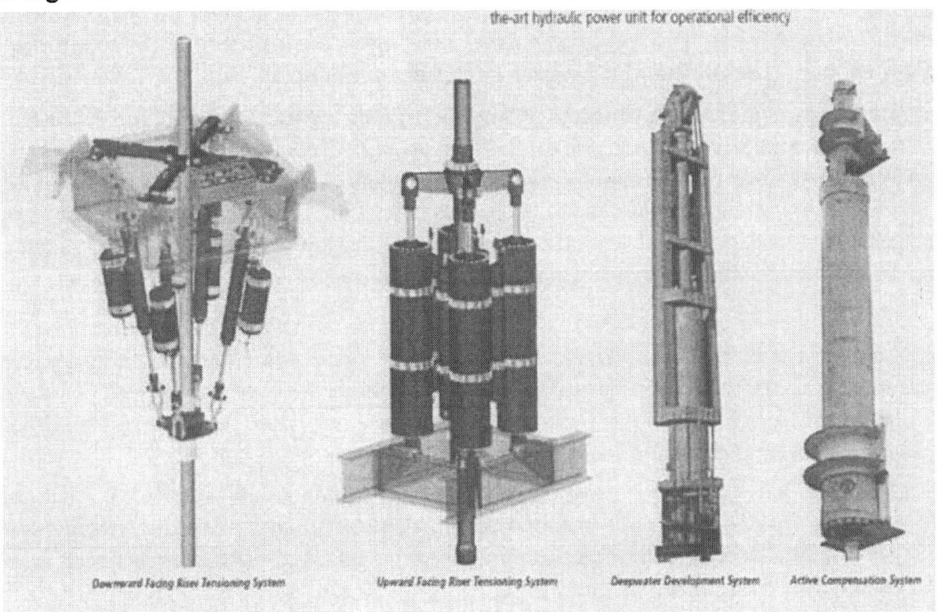

Figure 30.5 Motion compensation equipment (The Composite Catalog ® of Oilfield Equipment & Services, 2002/03, 45th Edition).

30.2.3 Choke and Kill Lines and Drill String

Choke and kill lines are attached to the outside of the main riser pipe, see Figure 30. 6. They are used to control high-pressure occurrence. Both lines are usually rated for 15 ksi. High pressure is circulated out of the well bore through the choke and kill lines by pumping heavier mud into the hole. Once the pressure is normal, the BOP is opened and drilling resumes. If the pressure can't be controlled with the heavier mud, cement is pumped down, the kill line and the well is killed.

The drill string permits the circulation of drilling fluid or liquid mud. Some function of this mud is:

- to cool the bit and lubricate the drill string;
- to keep the hole free of cuttings by forced circulation to the top;
- to prevent wall cave-ins or intrusions of the formations through which it passes;
- to provide a hydrostatic head to contain pressures that may be present.

Figure 30.6 Complete riser joint ("Offshore" Mgazine, May 2001).

30.3 Key Components of Subsea Production Systems

30.3.1 Subsea Wellhead Systems

The foundation of any subsea well is the subsea wellhead. The function of the subsea wellhead system is to support and seal the casing string in addition to supporting a blowout preventer (BOP) stack during drilling, and the subsea tree under normal operation.

Installation of equipment to the seabed is generally done by one of two methods: 1) with the use of tensioned guidelines attached to guide sleeves on the subsea structure orienting and guiding equipment into position, 2) guidelineless, using a dynamic positioning reference system to move the surface vessel until the equipment is positioned over the landing point where after the equipment is lowered into place. Regardless of the guiding system, the procedure in which the wellhead system is installed is as follows:

1. The first component installed is the *temporary guidebase*. The temporary guidebase serves as a reference point for the installation of subsequent well components compensating for any irregularities of the seabed. For guideline systems the temporary guidebase also act as the anchor point for guidelines.
2. The *conductor housing* is essentially the top of the casing conductor. The casing conductor and housing are installed through the temporary guidebase, either by piling or drilling, and provides an installation point for the permanent guidebase and a landing area for the wellhead housing.
3. The *permanent guidebase,* which is installed on the conductor housing, establishes structural support and final alignment for the wellhead system. The permanent guidebase provides guidance and support for running the BOP stack or the subsea tree.
4. The *wellhead housing* (or high-pressure housing), which is installed into the conductor housing, provides pressure integrity for the well, supports subsequent casing hangers, and serves as attachment point for either BOP stack or subsea tree by using a wellhead connector.
5. To carry each casing string a *casing hanger* is installed on top of each string, the casing hangers are supported in the wellhead housing which thus supports the loads deriving from the casing. To seal the inside annuli an *annulus seal assembly* is mounted between each casing hanger and the well housing.

30.3.2 BOP

The marine riser is used as a conduit to return the drilling fluids and cuttings back to the rig and guide the drill and casing strings and other tools into the hole. Geiger and Norton (1995) gave a short and relevant description on floating drilling.

The well is begun by setting the first casing string known as the conductor or structural casing – a large-diameter, heavy-wall pipe – to a depth depending on soil conditions and strength/fatigue design requirements. Its primary functions are;

- to prevent the soft soil near the surface from caving in;

- to conduct the drilling fluid to the surface when drilling ahead,
- to support the blowout preventer (BOP) stack and subsequent casing strings, and
- to support the Christmas tree after the well is completed.

The depth and size of each drilling string is determined by the geologist and drilling engineer before drilling begins. When drilling from a semi-submersible or drillship, the wellhead and BOP must be located on the seabed.

The BOP stack is used to contain abnormal pressures in the well bore while drilling the well. The primary function of the BOP stack is to preserve the fluid column or to confine well fluids or gas to the borehole until an effective fluid column can be restored.

At the lower end of the riser is the lower flexjoint. After the hole has been drilled to its final depth, electric logs are run to determine the probable producing zones. Once it is determined that sufficient quantities of oil and has exist, the production tubing is then run to the zone determined to contain oil or gas. Only after this takes place is the well "completed" by removing the BOP stack and installing the fittings used to control the flow of oil and gas from the well head to the processing facility.

30.3.3 Tree and Tubing Hanger System

In order to complete the well for production a tubing string is installed in and supported by a tubing hanger. The tubing hanger system carries the tubing and seals of the annulus between casing and tubing. To regulate flow through the tubing and annulus a subsea tree is installed on the wellhead. The subsea tree is an arrangement of remotely operated valves, which controls the flow- direction, amount, and interruption.

30.4 Riser Design Criteria

30.4.1 Operability Limits

Table 30.1 presents typical criteria to be used for determination of the operability limits for the drilling riser, defined mainly based on API 16Q.

Table 30.1 Criteria for drilling riser operability limits.

Design Parameter	Definition	Drilling Conditions	Non-Drilling Conditions
Low Flex Joint Angle	Mean	1°	NA
	Max	4°	90% of Capacity (9°)
Upper Flex Joint Angle	Mean	2°	NA
	Max	4°	90% of Capacity (9°)
Von Mises Stress	Max	67% of σ_y	80% of σ_y
Casing Bending Moment	Max	80% of σ_y	80% of σ_y

In general, DNV F2 curve is used for the weld joints and DNV B curve for the riser connectors (coupling). Two stress concentration factors (SCFs) are used in fatigue analysis, one is 1.2 for the girth welds, and the other is roughly 2.0 for riser connectors, depending on the type of risers. In recent years, fatigue qualification testing is performed to determine the actual S-N curve data. An Engineering Criticality Assessment (ECA) analysis is conducted to derive defect acceptance criteria for inspection.

For the drilling riser, the safety factor on fatigue life is 3 because the drilling joints can be inspected. The fatigue calculations are to account for all relevant load effects, including wave, VIV and installation induced fatigue. In some part(s) like the first joints nearest to the LFJ joint, the fatigue life could be less, in which case the fatigue life will determine the inspection interval.

30.4.2 Component Capacities

For strength check, various component capacities need to be defined such as:

- Wellhead connector;
- LMRP connector;
- Lower flexjoint;
- Riser coupling and main pipe;
- Peripheral lines;
- Telescopic joint;
- Tensioner/Ring;
- Active heave drawworks;
- Hard hang-off joint;
- Soft hang-off joint;
- Spider-gimbal;
- Riser running tool.

30.5 Drilling Riser Analysis Model

30.5.1 Drilling Riser Stack-up Model

A typical drilling riser stack-up is schematically shown in Figure 30.2. The weight in air and seawater for the telescopic joint, flex joints, LMRP and BOP need to be defined for riser analysis.

The submerged weight and dimensions (length x width x height) for the production trees, manifolds and jumpers are required for the dual activity interference analysis. The properties of the auxiliary rig drill pipe and wire rope will also be used in the interference analysis.

The recommended tensioner forces used in the analysis are calculated based on the mud weight. In the analysis, the drilling mud densities are typically assumed to be 8.0 ppg, 12.0ppg, and 16.0ppg, etc. The maximum allowable bending moment in the casing may be determined assuming the allowable stress is 80% of yield strength.

The hydrodynamic coefficients to be used in the analysis include the normal drag coefficient and the associated drag diameters for the bare and buoyancy joints. The tangential drag coefficient may be taken from API RP 2RD, section 6.3.4.1, equation 31. For the LMRP and the BOP, the vertical and horizontal drag areas and coefficients may be provided by suppliers.

The red alarm is typically 60 seconds before disconnect point, and yellow alarm is roughly 90 seconds before red alarm.

30.5.2 Vessel Motion Data

The required vessel motion data include the following:

- The principal dimensions of the vessel;
- The mass and inertia properties at maximum operation draft;
- Reference point locations for RAOs;
- The survival draft RAOs for various wave directions;
- The maximum operating draft RAOs for various wave directions;
- The transit draft RAOs for various wave directions.

In addition, wave drift force quadratic transfer functions for surge, sway, and yaw are required to conduct irregular wave force calculation in a drift-off analysis. Wind and current drag coefficients for the vessel are also required.

30.5.3 Environmental Conditions

Generally angles denote the direction "from which" the element is coming, and they are specified as clockwise from the true North.

Tidal variations will have a negligible effect on the loads acting on deepwater risers and may be negligible in the design.

The environmental conditions include:

- Omni-directional hurricane criteria for the 10-year significant wave height and associated parameters;
- Omni-directional winter storms criteria for 10-, and 1-year return periods;

- The condensed wave scatter diagram for the full population of waves (operational, winter storm, and hurricane);
- Loop/Eddy normalized profiles.
- The 10-, and 1-year loop/eddy current profiles along with the associated wind and wave parameter;
- The bottom current percent exceedance and the normalized bottom current profile;
- The combined Loop/Eddy and Bottom Current normalized current profile as a fraction of the maximum;
- The combined Loop/Eddy and Bottom Current profiles for 10-year eddy + 1-year bottom, and 1-year eddy + 1-year bottom;
- 100-year submerged current, probability of exceedance and profile duration.

Background current is the current that exists in the upper portion of the water column when there is no eddy present. Mean values of the soil undrained shear strength data, submerged unit weight profile, and ε_{50} profiles are used along the soil column to calculate the equivalent stiffness of the soil springs, for analysis of the connected riser.

30.5.4 Cyclic P-y Curves for Soil

The methodology for deriving p-y curve of soft clay for cyclic loading was developed by Matlock (1975). A family of p-y curves will be required to model the conductor casing/soil interaction at various depths below mudline.

30.6 Drilling Riser Analysis Methodology

Some key words for drilling riser design and analysis are shown in Figure 30.7. From structural analysis point of view, a drilling riser is a vertical cable, under the action of currents. The upper boundary condition for the drilling riser cable is rig motions that are influenced by rig design, wave and wind loads. One of the key technical challenges for deepwater drilling riser design is fatigue of VIV due to (surface) loop currents and bottom currents.

Figure 30.8 shows a typical finite element analysis model for C/Wo risers. It illustrates the process of running riser and landing, the riser being connected or disconnected or in hang-off mode.

30.6.1 Running and Retrieve Analysis

The goal of running/retrieval analysis is to identify the limiting current environment that permits this operation. During this operation, the riser could be supported by the hook at 75 ft above the RKB, or it could be hanging in the spider. The critical configuration is the hook support because of greater potential for contact between the joint and the diverter housing. The BOP is on the riser for deployment, and may not be on the riser if the riser is disconnected at the LMRP, in which case only the LMRP is on the riser.

The hook is considered to be a pinned support with only vertical and horizontal displacement restraints. The riser may rotate about the hook under current loading. The limiting criterion is contact between a riser joint and the diverter housing.

Static analysis is used to evaluate the effects of the current drag force. Wave dynamic action on the riser's lateral motions is not considered.

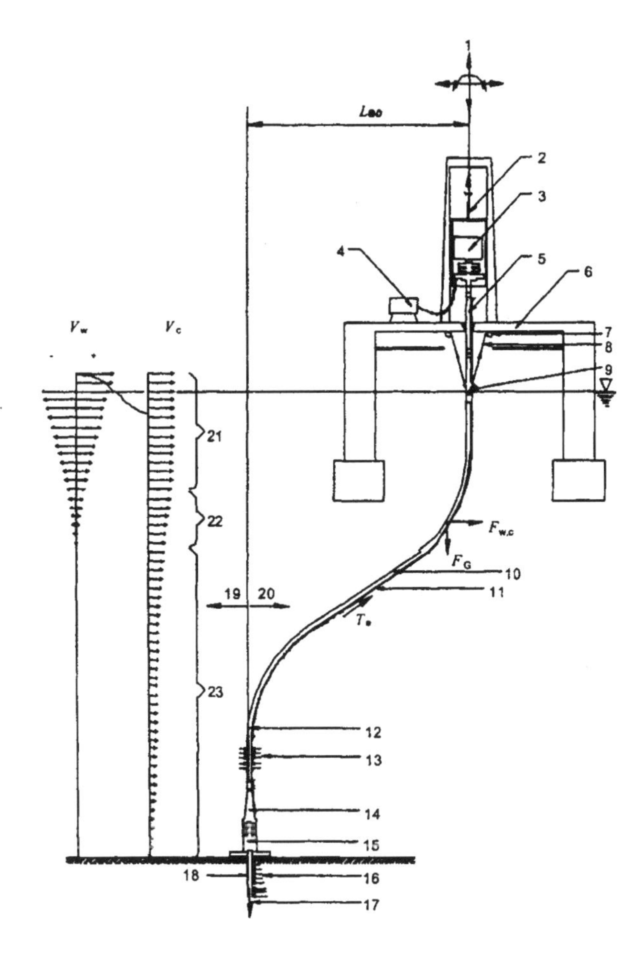

Key

1	Wave motions due to first order wave motions	11	Riser joints	21	Excitation zone
2	Draw works tension and stroke	12	Bending stiffness	22	Shear zone
3	Surface equipment	13	External pressure	23	Damping zone
4	Surface pressure (Choke or mud-pump)	14	Stress joint	$F_{w,c}$	Wave and current forces
5	Slick joint	15	Subsea equipment	F_G	Gravity forces
6	Drill floor	16	Soil restraint	T_e	Effective tension
7	Tensioner sheaves	17	Tool	V_w	Wave velocity
8	Tensioner tension and stroke	18	Conductor bending stiffness	V_c	Current velocity
9	Tensioner joint	19	Upstream	L_{so}	Vessel offset (+)
10	Outside diameter	20	Downstream		

Figure 30.7 Principal parameters involved in C/WO riser design and analysis (ISO 13628-7, 2003 (E)).

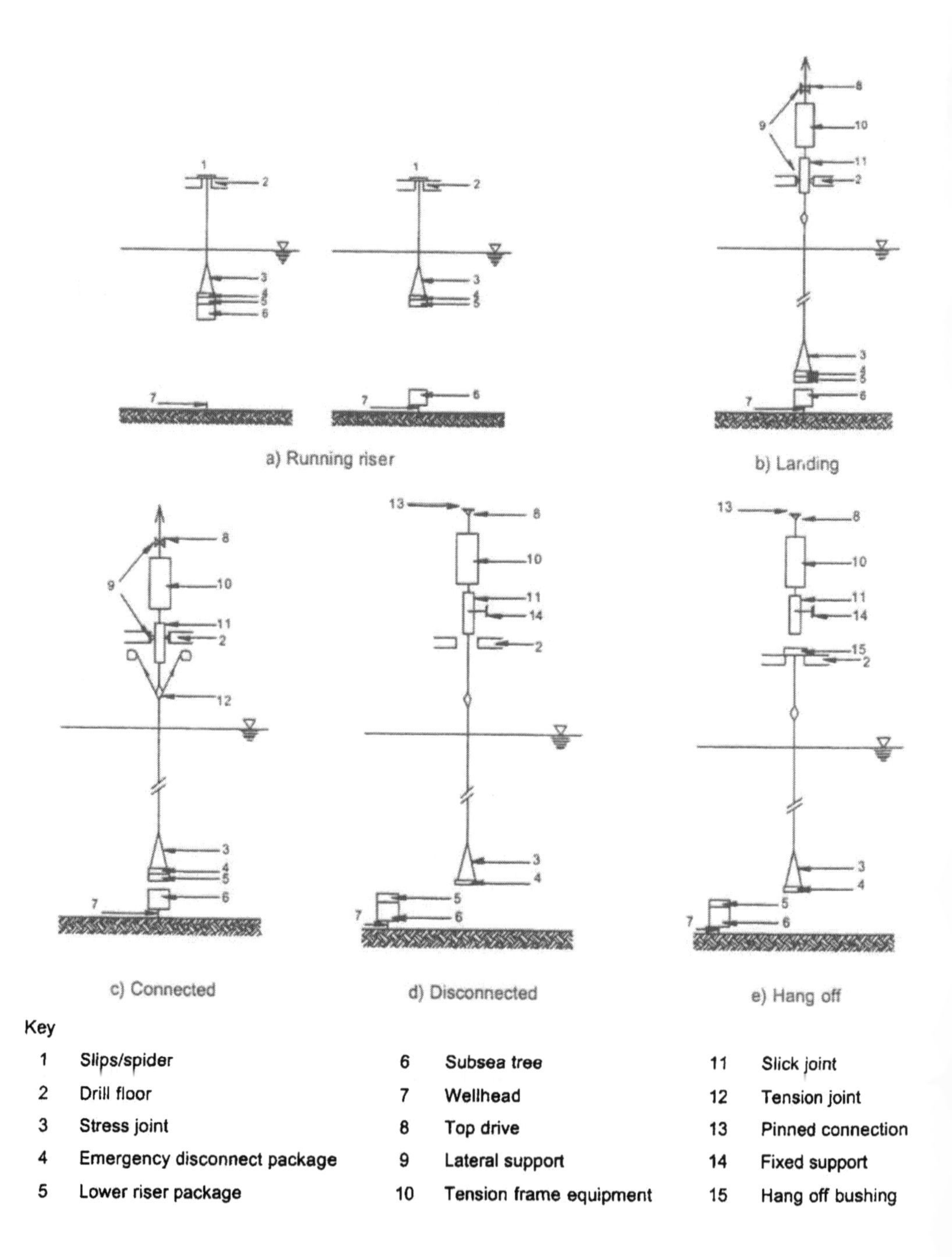

Figure 30.8 Typical finite element analysis model for C/WO risers (ISO 13628-7, 2003 (E)).

30.6.2 Operability Analysis

The objective of the operability analysis is to define the operability envelope, for various mud weights and top tensions, as per the recommendations of API RP 16Q.

The operability envelope for the limiting criteria is computed using both static and dynamic wave analysis. The static analysis involves offsetting the rig upstream and downstream under the action of the current profile to find the limiting up and down offsets at which one limiting criterion is reached. Two current combinations are typically considered: Background + Bottom and Eddy + Bottom. Typically three mud weights are modeled with their respective top tensions.

The procedure for the dynamic analysis is the same as that for the static analysis except that wave loading is added, and the analysis is carried in the time domain, using regular waves based on H_{max}, and for at least 5 wave periods. The dynamic analysis predicts the maximum LFJ and UFJ angles which should be checked against their limiting values.

The limiting conditions for the flex joint angles are typically as follows:

(1) Connected Drilling for dynamic analysis:

- Upper flex joint mean angle < 2 degrees, and 4 degree maximum;
- Lower flex joint mean angle < 1 degree, and 4 degrees maximum

(2) Connected Non-Drilling

- Upper flex joint max angle < 9 degrees;
- Lower flex joint max angle < 9 degree.

Note that the allowed limit for the upper and lower flex joint angles is 1 degree for static analysis of connected drilling risers.

Other limitations on the dynamic riser response are as follows:

- Riser von Mises stresses < 0.67 yield stress for extreme conditions;
- Riser connector strength;
- Tensioner and TJ Stroke limit

Limitations may also arise from loading on the wellhead and conductor system, which are as follows:

- LMRP connector capability;
- BOP flanges or clamps;
- Wellhead connector capacity;

- Conductor Bending Moment (0.8 yield stress).

For drilling, usually it is the mean angles of the LFJ (1 degree) and the UFJ (2 degrees) that determine the envelope. For non-drilling conditions, usually it is the maximum dynamic bending moment in the casing that controls the envelope.

30.6.3 Weak Point Analysis

Weak Point Analysis forms a part of the design process of a drilling riser system. The objective of the Weak Point Analysis is to design and identify the break point of the system under extreme vessel drift-off conditions, should the LMRP fail to disconnect. The riser system should be designed so that the weak point will be above the BOP.

The basic assumption here is that all equipments in the load path are properly designed as per the manufacturer's specifications. The areas of potential weakness in a drilling riser system are typically:

- overloading of the drilling riser;
- overloading of connectors or flanges;
- stroke-out of the tensioner;
- exceedance of top and bottom flex joint limits;
- overloading of the wellhead.

Evaluation criteria for Weak Point Analysis are derived for each potential weak point of the drilling riser system. The weak point criteria determine failure of the system. The evaluation criterion for stroke-out of the tensioners is typically the tensile strength (rupture) of the tensioner lines for each line. The maximum capacity of a padeye will be based on the load that causes yield in each padeye.

The failure load capacity for a flex joint typically corresponds to the maximum bending moment and tension combination that the flex joint can withstand. This typically relates to additional loading following angular lock-out.

The failure load capacity for standard riser joints and conductor joints is typically taken as the maximum combined tension and bending stress that the joint can withstand before exceeding the yield stress of the riser material.

In order to eliminate the uncertainty, a full time-domain weak point computer analysis may be conducted, as follows:

- Perform dynamic regular wave analyses for selected combination of wind, waves, and currents and to establish the dynamic amplification of the loads generated at potential weak points, especially at the wellhead connector and at the LMRP connector.

- Sensitivity analyses are typically performed to determine the effect on the weak point of varying the critical parameters such as mud weight and soil properties.
- Vessel offsets should range from the drilling vessel in the mean position to extreme vessel offsets downstream, as determined by the coupled mooring analysis.
- Following the offset analysis of the drilling riser system, the results will be post-processed to extract the forces and moments generated by the offset position. These are then compared with the corresponding evaluation criteria at the potential weak points along the drilling riser system.

If the weak point is below the BOP, failure would have severe consequences in terms of well integrity, riser integrity or cost. Then further analysis should be conducted to re-locate the weak point to a position with less onerous consequence of failure. One option which might be considered in this context is to re-design the capacity of hydraulic connectors or bolted flanges/bindings in the drilling riser system such that Weak Point occurs at one of those locations.

In a mild environment, slow drift-off generates low static and dynamic moments on the wellhead because of the mild current and the low wave height. In fast drift-off environment, the lower riser straightens-out quickly before wave action magnifies the wellhead connector static moment when the tensioner strokes-out. The suggested critical environment would be a combination of high current to generate a high static moment at the wellhead connector, high waves to cause high dynamic moments, and slow wind to generate slow drift.

30.6.4 Drift-off Analysis

Drift-off analysis is a part of the design process for a drilling riser system on a dynamically positioned (DP) rig. The objective of a drift-off analysis is to determine when to initiate disconnect procedures under extreme environmental conditions or drift-off/drive-off conditions. The analysis is performed for the drilling and the non-drilling operating modes. In each mode, the analysis will identify the maximum downstream location of the vessel under various wind and current speeds and wave height/period.

The first task in a drift-off analysis is to determine the evaluation criteria by which the disconnect point will be identified. These criteria are based on the rated capacities of the equipment in the load path:

- conductor casing –based on 80% of yield;
- stroke-out of the tensioner/telescopic joint;
- top and bottom flex joints limits;
- overloading of the wellhead connector;
- overloading of the LMRP connector;
- stress in the riser joint (0.67 of yield).

Coupled system analysis is used where the soil and casing, wellhead and BOP stack, riser, tensioner, and the vessel are all included in one model. Combinations of environmental actions (wind, current, and waves) are applied to the system, and the dynamic time-domain response is then computed. In this coupled vessel approach, the vessel drift-off (or vessel offset) is an output of the analysis. This approach, which accounts for soil-casing-riser-vessel interaction is more accurate than the uncoupled approach where the vessel offset is computed separately, and then applied to the vessel drift curve to the riser model in a second analysis.

Following the static and dynamic analyses of the drilling riser system, the disconnect point of the system is identified as follows.

- The vessel offset, for the specified environmental load conditions, which generates a stress or load equal to the disconnect criteria of the component is the allowable disconnect offset for that particular component.
- The allowable disconnect offset should be determined for each of the key components along the drilling riser system.
- Then the point of disconnect (POD) corresponds to the smallest allowable disconnect offset for all critical components along the drilling riser system.
- Once the vessel offset at which the riser must be disconnected has been determined, the offset at which the disconnect procedure must be initiated (red limit) will typically be based on 60 seconds. This is the EDS time.
- For non-drilling, the disconnect initiation offset is adjusted by 50 ft before the EDS time. This is the modified red limit for non-drilling.
- For drilling, the disconnect initiation offset is typically 90 seconds before EDS.

30.6.5 VIV Analysis

The objectives of performing VIV analysis of the drilling risers are as follows:

- Predict VIV fatigue damage;
- Identify fatigue critical components;
- Determine the required tensions and the allowable current velocity

Following the modal solution, the results are prepared for input to Shear7. Parameters that remain user defined are as follows:

- Mode cut-off value;
- Structural damping coefficient;
- Strouhal number;
- Single and multi-mode reduced velocity double bandwidth;
- Modeling of the straked riser section with VIV suppression devices

In a VIV analysis of the drilling riser, the vessel is assumed to be in its mean position. The analysis includes:

- Generate mode shapes and modal curvatures for input to VIV analysis using F.E. modal analysis program;
- Model the riser using Shear7 based on the tension distribution determined from initial static analysis;
- Analyze VIV response of the riser for each current profile using Shear7;
- Evaluate the damage due to each current profile;
- Plot the results in terms of VIV fatigue damage along the riser length for each current profile.

30.6.6 Wave Fatigue Analysis

A time domain approach is adopted for motion induced fatigue assessment of the drilling riser. No mean vessel offsets or low frequency motions are considered for motion fatigue analysis of the drilling riser.

The procedure for performing a fatigue analysis is described below:

- Perform an initial mean static analysis;
- Apply relevant fatigue currents statically as a restart analysis;
- Perform dynamic time domain analyses for the full set of load cases, applying the relevant wave data for each analysis;
- Post-process the results from the time domain analyses to estimate fatigue damage of the drilling riser at the critical locations.

30.6.7 Hang-off Analysis

Two hang-off configurations are assumed as follows: a "hard hang-off" in which the telescopic joint is collapsed and locked, thereby forcing the top of the riser to move up and down with the vessel; and a "soft hang-off" in which the riser is supported by the riser tensioners with all APVs (air pressure vessels) open and a crown-mounted compensator (CMC), providing a soft vertical spring connection to the vessel.

Time-domain analysis is conducted using random wave analysis and a simulation time of at least 3 hours simulations. The hard hang-off cases are the 1-year WS, 10-year Winter Storm, and 10-year Hurricane. The soft-hang-off cases are the 10-year WS and the 10-year Hurricane. The goal of the time-domain dynamic analysis is to investigate the feasibility of each mode.

In a hang-off configuration modelling, the riser is disconnected from the BOP, and only the LMRP is on the riser. For the hard hang-off method, only the displacements are fixed. The rotations are determined by the stiffness of the gimbal-spider. The trip saver is at the main deck.

For the soft hang-off method, the riser weight is shared equally by the tensioner and the drawworks. The drawworks have zero stiffness. The tensioner stiffness may be estimated based on the weight of the riser supported by the tensioners and the riser stroking from wave action.

The evaluation criteria for soft and hard hang-off analysis are as follows:

- For soft hang-off, use the stroking limit for the tensioner and slip joint;
- Minimum top tension to remain positive to avoid uplift on the spider;
- Maximum top tension: rating of substructure and the hang-off tool;
- Riser stress limited to 0.67Fy;
- Gimbal angle to prevent stroke-out;
- Maximum riser angle between gimbal and keel to avoid clashing with the vessel.

30.6.8 Dual Operation Interference Analysis

Dual operation interference analysis will evaluate the different scenarios proposed for having the drilling riser in place and connected on the main rig while performing deployment activities on the auxiliary rig. The goal of this analysis is to identify limiting currents and offsets where these activities can take place without causing any clashing between the drilling riser, the suspended equipment on the auxiliary rig or the winch. The distance between the main riser and the auxiliary rig and between the main rig and winch is an important design parameter.

Note that clashing of the main riser with moonpool, vessel hull or bracing will need to be assessed separately prior to finalizing the stack-up model.

A static offset will be applied according to supplied information on the dual operation activity and subsequently another static offset of the vessel due to current loading. Finally the current loading will be added and then the system will be evaluated for minimum distance between the drilling riser, the dual operation equipment, and the vessel.

A drag amplification factor will be applied to the completion riser (off of the auxiliary rig) to account for VIV induced drag. No drag amplification will be added to the drilling riser (off of the main rig) to conservatively estimate its downstream offset due to the current. The auxiliary rig equipment will be considered deployed at 10%, 30%, 60% and 90% of water depth and upstream, while the main drilling riser will always be considered to be downstream and connected.

30.6.9 Contact Wear Analysis

The contact between the drill string and the bore of the subsea equipment may result in wearing of both surfaces due to the rotation and running/pulling of the drill string. The softer bore of the subsea equipment will experience more wear than the drill string, and therefore, it is the subject of this study. The wear volume estimation is based on the work of Archard (1956) and others. The expression for wear is given by:

$$V_{wt} = (K/H)N\,S \quad (30.1)$$

Where,

V_{wt} : total wear volume from both surfaces (in^3);

K : material constant,

H : material hardness in BHN,

N : contact force normal to the surfaces (lbs),

S : sliding distance (in).

This result is based on several hundred experiments which included a wide range of material combinations. The experimental result demonstrates that the wear rate, V_w/S, is independent of the contact area and the rate of rotation or sliding speed, as long as the surface conditions do not change. Such a change can be caused by an appreciable rise in surface temperature. The H value for 80ksi material is 197 BHN. For the flex joint wear ring and wear sleeve, H is 176 BHN.

The normal force, N, is obtained from the contact analysis of the drill string for the load cases. The sliding distance, S, is related to the string RPM as follows:

$$S = \pi d(RPM)t \quad (30.2)$$

Where d is the diameter of the drill pipe/the tool joint, and t is the time in minutes. Substitution of Equation (30.1) into Equation (30.2) and solving for t as a function of V_w gives:

$$t = (H/K)(V_w/N)(\pi dRPM) \quad (30.3)$$

The drilling fluid provides lubrication with reduction in wear by comparison to the dry contact conditions. Therefore, the results of this study, which are based on unlubricated wear, will be conservative. The wear volume, V_w, can be further related to the wear thickness, t_w, by the wear geometry as discussed in the next section.

Since the goal is to find the wear thickness, the wear geometry should be considered. The wear area is the crescent bounded by the bore and the OD of the tool joint or the drill pipe. The following are the possible contact cases:

- Tool joint contact with casing;
- Tool joint contact with BOP-LMRP;
- Tool joint contact with riser joint;
- Tool joint contact with flex joint;
- Drill pipe contact with riser joint;
- Drill pipe contact with flex joint.

For each tension, and each position of the tool joint, the flex joint angle is increased between 0 and 4 degrees at increments of 0.1 degrees. The reaction forces at each increment are reported. As long as the drill string tension is maintained at a given angle which is greater than zero, the wear process will continue under the reaction forces. Because of the large scale of the problem, these reaction forces remain unchanged for wear thicknesses of up to 1". So the question becomes: how much time does it take to wear-out a certain thickness.

The first step in calculating wear is to estimate the drill string tension near the mudline since the contact reaction forces depend on this tension. In order to simplify the wear calculations, a conservative approach is implemented where the reaction forces for each contact location are normalized with respect to the drill string tension for the 5 positions of the tool joint.

A typical wear calculation procedure could be as follows:

1. Determine as input the following:
    - angle of drilling,
    - material Brinell hardness,
    - tension range,
    - RPM
2. Calculate the normal force from the reaction envelopes from the tension;
3. Obtain the sliding distance S From the $t_{w'} - V_{w'}$ values;
4. Obtain the time in minutes, t, for each $t_{w'}$.

30.6.10 Recoil Analysis

The objectives of conducting recoil analysis are to determine recoil system settings and vessel position requirements which ensure that during disconnect the following are achieved:

- The LMRP connector does not snag;
- The LMRP risers clear of the BOP;
- The riser rises in a controlled manner

Recoil analysis is not required for every specific application if the vessel has an automatic recoil system. The criteria to be considered at each stage of recoil are as follows:

- Disconnect – the angle of the LMRP as it leaves the BOP should not exceed the allowable departure angle of the connector. This may limit the possibility of reducing tension prior to disconnect.
- Clearance – the LMRP should rise quickly enough to avoid clashing with the BOP as the vessel heaves downwards.
- Speed – the riser should not rise so fast that the slip joint reaches maximum stroke at high speed.

Requirements for modeling the riser during recoil are the same as those needed for hang-off. In addition, it must be possible to account for the nonlinear and velocity dependent characteristics of the tensioner system. Time-domain riser analysis program may be used alone, or in conjunction with spreadsheet calculations from which tensioner characteristics are derived. The analysis sequence is as follows:

- Conduct analysis of the connect riser;
- Release the base of the LMRP, to reflect unlatching, and analyze the subsequent response for a short period of time;
- Change tensioner response characteristics to simulate valve opening or closure and analyze subsequent riser response for a number of wave cycles.

The analysis is repeated to determine the necessary time delay between operations. The upstroke of the riser must be monitored to detect whether top-out occurs and at what speed. If the riser is to be allowed to stroke on the slip-joint during hang-off, vertical oscillation of the riser following disconnect must also be monitored to ensure clashing with the BOP does not occur.

30.7 References

1. AMJIG (2000), "Deepwater Drilling Riser Integrity Management Guidelines," Revision 2, March 2000.
2. API Specification 5CT (1998), "Specification for Casing and Tubing," 6th Edition, October 1998.
3. Archard, J. F. (1953) "Contact and Rubbing of Flat Surfaces," *Journal of Applied Physics*, vol. 24, No. 8, pg 981.
4. J. F. Archard and W. Hirst (1956), "The Wear of Metals under Unlubricated Conditions," Proceedings of The Royal Society of London, Series A, Vol. 236, pg397, April 1956.
5. ASTM Standard E140-02, "Standard Hardness Conversion Tables for Metals Relationship among Brinell Hardness, Vickers Hardness, Rockwell Hardness, Superficial Hardness, Knoop Hardness, and Scleroscope Hardness".

6. API RP 16Q (1993), "Recommended Practice for Design, Selection, Operation and Maintenance of Marine Drilling Riser System," 1st Edition, November 1993.
7. API RP 2A-WSD (2000)," Recommended Practice for Planning, Designing and Constructing Fixed Offshore Platforms - Working Stress Design," Edition: 21st, December 2000.
8. API RP 1111 (1999), "Design, Construction, Operation, and Maintenance of Offshore Hydrocarbon Pipelines (Limit State Design)," 3rd Edition 1999.
9. Bharat Bhushan (1999), "Principles and Applications of Tribology," Wiley-Interscience Publishing, 1999.
10. E.R Booster (1985), "Handbook of Lubrication, Theory and Practice of Tribology," Vol. II, p201, 1985.
11. DnV Classification Notes (1984), "Fatigue Strength Analysis for Mobile Offshore Units," Note No. 30.2, August 1984.
12. Geiger, P.R. and Norton, C.V. (1995): "Offshore Vessels and Their Unique Applications for the Systems Designer", Marine Technology, Vol. 32(1), pp.43-76.
13. Matlock, H. (1975), "Correlations for Design of Laterally Load Piles in Soft Clay," paper OTC2312 presented at the Seventh Offshore Technology Conference, Houston, Texas, 1975.
14. MCS (2001), "DEEPRISER Reference Manual," Version 1.3, 2001.
15. Vandiver, K. and Lee, L., (2001), "User Guide for Shear7 Version 4.1,", Massachusetts Institute of Technology, March 25, 2001.
16. World Oil (2002/03): "The Composite Catalog ® of Oilfield Equipment & Services", the 45th Edition.

Part IV

Riser Engineering

Chapter 31 Integrity Management of Flexibles and Umbilicals

31.1 Introduction

This Chapter deals mainly with risk assessment and integrity management of flexible pipes and steel tube umbilicals. A recognized methodology for formulating an integrity management plan involves carrying out a risk assessment and determining the risks inherent for the flexible pipe. Once risks are determined, specific integrity management measures can be identified to mitigate these risks.

Figure 31.1 shows field layout for two 6" flexible pipes and umbilicals in the Marin field, connecting the Marlin subsea development to the Marlin TLP one mile away, in 3400 ft of water. The increased use of flexible pipes in deepwater and high pressure / high temperature applications demands operators to adopt an integrity management program in order to safeguard their assets and to avoid reduction of production. The integrity management programs may prolong the operating life through the use of preventive maintenance procedures. The data recording of in-service conditions may be used in re-assessments of flexible pipe. One must also keep in mind that the cost of a failure is many times over the cost of the implementation of an integrity management program.

Flexible pipe integrity management programs are well established in the industry, and definite methodologies for integrity management exist. One of the main standards for flexible pipe integrity management has been developed by MCS and details of this methodology may be found in various publications, e.g. UKOOA (2001 & 2002).

A high risk generally requires the implementation of some form of predictive inspection or monitoring measure. A medium risk generally requires a detective inspection or monitoring measure in order to detect any signs of initiation of failure due to the particular failure mode, or to ascertain that no significant defect leading to this particular failure is present in the pipe. Low risk levels do not require any quantitative inspection or monitoring procedures or any specific integrity management procedures.

In this Chapter, failure statistics are discussed in Section 31.2 and risk management methodology is presented in Section 31.2. Section 31.3 provides a review of failure drivers and failure modes for a flexible pipe. Sections 31.4 – 31.7 illustrate a number of industry-standard integrity management measures that can be applied to manage risks for flexible pipes.

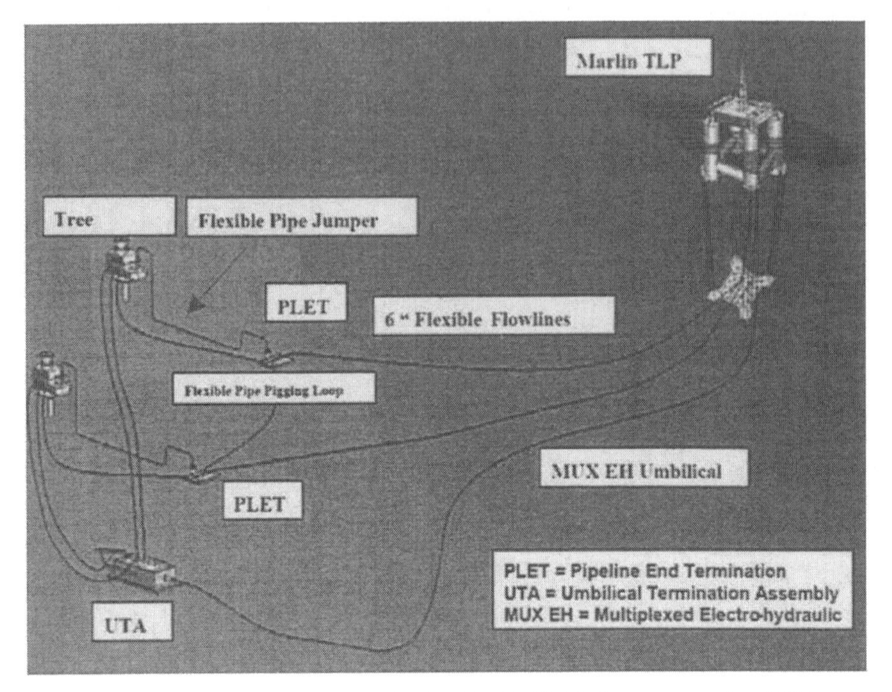

Figure 31.1 Flexible flowlines and umbilical in the Marlin Field (Lecomte et al., 2002).

31.2 Failure Statistics

It is a useful exercise to determine actual failure mode statistics from operational experience. Figure 31.2 illustrates damage and failure mechanisms that have been correlated by UKOOA (2001). From a total of 106 flexible pipe failure / damage incidents reported by UKCS operators (not including flooded annuli), it is found that 20% of flexible pipes have experienced some form of damage or failure. In this context, failure / damage incidents are defined as follows:

- Failure – an incident resulting in the loss of containment of the flexible pipe and requiring the pipe to be replaced.
- Damage – an incident resulting in damage to the flexible pipe where the flexible pipe can remain in service following mitigation measures taking place.

Of this 20%, two-thirds have occurred during installation and one-third during normal operation. Of these 106 failure / damage incidents, there were a total of 32 incidents that required the flexible pipe to be replaced. Tables 4.1 and 4.2 of UKOOA (2002) show the riser operational failure/damage incidents and riser installation & commissioning failure/damage incidents, respectively.

Figure 31.3 obtained from PARLOC (2001) shows a comparison between steel and flexible pipe statistics.

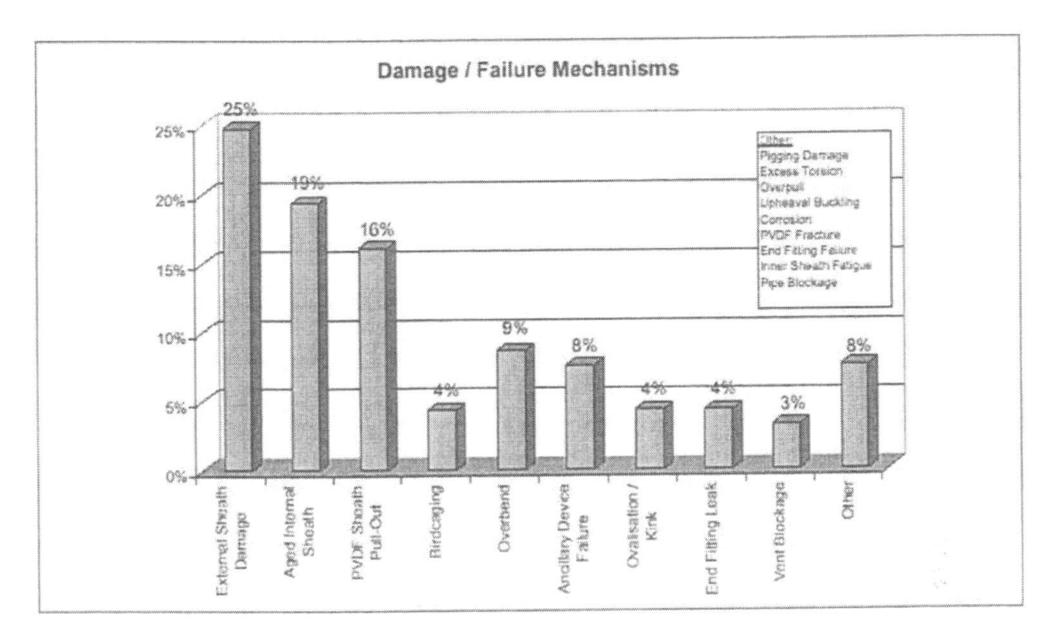

Figure 31.2 System failure mechanisms (UKOOA, 2001).

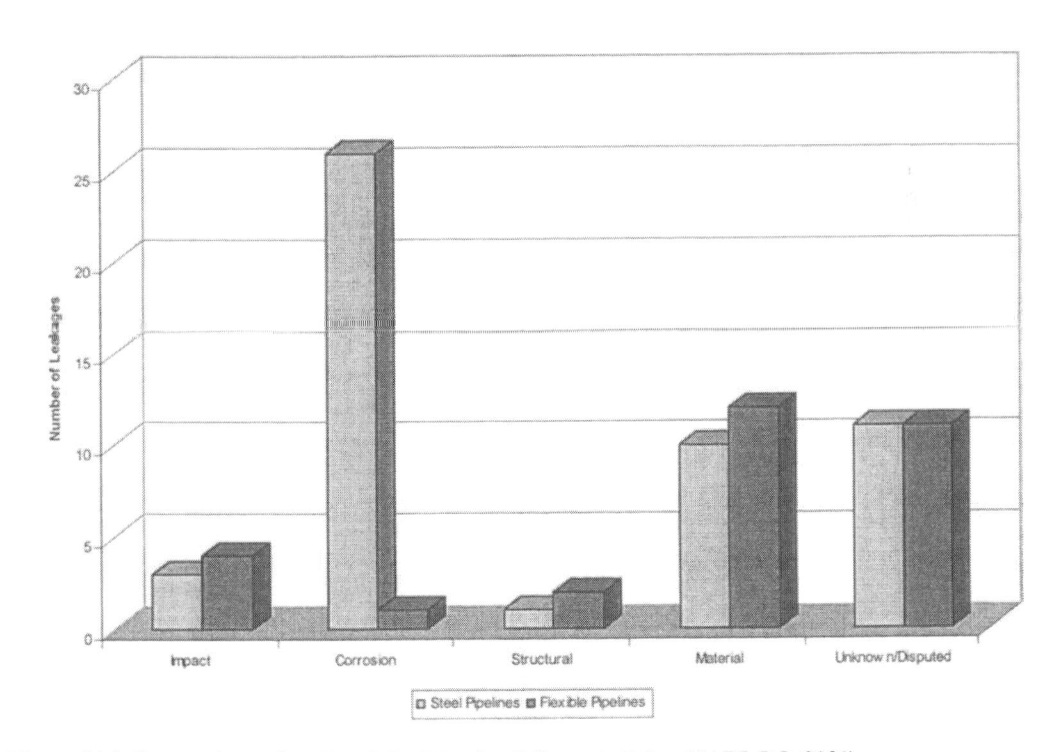

Figure 31.3 Comparison of steel and flexible pipe failure statistics (PARLOC, 2001).

31.3 Risk Management Methodology

Risk is often quantified as an integer that is the product of two other values, which are known as a Probability of Failure and a Consequence Rating. A risk management shall take all possible failure modes into consideration, through an analysis of failure drivers (such as temperature, pressure, product fluid composition, service loads, pipe blockage or flow restriction) and general failure modes (such as fatigue, corrosion, erosion, accidental damage, ancillary equipment).

The Probability of Failure is a scale that determines the probability of a failure mode occurring. It is estimated through in-service data and statistics of past failure occurrences. The Consequence Rating is an integer that describes the severity of a failure occurring due to a particular failure mode. For example, the rupture of a hydrocarbon-carrying pipe next to a platform would be considered to have a high consequence rating due to the potential for pollution, and danger to personnel and the platform. The rupture of an umbilical at a subsea manifold, though still a failure that could be costly and lead to the loss of production, would not have a high consequence rating due to a lack of pollution and danger to personnel or other equipment.

Examples of Probability of Failure Ratings, Consequence Ratings, and a combined risk matrix are:

- Risk assessment and Integrity Management Strategy, see Picksley et al. (2002).
- Evaluation of Probability of Occurrence Rating [POR] for Pipe Failure, see Picksley et al. (2002).
- Consequence ratings in terms of health and safety, environmental damage, operability, loss and disruption, see Picksley et al (2002).
- Strategic Levels of Inspection and Monitoring based on probability (level 1 to 5) and Consequence (level 1 thru 5), see Picksley et al. (2002).

31.4 Failure Drivers

31.4.1 Temperature

The most likely failure mode with temperature as a failure driver is cracking or ageing of the internal polymer sheath. Polyamide PA-11 (Rilsan) is a workhorse polymer within the flexible pipe industry, performing many important functions in inner sheath, outer sheath and other applications. According to API RP 17B, the maximum allowable temperature of PA-11 is 90°C if the water cut is 0 percent and 65°C if any water is present. Repeated temperature cycling will also cause increased stress in the polymer material. More detailed guidance on the use and limitations of PA-11 are given by the Rilsan Use Group in API TR17RUG (Groves et al., 2002).

Figure 31.4 shows a temperature management system adopted for HPHT Kristin field development.

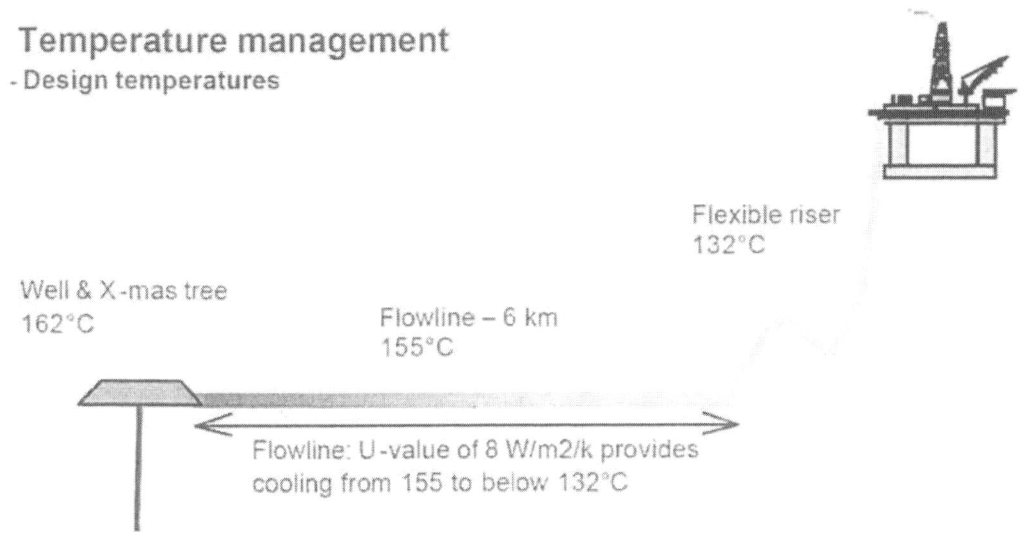

Figure 31.4 Temperature management system on HPHT Kristin field (Hundseid et al., 2004).

31.4.2 Pressure

Excessive pressure in the pipe bore will cause failure of the pressure armor wires. However pressure is usually carefully monitored and controlled and unless it exceeds the design limits, failure due to pressure is not considered a high risk. Thermal and pressure cycling could however play a significant role in fatigue damage of the pipe, and could also cause significant upheaval buckling in a buried flexible flowline. The risk of pressure surges always exists in case of well shut-down, and these pressure surges need to be carefully monitored and assessed to ensure they do not cause an unexpected failure of the pressure armor wires.

Hydrostatic collapse could also be a potential failure mode for pipes. However the pipes are designed with a factor of safety so it is unlikely for pipes to suffer hydrostatic collapse, especially if the external sheath remains intact. Should the external sheath be breached, water could enter the pipe annulus, and the pressure in the annulus could cause the internal polymer sheath to collapse. The presence of a carcass provides much more safety against this failure mode.

Another failure mode related to pressure is caused by gases diffusing from the pipe bore through the internal polymer sheath and into the annulus. The gases cause a pressure build-up in the annulus. In the case of rapid depressurization and evacuation of the pipe bore (for example through an emergency shut down), the pressure in the annulus could exceed the pressure in the pipe bore. This could lead to internal pressure sheath collapse in pipes with no carcass.

Figure 31.4 shows a pressure protection system for the HPHT Kristin field.

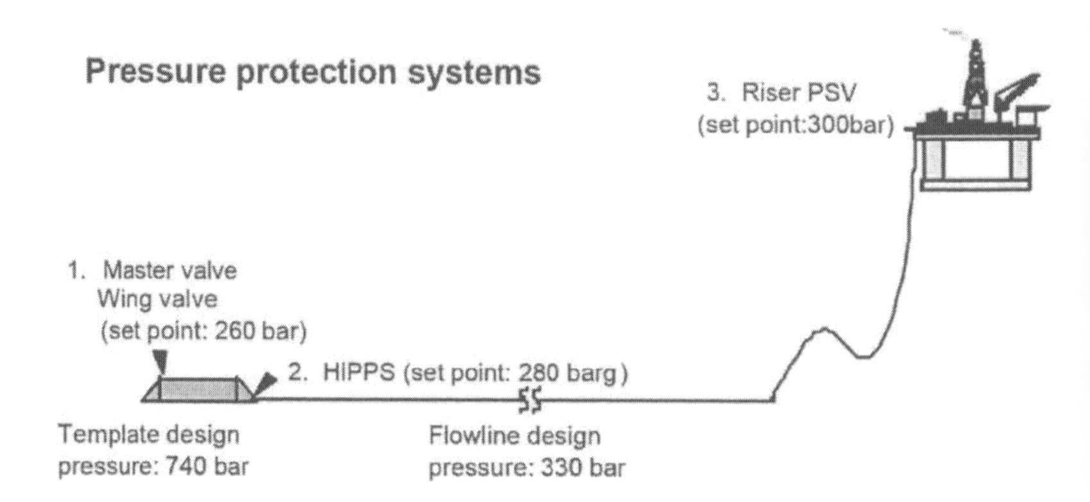

Figure 31.5 Pressure management system on HPHT Kristin field (Hundseid et al., 2004).

31.4.3 Product Fluid Composition

The products carried in the pipe bore could have a deleterious effect on the pipe wall materials. The presence of water, CO_2 and H_2S causes general, pitting, hydrogen induced cracking (HIC) and sulfur stress cracking (SSC) corrosion. These chemicals do not only attack the carcass, but they also diffuse through the internal polymer sheath into the annulus and can cause corrosion of the pressure and tensile armor steel wires. The diffusion of water into the annulus also has a negative effect on the fatigue life of the steel armor wires.

The products transported in the pipe bore (including inhibitors and acids) could cause the accelerated degradation of the polymer material of the internal sheath. The presence of sand could also cause erosion of the carcass and internal polymer sheath, particularly in curved sections of the pipe.

31.4.4 Service Loads

Excessive tension in a flexible riser could cause tensile armor wire failure, leading to collapse of the pipe. The loads experienced during installation are often the highest loads that a riser sustains during its lifetime.

Another mode of failure is over-bending of the pipe. This causes unlocking of the pressure armor layer and leads to collapse of the pipe wall. Bend stiffeners, bellmouths and bend restrictors are used to protect a pipe from over-bending in its most critical regions.

A failure mode that can occur at the touch down point of a riser is compression of the riser. This occurs when the effective tension in the pipe wall becomes negative. A high value of compression can result in buckling of the pipe wall.

31.4.5 Ancillary Components

Ancillary components such as bend stiffeners, bend restrictors, buoyancy modules and hold down tethers are all designed to support the flexible pipe configuration and prevent severe loads or over-bending of the pipe. A failure to any ancillary component could easily be followed by failure of the pipe itself, since the protection provided by the ancillary component would be lost. A loss or damage of an ancillary component could be due to a number of reasons, such as impact damage with another object or corrosion.

31.5 Failure Modes

31.5.1 Fatigue

Wave and riser motions will cause fatigue damage to the steel armor wires. A detailed fatigue life analysis is always carried out before pipe installation and the pipe manufacturer needs to prove that the fatigue life of the pipe is 10 times the pipe service life. Due to a high damping factor, flexible risers are not susceptible to vortex-induced vibrations (VIV). However, manufacturing and installation are also likely to introduce a certain amount of fatigue damage.

The presence of water in the annulus will accelerate fatigue of the steel armor wires. Small amounts of water will diffuse from the pipe bore through the internal polymer sheath into the annulus. A more serious form of corrosion fatigue damage could take place in case of a tear in the external sheath of a flexible riser (as shown in Figure 31.6). Such a hole would permit the ingress of seawater into the annulus, which would cause an accelerated rate of corrosion and a reduction in the fatigue life of the steel armor wires.

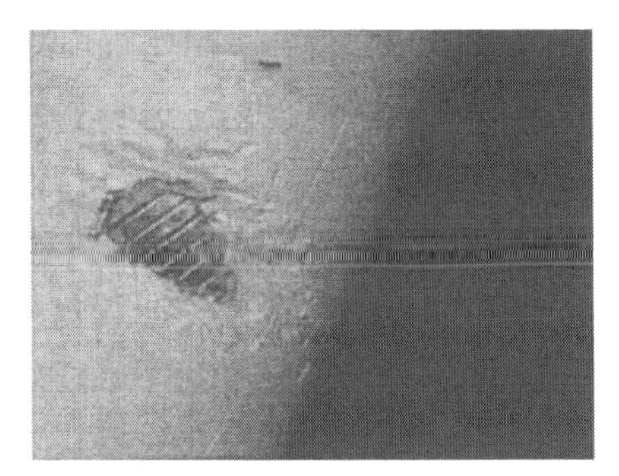

Figure 31.6 Hole in outer sheath – example of installation damage to external sheath (Picksley et al., 2002).

31.5.2 Corrosion

The steel carcass and steel armor wires of a flexible pipe are susceptible to corrosion due to the presence of water, CO_2 and H_2S.

31.5.3 Erosion

Carcass or internal polymer sheath erosion takes place when sand particles impact against the material. This failure mode is relevant for pipe sections that are curved.

31.5.4 Pipe Blockage or Flow Restriction

Hydrocarbon-carrying pipes are susceptible to the formation of wax or hydrate deposits. This normally occurs in cases of low temperature in the pipe bore. The formation of wax or hydrate deposits could cause a blockage in the pipe, restrict the flow of fluid, and lead to a pressure increase in the pipe bore. If left untreated, this could eventually cause rupture of the pressure armor layer and collapse of the pipe.

31.5.5 Accidental Damage

Pipes are susceptible to impacts due to dropped objects from vessels, anchors, interference with other pipes, and trawl boards. Excessive motions of a pipe could also cause impacts on the seabed. If these impacts are severe, they could cause damage to the external sheath of the pipe and ingress of water into the pipe annulus. An extremely severe impact to a pipe could also cause unlocking of the pressure armor layer or rupture of the tensile armor wires.

31.6 Integrity Management Strategy

This Section explains a number of integrity management procedures that are used to safeguard the integrity of flexible pipes. A complete integrity management strategy also includes a number of inspection procedures, monitoring measures, and testing and analysis methods that are listed in Sections 31.7 to 31.9 respectively. It is important that a risk assessment and an integrity management strategy are defined in the design stage of a flexible pipe, to ensure that the pipe does not suffer any unnecessary damage during all stages of its life, including manufacturing, qualification testing and installation.

31.6.1 Flexible Pipe Integrity Management System

Figure 31.7 shows an integrity management system applied on the Asgard field in Norway. The integrity management system establishes and maintains a database for design data and field operation data, including:

- Design basis and main design findings;
- Manufacturing data relevant for re-assessment of risers;
- Operating temperature and pressure with focus on temperature fluctuations;
- Fluid compositions, injected chemicals and sand production;
- Riser annulus monitoring, polymer coupons test results;
- Seastate conditions, vessel motions and riser response.

Figure 31.8 shows the management of operational data, i.e., handling of the data between the host vessel, operator's onshore team and the flexible pipe integrity management system. The database "EPVIEW" is developed to contain signals from the instruments related to operation of the marine systems and risers, including data from pressure and temperature transmitters and annulus vent rate monitoring.

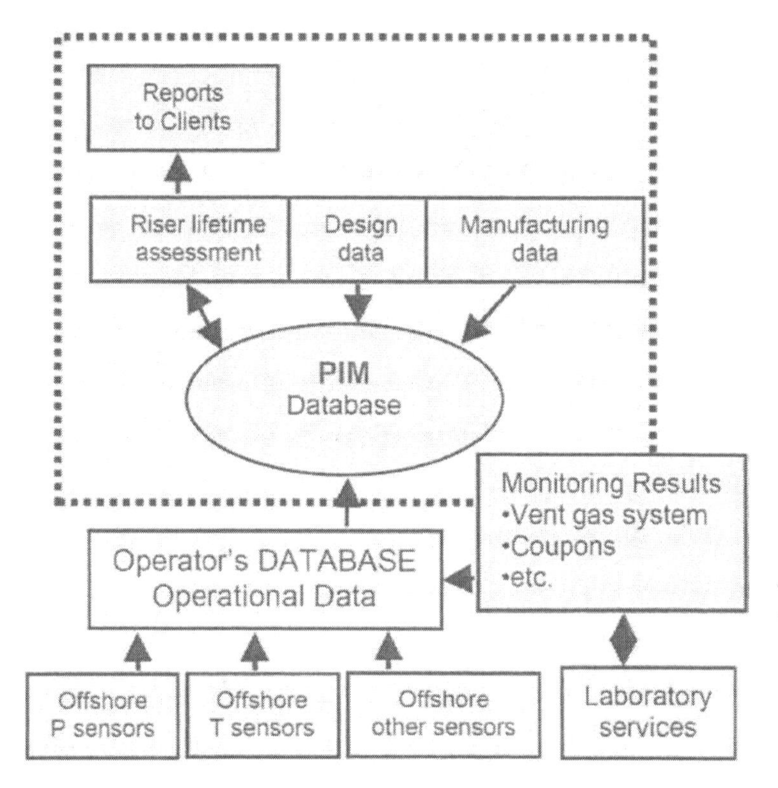

Figure 31.7 Flexible pipe integrity management system (Benit et al., 2003).

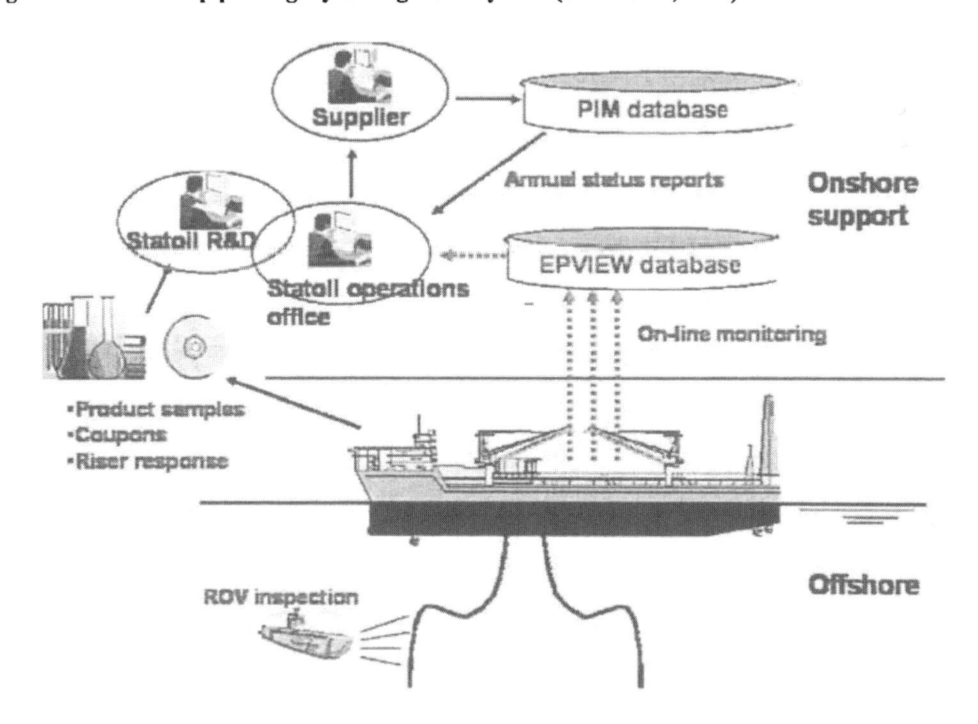

Figure 31.8 Flexible pipe integrity management system (Benit et al., 2003).

31.6.2 Installation Procedures

Installation has the potential of damaging the flexible pipe through over-bending, excessive tension loading, and clashing that might damage the external sheath. Hence strict installation procedures need to be adhered to in order to prevent damage to the flexible pipes before operational use. A number of installation incidents have been documented, and in some cases expensive mitigation measures had to be taken following these incidents to enable the pipes to remain fit for service.

31.6.3 Gas Diffusion Calculations

These calculations are required to predict the gas composition of the pipe annulus. This composition has an effect on the use of 'sweet' or 'sour' service wires in the annulus, the rate of corrosion of these wires, and also their fatigue life.

31.6.4 Dropped Object Reporting /Deck Lifting & Handling Procedures

Dropped objects have the potential of hitting a flexible pipe and causing damage such as a hole in the external sheath, or even pressure armor wire unlocking or tensile armor wire rupture. Hence a strict Dropped Object Reporting Protocol is required for any installation or vessel operating in the vicinity of flexible pipes. A dropped object must always be reported and an ROV deployed to determine the location of the dropped object on the seabed and examine the surrounding pipes for any potential damage. Deck Lifting and Handling Procedures are intended to prevent dropped object incidents from occurring in the first place.

31.6.5 Vessel Exclusion Zone

The fewer vessels operating in the vicinity of flexible pipes means the lower is the risk of a dropped object incident. Fishing vessels have also been documented to cause damage to pipes due to impacts between trawl boards and the pipes. Hence a vessel exclusion zone extending beyond the touch-down regions of all the risers is usually enforced.

31.6.6 Fatigue Life Re-analysis of Pipes

Fatigue life re-analysis should be undertaken at regular intervals during a pipe's lifetime to ensure that the fatigue life has not deteriorated significantly from design assumptions.

31.6.7 High Integrity Pressure Protection System (HIPPS)

This system is recommended for flexible pipes that could be prone to over-pressure, pressure surges, or uncertainty in the pressure profile throughout their lifetime. This system includes a series of sensors and monitors for automatic shut-down and reduction of pressure in the pipes to prevent any damage from occurring.

31.7 Inspection Measures

31.7.1 General Visual Inspection (GVI) / Close Visual Inspection (CVI)

This measure is one of the most important and commonly-used measures for ensuring the integrity of flexible pipes. Visual inspection is used both as a topsides measure and also at subsea. As a measure at topside, the end-termination and vent valves of the flexible risers are inspected. Bend stiffeners are inspected for any evidence of damage, and the general condition of the hang-off basket is documented.

Subsea inspection has the aim of determining any damage or anomalies to the flexible pipes such as buckling, kinking, and holes or tears in the external sheath. The touch-down zone should be examined thoroughly to determine the extent of trenching and any evidence of damage or pipe configuration outside the design limits. Evidence of clashing with other risers or other objects, including the seabed, should also be documented. Excessive marine growth should be controlled and cleaned if necessary. Subsea ancillary equipment should similarly be examined for any signs of damage. An important point to remember is that small holes in the external sheath of operational flexible pipes are difficult to detect by visual inspection, and hence other testing and inspection measures are employed to complement visual inspection and lessen the risk of damage being undetected.

31.7.2 Cathodic Protection Survey

This measure is normally employed as part of a general visual inspection. The condition of the anodes should be examined at regular intervals to ensure that adequate cathodic protection is available at all times for the flexible pipes and other ancillary equipment.

31.8 Monitoring

31.8.1 Inspection and Monitoring Systems

A typical inspection and monitoring system includes (Benit et al., 2003):

- Polymer monitoring: online, offline, topside and subsea;
- Annulus monitoring: vent gas rate, annulus integrity;
- Riser dynamics: tension, angle and curvature;
- Steel armour: inspection method, magnetic or radiograph;
- Use of existing sensors, pressure and temperature sensotrs.

31.8.2 Bore Fluid Parameter Monitoring

Various bore fluid parameters should be continuously monitored because changes in these parameters could exceed certain design limits and cause damage to the flexible pipe. The parameters that require monitoring are temperature, pressure, volume flow-rate and the bore fluid composition.

Temperature is monitored to ensure it does not exceed the internal polymer temperature limit and cause accelerated ageing of the internal sheath. Temperature cycling could also cause ratcheting of pipes, and temperature variations could cause a different gas diffusion rate from the pipe bore to the annulus, and hence alter the corrosion and fatigue life of the annulus armor wires.

Pressure is monitored because excessive pressure or pressure surges could cause damage to the pressure armor layer. Rapid depressurization and evacuation of the pipe bore could cause pipe collapse. Pressure variations could have an effect on the fatigue life of the pipe, and cause ratcheting or buckling of the pipe. The volume flow-rate is also monitored to give an indication of a depressurization event, and also to ensure that erosion calculations performed during the design stage remain valid throughout the lifetime of the pipe.

The bore fluid parameters are monitored to ensure that gas diffusion calculations are based on valid data.

31.9 Testing and Analysis Measures

31.9.1 Coupon Sampling and Analysis

This testing method involves the use of coupon samples that are ideally obtained from the same polymer extrusion run as the internal polymer sheath of the flexible pipe. The coupons are placed in a holder in-line with the transported bore fluid. The coupons are retrieved at regular intervals, and are tested to ensure that they do not suffer a worse rate of degradation than has been calculated for the internal polymer sheath during design calculations.

A more reliable method for determining the degradation of the internal polymer sheath is through the use of Frequency Dependent Electromagnetic Sensing (FDEMS). FDEMS involves keeping the coupon samples in their holder and having an electromagnetic online system for monitoring the degradation rate of the coupons. This method is more expensive to implement than coupon sampling and analysis, and would normally only be justified if the rate of degradation of the internal polymer sheath is estimated to be high and the risk rating for internal pressure sheath failure is high.

31.9.2 Vacuum Testing of Riser Annulus

Vacuum testing of a riser annulus is the most reliable method in the industry to determine the presence of water in the riser annulus.

Vacuum testing normally involves two separate procedures. Initially a vacuum is drawn from the vent valves at the end termination of the riser. If the vacuum is stabilized, this is usually a good indication that there is no breach of the outer sheath. However, a further procedure is carried out following the drawing of a vacuum. This next procedure involves the injection of a known volume of nitrogen gas, which creates an inert atmosphere inside the annulus and hence has a protective effect on the steel armor wires. The volume of nitrogen gas injected into the annulus is carefully measured and since the free volume of the annulus would be known from pre-installation tests, the free volume of the annulus at the time of the test can be examined.

31.9.3 Radiography

Radiography is used to monitor the condition of a PVDF internal sheath at the end termination.

31.10 Steel Tube Umbilical Risk Analysis and Integrity Management

31.10.1 Risk Assessment

Steel Tube Umbilical Risk Analysis is undertaken using the same methodology as for flexible pipes. A number of failure modes are evaluated and each failure mode is given a risk rating, which depends on a Probability of Failure and a Consequence Rating. A brief overview of the major failure modes for a steel tube umbilical is presented here:

Fatigue: A detailed fatigue analysis is required for a steel tube umbilical during its design phase. This fatigue analysis has to consider the fatigue damage caused during manufacture, installation and in-place wave fatigue damage. It is also to be noted that umbilicals are susceptible to VIV, and significant fatigue damage can result from this. The hang-off angle of the umbilical is a parameter that often significantly affects the occurrence of VIV. Strakes and fairings can be used to mitigate VIV in an umbilical.

Corrosion: Due to the fact that the spaces between the steel tubes in an umbilical are often flooded with seawater, the steel material throughout the umbilical needs to be corrosion-resistant to seawater.

Service Loads: Wave and current induced motions are liable to introduce high stress in the umbilical, particularly at the touch down zone and the hang-off region. Bend stiffeners are often employed to reduce the bending of the umbilical at the hang-off region. High stress in the touch-down region can be reduced by installing the umbilical with buoyancy modules, rather than in a free-hanging configuration. Compression at the touch-down point is also avoided as much as possible. If the umbilical motions remain within design limits, tube rupture due to excessively high service loads is not a high risk.

Accidental Damage: Umbilical impact with the seabed, other pipes, dropped objects or trawl boards can result in steel tube rupture and failure of the umbilical.

Ancillary Components: A bend stiffener is often employed to restrict umbilical motions in the hang-off area. Buoyancy modules, strakes and fairings, and other equipment such as hold-down tethers could also be used to protect the umbilical from excessive bending and stress. The loss of any of this equipment would mean that the protection afforded to the umbilical by these components would also be lost. Hence the loss of an ancillary component, if not immediately detected, could lead to tube rupture and total failure of the umbilical.

31.10.2 Integrity Management Strategy

This Section provides a brief overview of integrity management procedures that could be used to safeguard the integrity of steel tube umbilicals throughout their lifetime.

Fabrication and Installation Procedures: Significant fatigue damage is liable to be induced in a steel tube umbilical during fabrication and installation. Over-bending of the tubes and impact damage are the two most common forms of damage that could result during these stages. Hence procedures should be in place to reduce the risk of this form of damage, and these procedures need to be strictly adhered to.

Dropped Object Reporting / Deck Lifting and Handling Procedures / Vessel Exclusion Zone: As already discussed in Section 31.4, these procedures reduce the risk of dropped object and trawlboard impacts on the umbilical. Such impacts are liable to cause significant damage to the steel tubes.

Fatigue Life Re-Analysis: Fatigue life re-analysis should be undertaken at regular intervals throughout the lifetime of the umbilical.

General Visual Inspection: This measure is one of the most important integrity management procedures that could be undertaken to safeguard the integrity of a steel tube umbilical. A visual inspection does not detect any damage to the umbilical before this occurs. However it could be very useful in detecting any anomalies that could worsen and cause a total failure of the pipe. Hence a visual inspection should concentrate on detecting anomalies such as excessive marine growth, buckling of the pipe, any signs of damage to the steel tubes, over-bending, and any signs of an impact to the umbilical.

Umbilical Motion Monitoring: Vessel excursions, weather conditions (including currents) and the umbilical motions should be monitored throughout the lifetime of the umbilical. Through this monitoring, any motions that might exceed design limits could be detected and further investigation carried out to ensure that no damage occurs to the umbilical.

31.11 References

1. Binet, E. Tuset, P. and Mjoen, S. (2003), 'Monitoring of Offshore Pipe', OTC 15163, Offshore Technology Conference.
2. Energy Institute, London (2003), 'PARLOC 2001: The Update of Loss of Containment Data for Offshore Pipelines'.
3. Grove S. et al. (2002), "The Rilsan User Group and API TR17RUG", OTC 14062.
4. Lecome, H., Hogben, S., Smith, J., Bednar, J. and Palmer, M. (2002), "BP Marlin: First Flexible Pipelay with Newbuild Deepwater Pipeline Vessel", OTC 14185.
5. OTO 98019 (1998), 'Guidelines for Integrity Monitoring of Unbonded Flexible Pipe'.
6. Picksley, J.W., Kavanagh, K., Garnham, S. and Turner, D. (2002), 'Managing the Integrity of Flexible Pipe Field Systems: Industry Guidelines and Their Application', OTC 14064, Offshore Technology Conference.
7. UKOOA (2001), 'State of the Art Flexible Riser Integrity Issues', prepared by MCS, Available online at www.oilandgas.org.uk/issues/fpso/studies.htm#pdfs.
8. UKOOA (2002), 'Guidance Note on Monitoring Methods and Integrity Assurance for Unbonded Flexible Pipe', prepared by MCS, Oct. 2002. Available online at www.oilandgas.org.uk/issues/fpso/studies.htm#pdfs

PART V: Welding and Installation

Part V

Welding and Installation

Chapter 32 Use of High Strength Steel

32.1 Introduction

The research and development of new steels are for both sour and non-sour service. The materials being developed for subsea pipelines and risers are grades X70 and X80 for non-sour service and grades X65 and X70 with a wall thickness of up to 40 mm for sour service. Figure 32.1 shows, by way of example, the distribution curves determined on a production lot of grade X65 pipe intended for sour service. As can be seen, the distributions for the transverse direction are shifted to the right relative to those for the longitudinal direction. In this Chapter, we shall review use of high strength steel for subsea pipelines, technological challenges and solutions.

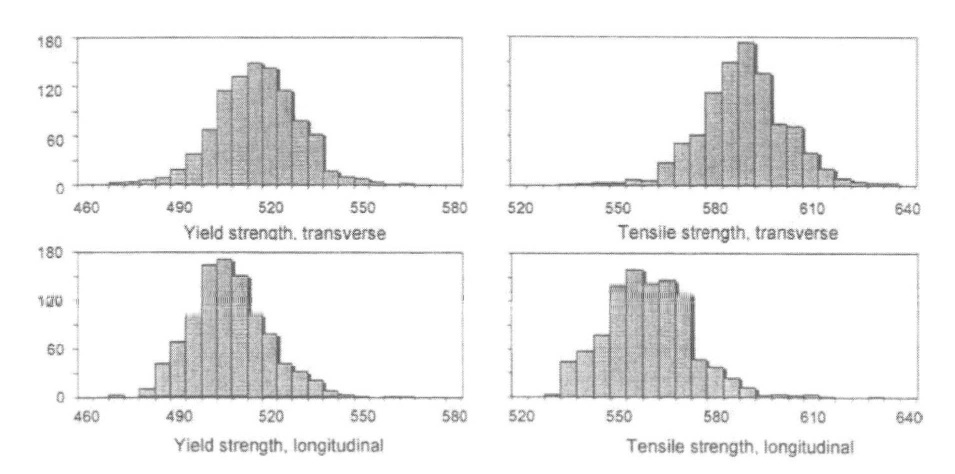

Figure 32.1 Results on 610 mm OD x 14.3 mm W.T. X65 production linepipe for sour service.

32.2 Review of Usage of High Strength Steel Linepipes

32.2.1 Usage of X70 Linepipe

32.2.1.1 General

For offshore pipelines, the current trend is towards linepipe in grade X70 with a wall thickness up to 40 mm. Fulfillment of the requirements for DWTT transition temperature will be progressively difficult as the wall thickness increases. For wall thickness in excess of 30 mm, low transition temperatures can only be achieved by means of highly expensive rolling processes. Until now, there has been only limited offshore use of X70 material. The main installation contractors have completed three projects with X70 and have two planned until 1997.

Mechanical properties	Requirement	Average $\bar{x}$	Stand. dev. s
Transverse yield strength $R_{t0.5}$ [MPa]	>482	525	12.6
Tensile strength R_m [MPa]	>565	635	14.0
Yield-to-tensile ratio $R_{t0.5}/R_m$ [%]	<92	82.6	1.6
Elongation A2" [%]	>25	34.8	1.6
CVN toughness, -20 °C			
Weld metal [J] *)	>50	172	36
HAZ [J]	>50	146	64**)
Base metal [J]	>80	220	45

*) MnMoTiB alloyed
**) No normal distribution

142 heats

97.14255

EUROPIPE

Production results on 682.4 mm O.D. x 15.9 mm W.T., API grade X70 line pipe

Figure 32.2 Production results on 682.4 mm OD x 15.9 mm W.T., API Grade X70 linepipe.

A large offshore project in grade X70 is the pipeline in the North Sea operated by Statoil, connecting Karstø, Norway, with Dornum, Germany. This pipeline has a length of 600 km and it is built of pipe 42" x 25 to 30 mm WT.

In the 1990's Europipe completed the development of grade X80 pipe 48" in OD and 18.3 mm to 19.4 mm in wall thickness for onshore pipelines. It has been demonstrated that it is feasible to manufacture commercially large diameter X80 pipe consistently for long transmission pipelines, see Gräf and Hillenbrand (1995).

As regards offshore applications, a series of pipes have been supplied for qualification testing with respect to pipelaying. Use of X80 linepipe for export pipelines was qualified by a joint industry project called EXPIPE.

For low-alloy steel pipelines operating in sour service, X65 is currently the established material. Special treatment in the steelmaking shop and fulfillment of special requirements for chemical composition help prevent the formation of nucleation sites for hydrogen induced cracking (HIC). Production trials show big potential for the development of higher grades up to X80 for slightly sour conditions, see Gräf and Hillenbrand (2000).

For high pressure transmission land lines, Grade X70 is now widely used for high pressure transmission lines in many countries. The supplier reference lists summarized in Table 32.1 provides 94 project references for 4 suppliers. This list is indicative rather than comprehensive, as other manufacturers have supplied this grade of material. A pipeline project installed in July 1997 for BP in the North Sea involves the laying of a grade X70, 24-inch diameter pipeline with a wall thickness of 25.8 mm. The reference list also shows only limited subsea use of X70 material, refer to Table 32.2.

32.2.1.2 Oman-India Gas Pipeline

In June 1993 a study was initiated to establish the feasibility of installing a subsea pipeline to connect the gas reserves in Oman to markets in India. The preliminary route of over 1,100 km would provide a direct link between Oman and India across the Arabian Sea with water depths up to 3,500 m. The Oman-India Gas Pipeline (O-IGP) project is currently on hold and design has not progressed past the preliminary stages.

The recommended pipe grade for the Oman-India Gas Pipeline is X70 for a 24-inch pipeline with constant internal diameter. Calculations have shown that the wall thickness along the majority of the route is predominately dictated by the prevention of external pressure collapse, (Refer to Table 32.3). For details of the development of design methods for hydrostatic collapse in deep water, refer to a paper by Tam et al. (1996).

Table 32.3 Required wall thickness based on collapse of the pipeline.

Water Depth (m)	Wall Thickness (mm)		
	API 5L X65	API 5L X70	API 5L X80
3500 – 3000	44.0	41.0	38.0
3000 – 2500	39.0	37.0	36.0
2500 – 2000	35.0	34.0	33.0
2000 – 1500	31.0	30.0	29.0
1500 – 1000	27.0	26.5	26.0
1000	22.7	22.0	22.0

32.2.1.3 Britannia Pipeline

The Britannia Field is a gas condensate reservoir in the Central North Sea approximately 200 km north-east of Aberdeen and 45 km north of Forties. Britannia Operator Ltd. (BOL) is a joint venture established by Chevron and Conoco for the Operatorship of Britannia on behalf of the Co-venturers.

Dry gas will be exported in dense phase mode through a pipeline to an extension of the Mobil SAGE Terminal at St Fergus. At the terminal, the gas will be processed for delivery into the British Gas National Transmission System. Offshore condensate will be delivered to the Forties Pipelines System through a condensate export pipeline from the Britannia Platform to the Forties Unity Platform.

The Gas Export Pipeline is nominally 28-inch diameter, 186 km in length with a bore of 650.6 mm. The pipeline design pressure is 179.3 barg and the design life of the pipeline is 30 years. The pipe grade is X70. The 14-inch Condensate Pipeline is 45 kilometers in length. The Britannia pipelines were completed in 1997.

Onshore lines are specified on the basis of transverse yield strength. The method of manufacturing these steels (TMCP, UOE) means that the axial yield strength will be around 4 - 5 ksi (~30 Nmm^{-2}) lower.

32.2.2 Usage of X80 Linepipe Onshore

High strength large diameter pipes are available from steelmakers e.g. Europipe for pipe diameter 20 – 60 inches and wall thickness of 12 – 32 mm, see Graf and Hillenbrand (1997). Five onshore projects have been identified in which X80 pipe has been used.A period of seven years elapsed before Ruhrgas AG in Germany began in 1992 to place an order for linepipe for the construction of the world's first ever grade X80 pipeline.

The 260 km 48-inch Ruhrgas pipeline from Schlüchtern to Werne in Germany was designed and built entirely to X80 capabilities and requirements. This pipeline, installed in 1992-1993, connects existing pipelines in new federal states in the former East Germany and started operations in late 1993. Considerable information has been published about this pipeline (Graf (1993), Chaudhari et al. (1995) and Behrens et al.).

Europipe GmbH, Ratingen, Germany, supplied the entire linepipe for the project. The material, specified as GRS 550 TM by Mannesmannroehren-Werke AG (MRW), Muelheim, Germany, has a specified minimum yield stress (SMYS) of 550 Nmm^{-2} and a minimum tensile strength of 690 Nmm^{-2}. The comparable API 5L X80 grade has an SMYS of 551 Nmm^{-2} and a minimum tensile strength of 620 Nmm^{-2}.

A test program was undertaken to determine the properties of the pipe steel and the weldment. The specified minimum values of yield and tensile strengths were exceeded in the tests. The impact energy values measured on the base material exceeded 95 J, thereby exceeding the minimum value for crack arrest recommended by the European Pipe Research Group (EPRG). The ductile-brittle transition temperatures measured on the drop-weight tear test (DWTT) specimens were well below the specified test temperature of 0°C. The impact energy values of the longitudinal weld metal measured at 0°C, the commonly specified test temperature in Germany, varied between 100 and 200 J. The average values of the impact energy for the base material and weld metal were 190 J and 158 J respectively (Chaudhari et al., 1995).

The strength of the seam weld was checked by means of flattened transverse weld specimens with the weld reinforcement removed by machining. For all specimens, failure occurred in the base metal, outside the weld region. The field welding for GRS 550 TM required the development of a new concept in order to achieve the mechanical-technological properties for the welding metal and welding joint. For this project, it proved necessary to implement a combined manual welding technology using cellulose-coated electrodes for root and hot pass welding and lime-coated (basic) electrodes for filler passes and cap pass welding.

Table 32.1 Supply record of major linepipe producers.

Supplier	Grade & location	No of References	OD range (in)	Thickness range (mm)	Comments
Europipe	All X70	35	30 – 48	11 – 25	All non-sour gas in 9 countries
	X80	2	48	18.3	For Ruhrgas, Germany, non-sour gas
Sumitomo	X70 land pipeline	33	20 – 56	7.9 - 35	Gas lines: USSR 17, USA: 4, Canada 2, Malaysia 1, Bangladesh 1.
	X70 subsea p/line	2	18 – 24	22 - 32	Both for water injection
	X70 other/misc	2	28	31	TLP tethers for Shell USA
	X80	1	21	22	Drilling riser for Vetco/Shell.
Nippon Steel (note 1)	X70 land pipeline	6	24 – 42	7.4 - 20.6	Major orders for USA, UAE, Columbia, Malaya; and other minor orders.
	X70 subsea p/line	1	32	14	Small order only
	X70 other/misc	1	44	38	Conoco Norway, 5000 tonnes, use not stated
	X80	None			
NKK (note 2)	All X70	8	24 – 56	9 - 34	
	X80	3	26 - 48	10.5 - 16	42 & 48" orders for Canada, 26" for Vetco Gray.

Notes:

1. Nippon Steel references are hard to interpret. Russian orders omitted as Grade not known. Structural steel orders also omitted.
2. All NKK references are believed to be land pipeline.

Table 32.2 X70 Subsea pipeline projects in 1997.

Date	Location	Onshore/ Offshore	Length (km)	OD (in)	WT (mm)	Steel Source, Type	Welding Method	Notes
1997	Shell Oil Mensa Gulf of Mexico	Offshore	100	12	19,21,32	-	Phoenix mechanized GMAW with shaw mechanized UT	2
August '97	BP ETAP, UK North Sea	Offshore	74	24	25.8	UOE	Passo GMAW	3
July '97	Britannia, UK North Sea	Offshore	190	28	17.5	-	Passo GMAW	3
1997	Norfra Pipeline, Norway to Franl[illegible] North Sea	Offshore	840	42	-	-	-	4

* Notes:

1. Information shown is based on available data.
2. Completed by Allseas.
3. Completed by EMC.
4. Refer to Thorbjornsen et al. 1997) for details.

32.2.2.1 NOVA Pipeline Projects

Pipe supplied to the two Canadian projects were to CSA Z245.1, typical compositions are again given in Table 32.4. The first Canadian project was a short (126 welds) cross over section of the 42-inch diameter pipeline at the Express East Compressor Station in Alberta, Canada, completed in 1990. A Japanese steel mill supplied the pipe. The second Canadian project was 53.8 km of 48-inch diameter pipeline for the Mitzihwin project in Alberta, Canada, completed in 1994. A Canadian steel company supplied the pipe.

32.2.2.2 Conclusions

These three projects have demonstrated that large diameter X80 pipe can be manufactured consistently for long pipelines. The approach to the X80 projects was significantly different when the welding procedure and consumables were selected. The field welding of the X80 pipe did not present any difficulty for the Ruhrgas and Mitzihwin projects. These projects demonstrated that conventional mechanized welding using the GMAW process could produce consistent, high quality welds for onshore pipelines.

Table 32.4 Chemical compositions for X80 linepipe and induction bends.

Typical values as weight %	**Ruhrgas 48" Schlüchtern to Werne**		**Empress East Compressor Station, Canada**	**Mitzihwin Project, Canada**
Element	Linepipe TMCP (Reference Chaudhari et al. (1995))	Bends Q&T (Reference Graf et al. (1993))	Japanese 42" OD UOE Linepipe (Reference Laing et al. (1995))	Canadian 48" DSAW spiral linepipe (Reference Laing et al. (1995))
C	0.09	0.12	0.06	0.04
Si	0.04	0.45	0.3	0.35
Mn	1.91	1.75	1.81	1.77
P	0.016	0.015	0.008	0.014
S	0.0009	0.003	0.003	0.005
Cu	0.04		0.16*	0.38*
Cr	0.05		0.02*	0.06*
Ni	0.04		0.09	0.15
Mo	0.01	0.22	0.18	0.26
V	-	0.06	0.08	0.00
Nb	0.042	0.035	0.03	0.09
Ti	0.018		0.01	0.03
Al	0.036	0.04	0.026	0.032
N	0.0035			
B	0.0003			
CE (IIW)	0.43	0.48		

* Note: The original reference has a typographical error, these values are all given as Cr so they are unreliable.

32.2.3 Grades Above X80

Higher grades are currently under active development. X100 grades are being actively developed by several companies (Nakasugi et al. (1990), Hillenbrand et al. (1997), Terada et al. (1995), Tamehiro (1996) and Kushida et al. (1997)) but at the present time no project use has been identified/indicated. Views of the future developments towards high strength steel, up to X100, are given by a consortium of companies and documented in Graf and Hillenbrand (1995). The supply capabilities of UOE linepipe as per 1997 are listed in Table 32.5.

32.3 Potential Benefits and Disadvantages of High Strength Steel

32.3.1 Potential Benefits of High Strength Steels

Potential Cost Reduction

The cost reduction is based on the premise that increasing material yield strength reduces the wall thickness required for internal (or external in the case of deep waters) pressure containment and hence the overall quantity of steel required. Price (1993) considered both direct and indirect consequences of using a high strength steel, and estimated a 7.5% overall project saving for a 42-inch offshore line laid with X80 instead of X65. Although the X80 pipe cost 10% more per tone, it was 6% less per meter. Further savings were identified for transportation, welding consumables, welding equipment rental and overall lay time.

On the recently completed Britannia gas pipeline, cost studies during detailed engineering showed that by increasing the linepipe material grade from X65 to X70, an approximate cost reduction of US$ 3.5 million could be achieved. The project CAPEX is approximately US$ 225 million. Although not directly related to the use of high strength material, other potential cost savings identified include:

- Tighter than normal (API 5L) definition of dimensions. Consideration should be given to reducing tolerances on ovality and wall thickness from API 5L requirements. The cost of reducing tolerances should be compared to the expected increase in pipeline construction rates and wall thickness reductions for mechanical design.

- Use of ECA based acceptance criteria for determination of maximum allowable defect sizes in pipeline girth welds. Traditionally, the acceptance criteria for weld defects are based on workmanship standards. ECA procedures typically rely on the application of Crack Tip Opening Displacement (CTOD) test results. The values of defect length are founded upon plastic collapse calculations which are based on assumptions regarding the flow stress and the yield/tensile strength ratio of girth and parent metal welds.

- Non-standard pipeline diameters should be considered. Optimization of the pipe ID based demonstrated that the linepipe cost could be reduced by procuring pipe of the exact ID required as opposed to selecting the larger standard size.

- Elimination of mill hydrostatic test with appropriate increased NDE.

Table 32.5 UOE linepipe supply capabilities in 1997.

SUPPLIER	Max single joint length available (ft/m)	OD RANGE (ins) at max thickness (note 1)	MAX THICKNESS BY GRADE (mm, rounded)					SUPPLY HISTORY (pipelines)	
				X60	X65	X70	X80	X70	X80
British Steel	45 / 13.7	30 - 42	49 (X52)		37	35	32	No	No
Europipe	60 / 18.3	20 - 64	40 (X52)	36	36	34	30	Yes	Yes
Sumitomo	60 / 18.3	30 - 48			38	38	32	Yes	No
Nippon Steel	60 / 18.3	29 - 56	40 (GrB)		38	36	33	Yes	No
Kawasaki	60 / 18.3	20 - 64		38	36	33	24	Not provided	
NKK (note 2)	60 / 18.3	16 - 56		33 / 29	33 / 27	32 / 26	29 / 26	Yes	Yes

Notes:

1. OD range may vary with grade, value is for X65.
2. Wall thickness given are for 18 m lengths first, then for shorter lengths down to 13 m

Pipeline welds are traditionally inspected using visual examination and radiography. Radiography systems are available which produce a real-time image of the weld being inspected. Normally, a radiograph of the weld is produced by exposing a suitable piece of film. The film is then processed and developed prior to viewing for interpretation. The real-time systems produce the image of the weld on a screen which can be viewed without the need for film processing. The radiographic image is stored on digital laser disc as a permanent archive and offers instant retrieval. The time to inspect each weld is reduced compared to traditional methods.

As an alternative to radiography, high speed ultrasonic inspection is available. This method has become a standard NDT method for inspecting GMAW (onshore) pipeline girth welds in Canada. Currently available high speed ultrasonic equipment is capable of inspecting a 40-inch diameter girth weld in 90 seconds. The inspection can be performed immediately on completion of production welds. A limitation of this technique is that it is not reliable for wall thickness below 10 mm. For project wall thickness above 10 mm ultrasonic inspection is a viable option. The use of automated ultrasonic inspection for onshore and offshore pipeline welding may reduce construction costs.

Wall Thickness and Construction

Given two similar design conditions, increasing the grade of linepipe in simplistic terms will correspondingly decrease the wall thickness and therefore provide cost benefits. In addition to this, a thinner wall thickness will also have various impacts on construction activities. A thinner wall thickness will require less field welding and therefore, in theory, has the potential to reduce construction/lay time.

By increasing the material grade, it is possible to lay pipeline in deeper waters. A thinner wall thickness has a direct impact on this installation method since the requirements for lay barge tensioners is related to the water depth and weight of pipe. For the Oman-India Gas Pipeline project, the question is how this pipeline can be laid with a massive top tension of 10,600 kN, during normal laying operation, necessary for controlling the catenary. It is recommended that a laybarge that has a tension capability of at least 26,700 kN is used. This requirement is dictated by a wet buckle abandonment/recovery scenario, that is, a buckle together with rupture leading to pipeline flooding. J-lay techniques may be used but lay-rate can be low.

Weldability

Thick wall thickness creates additional problems related to weldability. As the wall thickness of the linepipe increases, the cooling rate of the weld increases leading to possible problems with hardness, fracture toughness and cold cracking (when non-hydrogen controlled welding processes are used). A thinner wall thickness due to increase in material strength means that the cooling rate of the weld will also decrease.

Pigging Requirements

The thicker walled sections of the pipeline in deeper waters may restrict the full capabilities of intelligent pigging. There is a limitation on the wall thickness depending on the type of pigging tool used.

32.3.2 Potential Disadvantages of High Strength Steels

Increase in Material Costs per Volume

Generally an increase in material grade will equate to an increase in cost of material. Refer to Figure 32.3. However, it is also interesting to note that for a given design case, an increase in the material grade equates to a slight decrease in cost per meter.

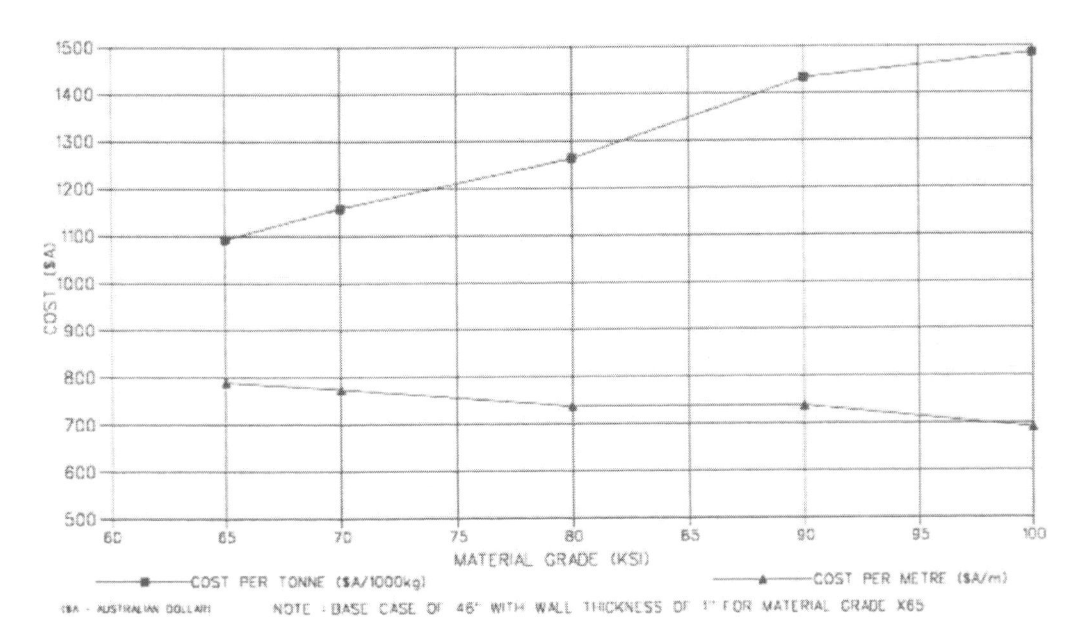

Figure 32.3 Cost variation of high grade line pipe.

Limited Suppliers

The worldwide availability of proven suppliers for material grades above X70 is still relatively limited.

Welding Restrictions

With regards to the weldability of X80 steel, there is a medium risk of schedule extension and cost increase since it has only been used on a small number of onshore projects and there is no experience offshore. Welding to the required quality may be slowed by more process restrictions and more complex controls. Due to the limited worldwide experience of welding X80 linepipe, certain key welding issues will have to be addressed in further studies, particular that of welding consumables.

Limited Offshore Installation Capabilities

The number of offshore pipelay installation contractors with proven experience of welding X70 steel linepipe is limited. Additionally, the experience of laying deepwater pipelines by the J-lay method is limited to relatively small diameter pipelines.

Repair Problems

Repair techniques for any pipeline is largely dependent on the water depth. At diverless water depths, (that is, at water depths without the use of divers), excluding the use of diverless hyperbaric welding systems, (that is, diverless subsea welding systems), the current state of the deep water repairs involves the use of mechanical connectors. These connectors are attached to the open end of a pipeline by relying on a metal to metal sealing arrangement.

Repair by hyperbaric welding, whether at diverable or diverless water depths, for material grade of X70 or above has not been undertaken and therefore there is currently no information regarding its behavior under hyperbaric conditions. Research programs should be monitored and initiated to develop understanding in this area.

An alternative repair method is to use the hot tap technique to bypass the area of pipeline damage. However, for offshore use this experience is limited and certainly unproven in high strength material pipelines.

Hot tap repairs are regularly performed onshore for API 5L X65 pipe grades and lower. BS 6990 states that hot tap welding of material above X65 yield strength should not be performed without welding trials being performed. The inferior weldability of high grade linepipe combined with the high cooling rates experienced during welding onto a live pipeline increase the safety risks associated with hot-tapping operations. For linepipe grades above API 5L X70, it is recommended that hot tapping is not performed unless extensive weld testing can be conducted.

Additionally, the subsea hot tap technique is limited to a maximum size of 24/36-inch (i.e. 24-inch bypass into 36-inch pipeline) at a limited water depth of 100 m for relatively low pressure lines (1,000 psi). This technique needs to be further evaluated.

32.4 Welding of High Strength Linepipe

32.4.1 Applicability of Standard Welding Techniques

The range of welding techniques used for pipeline construction includes Shielded Metal Arc Welding (SMAW), Gas Metal Arc Welding (GMAW), Submerged Arc Welding (SAW), Flux Cored Arc Welding (FCAW) and Gas Tungsten Arc Welding (GTAW). All of these techniques have been applied successfully to API 5L X65 linepipe and lower in accordance with internationally recognized pipeline construction codes and standards.

When welding higher strength grades of linepipe (X70 and above), special techniques are generally specified to avoid defects in high strength welds. Some of the additional measures that are necessary include:

- control of joint preparation and line-up;
- using adequate preheat;

- additional inter-run griding;
- careful selection of electrical characteristics;
- no movement of the pipe until completion of the root pass.

The specific application of standard welding technology to onshore and offshore pipeline construction is discussed in the following sections.

Onshore

The SMAW process is the standard welding methods for onshore pipelines. Low hydrogen SMAW has been used on pipelines up to API 5L X80 grade. Cellulosic SMAW is not generally used on linepipe above API 5L X70 strength due to problems with hydrogen cracking. For onshore pipelines above API 5L X70 grade, low hydrogen processes such as GMAW or FCAW are required. Refer to Section 32.3.2 below for details of project welding experience.

Offshore

The semi-automatic GMAW process is used extensively on laybarges for offshore pipelay. GMAW is sensitive to changes in carbon equivalent of the material. Generally, the carbon equivalent of linepipe increases as the grade is increased. GMAW is also sensitive to Boron alloying in the linepipe, however control of Boron in Japanese and European mills is very good and hence this is not considered to be an issue provided high quality linepipe is used. It is possible that development of GMAW procedures will take significantly longer for linepipe above API 5L X70 strength.

SAW is used on third generation lay barges for double jointing. SAW is a high heat input, high dilution process. Therefore, the chemistry of the linepipe being welded has a large influence on the properties of the final weld.

Welding API 5L X70 and X80 linepipe with SAW will require careful control of alloying elements to ensure that the final properties of the weld will be satisfactory. There have been problems with poor root toughness of SAW welds due to pick up of elements such as aluminium from the linepipe. The construction contractor should be given the opportunity to review chemistry requirements prior to linepipe manufacture in order to ensure compatibility with proposed SAW procedures.

FCAW is currently used for structural welding and for performing certain types of repairs on pipeline welds. Properties of FCAW welds are generally good, however, there have been historical problems in obtaining consistent weld toughness. FCAW consumables have been developed for welding linepipe up to API 5L X80 grade.

GTAW produces very high quality welds with excellent properties. However, the process is slow and is not generally used offshore (with the exception of hyperbaric welding and welding of Corrosion Resistant Materials).

In principle all the standard pipeline welding methods (with the exception of cellulosic SMAW) should be suitable for welding API 5L X70 and X80 linepipe provided additional time is allocated for weld procedure and consumable development.

32.4.2 Field Welding Project Experience

Manual Welding

The quality requirements of the Megal II and Ruhrgas lines required development of a welding procedure to overcome concerns over cold cracking of the high strength weld metal during conventional vertical-down welding with cellulosic electrode. The technique adopted used conventional cellulosic electrodes for the root and hot passes and basic electrodes for the fill and cap passes. The root was welded with an under-matched consumable, whilst over-matched consumables were used for the fill. All welding was downhill.

Pass	Consumable		
	Type	AWS Designation	Diameter (mm)
Root pass	Cellulosic	E6010	4
Hot pass	Cellulosic	E9010-G	5
Filler passes	Basic	E10018-G	4, 4.5
Cap passes	Basic	E10018-G	4

Note: Above data is as per Graf et al. (1993), Chaudhari et al. (1995) gives root and hot pass consumables differently as E7020-A1.

It should be noted that downhill welding is the norm for pipelines, at least outside of Japan, because it is fastest overall. Downhill is conventionally used with cellulosic electrodes which have a finite moisture content and are therefore not 'low hydrogen' but can be used on conventional linepipe steels when other suitable precautions are taken to prevent hydrogen cracking. Apart from pipelines, downhill welding is regarded as a poor practice for high quality welding and so it appears that the Japanese uphill practice is more cautious.

High strength steels and weld metals are more sensitive to hydrogen cracking. They cannot be reliably welded with cellulosic electrodes and so 'low hydrogen' consumables are required such as basic electrodes which are normally used in the uphill practice as per Japanese practice. It appears that cellulosic electrodes were used vertical down on the Ruhrgas line but only after 2 weeks special training of welders.

This approach allowed conventional welding of the first two passes without loss of productivity or risk of cold cracking. Chaudhari et al. (1995) states that the use of basic electrodes caused only a small loss of productivity for the subsequent passes. This is based on an overall welding cycle time of 5 - 6 hours which includes 3.3 hours for moving equipment between joints, setting up, etc. If only the welding time is considered Chaudhari et al. (1995) shows the time to complete a joint was 103 minutes using cellulosic electrodes (for all passes) compared with 137 minutes using basic, low hydrogen electrodes. At 33%, the increased welding time is significant and a consequence of requiring the improved mechanical properties obtainable from the basic electrodes. The increased time was due to more 'arc off' time for removal of the basic slag between passes.

The repair rate for manual field welding is reported to have been less than 3%. Maximum hardness of 350 HV10 are reported in the cap HAZ.

Mechanized Welding

A general discussion of mechanized welding of X80 is provided in Price (1993). Experience with the use of mechanized welding on three projects is presented in Chaudhari et al. (1995), Laing et al. (1995). The CRC Evans GMAW mechanized system was used in all three cases.

- An internal root pass was used in conjunction with external passes deposited into a narrow gap bevel.
- The Empress Project used pulsed GMAW for all external passes which resulted in improved weld metal and HAZ toughness compared to conventional GMAW.
- All three projects used titanium treated wire, the Thyssen wire used in Germany contained 1% Ni, 0.4% Mo which was not used in Canada.

The Mitzihwin project achieved an average rate of 103 butts at 48-inch OD x 12.1 mm WT in an 8 hour day though the repair rate was considered high at 6%, compared with 4% achieved on the other two projects (Laing et al. 1995). It is stated that repair rates have been less than 1% in comparable subsequent projects.

Properties of Field Welds

A detailed review of the inter-relation of welding process and properties is beyond the scope of this study. In the present context, the main point to be noted is that project specifications for weld quality, strength and toughness were met in all cases for X80 with wall thickness in the range 10.6 - 18.3 mm and that techniques have been developed sufficiently to allow consideration of X80 for both land and offshore pipelines.

32.5 Cathodic Protection

Subsea pipelines require compatibility with CP in sea water. High hardness steels are at risk of brittle failure caused by hydrogen embrittlement. Compatibility is conventionally satisfied by hardness values below 350HV10. The limit applies to parent metal and all weld zones. Chaudhari et al. (1995) and Laing et al. (1995) report maximum values of 350HV10 for manual welding (Ruhrgas project) and 303HV for mechanized welding (three projects, test load not given). The value of 350HV10 (10 for 10 g load in Vickers Hardness test) has been shown to be an acceptable maximum hardness for avoiding hydrogen embrittlement of structural steels and welds under CP in seawater (to minimum negative potential, maximum polarization's) of conventional sacrificial anodes. In all cases maxima were in the HAZ. This data indicates that X80 can be welded within the conventional limit for compatibility with CP.

In the context of future developments beyond X80, it is worth noting two points:

1) Marine sacrificial CP systems are available with potential control (as opposed to full open circuit potential capability of normal systems) to allow the use of steels with higher hardness values. Open circuit is the condition of maximum negative potential (or polarization) of protected steel from a conventionally mounted sacrificial anode when no current flows as can (almost) occur in practice at low current demands. This condition is the worst for hydrogen evolution and consequent hydrogen cracking. Steels conventionally need to be compatible with this potential which is more negative than that required for corrosion protection. Smart CP systems now exist which have local, potential sensing devices to control the applied potential only to the value required for corrosion protection, thus risks of hydrogen cracking are minimized. These systems have been used on high strength steels of jack-up rigs which previously have been known to crack due to hydrogen uptake.

2) Developments of linepipe for sour service will impose lower hardness limits, typically 250 - 275HV10.

Corrosion fatigue in the presence of CP is a secondary consideration in that pipelines would not normally be designed against a specified fatigue life. However fatigue concerns may arise in the event of spanning of subsea pipelines and so it is prudent to confirm that candidate materials do not have degraded fatigue properties relative to established grades. The concern arises from the unwanted uptake of hydrogen under the influence of CP. Hydrogen uptake adversely influences toughness and fatigue crack growth rates. Healy and Billingham (1993) indicates that fatigue properties of high strength grades under CP are comparable to conventional steels but information should be obtained that is specific to candidate linepipe steels.

Pipelines on land similarly require compatibility with CP and the above hardness criteria are also conventionally applied. Occurrences of external stress corrosion cracking (SCC) do not correlate with steel grade. Hydrogen embrittlement is associated with hydrogen uptake, normally in seawater. External SCC is fundamentally different and is a known risk for land pipelines and can be potentially a problem for all lines.

32.6 Fatigue and Fracture of High Strength Steel

It is recommended to obtain fatigue data for the proposed materials and apply the data to mechanical design. Fatigue life is used as the basis for many of the limits placed on offshore pipeline strength design. These limits have often been established based on empirical data from tests on low strength steels, with a safety margin applied. In general, the ability of steels to resist fatigue failure increases with increasing yield strength. Fatigue analysis data from linepipe manufacturers can be used to challenge the requirements of pipeline codes in the areas of thermal buckling analysis, freespan and pipeline stability analysis.

As the strength of linepipe increases, weld metals of increased strength and sufficient toughness are required to ensure overmatching behavior of girth welds.

32.7 Material Property Requirements

32.7.1 Material Property Requirement in Circumferential Direction

Necessary CTOD value requirements for Heat Affect Zone (HAZ) and weld metal are to be established that are relevant for the specific design conditions with regard to type and extent of longitudinal weld defects likely to exist. Typically, the required CTOD value is established through ECA (Engineering Criticality Assessment) using British Standard PD 6493.

The extent of longitudinal weld defects that likely to exist, is defined in the operators' welding qualification specifications. Typical values are: depth 3 mm and width minimum of 25 mm and pipe wall-thickness.

Practical experience from field use of the line pipes have, demonstrated that there has been very little structural failure due to lack of CTOD value in hoop direction for line pipes. Similar observations may be made on the CTOD requirements for the longitudinal direction. It is therefore suggested to closely evaluate the following:

- CTOD testing methods, scatters and statistical evaluation of scatters;
- Possibility to reduce the number of CTOD tests;
- Safety factors used in ECA determination of CTOD requirements;
- ECA design equations and analysis methods.

It is likely that fracture occurs in the weldment. Weldability of the pipe is a more important parameter than CTOD value.

32.7.2 Material Property Requirement in Longitudinal Direction

The CTOD value for line pipes in longitudinal direction is influential for fracture limit-state when ECA such as PD 6493 is applied to calculate the limiting loading condition to avoid fracture.

The CTOD value needed to avoid fracture depends on the extent of girth weld defects likely to exist and the applied load. For a defect depth of 3 mm, a wall thickness of 25.4 mm and loading up to 0.5% total strain a defect length of 177 mm (7 x wall thickness) was shown to be safe when CTOD is minimum 0.10 mm, see Knauf and Hopkins (1996). The discussions on unstable fracture and CTOD for hoop direction are also valid for longitudinal direction.

The fact is that the yield stress in longitudinal direction does not significantly affect pipe strength as long as strain-based design is applicable to the design situation. The reasoning for this statement is that strain acting on pipelines in operating condition is typically as low as 0.2% unless the pipeline is under a high pull-over load.

With exception of some special material problems, the Y/T (SMYS/SMTS) ratio requirements can be replaced by introducing strain-hardening parameters such as σ_R and n used in a Ramberg-Osgood equation. The level-2 and level-3 failure assessment diagrams in PD 6493 do also account for strain-hardening effects.

32.7.3 Comparisons of Material Property Requirements

Which material properties are dominant in local buckling/collapse? The answer is dependent on loads as the following:

- For internal pressure containment, hoop SMTS;
- For external-pressure induced buckling, hoop SMYS;
- For bending collapse, longitudinal SMYS;
- For combined internal pressure and bending, hoop SMTS;
  Longitudinal SMYS & SMTS;
- For combined external pressure and bending, hoop SMYS;
  Longitudinal SMYS & SMTS.

Raising hoop SMYS will directly result in a proportional reduction of the required wall-thickness of the line pipe for water depth shallower than 350 mm. As a conclusive remark on materials property requirements, it is believed that:

- The minimum CTOD values in both hoop and longitudinal directions typically should be 0.1 mm; the applicability of lower CTOD values can be validated by ECA methods.
- It is economically beneficial and technically justifiable that for pipe grades X60 to X80 yield and tensile strength in longitudinal direction can be lower by up to 10% than those in the transverse direction for water depths shallower than 450 m.
- For fracture and local/buckling failure modes, the Y/T value requirement can be removed if the strength analysis explicitly account for the difference of strain-hardening whose parameters (σ_R and n) are a function of SMYS and SMTS as the equations given in Bai et al. (1994).

As a further study, it is proposed to compare the Y/T ratio requirements from alternative codes (e.g. 0.93 from API for onshore pipelines, 0.85 from EPRG). It is perhaps possible to find some other rational criteria that can replace the Y/T ratio requirement in strength design. In order to develop alternative criteria, it is necessary to understand the reasoning of using Y/T ratio as a design parameter.

32.8 References

1. API 5L (1995) "Specification for Line Pipe", 41st Edition.
2. Bai, Y., Knauf, G. and Hillenbrand, H.G. (2000) "Materials and Design for High Strength Pipelines", Proc. of ISOPE'2000.
3. BSI: PD6493, Guidance on methods of assessing the acceptability of flaws in fusion welded structures, British Standards Institute, (1991)
4. Chaudhari, V., Ritzmann, H.P., Wellnitz, G., Willenbrand, H.G. and Willings, V., (1995) "German gas pipeline first to use new generation linepipe", Oil and Gas Journal, January, 1995.
5. Graf, M.K., Hillenbrand, H.G. and Niederhoff K.A., (1993) "Production of Large-diameter Linepipe and Bends for the World's First Long Range pipeline in Grade X80 (GRS 550)" PRC/EPRG Ninth Biennial Joint Technical Meeting on Linepipe Research, Houston, Texas, May 11-14th, 1993.
6. Graf, M. and Hillenbrand, H. G., (1995) "Production of Large Diameter Linepipe - State of The Art and Future Development Trends" Europipe GmbH 1995.
7. Healy, J. and Billingham, J., (1993) "Increased Use of High Strength Steels in Offshore Engineering", Welding & Metal Fabrication, July 1993.
8. Hillenbrand et al. (1995) "Manufacturability of Linepipe in Grades up to X100", TM Processed Plate HG Pipeline Technology, Volume II 1995.
9. Knauf, G. and Hopkins, P. (1996) "The EPRG Guidelines on the Assessment of Defects in Transmission Pipeline Girth Welds", 3R international (35), heft 10-11/1996, pp. 620-624.
10. Laing, B.S., Dittrich, S. and Dorling, D.V., (1995) "Mechanized Field Welding of Large Diameter X-80 Pipelines". Pipeline Technology, Proceedings 2nd Int-Conf. Sept 1995. Elsevier. ISBN 0-444-82197-X Vol 1, p505-512.
11. Nakasugi, H., Tamehiro, H., Nishioka, K., Ogata, Y. and Kawada, Y., (1990) "Recent Development of X80 Grade Linepipe", Welding-90, Hamburg, F R Germany, October 22-24, 1990.
12. Price, C., (1993) "Welding and Construction Requirements for X80 Offshore Pipelines", 25th Annual Offshore Technology Conference, Houston ,Texas, May 3-6, 1993.
13. Tam, C. et al., (1996) "Oman-India Gas Pipeline: Development of Design Methods for Hydrostatic Collapse in Deep Water", OPT '96.
14. Tamehiro, H., (1996) "High Strength X80 and X100 Linepipe Steels", Nippon Steel Corporation. Int. Convention 'Pipelines: The Energy Link', Australia 26-31 October.
15. Thorbjornsen, B, Dale, H. and Eldoy, S. (1997) "The NorFra Pipeline Shore Approach: Engineering Environmental and Construction Challenges", 7th International Offshore and Polar Engineering Conference, Honolulu, USA.

Part V

Welding and Installation

Chapter 33 Welding and Defect Acceptance

33.1 Introduction

During offshore pipeline installation, the occasional weld defect repair may be carried out just after the stern-most tensioner and before the next roller support in the repair station. The weld defect is removed by grinding and subsequent re-welding. Aft of the tensioner, the pipe undergoing excavation is highly loaded by bending and tension. The local stresses at the weld repair are intensified during the weld excavation process. Defects in the girth weld are located and measured by radiography or ultrasonics. Defects exceeding the project criteria are thoroughly removed, however, the removal of excess material should be minimized so as to minimize local stresses. The determination of safe weld excavation sizes for repair is one of the more difficult evaluations in pipeline installation engineering. This subject has attracted much attention from pipeline owners, installation contractors, and operating companies because of personnel safety and also the risk of the pipeline parting or buckling on the ramp during excavation. This paper presents an analytical method of determining safe weld excavation lengths preventing both plastic collapse (buckling) and fast fracture of the pipe during weld defect repair.

33.2 Weld Repair Analysis

The weld excavation and repair procedure have been analyzed and performed successfully for the major GoM operators and many contractors in the past years due to their awareness of risk and increased focus on safety. There are many factors affecting the safe weld repair at the lay ramp, for example, plastic collapse of section, brittle fracture, or cracking growth and fatigue. Fatigue is an extension of the present subject and is treated by the fracture mechanics approach. This topic is the subject of a separate paper. Plastic collapse and fast fracture are both considered in the following weld repair analysis:

1. Stress analysis is performed on the excavated pipeline weld under the most severe tension and bending moment loading conditions experienced during repair on the lay ramp. This analysis is performed to determine the maximum allowable excavation lengths and depths to avoid plastic collapse.

2. Given that the allowable excavation length and depth have been established by stress analysis, fracture mechanics analysis is performed. This ensures that the weld and heat affected zone (HAZ) will not fast fracture, given the materials properties of the weld region.

33.2.1 Allowable Excavation Lengths for Plastic Collapse

This part of the analysis ensures that the pipe will not suffer plastic collapse of the section during repair. The maximum allowable weld repair length and depth is established to ensure that the stress at the root of the excavated area is kept below the yield strength of the pipe during the repair process. API 1104 and BS 4515 are typical guidelines required for welding and weld acceptance standards for construction of offshore pipelines and flowlines. In accordance with British Standard 4515, which considers welding of steel pipelines on land and offshore, the following criteria should be considered during repairing:

1. For a full penetration repair the maximum weld repair length is 20% of the girth weld length.
2. For a partial penetration repair the maximum allowable repair length is 30% of the girth weld length.

These guidelines begin to address the issue under discussion but do not necessarily speak to a particular loading condition for a specific excavation or material toughness.

When pipeline repairs are made between the stern-most tensioner and the next roller support on the lay ramp, the pipe butt weld is excavated under high loads caused by the curvature bending moment and lay tension. With elastic stress analysis, the stress distribution in the pipe may be calculated by,

$$\sigma = \frac{N}{A} + \frac{M_x}{I_x} y + \frac{M_y}{I_y} x \tag{33.1}$$

where the cross-section of pipe is assumed to be in an x-y plane and σ is stress; N, tension force; A, the pipe cross section area; M, bending moment and I, the second moment of area of the cross-sectional area about the neutral axis.

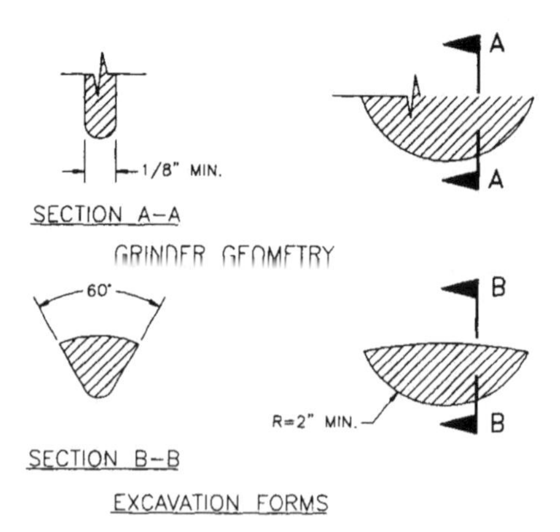

Figure 33.1 Geometries of grinder and excavation.

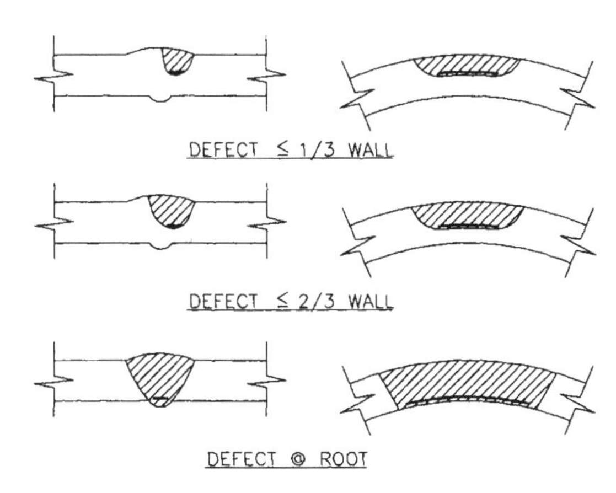

Figure 33.2 Defect positions and its corresponding excavation forms.

For a simple excavation geometry, the stress estimate may be calculated basing on the formula [5]. For real-life repair geometries, however, the excavation profile is too complicated to obtain an accurate result of stress distribution using Formula (1). Figure 33.1 shows the geometries of typical grinder forms used in the excavation. The radii may be as large as 6 inches and as small as 2 inches. The general defect position and its corresponding excavation forms are shown in Figure 33.2.

Finite Element Analysis (FEA) is a powerful tool to analyze the complex stress map for this kind of problem. The excavated pipelines are modeled and analyzed with finite elements to provide an accurate assessment of the stresses due to the applied bending and tensile loads of the pipe cross-section for the conditions of varying excavation lengths and depths.

The loads of bending moment and tension force during repair are chosen from the installation simulation with installation software widely used such as OFFPIPE. A dynamic factor of 1.20 is included with the static analysis results to account for the effects of vessel and pipe motion. This factor is dependent upon the vessel motion, pipe properties and the seastate acceptable for pipeline installation. The seastate can be resolved by utilizing dynamic OFFPIPE or similar software sensitivity analyses. Based on the stress distribution from FEA, the allowable excavation lengths for plastic collapse at various positions and depths of excavation are determined directly.

33.2.2 Allowable Excavation Lengths Using Different Assessments

The stresses resulting from the tensile and bending load and the residual weld stresses, combined with an excavation in the weld-zone, is of a critical value with respect to the fracture failure mode. Allowable flaw length of offshore pipelines may be assessed by using different methods, as shown below.

Level 1 Assessment - Workmanship Standards

Pipeline welding codes in the U.S. and elsewhere establish minimum weld quality standards based on the inspection of a welder's workmanship. The initial flaw acceptance criteria evolved through industry experience over many decades. Most workmanship standards are similar, though not identical, in terms of imperfection types and sizes. The advantage of workmanship standards is that they are time-tested, compatible with normal levels of NDE quality, easy to apply, and they do not require material strength or weld toughness data. However, it has been recognized that some rejectable flaws may not pose a real threat to pipeline integrity but still are rejected solely on the basis of workmanship standards. That is, the workmanship standards tend to be conservative. A girth weld with a defect assessed unacceptable by a workmanship standard may be extremely costly to repair or replace and yet with additional assessment may be shown to be safe and fit for service. Still the basic workmanship standard approach is time tested and has been proven over more than thirty years, as shown by applications on the Alaskan pipeline.

The principal workmanship standards that are recognized by U.S. gas pipeline regulations are those contained in API Standard 1104. Weld workmanship standards are also contained in other specifications, e.g., ASME B&PV Code, CAN/CSA-Z184, and BS 4515, but there are not recognized by U. S. pipeline regulations.

Level 2 Assessment - Alternative Acceptance Standards

Alternative acceptance standards were developed to facilitate acceptance of flaws that do not meet workmanship standards. Incentives for alternative standards are usually economic, arising due to the inaccessibility or quantity of welds that would otherwise require repair. Alternative standards recognize that the true severity of a flaw is dependent on material toughness and applied stress levels, and can only be determined using fracture mechanics principles. The crack tip opening displacement (CTOD) test is most commonly used as the weld toughness assessment method when alternative acceptance methods are utilized.

Three alternative criteria that are recognized by their respective national regulating agencies and that are often cited are the Appendix A of API Standard 1104, Appendix K of CSA-Z184, and BSI PD 6493 (PD 6493, Level 1 and revised BS 7910). All three standards are based on the CTOD Design Curve approach developed by The Welding Institute, and they extend linear elastic fracture mechanics (LEFM) concepts into the elastic-plastic regime. In spite of their common origins, they differ in their treatment of residual stresses, summation of stress components, minimum toughness level, and factors of safety.

Level 3 Assessment - Detailed Analysis using fracture mechanics

Flaws that are not permitted by Level 2 assessment may be further evaluated by detailed fracture mechanics analysis. PD 6493 provides an appropriate Level 3 procedure based on R-6 Failure Assessment Diagram (FAD) methodology. This diagram is broadly used, however it is generally more representative to plot defect length versus stress field.

The allowable excavation length for weld excavation in this paper is determined by using the British Standard PD 6493, Guidance on Methods for Assessing the Acceptability of Flaws in

Fusion Welded Structures, or its revision, BS 7910, Guide on Methods of Assessing the Acceptability of Flaws in Structures. Using fracture mechanics, a CTOD based design curve is generated to ensure that failure by unstable fracture will not occur for a given defect size, applied stress and critical CTOD toughness. These codes involve a two-parameter assessment depicted by a Failure Assessment Diagram (FAD), which considers the independent possibility of plastic collapse and fast fracture.

FractureGraphic software is a typical software which uses the analytical approach that is detailed in the BSI PD 6493 and BS 7910 codes. As discussed in BS 7910, both the primary and secondary stresses must be considered in determining the allowable crack length. The primary stresses result from the tensile and bending load and the secondary stresses result from residual weld stresses. FractureGraphic calculates the secondary weld residual stresses based on the material strength and the weld heat input. The relationship between the permissible crack depth and length as a function of applied stress and calculated secondary stress is obtained. Based on these results, the allowable excavation lengths for fracture mechanics at various positions and depths of excavation are determined.

33.3 Allowable Excavation Length Assessment

33.3.1 Description of Pipeline Being Installed

A 12.75-inch diameter by 0.625-inch wall thickness X52 pipeline was chosen as an example to evaluate the allowable excavation length in weld repair during laying. The OFFPIPE installation analysis provided the worst stress case during installation resulting from tension forces and moments including dynamic effects. For this example, tension force and moment were 29.35 kips and 81 ft-kips, respectively. It was assumed that any defects would be removed by grinding using a 4 inch disc, or slightly larger. A CTOD value of 0.008 inches was used for the fracture mechanics analysis.

33.3.2 Analysis Method

Finite Element Analysis

FEA analysis was performed using ALGOR software. The pipe under consideration was located aft of the tensioner on the lay barge, and was subjected to a bending moment of 81 ft-kips with a tensile load of 29.35 kips. The analysis considered defects removed at 1/3, 2/3 and through–wall thickness. These conditions are shown in Figure 33.3. The excavation profile following defect removal had a combined angle of 60 degrees with a root radius of 0.125 inches. The bevel that was modeled is the minimum size necessary for weld repair following removal of the defect. The length of the groove was taken as just slightly larger than the length of the defect. In addition to the length of the defect, the excavated groove includes a curved section required to accommodate the grinder.

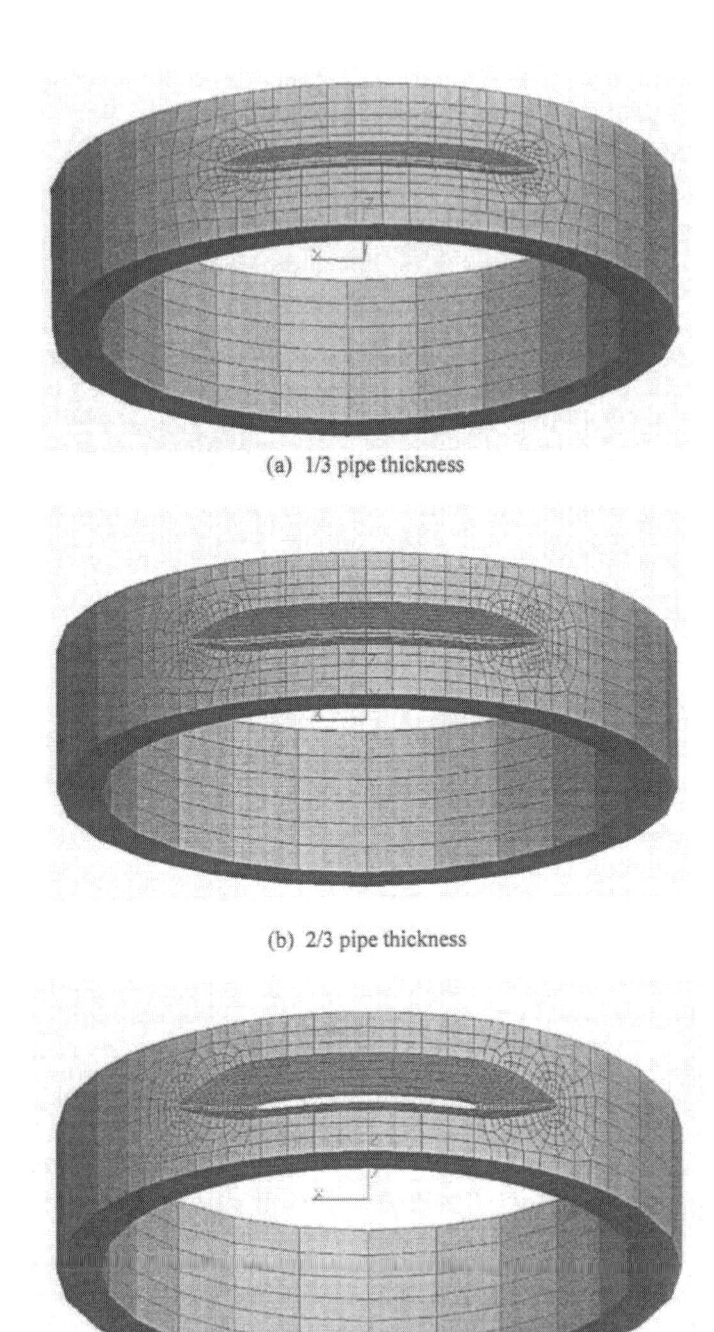

(a) 1/3 pipe thickness

(b) 2/3 pipe thickness

(c) through pipe thickness

Figure 33.3 Groove shapes and mesh distribution for FEA

In addition to varying the depth of the defect, the length of the defect was altered, along with the position on the circumference of the pipe. As the circumferential position varies, the pipe stresses vary. The bending stress at top center results in a pure tensile load and at the bottom center normally a pure compressive load since tension stresses are much smaller than bending stresses. Four circumferential positions are shown evaluated. Top center, 45° off of top center, 90° off of top center, and bottom center. A fifth position, 45° above bottom center, is shown analyzed by interpolation between the two adjacent locations. Each change in a variable requires a new model to be created for stress analysis.

Fracture Mechanics Analysis

The fracture mechanics investigation is demonstrated using the FractureGraphic software, with manual checking using BSI PD 6493 and BS 7910. The defects with depth of 1/3 wall thickness, 2/3 wall thickness, and through-wall that were considered are shown in Figure 33.2. A CTOD value of 0.008 inches was maintained for all positions for these analyses.

Defects may be surface breaking or mid-wall defects. Surface breaking defects are more severe from a fracture standpoint. A mid-wall defect behaves the same as a surface-breaking defect once the top ligament is removed. The limiting length of the crack for this part of the evaluation is the actual crack length measured by ultrasonics or by radiography.

33.3.3 Analysis Results

Typical FEA results are shown in Figure 33.4, which are presented as Von-Mises stresses and displacement distributions. The stresses as a function of crack depth and pipe position for all cases are shown in Table 33.1.

The fracture mechanics analyses results are shown in the form of permissible crack depth and length as a function of applied stress. Figure 33.5 shows the results for the applied stress levels of 25, 30, 35, 40, 45, and 50 kips respectively, where the stresses are perpendicular to the excavation.

The FEA results in Table 33.1 are compared to the fracture mechanics analyses results to determine the governing condition. The failure condition limit, either plastic collapse or fast fracture, is determined by the least loading condition and determines the maximum allowable amount of material that can be excavated. The summary of the allowable length excavation for the example is presented in Figure 33.6.

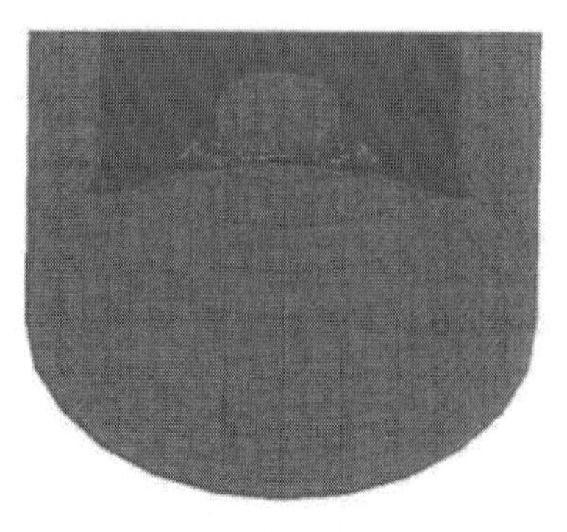

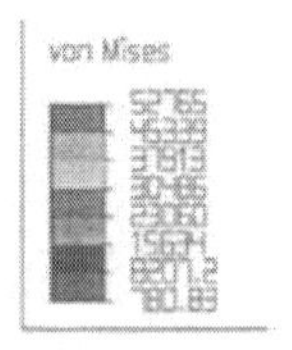

(a) Von Mises stress distribution

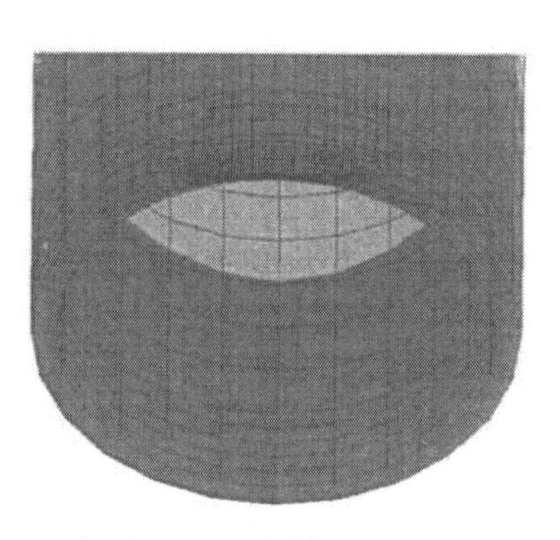

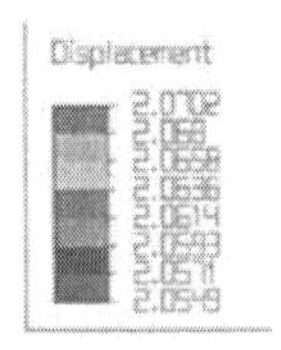

(b) Displacement distribution

Figure 33.4 Stress and Displacement distributions for top center case.

Table 33.1 Variations of maximum Von Mises stress with Groove's position and depth

Position \ Depth	1/3t	2/3t	1t
0 degree	L=12 in σ_{max}=42.9ksi	L=2 in σ_{max} = 52.7 ksi	L=1 in σ_{max} = 54.0 ksi
45 degree	L=12 in σ_{max} = 44.4 ksi	L=2 in σ_{max} = 52.2 ksi	L=1 in σ_{max} = 54.0 ksi
90 degree	L=12 in σ_{max} = 43.4 ksi	L=3 in σ_{max} = 19.5 ksi	L=2 in σ_{max} = 45.3 ksi
180 degree	L=12 in σ_{max} =32.9 ksi	L= 3 in σ_{max} = 51.9 ksi	L=2 in σ_{max} = 52.6 ksi

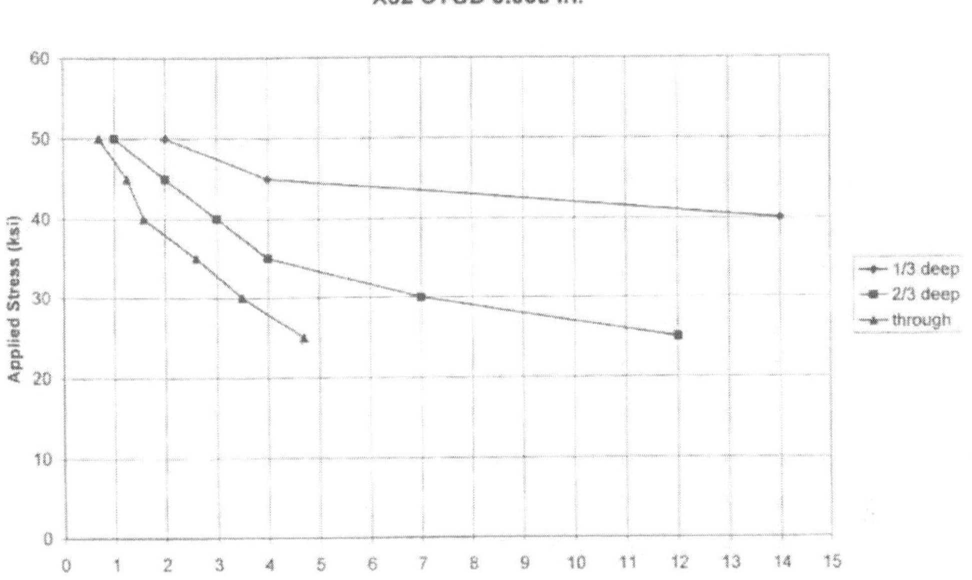

Figure 33.5 Permissible crack depth vs. length for fracture mechanics analysis

It is noted that in general, the limiting failure mechanism is fast fracture rather than plastic collapse. For a through thickness crack at top center, a crack length of only 1/2 inch can be tolerated for this example. Serious consideration is warranted if planning any repair of through thickness or root defects after the tensioner. If such repairs are attempted, they should be performed only under close supervision. Means of modifying the ramp configuration for a 'repair area' can be accommodated on larger vessels to reduce or minimize bending loads. In other cases, only the material and weld toughness can be increased to permit larger excavation.

When the defect length is found to be more than the allowable groove length at that defect depth, either a multi-stage repair can be undertaken or the joint may be backed up to the front of the tensioner. Backing up is a costly operation offshore. In a multiple repair situation, the first weld repair groove is excavated from one end of the defect up to the allowable length. This section is then weld repaired. The subsequent weld repair grooves are then excavated and repaired without exceeding the allowable repair groove length. The procedure continues until the entire defect length is repaired. Radiographic procedures are performed to demonstrate that the acceptance criteria are met.

33.4 Conclusions

A procedure for repairing offshore pipeline welds after the tensioner is presented. This procedure has been utilized many times in actual practice. Within the discussion, broadly used and available software are identified, and steps for problem resolution are described. The example problem is taken from an actual project case study in order to demonstrate defect size limitations.

MAXIMUM WELD EXCAVATION LENGTH
VERSUS DEPTH
VERSUS PIPE POSITION
CTOD = 0.008 in.

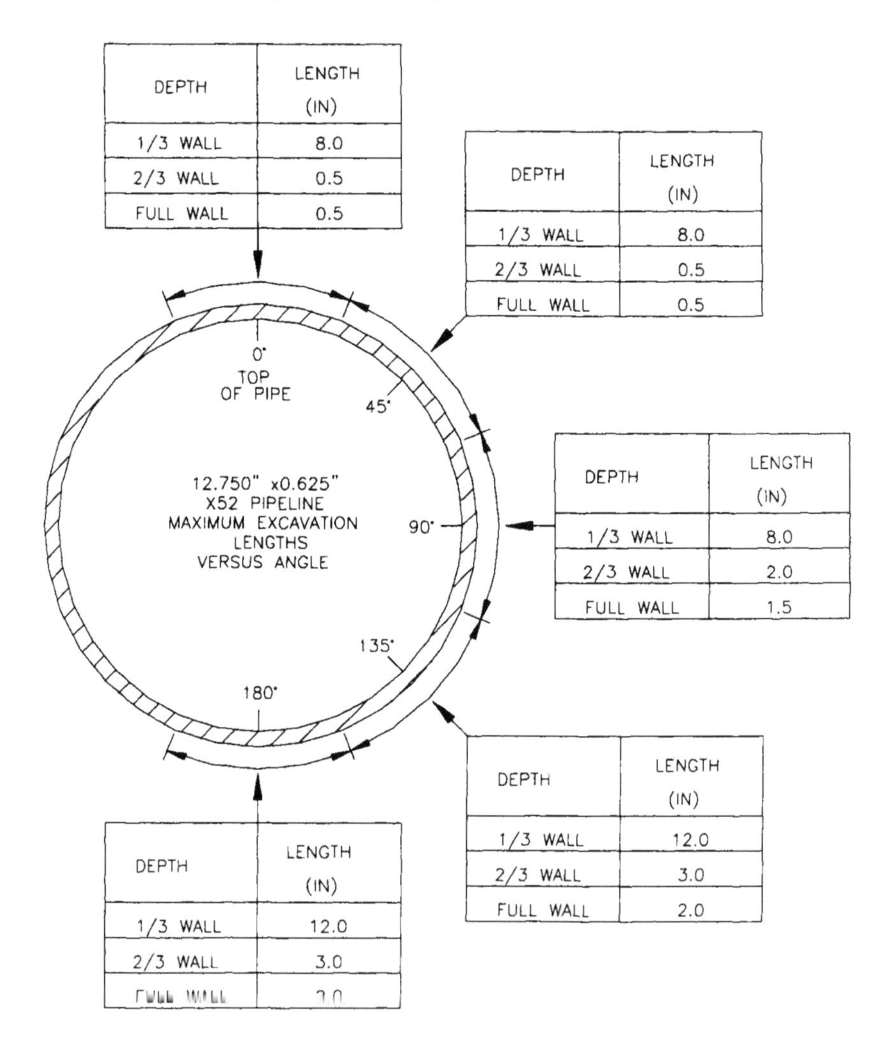

NOTES:
1. 1/3 WALL MEANS EXCAVATION DEPTH FROM 0 TO 3/16 DEEP.
2. 2/3 WALL MEANS EXCAVATION DEPTH FROM 3/16 TO 3/8 DEEP.
3. FULL WALL MEANS EXCAVATION DEPTH TO 9/16 DEEP.
4. BETWEEN 2/3 AND FULL WALL LINEAR INTERPOLATE.
5. LENGTHS NOTED ARE CENTERED AT ANGLE SHOWN.
6. 12-INCH LENGTH IS SET AS A MAXIMUM LIMIT FOR ALL CASES.
7. EXCAVATION LENGTH DOES NOT EQUAL DEFECT LENGTH.
   LENGTHS SHOWN ARE EXCAVATION LENGTHS MEASURED AT PIPE SURFACE.
8. USE A 4-INCH GRINDING DISK OR CLOSEST POSSIBLE SIZE TO MINIMIZE METAL REMOVAL.
9. NO FULL WALL REPAIRS CAN BE MADE POST TENSIONER.

Figure 33.6 Summary of allowable length and depth of excavation for 12" pipeline.

33.5 References

1. Bai, Q., Haun, R, and Sumner, K., (2003), "Assessment of Allowable Pipeline Weld Excavation Lengths on the Lay Ramp", OPT 2003, Houston.
2. API 1104, (1999), "19th.Standard for Field Welding Pipelines and related Facilities", American Petroleum Institute.
3. BS 4515, (1996), "Specification for Welding of Steel Pipelines on Land and Offshore", British Standards Institute, London.
4. BSI-PD 6493, (1991),"Guidance on Some Methods for Assessing the Acceptability of Flaws in Fusion Welded Structures", British Standards Institute, London.
5. BS 7910, (1999), "Guide on Methods for Assessing the Acceptability of Flaws in Fusion Welded Structures", British Standards Institute, London.
6. Young, W.C. and Budynas, R. D., (1989), Roark's Formulas for Stress and Strain, 7th edition, McGraw-Hill.

Part V

Welding and Installation

Chapter 34 Installation Design

34.1 Introduction

Marine pipeline installation is performed by specialized lay-vessel. There are several methods to install a pipeline, the most common methods being S-lay, J-lay and reeling.

Depending on the method, a marine pipeline is exposed to different loads during installation from a lay-vessel. These loads are hydrostatic pressure, tension and bending. An installation analysis is conducted to estimate the minimum lay-tension for the pipeline for a given radius of curvature to ensure that the load effects on the pipeline is within the strength design criteria.

A commonly used FEM computer program for installation analysis is OFFPIPE. This program can give indicative global results for most situations but not the effects of stress/strain concentration and point loads due to change in stiffeners.

This chapter also describes a finite element model for pipeline installation analysis. The model should be able to compute static load effects on a pipeline during installation, based on the layramp geometry, pipeline design data and water depth for the pipeline to be installed.

The established model should be a tool for analyzing the static configuration of a pipeline during installation. The static configuration of the pipeline is the shape of the pipeline from the lay-vessel to the seabed when it is in static equilibrium. The model should also be capable of analyzing the load effects on the pipeline when a section like a valve is installed. The model should also be capable of letting the pipeline slide over the stinger. A pipeline cross section will then move from the lay-vessel, over the stinger and through the sagbend to the seabed.

The purpose of developing this finite element model for pipeline installation is to calculate the load effects on a pipeline during the installation of an in-line valve. These analyses do not involve the response due to environmental loads.

34.2 Pipeline Installation Vessels

Pipeline installation methods have significantly changed over the last twenty years. This is pertinently enforced by the recent replacement of the BP Forties 170km trunkline. When it was first installed in 1974 it took two lay barges more than two summers, and each lay barge suffered 60% downtime due to weather. In 1990 it took one (relatively old) pipelay vessel to install the replacement pipeline (and the pipewall was significantly thicker- 28.5mm compared to the original 19mm). The significant increase in layrate is due to a combination of factors including:

- Improved welding techniques;
- Improved survey capabilities;
- Improved anchor handling techniques;
- Improved procedure.

The methods available to install pipelines are discussed under the following headings (see Figures 34.1 to 34.4 from Langford and Kelly (1990)). Different vessel types are used depending on the pipelay method and site characteristics (water depth, weather etc.).

- S-lay/J-lay semisubmersibles;
- S-lay/J-lay ships;
- Reel ships;
- Tow or pull vessels.

34.2.1 Pipelay Semi-submersibles

Pipelay semisubmersibles were developed as a direct response to the large weather downtime being experienced by the monohull pipelay barges (especially in the North Sea). These vessels have excellent weather capabilities and can provide a stable platform for pipelaying in seas experiencing Beaufort force 8 conditions. It is usually the limitations of the anchor handling vessels which prevent the semi-submersible from operation in rough weather.

There are presently several such vessels operating in the North Sea (see Figure 34.2 for typical vessel- Semac).

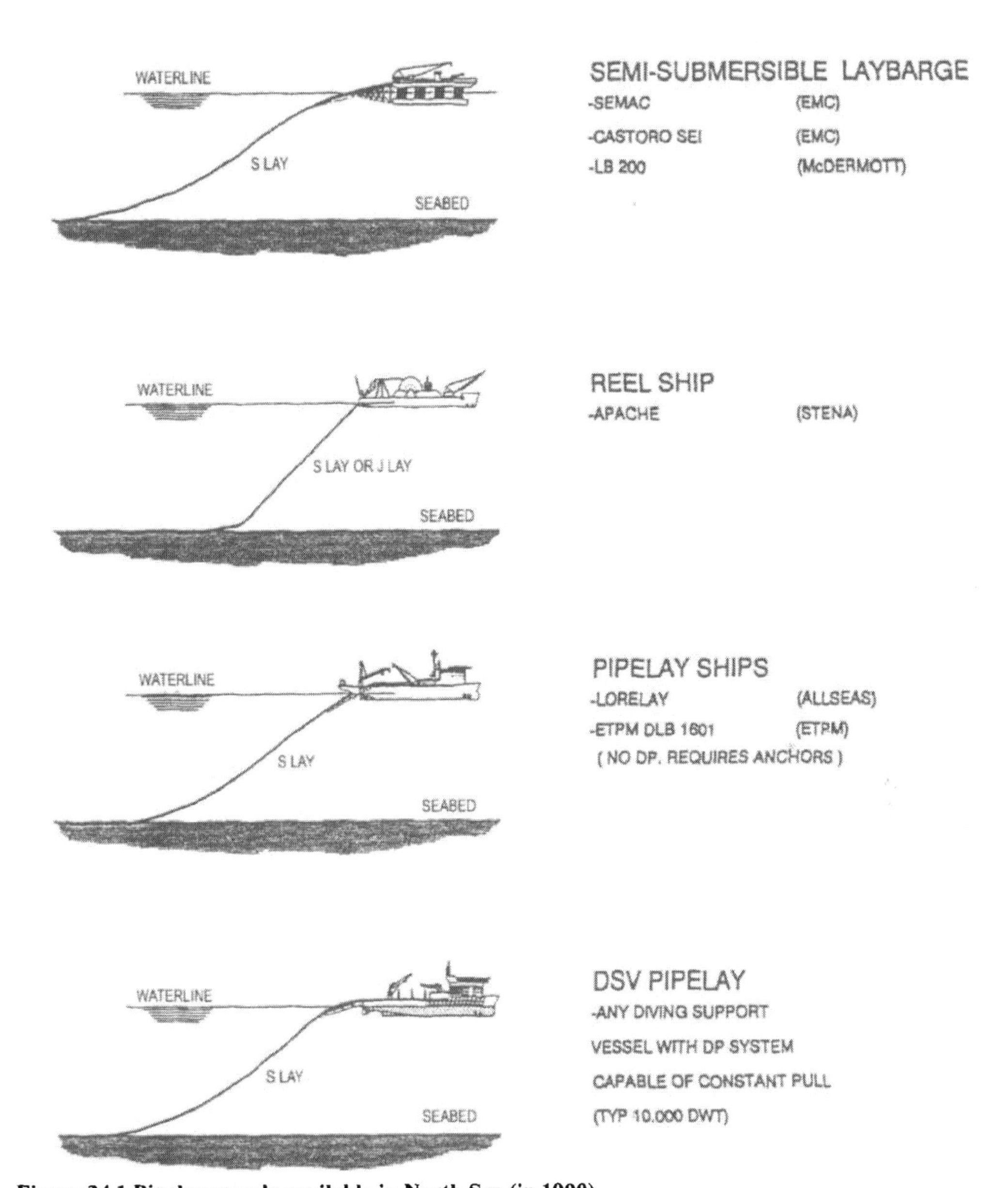

Figure 34.1 Pipelay vessels available in North Sea (in 1990).

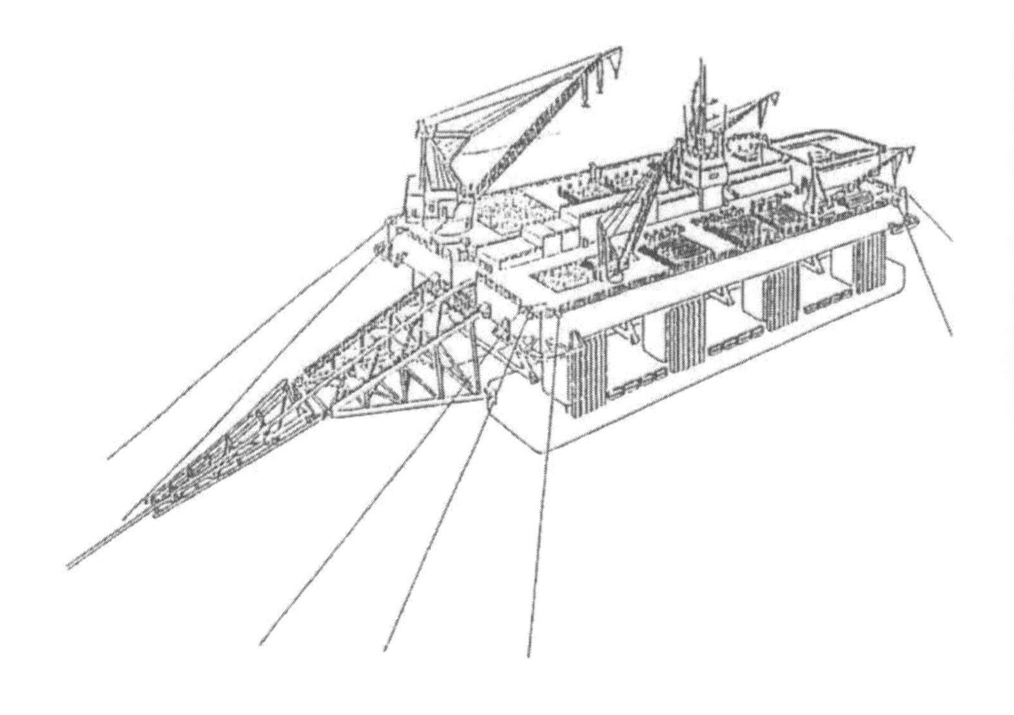

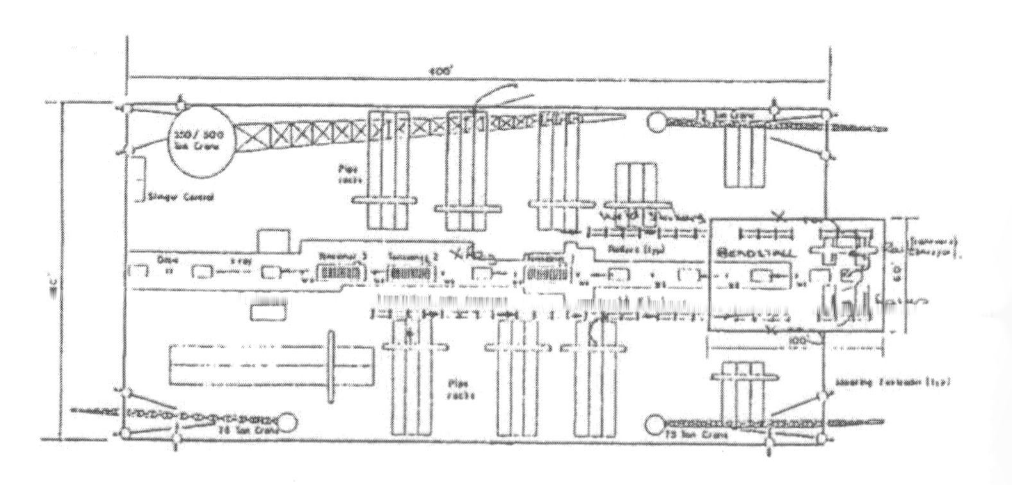

Figure 34.2 Pipelay semi-submersible.

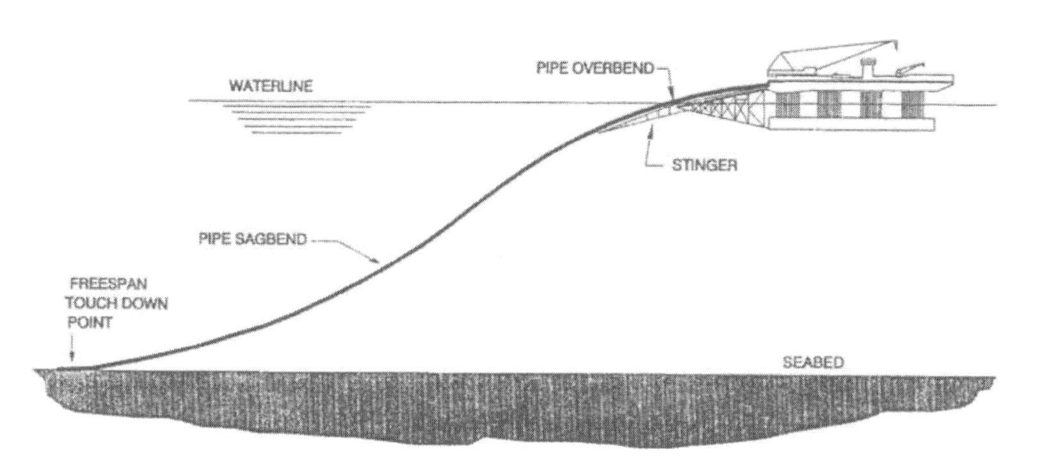

Figure 34.3 Typical pipe configuration during installation.

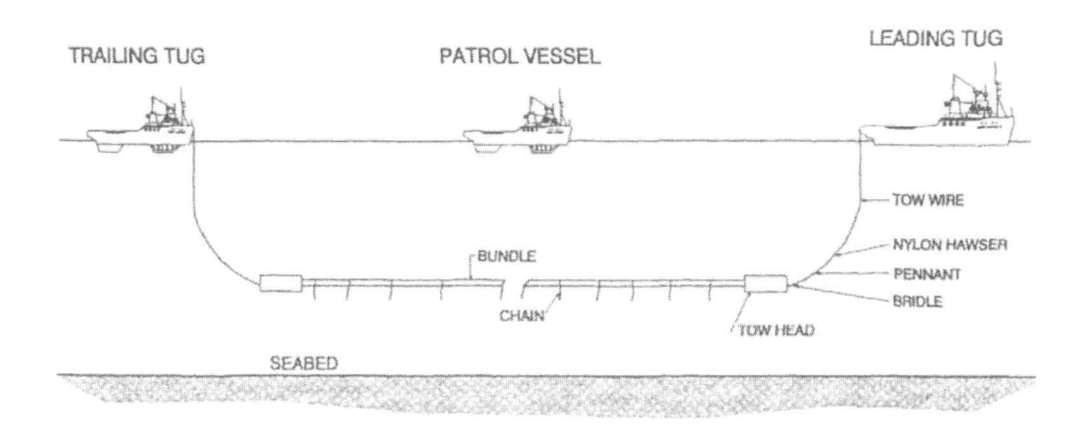

Figure 34.4 Flowline tow method.

General Principle

Pipelay semisubmersibles are effectively a floating factory which weld line pipe joints together and installs the pipe accurately on the seabed. Pipelay barges used originally in the Gulf of Mexico experienced difficulty in the North Sea in laying pipelines quickly and without damage. Consequently, pipelay semisubmersibles were developed. These vessels perform the pipelaying in the following sequence (see Figure 34.2).

- Handle 12 m pipe joints onto deck, using vessel cranes;
- Handle joints onto conveyors for beveling and joining joints into pairs (double joints);
- Storing double joints;
- Join double joints onto main firing line;
- Lay pipeline onto seabed without overstressing the line.

To perform these tasks the vessel should exhibit the following capabilities:

- Stable platform and constant tension tensioners to permit the line to be "S" laid into the sea (see Figure 34.3). Should the vessel move too much (i.e. due to weather) the pipeline may overstress and possibly buckle.
- Method of handling joints quickly. These vessels can install up to 5km of pipeline a day (one joint every 3.5 minutes);
- Method of welding and nondestructive testing of joints with sufficient speed to average 3.5 minutes per joint. The time given to weld a joint is assisted by the double joints, and having up to 4 welding stations on the firing line.

This means that each weld can actually take (3.5 x 4 x 2) 28 minutes to complete.

Installation Capabilities/Constraints

Pipelay semisubmersibles can install in a wide range of diameter pipelines (6" to 40") in water depths from 10 to 1500m deep, the deepest to date is 600m.

The main constraints of the pipelay vessels is the cost, they typically require 400 personnel, two anchor handling vessels, a survey vessel and the supply vessels for transporting linepipe. The total cost of the spread varies annually depending on the equipment rates etc.

34.2.2 Pipelay Ships and Barges

General Principle

Pipelay ships install pipelines in the same manner as the pipelay semisubmersibles. The principal difference is that these vessels are monohulls, and hence do not have as good sea keeping abilities as the semisubmersibles. Flat barges have worse sea keeping abilities than the ships and are used only in the calmer wave climates.

Apart from this the handling, welding and lay down of pipe is performed in the same manner as discussed for semisubmersibles (see Figure 34.1).

Installation Capabilities/Constraints

Pipelay ships have very similar installation capabilities as pipelay semisubmersibles. This includes the wide range of pipeline diameters in water depths from 15m to over 1000m.

The monohull pipelay ships have poorer seakeeping capabilities than the semisubmersibles. This results in greater periods of downtime and reduces the total time per season during which pipe can be installed.

The main advantage of the pipelay ship is the cost: the relatively smaller, dynamically positioned ships can operate without anchor handling vessel assistance. Presently none of the existing dynamic positioned semisubmersibles are equipped for pipe laying.

If the work could be confined to the summer season then a small dynamically positioned ship would provide a more economical means for installation of short pipelines than would any of the large existing semisubmersibles. If smaller dynamically positioned semisubmersibles were equipped for pipelaying purposes, however, the relative cost effectiveness between the application of a ship and a semisubmersible could be different, also in the summer season.

34.2.3 Pipelay Reel Ships

There is presently one reel ship in the UK sector of the North Sea (Stena Apache). This vessel has provided an economical tool for installing short, small diameter pipelines (see Figure 34.1).

General Principle

The pipe reeling method is applied for line sizes up to 16-inch. The pipeline is made up onshore and is reeled onto a large drum on a purpose built vessel. During the reeling process the pipe undergoes plastic deformation on the drum.

During the installation the pipe is unreeled and straightened using a special straight ramp. The pipe is then placed on the seabed in a similar configuration to that used by the laybarge (S-lay) although in most cases a steeper ramp can be used and overbend curvature is eliminated, (i.e. J-lay). Using the J-lay method very deep water depths can be achieved.

The analysis of reeled pipelay can be achieved using the same techniques as for the laybarge. Special attention must be given to the compatibility of the reeling process with the pipeline steel grade and the welding process used. Recent tests have indicated that the reeling process can cause unacceptable work hardening in higher grade steels.

A major consideration in pipeline reeling is that the plastic deformation of the pipe must be kept within limits specified by the relevant codes. The existing reelship reflects such code requirements.

Installation Capabilities/Constraints

Due to the requirement to reel the pipeline onto a small diameter drum, the pipeline experiences some plastic strain. The permissible amount of strain (and ovalisation of the pipe) limits the maximum diameter of the pipeline that can be installed using this method. Usually, depending on the wall thickness, the maximum diameter is 16-inch. Also, due to the limited size of drum, only short lengths of pipe can be laid (usually 3-15 km depending on pipe diameter). However, it is possible to install larger lines if more drums of pipe are available.

Accepting the constraints of this method, the reel method of installation has been proven to be a reliable and economical method of installing pipelines. The main advantages of the system are:

- Short offshore installation duration;
- Minimum offshore spread (no anchor handlers).

34.2.4 Tow or Pull Vessels

This method is applied to short lines, usually less than 4 km, (7 km has been laid), which would prove difficult, impossible, or more expensive to install from a pipelay vessel.

General Principle

The pipeline is fabricated onshore, and towed into the sea when it is completed (see Figure 34.4). The buoyancy of the line is selected and designed to verify that a controlled depth tow can be performed. Usually two rigs tow the pipeline to location; one leading and one trailing. On location the pipeline is positioned and flooded.

This method of installation is usually used when several flowlines are fabricated together (i.e. a bundle).

Installation Capabilities/Constraints

The main advantages of fabricating a pipeline onshore are:

- Low equipment costs compared to fabricating offshore;
- Long fabrication duration's permitting more difficult fabrication techniques to be applied. Some fabrication techniques, such as using bundling (several pipelines tied together and sometimes installed into a carrier pipe), cannot be performed by offshore vessels.
- Pipeline fabrication is not prone to weather interruptions.

However, the constraints of using the tow/pull methods are:

- Limited fabrication length of pipestrings due to the size (length) of the fabrication yard and the difficulty of controlling a long line during tow out. The maximum line length to date using this method is 7 km;
- The line should be installed in a straight line. A substantial amount of intervention work is required to install bends in the system.

34.3 Software OFFPIPE and Code Requirements

34.3.1 OFFPIPE

For pipeline installation analysis the fit-for-purpose computer program OFFPIPE may be used. OFFPIPE is a finite element method program specifically developed for the modeling and analysis of non-linear structural problems encountered in the installation of offshore pipelines.

The static analysis carried out in this chapter considers the following 2-dimensional functional external loads:

- Tension at lay barge tensioners;
- Buoyancy uniformly distributed;
- External hydrostatic pressure;
- Reaction forces from the lay barge rollers;
- Vertical seabed reaction (assumed continuous elastic) foundation.

The material modeling used by the OFFPIPE computer program is a Ramberg-Osgood material model. This Ramberg-Osgood material model used in OFFPIPE is expressed as follows:

$$\frac{\kappa}{K_y} = \frac{M}{M_y} + A\left(\frac{M}{M_y}\right)^B \tag{34.1}$$

where:

κ : pipeline curvature;

M : pipeline bending moment;

K_y : $2 \cdot \sigma_y / (E \cdot D)$;

M_y : $2 \cdot I_c \cdot \sigma_y / D$;

E : modulus of elasticity of the pipe steel;

D : diameter of the pipe steel;

I_c : cross sectional moment of inertia of the pipe steel;

σ_y : nominal yield stress of the pipe steel;

A : Ramberg-Osgood equation coefficient;

B : Ramberg-Osgood equation exponent.

The method described above is for typical standard S-lay, J-lay or reeling method installation of an offshore pipeline. The analysis can be carried out both by static analysis or dynamic in order to determine the effect of the weather conditions. For special consideration of local constraints on the pipeline in terms of structures or similar other simulation tools may be used in terms of more generalized computer programs (ANSYS, ABAQUS). More generalized computer software tolls may also be used if special installation methods should be used, where OFFPIPE is not found to be applicable.

34.3.2 Code Requirements

For pipeline installation analysis code requirements may be related to the pipeline curvature on the stinger and in the sagbend, for S-laying. A typical code is the Statoil Specification F-SD-101. For a carbon steel material complying with API-5L- X65, the code requirements are listed below:

- Pipeline overbend (stinger) 0.20 %
- Pipeline sagbend (spanning section) 0.15 %

In line with the tendency of allowing higher strains a level of 0.23 % may be used for the pipeline overbend. This is based on the recommendation in Statoil F-SD-101, Amendment 1.

It should be indicated the allowable strain for installation may be developed using limit-state based design as discussed in Chapter 4.

34.4 Physical Background for Installation

34.4.1 S-lay Method

Different technologies and equipment are adopted to install pipelines offshore. One of these methods is the S-lay method.

The lay-vessel can be either a normal vessel or a semi-submersible vessel. What makes the lay-vessel special is that it has a long ramp extension or "stinger" at the stern. At the vessel there is a near horizontal ramp. This ramp includes equipment like welding stations and tension machines. When the pipeline is welded the pipeline is fed into the sea by moving the vessel forward on its anchors. A number of rollers are placed at the stinger and vessel. These rollers support the pipeline when it moves from the vessel and into the sea. The rollers placed on the stinger and the vessel, together with the tension machines, create a curved support for the pipeline. The pipeline is bend over the curved support on its way into the sea and this part of the pipeline is named "overbend", see Figure 34.5. The stinger radius controls the overbend curvature.

Number of tension machines, the positions of them and the capacity is different for each vessel. The last tension machine is normally placed at the stern of the vessel, close to the stinger. The first tension machine is placed somewhere further forward on the horizontal ramp. The purpose of applying tension to the pipeline through these tension machines is to control the curvature of the sagbend and the moment at the stinger tip through supporting the submerged weight of the suspended part of the pipeline, see Figure 34.5. The tension capacity for the vessel depends on the capacity of each tension machine and the number of tension machines.

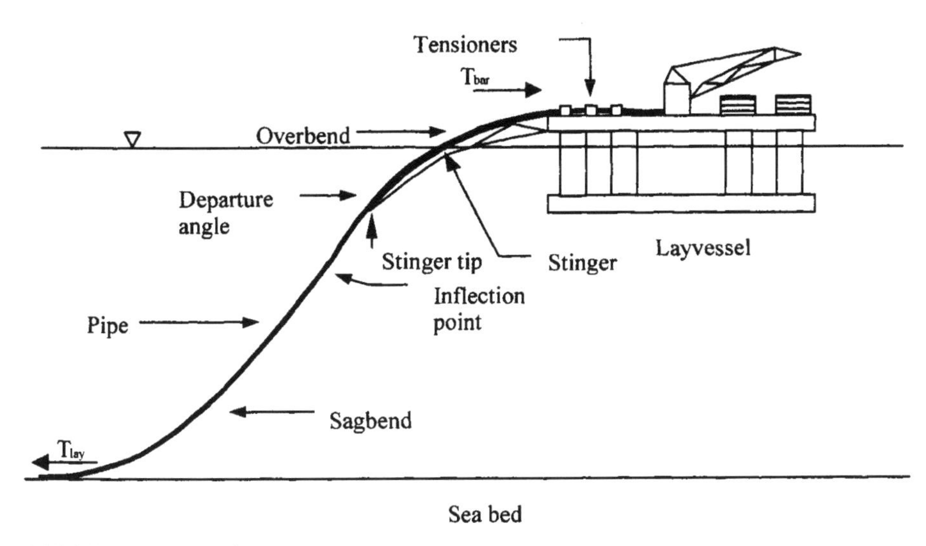

Figure 34.5 S-lay configuration.

The required tension depends on the water depth, the submerged weight of the pipeline, the allowable radius of curvature at overbend, departure angle and the allowable curvature at the sagbend.

The stinger is normally made up of more than one section. Different set-ups can be made through moving the sections relative to the vessel and each other. The position of the rollers relative to the section they belong to can also be changed. This means that a vessel can be configured for a number of different radiuses of curvature.

The stinger on a lay-vessel has limitations both for minimum and maximum radius of curvature. These limits are different for each lay-vessel. Because of this, each lay-vessel also have an upper and lower limit for the angle the pipelines can departure from the stinger. Through trim of the vessel, small changes can be made to the departure angle for a specific radius of curvature. The necessary lay-tension is very influenced by the departure angle from the stinger.

The curvature of the support for the pipeline is very often referred to as stinger radius. This doesn't mean that the stinger has a constant radius equal to this value. It is more like an average value for the radius of curvature that are made of the rollers at the stinger and vessel. A roller/support is normally build up of some wheels, see Figure 34.6.

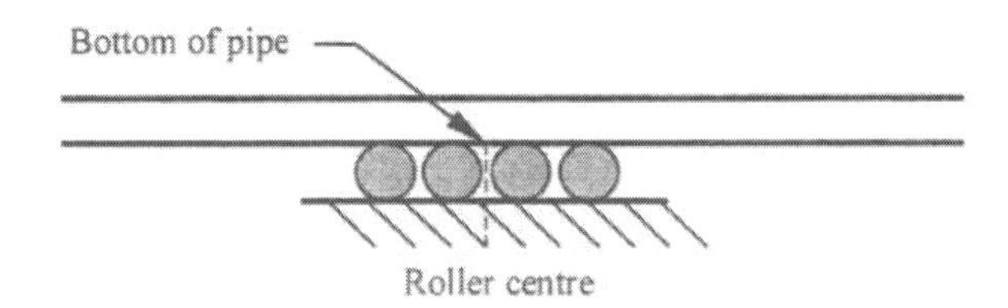

Figure 34.6 Typical roller/support for pipeline.

The tensioners normally consist of an upper and lower track loops. Wheels within the track loops apply squeeze forces to the tracks, which in turn grip the pipeline, see Figure 34.7.

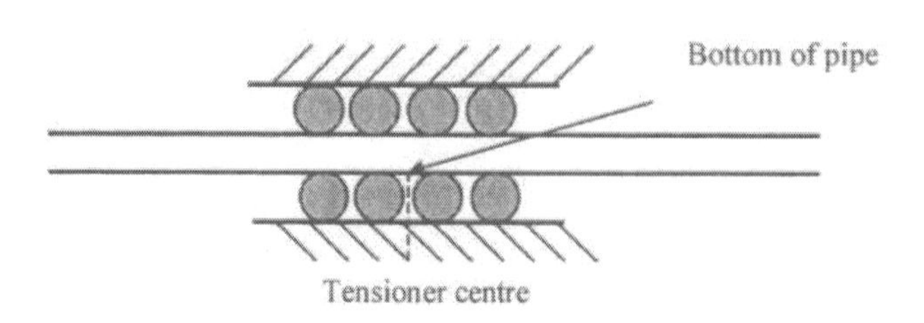

Figure 34.7 Typical tensioner support.

34.4.2 Static Configuration

During installation, the pipeline will experience a combination of loads. These loads are: tension, bending, pressure and contact forces perpendicular to the pipe axis at the supports on the stinger and at the seabed.

The static configuration of the pipeline is governed by following parameters:

- tension at the lay-vessel;
- radius of curvature for the stinger;
- roller positions;
- departure angle from stinger;
- pipe weight;
- pipe bending stiffness;
- water depth.

34.4.3 Curvature in Sagbend

Under the action of tension and pipe weight, the pipeline will exhibit large deflection from its stress free state. The curvature of the pipeline in the sagbend is governed by the applied axial tension. The simplest model for the calculation of the relationship between tension and curvature is the catenary model. The catenary model ignores the flexural rigidity of the pipeline. The horizontal component of the tension (T_h) is constant from the point where the pipeline touches the seabed and up to the stinger tip. The vertical component of this force (T_v) increases from the touch down point on the seabed and up to the stinger, because of the submerged weight of the suspended part, see Figure 34.8.

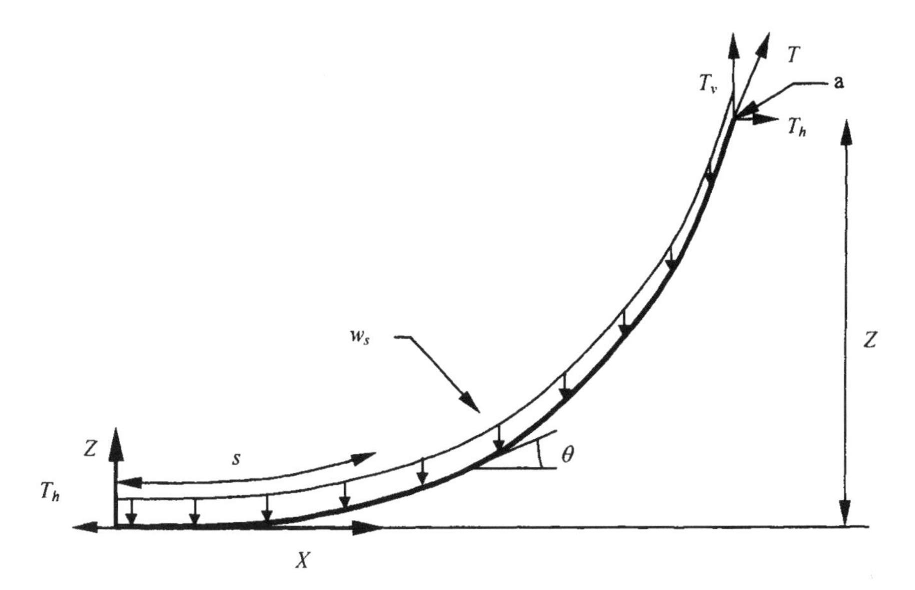

Figure 34.8 Catenary model.

Shape of the catenary can be expressed as:

$$z = \frac{T_h}{w_s}\left(\cosh\frac{x\,w_s}{T_h} - 1\right) \tag{34.2}$$

where:

x	: horizontal distance from touch down point
z	: height above seabed
T_h	: horizontal force at seabed
w_s	: submerged weight pr. unit

Curvature is then:

$$\frac{d\theta}{ds} = \frac{d^2 z}{dx^2}\cos\theta = \frac{w_s}{T_h}\ \cosh\frac{xw_s}{T_h}\ \cos\theta \tag{34.3}$$

where:

θ	: angle to the x-axes
s	: arclength

The greatest curvature is at the touch down point:

$$\frac{1}{R} = \frac{w_s}{T_h} \tag{34.4}$$

The relationship between curvature and strain for the pipe is:

$$\varepsilon = \frac{r}{R} \tag{34.5}$$

The vertical component Tv is equal to the weight of the suspended part of the pipeline:

$$T_v = w_s \ s \tag{34.6}$$

Where s is the length of the suspended part of the pipeline and can be expressed as:

$$s = z\sqrt{1 + 2\frac{T_h}{zw_s}} \tag{34.7}$$

The angle between the pipeline and the x-y plane is:

$$\tan\theta = \frac{T_v}{T_h} \tag{34.8}$$

T_h can be expressed through θ, w_s, and z by setting T_v into the expression for tan θ.

$$T_h = \frac{zw_s}{\tan^2\theta}\left(1 + \sqrt{1 + \tan^2\theta}\right) \tag{34.9}$$

The departure angle and the height above seabed at stingertip are known for a specific lay-vessel and stinger radius, while the location of the inflection point is unknown. At deep water is it reasonable to say that the departure angle from stinger tip and the angle in the inflection point are approximately the same. The inflection point in Figure 34.1 is the same as point a in Figure 34.8. The horizontal tension can therefore be estimated using Eq. (34.9). Since the inflection point and its location are unknown the tension can be estimated through using the departure angle and height above seabed at the stinger tip. The predicated tension is overestimated because θ is smaller and z is greater at the stinger tip than in the inflection point. The tension is also overestimated because the flexural rigidity of the pipeline are neglected. The calculated curvature and strain in the sagbend will be conservative because the flexural rigidity of the pipeline are neglected.

To get an accurate model the flexural rigidity of the pipeline has to be included in the analyses. This is done in the finite element model. The finite element method deals with the large deflection effects at a global level by stiffness and load updates, i.e. re-calculating stiffness and loads at the deflected shape and iterate until convergence.

34.4.4 Hydrostatic Pressure

The pipeline is exposed to hydrostatic external pressure when it is submerged. There is no internal pressure during installation. The external pressure has an effect on the pipeline response. The radial pressure will induce an axial strain via the Poisson's ratio effect.

$$\varepsilon_{xx} = -\frac{\nu}{E}(\sigma_h + \sigma_r) \tag{34.10}$$

where:

ε_{xx} : Axial strain;

ν : Poisson ratio;

σ_η : Hoop stress;

σ_r : Radial stress;

E : Young's modulus.

The hoop and radial stresses are given by the Lame's equation. If the pipe ends are free, the strain will not introduce any stress. However, if the ends are constrained, axial force will develop. This effect is similar to thermal loads.

When the pipe ends are capped, a force will be induced:

$$T_p = p_0 A_0 - p_i A_i \tag{34.11}$$

where:

p_0 : external pressure;

p_i : internal pressure;

A_0 : outside cross-sectional area;

A_i : inside cross-sectional area.

The distributed pressure on a deflected pipeline will alter the tension-stiffening effect and indirectly affect the pipeline curvature.

The effective axial tension T_e in the pipeline is defined as, see Figure 34.9.

$$T_e = T_a + T_p \tag{34.12}$$

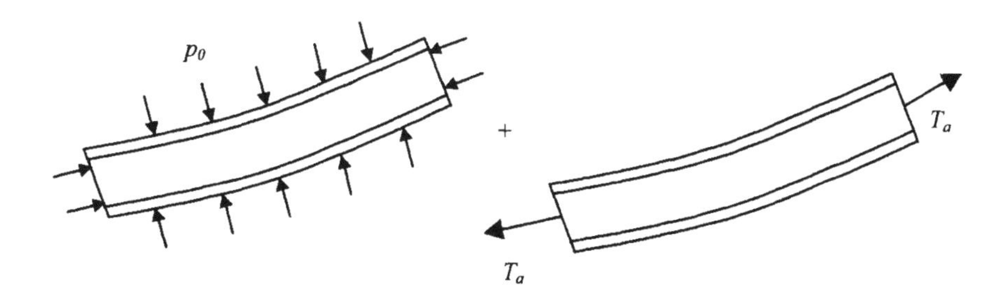

Figure 34.9 Effective axial tension.

The true tension is an integration of stress over the cross-section of the steel wall. In deep water, T_p usually are greater than T_a. The result of this is that T_e becomes negative and the pipe section, as a beam, will be in compression instead of tension. The force T_p is a function of the water depth so T_e will always be positive at the sea surface and be positive or negative at the seabed depending on the relationship between T_a and T_p.

34.4.5 Curvature in Overbend

The part of the pipeline that is supported by the layramp that are made up of the rollers placed on the stinger and the vessel will have the same curvature as the layramp. A target for installation analysis is to find the best layramp configuration for the pipeline that is going to be installed.

The layramp consists of the lay-vessel and the stinger. The function of the layramp is to provide a curved support with an overall radius of curvature. The result of this is a bending moment in the pipeline and strain. This curve is created by placing out a number of rollers at the barge and at the stinger. The location of these rollers depends on which radius of curvature that is needed to control the overbend-strain in the pipeline within acceptable level.

The configuration and curvature of the pipe section are displacement controlled at the stinger. This means that the pipeline displacement is governed by the stinger and roller properties.

The stinger and the vessel do not create a support with a constant curvature. This means that the bending moment is not constant along the pipeline on the stinger. The rollers/supports don't create a continuous support for the pipeline. The result of this is peaks in the moment level at every roller NOU (1974) and Igland (1997). The moment distribution over the stinger will therefore in principle be like illustrated in Figure34.10.

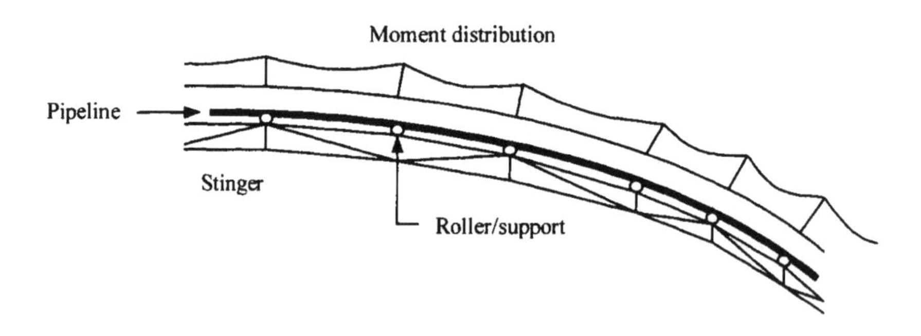

Figure 34.10 Moment distribution over stinger.

It is therefore very important to represent the stinger geometry as accurate as possible in the finite element model.

34.4.6 Strain Concentration and Residual Strain

Offshore pipelines are usually coated with concrete in order to counteract buoyancy and through this ensure on-bottom stability. The pipeline is also covered with corrosion coating. The effect of the coating weight may be easily accounted for in analysis. The concrete also has an effect on the pipe stiffness. The concrete has high compressive strength and low tensile strength. There is a discontinuity in the concrete coating on the pipeline. The most important effect of this is the occurrence of strain concentration at field joints during bending of the pipeline. The effect the concrete has on stiffness and strain in the pipeline is not accounted for in the model.

During installation, the pipeline is exposed to plastic strains when the pipeline passes over the stinger and exceeding a certain curvature. This means that the pipeline leaves the stinger with a residual curvature. When passing the inflection point, the bending of the pipeline is reversed; i.e. the residual curvature has to be overcome. This occurs partially through bending and partially through twisting. The pipeline will have residual strain when it is installed at the seabed because it has been exposed to plastic strains (Endal et al., 1995).

34.4.7 Rigid Section in the Pipeline

A valve has bigger outer diameter and is more rigid than the adjacent pipeline. Both these facts have an effect on the pipeline response. The result of a more rigid section in the pipeline is a higher bending moment. The increase in bending moment induces higher strains in the adjacent pipeline. The increase in bending moment because of the fact that the valve is more rigid will occur both in overbend and in the sagbend. To reduce the bending moment in the sagbend a higher lay tension can be applied to the pipeline. The lay tension will then have to be higher than normal as long as the valve is located in the sagbend.

Reducing the bending moment in overbend can be more complicated. When the valve is located at a support, the pipeline configuration will be lifted locally because the valve has a bigger outer radius than the pipeline. The bending moment is greatest when the valve is located at a support, because the pipeline then is lifted. The distance the pipeline is lifted is here named offset, see Figure 34.11.

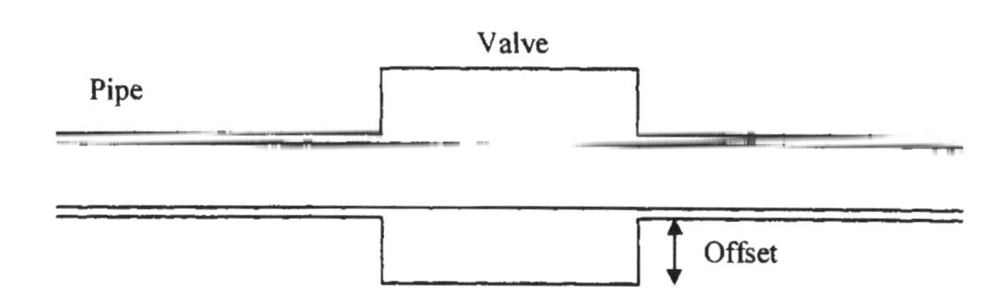

Figure 34.11 Offset of pipeline when valve located at support.

One way to reduce the bending moment in overbend is to increase the stinger radius. The lay tension for the entire pipeline will be higher if the stinger radius is increased to reduce the moment. Increasing the stinger radius may not reduce the moment enough in overbend. Keeping the strain in the adjacent pipeline at an acceptable level may require strapping of wood timber (or similar) onto the adjacent pipeline sections. This strapping of wood timber (or similar) are here named tapering. The tapering can be strapped to the underside of the pipeline on each side of the valve, see Figure 34.12. This tapering can have different shapes. It can be linear, parabolic or have other shapes.

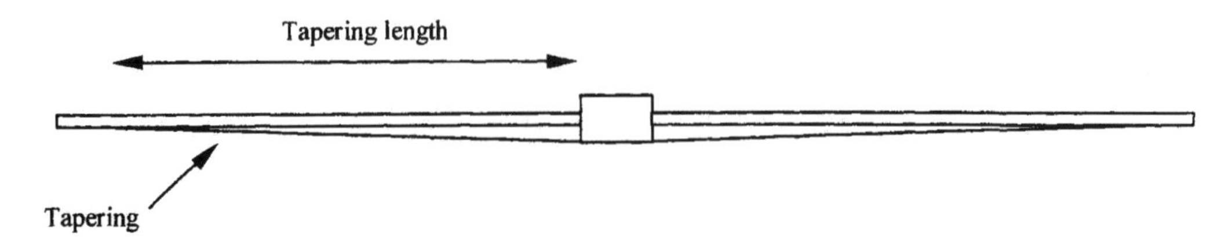

Figure 34.12 Principle for tapering of pipeline with valve.

The positive effect from this tapering is a reduction of the effect the outer radius has on the moment due to the fact that there will be an offset of the pipeline when the valve are located at a support. The tapering has to be able to withstand the loads between the pipeline and the support. If the pipeline is not tapered, the reaction force normal to the valve will be very high locally. By tapering the pipeline, the reaction force will be better distributed to the adjacent supports. The local change in pipe curvature will be less as a result of this.

34.4.8 Dry Weight/Submerged Weight

To be able to calculate the response of the pipeline we need the dry weight and the submerged weight of the pipeline. In ABAQUS, one way to represent the dry weight of the pipeline is through the density of the steel cross-section. The dry weight is then calculated as a function of the pipe external diameter, the wall thickness, the density of the steel (material) and the acceleration due to gravity.

The pipeline is represented with a mass when the weight of the pipeline is modeled this way instead of as a distributed load. This makes it possible to perform dynamic analysis. Dynamic analysis will be performed after further development of the present model.

In the "real life" the steel pipe is covered with corrosion coating and concrete coating. The steel pipe, corrosion coating and concrete coating have different density. In the analyses, the bare steel pipe is used to represent the pipeline.

The total weight of the pipeline has to be represented by the steel pipe because just the bare steel pipe is used in the analyses. Another density has to be used for the steel (material) than the steel density from the design data to be able to represent the total weight of the pipeline by the steel pipe. The density used for the cross-section to the pipeline is therefore calculated as explained below.

$$A_s = \frac{\pi}{4}\left((D_i + 2t_s)^2 - D_i^2\right) \tag{34.13}$$

$$A_{con} = \frac{\pi}{4}\left((D_i + 2(t_s + t_{cor} + t_{con}))^2 - (D_i + 2(t_s + t_{cor}))^2\right) \tag{34.14}$$

$$A_{cor} = \frac{\pi}{4}\left((D_i + 2(t_s + t_{cor}))^2 - (D_i + 2t_s)^2\right) \tag{34.15}$$

$$\rho_{inp} = \frac{A_s\rho_s + A_{cor}\rho_{cor} + A_{con}\rho_{con}}{A_s} \tag{34.16}$$

where:

A_s : Cross section area of steel [m];
A_{cor} : Cross section area of coating [m];
A_{con} : Cross section area of concrete [m];
D_i : Pipeline internal diameter [m];
t_s : Pipeline steel wall thickness [m];
t_{co} : Corrosion coating thickness [m];
t_{con} : Concrete coating thickness [m];
ρ_s : Steel density [Kg/m^3];
ρ_{cor} : Corrosion coating density [Kg/m^3];
ρ_{con} : Concrete coating density (with 4% water [Kg/m^3];
ρ_{inp} : Density for input [Kg/m^3].

This density together with the steel pipe outer diameter and wall thickness as input makes ABAQUS able to calculate the dry weight of the pipe. The dry weight of the pipe is here thought of as the weight of the pipe in air.

During the installation, a part of the pipeline will be above the sea-surface and rest of the pipeline will be under the sea-surface. From a point at the stinger, the pipeline will be into the water. The pipeline will then be exposed to a buoyancy force and hydrostatic pressure.

This is applied to the pipeline in ABAQUS by using a command named PB. This command applies a distributed pressure load and a distributed buoyancy load to the submerged part of the pipeline.

When computing the distributed buoyancy loads (load type PB) ABAQUS assumes closed-end conditions. The pressure field varies with the vertical co-ordinate z. For the hydrostatic pressure the dependence on the vertical co-ordinate is linear in z,

$$p = \rho\, g(z_0 - z) \tag{34.17}$$

Here z_0 is the vertical location of the free surface of the fluid, ρ is the density of the water and g is acceleration due to gravity.

The calculation of the pressure load and buoyancy load on the pipeline is based on the outer diameter of the pipeline. This is a problem since the outer diameter for computing the buoyancy load has to be different from the outer diameter that is used for computing the right pressure load on the pipeline. This is because the pipeline is covered with concrete and coating which contribute to the buoyancy load but not to the pressure load. The outer diameter of the steel pipe is defined to give the right pressure load. This means that the buoyancy load is too small when the command PB is used to specify the buoyancy and pressure load on the pipeline. A User Subroutine is used to specify a distributed load to the pipeline to get the right submerged weight. This load is applied to the pipeline at the same time as the buoyancy load computed by PB. The magnitude of this load is equal to the difference in buoyancy load caused by the fact that the outer diameter of the steel pipe is smaller than the outer diameter for a pipe covered with concrete and coating. The magnitude of the distributed load specified in the user subroutine is computed as explained below.

$$b_d = b_a - b_{pb} \tag{34.18}$$

$$b_a = \frac{D_{oc}^{\ 2}\pi}{4}\rho\, g \tag{34.19}$$

$$b_{pb} = \frac{D_{os}^{\ 2}\pi}{4}\rho\, g \tag{34.20}$$

where:

D_{oc} : Outer diameter of pipe with concrete and coating

D_{os} : Outer diameter of steel pipe

b_a : Actual buoyancy load/m

b_{pb} : Buoyancy load (PB)/m

b_d : Distributed load from user subroutine/m

The User Subroutine DLOAD has been used. The subroutine can be used to define the variation of a distributed load magnitude as a function of position, time, element number, etc. This subroutine is made such that the calculated distributed load only will be applied to elements beneath the still water surface.

34.4.9 Theoretical Aspects of Pipe Rotation

Severe pipe rotation has been experienced during deepwater pipelaying, but the reasons causing the phenomenon are not understood in the industry. While analytical models have demonstrated the influence of residual curvature on pipe rotation, 3D FE simulations of the pipelay process are needed to predict rotation.

This section, which is taken from Damsleth et al. (1999), deals with the consequences of the plastic strain that can occur in the outer fibres of the pipe wall as it passes over the stinger during laying. Endal et al. (1995) have shown that the pipe twists, i.e. it rotates around its axis. They also show that, provided the plastic strain is small, the on-bottom configuration is straight and flat as for an entirely elastic process. Thus the main consequence is the rotation during pipelaying. They also state that pipeline twist acts only in the elastic sagbend (or underbend) section and characterize it as a typical instability phenomenon. These aspects will be reviewed here as we elaborate their theoretical approach, the main modification being the inclusion of the gravitational potential energy.

During installation, the pipe extends from the horizontal tension machine, bends over the stinger and, while sloping downward through the water, bends gradually in the opposite direction onto the horizontal seabed. The tensioner provides the upper support for the pipe while the seabed provides the lower support where residual tension is balanced by friction. Customary terms used to describe this S-lay pipe configuration are overbend, inflection point and underbend, or sagbend.

A structural analysis of this system requires the specification of the properties of the pipe (stiffness, weight), the configuration of the stinger, any environmental loads and the contact conditions on the seabed and the stinger. For the purpose of this discussion the dynamic effects that arise from lay vessel motions, waves and the motion of the pipe will be neglected.

In order to predict the equilibrium configuration, the principle of virtual work is used:

$$\int_V \sigma\delta\varepsilon \; dV = \int_S \delta\vec{u} \cdot \vec{t}\, dS + \int_V \delta\vec{u} \cdot \vec{f} dV \tag{34.21}$$

Customary symbols represent stress, strain and displacement. Here **t** represents the surface tractions and **f** the body forces. Notice that the integrals are over the volume of the pipe and over the surfaces of the pipe, both inner and outer.

If the deformation of the cross-section is neglected, the integrals simplify considerably and they can be transformed into line integrals. We will assume this has been done. Let us consider the different terms.

The body force in the pipe-lay problem is the weight of the pipe and its contents. The energy related to this virtual work term is the potential energy, which is:

$$-\int z(\lambda g)ds \tag{34.22}$$

where z represents the vertical coordinate of the center line, λ the mass per unit length, g the acceleration of gravity and ds the length of the line element along the center line.

The surface integral arises from the surface tractions, which are of diverse origins. The pressure of the contents, the water pressure and the water-motion tractions on the wetted part of the outer surface are the most obvious. Contact stresses arise on the seabed and on the stinger. The pressure integrals are easily integrated and give rise to pressure terms in the effective force and the buoyancy.

The contact surface tractions are important in connection with pipe twist. An example is illustrative: suppose the touch-down point of the pipe on the seabed is forced sideways. Friction forces arise due to the transverse displacement, and these forces will tend to twist the pipe around the centerline because the outer radius acts as the lever of the force. The internal virtual work must be evaluated taking into account the stress-strain history of each material point, so that a correct plastic state is maintained. For pipelay, plastic flow is only allowed over the stinger while elastic conditions are required in the underbend. We can, therefore, formulate the strain energy as:

$$\int \left(c_1 T^2 + c_2 M_t^2 + c_3 M_b^2\right) ds \tag{34.23}$$

where the three constants represent section-factors and T = axial force, M_b = bending moment and M_t = twisting moment.

There is an interesting difference between the contribution of the two moment terms along the centerline. From the solution of the elementary problem of a horizontal long beam tensioned axially in the gravitational field, it follows that only at distances less than:

$$\sqrt{\frac{EI}{T}} \tag{34.24}$$

from the ends does the bending moment differ from a constant term. The rest behaves as a catenary. For pipelay, this distance is normally small compared to the length of the free span. Therefore, we can say that the boundary conditions only have a local influence in bending. The twist is entirely different: A rotation of the pipe at the tensioner is immediately felt at the touch-down point. Twist acts over long distances, as does the gravity force.

How does the residual strain in the overbend change the value of the potential energy? An example will illustrate the point: Consider first that the suspended pipe is entirely in the vertical plane. Assume two pipelay scenarios that only differ because one material remains completely elastic, whereas the other experiences plastic strains in the overbend section on the stinger. In the underbend section, the pipe with plastic strain hangs higher than the elastic one because its natural (unloaded) shape has become convex. This means that the potential energy is higher for the plastically deformed pipe than for the elastic one. Allowing for a 3D deformation, the bent pipe can reduce its potential energy through twist. The elastic pipe is already at its lowest potential energy and so it is stable.

It is reasonable to conclude from this argumentation that the reduction of potential energy is the mechanism that underlies pipeline rotation during pipelaying. The theory of large deflection of beams is found in classic texts, e.g. Landau or Love. A non-linear 3D finite element program can solve the virtual work equation with very few approximations. Three simple models will illustrate the main point of interest. All represent a pipe of length 1218 m and with D/t=36. They are all fixed in one end and pinned in the other where a sliding condition is specified. Both ends are at the same elevation and the body force is equal to the submerged weight. In order to produce elastic strains below 0.035% an appropriate horizontal force is applied in the pinned end to represent the lay tension.

A 3D load-case is created by means of a horizontal force corresponding to a sea current of 0.5 m/s that is applied normal to the plane of the equilibrium configuration. First the horizontal force is applied, and then the submerged weight. Before the application of the horizontal force, the pinned end is locked in all translational degrees of freedom at their current values. The models are:

1. Straight pipe;
2. Pre-curved "overbend" pipe, R=571 m;

3. Pre-curved "underbend" pipe, R=571 m;

The displacement and rotation of a point in the middle of the span will be studied for each of the models. The equilibrium configurations shown are similar to that of the underbend during pipelay where the pipe is subject to its submerged weight and axial tension. The pre-curved overbend pipe represents a pipe that has been plastically deformed on the stinger to give 0.1% residual strain while the pre-curved underbend pipe illustrates a naturally stable case from the gravitational viewpoint. The reference system has its x-axis origin at the fixed point with positive direction towards right in the figures below and positive z is upward.

The models are shown in the two figures below, where both unloaded shapes and equilibrium configurations under tension and weight are included. One important observation can be made from the equilibrium of the overbend curve: since the midpoint is below the horizontal plane, this pipe would have ended up flat if laid on the seabed, as found by Endal et al. (1995).

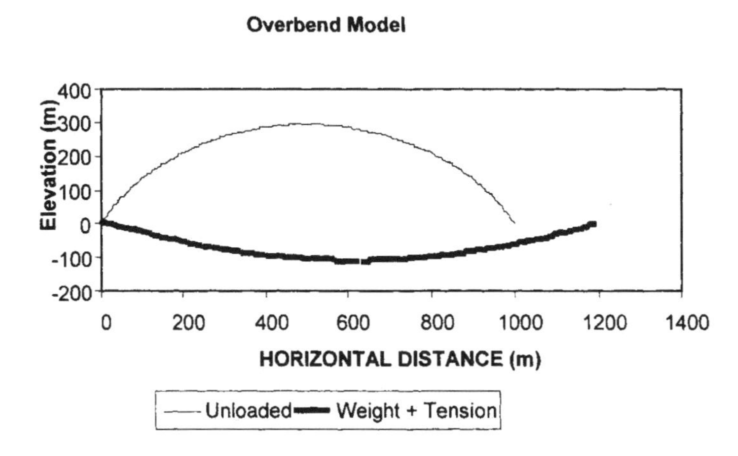

Figure 34.13 Pre-curved "overbend" model in its free and loaded conditions.

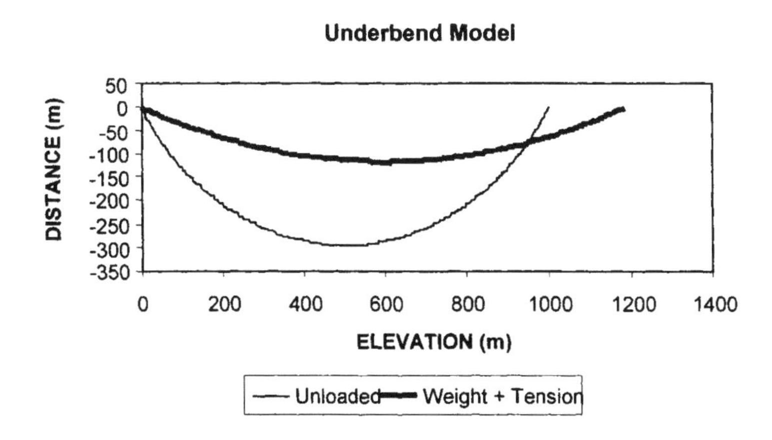

Figure 34.14 Pre-curved "underbend" model in its free and loaded conditions.

The next figure shows the lateral deflections under horizontal loading. All three curves show practically the same deflections as expected.

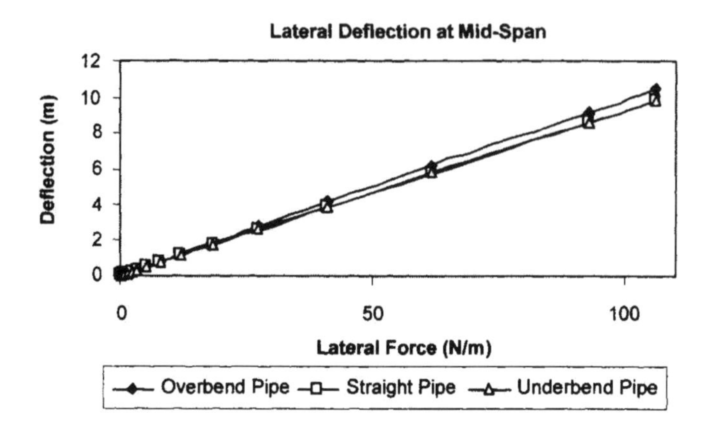

Figure 34.15 Displacement at mid-span subject to a lateral force.

The twist of the midpoint is exhibited in the figure below. The horizontal axis represents the evolution of the rotation as the applied horizontal force raises from zero to full value. First consider the straight model. It has negligible twist around the x axis. The underbend curve rotates at maximum one degree in the direction of a pendulum. However, the overbend curve shows a rotation of 17 degrees in the negative direction. When it rotates it tries to tip over from the overbend shape into the underbend shape to reduce its potential energy. At the same time twist energy starts to increase, reaching equilibrium when the two energy changes are equally large.

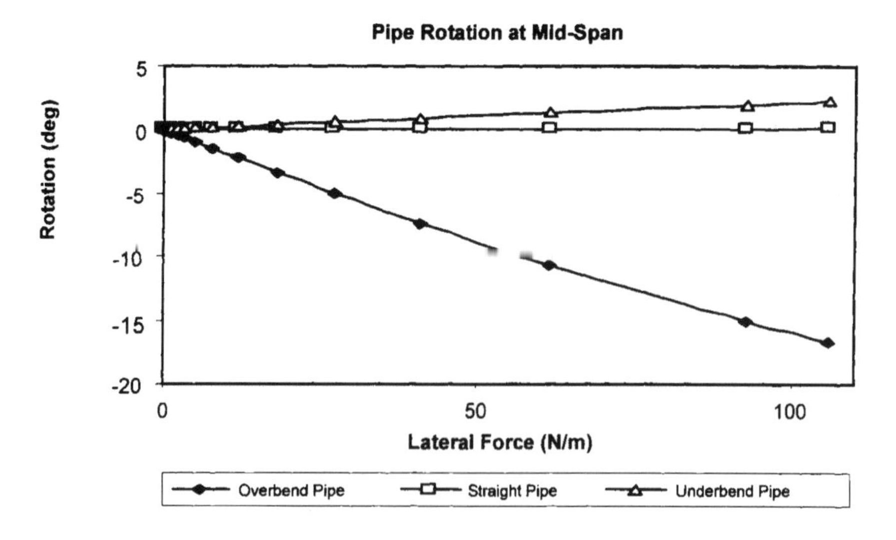

Figure 34.16 Rotation at mid-span subject to a lateral force.

While twist cannot occur in a 2D case, the simple models demonstrate that 3D is a necessary, but not sufficient, condition for twist. Endal et al. (1995) characterize the twist as an instability as shown in their figure 4. We have repeated their analysis with a stiff pipe (D/t=36) where the twist emerges much more slowly. In a model of the pipelay process, the initiation of the lay with plastic strain over the stinger shows that the twist occurs even more slowly. After a few kilometers of pipelay, the twist in one joint has a constant angular velocity as the joint leaves the stinger and descends toward the seabed. The nature of the twist phenomenon is thus, in general, not an instability.

A last observation deals with the modeling of pipelay process by means of beam elements in a general-purpose finite-element program. These elements represent an elastic line and therefore they have zero section radii. As a consequence, they cannot represent the coupling between transverse motion and twist as discussed above. Special-purpose elements have to be used.

The examples illustrate elastic twist behavior that reduces the overbend pipe's potential energy when subject to out-of-plane loads due to current or lateral displacements.

34.4.10 Installation Behaviour of Pipe with Residual Curvature

Pipelay vessels have gradually adapted to the technical challenges of deepwater projects by increasing tension capacity and stinger length. The larger lay vessels have reached physical limitations where further increase in their capacity would, in principle, be too costly for a low oil price scenario. Increasing the utilization of the pipe strength capacity by curving the stinger more sharply to obtain steeper departure angles is a cost-effective alternative. Since the tension required to install the pipe will be lower, it brings the added benefit of reducing the seabed intervention needed for freespan support. See Damsleth et al. (1999).

Today's larger S-lay vessels are fitted with total tension capacity of 300 to 500 tonnes. The stingers are 60 to 100 m long to cope with installing pipelines in 300m to 700m water depths. But the present 45° to 55° stinger departure angles result in about half the lay tension remaining with the pipe on the seabed. In areas where the seabed is uneven, the high residual tension develops both larger and more frequent freespans. In order to obtain the lowest residual tension, the stinger must provide as steep departure angle as possible.

The stingers of most of the larger pipelay vessels have already been extended to install increasing pipe sizes in deeper water. Extending them further would make them more vulnerable to environmental loads and increase weather downtime. To install large diameter pipe in very deep water (1500m to 2500m) with the present tension capacity requires stingers with up to 90-degree departure angles. The present stinger arc lengths can be maintained while the curvature is increased. Depending on the D/t of a given pipe size, a permanent curvature in the overbend may develop causing eventual pipe rotation.

While the controlled curvature of the stinger permits the use of strain criteria, deeper water installation demands stinger curvature leading to greater plastic deformation of the pipe in the overbend. Detailed structural analysis can be used to develop project-specific strain criteria for installation (Bai et al. 1999) that allows plastic strain in the overbend. However, it has been demonstrated that permanent curvature of the pipe can potentially lead to unacceptable rotation where Tees and other fixtures are to be installed in the line. This phenomenon need not become an installation problem provided the rotation can be predicted and controlled.

It is difficult to quantify pipeline twist for the construction phase since the behavior of the pipeline during installation is specific to the pipe characteristics and the installation configuration. While design codes provide criteria for maximum overbend strain to avoid pipeline twist, the resulting lay configuration may be too costly. Or, strain concentrations due to coatings, under-matched welds, buckle arrestors and other in-line components may produce permanent overbend curvature that could cause pipe rotation.

Therefore, non-linear 3D FE models using elasto-plastic beam and friction/contact elements are used to analyze the load history of the pipeline during the pipelay process that accounts for the complex interaction between constant as well as time and position varying loads involving all 6 degrees of freedom. The FE model can simulate pipeline rotation to determine whether control measures are necessary as well as demonstrate the effectiveness of correction measures.

The three figures below illustrate the twist phenomenon during laying of a 2.4 km section of deepwater pipeline with a 0.5 m/s lateral current. Pipelay initiation was by dead man anchor so that the end was free to rotate. Rotational friction on the seabed is ignored in this case.

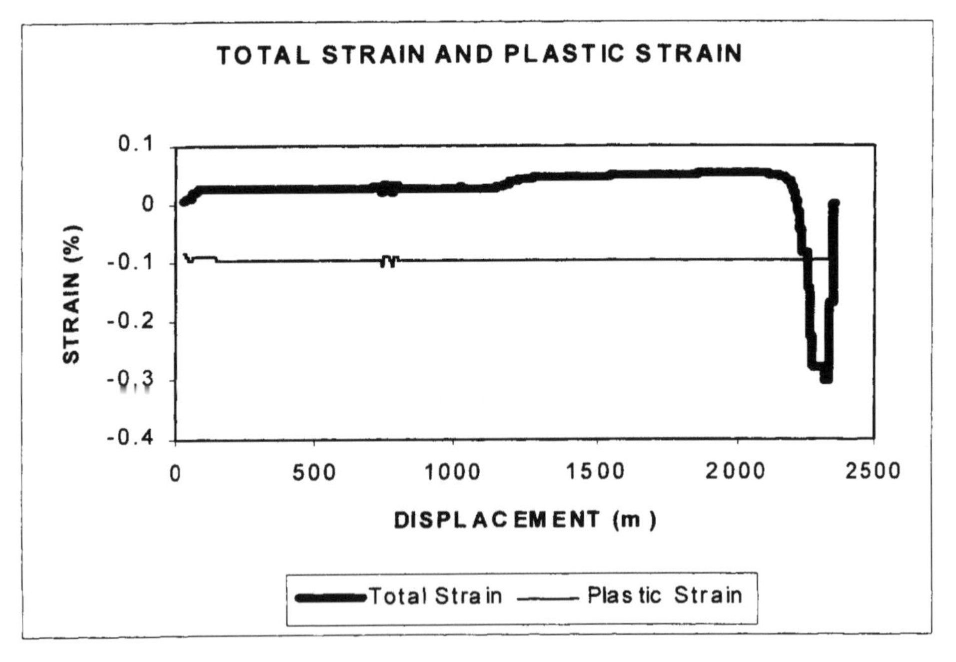

Figure 34.17 Total axial strain and plastic strain in an elasto-plastic pipe from the free end on the seabed to the tensioner on the laybarge.

Figure 34.17 shows the situation after 2.4 km of pipelay indicating total strain and permanent strain in the vertical plane after the pipe has been subject to elasto-plastic bending over the stinger during laying. Figure 34.18 shows a net rotation of 60 degrees of the free end due to the twist effect for an elasto-plastic pipe material and 0 degrees for a completely elastic pipe. Figure 34.19 shows the resulting rotational moment along the elasto-plastic pipe that compared to the near-zero moment of the elastic pipe. The only difference between the curves is due to the 0.1% residual strain of the elasto-plastic pipe material, demonstrating that plastic strain, combined with a lateral disturbing force, is the source of the pipe rotation.

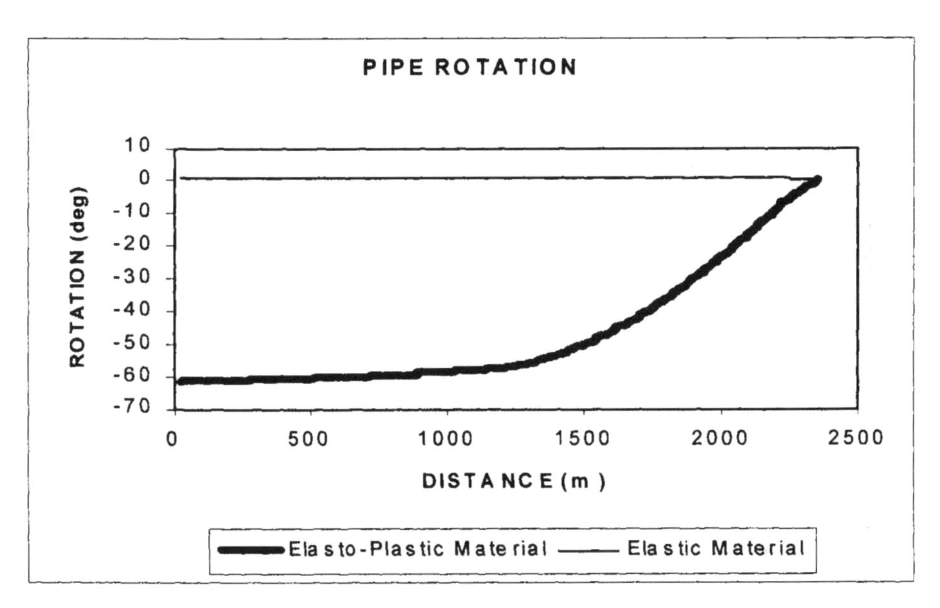

Figure 34.18 Axial rotation of the pipeline from the free end.

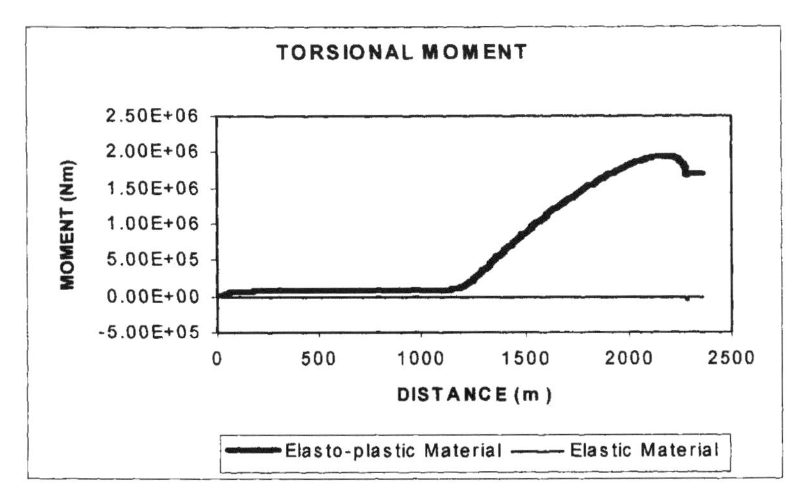

Figure 34.19 Torsion moment in the pipeline from the free end on the seabed to the tensioner on the laybarge.

With 0.1 % residual strain, this pipe lays flat on the seabed. The 60° rotation along the free-span and seabed does not pose a problem unless in-line components requiring access are present. In this case, the most cost-effective solution is to design the pig launcher for multiple point access because the pipe may continue rotating as the lay vessel moves forward until friction builds up.

The presence of an in-line Tee with +/-15° access limitations demands that the pipeline rotation is controllable or predictable. A simulation of the pipelay process as shown above demonstrates that unacceptable rotation is expected for the given pipe configuration. Further simulations can demonstrate the effectiveness of various corrective measures such as added buoyancy, pre-rotating the Tee, increasing tension, current monitoring and vessel offset. The simulation results will allow operators to develop contingency strategies to avoid unacceptable rotation, depending on available means and equipment.

34.5 Finite Element Analysis Procedure for Installation of In-line Valves

34.5.1 Finding Static Configuration

A Finite Element Analysis by Martinsen (1998), a M.Sc. Thesis supervised by the author, is given below.

The initial configuration for the pipeline is a straight line when starting to find the static configuration for the pipeline. The pipeline is stress free in its initial configuration. All nodes in the pipe are fixed against displacement in y-direction and against rotation about the x-and the y-axes. All degrees of freedom are fixed at the pipe end located at the lay-vessel (first tensioner). Load step 1 is to apply a horizontal concentrated force at the other end, see Figure 34.20. An estimate of the necessary force can be calculated with Equation (34.9).

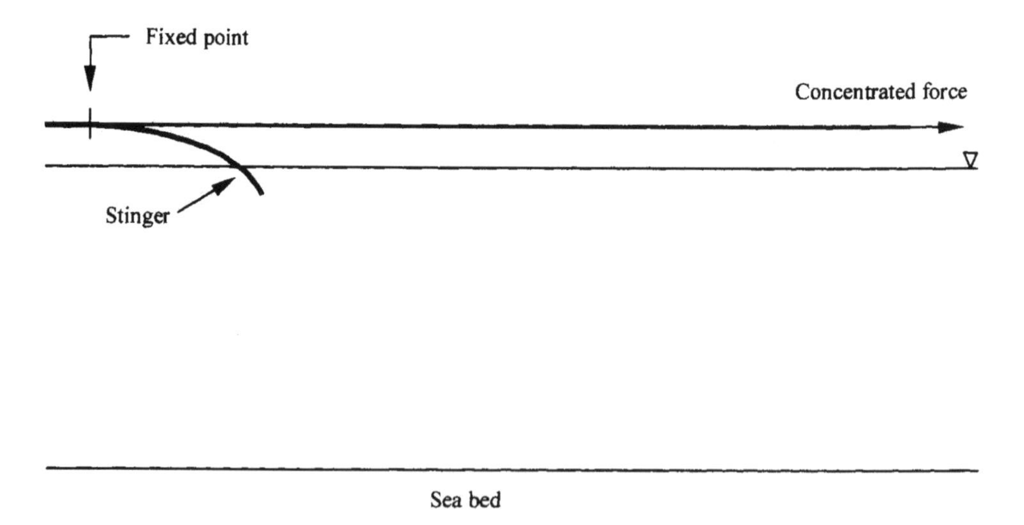

Figure 34.20 Initial configuration and load step 1.

Load step 2 is a prescribed displacement in vertical direction of the node where the concentrated force is applied, see Figure 34.21. The displacement of this node is equal to the initial distance between the end node and the seabed. The prescribed displacement induces a displacement of the node in x-direction and rotation about the z-axes. The concentrated force is a follower force. This means that the direction of the force rotate with the rotation of the node. During this step a part of the pipeline will encounter the stinger. This part of the pipe is bend. The rest of the pipe is almost straight.

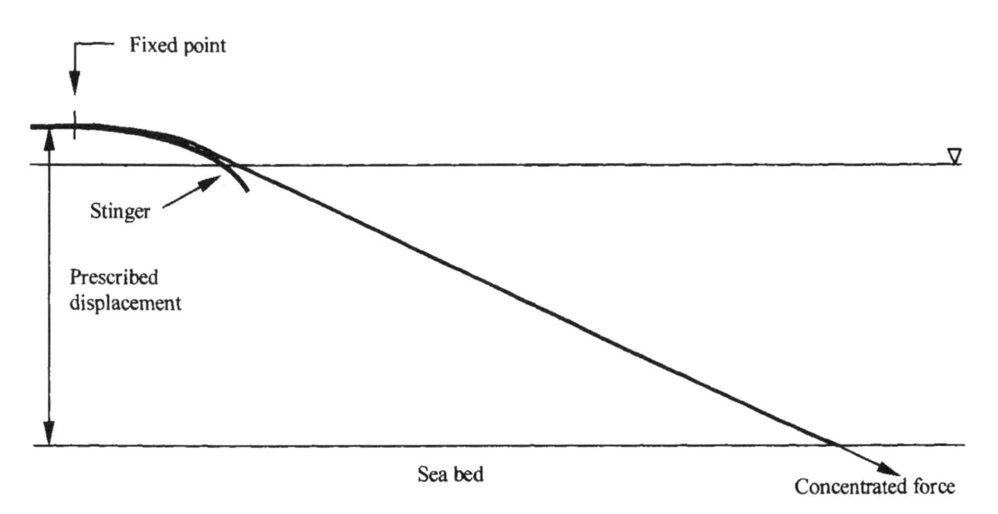

Figure 34.21 Configuration after load step 2.

As the last load step, load step 3, the dry weight, buoyancy, pressure and the distributed load specified in the user subroutine are applied. The node where the concentrated force is applied moves left until the pipeline has found static equilibrium, see Figure 34.22. The pipeline has to be long enough in its initial configuration so a part of the pipeline is lying horizontal and slides on the seabed when the static configuration are computed.

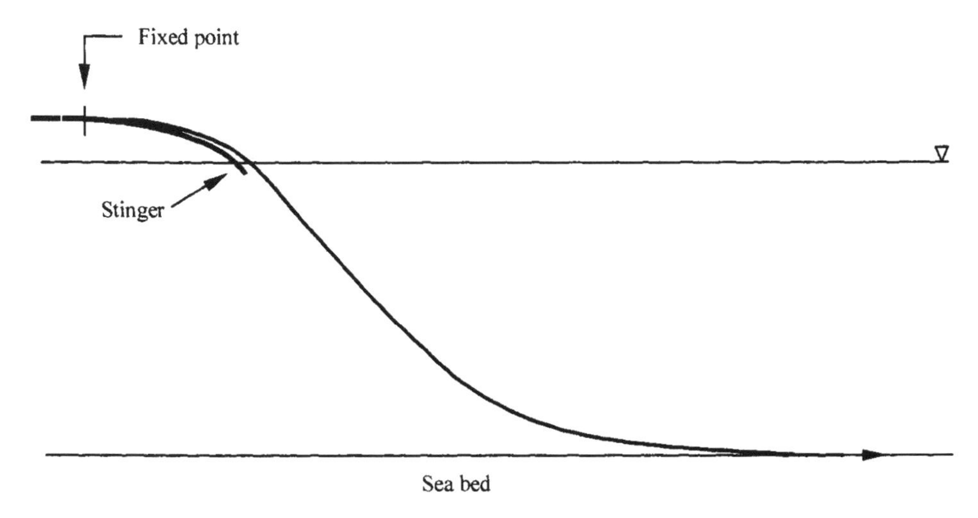

Figure 34.22 Configuration after load step 3.

The required tension at the laybarge for installation of the pipeline is the reaction- force at the fixed node. The load effects on the pipeline for the applied tension has to be checked against design criteria's. The result of this check may be that the applied tension has to be changed. The pipeline has found its static configuration but the design criteria tell us that the tension has to be increased or that it can be decreased. If Equation (34.9) is used to calculate the tension, the tension normally can be decreased. One way is to change the applied tension in the original input file and run the analysis all over again. To save computing time the file can be restarted. An analysis is restarted (continued) by including the RESTART, READ option in an input file. This option will read the result file created by the original analysis. A file including the RESTART, READ option with a load step similar to load step 1, but with a new concentrated load, has to be created. The analysis will then continue from the last increment at the last load step in the original file. This procedure can be done repeatedly until a load that satisfies the design criteria's has been applied. For each new try, there is enough to change the concentrated load in the restart file and read from the original result file.

It is possible that the pipeline has been overloaded during the process finding the necessary tension. The material is not perfect elastic and an overload will then have an effect on the result. The tension that is found to satisfy the design criteria has to be used as input in the original file as a last check.

The same basic procedure (the three load steps) is used for a pipeline with valve. The stinger configuration has to be changed to account for the thickness of the valve and the tapering. The valve has to be represented by more rigid section than the adjacent pipeline.

The first case is to find what lay-tension that is necessary to satisfy the design criteria's when the valve is located in the sagbend.

The second case is to place the valve at the support on the stinger that results in highest strains for a pipeline without valve.

A third case can be to place the valve at the support where the distance to the adjacent supports is greatest.

Analyses with different lengths and types of tapering have to be performed for the cases where the valve is located at the stinger. The result of this is that a lot of analyses have to be performed for each case. This is the reason for making a spreadsheet that computes some of the input to the ABAQUS file that will change from case to case.

34.5.2 Pipeline Sliding on Stinger

Pipeline with a length equal to the length that is going to be installed at the seabed has to be specified in front of the first tensioner (fixed point). A horizontal surface is also specified in front of the first tensioner to support this part of the pipeline. The pipeline is fixed at two nodes on the layramp. These nodes are the one located at the same place as the first tensioner and the end node of the pipeline located at the "vessel". The initial configuration of the pipeline will then be as in Figure 34.23.

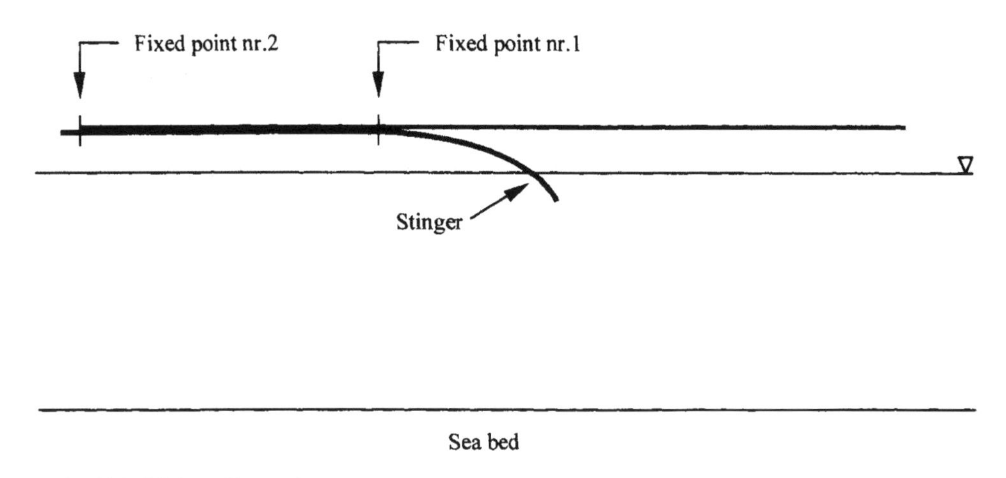

Figure 34.23 Initial configuration.

The first three load steps are the same as used for finding the static configuration explained earlier. Weight, buoyancy and pressure are applied to the entire pipeline in step 3. This means that weight is applied to the entire pipeline and pressure and buoyancy to the part of the pipeline that are submerged. Next step is to change the boundary conditions in fixed point nr.1. The node in this point is then released. This is step no. 4. The pipeline separates 0.01-0.1 m from the stinger at this point when the node is released. The result of this is a small change in the static configuration for the pipeline and the contact forces between the pipeline and the stinger.

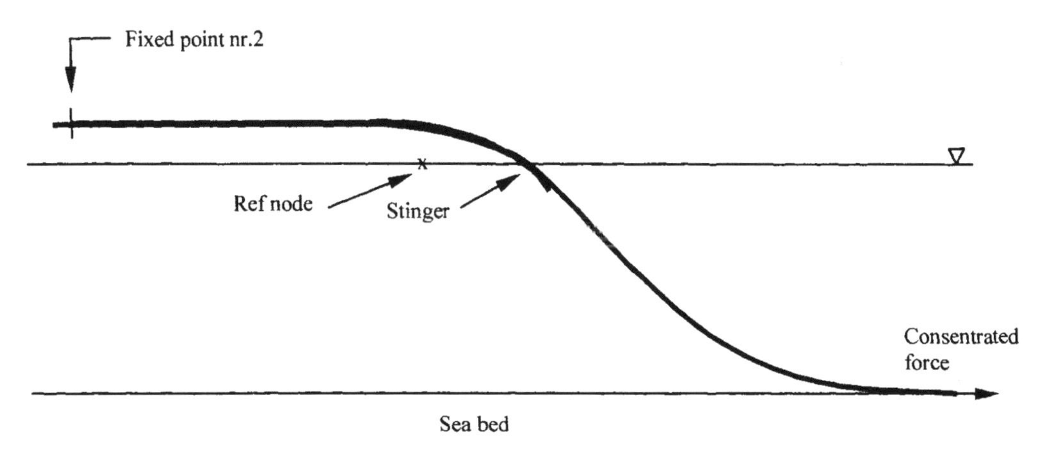

Figure 34.24 Pipeline configuration before the pipeline slides on the stinger.

The next and last step is to move the surface representing the lay-vessel/stinger towards left. The whole lay-vessel/stinger has one single reference node. The lay-vessel/stinger is then moved through moving the reference node in Figure 34.24. Point nr.2 remains fixed. A pipeline cross section will then move from the vessel, over the stinger and through the sagbend to the seabed.

Convergence problems often occur when the node in fixed point nr.1 is released. The longer the distance between fixed point nr.2 and fixed point nr.1 is, the more difficult it is to make the model converge.

34.5.3 Installation of In-line Valve

The purpose with this design example is to illustrate the effect an in-line valve has on the strain level in the pipeline. Analyses will be performed for valve located in sagbend and valve located at a support on the stinger. What effect tapering of the pipeline can have on the strain level will also be illustrated.

The results will also be compared with design criteria's regarding the allowable strain level in sagbend and overbend, as defined in Statoil specification F-SD-101.

- Pipeline overbend (stinger) 0.23%
- Pipeline sagbend (suspended part) 0.15%
- No contact between pipeline and last support

Less conservative criteria may be defined based on the principles presented in Chapter 4.

The problem with installation of an in-line valve is the increase in bending moment and strain locally because the valve has stiffness larger than the adjacent pipeline. The increase in bending moment because of the fact that the valve has a larger stiffness will occur both in overbend and in the sagbend. The increase in bending moment induces higher strains in the adjacent pipeline. If the strain in the adjacent pipeline in the sagbend exceeds the design criteria a higher lay-tension can be applied to the pipeline to reduce the strain.

When a valve is placed at a support, the adjacent pipeline will be lifted as a result of the contact between the valve and the support. This also leads to an increased bending moment locally. The result of these two effects is that the strain in the adjacent pipeline increases. To reduce the increased bending moment because the pipeline is lifted, the pipeline can be tapered.

An example design analysis was performed by Martinsen (1998).

34.6 Two Medium Pipeline Design Concept

34.6.1 Introduction

The design and construction of pipelines and flowlines is one of the key issues for the development of deepwater production and transportation facilities. The installation of large diameter trunklines has been limited to around 600m (Rivett, 1997). Smaller diameter flowlines have been installed in as much as 1000m depth. New challenges presented by projects currently undertaken in even deeper water are challenging the present pipeline technology and have stimulated the development of new concepts (Damsleth and S. Dretvik, 1998, Walker and Tam, 1998).

It is known that linepipe material cost takes a large portion of the CAPEX of pipeline projects. Using present technology, installation design for external pressure would govern wall thickness selection for deepwater pipelines. There is a need to develop new design concepts to avoid this situation (Palmer 1997) and make deepwater pipelines as commercially competitive as their shallow water counterparts.

Until a few years ago, pipeline design has based on simplified capacity equations and some special purpose computer programs for installation and on-bottom stability design. Recently, use of nonlinear finite element simulations and limit-state design has become acceptable practice (Bai and Damsleth, 1997, 1998) in situations where design criteria has significant cost impact. The technological advances in finite element simulation have permitted project specific optimizations that have saved up to 16% of the pipeline CAPEX development (Home, 1999) for pipelines in water depths of 350m. The potential for optimization can be even greater for deeper water pipelines.

This section presents a new design concept for deepwater installation, which is called Two Medium Pipeline (Bai et al., 1999).

34.6.2 Wall-thickness Design for Three Medium and Two Medium Pipelines

Subsea pipelines have historically been designed for three different mediums: air (during installation), water (during pre-commissioning) and finally the product (gas/oil). In shallow water, air-filled pipelines at near atmosphere pressure do not cause particular difficulty because the wall-thickness is sized for the internal pressure of the product or the pressure test. In deep water, provided the same installation and operation approach is adopted, the pipeline will be sized for external pressure (collapse/local buckling) for the installation phase. This phenomenon is clearly illustrated in Figure 34.25 and 34.26 that shows how the operation, testing and installation phases dictate the pipeline wall-thickness requirements for increasing water depths.

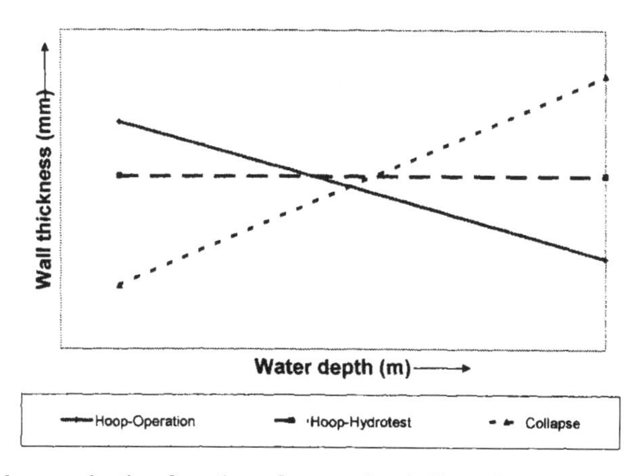

Figure 34.25 Wall thickness selection function of water depth (3 mediums).

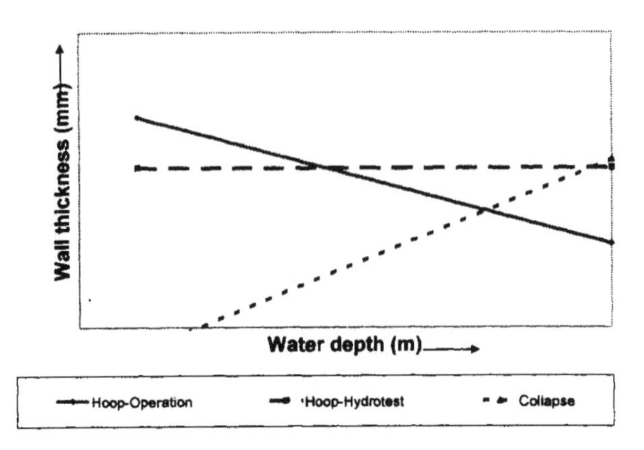

Figure 34.26 Wall thickness selection function of water depth (2 mediums).

Should the pipeline be designed to carry two mediums, water (during installation/testing) and product (gas/oil during operation) then the wall-thickness requirements can be drastically reduced for the deepwater pipeline. This approach (Figure 34.25) is not limited by collapse resistance until a significantly deeper water depth is reached (Figure 34.26). The densities of both of the two mediums (water and oil/gas) then the wall-thickness requirements for hoop stress/collapse will converge. Hence the wall-thickness requirements should be less for dense oil than for gas.

34.6.3 Implication to Installation, Testing and Operation

What are the implications of changing from a 3 medium to a 2 medium pipeline? All phases of the pipeline installation, testing and operation will be affected by providing a limitation to the contents density and the minimum pressure. The principal issues for each phase are:

Installation: The pipeline will need a facility to free flood during installation and cancel any differential pressure in this phase. Although the pipeline wall-thick-ness may be significantly reduced, the submerged pipe weight is still increased and the pipelay tensions can be-come too high for the present laybarge capacity. This phenomena is true in comparison with lines installed dry in shallow water but not deepwater.

Testing: The pipeline internal pressure must stay above a design minimum, so although the pressure testing will be unaffected the drying the pipeline with ambient pressure hot air and/or vacuum drying will not be possible. Pre-commissioning would have to employ methanol slugs or similar followed by the product at operational pressure.

Operation: Throughout the production life, a minimum operating pressure must be maintained otherwise the pipeline will collapse. In practice this would require a minimum pressure assurance system, such as having isolation valves, which would prevented the pipeline pressure dropping below a specified minimum.

The implications for both the testing and operation phases, although significant, are not insurmountable with existing technology and practices. On the other hand, the impact on installation is significant and is what this section will focus on.

34.6.4 Installing Free Flooding Pipelines

Installing pipelines dry has been logically adopted as the lay tensions can be kept relatively low and there is a large margin to be gained between with the increased submerged weight during operation (for stability purposes). This logic is sound in shallow water but can not be extrapolated to depths in excess of 1000m. The required wall-thickness of an airfilled pipeline becomes so large that the associated submerged weight will require lay tensions significantly greater than present lay barge capacity.

To illustrate this phenomenon Figure 34.27 and 34.28 illustrates the required pipeline wall-thickness for a range of pipeline diameters. For the purposes of comparison it is assumed the pipeline would be carrying oil at a density of 800 kg/m3 at a pressure of 200 barg. Figure 34.27 illustrates the wall-thickness for a pipeline installed while empty and Figure 34.28 illustrates the associated wall-thickness when the pipeline is installed flooded.

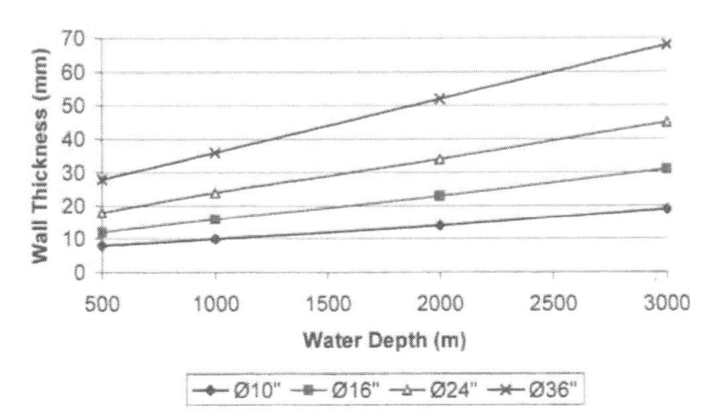

Figure 34.27 Pipeline wall-thickness, pipeline installed empty.

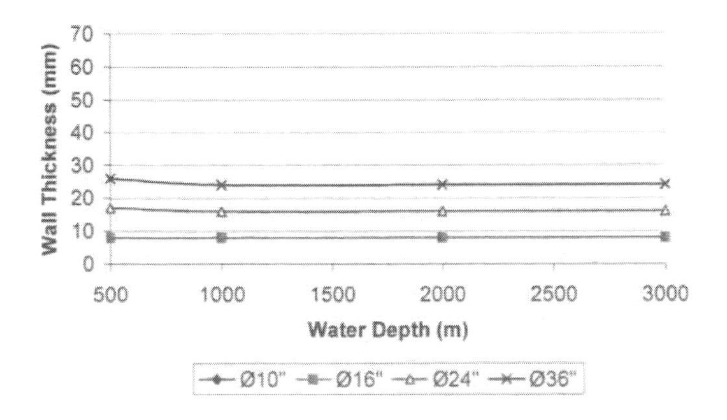

Figure 34.28 Pipeline wall-thickness, pipeline installed flooded.

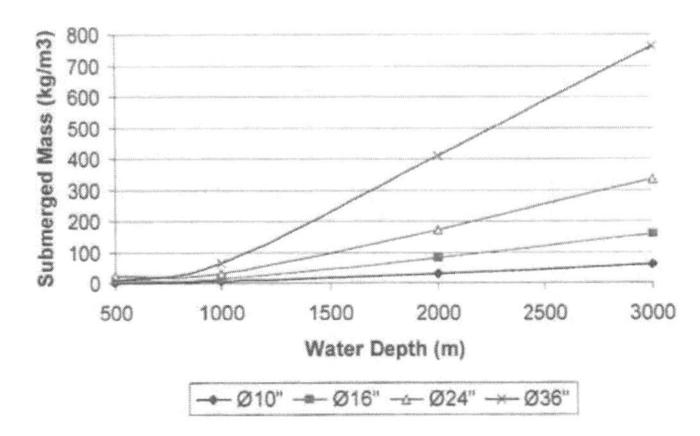

Figure 34.29 Installed submerged weight, pipeline installed empty.

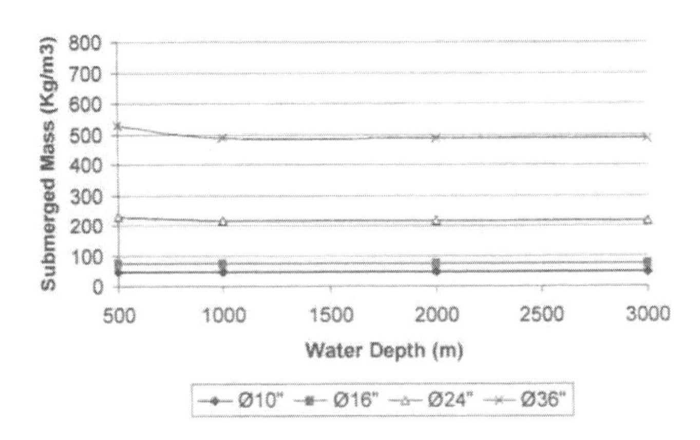

Figure 34.30 Installed submerged weight, pipeline installed flooded.

When the line is installed empty a direct consequence of the wall-thickness required in deep water is the large submerged weight. This becomes significant when water depth is deeper than 1000m where the submerged weight doubles every 1000m (Figure 34.29). As would be expected, the submerged weights are still lower than having a flooded line - until water depths of circa 2000m are reached. At 2000m the submerged weight of a flooded line can be less than an empty line because hydrostatic collapse is not a failure mode.

Figure 34.30 illustrates the associated pipeline submerged weights for a range of pipeline diameters when installed wet. The on-bottom stability requirements benefit from the increase in the submerged weight due to the heavier wall thickness. This example has not accounted for thermal insulation coating that would reduce the submerged weights while still satisfying stability requirements.

34.6.5 S-lay vs. J-lay

The offshore pipeline industry is familiar with experienced in installing air-filled pipelines by the S-lay method. An indication of the absolute minimum lay tensions are illustrated in Figure 34.31, which is generated on the basis that no additional weight coatings are required for

for stability purposes. On the basis that existing spreads have a maximum lay tension capacity of between 400 and 500 tones then the deepest a Ø16" line can be installed is 2000m (Figure 34.31).

It is interesting to note that the lay tension in 2000m depth would be the same (or less) to install the Ø16" pipeline flooded as opposed to dry (Figure 34.32), but the associated cost would be less as the required pipe steel would be approximately half of that installing the line dry. The difference is even more dramatic for a Ø10" line, which can be installed in over 3000m of water with existing spreads (when flooded) compared to 2500m when dry.

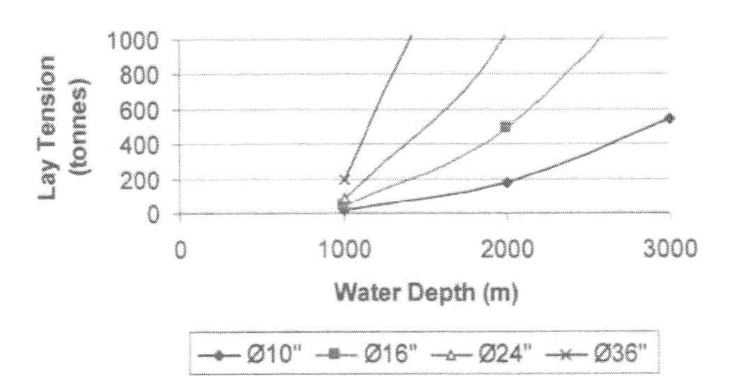

Figure 34.31 S-Lay of dry pipeline.

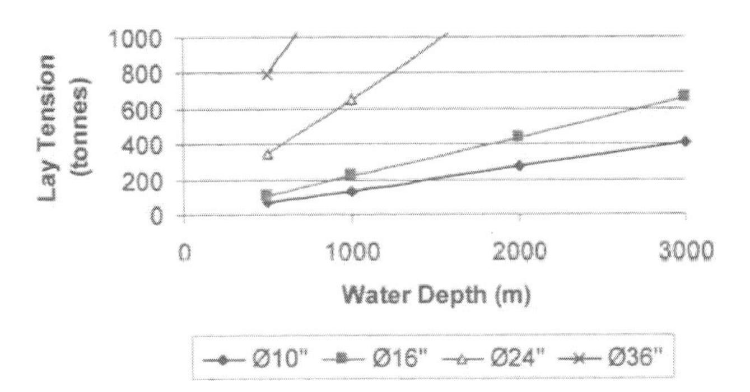

Figure 34.32 S-Lay of wet pipeline.

The difference becomes even more noted when comparing J-Lay capabilities. To install the same Ø16" pipeline in 2000m by J-Lay will require a tension of 200 tones, whether installed wet or dry (Figure 34.33 and 34.34). The big benefit from the wet installation, apart from requiring only half the material, is that existing equipment can install a Ø16" pipeline in over 4000m compared to over 2000m in the dry state. An observation from this study is that installation of a Ø24" pipeline still requires very large lay tensions – even with J-Lay.

There is scope to reduce the submerged weight of the line by addition of buoyant insulation, with this and flooding the line there is the potential to reduce lay tensions in 3000m to more achievable lay tensions.

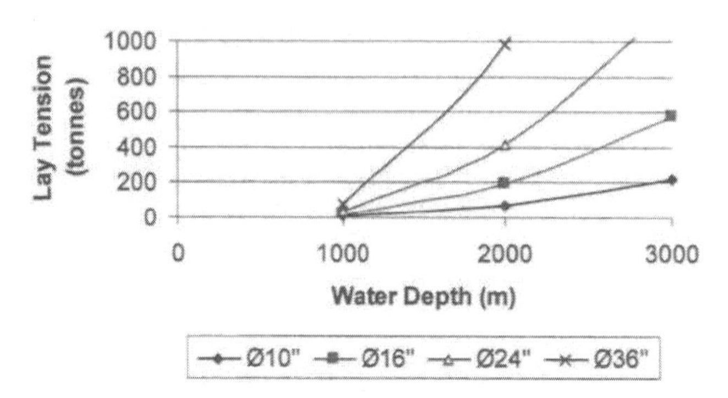

Figure 34.33 J-Lay of dry pipeline.

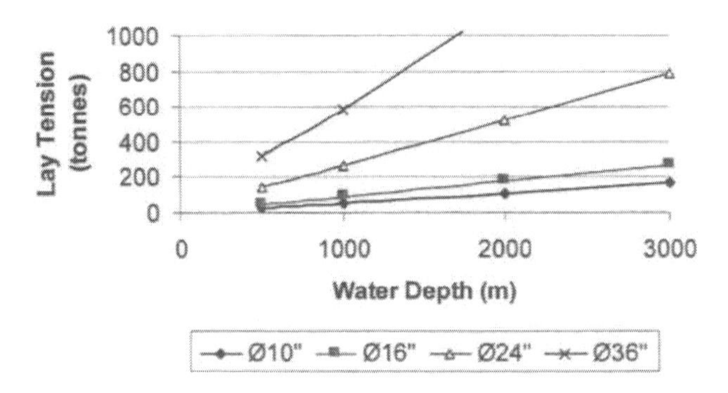

Figure 34.34 J-Lay of wet pipeline.

S-Lay installation tension is limited by a more horizontal departure angle at the stinger tip. The present stingers on the larger vessels have already been extended for 400 to 600m and are designed to provide departure angels of up to about 60 degrees. The required angle for ultra deep water would be the equivalent of J-Lay, or virtually 90 degrees. To keep stinger lengths within the present size (max 100 m arc length), it is necessary to increase the curvature This will plastically deform the pipeline in the overbend providing a permanent residual strain in the pipe on the seabed. The effect of residual strain is not well documented but two phenomena are identified. The first is the tendency of the pipe to twist due to instability in the sag bend introduced by the reverse plastic strain. The second is that the pipe may adopt a "corkscrew" configuration on the seabed. If the plastic strain is not severe then these effects can be avoided or be used to benefit the installation operation.

34.6.6 Economic Implication

What are the economic implications of installing a waterfilled pipeline?

Pipeline project CAPEX can be broken down into the following main areas:

• Management and design	5%
• Materials and fabrication	55%
• Installation	29%
• Commissioning	1%
• Insurance and miscellaneous	2%

The cost impact is discussed for each main area.

Management and design: The approach will have no direct commercial impact on the management/design process. However the design should address all the potential limitations of a 2 medium pipeline to ensure that they are acceptable to the operations phase. One area that must be addressed is a system of assuring that pressure in the line never drops below a prescribed minimum. One approach is to have isolation valves at the ends of the end, which are activated on detection of pressure drops as a HIPPS is applied to HP lines;

Materials and fabrication: The required wall-thickness is significantly reduced. Some reduction in wall-thickness is achieved for water depths less than 1000m. But in water depths of 2000m or more, the reduction is at least 50%;

Installation: A addressed above, for pipelines in excess of 2000m there is a significant reduction in lay tension requirements. An added benefit is that lay rates should be faster as the wall-thickness of the pipeline is significantly reduced. It is envisaged that there will be no adverse commercial impact to this phase. If conventional installation methods can be used, the cost is reduced further;

Commissioning: As conventional approach of de-watering and drying the pipeline will not be possible. However pre-commissioning techniques using liquids (methanol) at the required pressure could be implemented. The approach may have a commercial impact;

Insurance etc: Although this is not a well-proven technique it is technically feasible. It has been used where the pipeline is too light to be stable on the seabed, but not in ultra deep water. Insurance would be related to repair cost rather than risk of damage.

In summary, installation of a wet pipeline in 1000 – 3000m water depths is technically feasible and it could reduce the pipeline CAPEX (due to material savings) by up to 27%. There is a greater emphasis on the design aspects but with modern analysis methods and tools, the engineering can be performed reliably and efficiently.

34.7 References

1. Bai, Y. and Damsleth, P.A. (1997), "Limit State Design of Offshore Pipelines", Proc. of OMAE'97.
2. Bai, Y. and Damsleth, P.A. (1998), "Design Through Analysis Applying Limit State Concepts and Reliability Methods", Proc. of 8th International Offshore and Polar Engineering Conference, ISOPE-98.
3. Bai, Y. and Damsleth, P.A. and Langford, G. (1999), "Strength Design of Deepwater Pipelines", Proc. of 2nd International Conference on Deepwater Pipeline Technology, DPT-99.
4. Damsleth, P.A. and Dretvik, S. (1998), "The Åsgard Flowlines – Phase 1 Design and Installation Challenges", OPT'98.
5. Damsleth, P.A., Bai, Y., Nystrøm, P.R. and Gustafsson, C. (1999), "Deepwater Pipeline Installation with Plastic Strain" Proc. of OMAE'99.
6. Endal, G., Ness, O.B., Verley, R., Holthe, K. and Remseth, S., (1995) "Behaviour of offshore pipelines subjected to residual curvature during laying", OMAE'95.
7. HKS, (1998) "ABAQUS User manuals, version 5.7", Hibbit, Karlsson og Sorensen
8. Holme, R., Levold, E., Langford, G. and Slettebø, H. (1999), "Åsgard Transport – The Design Challenges for The Longest Gas Trunkline in Norway", OPT'99.
9. Igland, R.A., (1997) "Reliability analysis of pipelines during laying, considering ultimate strength under combined loads", Doktor ingeniøravhandling 1997:80, Institutt for marine konstruksjoner, NTNU Trondheim.
10. Langford, G. and Kelly, P.G., (1990) "Design, Installation and Tie-in of Flowlines", JP Kenny Report Job Nr. 4680.1.
11. Malahy, R.C.Jr., "OFFPIPE user's guide version 2.05".
12. Martinsen, M., (1998) " A Finite Element Model for Pipeline Installation Analysis", A M.Sc. Thesis Performed at Stavanger University College for JP Kenny A/S, 1998.
13. NOU 1974:40, Rørledninger på dypt vann.
14. Palmer, A. (1997), "Pipelines in Deep Water, Interaction between Design and Construction", Proceedings of Workshop on Subsea Pipelines, Edited by S.E. Estefen et al, Federal Univ. of Rio de Janeiro.
15. Rivett, S.M., Raven, P.W.J. and Baker, D.W. (1997), "Pipeline Design and Construction in Deep and Ultra Deep Water – An Overview", Deep Water Drilling and Production Technology Symposium, Black Sea.
16. Walker, A. and Tam, C.K.W. (1998), "Deepwater Pipeline Design", Deepwater Pipeline Technology (DPT) Conference, March 9-11.

Part V

Welding and Installation

Chapter 35 Route Optimization, Tie-in and Protection

35.1 Introduction

Over the last twenty years the installation equipment used has been developed to meet the needs of the industry and the harsh environmental conditions. The available equipment and its associated capabilities and limitations play a major role in the design of all offshore installations, including pipelines.

This section outlines some of the available equipment and discusses their capabilities and constraints. The installation equipment is discussed in the following format (Langford and Kelly, 1990):

- Route Optimization;
- Pipeline tie-ins;
- Pipeline trenching/burying;
- Pipeline rock dumping.

35.2 Pipeline Routing

35.2.1 General Principle

Route selection is a complex procedure, which can be governed by several variables. Clearly, the shortest distance between the terminal points is likely to be the most economic from a material standpoint, but possible overriding factors must be considered. Typically the route selection will be affected by:

- End point locations;
- Water depths;
- Presence of adverse environmental features such as high currents, shoaling waves;
- Presence of other fields, pipelines, structures, prohibited zones (e.g. naval exercise areas, firing ranges);
- Presence of unfavorable shipping or fishing activity;
- Suitability of landfall sites, where applicable.

35.2.2 Fabrication, Installation and Operational Cost Considerations

A significant proportion of the total cost to install a pipeline which is directly affected by the chosen route is incurred during fabrication and installation. The associated activities are:

1. Length of fabricated pipeline pipe (coated);
2. Pre sweeping of route;
3. Pre lay installed freespan correction supports;
4. Post lay installed freespan correction supports;
5. Trenching, burying or rock dumping.

Some or all of these activities will be present within the selected pipeline route. As a general rule the design should be performed to:

- Minimize length of pipeline required;
- Avoid requirement for presweeping;
- Avoid pre-lay installed freespan correction supports;
- Minimize post-lay freespan correction supports;
- Minimize trenching, burying and rock dumping.

35.2.3 Route Optimization

Optimization of pipeline routing is usually not performed as the route probably has no obstruction, is in an accessible water depth and the seabed topography is flat: Hence a straight line between the two termination points would suffice. However, on seabeds with onerous terrain significant savings on fabrication and installation costs can be made if route optimization is performed.

To perform a route optimization, reasonably accurate costs for the following activities are required:

- Supply of additional pipeline pipe/unit length;
- Presweeping a corridor/unit length, including cost of reduced lay rate due to a smaller lay corridor;
- Prelay freespan correction supports (each), again including cost of reduced layrate due to smaller lay corridor;
- Post lay freespan correction supports (each);
- Trenching, burying and rockdumping/unit length (for each).

Based on the derived costs, a total cost for each route can be derived.

It is worth noting that the optimization cannot be completed until all the pipeline design parameters are finalized (for instance the number of freespan correction supports will not be known until the allowable freespan has been determined).

35.3 Pipeline Tie-ins

It might be natural to assume that each pipeline has two tie-ins, one at each end. This is however, not always the case. Where the installation method is only suitable for limited lengths of pipeline, midline tie-ins may be required.

The methods of pipeline tie-in are discussed under the following headings (see Figure 35.1).
- Spoolpiece;
- Lateral pull;
- J-tube;
- Connect and lay away;
- Stalk on.

Based on the review of the above tie-in methods it may appear that there is a good choice of tie-in methods, but closer review shows that this is not the case. Only certain combinations of installation and tie-in methods are practical and other factors limit the selection. These are discussed in the following subsections.

35.3.1 Spoolpieces

General Principle

This method is probably the most popular method of tie-in for flowlines/pipelines. Divers measure and then assist the installation of a piece of pipe to fit in between the two ends of flowline to be tied together.

Installation Capabilities/Constraints

This method is popular because of the flexibility of the method. Misalignment of the two pipes can be accommodated by installing bends into the spool, and inaccuracies in placing the pipelines down can be accommodated when the spool is made up (after diver measurements).

The connection method can either be by flanges or welding. The welding method requires a hyperbaric habitat. From a design viewpoint, should there be large flowline expansion, then this can also be accommodated by incorporating a dogleg in the spool. This will permit expansion of the pipe without transmitting high loads into the adjacent pipe.

The obvious disadvantages are that remedial work is required after flowline installation to tie the line in. Additional work requires divers. This could limit this option should the tie-in be required in very deep waters.

35.3.2 Lateral Pull

General Principle

Lateral deflection involves positioning the flowline end to one side of the target structure and then pulling it laterally into position. This has two disadvantages compared with a direct pull-in.

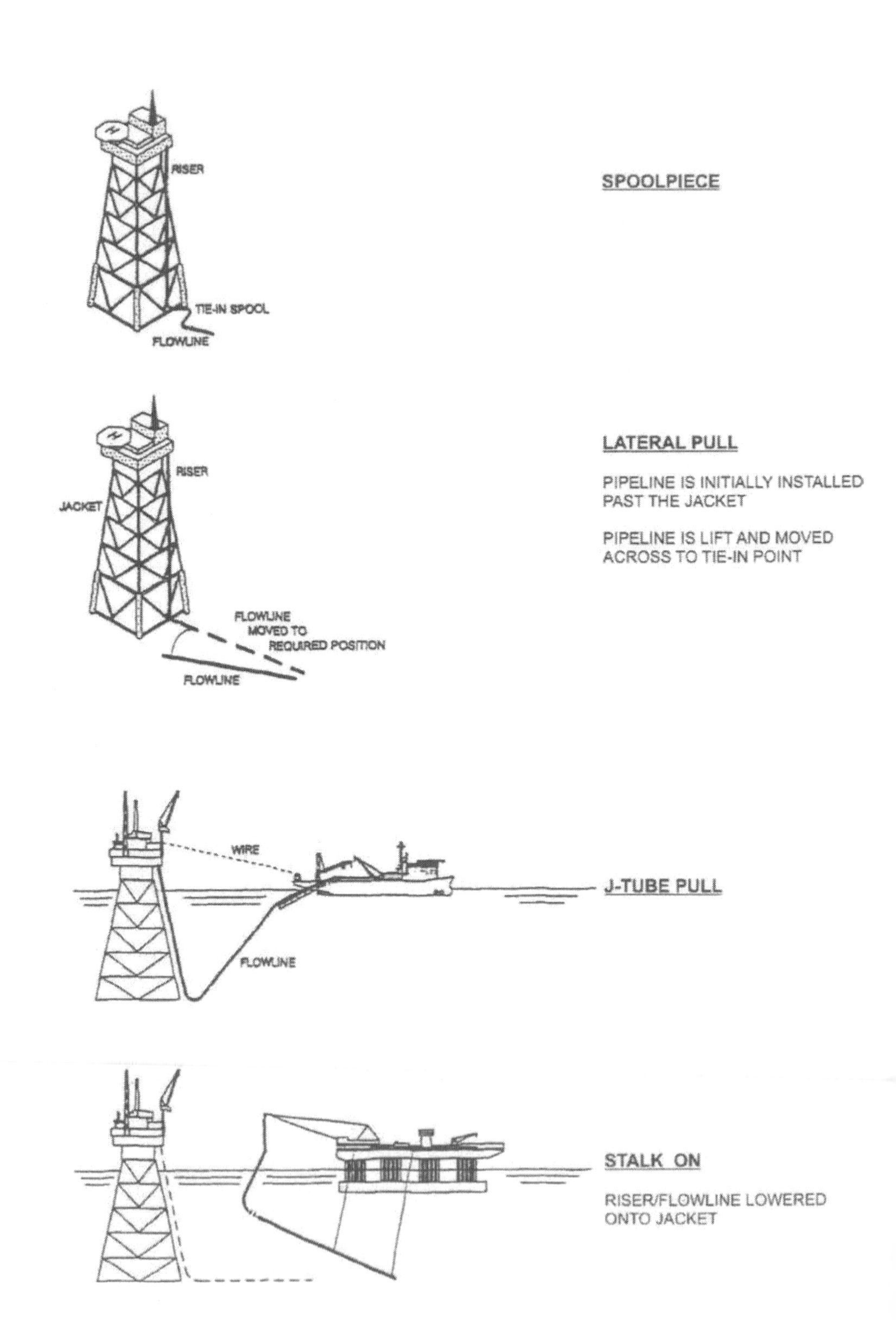

Figure 35.1 Tie-in methods.

- Alignment is more difficult to achieve;
- A clear (swept) area is required to one side of the tie-in site.

The flowline may be pulled towards the target by a single wire, or a series of wires may be developed through dead-man anchors to give greater control of alignment. A bell mouth or stab-in guides usually assist final alignment.

For large diameters such as export lines or bundles, it is necessary to make a length of pipeline neutrally buoyant. This gives greater flexibility and reduces the pull forces, but can expose the pipe to large current forces.

One development of this technique is the use of vertical deflection rather than lateral deflection. The required initial shape could be attained by local adjustments to buoyancy, pull-in being again by a system of wires. The principal advantage of this method is that it does not require the same amount of seabed space.

In addition, it should be possible to devise initial configurations which it would be difficult to create laterally by laying or towed installation. Direct pull for second end tie-ins may then become available by creating a vertical slack loop behind the pullhead.

Installation Capabilities/Constraints

This method is usually utilized when direct pull-ins (i.e. J-tube, connect and lay away or stalk-on) are not feasible options. However, this method is frequently used as direct pull-ins are usually not possible. For instance J-tube pull-in and connect and lay away can only be performed by a vessel laying away, and not when a vessel lays down. Stalk-ons can only be performed in shallow water.

The main disadvantages of this method are:

- Requires extensive diver intervention;
- Difficult operation to perform; several experienced diving operators have buckled flowlines using this method;
- If connected directly to the tie-in point then all the pipeline expansions will be fed into that point. The tie-in point must either take high axial loads or large deflections.

35.3.3 J-tube Pull-in

General Principle

This method requires the flowline to have the capacity to easily move axially over a relatively large distance. This limits the option to pulling the flowline directly from the pipelay vessel.

The method of J-tube pull-in is to connect the flowline to a wire and, by pulling the wire, to pull the flowline through a riser (J-tube) up to the topsides of the platform. This method requires the J-tube to be of a reasonably large diameter compared to the flowline. This method is discussed in further detail by Ellinas (1986).

Installation Capabilities/Constraints

The principal advantage of this system is that the flowline is tied in directly to the jacket topsides, so avoiding subsea tie-in work. However, the main disadvantages are:

- Normally used for small diameter lines; the forces involved with large diameter lines become too high;
- The line is directly tied into the jacket, with no system to accommodate the pipeline expansion. Large deflections and/or axial forces will be fed into the J-tube during operation.

This method is very popular for small diameter flowlines, when the pipelay starts at the jacket.

35.3.4 Connect and Lay Away

General Principle

This method is very similar to the J-tube pull-in method with the exception that the tie-in is performed subsea. This method is usually applied in diverless operations, where a mechanical connecting system will be utilized to perform the connection.

Two examples of diverless pull-in and connection tools, McEVOY and FMC, are presented in Figures 35.2 and 35.3 respectively. There are also diverless connection systems for bundled lines, two such systems, VETCO and CAMERON are presented in Figures 35.4 and 35.5 respectively. Please see Phillips (1989) for a more detailed description of these connection systems.

Installation Capabilities/Constraints

This system is mainly used at subsea manifolds/wellheads, where the water depth prohibits the use of divers. This is the only system developed for performing diverless connection of pipelines.

The principal advantage is that this system can be adapted to perform diverless connections.

The main disadvantages are:

- Expensive technology to perform diverless work;
- The connection will be subjected to the pipeline axial loads, when it can not accommodate expansion.

35.3.5 Stalk-on

General Principle

The stalk-on method is primarily used in shallow water applications (less than 40 m) and hence would only be applicable in the Southern North Sea. The method involves laying the flowline down adjacent to the jacket it shall be tied into. The vessel maneuvers over the flowline, lifts it up and welds on (or flanges on) the jacket riser. The pipeline and riser are then lowered onto the seabed/jacket. The jacket clamps are subsequently closed around the riser. See Figure 35.1 for illustration.

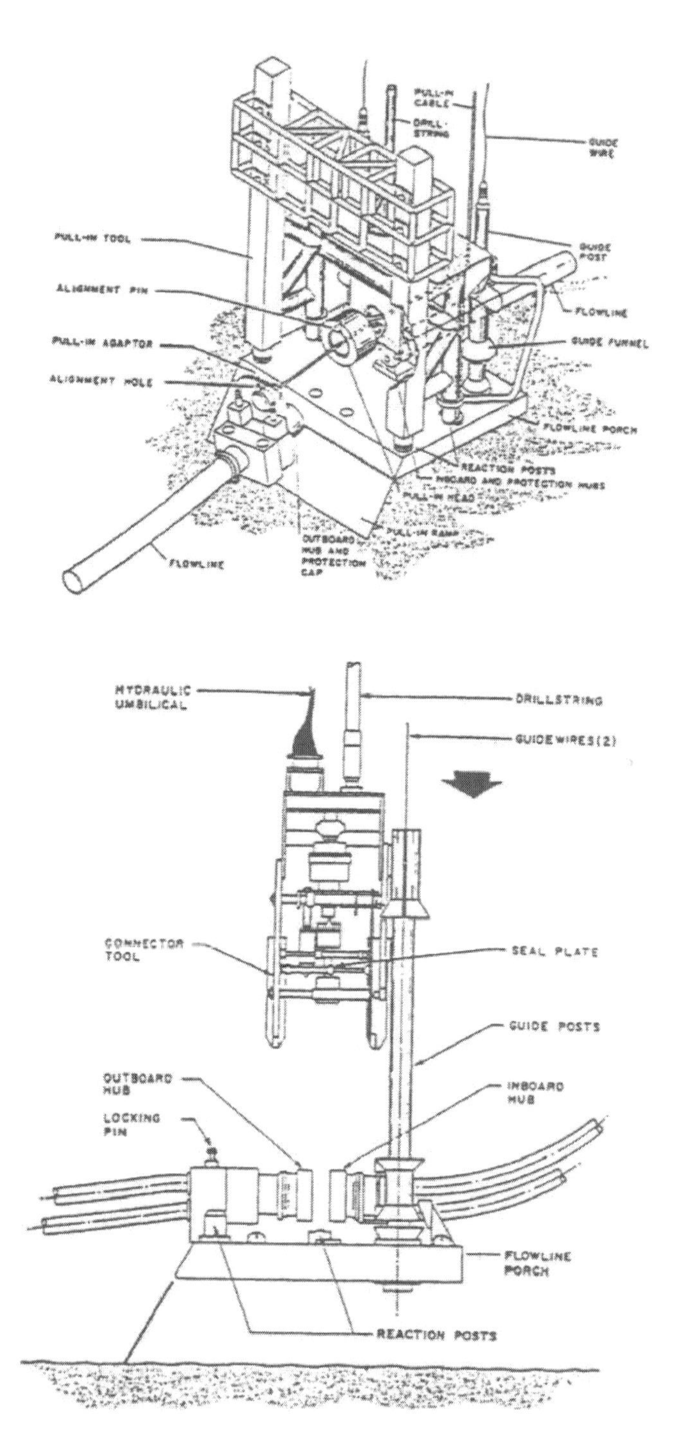

Figure 35.2 McEvoy flowline connection system.

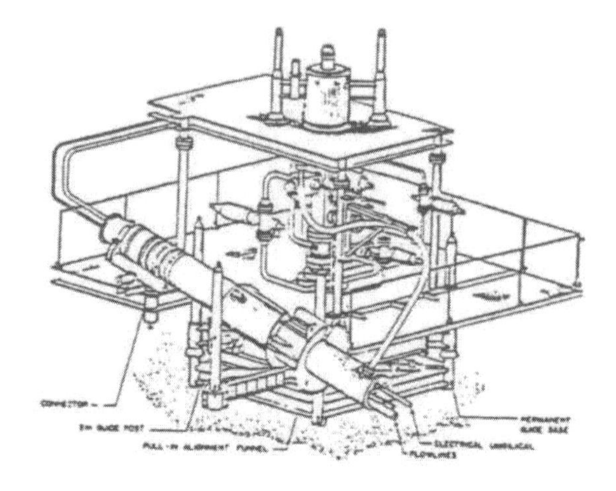

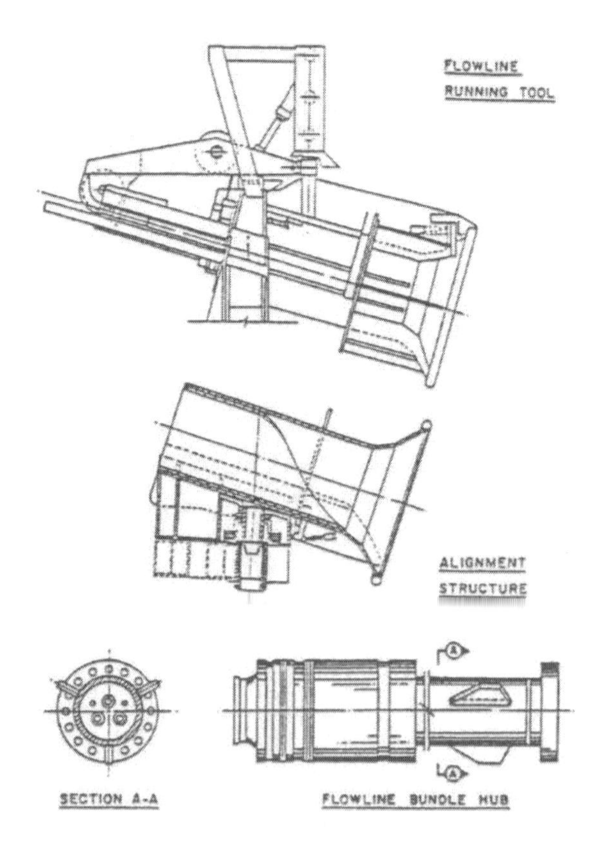

Figure 35.3 FMC diverless flowline connector system.

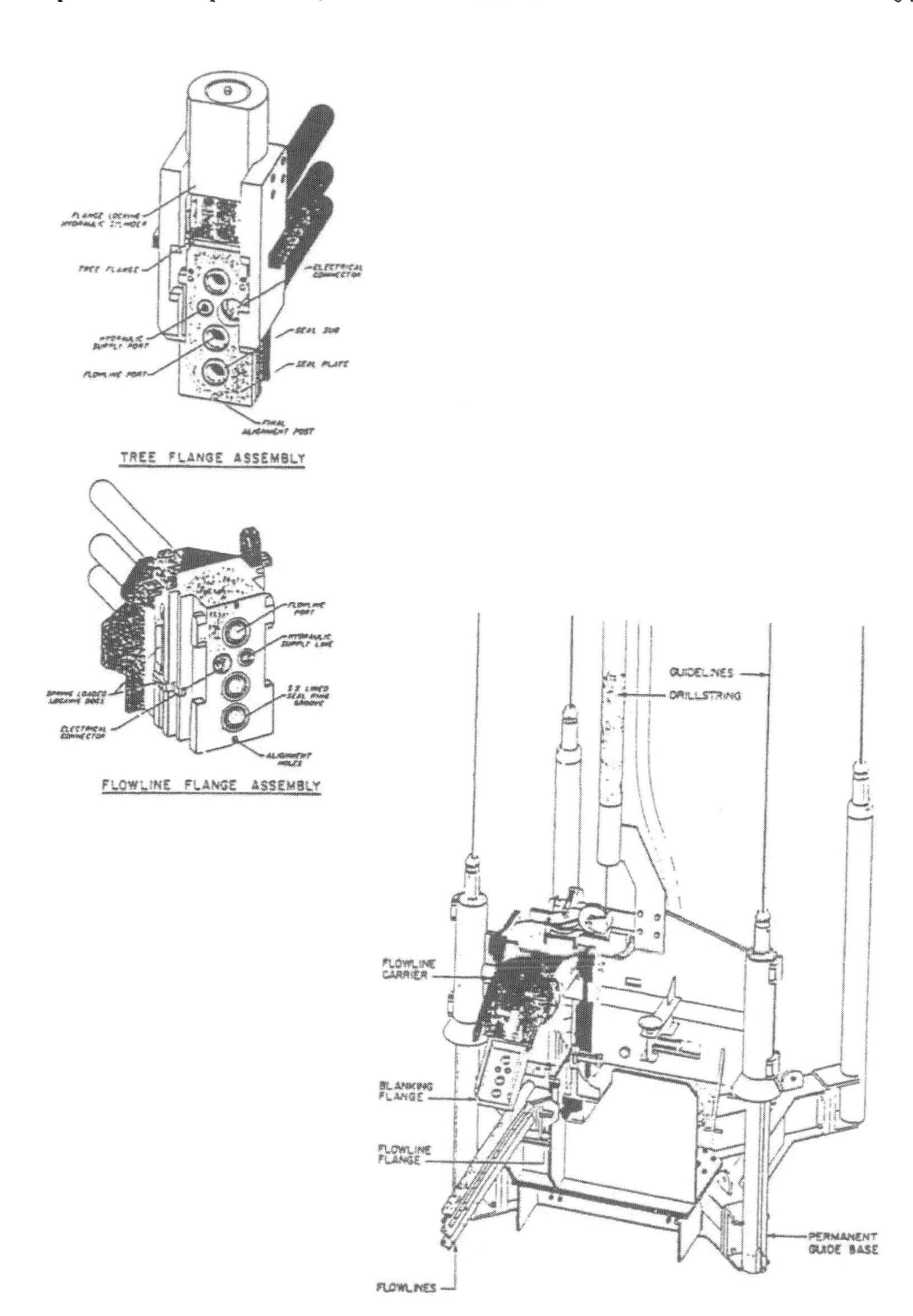

Figure 35.4 VETCO diverless flowline connection tool.

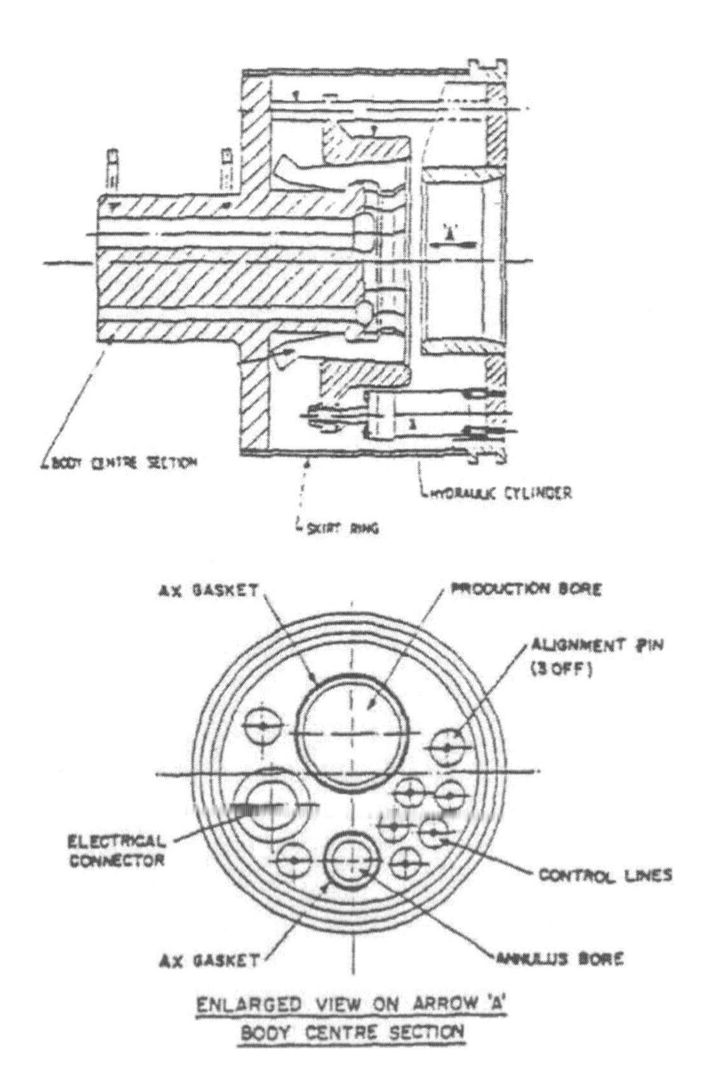

Figure 35.5 CAMERON multibore connector system.

Installation Capabilities/Constraints

The primary advantage of this method is that the same vessel that installs the flowline can also perform the stalk-on. However, the disadvantages are:

- The riser will be subjected to the expansion of the pipeline, as no expansion, as no expansion spool is used;
- The operation can only be used in shallow waters.

35.4 Flowline Trenching/Burying

The development of trenching/burying equipment has, like the flowline installation equipment, changed significantly over the last twenty years. The trend has been to move away from dedicated trenching vessels to equipment that can be used from Diving Support Vessels (DVS's). The flowline trenching/burying equipment is discussed under the following headings:

- Jet sled;
- Ploughing;
- Mechanical cutter.

35.4.1 Jet Sled

This method is the traditional method of trenching a pipeline. Dedicated vessels with turbine engines were built to provide jet sleds that would trench through most soils (see Figure 35.6).

General Principle

The jet sled works on two principles:

1. High pressure jet nozzles power water to break up the soil;
2. Air is pumped into pipes which generates lift, this lifts the broken soil away from the location.

Using these two principles a jet sled is able to trench.

Installation capabilities/Constraints

The size of jet sleds (and associated costs) varies considerably. The largest can weigh up to 80 tones and are controlled by dedicated vessels, and the smallest can fit onto a DSV and weigh up to 0.5 tone. The associated capabilities also vary. The large ones can trench through sand, silt and clay and even through soft rock (sandstone), the trench rates vary depending on the soil conditions. The small jet sleds are only suitable for sand, silt and soft clay.

The main constraint of the jet sleds is that they cannot bury the flowlines. They can excavate a hole for the flowline to sink into, but they cannot backfill the hole. Jet sleds are still a popular means of trenching pipes as the method is well proven and little damage to the pipeline occurs compared to that caused by damage by other methods. This system usually requires divers but may be operated diverless.

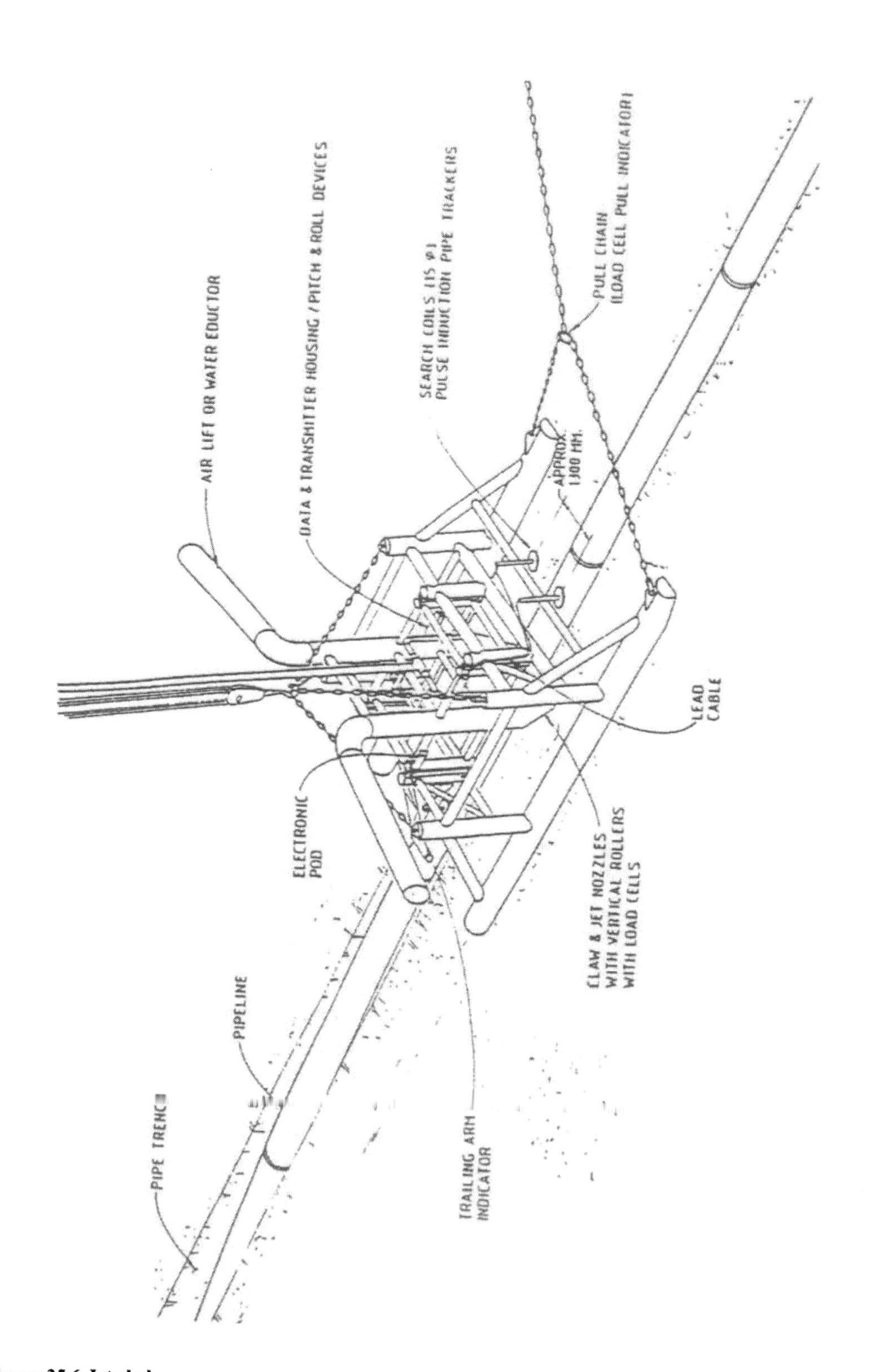

Figure 35.6 Jet sled.

35.4.2 Ploughing

Ploughing was first developed in 1980 for the North Sea in order to provide a cheaper alternative to trenching of pipelines. Since then it has become a popular method of pipeline trenching. (see Figure 35.7).

General Principle

The general principle of pipeline ploughing has been adapted from the technique used in agriculture to plough fields. The pipeline plough consists of a very large "share" which the pipeline rests on top of. The pipeline is pulled along (usually by the surface vessel) and as the ploughshare passes the flowline settles in the trench. Should a backfill plough also be employed, this will reverse the process by pushing the spoil back into the trench, so burying the flowline.

Installation Capabilities/Constraints

The main advantage of this system is that it can trench a large range of flowline sizes (up to 24-inch diameter) operated from a DSV. The trench rates can be very high depending on the soil conditions.

The system is probably the only system that can bury flowlines in one operation, (should it be required). It should be noted, however, that some operators prefer rock or imported material to be used as backfill.

The main disadvantage of this system is that it has a limitation on the depth which can be excavated. To date, the maximum trench depth is 1.5m. An additional disadvantage is that the plough system can cause damage to flowlines, especially on those lines not protected by concrete coating. However, this system is better than most. This system usually requires divers for plough placement and retrieval, but in some cases it can be performed without divers.

35.4.3 Mechanical Cutters

Mechanical cutters have been developed as a diverless option to trenching for small diameter flow lines (see Figure 35.8).

General Principles

There are many varied and different types of mechanical diggers available for subsea flowline trenching. However, the methods are all based on the same basic principle. The controls and power source is onboard a surface vessel, which via an umbilical powers a subsea machine. This machine moves along the seabed on tracks.

Installation Capabilities/Constraints

These machines can usually handle only small diameter flowlines (and preferably flexible). Since they provide their own traction the machines require reasonably firm soil. They cannot trench in very soft soil or very hard clay/rock.

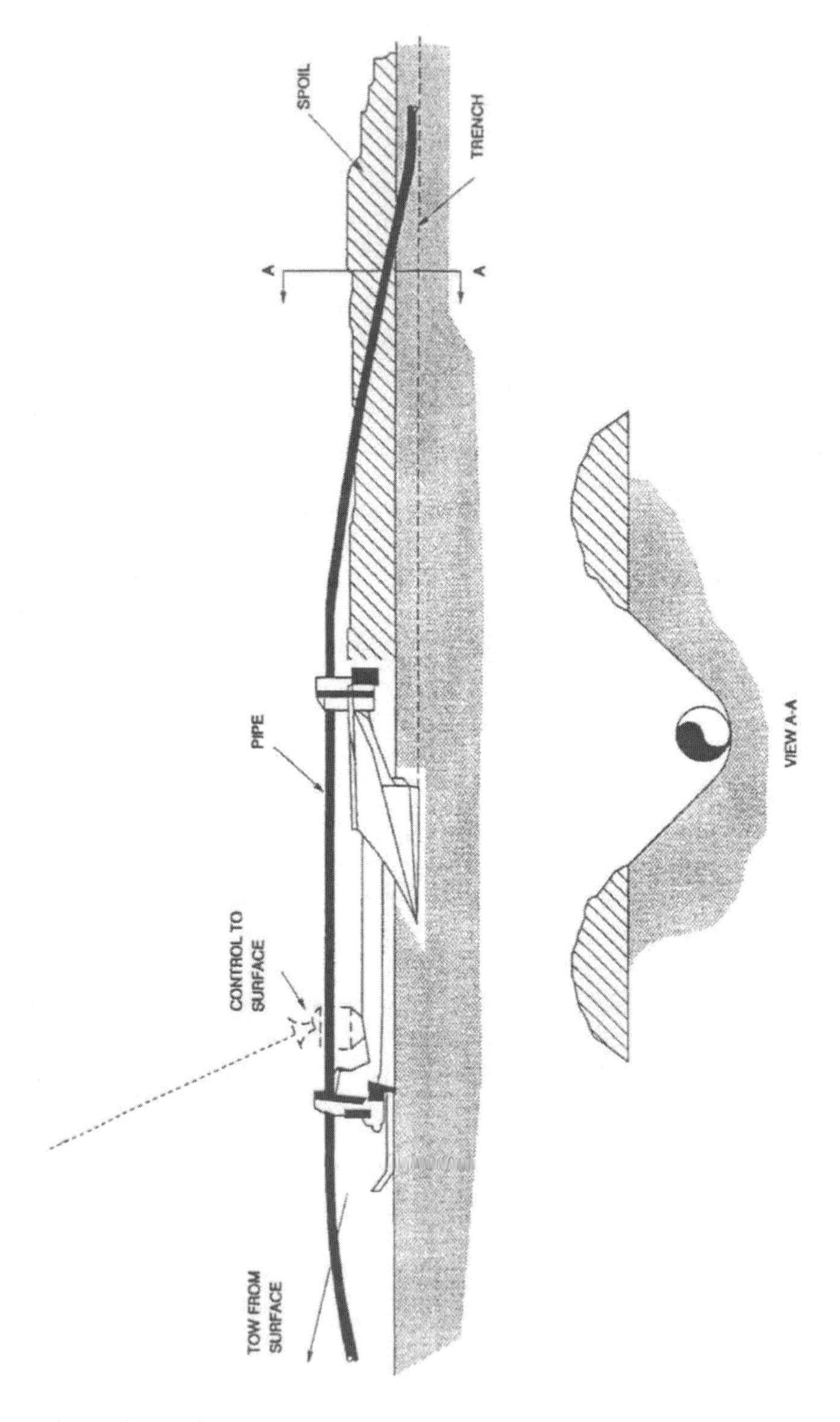

Figure 35.7 Pipeline plough.

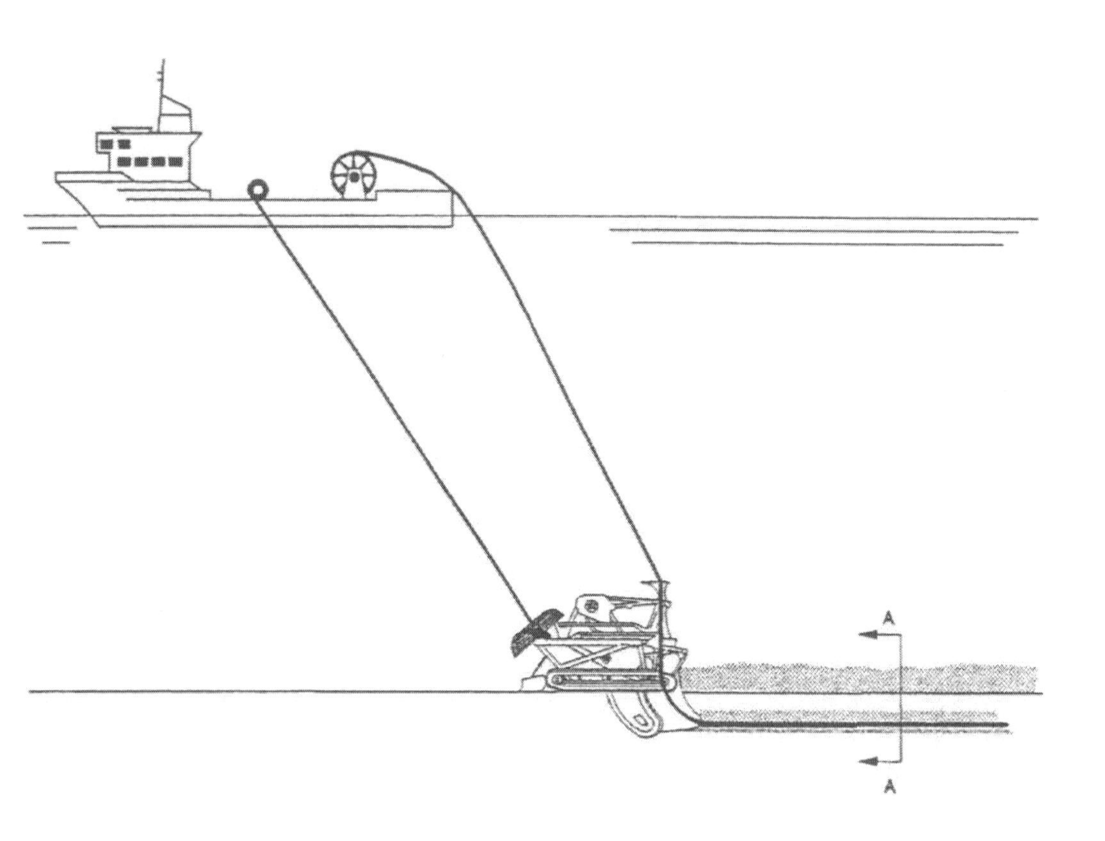

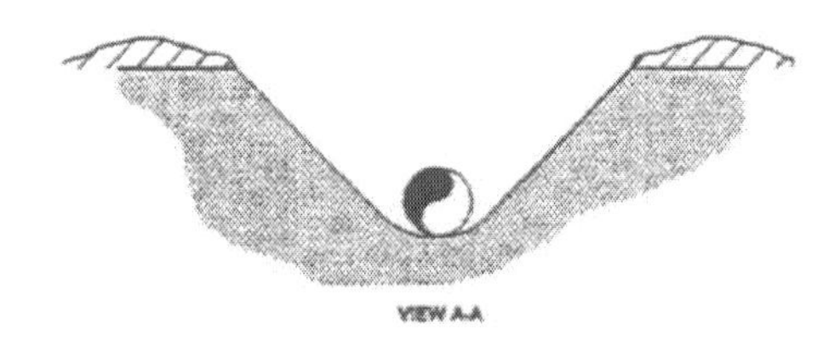

Figure 35.8 Mechanical cutter.

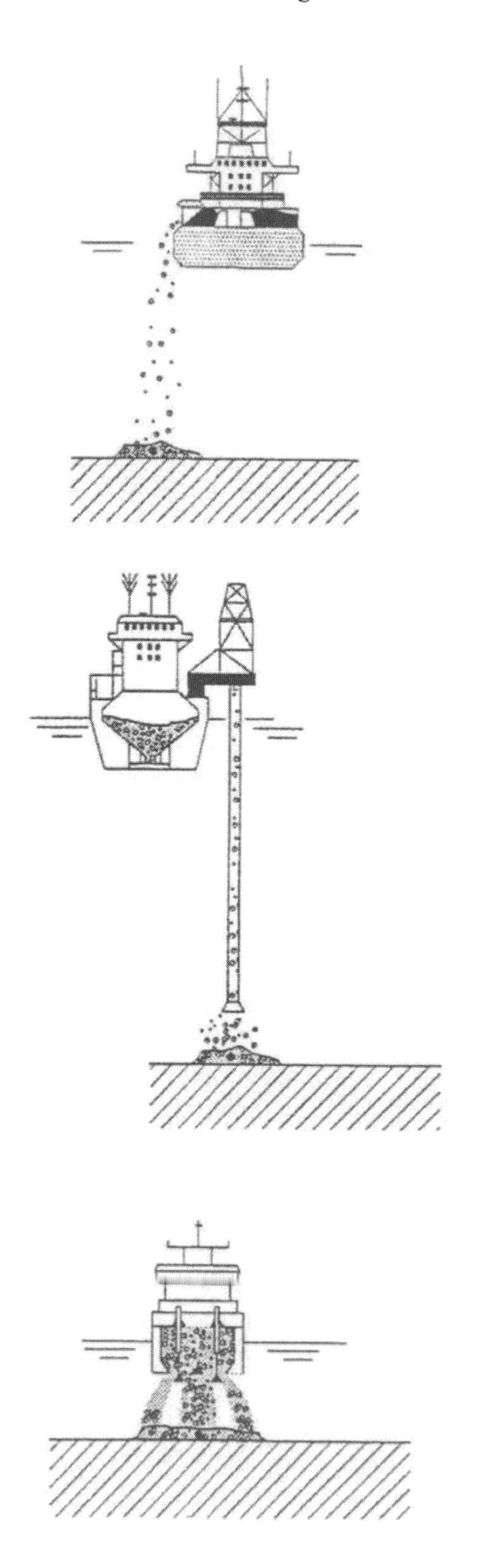

Figure 35.9 Rockdumping methods.

35.5 Flowline Rockdumping

Rockdumping, like the other installation activities for offshore, has become more specialized in the last 17 years. The rockdumping vessels were designed to deposit large quantities of rock in localized areas. Along with the requirement of small quantities of rock being placed over pipelines, new vessels have been developed (see Figure 35.9).

The three main rockdumping techniques are:

- Side dumping;
- Fall dumping;
- Bottom dropping.

35.5.1 Side Dumping

General Principle

This method involves loading selected stone onto a flat decked ship, positioning the ship over the required location to dump rock, and pushing rock over the side is by hydraulic rams which clear the rock from the center line of the vessel and outboard.

Installation Capabilities/Constraints

This method of rockdumping is very efficient at dumping large quantities of rock in short lengths. This is suitable for protecting bases of jackets or subsea manifolds, but is wasteful of rock for dumping of flowlines.

35.5.2 Fall Pipe

General Principle

The method is based on loading the selected stone onto the vessel, mobilizing offshore to the selected location to rock dump, and dropping the rock down through a tube to the location. To provide further accuracy the "fall pipe" has a remote operated vehicle at the end so the location of the rock can be monitored and controlled.

Installation Capabilities/Constraints

This method of rockdumping was developed for dumping of rock on pipelines/flowlines. It provides accurate dumping, which minimizes wastage and permits long stretches of rock to be dumped during one trip.

35.5.3 Bottom Dropping

General Principle

There are two methods of bottom dropping. One method incorporates ports which open at the bottom of the hold and the second is to apply a split barge which drops all the rock at once.

Installation Capabilities/Constraints

Both methods are again suitable for dropping large quantities of rock, when great accuracy is of less importance. This method is not suitable for rock dumping on flowlines.

35.6 Equipment Dayrates

The costs for installation equipment vary each season, generally depending on its availability. These costs play an important part in the selection of the chosen method of installation.

It is recommended that respective installation contractors be contacted should a costing exercise be conducted.

35.7 References

1. Ellinas, C.P., "J-Tube Method of Riser Installation", Offshore Pipeline Engineering, Park Lane Hotel, April 1986.
2. Langford, G., Kelly, PG., (1990) "Design, Installation and Tie-in of Flowlines", JPK Report Job No. 4680.1.
3. Phillips, P.W.J., "The Development of Guidelines for the Assessment of Submarine Pipeline Spans, Overall Summary Report", HMSO, 1986.

Part V

Welding and Installation

Chapter 36 Pipeline Inspection, Maintenance and Repair

36.1 Operations

36.1.1 Operating Philosophy

A pipeline operations philosophy needs to be developed and incorporated into the Operations Manual. The philosophy should address the overall issues that dictate gas operation such as:

- Maximum and minimum design and operating limits on gas flowrate, pressure and temperature;
- Sales contract requirements;
- Utilization of line pack to satisfy fluctuations in demand;
- Gas delivery requirements at third party tie-ins;
- Actions to be taken in the event of planned or unplanned shutdowns of the compressor station, e.g. allow gas delivery to continue via line pack inventory until minimum delivery pressure limits are reached;
- Actions to be taken in the event of planned or unplanned shutdowns of the delivery station, e.g. continue gas pumping until maximum pipeline pressure limits are reached.

36.1.2 Pipeline Security

Certain control systems must be provided so that the pipeline may be operated safely. The following functions are the minimum to be provided.

Emergency Shutdown

A means of shutting down the pipeline must be provided at each of its initial and terminal points. The emergency shut-down systems must be equipped so that any shut down will register at the control center and a positive alarm system will draw the attention of the person in charge of the control center to the event. The response time of an emergency shut down (ESD) valve should be appropriate to the fluid in the pipeline (type and volume) and the operating conditions.

Pressure, Temperature and Flow Control

Instrumentation must be provided at the control center to register the pressure, temperature and rate of flow in the pipeline. Any variation outside the allowable transients must activate an alarm in the control center.

To ensure protection to the pipeline against over (and under, for example, when there is leakage) pressurization and excessively high temperatures, automatic primary and secondary trips should be installed at the compressor station. Details as to their location and their high / low pressure and high temperature settings are required as input into the Operations Manual.

Relief Systems

Relief systems such as relief valves, are typically required to ensure the maximum pressure of the pipeline does not exceed a certain value. Relief valves must be correctly sized, redundancy provided, and they must discharge in a manner that will not cause fire, health risk or environmental pollution.

High Integrity Protective Systems (HIPS) may be considered when the conventional relief methods are unsuitable for ultimate plant protection. However, the application of a High Integrity Protective System must be justified and its design must be agreed with the relevant Regulatory Authority. The following main principles apply:

- A clear economic advantage must be demonstrated over the conventional approach to justify the increased complexity and dependence on rigorously controlled maintenance associated with HIPS;
- HIPS must be designed with appropriate redundancy and testing frequency to ensure higher reliability than conventional protection systems;
- Economic comparisons should take into account life cycle maintenance and testing costs;
- HIPS must respond quickly enough to prevent over pressure if downstream systems can be suddenly blocked-in. This is one reason why HIPS lend themselves to protection of large volume systems, including pipelines, rather than small sections of plant;
- HIPS isolation valves must have a tight shut-off. Otherwise, partial capacity relief valves will be needed after the HIPS isolation valves to accommodate leakage rates should the HIPS isolation valves fail.

Leak Detection

The pipeline must have an integrity monitoring system capable of detecting leak. A leak detection system in itself has no effect on the leak expectancy of a pipeline and will only make the operator aware of the occurrence of a leak, enabling him to take remedial actions in order to limit the consequences of the release. The leak detection system requirements will vary depending on the pipeline system in question (e.g. offshore or onshore, length etc.) however, the following should be considered at the design stage and/or implemented during operation.

On-Line Leak Detection

- continuous mass balance of the pipeline;
- continuous volumetric balance corrected for temperature and pressure of the pipeline;
- continuous monitoring of rate of change of pressure;
- continuous monitoring of rate of change of flow;
- low pressure alarms;
- high pressure alarms;
- high flow alarms;

Off-Line Leak Detection

- visual inspection of the pipeline route;
- running of a leak detection pig (see Chapter36.3.3);
- methane-in-water sensing by Remotely Operated Vehicle (ROV).

Several other methods of on-line leak detection are available, some of which will also indicate the location of a suspected leak. However, in general a good deal of intermediate pressure, temperature and flow information is required with attendant telemetry and for this reason such methods are not generally suitable for offshore use.

36.1.3 Operational Pigging

The conflicting balance of sensitivity to leaks and false alarms will determine the sensitivity of an on-line leak detection system. Large leaks can normally be detected more rapidly than small ones. To maintain the user's confidence in the system, avoiding false alarm should have a higher priority than attempting to shorten the leak detection time or reducing the minimum detectable leak rate.

Operational pigging is performed to maintain pipeline integrity. With regular operational pigging the pipeline should be maintained at its optimum throughput capacity and a higher efficiency will be achieved. Typically the following purposes will be served with regular pigging:

- prevention of scale build-up;
- cleaning of the pipe wall;
- removal of internal debris;
- removal of liquids (condensate and water);
- enhancement of the performance of corrosion inhibitors;
- provision of a means to verify the occurrence of corrosion.

Pig Type and Frequency

Operational pig runs are performed in pipelines using cup or bi-directional pigs to remove water drop-out, soft wax, sand deposits, scale and other debris build-up. The operational pigging frequency is different for each pipeline and varies with changes in flow conditions, gas composition and corrosion condition in the pipeline. Depending on the results of the pigging evaluation and the corrosion monitoring assessment, the pigging frequencies will be reviewed and updated regularly. As an example, for the Balingian gas trunkline network operated by Sarawak Shell Berhad (SSB) in Malaysia, the following cleaning pigging frequency is required:

Pipeline	Pigging Frequency
Balingian oil (12, 16 & 18-inch)	1 in 3 months
Temana oil (8 & 12-inch)	1 in 3 months
Samarang oil (8, 12 & 18-inch)	1 in 2 weeks
Balingian gas (12 & 18-inch)	1 in 6 months
Loconia gas (30, 32 & 36-inch)	1 in 6 months

Several types of pigs can be used. The selection of pig type will depend on the purpose of the pigging run. The following gives a brief description of some of the main types of pig used in normal operation. Non routine or intelligent pigging is addressed within Section 36.3.

Cleaning pigs - cleaning pigs are available fitted with a number of sealing cups (omni-directional) or sealing discs (bi-directional). The cleaning devices attached to the pig body range from carbon or stainless steel wire brushes that are spring loaded to the pipe wall to oversized circular wire brushes interfering on the pipe wall. For internally lined pipelines, nylon bristle brushes can be used. There are also scrapers moulded to resemble plough blades in polyurethane, or for non-lined pipes, hardened steel blades profiled to suit the pipeline inner diameter. All pigs are designed for the brushes or blades to cover the circumference of the pipe surface.

Foam pigs - pipeline cleaning foam pigs are made of hard polyurethane and covered with abrasive coating or wirebrush bands. Pigs manufactured of soft open cell polyurethane foam are used for water absorption in swabbing and drying service.

Spheres - spherical moulded tools made of polyurethane or neoprene of which the larger sizes are inflatable. The larger diameter spheres have facilities (inlets) whereby the spheres may be inflated to slightly greater diameters. Main application is in pipelines that have not been designed to accept standard pigs and / or in two phase pipelines to remove liquid hold-up and product separation.

Pigging Operations

The detailed procedures required for carrying out a routine pigging run should be contained or referenced within the Pipeline Operating Manual. Typically, the pipeline operations department should carry out the following activities:

a. Check whether the pig trap isolation valves have been leak tested in the previous six months. The six month durations is good common practice. If not then leak testing is recommended prior to commencing a pig run;

b. Ensure that the launcher has been correctly isolated, depressurized, vented and purged and is safe to open and ready to receive pigs;

c. Ensure that all valves on the pig route are or will be fully open;

d. Ensure that all pig indicators are correctly set and operational;

e. Inform the receiving station of the following:

- pig type;
- by-pass setting (that is, the pig has a by-pass facility in the case that the pressure build behind the pig is too great)
- time of launch;
- estimated time of pig arrival;
- inlet / outlet flow conditions at time of launch.

f. Keep track of the pig run by continuously monitoring the pressure and flow conditions at inlet and outlet. Remain in regular contact with the receiving station and exchange updates on estimated time of pig arrival.

g. Receiver station to notify launcher station when pig arrives in receiver and then isolate, depressurize, vent and purge receiver prior to removal of pig and inspection for damage and wear.

Data Monitoring

Each pig run shall be evaluated to determine the effectiveness of the operation. This information shall be used to enable a proper decision for future pigging runs and/or any other action to be taken. Typically, the following shall be evaluated.

- The actual pig arrival time compared with the estimated arrival time. In conjunction with known flow rates and associated flow conditions throughout the pigging run, an estimate of pig by-pass/pig slippage can reasonably be made;
- The wear on the pigs shall be determined and classified;
- The total weight of the debris received in the pig trap shall be measured. A sample shall also be taken for subsequent analysis;
- An estimation of the water volume swept ahead of the pig should be made if suitable equipment is available at the receiver station.

36.1.4 Pipeline Shutdown

A pipeline shutdown can be initiated in the following three circumstances:

- an emergency;
- major maintenance;
- production shutdown.

An emergency shutdown of the pipeline is achieved by closing the appropriate Emergency Shut-down (ESD) Valves. The ESDVs valves will be closed automatically by one of the following:

- fusible plug loops (which can be tripped) in case of fire;
- low pressure trips in case of a pipeline leak;
- high pressure trips in case of high pressure in the pipeline;
- low instrument air supply;
- terminal ESD valve as per the shutdown sequence.

A pipeline ESD valve should also be able to be closed manually at the control room and locally at the valve itself. The closing of the ESD valve should be linked to the prime mover shutdown.

36.1.5 Pipeline Depressurization

For most pipelines, in the event of pipeline rupture, depressurization of the line must be carried out immediately in order to reduce the amount of escaped gas. For onshore pipelines, closure of line sectioning valves, each side of the rupture, may further limit the amount of product inventory escaping.

The time taken to fully depressure a pipeline to atmospheric pressure will depend on several factors, not least of which will include, the size and type of pipeline inventory, the operating pressure at time of rupture, the rate of flow escaping and the maximum vent rate at the end station.

For long, large diameter gas trunklines the time taken to fully depressure a line can easily be in the order of several days.

The procedures for emergency depressurization are an essential part of the Pipeline Operating Manual and should state, along with the actions required, the maximum achievable depressurization rate during emergency blowdown.

36.2 Inspection by Intelligent Pigging

36.2.1 General

In Europe, the use of intelligent pigs has increased from, on average, about 2% of the pipelines per year at the beginning of the eighties to about 8% in the nineties. The inspection capabilities of intelligent pig contractors have continuously improved by developments on sensor technology and on data processing, storage and analysis. Despite all the developments on the mechanical design of pigs and on the inspection technology, intelligent pigs should not be seen as being infallible. Each different tool has inherent limitations on inspection capabilities that should be realized. Various experiences within the industry whereby unsatisfactory inspection results were obtained emphasize this point. The main causes for unsatisfactory results have been; no appreciation of the limitations of the inspection tool, selection of the wrong technique and/or contractor, poor performance of the contractor and lack of expertise to interpret and analyze the inspection results. With regards to the frequency for intelligent pig inspection, there is no norm in the industry and the requirement for intelligent pigging depends on the operators inspection philosophy, and the nature and operational risks (Chapter 36.1.2) of the pipeline. Indeed there are many pipelines that are not designed to be or have never been intelligently pigged.

36.2.2 Metal Loss Inspection Techniques

General

Several techniques are available for the inspection of pipelines using pigging technology however, each different technique and tool has inherent limitations on inspection capabilities that should be realized. The type of pig chosen will depend on the purpose of the inspection and the nature of the inspection data required.

Although on occasions the objectives of pipeline inspection using an intelligent pigging tool may vary, in general it is the requirement to detect metal loss that concerns most operators of oil and gas pipelines.

Several techniques are applied in metal loss intelligent pigs, these are:

- Magnetic flux leakage
- Ultrasonics
- High frequency eddy current
- Remote field eddy current

Magnetic Flux Leakage

Principle

About 90% of all metal loss inspections are performed with magnetic flux leakage (MFL) pigs hence, the MFL technology can be regarded as the most important technique for metal loss inspections of pipelines.

The magnetic flux leakage technique is based on magnetizing the pipe wall and sensing the MFL of metal loss defects and other features. From the MFL signal patterns it is possible to identify and recognize metal loss corrosion defects, but also other features such as girth welds, seam welds, valves, fittings, adjacent metal objects, gouges, dents, mill defects, girth weld cracks and large non-metallic inclusions.

Magnetism

MFL pigs are equipped with large magnetic yokes to magnetize the pipe wall in the axial direction. The magnetic yoke consists of a backing bar, permanent magnets, pole shoes and brushes. The combination of the magnetic yoke and the pipe wall is called the magnetic circuit. The magnetic resistance called reluctance, in the magnetic circuit should be minimized in order to obtain a high magnetic flux density, also referred to as level of magnetism, through the pipe wall. Minimization of the magnetic reluctance is achieved by optimizing the design of the magnetic yoke and by using steels with a high magnetic permeability. The magnetic power is given by the strength of the permanent magnets. The strongest permanent magnets applied today are made of NdFeB. Alternatively, an electromagnet can be applied as the magnetic power source instead of employing permanent magnets.

Pipe wall magnetism is dependent on wall thickness, tool velocity, and pipe material, beside the design of the magnetic yoke. The minimum pipe wall magnetism required in order to obtain good flux leakage signals is 1.6 Tesla. Lower pipe wall magnetism levels will make the measurement sensitive to all sorts of disturbances. The best performance is achieved at higher magnetization levels, i.e. in excess of 1.7 Tesla. A magnetic field moving through a pipeline will induce eddy currents in the pipe wall. At high velocities these eddy currents lead to a lower pipe wall magnetization and a distorted MFL field from a defect. At thick walled pipe and/or high tool speed there comes a point where the pipe wall is no longer sufficiently magnetized.

Measurement errors can occur when the level of magnetization in the pipe wall deviates from expectation. This has a higher probability to occur at lower D/t ratios (D/t <30), higher tool velocities (above 3 m/s) and lower steel grades.

Sensors and Resolution

Two types of sensors are applied to sense the magnetic flux leakage fields. In the past mostly coil sensors were used since they could be shaped in all geometry's and do not need power. Disadvantages are that they require a minimum tool speed and that a time differential signal of the absolute flux leakage fields is obtained which requires integration

Nowadays more and more MFL pigging contractors apply Hall effect sensors which have the advantage that they measure absolute magnetic field, are sensitive and small (i.e. make a point measurement) and do not have a limit on minimum tool speed. The major disadvantage of Hall effect sensors is that they require power.

A measurement grid is made over the pipeline, both in the circumferential and axial directions. The resolution of the grid plays an important role on the detectability and sizing performance of small defects; hence the best performance can only be obtained with a fine grid. The grid spacing circumference is determined by the circumferential sensor spacing and in the axial direction by the sampling frequency. The sensor spacing varies between 8 mm and 100 mm for the various MFL pigs. The axial sampling distance varies between 2.5 mm and 5 mm. The smallest defect to be detected and properly sized has a width equal to the sensor spacing and a length equal to about three times the axial sampling distance.

Within the intelligent pigging industry, a distinction is made between low resolution and high resolution MFL pigs referring to the quality of measurement. However, it should be noted that a proper definition on low and high resolution is non-existent. Therefore the fact that an MFL pig is called high resolution does not guarantee a good performance.

Many MFL pigs contain additional sensors to discriminate between internal and external defects and to get a measure of wall thickness changes. Internal/external discrimination is done by means of sensors that are only sensitive to internal defects. Most contractors apply weak magnets combined with a magnetic field sensor placed in a second sensor ring outside the magnetic yoke that measure the decrease in magnetic field when the lift off distance of the magnet to the pipe wall increases by internal metal loss defects. Some contractors make use of eddy current proximity probes that may be placed within the magnetic yokes.

A measure of the wall thickness is obtained by measuring the axial background magnetic field by means of Hall effect sensors. The axial background magnetic field is related to pipe wall magnetization and thus pipe wall thickness.

Data Analysis

MFL pigs record a large amount of data that needs to be analyzed. Most contractors have developed software that automatically analyze the data and detect the relevant features. However, manual analysis and data checks are still necessary to obtain the most accurate defect data.

The relation between MFL signals and defect dimensions is indirect and non linear. Consequently good data analysis algorithms are of importance. Defect length can be accurately determined from the start and end of the MFL signals. Defect width can be determined with limited accuracy from the circumferential signal distribution as measured by adjacent sensors. Defect depth is related to the integrated signal amplitudes but corrections have to be made for defect length and length / width aspect ratios. For defects with a length above 3t (t = wall thickness) or 30 mm, this relation tends to become linear. The relationship between metal loss defect depth and MFL signals becomes more non linear and length dependent below a defect length of 3t or 30 mm for which reason defect sizing accuracy will be of lesser quality.

Capabilities and Limitations

Defect detectability levels are highly dependent on the magnetization level in the pipe wall, the MFL noise as generated by the pipe and the geometry metal loss defect

The pipe material make influence magnetic noise levels. In particular seamless pipe creates a high magnetic noise level whilst on the other hand the ERW manufacturing process gives relatively low MFL noise levels. In addition the quality of the line pipe steel in terms of the number of non-metallic inclusions also influences magnetic noise levels.

The geometry of the defect plays an important role on defect detectability. Mainly the defect depth and width, i.e. the cross sectional area of metal loss normal to the pipe axis, have a strong influence on detectability. Defect length has a secondary effect on defect detectability. In general, the detectability and sizing performance reduce for very short defects (pinhole pitting, circumferential cracks) and for very long smooth defects (axial grooves, general corrosion). Hall effect sensors that measure the absolute axial magnetic field are better suited to measure smooth grooves than coil sensors.

Under optimal conditions, the MFL pigs can detect pits as small as 5% wall thickness loss however, most MFL pigging contractors specify pit detectability between 10% and 40% wall loss whereby the large influence of pipe wall magnetization and line pipe manufacturing process has been taken into account.

Under optimal circumstances, the depth sizing accuracy of general and pitting defects will be about 10% of the pipe wall thickness at 80% confidence.

Depth sizing of axial pits and grooves requires a good length/width correction factor on data analysis and an accurate measurement of defect width. In general depth sizing of axial pits will be less accurate. It has been found that the depth of defects with a length / width aspect ratio above 2 and a width smaller than the sensor spacing can be severely undersized. Under optimal conditions, the accuracy of depth sizing of axial pits will be +10% and -20% of pipe wall thickness at 80% confidence.

Depth sizing of circumferential pits and grooves requires a good length/width correction factor on data analysis. Under optimal conditions the sizing accuracy can be as good as that of general and pitting defects.

It should be realized that defect sizing of bottom-of-the-pipe corrosion whereby general and localized corrosion interacts is more complex. Often only the localized defects are measured

Applicability

MFL pigs can be used under the following conditions:

- Up to velocities of 5 m/s but preferably between 0.5 and 3 m/s;

- D/t > 15, but in case D/t <30 precautions may be required to ensure sufficient magnetization and good reliability of measurement;
- Pipe diameter range from 4-inch to 60-inch;
- All sorts of product.

Ultrasonics

Principle

Ultrasonic pigs utilize ultrasonic transducers that have a stand-off distance to the pipe wall. A fluid coupling is required between the transducer and pipe wall. The transducers emit sound pulses which are reflected at both the inner and outer surface of the pipe wall. The time elapsed detection of these two echoes gives a direct measure of the remaining wall thickness of the pipe. The time elapsed between pulse emittance and the first echo is used to determine the stand-off distance. Any increase in stand-off distance in combination with a decrease in wall thickness indicates internal metal loss. A decrease in wall thickness, while the stand-off distance keeps constant, indicates external metal loss, laminations or inclusions. The outer wall echo cannot be distinguished from the inner wall echo for too thin (remaining) wall thickness.

Sensors

Ultrasonic pigs utilize piezoelectric ultrasonic transducers that emit 5 MHz sound pulses. The transducers are placed in a stand-off distance to the pipe wall. Normally the transducer and stand-off are chosen such that the ultrasonic beam at the pipe wall has a spread of below 10 mm. The circumferential sensor spacing of the state-of-the-art ultrasonic pigs is a little under 10 mm. Consequently the smallest detectable pits have a diameter of about 10 mm. A number of measurements, about 4 or 5, must be made in the axial direction for a pit to be recognized. The sampling frequency depends on the firing frequency of the ultrasonic transducers and the speed of the pig. Under optimal circumstances, the axial sampling distance is about 3 mm.

For accurate metal loss monitoring in heavy wall pipelines the ultrasonic technique is better suited than the MFL technique. In gas or multiphase lines this can be achieved by running the ultrasonic tool in a batch of liquid such as glycol. In view of the maximum allowable speed of an ultrasonic tool the velocity excursions of the gas driven pig-slug train needs to be properly controlled. The dynamics of a pig-slug train in a gas pipeline has been extensively studied to determine the optimum parameter settings in order to avoid the pig-slug train from stopping during the survey and subsequently shooting off at high velocities. The maximum allowable speed of the ultrasonic tool is determined by the firing frequency of the ultrasonic sensors and was in the past limited to about 1 m/s. However, due to the improved electronics the firing frequency has been increased which now allows a maximum velocity of around 3 m/s.

Data Analysis

Interpretation of ultrasonic signals is more straight forward than MFL signals. The stand-off and wall thickness signals give a direct mapping of the pipe wall, showing all corrosion defects. A rough surface and internal debris may lead to loss of signal and can be recognized as such. In addition laminations, inclusions, girth welds, valves and tees can be easily recognized. Nowadays defect detection and sizing is fully automated however, the data is still often checked manually.

Capabilities and Limitations

Ultrasonic pigs have the advantage that they provide a better quantification of the defect sizes than MFL pigs. Detection of defects starts at lengths of 10 mm. The probability of detection becomes high at surface lengths of about 20 mm. Depth sizing accuracy of the remaining wall thickness is in the order of +/- 1 mm for pits and +/- 0.5 mm for general corrosion at a confidence level of about 80%. Small pits can be missed. This performance is achieved by the state of the art tools.

The depth sizing error is absolute and independent of nominal pipe wall thickness. The relative error however, will increase significantly for smaller wall thickness. Most pipeline operators conclude that ultrasonics is more suited for thick wall pipe than for thin wall pipe. A threshold wall thickness of 7 mm is generally chosen below which ultrasonic pigs are not recommended for use.

The amplitudes of the inner and outer wall echoes must exceed pre-set threshold values to be detected. The echo signal can be attenuated by fouling, roughness of surfaces, tilting of probe and curvature of surface profile. Dirt at the bottom of the line during a survey may mask the most critical defects.

A rough internal pipe surface, e.g. due to corrosion, may result in a double inner wall reflection causing the tool to ignore the second reflection coming from the outer wall. When this shortcoming is not realized the metal loss is reported to be external with a completely wrong depth.

Applicability

Ultrasonic pigs can be applied under the following conditions :-

- Diameter range from 6-inch to 60-inch;
- Velocities from 1 m/s through to 3 m/s;
- For pipe wall thickness above 7 mm;
- Only for liquid products unless the tool is run in a batch of liquid.

High Frequency Eddy Current

Principle

High Frequency Eddy Current (HFEC) technology has been developed for monitoring internal corrosion in heavy wall, small diameter pipelines.

HFEC proximity sensors are mounted on a polyurethane sensor carrier and applied for two different types of measurement so called global and local. The local sensors measure the distance from the sensor to the pipe wall. The global sensor is used to measure the distance from the center of the carrier to the local sensors. The combination of the measurements from local and global sensors provides the internal profile of the pipeline by which both internal pitting and general corrosion can be determined.

The principle of eddy current is based on the phenomenon that an alternating current in a transmitter coil induces alternating currents or eddy currents in any nearby conductor through inductive electromagnetic coupling. The eddy currents in the conductor will in turn induce currents in other nearby conductors, establishing an indirect electromagnetic coupling from the transmitter coil via the first conductor to the second conductor. Hence, a receiver coil can be indirectly coupled to a transmitter coil via the pipe wall. By designing the receiver coil in a figure eight shape, the direct electromagnetic coupling between transmitter and receiver coil is canceled out and the receiver coil is only responsive to the indirect electromagnetic coupling via the pipe wall. The phase and amplitude of receiver coil signal are highly sensitive to the distance between the coils and the pipe wall. By a proper selection of frequency and phase of the eddy currents, the signals have been made insensitive to pipe wall material properties.

Capabilities and Limitations

The sensor geometry has been optimized so that internal pitting and general corrosion with a length exceeding 10 mm and a depth exceeding 1 mm should be detected and sized with an accuracy of +/- 1 mm up to a maximum depth of 8 mm. Furthermore, the technique can accurately measure ID reductions such as dents and ovalities

The HFEC technique can only measure internal defects, no measurement is obtained from external defects. The measurement is insensitive to the pipeline product and to debris.

Applicability

HFEC pigs can be applied under the following conditions :-

- Diameter range from 6-inch to 12-inch;
- Velocities up to 5 m/s;
- All sorts of products;
- When only internal corrosion is of concern.

Remote Field Eddy Current

The Remote Field Eddy Current (RFEC) dates back to the 1950's (well bore inspection) but use of the technique for pipeline inspection has not passed the experimental stage.

The RFEC technique utilizes a relatively large solenoidal exciter coil, internal to and coaxial with the pipe, which is energized with a low frequency alternating current to generate eddy currents in the pipe wall. At two to three pipe diameters distance (remote field) one or more receivers are located detecting those eddy currents which have penetrated the pipe wall twice (outward at exciter, inward at receiver). Both amplitude of the received signal and phase lag between remote field and exciter field provides information on wall loss and changes in material properties (electrical conductivity and magnetic permeability). Because of the double wall transit the RFEC technique has equal sensitivity to internal and external wall loss.

Detection and sizing performance are dependent on pipeline diameter, wall thickness, magnetic permeability and tool speed. Tool speed is limited to less than 0.5 m/s due to the low frequency applied to generate the eddy currents. The maximum wall thickness that can be inspected with a RFEC tool depends on test frequency in combination with pipe magnetic permeability. For carbon steel pipes the maximum inspectable thickness is approximately 10 - 12 mm.

36.2.3 Intelligent Pigs for Purposes other than Metal Loss Detection

General

If one excludes metal loss detection then, broadly speaking, pipeline inspection by intelligent pigging can be categorized into the following five groups of inspection capability:-

- Crack detection
- Calipering
- Route surveying
- Freespan detection
- Leak detection

The purpose of this section is to briefly describe the tools and techniques that are currently available with respect to the above inspection requirements.

Crack Detection

British Gas have developed a crack detection pig based on ultrasonic wheel probes. This pig is called the Elastic Wave Inspection Vehicle and can be operated in both gas and liquid pipelines. The first prototype was a 36-inch pig which contained 32 wheel probes. In addition a 30-inch pig has been built. Main difficulties with this technology has been on data interpretation with regards to minimizing the rate of false calls. However, in recent years much work has been carried out by British Gas on data analysis algorithms to discriminate real cracks from spurious indications. British Gas claim that the number of false calls has decreased significantly by their recent improvements on data analysis.

PTX have developed an ultrasonic crack detection pig that aims to detect both internal and external longitudinal cracks in clean liquid pipelines. The tool can also detect potential fatigue cracks in the longitudinal weld seam. Note that this pig cannot be run in gas pipelines unless this is done in a liquid slug. The key in the concept is the complete coverage of the pipe by a large number of ultrasonic piezoelectric transducers (512 for a 24-inch pig).

Calipering

Caliper pigs measure internal profile variations like dents, ovality and internal diameter transitions with the primary objective being to detect mechanical damage and/or ensure that a less flexible metal loss inspection pig can pass through the pipeline. Caliper pigs are normally designed to be flexible and can pass 25% ID reductions.

Most of the Caliper pigs are equipped with mechanical sensors (fingers) that follow the inner profile of the pipe wall. Typically, these pigs can detect dents and ID reductions of between 1% and 2% of the pipe diameter. A drawback of the mechanical caliper pig is that false readings can be obtained from debris or solid wax. Established contractors that offer services with mechanical caliper pigs are Pipetronix, Enduro Pipeline Services and TD Williamson (TDW). Some tools have the additional capability to measure the bend radii.

H Rosen Engineering (HRE) offers a service with a caliper pig that uses eddy current proximity probes and which is called the Electronic Gauging Pig (EGP). The 8 probes are mounted in a conical nose at the front or rear of the pig. This pig has the advantage that the pig is very rugged and insensitive to debris or wax. When required the EGP can be mounted with a larger cone by which the sensitivity can be increased from about 1.5% ID reduction to about 0.5% ID reduction, at the expense of the pig's flexibility.

Route Survey

The Geopig of BJ Pipeline Services (formally Nowsco) is the market leader for route surveying. The Geopig was developed by Pulsearch, Canada in the mid eighties with the aim to measure subsidence in the "Norman Wells" pipelines in Canada which lie in an active permafrost region. The Geopig is capable of determining the latitude, longitude, height, bend location and curvature and center point of a complete pipeline in a single run. The heart of the Geopig is a strapdown inertial measurement unit giving an accuracy on location of 0.5 m/km and a curvature with a radius up to 100m. Two fixed rings with ultrasonic probes are mounted to measure the internal profile of the pipeline. In liquid pipelines undamped and unfocused 2.5 MHz transducers are used. The sensors for gas service operate at 250 KHz and require a minimum internal pressure of 10 bar. A footprint of the sonar on the wall has a diameter of 10mm. The accuracy of the sonar to measure dent depths is +/- 2.5 mm.

Some pipeline operators have found good use of the Geopig to assess the pipeline profile for upheaval buckling and the necessity for rock dumping.

An alternative to the Geopig is offered by Pipetronix in the form of their Scout pig, which uses inertial navigation by means of built-in gyroscopes.

Freespan Detection

British Gas has developed the Burial and Coating Assessment (BCA) pig based on neutron backscattering, that aims to detect freespans. However, the BCA pig has not become a commercial success because of its limited competitiveness with respect to remotely operated vehicle (ROV) inspection.

HRE have recently developed a freespan detection pig based on gamma ray technology.

BJ Pipeline Services claim that their Geopig (see previous section) can detect freespans by measuring vibrations of the pipeline when the pig passes an unsupported section however, this capability has not yet been field proven.

Leak Detection

Two types of pig are available for leak detection.

The first type aims to acoustically detect leaks in on-stream liquid pipelines by means of the escaping noise. Acoustic pigs are offered by Maihak and recently by TUV Osterreich. With this type of pig it is considered feasible to detect leaks at a leak rate of about 10 liters per hour.

The second type of pig aims to detect leaks in shut-in pipelines by measuring the flow or differential pressure over the pig. Service with this type of pig is offered by Pipetronix and H Rosen Engineering.

36.3 Maintenance

36.3.1 General

The principle function of maintenance is to ensure that physical assets continue to fulfil their intended purpose. The maintenance objectives with respect to any item of equipment should be defined by its functions and its associated standards of performance.

Prior to setting out to analyze the maintenance requirements of equipment it is essential to develop a comprehensive equipment register. In general terms the equipment included will relate only to onshore pipelines (or onshore sections) since maintenance work on subsea pipelines is not foreseen, that is, all subsea equipment should be designed to be maintenance free throughout the design life expectancy of the pipeline. This is not to say that remedial work on a subsea pipeline will never occur, but only that it should not be a planned occurrence. However in the case of subsea pipeline repairs, it is prudent for most operators to keep a set (or to share a set) of emergency pipeline repair equipment on stand by. This may include repair equipment such as pipeline repair clamps and full hyperbaric welding spreads. This equipment should be maintained along with onshore pipeline equipment.

Generally preventive maintenance is carried out on onshore pipeline equipment with dominant failure modes (e.g. wear out of pump impellers) at pre-determined intervals or to prescribed criteria, with the intent to reduce the probability of failure or the performance degradation of the item. It should go without saying that all maintenance work should attempt to minimize the effect to normal production operations. (e.g. schedule critical activities to coincide with a planned pipeline shutdown).

Maintenance should be carried out on all pipeline associated equipment (e.g. pipeline valves and actuators, pig traps, pig signalers and other pipeline attachments). Maintenance procedures and routines should be developed with account taken of previous equipment history and performance.

36.3.2 Pipeline Valves

Pipeline valves should be lubricated and functionally operated at least once annually and in accordance with the valve manufacturers recommendations. Functional operation of subsea valves should also be carried out annually. However, where valves are located in unfavorable conditions (e.g. valve pits subject to flooding or general dampness) it may be advisable to increase the maintenance frequencies to account for these conditions.

All valve actuators whether they be manual, pneumatic, hydraulic or electrical should be functionally tested at least once per year and in accordance with the actuator manufacturers recommendations.

In developing maintenance routines, account should be taken, where applicable, of the requirement to test the equipment by remote operation or by simulating line-break conditions. Operations involving the closure of block valves should be a co-ordinated exercise with all the relevant parties.

36.3.3 Pig Traps

Pig trap maintenance shall be carried out strictly in accordance with the manufacturers guideline for the type of pig launcher and receiver facilities used, and these guidelines incorporated in the maintenance routine. However, as a minimum a full inspection and survey of the condition of the pig traps should be conducted annually, and should include:

- Condition of launcher / receiver barrel;
- End closure seals;
- Bleed locks and electrical bond;
- Locking rings;
- Pig signalers;
- Associated valves and pipework.

36.3.4 Pipeline Location Markers

Aerial markers and pipeline markers should be maintained on an ongoing basis with the information contained on the marker posts verified and updated annually.

Above ground crossing points should be examined at least once per year for condition of supports and associated structures, including paintwork and protective wrap, and refurbished where necessary.

36.4 Pipeline Repair Methods

36.4.1 Conventional Repair Methods

Damage to a submarine pipeline can be repaired in different ways depending on the water depth and on the type and extent of the damage. This section describes the various types of conventional repair methods currently available for repairing a damaged subsea pipeline in water depths of less than 300 m. This maximum depth limitation is one that is realistically imposed as a result of diver constraints. Non-conventional pipeline repairs are considered to be those carried out diverless and in water depths exceeding 300 meters, as discussed in Table 36.1 summarizes the various repair methods and their applicable water depths. The various types of conventional repair methods can be summarized as follows:

- Non-critical repair work;
- Minor repair requiring the installation of a pin hole type repair clamp;
- Medium repair requiring the installation of a split sleeve type repair clamp;
- Major repair requiring the installation of a replacement spool.

Non-critical intervention work such as freespan correction, retrofitting of anode sleds and rock dumping, can usually be considered as planned preventive measures to reduce the risk of an emergency occurring. For the localized repair of non-leaking minor and intermediate pipeline damage, repair clamps are likely to be utilized and without the necessity of an emergency shutdown to the pipeline system. For major pipeline damage resulting in, or likely to result in, product leakage, immediate production shutdown and depressurization is invariably required allowing the damaged pipe section to be cut out and replaced by a spool using surface/hyperbaric welding techniques or mechanical connectors. In the case of surface welding, this procedure relies on the requirement for the damaged section of the line to be located in shallow water, typically not greater than 30 meters. This should allow the pipeline to be raised to the surface using suitably equipped attendant vessels and thereby permit the repairs to be performed in a dry environment.

Table 36.1Repair methods vs. applicable water depths.

Repair Method	Water Depth		
	0-50m	50m-300m	>300m
Repair Clamp	➽	➽	➽ (note 1)
Hyperbaric Welding	➽ (note 2)	➽	N/A
Mechanical Connectors	➽	➽	➽ (note 1)
Surface Welding	➽ (note 3)	N/A	N/A

Notes:

1. Technology exists for the diverless installation (by ROV) and the diverless installable hardware such as repair clamps and mechanical connectors.
2. Hyperbaric welding in water depths less than 20 m is not practical and other repair solutions are required.
3. Water depth limitation for surface welding is governed by size of pipeline, weight of pipeline and vessel lifting capabilities.

36.4.2 General Maintenance Repair

This section deals with those non-critical repairs which in the short term will not jeopardize the safety of the pipeline and hence, can form part of a planned maintenance program. Examples include:

- Corrosion coating repair;
- Submerged weight rectification;
- Cathodic protection repair;
- Span rectification procedures;
- Installation of an engineered backfill (rock dumping).

Corrosion Coating Repair

Repairs carried out on the corrosion coat of a submarine pipeline may be undertaken under two differing environments. They are:

- Marine conditions - coating applied in seawater.
- Hyperbaric conditions - coating applied in dry conditions inside a habitat.

The need for any major repairs at a particular site usually dictates the conditions in which the coating repair is carried out. Repairs to a subsea pipeline that involve only repairs to the corrosion coating is unlikely.

Submerged Weight Rectification

In a submerged pipeline system the concrete weight coating provides negative buoyancy. If a loss of concrete weight coating occurs at locations where a pipeline is exposed on the seabed, the stability and structural integrity of the system may be affected. If the condition worsens it may be that some rectification measures are necessary to stabilize and protect the pipeline system. These remedial measures may include:

- Installation of concrete sleeves;
- Installation of engineered backfill;
- Installation of sand or grout bags;
- Installation of stabilization mattresses or saddles.

For each situation which arises the requirements for stabilization and protection of the pipeline due to its exposure or loss of weight coating should be analyzed to assess its weight rectification requirements.

Installation of Concrete Sleeves

If concrete sleeves are utilized, the damaged concrete weight coating may be replaced in-situ. Fabric sleeves, which are prefabricated, may be zipped and strapped around the damaged section of pipe and subsequently pumped full of grout via the relevant facilities located on board the surface vessel. Refer to Figure 36.1.

The sleeves may be manufactured to suit the pipe size and coating and provide sufficient flexibility to adapt to uneven surfaces of the pipe. Typically, they may be provided in lengths of up to 6 meters. The underside of the pipe has to be made accessible to enable the installation of the sleeve. This option, may be used for local or one-off type repair, but is expensive for more extensive repair requirements.

Installation of Engineered Backfill

If this method is adopted, the engineered backfill material is positioned so as to bury completely the damaged section of weight coating and thus provide the requisite protection and stability. Refer to Figure 36.2.

Installation of Sand or Grout Bags

Sand or grout bags may be employed in a similar manner to the engineered backfill to provide local cover and burial of the damaged section of the pipeline. Divers are used to place the bags around the pipeline system. Refer to Figure 36.1. Comparatively the operation is more labor intensive than a similar operation using engineered backfill hence, the financial ramifications may be restrictive for extensive repairs to the pipeline weight coating.

Methods similar to these are frequently used as a integral part of localized span rectification.

Installation of Stabilization Mattresses or Weight Saddles

When this method is employed, flexible mattresses or concrete saddles are positioned over the pipeline system to provide the required stability and protection. In each case the actual positioning operation is usually completed using a subsea handling frame located over the exposed pipeline. In general, the flexible mattresses are considered to be more suitable than the concrete saddle due to their greater ability to adapt to transient seabed conditions. Refer to Figure 36.3.

This option may be used for a considerable number of situations and provides a versatile facility for one-off or the more extensive type of repair.

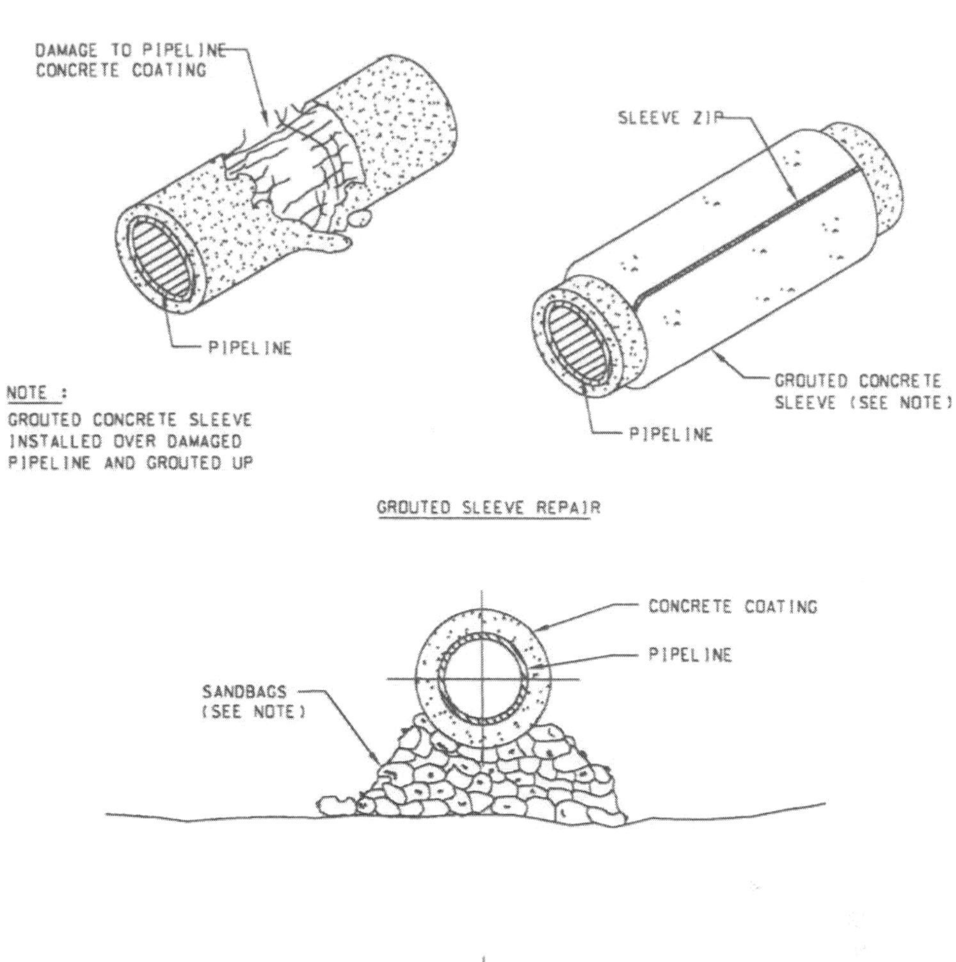

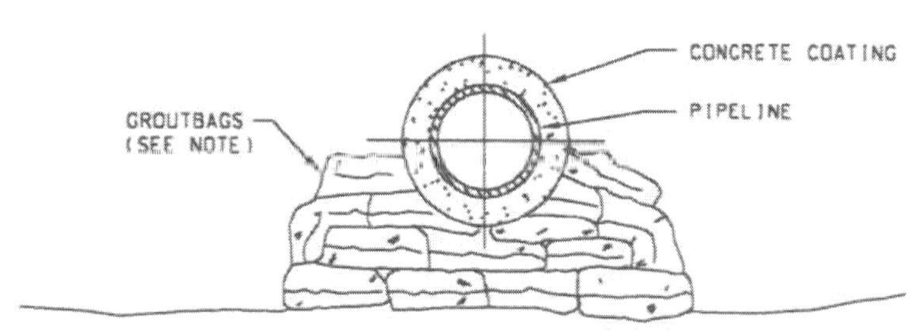

NOTE :
RANDOMLY PLACED INDIVIDUAL BAGS PRODUCING PIPELINE SUPPORT OR COVER

GROUTBAG OR SANDBAG REPAIR

Figure 36.1 Typical methods of concrete sleeves groutbags and sandbags.

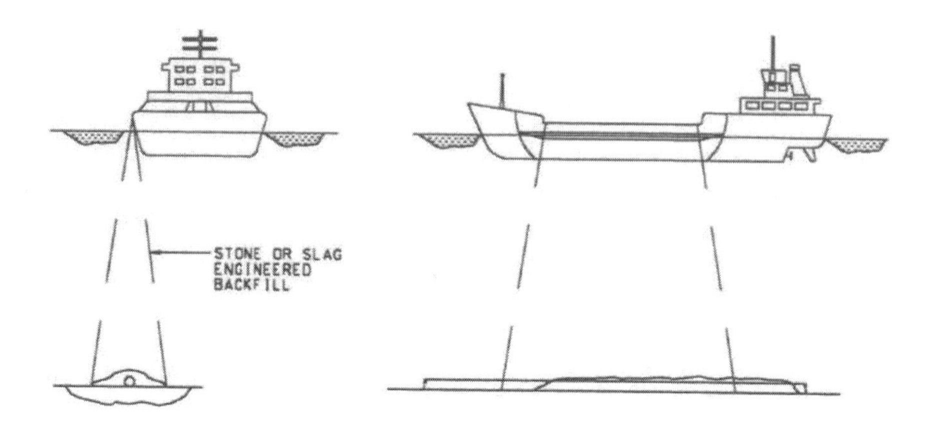

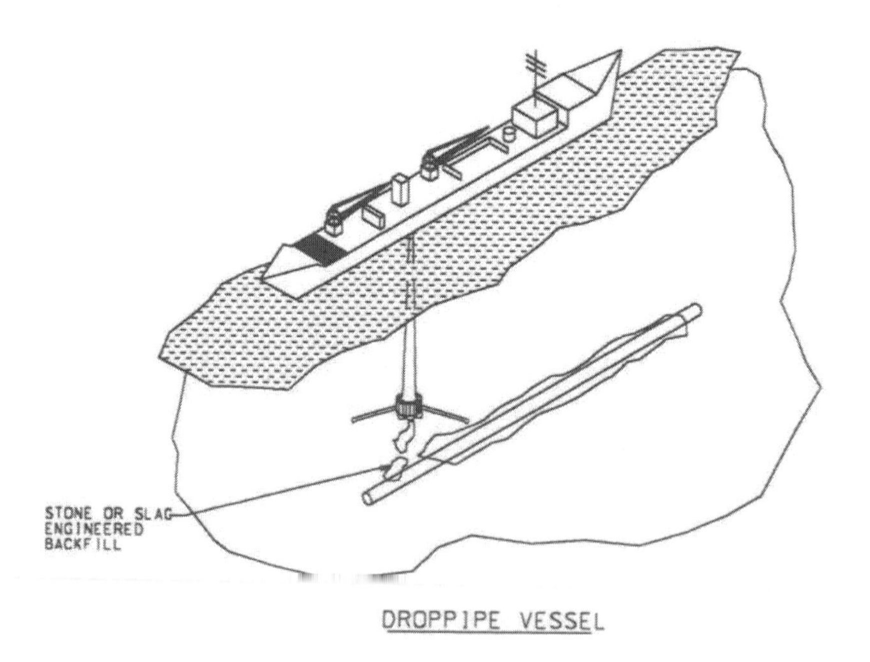

Figure 36.2 Typical methods of rock dumping.

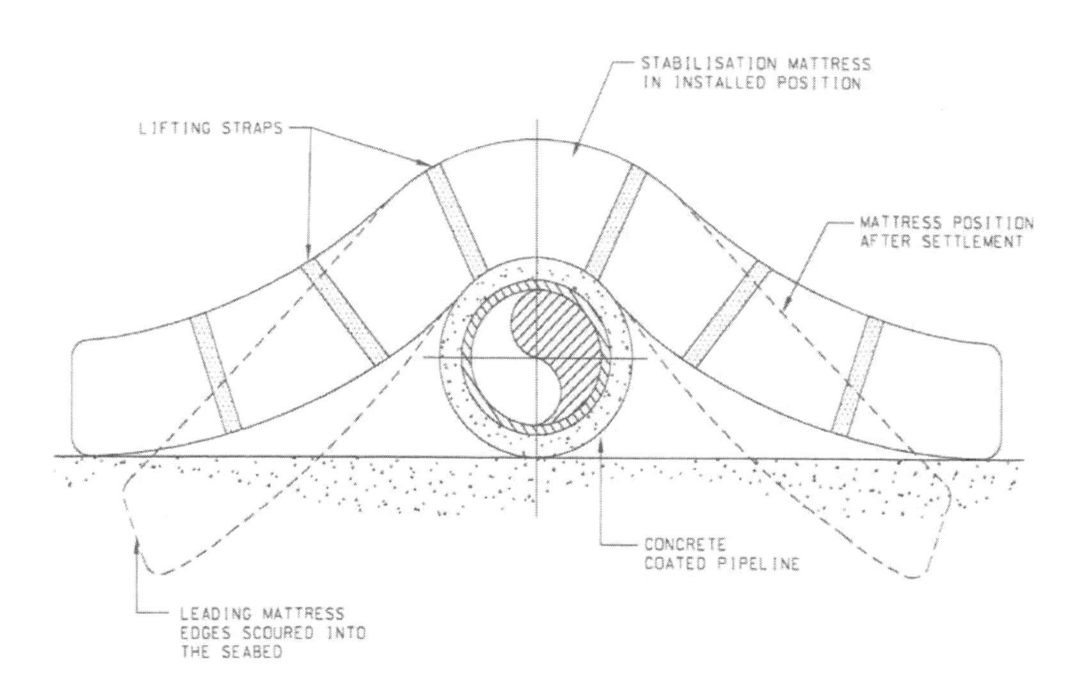

Figure 36.3 Stabilization mattress type stability method.

Cathodic Protection Repairs

The cathodic protection facilities of the pipeline system may need to be repaired or enhanced if the system performance is shown to be inadequate This ineffectiveness may be due to a the anodes being damaged or being prematurely depleted as a result of bad CP design or unexpected and severe corrosion coating breakdown.

The introduction and connection of anode "sledges" may be utilized to achieve extra cathodic protection. These anode "sledges" are connected at specified intervals along the pipeline system and at a minimum stand-off distance from the line, both requirements being optimized for a given situation.

Electrical connection between the end of the anode "sledge" cable and the pipeline is typically achieved by employing mechanical screws or by "wet" welding onto an in-situ doubler plate from an original anode. The use of screwed connections, although simpler in concept, have been known to loose their electrical contact over time. The technique of "wet" welding onto an in-situ doubler plate or strap is therefore recommended as the preferred method of providing electrical contact.

Pipeline Span Rectification

Within the pipeline system's design life unacceptable freespans may develop due to a number of factors which include scouring action or the passage of sand waves. It is usual practice, during the pipeline design phase to calculate the permitted spans of the system for all phase of installation and operation. With the pipeline full of water, air or gas allowable spans are calculated for the both static and dynamic conditions. Accordingly, a "worst case" envelope can be developed, which may be used as a basis for designating the allowable span criteria.

Any spans which exist may be detected by subsequent regular inspection program. The span assessment and method of support should also take into account any proposed changes in the submerged weight of the pipeline.

Span rectification measures will have to be employed if the pipeline span exceeds the allowable span criteria. Generally span rectification measures will take the form of installing discrete supports within the length of the unacceptable pipeline span, thus reducing the actual freespan length. The installation of an engineered backfill may also be necessary to fill in the voids between the supports and to ensure a smooth contour over the pipeline system.

Before the surface support vessel is mobilized, the repair contractor should, in consultation with the Company, propose a design for span supports and the method of installing them. Design calculations should be undertaken in order that the supports conform to the following requirements:

- The supports are positioned such that all relevant spanning conditions of the pipeline are satisfied;
- Realistic installation tolerance is to be included for the horizontal positioning of the calculated spacing of the supports;
- The supports are stable and fully support the pipeline over its remaining design life period;
- The support system is not susceptible to scouring action;
- Lateral movement of the pipeline is prevented by the support installation.

Supports may be developed by placing numerous individual sand or grout bags under the pipeline. An alternative to this is to install an empty fabric form-work under the pipeline and subsequently fill it with grout. This technique is considered to provide a more reliable and complete structural support than by using sand or grout bags and for larger supports may be comparatively faster to install. Refer to Figure 36.4. The grouted fabric form-work may be shaped to match the contours of the pipe and may be provided with straps to ensure a permanent connection with the pipeline. Additionally, these units may be designed such that during the grouting operation the injection pressure may provide an upward lifting mechanism to the pipeline. This feature may provide a useful facility for stress relief in the pipeline span if they are out with acceptable limits. Alternatively, if required, other equipment may be installed to temporarily lift the pipeline during the support installation.

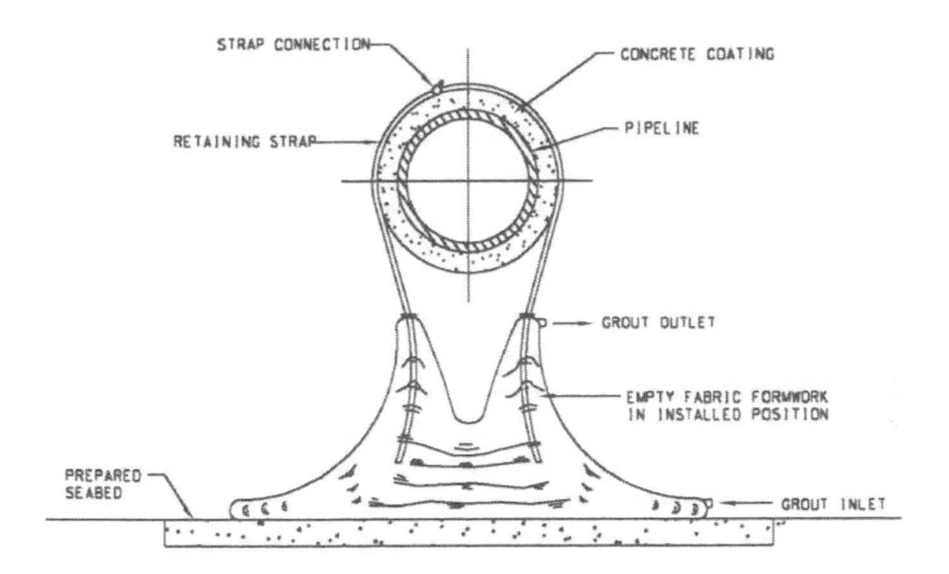

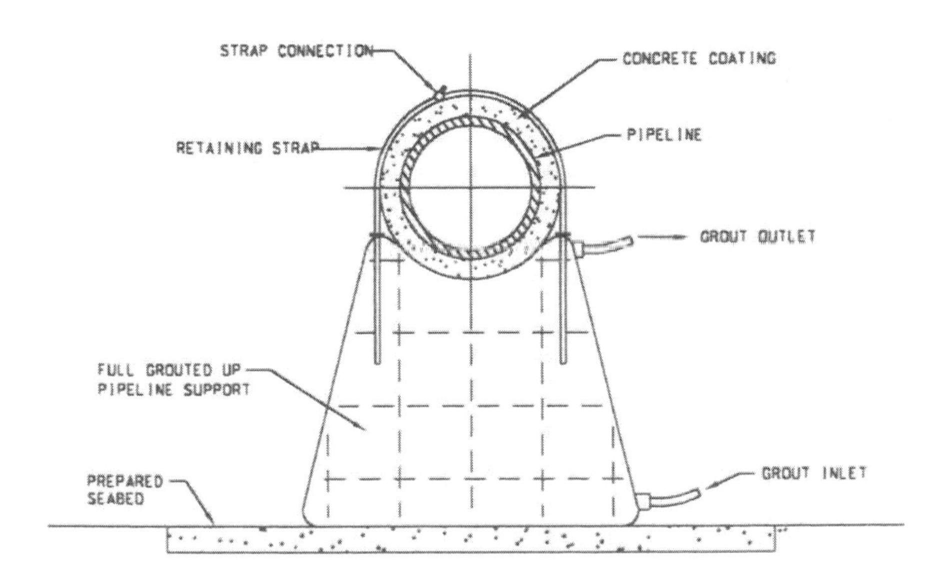

Figure 36.4 Typically methods of using formwork for grouting.

36.5 Deepwater Pipeline Repair

36.5.1 General

In the last decade the world's hydrocarbon industry has moved into deep waters and the underwater pipeline repair technology is continuously developing to keep pace. In general, a well proven capability exists to conduct repairs on pipelines up to a water depth of about 300 m, beyond which divers cannot realistically work in saturation. However, recently the use of robotics has undergone significant advancement which together with experience gained in the past few years in the field of pipe repair in deep waters (to 600 m) suggests that there is now such a thing as deepwater pipeline repair technology, although improvements would be necessary for specific scenarios.

Typically any deepwater repair procedure requiring the replacement of a pipe section will be based on the concept of a spoolpiece installation using diverless mechanical connectors to attach onto the free ends of the pipeline. End connector hardware capable of being installed without divers has been developed by Hydratight of UK and HydroTech of USA. Refer to vendor details contained in Attachments. The basic concept remains the same regardless of whether divers are employed to carry them out as in more conventional repair operations (refer to Figure 36.5). Unfortunately, the problems associated with physically accomplishing each task as a diverless operation, remain significant.

Notwithstanding the above, there is a growing consensus that various ROV contractors could collectively perform virtually all the tasks required with a minimum amount of special support equipment having to be constructed.

This section outlines the progress made in the art of deepwater repair, presents guidelines for new repair technology and discusses different ways to approach and solve a diverless repair task.

36.5.2 Diverless Repair Research and Development

Diverless repair systems had been considered since 1971 with two significant studies being performed as Joint Industry Studies, one sponsored by Exxon Production Research and the other by Shell. Aims of these studies were twofold; firstly to allow pipeline repairs at water depths beyond diver capabilities and secondly to have a cost effective diverless repair system that could compete with diver assisted repair systems. Some of the earlier studies were a little too ambitious in that they attempted, optimistically, to solve all problems for both small and large diameter pipe sizes and in water depths reaching 1300 m. As a result, although the studies identified many of the major problem areas, they did not lead to the development of actual repair capabilities since, at that point in time, the conclusions and recommendations were considered to be either impractical or too expensive to implement. Also these earlier studies were prompted by the industry anticipating in the very near future (at that time) the need to repair large diameter, concrete coated pipelines in water depths to 1300 m. As we now know, this did not materialize. This again contributed to the fact that the early studies did not result in any repair system.

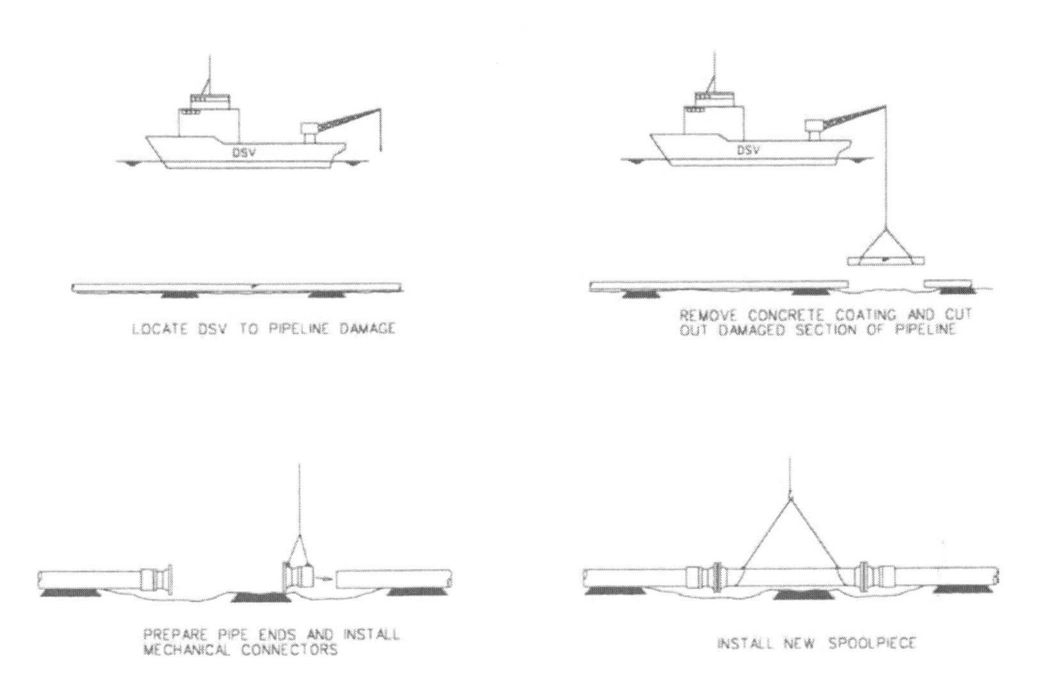

Figure 36.5 Replacement of pipe section.

36.5.3 Deepwater Pipeline Repair Philosophy

In view of the increasing global trend towards deepwater developments, greater emphasis will be placed on the development of reliable and cost effective deepwater pipeline repair systems. However, given that it is considered highly unlikely that any one system will be all encompassing in terms of its ability to repair all forms of damage for all types of pipeline, each operator must carefully evaluate his own requirements.

There are basically two different possibilities to establish a repair strategy for each specific pipeline scenario. The first option maybe to make a joint agreement among a pool of pipeline operators, whereby the total investment and ongoing maintenance costs can be shared among all the involved partners. In this case the repair system should be able to repair pipelines with different characteristics and each partner has to foresee purchasing of his dedicated connectors and connection tools. For example, the Pipeline Repair Service (PRS) is a joint venture between Statoil and Norsk Hydro in Norway. The PRS is a collection of dedicated standby equipment and a pool of services for the rapid deployment of repair equipment, including hyperbaric facilities, for pipeline repair. The disadvantage of this approach is that there may not be other pipeline operators in the region who have a need for a similar deepwater repair capability. An alternative would be for the individual operator to establish his own dedicated repair contingency.

Regardless of which option is chosen, it should be noted that the substantial investment and operating costs of a repair system are completely sustained by the Pipeline Operator or by the pool of Operators and will never be recovered even if considerable saving of money is involved when comparing the selected repair method with alternatives. In addition if no damage occurs to the pipeline, as is the desirable intent, then the equipment will never be used. Also, based on the results of a risk assessment/occurrence probability analysis, the total investment cost for a repair using a diverless repair system could be compared with the cost of a repair performed by relaying a section of pipeline and performing the necessary tie-ins in a water depth reachable by divers. For these and other reasons it is absolutely essential to choose a repair strategy that can solve all the envisaged repair tasks in a reliable and efficient manner and at the lowest possible cost.

36.6 References

1. "Diverless Pipe Repair System Set for Deepwater Trials", Offshore Journal, August 1995.
2. Jackson L. and Wilkins R. "The Development and Exploitation of British Gas Pipeline Inspection Technology", Institution of Gas Engineers 55th Autumn Meeting, 1989
3. Kiefner, J.F., Hyatt, R.W. and Eiber, R.J. (1986) :"NDT Needs for Pipeline Integrity Assurance," Battelle/AGA.
4. Manelli, G. and Radicioni, A. (1994), "Deepwater Pipeline Repair Technology: A General Overview", OMAE'1994.
5. South East Asia Oil Directory 1997 Produced by Oil & Gas Journal, Published by Penn Well Publishing Company.

PART VI: Integrity Management

Part VI

Integrity Management

Chapter 37 Reliability-based Strength Design of Pipelines

37.1 Introduction

37.1.1 General

In principle, reliability-based design of offshore pipelines involves the following aspects:

- Identification of failure modes for specified design cases;
- Definition of design formats and Limit State Functions (LSF);
- Uncertainty measurements of all random variables;
- Calculation of failure probability;
- Determination of target reliability levels;
- Calibration of safety factors for design;
- Evaluation of design results.

37.1.2 Calculation of Failure Probability

Generally, limit-state function (LSF) is introduced and denoted by g(Z) where Z is the vector of all uncertainty variables. Failure occurs when g(Z)≤0. For a given LSF g(Z), the probability of failure is defined as:

$$P_f(t) = P[g(Z) \leq 0] \tag{37.1}$$

The results can also be expressed in terms of a reliability index β, which is uniquely related to the failure probability by:

$$\beta(t) = -\Phi^{-1}(P_f(t)) = \Phi^{-1}(-P_f(t)) \tag{37.2}$$

where Φ(.) is standard normal distribution function.

Two general approaches are available to solve Equation (37.1) namely analytical and simulation methods respectively, see Bai (2003).

37.2 Uncertainty Measures

Considering uncertainties involved in the design format, each random variable X_i can be specified as:

$$X_i = B_x \cdot X_C \tag{37.3}$$

where X_C is the characteristic value of X_i, and B_X is a normalized variable reflecting the uncertainty in X_i.

37.2.1 Selection of Distribution Functions

Usually, the determination of the distribution function is strongly influenced by the physical nature of the random variables. Also, its determination may be related to a well-known description and stochastic experiment. Experience from similar problems is also very useful. If several distributions are available, it is necessary to identify by plotting of data on probability paper, by comparisons of moments, statistical tests, etc. Normal or lognormal distributions are normally applied when no detailed information is available. For instance, resistance variables are usually modeled by normal distribution, and lognormal distribution is used for load variables. The occurrence frequency of a damage (e.g. an initial crack), is described by Poisson distribution. Exponential distribution is used to model the capacity of detecting certain damage.

37.2.2 Determination of Statistical Values

Statistical values used to describe a random variable are mean value and coefficient of variation (COV). These statistical values shall normally be obtained from recognized data sources. Regression analysis may be applied based on methods of moment, least-square fit methods, maximum likelihood estimation technique, etc.

37.3 Calibration of Safety Factors

37.3.1 General

One of the important applications of structural reliability methods is to calibrate safety factors in design format in order to achieve a consistent safety level. The safety factors are determined so that the calibrated failure probability, $P_{f,i}$ for various conditions is as close to the target reliability level P_f^T as possible:

$$\sum f_i \left(P_{f,i}(\gamma) - P_f^T\right)^2 = \min imum \tag{37.4}$$

where f_i is the relative frequency of the design case number i.

37.3.2 Target Reliability Levels

When conducting structural reliability analysis, target reliability levels in a given reference time period and reference length of pipeline should be selected. The selection is based on consequence of failure, location and contents of pipelines, relevant rules, access to inspection and repair, etc. The target reliability levels have to be met in design to ensure that certain safety levels are maintained.

The following safety classes are proposed:

Low safety class: where failure implies no risk of human injury, minor environmental damage and economic consequences.

Normal safety class: classification for temporary conditions where failure implies risk of human injury, significant environmental and economical consequences.

High safety class: classification for operation conditions where failure implies risk of human injury, significant environmental and economical consequences.

Target reliability levels may be specified by the operator guided by authority requirements, design philosophy and risk attitude in terms of economics. The target reliability level for damaged pipelines should be defined in the same level as intact pipeline. The target reliability level needs to be evaluated considering the implied safety level in the existing rules and codes. Sotberg et al. (1997) proposed target reliability levels as below:

Table 37.1 Target reliability levels (Sotberg, et al., 1997).

Limit States	Safety Classes		
	Low	Normal	High
SLS	10^{-1}--10^{-2}	10^{-2}--10^{-3}	10^{-2}--10^{-3}
ULS	10^{-2}--10^{-3}	10^{-3}--10^{-4}	10^{-4}--10^{-5}
FLS	10^{-3}	10^{-4}	10^{-5}
ALS	10^{-4}	10^{-5}	10^{-6}

37.4 Reliability-based Determination of Corrosion Allowance

37.4.1 General

This Sub-section is based on Nødland et al. (1997).

In order to calculate the pipeline reliability accurately, sufficient statistical data must be available as a basis for describing key input parameters in the form of probability distribution functions. In addition to this, an element of engineering judgment must be included. The reason for this is that experiences from a particular pipeline or from the laboratory are never directly transferable to a new pipeline because differences will always exist, such as amount and chemical composition of water, flow regimes, condensation rates etc. Furthermore, the methods available to calculate corrosion rates are empirical or semi-empirical, and as such valid only for a narrow band of operating parameters (i.e. flowrate, temperature, pH and pressure). The picture can be further complicated by possible formation of scales or deposits on the pipe wall. In addition, the corrosion rate model which is normally used, i.e. the deWaard-Milliams or Shell formula deWaard et al. (1993, 1995) is based on laboratory data only and is generally recognized to be very conservative when applied to "real life".

At present, statistical data which can be used in a reliability analysis are limited. Some data exist to give a reasonable representation of the uncertainty in defect lengths in gas lines and in pipe wall thickness. When it comes to the calculation of corrosion rates, very little work has been carried out to provide a statistical basis to work from. Therefore, the probability functions are estimated primarily from engineering judgment. Hence, the reliability found from the calculated probabilities of failure (or unacceptable corrosion depth) should not be taken as an exact value, but rather as a subjective evaluation of pipeline reliability.

The calculated reliability can then be used in life cycle cost (LCC) analyses, where the combination of corrosion allowance and operating parameters which give the highest net present value (NPV) will represent the optimum corrosion allowance. Following this approach, the relative effect of different design and operating parameters on the pipeline reliability may be evaluated.

The scope of this Section is to describe a method to calculate the reliability of a corroded pipeline for different design and operating parameters. Examples are included to illustrate the effect of different parameters on the reliability.

37.4.2 Reliability Model

Corrosion in a pipeline becomes unacceptable when the corrosion depth exceeds the allowable corrosion depth. This can be described by the limit state function, e.g, Edwards et al (1996)

$$g(\overline{X}) = d - CR \cdot t \tag{37.5}$$

Where:
- d : Maximum allowable corrosion depth;
- CR : Corrosion rate;
- t : Duration of wet service;
- $\overline{X}$: vector containing all the basic uncertainty variables.

Because the parameters in g cannot be determined precisely, probability distributions are used to describe them. This is shown in Figure 37.1.

The corrosion is unacceptable in the shaded region, representing a high corrosion rate and a low allowable corrosion depth. This corresponds to $g(\overline{X}) < 0$, which is the criterion for failure. The probability of failure will therefore be:

$$P(g(\overline{X}) < 0) = \int_V f_{\overline{X}}(\overline{x})d\overline{x} \tag{37.6}$$

Where:
- V : failure domain = $\{\overline{x} | g(\overline{X}) < 0\}$
- $f_{\overline{X}}(\overline{x})$: joint probability density function for $\overline{X}$
- $\overline{x}$: Realization of $\overline{X}$ in the basic variable space

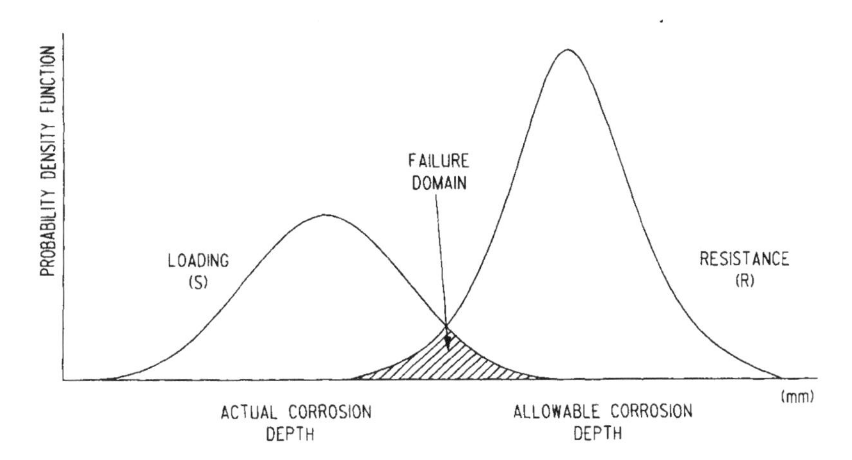

Figure 37.1 Probabilistic limit state function.

The reliability is expressed as (1-probability of failure). The calculation of the probability of non acceptance is carried out using the proprietary software STRUREL (RCP, 1996). The Second Order Reliability Method (SORM) is used (Madsen, et al 1986).

Determination of corrosion rate

Corrosion rates (CR) can be based on empirical equations (e.g. Shell models deWaard et al (1993, 1995) or on experience/measured data. In a pipeline design a combination of the above and sound engineering judgment from a qualified corrosion engineer is necessary. In the examples below, a deterministic corrosion rate (based on deWaard et al (1995)) is multiplied by a probabilistic variable, x_m, called the corrosion rate uncertainty factor. Engineering judgment and operational experience is used to determine x_m.

Determination of maximum allowable defect depth, d.

Formulas for calculating d are available from codes or from the literature (ASME, 1991). In the examples presented below, d is based on ASME B31G (ASME, 1991). This means that the calculated reliability is not based on a failure probability but rather on a probability of not satisfying the acceptance criteria of ASME B31G.

37.4.3 Design Examples

Two examples are given to show how the reliability of a pipeline may be calculated, and how changing the corrosion allowance will affect the calculated reliability. In each example a sensitivity study is done to demonstrate the effect of future operation on the pipeline reliability. The results from the sensitivity study can be used to evaluate the most efficient method to increase reliability; adding wall thickness or changing procedures for operation.

The examples are:

- A dry gas line which is occasionally wetted due to production upsets. Corrosion will only occur for a limited time following an upset. Assumed $L \leq 4.48 \cdot \sqrt{D \cdot t}$. Sensitivity parameter: Number of upsets per year.
- A wet line transporting liquids. Corrosion will occur continuously over the whole lifetime. Assumed $L > 4.48 \cdot \sqrt{D \cdot t}$. Sensitivity parameter: Inhibitor performance.

1. Dry gas line

Design data

The deterministic design data for the line are presented in Table 37.2 and the probabilistic parameters are shown in Table 37.3. Two values of the corrosion allowance are considered;

- 0 mm
- 3 mm (commonly used in pipeline design, based on experience)

Table 37.2 Deterministic design parameters.

Parameter	Unit	Dry gas line	Wet liquid line
D	mm	1016	508
t_{-c} [1)]	mm	26.6	13.5
Pressure	bar	200	130
Temperature	°C	50	50
CO_2 content	mole %	4.5	0.5
Design Life	years	50	20
F_{scale}, F_{pH}, i	-	1	1

NOTE:

1): t_{-c} : Total wall thickness - corrosion allowance

Table 37.3 Probabilistic design parameters and uncertainty factors, dry gas line.

Parameter	Distribution	Mean value	Std. dev.	Reference
Corrosion rate uncertainty factor[1)]	Normal (Gauss)	1.5	0.3	See text below
Duration of wet service	Normal (Gauss) (see below)	$3 \cdot n$	$2 \cdot \sqrt{n}$	See text below
Defect length	Lognormal	30 (mm)	20 (mm)	Emden gas line
Wall thickness uncertainty factor[1)]	Normal	1.04	0.04	Engineering judgment

NOTE:

1). In the limit state function corrosion rate and wall thickness values are multiplied by their respective uncertainty factors.

2). n = #upsets/lifetime

The conservatism in the corrosion rate model is often explained by the model being based on laboratory experiments which reflect "real life" rather poorly. However, condensation of water in a gas line may result in "fresh", unsaturated water which is known to give a more corrosive environment than water saturated with corrosion products. Also, in a long pipeline it is appropriate to consider extreme value statistics, as the probability of seeing "worst case" corrosion rates increase with length. The mean value of the corrosion rate uncertainty factor is therefore > 1. There is also a large scatter in the laboratory data upon which the Shell model is based deWaard et al (1995), which is reflected in the large standard deviation.

It has been assumed that three operational upsets will occur per year (Base Case). Sensitivity cases are shown to evaluate the effect of reducing this number to one or two upsets per year. In this context, an upset is defined as water or wet gas ingress into the pipeline. It has further been assumed that the time from the ingress starts until the upset is detected and corrected and the line has returned to the dryness before the upset, can be described by a lognormal probability distribution function with a mean value, m, of 3 days and a standard deviation, s, of 2 days. It has also been assumed that the upsets are independent occurrences and that the line will be dried after the previous upset before a new upset occurs. The total duration of wet service, t_w, throughout the lifetime of the field can then be described by a normal distribution with the following parameters:

$$m_{t_w} = n \cdot m \tag{37.7}$$

$$s_{t_w} = \sqrt{n} \cdot s \tag{37.8}$$

where:

m_{tw}	:	mean value for total duration of wet service;
s_{tw}	:	standard deviation for total duration of wet service;
n	:	total number of upsets during lifetime (should not be less than approx. 20 for the assumption of the normal distribution to be true);
m	:	mean value for drying time, single upset;
s	:	standard deviation for drying time, single upset.

The cases are summarized in Table 37.4. The results are presented in Figure 37.2.

Table 37.4 Sensitivity cases, duration of wet service.

Case	n	m_{tw}	s_{tw}	Comment
Base Case	150	450	24.49	3 single upsets/year
S1	100	300	20	2 single upsets/year
S2	50	150	14.14	1 single upset/year

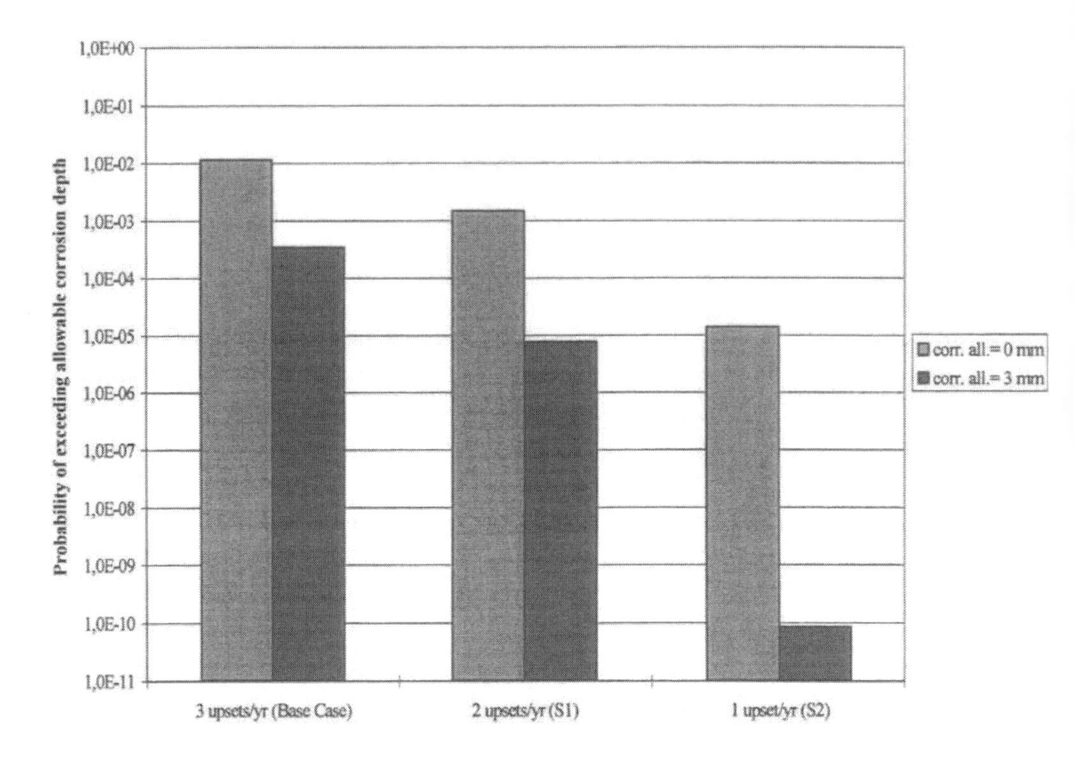

Figure 37.2 Calculated probabilities for a gas pipeline with operational upsets.

2. Wet liquid line

Design data

The deterministic design data for the line are presented in Table 37.2 and the probabilistic parameters are shown in Table 37.5.

Table 37.5 Probabilistic design parameters and uncertainty factors, wet liquid line.

Parameter	Distribution	Mean value	Std. dev.	Reference
Shell '95 model uncertainty factor[1)]	Normal (Gauss)	0.04	0.25	See below
Inhibitor efficiency, i	Beta	see below	0.03	See below
Wall thickness uncertainty factor[2)]	Normal	1.04	0.04	Bai et al. (1997) & engineering judgment

NOTE: 1). corrosion rate uncertainty factor, x_m = 1+ Shell '95 model uncertainty factor.

2). in the limit state function corrosion rate and wall thickness values are multiplied by their respective uncertainty factors.

Three values for the corrosion allowance are considered;

- 5 mm,
- 6 mm (calculated from conventional, deterministic approach)
- 7 mm

The rationale behind establishing the corrosion rate uncertainty factor is the same as for the dry gas line. However, because the pipe wall in a liquid or multiphase line is washed with liquids, "fresh" water from condensation is assumed not to be present. Also, the corrosivity of the water phase is limited by the presence of hydrocarbons, corrosion products and inhibitor. The "worst case" corrosion rate is therefore considered to be less in a liquid line compared to a gas line. The mean value of x_m is accordingly assumed to be less in a liquid line than in a gas line. Based on this, the corrosion rate model uncertainty factor, x_m is taken to be (1+Shell'95 model uncertainty factor), i.e. it is assumed that the Shell model is a reasonable presentation of the extreme corrosion rates.

Three cases are considered for the inhibitor efficiency. These are described in Table 37.6. The cases are meant to represent different levels of commitment to prudent operation of the pipeline. In this context, commitment comprises proper selection of inhibitors, monitoring of inhibitor performance and ability to execute corrective actions if the inhibitor protection becomes inadequate.

Table 37.6 Sensitivity parameters, inhibitor performance.

Case	Distribution	Mean	Std. deviation
1	Beta; Range 0.7 to 0.99	0.9	0.02
2	Beta; Range 0.7 to 0.99	0.9	0.03
3	Beta; Range 0.7 to 0.99	0.85	0.035

The basis for the suggested inhibitor efficiency distributions is the deterministic value of 85% as described in NORSOK MD-P-001. The rationales behind the selection of distributions are as follows:

Case 1: Very high level of commitment to operation. Probability of seeing values of inhibitor performance < 85% is very low (~ 1%)

Case 2: High level of commitment to operation. Probability of seeing values of inhibitor performance < 85% is low (~ 10%)

Case 3: Low level of commitment to operation. Probability of seeing values of inhibitor performance < 85% is high (~ 50%)

An upper efficiency limit of 99% and a lower limit of 70 % is assumed. It is assumed that measures will be taken during operation to keep the efficiency within this band.

Note that the inhibitor efficiency is an average value for the whole life of the line, as the limit state function is considering the average corrosion over the total life, i.e. it does not give a "day to day" picture of the corrosion in the line. From this follows that values well below 85% may still be acceptable for short time periods.

The results are presented in Figure 37.3.

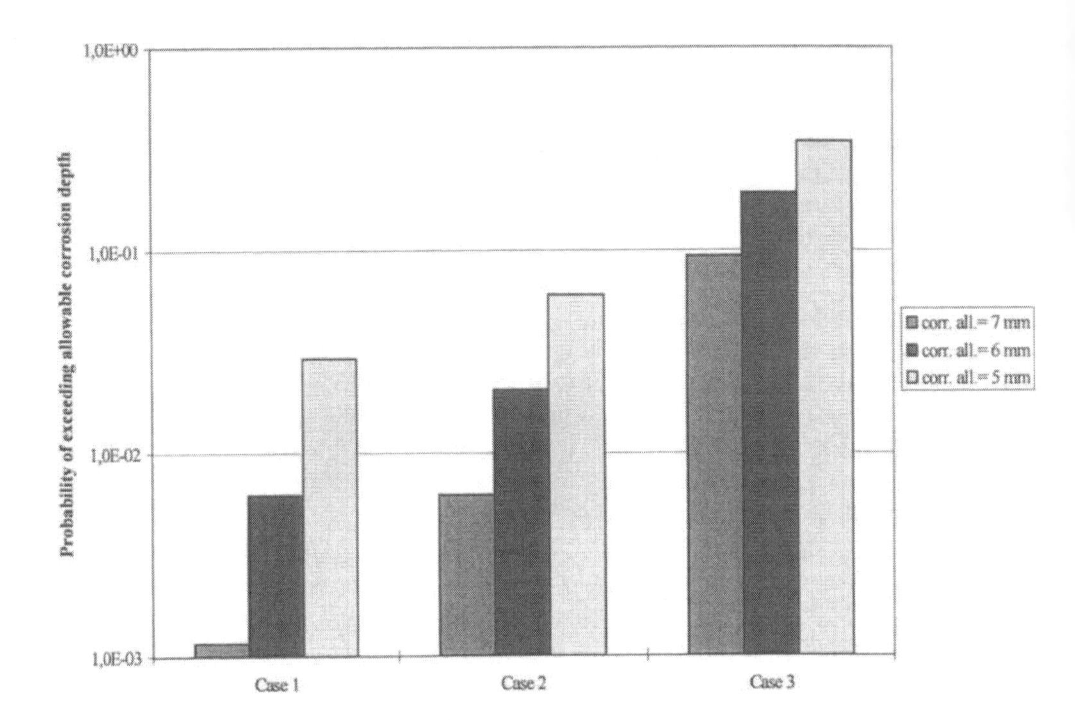

Figure 37.3 Calculated reliability of wet liquid pipeline with different inhibitor efficiencies.

37.4.4 Discussions

The examples illustrate how the uncertainties in the design parameters may be addressed and how the reliability of the pipeline can be subjectively evaluated. The results show that the reliability of a pipeline is related to both the future operation of the line and to the size of the corrosion allowance.

An important trend can be seen from the results; The effect of the corrosion allowance on increasing the reliability decrease as the corrosivity of the line increase. This means that if the corrosion rate is relatively high, a corrosion allowance will be a poor way of increasing the reliability compared to reducing the corrosion rate, e.g. by changing the operating parameters.

The reliability is expressed as (1-probability of exceeding acceptable corrosion depth). Reaching the maximum acceptable depth does not mean that the line will fail, due to the inherent safety level in the B31G code. Furthermore, the corrosion depth is likely to increase over a long period of time, which means that the defects are likely to be detected by inspections before they reach a critical size. Cost of repair and/or reducing the capacity of the line is therefore the most likely consequence of exceeding the maximum depth.

The optimum corrosion allowance is selected by minimizing the life cycle cost (LCC), where capex, opex and risk (probability of failure x cost of failure) are considered for each candidate corrosion allowance.

37.4.5 Recommendations

Better statistical presentation of the input parameters will increase the confidence in the calculated reliability and consequently also the life cycle cost. Data taken from operational experience is particularly in demand as this will reduce the gap between the existing laboratory models and "real life". Similarly, corrosion rate models are needed which can predict field corrosion rates with better accuracy.

Reliability based methods are being developed to predict the maximum allowable corrosion depth (Bai et al, 1997). By merging these methods with the method presented here, the probability of failure (or bursting) can be calculated instead of the probability of exceeding an allowable depth. This will remove unnecessary conservatism, but it will also affect the acceptable reliability because factors such as environmental damage and human safety will have to be considered in addition to cost.

37.5 References

1. ASME (1991) "B31G, Manual for Determining the Remaining Strength of Corroded Pipelines", American Society of Mechanical Engineers.
2. Bai, Y. (2003) "Marine Structural Design", Elsevier Science.
3. deWaard, C., Lotz, U. (1993) "Prediction of CO_2 corrosion of carbon steel", CORROSION/93, paper no. 69.
4. deWaard, C., Lotz, U. and Dugstad A. (1995) "Influence of liquid flow velocity on CO_2 corrosion: a semi-empirical model", CORROSION/95, paper no. 128.
5. DNV (1996) "Rules for Submarine Pipelines", Det Norske Veritas.
6. Edwards, J.D., Sydberger, T. and Mørk, K.J. (1996)" Reliability based design of CO_2 - corrosion control", CORROSION/96, paper no.29.
7. Guedes Soares, C., (1997) "Quantification of Model Uncertainty in Structural Reliability", in Probabilistic Methods for Structural Design, edited by C. Guedes Soares. Kluwer Academic Publishers.
8. ISO/DIS 13623(1997) "Petroleum and Natural Gas Industries; Pipeline Transportation Systems", International Standard Organisation.
9. Jiao, G., Sotberg T., Bruschi, R. and Igland, R.T., (1997) "The SUPERB Project: Linepipe Statistical Properties and Implications in Design of Offshore Pipelines", Proc. of OMAE'97.
10. Kiefner, J.F. (1974) "Corroded pipe: Strength and repair methods", 5th Symposium on Line Pipe Research.

11. Nødland, S., Bai, Y., Damsleth, P.A. (1997) "Reliability approach to optimise corrosion allowance", Risk based & limit state design & operation of pipelines, IBC, Aberdeen 21/22 May 1997.
12. Nødland, S., Hovdan, H. and Bai, Y. (1997) "Use of Reliability Methods to Assess the Benefit of Coorsion Allowance", EUROCORR '97, Trondheim, Norway 22-25 September 1997.
13. NORSOK MD-P-001, Materials Selection, Rev. 1 December 1994.
14. NORSOK Standard Y-002 (1997) "Reliability-based Limit-state Principles for Pipeline design".
15. RCP (1996) "SYSREL 9.0", RCP Consulting, Munich.
16. Sotberg, T., Moan, T., Bruschi R., Jiao, G. and Mørk, K.J., (1997) "The SUPERB Project: Recommended Target Safety Levels for Limit State Based Design of Offshore Pipelines", Proc. of OMAE'97.

Part VI

Integrity Management

Chapter 38 Corroded Pipelines

38.1 Introduction

The purpose of this chapter is to develop corrosion defect prediction models and reliability-based design and re-qualification criteria for assessing corroded pipelines. This evaluation focuses on three interrelated issues:

- Corrosion defect growth
- Check burst strength (allowable versus maximum internal service pressure).
- Check bending capacity (allowable versus maximum external service pressure, bending moment, and axial load, see Chapter 4).
- Check adequacy of residual corrosion allowance for remaining service life.
- Corrosion defect inspection
- Updated inspection and maintenance program.

38.2 Corrosion Defect Predictions

The scope of the assessment for corrosion defects consists of a proper characterization of defects by thickness profile measurements and an initial screening phase to decide whether detailed analysis is required.

38.2.1 Corrosion Defect Inspection

The assessment of a single isolated defect is to be based on a critical profile defined by the largest measured characteristic dimensions of the defect (e.g., depth, width, length) and properly calibrated safety/uncertainty factors, in order to account for uncertainties in the assessment and thickness measurements.

A distance equivalent to the nominal pipe wall thickness may be used as a simple criterion of separation for colonies of longitudinally oriented pits separated by a longitudinal distance or parallel longitudinal pits separated by a circumferential distance. For longitudinal grooves inclined to pipe axis:

- If the distance x, between two longitudinal grooves of length L_1 and L_2, is greater than either of L_1 or L_2, then the length of corrosion defect L is L_1 or L_2, whichever is greater. It can be assumed that there is no interaction between the two defects;
- If the distance x, between two longitudinal grooves of length L_1 and L_2, is less either of L_1 and L_2, it is to be assumed that the two defects are fully interacted and the length of the corrosion defect L is to be taken as $L = L_1 + L_2 + x$.

38.2.2 Corrosion Defect Growth

The corrosion defect depth d after the time T of operation, may be estimated using an average corrosion rate V_{cr}:

$$d = d_0 + V_{cr} \cdot T \tag{38.1}$$

where d_0 is defect depth at present time.

The defect length may be assumed to grow in proportion with the depth, hence:

$$L = L_0 \left(1 + \frac{V_{cr} \cdot T}{d_0} \right) \tag{38.2}$$

where L and L_0 are defect lengths at the present time and the time T later.

38.2.3 CO_2 Corrosion Defects

38.2.3.1 CO_2 Corrosion Defect Prediction

CO_2 corrosion rates in pipelines made of carbon steel may be evaluated using industry accepted equations that preferably combine contributions from flow independent kinetics of the corrosion reaction at the metal surface, with the contribution from flow dependent mass transfer of dissolved CO_2.

The corrosion rate V_{cr} in mm/year, can be predicted by:

$$V_{cr} = \frac{1}{\frac{1}{V_r} + \frac{1}{V_m}} \tag{38.3}$$

where V_r is the flow independent contribution, denoted the reaction rate, and V_m is the flow dependent contribution, denoted the mass transfer rate. The reaction rate V_r can be approximated by:

$$\log(V_r) = 4.93 - \frac{1119}{T_{mp} + 273} + 0.58 \cdot \log(p_{CO_2}) \tag{38.4}$$

where the temperature T_{mp} is to be given in °C, and the partial pressure p_{CO2} of CO_2 in bar. The partial pressure p_{CO2} can be found by:

$$p_{CO_2} = n_{CO_2} \cdot p_{opr} \tag{38.5}$$

where n_{CO2} is the fraction of CO_2 in the gas phase, and p_{opr} is the operating pressure in bar.

The mass transfer rate V_m is approximated by:

$$V_m = 2.45 \cdot \frac{U^{0.8}}{d^{0.2}} \cdot p_{CO_2} \tag{38.6}$$

where U is the liquid flow velocity in m/s and d is the inner diameter in meters.

38.2.3.2 CO_2 Corrosion Models Comparison

The corrosion caused by the incidences of CO_2 represents the greatest risk to the integrity of carbon steel equipment in a production environment and is more common than damage related to fatigue, erosion, or stress corrosion cracking. NORSOK, Shell, as well as other companies and organizations have developed models to predict the corrosion degradation.

Shell's prediction model for CO_2 corrosion is mainly based on the De Waard's equation published in 1991. Starting from a "worst case" corrosion rate prediction, the model applies correction factors to quantify the influence of environmental parameters and corrosion product scale formed under various conditions.

NORSOK's standard M-506 may be used to conduct the CO_2 corrosion rate calculation, which is an empirical model for carbon steel in water containing CO_2 at different temperatures, pH, $CO_{2\ fugacity}$ and wall shear stress. The NORSOK model covers only the corrosion rate calculation where CO_2 is the corrosive agent. It doesn't include additional effects of other constituent, which may influence the corrosivity, e.g. H2S, which commonly appears in the production flowlines. If such constituent is present, the effect must be evaluated separately. None of the De Waard models includes H_2S effect.

38.2.3.3 Sensitivity Analysis for CO_2 Corrosion Calculation

Sweet services are due to the presence of carbon dioxide (CO_2) and free water in the production system. Sour services are defined by hydrogen sulphide (H_2S) content, which is above a critical concentration defined by as a partial pressure greater than 0.345 kPa. In this environment, the pipeline is prone to corrosion by the formation of pitting and sulphide films, sulphide stress corrosion cracking (SSC) and hydrogen induced cracking (HIC).

Table 38.1 presents the base case for the following sensitivity analysis. These data are based on the design operating data for 10 inch production flowline.

Table 38.1 Base case for sensitivity analysis

Parameter	Units	Base Case
Total Pressure	bara	52
Temperature	°C	22.5
CO_2 in Gas	Mole %	0.5
Flow Velocity	m/s	2.17
H_2S	ppm	220
pH		4.2
Water Cut		50 %
Inhibitor Availability		50 %

1. Total System Pressure & CO_2 Partial Pressure

An increase in total pressure will lead to an increase in corrosion rate because p_{CO2} will increase in proportion. With increasing the pressure, the CO_2 fugacity f_{CO2} should be used instead of the CO_2 partial pressure p_{CO2} since the gases are not ideal at high pressures. The real CO_2 pressure can be expressed as:

$$f_{CO2} = a^* \; p_{CO2} \qquad (38.7)$$

Where a is fugacity constant which depends on pressure and temperature:

$$a = 10^{P(0.0031-1.4/T)} \qquad \text{for } P \leq 250 \text{ bara}$$

$$a = 10^{250(0.0031-1.4/T)} \qquad \text{for } P > 250 \text{ bara}$$

The following figures present the effect of total pressure and CO_2 partial pressure on the corrosion rate. With increasing the total pressure and CO_2 partial pressure, the corrosion rate is greatly increased.

2. System Temperature

Temperature has the effect of the formation of protective film. At lower temperatures the corrosion product can be easily removed by flowing liquid. At higher temperature, the film becomes more protective and less easily washed away. Further increase in temperature results in lower corrosion rate and the corrosion rate goes through a maximum (De Waard et al., 1991). This temperature is referred as the scaling temperature. At this temperature, pH and Fe++ concentration formed at the steel's surface. At temperature exceeding the scaling temperature, corrosion rates tend to decrease to close to zero, according to De Waard. Tests in IFE Norway revealed that the corrosion rate is still increasing when the design temperature is beyond the scaling temperature (Dugstad et al., 1994).

The following figure shows the effect of temperature on the corrosion rate, where the total pressure is 48 bara and pH is equal to 4.2. Corrosion rate increases with increasing the temperature, when the temperature is lower than the scaling temperature.

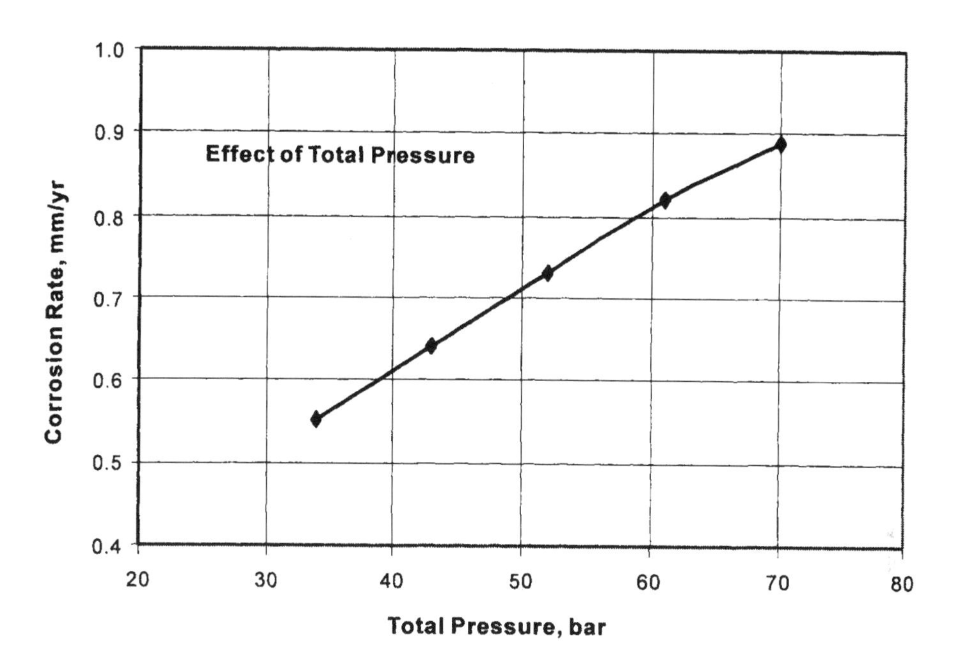

Figure 38.1 Effect of total pressure on corrosion rate.

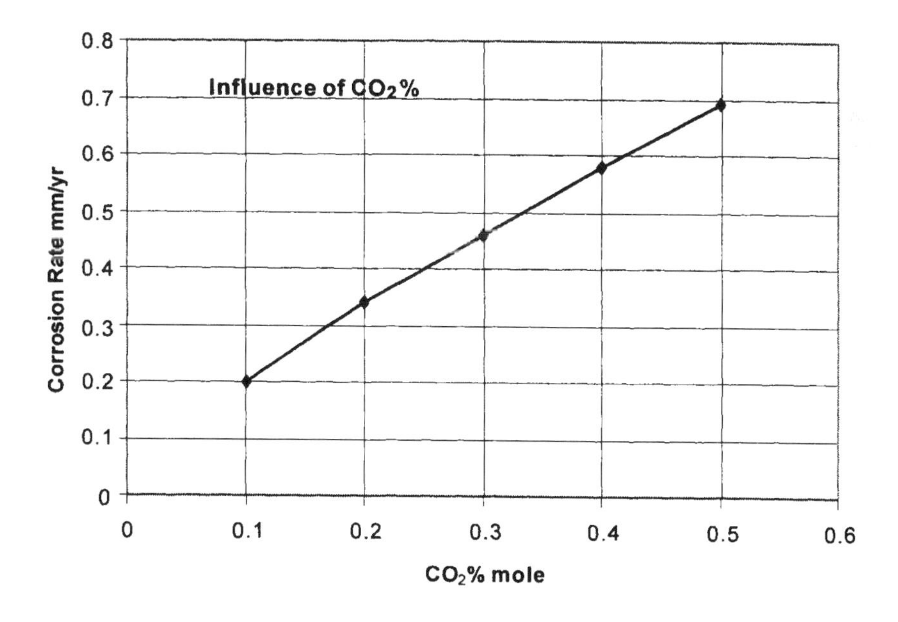

Figure 38.2 Effect of CO_2 on corrosion rate.

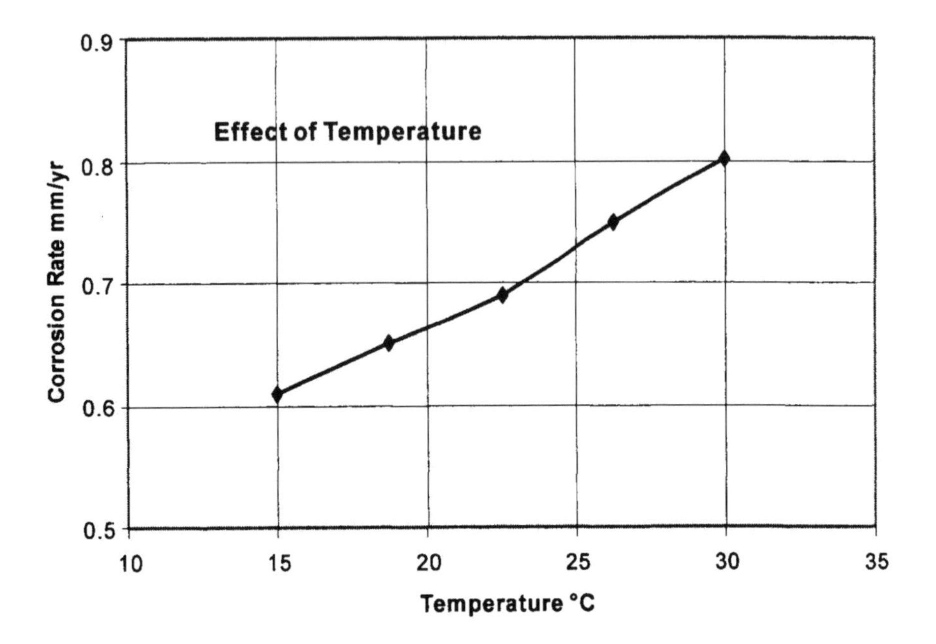

Figure 38.3 Effect of temperature on corrosion rate.

3. H_2S

H_2S can depress pH when it dissolved in a CO_2 aqueous solution. The presence of H_2S in CO_2/brine systems can reduce the corrosion rate of steel when compared to corrosion rate under conditions without H_2S at the temperature less than 80 °C, due to the formation of a meta-stable iron sulphide film. At higher temperature the combination of H_2S and chlorides will produce higher corrosion rates than just CO_2/brine system since the protective film isn't formed.

H_2S at levels below the NACE criteria for sulphide stress corrosion cracking (per MR0175, NACE publication) reduces general metal loss rates but can promote pitting. The pitting proceeds at a rate determined by the CO_2 partial pressure and therefore CO_2 based models are still applicable at low levels of H_2S. Where the H_2S concentration is greater or equal to the CO_2 value, or greater than 1 mol %, the corrosion mechanism may not be controlled by CO_2 and therefore CO_2 based models may not be applicable.

4. pH

pH affects the corrosion rate by affecting the reaction rate of cathode and anode therefore the formation of corrosion products. The contamination of the CO_2 solution with corrosion products reduces the corrosion rate. pH has a dominant effect on the formation of corrosion films due to its effect on the solubility of ferrous carbonate. An increase of pH slows down the cathodic reduction of H+.

The following figure presents the relationship between the pH and corrosion rate. In the solution with pH less than 7, corrosion rate decreases with increasing pH.

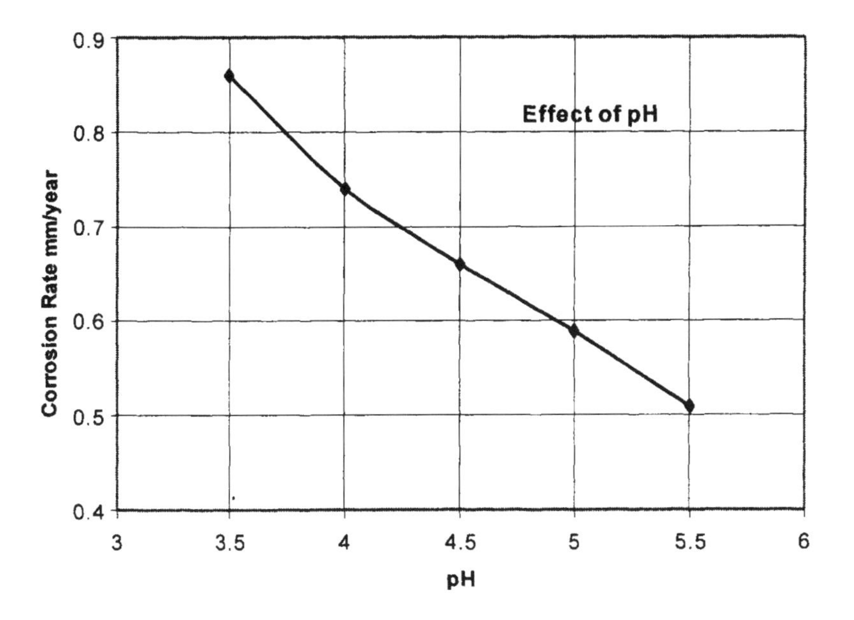

Figure 38.4 Effect of pH on corrosion rate.

5. Inhibitors and Chemical Additives

Inhibitors can reduce the corrosion rate by presenting a protective film. The presence of the proper inhibitors with optimum dosage can maintain the corrosion rate at 0.1 mm/year. Use of inhibitor can greatly decrease corrosion rate therefore increase pipeline life.

The impingement of sand particle can destroy the inhibitor film and therefore reduces the inhibitor efficiency. Inhibitors also perform poorly in low velocity lines particularly if the fluids contain solids such as wax, scale or sand. Under such circumstances, deposits inevitably form at the 6 o'clock position, preventing the inhibitor from reaching the metal surface. Flow velocities below approximately 1.0 m/s should be avoided if inhibitors are expected to provide satisfactory protection and this will be critical in lines containing solids.

6. Inhibitor Efficiency vs. Inhibitor Availability

When inhibitors are applied, there are two ways to describe the extent to which an inhibitor reduces the corrosion rate, the use of inhibitor efficiency (IE) and the use of the inhibitor availability (IA). A figure of 95% for IE is commonly used. However, inhibitors are unlikely to be constantly effective throughout the design life. For instance, increased inhibitor dosage or better chemicals will increase the inhibitor concentration. It may be assumed that the inhibited corrosion rate is unrelated to the uninhibited corrosivity of the system and all systems can be inhibited to 0.1 mm/year. The corrosion inhibitor is not available 100% of the time and therefore corrosion will proceed at the uninhibited rate for some periods.

The following equation is based on the assumed existence of corrosion inhibitors that are able to protect the steel to a corrosion rate CRmit (typically 0.1 mm/year) regardless of the uninhibited corrosion rate CRunmit, taking into consideration the percentage of time IA the inhibitor is available.

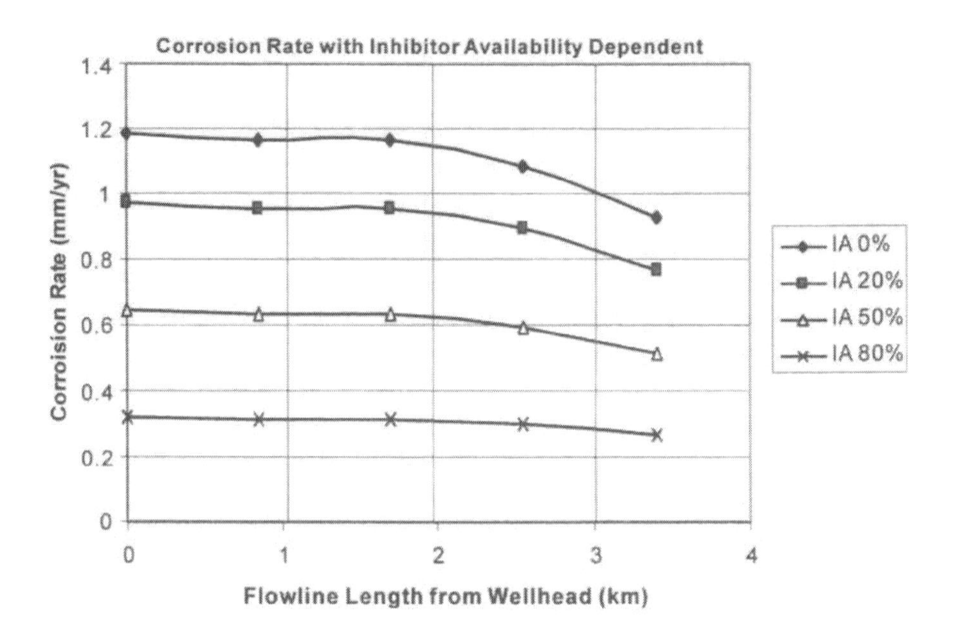

Figure 38.5 Inhibited corrosion rate under different inhibitor availability

7. Chemical Additives

Glycol (or methanol) is often used as the hydrate preventer on a recycled basis. If glycol is used without the addition of corrosion inhibitor, there will be some benefit from the glycol. De Waard has produced a glycol correction factor. However, if glycol and inhibitor are both used there will be little additional benefit from the glycol and it should be ignored for deign purpose.

Methanol is batch injected during start-up until flowline temperatures rise above the hydrate formation region and during extended shut down.

8. Single Phase Flow Velocity

Single-phase flow refers to a flow with only one component, normally oil, gas or water through a porous media. Fluid flow influences corrosion by affecting mass transfer and by mechanical removal of solid corrosion products. The flow velocity used in corrosion model is identified as the true water velocity. The following figure shows the corrosion rate increases consistently with increased flow rate at low pH.

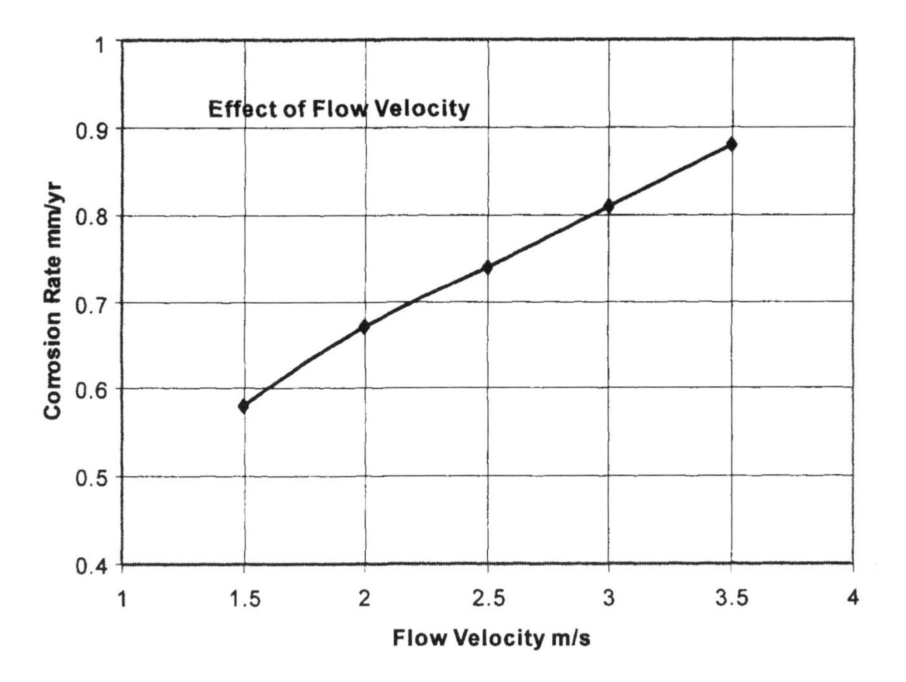

Figure 38.6 Effect of flow velocity on corrosion rate.

9. Multiphase Flow

Multiphase flow refers to the simultaneous flow of more than one fluid phase through a porous media. Most oil wells ultimately produce both oil and gas from the formation and often produce water. Consequently, multiphase is common in oil wells. The multiphase flow in a pipeline is usually studied by flow regime and corresponding flow rate. Because of the various hydrodynamics and the corresponding turbulence, multiphase flow will further influence the internal corrosion rate, significantly different from the influence of single phase flow in the pipeline on the corrosion.

10. Water Cut

Water cut means the ratio of water produced compared to the volume of total liquid produced. CO_2 corrosion is mainly caused by present water in contact with the steel surface. The severity of the CO_2 corrosion is proportional to the time during which the steel surface is wetted in water phase. Thus the water cut is an important factor to influence the corrosion rate. However, the effect of the water cut cannot be separated from the flow velocity and the flow regime.

11. Free Span Effect

Pipeline spanning can occur on a rough seabed or a seabed subjected to scour. The evaluation of allowable free-span length should be considered in order to avoid the excessive yielding and fatigue. The localized reduction of wall thickness influences the strength capacity of the pipeline therefore the allowable free span length. This is discussed in many reports and papers. It is not within the scope of work for this report to assess yielding and fatigue of free spans. Instead, a qualitative discussion will be given on possible development or acceleration of the development of corrosion.

Figure 38.7 shows at the middle point of the free spans, there are additional accumulated waters and marine organism that may accelerate corrosion development. The flow regime and flow rates will change. The corrosion defect depth in the region close to the middle point will be most likely deeper.

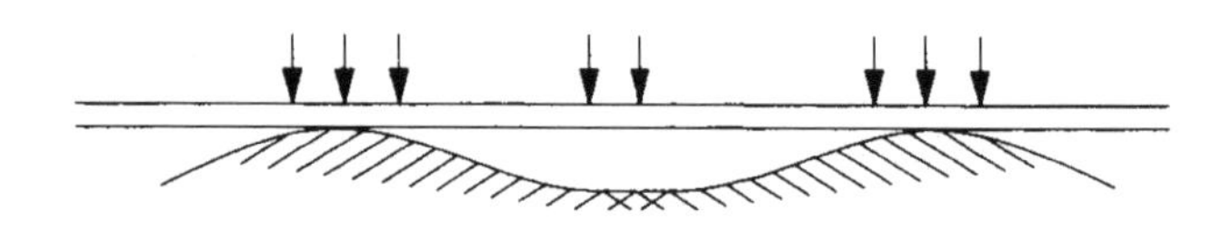

Figure 38.7 Effect of free-spans on corrosion defect development.

These three corrosion models were developed based on the results of tests using water-only, *i.e.* 100 % water cut, system in the laboratory. Therefore, the corrosion rate predicted with these models represents the worse case corrosion rates. For comparison, the corrosion rate under the flow condition with smaller water cut is generally lower than the worse case rate. Therefore, the corrosion rates prediction with these models is very conservative compared to the real corrosion rate in the field. With more corrosion data from pipeline pigging, the accuracy of corrosion rate prediction can be improved. However, the accuracy of corrosion rate prediction still cannot be exaggerated since the internal corrosion is influenced by numerous parameters discussed above. The combination of the corrosion rate prediction method and the pipeline pigging method can provide a benchmark to pinpoint the weakest links in the pipeline, predict the remaining life, and maintain the pipeline integrity.

38.3 Remaining Strength of Corroded Pipe

The design criteria for corroded pipelines are generally expressed as equations to determine the operating parameters:

1. Maximum allowable length of defects;
2. Maximum allowable design pressure for uncorroded pipelines;
3. Maximum safe pressure.

A number of criteria exist to determine these operating parameters.

38.3.1 NG-18 Criterion

The NG-18 criterion developed in the late 1960s and early 1970s is to evaluate the remaining strength of corroded pipe (Maxey et al. (1971)). It was developed for a pipe with a longitudinal surface flaw:

$$S_p = S_{flow} \frac{1 - AREA / AREA_0}{1 - M^{-1}(AREA / AREA_0)} \tag{38.8}$$

where:

S_p : predicted hoop stress level at failure;

S_{flow} : flow stress of the material;

AREA : area of through thickness profile of flaw;

$AREA_0 = L_t$

L : maximum axial extent of the defect;

t : nominal wall thickness of the pipe;

M : Folias factor which is determined by:

$$M = \sqrt{1 + \frac{2.51(L/2)^2}{Dt} - \frac{0.054(L/2)^4}{(Dt)^2}} \tag{38.9}$$

where:

D : nominal outside diameter of the pipe.

Equation (38.9) can be further simplified as (Kiefner, 1974):

$$M = \sqrt{1 + \frac{0.8L^2}{Dt}} \tag{38.10}$$

The calculation of AREA is simplified by assuming the shape of corroded area is parabolic for short corrosion and rectangular for long corrosion (Kiefner, 1974). The maximum allowable length L_{allow}, and the failure pressure P is solved from a formula which equates predicted bursting hoop stress S_P to 1.1 SMYS (Specified Minimum Yield Stress) assuming that the flow stress is 1.1 SMYS (Bai et al., 1994).

38.3.2 B31G Criterion

The B31G criterion (ASME 1993) is widely used to assess corroded pipelines for fitness for purpose evaluation. The main equations in the ASME B31G criteria (1993) can be summarized as follows.

Maximum Allowable Design Pressure, P

The maximum allowable design pressure in B31G criterion is expressed as:

$$P = \frac{2\,\text{SMYS}}{D} Ft \qquad (38.11)$$

where:

P : the maximum allowable design pressure;

SMYS: the Specified Minimum Yield Strength;

F : the design factor, which is normally 0.72.

Maximum Allowable Defect Length and Depth

In the B31G (ASME 1993), a criterion for the acceptable corroded length is given as below for a corroded area having a maximum depth 'd' in the range of $0.1 < d/t < 0.8$ where t is the nominal wall thickness

$$L_{allow} = 1.12B\sqrt{Dt} \qquad (38.12)$$

where:

L_{allow} : the maximum allowable axial extent of the defect;

$$B = \sqrt{\left(\frac{d/t}{1.1d/t - 0.15}\right)^2 - 1} \qquad (38.13)$$

The Maximum Allowable Operating Pressure (MAOP) is defined to be less or equal to the maximum allowable design pressure P given by Equation (38.11).

$$\text{MAOP} \le P \qquad (38.14)$$

Equating the Safe Maximum Pressure Level P' to the Maximum Allowable Operating Pressure (MAOP), the maximum allowable defect depth d_{allow} is:

a) For $A \le 4$

$$d_{allow} = \frac{3t}{2}\left[\frac{1 - \frac{\text{MAOP}}{1.1P}}{1 - \frac{\text{MAOP}}{1.1P\sqrt{A^2+1}}}\right] \qquad (38.15)$$

b) For $A > 4$

$$d_{allow} = \left[1 - \frac{\text{MAOP}}{1.1P}\right]t \qquad (38.16)$$

The Safe Maximum Pressure Level P'

The safe maximum pressure level P' for the corroded area is:

$$P' = 1.1P\left(\frac{1 - \frac{2}{3}\left(\frac{d}{t}\right)}{1 - \frac{2}{3}\left(\frac{d}{t\sqrt{A^2+1}}\right)}\right); \quad P' \le P,\ A \le 4 \qquad (38.17)$$

$$P' = 1.1P\left(1 - \frac{d}{t}\right); \quad P' \le P \text{ and } A > 4 \tag{38.18}$$

where:

$$A = 0.893\left(\frac{L}{\sqrt{Dt}}\right) \tag{38.19}$$

38.3.3 Evaluation of Existing Criteria

The existing criterion ASME B31G (1993) for corroded pipelines was established based on the knowledge developed over 20 years ago. This criterion is re-examined to develop an improved criterion based on current knowledge. This evaluation is conducted based on the corrosion mechanisms, parameters in the existing criterion and the applications which are not included in the existing criterion.

38.3.4 Corrosion Mechanism

Figure 38.8 shows the types of corrosion defects. For marine pipelines, internal corrosion is a major problem (Mandke (1990), Jones et al (1992)). Many forms of internal corrosion occur, e.g., (a) girth weld corrosion, (b) massive general corrosion around the whole circumference, and (c) long plateau corrosion at about 6 o'clock position. External corrosion, on the other hand, is normally thought of as being local, covering an irregular area of the pipe. However, when the protective coating is failed, the external corrosion may tend to be pattern of long groove.

The B31G criterion has several problems for corrosion defects in real applications. It cannot be applied to spiral corrosion, pits/grooves interaction and the corrosion in welds. For very long and irregularly shaped corrosions, the B31G criterion may lead to overly conservative results. It also ignores the beneficial effects of closely spaced corrosion pits.

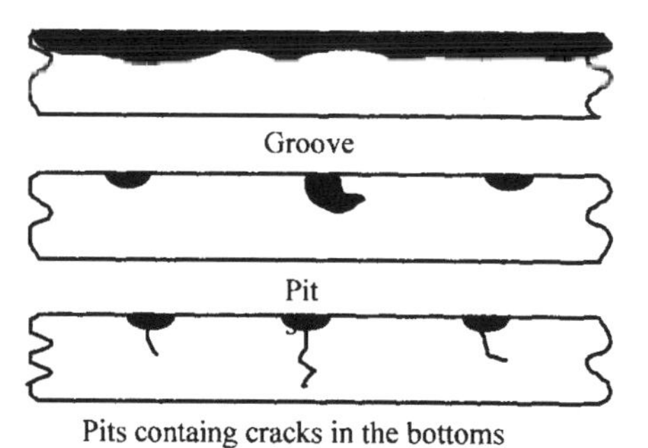

Figure 38.8 Type of corrosion defects.

Spiral Corrosion

For defects in other orientations, the B31G criterion recommends that the defect is projected on the longitudinal axis of the pipe to be treated as a longitudinal defect. This recommendation appears to be adequate for short defects. It is conservative for long spiral defects (Bai et al., 1994).

Mok (Mok et al., 1990, 1991) conducted extensive tests in the applicability of the B31G criteria to long spiral corrosion. For spiral defects with spiral angles other than 0 or 90 degrees, the study found that B31G underpredicted the burst pressure by as much as 50%. The effect of spiral angle is illustrated in Figure 38.9 (Mok et al., 1990).

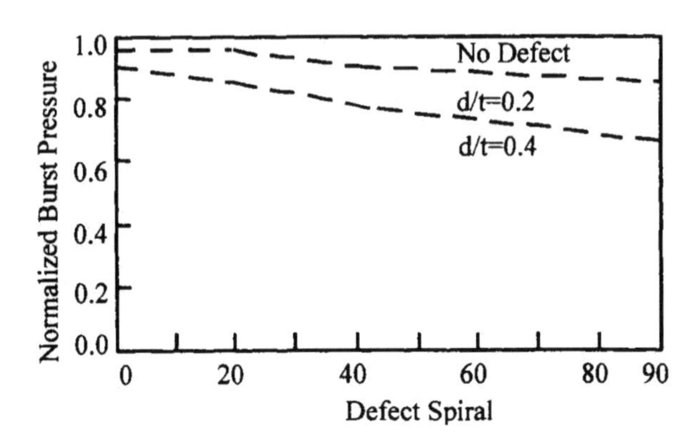

Figure 38.9 Effect of spiral angle.

Based on the experimental and numerical studies, Mok et al. (1990) recommended the spiral correction factor in determining the burst pressure for $W/t \le 32$ as:

$$Q = \frac{1-Q_l}{32}\frac{W}{t} + Q_l \tag{38.20}$$

in which W is the defect width, and coefficient Q_l is a function of the spiral angle φ ($\varphi = 90^0$ for longitudinal corrosion, $\varphi = 0^0$ for circumferential corrosion):

$$Q_l = \begin{cases} 0.2 & \text{for } 0^0 < \varphi < 20^0 \\ 0.02\varphi - 0.2 & \text{for } 20^0 < \varphi < 60^0 \\ 1.0 & \text{for } \varphi > 60^0 \end{cases} \tag{38.21}$$

for $W/t > 32$, the value of Q must be taken as 1.0.

Pits Interaction

Corrosion in pipelines often results in colonies of pits over an area of the pipe. For closely spaced corrosion pits, a distance of t (wall-thickness) is used as a criterion of pit separation for a colony of longitudinal oriented pits separated by a longitudinal distance or parallel longitudinal pits separated by a circumferential distance.

For circumferentially spaced pits separated by a distance longer than t, the burst pressure can be accurately predicted by the analysis of the deepest pits within the colonies of pits. For longitudinal oriented pits separated by a distance less than t, the failure stress of interacting defects can be predicted by neglecting the beneficial effects of the non-corroded area between the pits. For parallel longitudinal pits separated by a circumferential distance, experiments suggested that pits could be treated as interacting pits if the circumferential spacing is less than t (Bai et al., 1994).

Groove Interactions

For the interaction of longitudinal grooves, if the defects are inclined to pipe axis and the distance x between two longitudinal grooves of length L_1 and L_2 is larger than L_1 and L_2, the length of corrosion L is the maximum of L_1 and L_2. If the defects are inclined to pipe axis and the distance x between two longitudinal grooves of lengths L_1 and L_2 is less than L_1 and L_2, the length of corrosion L is the sum of x, L_1 and L_2, $L = L_1 + L_2 + x$.

Corrosion in Welds

One of the major corrosion damages for marine pipelines is the effects of the localized corrosion of weld on the fracture resistance. Figure 38.10 shows typical pattern of weld corrosion. The B31G criteria did not cover the assessment of corroded welds. The existing fracture assessment procedures (BSI PD6493) are recommended.

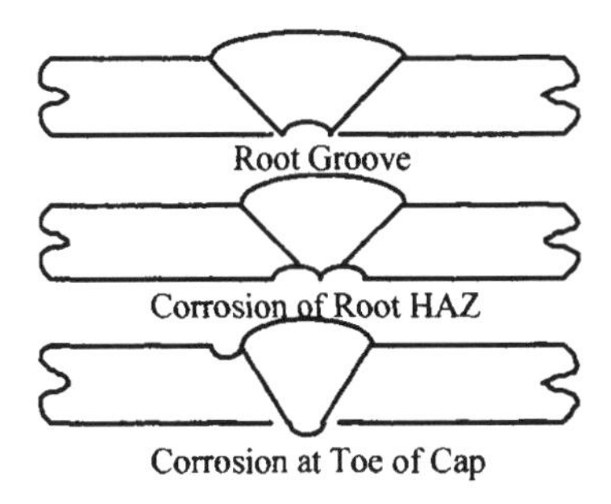

Figure 38.10 Typical patterns of weld corrosion.

Effect of Corrosion Width

Figure 38.11 shows the effect of defect width on burst pressure with a longitudinal defect (Mok et al., 1991), for the case of X52, OD = 508mm, t = 6.35mm, d/t = 0.4. It can be concluded in Mok's studies that the width effect is negligible on the burst pressure of pipe with long longitudinal defects.

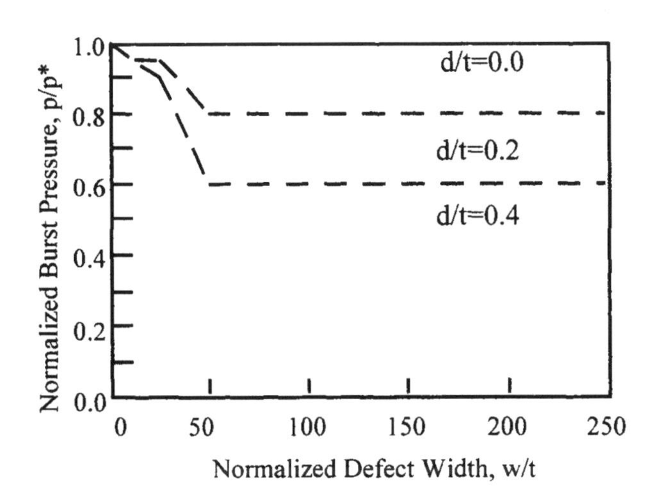

Figure 38.11 Effect of defect width.

Irregular Shaped Corrosion

The major weakness of the existing B31G criterion is its over-conservative estimation of the corroded area for long and irregular shaped corrosion (Bai et al. (1994), Kiefner and Vieth (1990), Hopkins and Jones (1992)). Therefore, the key to the irregularly shaped corrosion is the accurate estimation of the corroded area.

Two shapes were considered in the development of the original B31G criterion. One was the rectangle area method. The other was the parabola area method. Tests of Hopkins and Jones (1992) indicated that irregularly shaped corrosion could be conservative assessed using the B31G criteria when the accurate cross-sectional area of the corrosion defect was used.

We recommended two levels of AREA assessment. In the level 1, the AREA is estimated as:

$$\begin{aligned} &L/(Dt)<30 \quad AREA=\frac{2}{3}L\bullet d \\ &L/(Dt)>30 \quad AREA=0.85L\bullet d \end{aligned} \tag{38.22}$$

In the level 2, the exact area (AREA) of the corrosion profile is estimated by Simpson integration method.

38.3.5 Material Parameters

The major material parameters in the B31G criterion are flow stress, Specified Minimum Yield Stress (SMYS), Folias Factor M.

Flow Stress and SMYS

In the B31G-1993 manual, the flow stress was defined as 1.1 SMYS which is an appropriate value for the new pipelines. However, the flow stress is influenced by a number of factors, fabrication process (e.g. hot rolled versus cold expanded) and material aging. Furthermore, the flow stress used in burst strength criteria is influenced by possible cracks in the pit bottom due to corrosion fatigue. Therefore, specific attention should be made for accurate estimate of flow stress for aging pipelines. Many researchers (Hopkins and Jones (1992), Klever (1992), Stewart et al. (1994)) indicted that the flow stress for base material could be estimated as ultimate tensile stress. An approximation of the ultimate tensile stress is the Specified Minimum Tensile Stress, a statistic minimum of the ultimate tensile stress:

$$\sigma_{flow}=SMTS \tag{38.23}$$

The value of SMTS are available in some design specification (API 5L).

Folias Factor M

The Folias factor M is a geometric factor developed by Folias (1964) to account for the stress concentration effect of a notch in the pipes. Recent studies (Kiefner and Vieth, 1989) recommended the following expression to improve the accuracy of the Folias factor:

$$M = \begin{cases} \sqrt{1 + \dfrac{2.51(L/2)^2}{Dt} - \dfrac{0.054(L/2)^4}{(Dt)^2}} & \text{for } \dfrac{L^2}{Dt} \leq 50 \\ 0.032\dfrac{L^2}{Dt} + 3.3 & \text{for } \dfrac{L^2}{Dt} > 50 \end{cases} \tag{38.24}$$

38.3.6 Problems Excluded in the B31G Criteria

The ASME B31G criterion cannot be applied in some practical corrosion problems including corroded welds, ductile and low toughness pipe, and corroded pipes under combined pressure, axial and bending loads. Recent studies concluded that the corrosion in submerged-arc seams (longitudinal welds) should be handled in the same manner as corrosion in the body of the pipe. Corrosion in Electric Resistance Welds (ERW) or flash-welded seams should not be evaluated on the basis of the existing B31G criteria. It is recommended that Kastner's local collapse criteria (Kastner et al., 1981) is to be used to evaluate corrosion in (circumferential) girth welds.

A fracture mechanics approach (PD 6493) should be applied for assessing corroded welds, considering possible defects in the welds. The effect of material's fracture toughness (in ductile and low toughness pipe) is reflected by the critical fracture toughness of the material used in the fracture assessment criteria.

In the B31G criteria, the effect of axial load is not discussed. In general, tensile longitudinal stress may delay yielding and pipe bursting. On the other hand, compressive longitudinal stress may accelerate yielding and result in reductions in bursting pressure. Figure 38.12 shows the effect of axial load on collapse pressure (Galambos, 1988).

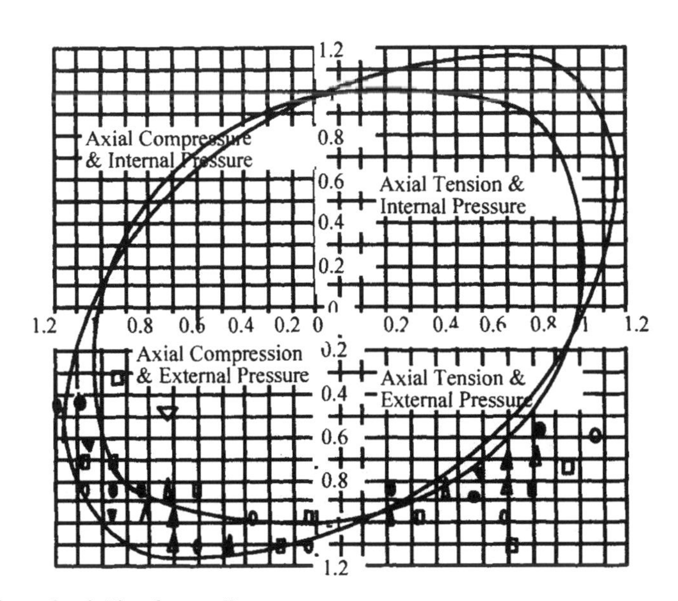

Figure 38.12 Effect of axial load on collapse pressure.

Figure 38.12 shows that:

- The internal burst pressure is largely reduced by axial compression
- The effect of axial tension is beneficial. The tension load is not significant when it is less than 60 percent of yield strength of the pipe section. This effect is significant when the axial tension is larger than 60% of the yield strength.

The dominant effect of bending stress, on the other hand, is the reduction of the hoop stress in the corroded region. Therefore, the B31G criteria of burst pressure that considers internal pressure alone may lead to unconservative results when large axial and bending stresses are coupled with corrosion.

38.4 New Remaining Strength Criteria for Corroded Pipe

38.4.1 Development of New Criteria

In this section, a new criterion is developed for longitudinally corroded pipelines. For longitudinally corroded pipe, pit depth exceeding 80% of the wall-thickness is not permitted due to the possible development of leaks. General corrosion where all of the measured pit depths are less than 20% of the wall-thickness is permitted, without further burst strength assessment. If the ratio of maximum pit depth and wall-thickness is between 0.2 and 0.8, the following equations are recommended.

The Maximum Design Pressure Level P

The maximum allowable design pressure in the new criterion is the same as that of the original B31G criteria:

$$P = \frac{2SMYS}{D} \bullet Ft \tag{38.25}$$

where F is the usage factor for intact pipe which is 0.72 according to the B31G criterion.

The Safe Maximum Pressure Level P'

$$P' = \frac{1}{\gamma} \frac{2\sigma_{flow}t}{D} \frac{1 - QAREA / AREA_0}{1 - M^{-1}AREA / AREA_0} \tag{38.26}$$

where:

P' : safe maximum pressure level

σ_{flow} : flow stress of the material

t : wall-thickness of the pipe

D : outside diameter of the pipe

$AREA$: Lt

$AREA_0$: original area prior to metal loss due to corrosion within the effective are which is Lt

L : defect length of corrosion profile

M : Folias factor
Q : Spiral correction factor
γ : Factor of Safety.

The predicted bursting pressure level P_b is

$$P_b = \gamma P' \tag{38.27}$$

Maximum Allowable Defect Area/Length

The Maximum Allowable Design Pressure P is

$$P = F\frac{2t}{D}SMYS \tag{38.28}$$

Equating the Safe Maximum Pressure Level P' to the Maximum Allowable Design Pressure:

$$\frac{1}{\gamma}\frac{2\sigma_{flow}t}{D}\frac{1-QAREA/AREA_0}{1-M^{-1}AREA/AREA_0} = F\frac{2t}{D}SMYS \tag{38.29}$$

The Maximum Allowable Effective Area $AREA_{allow}$:

$$\frac{AREA_{allow}}{AREA_0} = \frac{F\gamma\frac{SMYS}{\sigma_{flow}}-1}{F\gamma\frac{SMYS}{\sigma_{flow}}M^{-1}-Q} \tag{38.30}$$

in which is the safety factor used in the new criterion.

The Maximum Allowable Length L_{allow} is

For $M_{allow} \le 4.9$

$$L_{allow} = \sqrt{\frac{1.255}{0.0135}-\sqrt{\left(\frac{1.255}{0.0135}\right)^2+\frac{4\left(1-M_{allow}^2\right)}{0.0135}}}\sqrt{Dt}$$

and for $M_{allow} > 4.9$

$$L_{allow} = \sqrt{\left(M_{allow}-3.3\right)/0.032}\sqrt{Dt} \tag{38.31}$$

where M_{allow} is solved by equating the Safe Maximum Pressure to the Maximum Allowable Design Pressure as:

$$M_{allow} = \frac{F\gamma\frac{SMYS}{\sigma_{allow}}}{Q+\frac{AREA_0}{AREA}\left(F\gamma\frac{SMYS}{\sigma_{flow}}-1\right)} \tag{38.32}$$

Effective Area, AREA

Two levels of AREA assessment are recommended in Section 38.3.1.

Closely Spaced Corrosion Pits

A distance of t (wall-thickness) is used as a criterion of pit separation for a colony of longitudinal oriented pits separated by a longitudinal distance or parallel longitudinal pits separated by a circumferential distance.

Interaction of Longitudinal Grooves

For defects inclined to pipe axis, if the distance x, between two longitudinal grooves of lengths L_1 and L_2, is greater than either L_1 or L_2, then the length of corrosion L is the maximum of L_1 and L_2; if the distance x, between two longitudinal grooves of lengths L_1 and L_2, is less than L_1 and L_2, the length of corrosion L is the sum of x, L_1 and L_2, $L = L_1 + L_2 + x$. For two longitudinal grooves separated by a circumferential distance x, the wall thickness t is used as groove separation criterion.

Spiral Correction Factor

The spiral correction factor Q is determined as:

$$Q = \frac{1-Q_1}{32}\frac{W}{t} + Q_1 \tag{38.33}$$

in which W is defect width, and coefficient Q_1 is a function of the spiral angle φ ($\varphi = 90^0$ for longitudinal corrosion, φ (=0(for circumferential corrosion)

Flow stress

Consideration should be given to factors affecting flow stress, e.g., fabrication process (e.g. hot rolled versus cold expanded), material aging, possible size effect, installation process and possible crack in corrosion defect bottom. Use of the actual value of the flow stress is allowed provided the value has been obtained from a reliable approach (e.g., material testing of the pipe in situ. etc.).

If the ultimate tensile stress is known, the flow stress can be estimated as the ultimate tensile stress. For API 5L materials, SMTS (Specified Minimum Tensile Stress) is recommended as flow stress.

Folias Factor M

The Folias factor is estimated based on the following equations:

$$M = \begin{cases} \sqrt{1 + \frac{2.51(L/2)^2}{Dt} - \frac{0.054(L/2)^4}{(Dt)^2}} & \text{for } \frac{L^2}{Dt} \le 50 \\ 0.032\frac{L^2}{Dt} + 3.3 & \text{for } \frac{L^2}{Dt} > 50 \end{cases} \tag{38.34}$$

Corroded Welds

Corrosion in submerged-arc seams (longitudinal welds) should be handled in the same manner as corrosion in the body of the pipe. Corrosion in girth welds (circumferential) should be assessed using the Kastner's local collapse criterion. The level 2 (or level 3 analysis implemented in PD 6493 (1991) should be applied for assessing corroded welds. The corroded groove could be considered as a crack of the same depth and length. The effect of the material's fracture toughness (in ductile and low toughness pipes) could be taken into account in the assessment procedure of the material fracture toughness.

Safety Factor

Traditional safety factors are given based on engineering experience and judgment. Within Kiefner and Vieth (1989, 1993) studies, several modified B31G criteria were developed. In all cases, the safety factors are assumed to be 1/0.72=1.39, as the original B31G criterion. However, the safety factor for the new criteria is calibrated based on reliability methods. It is around 1.8 and dependent on the accuracy of inspection tools and corrosion depth.

38.4.2 Evaluation of New Criteria

The evaluation of the new criteria is conducted in this section to compare with the test data from AGA database, NOVA tests, British Gas tests, and Waterloo tests. In the comparison, a model uncertainty parameter X_M is introduced as:

$$X_M = \frac{X_{true}}{X_{pred}} \tag{38.35}$$

where X_{true} is the true strength in the tests and X_{pred} is the capacity predicted by a given criteria (existing or new). Table 38.2 is the statistical parameters for X_M (mean and COV). It is demonstrated in the table that the uncertainty of the new criteria is much smaller than that of the existing criteria.

Table 38.2 Statistics for different criteria and test data.

	B31G	NG18	New Criteria
Mean	1.74	1.30	1.07
COV	0.51	0.19	0.18

38.5 Reliability-based Design

38.5.1 Target Failure Probability

The target failure probability is developed based on the historical failure data and the safety level implied in the existing B31G criteria. The target safety level should be determined considering the consequence of failure as well as the effects of inspection, maintenance, and repair. The safety level to be applied in the new criteria should be the same level as the safety level in the existing B31G criteria. Based on the historical data, reliability analysis of the existing B31G criteria, and other factors, an annual target safety level of 10^{-4} is used in the development of the reliability-based criteria.

38.5.2 Design Equation and Limit State Function

For the sake of simplicity, only internal pressure is considered in the design equation. The LRFD approach leads to:

$$P_R \geq \gamma P_L \tag{38.36}$$

where, P_R is the characteristic strength of the pipe based on a criterion, P_L is the characteristic load (internal pressure), $\gamma = \frac{\gamma_L}{\varphi_n}$ is referred to as the partial safety factor.

A bias factor X is introduced to reflect the confidence in the criterion in prediction of burst strength:

$$X = \frac{\text{true burst strength}}{\text{predicted burst strength}} \tag{38.37}$$

Normalized random variables in the design equation are:

$$X_p = \frac{P_L}{\frac{2\,SMYSt}{D}} \tag{38.38}$$

$$X_f = \frac{\sigma_{flow}}{SMYS} \tag{38.39}$$

$$X_A = \frac{AREA}{AREA_0} \tag{38.40}$$

$$X_L = \frac{L^2}{Dt} \tag{38.41}$$

$$X_d = \frac{d}{t} \tag{38.42}$$

The B31G design equation for corroded pipelines is:

For $\frac{L^2}{Dt} \leq 20$,

$$0.72\frac{1.1SMYS}{SMTS}X_t\frac{1-\frac{2}{3}X_d}{1-(1+0.79745X_L)^{-1/2}\frac{2}{3}X_d} - X_p \geq 0 \tag{38.43}$$

For $\frac{L^2}{Dt} > 20$,

$$0.72\frac{1.1SMYS}{SMTS}X_t(1-X_d)-X_p \geq 0 \tag{38.44}$$

The limit state function is then expressed as:

For $\frac{L^2}{Dt} \leq 20$,

$$g(\bar{x}) = 0.72\frac{1.1SMYS}{SMTS}X_M X_t \bullet \frac{1-\frac{2}{3}X_d}{1-(1+0.79745X_L)^{-1/2}\frac{2}{3}X_d} - X_p \tag{38.45}$$

For $\frac{L^2}{Dt} > 20$,

$$g(\bar{x}) = 0.72\frac{1.1SMYS}{SMTS}X_M X_t(1-X_d)-X_p \tag{38.46}$$

The design equation for corroded pipelines, based on the new criteria is given by:

For $\frac{L^2}{Dt} \leq 50$

$$\frac{1-X_A}{1-\left(1+0.6275X_L-0.003375X_L^2\right)^{-1/2}X_A} \bullet \frac{1}{\gamma}X_{flow}X_t - X_p \geq 0 \tag{38.47}$$

For $\frac{L^2}{Dt} > 50$

$$\frac{1}{\gamma}X_{flow}X_t\frac{1-X_A}{1-(0.032X_L+3.3)^{-1}X_A} - X_p \geq 0 \tag{38.48}$$

The limit state function is

For $\frac{L^2}{Dt} \leq 50$

$$g(\bar{x}) = \frac{1}{\gamma} X_M X_{flow} X_t \bullet \frac{1 - X_A}{1 - \left(1 + 0.6275 X_L - 0.003375 X_L^2\right)^{-1/2} X_A} - X_p \tag{38.49}$$

For $\frac{L^2}{Dt} > 50$

$$g(\bar{x}) = \frac{1}{\gamma} X_M X_{flow} X_t \bullet \frac{1 - X_A}{1 - \left(0.032 X_L + 3.3\right)^{-1} X_A} - X_p \tag{38.50}$$

38.5.3 Uncertainty

Bias for Criteria, X_M

Model uncertainty X_M is introduced for the criteria to account for modeling and methodology uncertainties. It reflects a general confidence in the design criteria for a real life in-situ scenario.

The model uncertainty is calibrated from the 86 tests results in the AGA database (Kiefner and Vieth, 1989). A Hermit model is applied to simulate the four lower moments. The mean bias and COV for the existing and new criteria is listed in Table 38.2.

Bias for Normalized Pressure, X_p

The characteristic value of the normalized pressure X_p is obtained by substituting safety factors, characteristic values of the other parameters into the design equation. In general, the annual maximum operating pressure is higher than the nominal operating pressure. This is reflected by the mean bias in X_p. Sotberg and Leira (1994) assumed that the ratio of the annual maximum operating pressure to the design pressure followed a Gumble distribution. Its mean and COV is 1.07 and 1.5%. By further analyzing the data (Bai, 1994), a Gumbel distribution with a mean of 1.05 and a COV of 2% is used in this development.

Bias for Normalized flow stress, X_f

The X_f mainly reflects the material property. Uncertainty of X_f is largely dependent of the material grade. A log-normal distribution is assumed to fit the data in the existing database. From the data analysis, the mean value and COV are selected as 1.14 and 6%, respectively.

Bias for Normalized Area x_a

The normalized defect area x_a is the ratio of metal loss area and its original area. Two kinds of inaccuracy are possible:

- inaccuracy due to the calculation method for the area of metal loss.
- inaccuracy due to use of measurement instruments.

A log-normal distribution with mean 0.8 and COV 0.08 is recommended for x_a.

Bias for Normalized Depth x_d

The uncertainty in the corrosion depth is the combination of the uncertainties associated with pit separation, inspection, and future corrosion prediction. A log-normal distribution is thus assumed for x_d, and the mean value and COV are taken as 0.8 and 8% respectively based on the experimental data and expert judgment.

Normalized Length x_L

Similar to the discussion on corrosion depth, the uncertainty in normalized defect length x_L is the combination of the uncertainties associated with pit separation, inspections, and future corrosion. However, corrosion length is easier to measure in inspections. Normal distribution is used to fit the x_L. Its mean value and COV are taken as 0.9 and 5% respectively.

38.5.4 Safety Level in the B31G Criteria

Reliability methods are applied to estimate the implied safety level of the B31G criterion. The uncertainties described in the section 38.6.3 are used in the reliability analysis. The safety factor is taken as 1.4 in the B31G criteria.

The obtained implied safety level (safety index) of the B31G criterion is shown in Figure 38.13 for short corrosion $x_L = 10$ and Figure 38.14 for long corrosion $x_L = 200$, as functions of defect depth x_d and material grade (SMYS). Due to large model uncertainty in the B31G criteria, the implied safety level in the B31G criteria is quite low. It is found for short ($x_L = 10$) and shallow corrosion defects ($x_d < 0.4$), the implied safety level is lower than 10^{-3}. For long ($x_L = 200$) and deep ($x_d > 0.4$) corrosion defects, the implied safety level is between 10^{-3} and 10^{-4}.

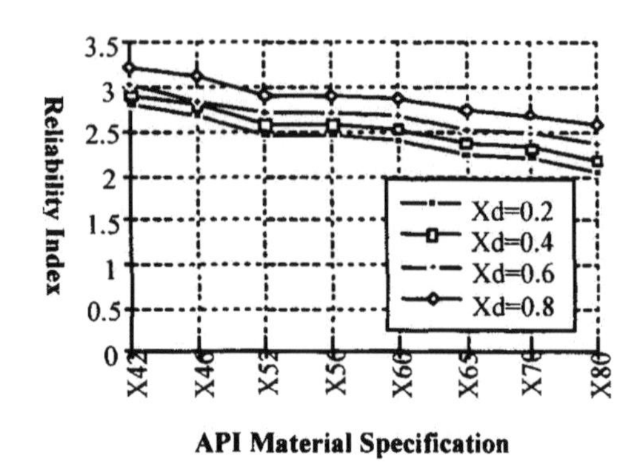

Figure 38.13 Implied safety level in the B31G criteria (short corrosion).

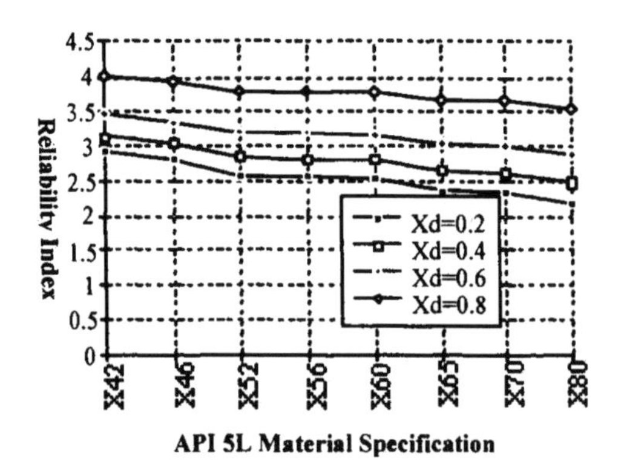

Figure 38.14 Implied safety level in the B31G criteria (long corrosion).

38.5.5 Reliability-based Calibration

It is proposed that the target safety level for the new criterion is set between 10^{-3} and 10^{-4}, based on the implied safety level in the B31G criterion. This ensures that the safety level in the new criteria is higher than or equal to the implied safety level in the original B31G criterion.

The relationship between the reliability index and the safety factor γ is shown in Figure 38.15 for short corrosion defects ($X_L = 10$) and Figure 38.16 for long corrosion defects ($X_L = 200$). The obtained reliability index for $X_A < 0.5$ was found to be close to the case $X_A = 0.5$.

Comparing Figure 38.15 with Figure 38.16, it is obvious that the reliability index β for a given set of X_A and safety factor γ is only slightly different for short corrosion detects ($X_L = 10$) and long corrosion defects ($X_L = 200$). The sensitivity analysis indicated that the model uncertainty of the criterion in questions was the dominantly important factor in the reliability analysis.

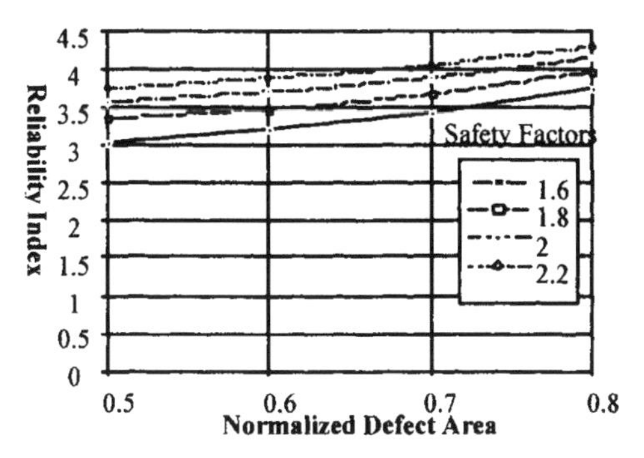

Figure 38.15 Reliability index for different defect area (short corrosion- new criteria).

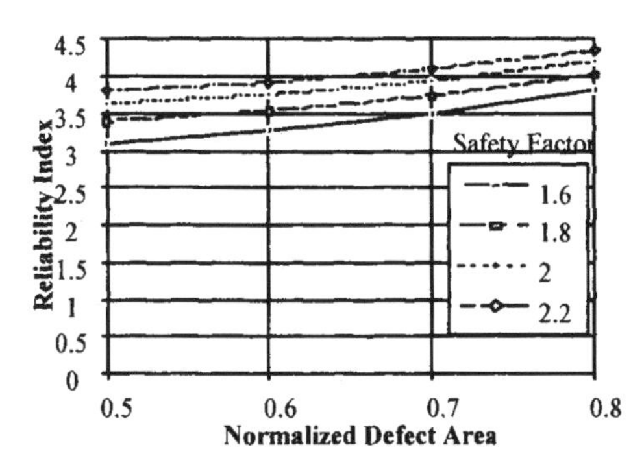

Figure 38.16 Reliability index for different defect area (long corrosion- new corrosion).

38.6 Re-qualification Example Applications

38.6.1 Design Basis

An example is presented to illustrate the application of the new criteria in the pipeline re-qualification. As a result of a corrosion detection pigging inspection of a 10 year old offshore gas pipeline, grooving corrosion was found in the pipeline. The re-qualification of this pipeline is divided into following steps.

Re-qualification Premises

Extensive groove corrosion has been observed in a gas pipeline after 10 years of service. The observed grooving corrosion results in a reduced rupture (bursting) capacity of the pipeline, increasing the possibility for leakage with resulting possible environmental pollution and unscheduled down time for repair

The intended service life

The gas pipeline is scheduled for a life of 20 years, resulting in residual service life of 10 years after the observation of the corrosion. There is no intended change in the service of the pipeline within the residual life.

Available Information

The design and operation parameters and their uncertainties for the pipeline are given in Table 38.3. It is assumed that gas pressure and temperature linearly vary over the entire pipeline length based on the conditions specified at the inlet and outlet point. The gas pressure varies over the service life. The gas temperature, on the other hand, is assumed to be constant.

Service History

The pipeline is routinely inspected on 5-year interval with a conventional corrosion detection pig. The pipeline inspection after the first 5 years of service did not bring up any observed corrosion.

Present Conditions

The inspection after the first 10 years service resulted in the detection of grooving corrosion. The maximum measured corrosion was detected at 0.6 km from the inlet point with a corrosion depth of 35% of the wall thickness, $d_{corr} = 0.35t$, and a detection accuracy is represented through a COV of 5% of the wall thickness.

Table 38.3 Uncertainty parameters in the analysis.

Var.	Description	Distribution
$X_{\Delta p,max}$	Annual max pressure ratio	G (1.05, 2%)
N	Hardening index	N(0.2, 6%)
X_f	Burst capacity model	N(1.0, 10%)
X_{SMTS}	Ult. Tensile Strength uncrt	N(1.09, 6%)
SMTS	Ultimate Tensile Strength	517.0 N/mm2
X_t	Wall thickness uncrt.	N(1.04, 10%)
X_{corr}	Corrosion model uncrt.	LN(0.2, 20%)
ϕ	Degree of circum. corr.	0.17
n_{CO_2}	mole fraction of CO_2	0.02
α	Influence of inhibitor	Beat(a, 50%)
d_{obs}	Observed relative corrosion	N(0.35, 14%)
Y_a	Normalized Area Uncrt.	LN(0.8, 8%)
X_d	Normalized Depth Uncrt.	LN(0.8, 8%)
X_L	Normalized Length Uncrt.	N(0.9, 5%)

Bursting Model

The burst strength formulation is expressed as

$$M_{burst}(t) = \Delta p_{grv}(t) - \Delta p_{max}(t) \tag{38.51}$$

where $\Delta p_{max}(t)$ is the annual maximum operating pressure, $\Delta p_{grv}(t)$ is the pressure resulting in bursting in year t.

The annual maximum occurring pressure in year t is expressed as a function of the operating pressure:

$$\Delta p_{max}(t) = X_{\Delta p,max} \Delta p_{oper}(t) \tag{38.52}$$

where, $X_{\Delta p,max}$ defines the relationship between annual maximum pressure and the average operating pressure.

The bursting capacity of the pipeline depends on the degree of grooving corrosion, and is modeled as:

$$\Delta p_{grv} = \frac{1}{\gamma} \frac{1 - QAREA / AREA_0}{1 - M^{-1} AREA / AREA_0} \Delta p_f \tag{38.53}$$

where Δp_f is the burst pressure for uncorroded pipe, $AREA_0$ is the original area prior to metal loss due to corrosion L_t. AREA is the exact area of the metal loss due to corrosion in the axial direction of through-wall thickness. γ is the factor of safety, M is the Folias factor, Q is the spiral correction factor.

The burst pressure Δp_f for uncorroded pipe is:

$$\Delta p_f = \frac{2\sigma_{flow} t}{D} \tag{38.54}$$

where D is the pipe diameter and t is the pipe thickness.

The flow stress is defined by Tresca or von Mises yield criterion as:

$$\sigma_{flow} = X_f \left[\left(\frac{1}{\sqrt{3}} \right)^{n+1} + \left(\frac{1}{2} \right)^{n+1} \right] \sigma_u \tag{38.55}$$

where, X_f is the model uncertainty for predicting the burst capacity, n is the hardening index, σ_u is the ultimate stress.

Corrosion Rate

The corrosion rate, or the annual degree of grooving corrosion, is estimated based on the empirical "deWaard & Milliams" formula that the influence of the operating pressure and temperature on the corrosion rate is defined:

$$\log(v(t)) = 5.8 - \frac{1710}{T(t)} + 0.67 \bullet \log_{10}(n_{co_2}(t) \bullet \Delta p_{oper}(t)) \tag{38.56}$$

where T is the temperature in Kelvin, n_{co_2} is the mole fraction of CO_2 in the gas phase and $\Delta p_{oper}(t)$ is the operating pressure (bar).

The estimated degree of corrosion over a time period, t, can be derived by integrating the corrosion rate over the time period:

$$d_{corr}(t) = X_{corr} \bullet \int_0^t \alpha(t) v(t) dt \tag{38.57}$$

where parameter $\alpha(t)$ expresses the influence of inhibitors and X_{corr} defines the model uncertainty associated with the empirical corrosion rate.

Basic Variables

The uncertainty defined in the Table 38.2 is introduced in the model, where the symbols N, LN, Beta and Gumbel indicated a normal, log-normal, Beta or Gumbel distribution. The first parameter is the mean value, the second is the COV, the third and fourth parameters are the lower and upper limits of the distribution.

38.6.2 Condition Assessment

The first stage of the re-qualification process is an evaluation of the present state of the system. If the system satisfies the specified constraints, the system will continue to operate as initially planned prior to the corrosion observation.

The specified constraints are summarized as:

- Acceptable level of safety within the remaining service, or at least until next scheduled inspection;
- The annual bursting failure probability is less than 10^{-3} within the next 5 years.

Three level analyses are conducted:

1. simplified analysis,
2. deterministic analysis
3. probabilistic analysis in the conditional assessment.

For the simplified analysis, the observed corrosion is compared with the corrosion allowance.

Estimated corrosion: 0.35t = 7.8mm

Corrosion allowance: 1.6mm

The observed corrosion is larger than the corrosion allowance.

For the deterministic analysis, the experienced degree of corrosion (stationary corrosion rate) is assumed to be valid over the remaining service life.

Corrosion after 15 years of service

corrosion rate: $\hat{v} = 0.35t / T = 0.78\text{mm / year}$

corrosion after 15 years: $\hat{d} = \hat{v}15 = 11.7\text{mm}$

Specified yield strength for X60 steel: $\sigma_y = 413\text{MPa}$

Acceptable hoop stress: $\sigma_{rules} = 297\text{MPa}$

Specified design pressure: $\Delta P = 13.4\text{Mpa}$

Hoop stress for uncorroded pipe:

$$\sigma_H = \frac{\Delta P \bullet D}{2 \bullet t} = \frac{13.4 \bullet 914.0}{2 \bullet 22.2} = 276\text{MPa}$$

Hoop stress for 11.7 mm corrosion:

$$\sigma_H = \frac{\Delta P \bullet D}{2 \bullet (t - \hat{d})} = \frac{13.4 \bullet 914.0}{2 \bullet (22.2 - 11.7)} = 583\text{MPa}$$

Based on the observed corrosion, the estimated stress after 15 years service is larger than the acceptable stress.

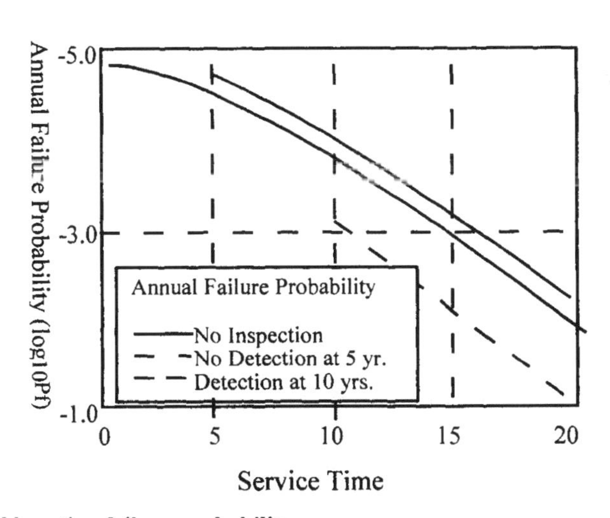

Figure 38.17 Annual bursting failure probability.

For the probabilistic analysis, the following approaches are applied:

- The corrosion rate is based on the de Waard & Milliams formula,
- The reduced burst capacity is estimated based on the new criteria,

- The design pressure for which the capacity model is to resist is developed over the service life as a function of the operating pressure.

Based on the capacity and loading model, the annual probability for bursting of the corroded pipelines is illustrated in Figure 38.17. It is shown that the estimated probability of failure increases slightly with time in spite of the reduced operating pressure due to the increase in the expected level of corrosion.

Evaluating of Repair Strategies

A minor repair/modification is recommended. The alternatives are summarized as:

- A reduction of the operating pressure, de-rating;
- Use of corrosion mitigation measures (inhibitors);
- Rescheduled inspection;
- Combination of the above alternatives.

The life-cycle cost of mitigation measures and lost income are set as the evaluation criteria. The constraint requirements are:

- Acceptable level of safety within the remaining service life, or at least until next inspection;
- The annual failure probability of the pipeline should be less than 10^{-3} with the remaining service life or until next inspection;
- Next inspection is scheduled for a service life of 15 years. Meanwhile, an early inspection can be recommended.

Two alternatives are studied in this example:

1. de-rating;
2. inhibitors.

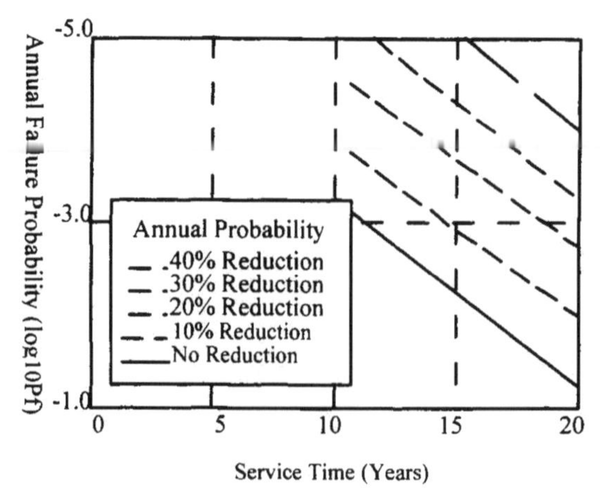

Figure 38.18 Annual failure probabilities for induced operating pressure.

De-rating

The reduced operation pressure reduces the annual maximum pressure as well as, to some extent, reduce the additional corrosion growth.

In Figure 38.19, the estimated annual bursting failure probability in the time period after the year 10 is shown as a function of the relative reduction in the operating pressure. It is illustrated in Figure 38.19 that the time period until probability of failure 10^{-3} is approximately 14, 17 and 21 years when the operating pressure is reduced with 10%, 20%, and 30% respectively.

Inhibitors

The use of inhibitors reduces the additional corrosion growth over the remaining service life and thereby reduces the annual failure probability over time. Inhibitors resulting in 50%, 60%, 70% and 80% corrosion reduction are considered in the example applications. As the mitigation effects are uncertain, the influence of the inhibitors is modeled as Beta distribution with a median (50%) value as the specified corrosion reduction effect and a COV of 50%.

The reduction in the degree of grooving corrosion due to the use of inhibitors is illustrated in Figure 38.20. The figure shows the expected corrosion depth over the time. The use of inhibitors greatly reduces the corrosion rate. Figure 38.21 shows the estimated annual bursting failure probability in the time period after the 10 years service. The use of inhibitors reduces the failure probability.

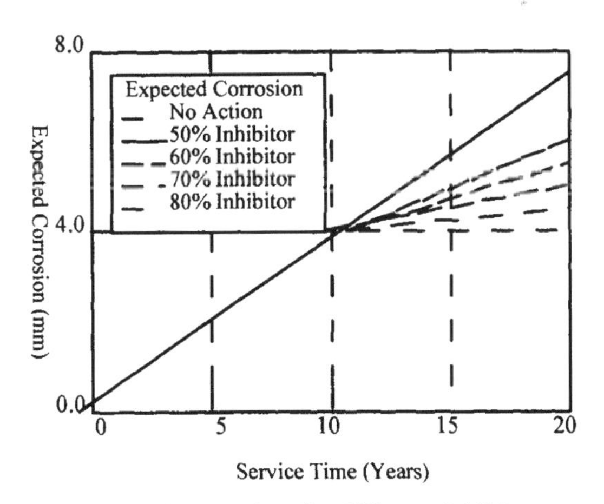

Figure 38.20 Expected corrosion depth over time for different inhibitors.

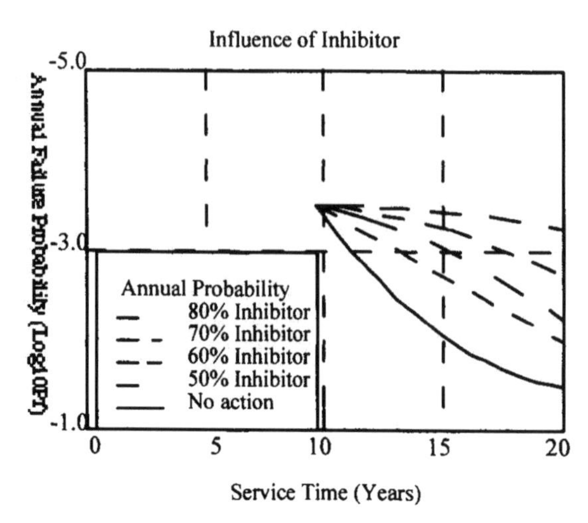

Figure 38.21 Annual failure probability for different inhibitors.

Evaluation of Alternatives

The selection of the minor repair/modification alternatives (de-rating or inhibitors) satisfies the constraints. Table 38.4 summarizes the combination effects. It summarizes the operating time after 10 years service until the target probability 10^{-3} is reached.

If the next inspection is not scheduled prior to 15 years of service, the combinations of de-rating and inhibitors in the shaded area of Table 38.4 are the realistic decision alternatives. The darker shaded area indicates the most attractive combination of use of inhibitor with specified effect and degree of pressure reduction. If the pipeline inspection is rescheduled, the alternatives of upper left corner in Table 38.4 are recommended. However, in the evaluation of the alternatives incorporating a reduction of the time period until next inspection, the likelihood of possible major repair/mitigate measures at an earlier period should be addressed in the decision process.

Table 38.4 Operating years inspection until the target failure probability.

	Effect of Inhibitors				
P Red.	0%	50%	60%	70%	80%
0%	1	3	4	6	12
5%	2	5	6	9	15
10%	4	7	9	12	18
15%	5	9	11	14	20
20%	6	11	13	17	23
25%	8	13	15	19	23
30%	9	15	17	21	29

38.6.3 Rehabilitation

A possible major repair alternative is replacement of a fraction of, or the whole pipeline. The major repair/modification can greatly reduce the estimated failure probability. However, as the observed damage can be effectively controlled by the minor repair/modifications, the major modification is not recommended in the re-qualification process of this pipeline.

38.7 References

1. ABS (2001), "Guide for Building and Classing Subsea Pipelines and Risers", American Bureau of Shipping., March 2001.
2. ASME (1996), "B31G - Manual for Assessing Remaining Strength of Corroded Pipes", American Society of Mechanical Engineers.
3. Bai, Y. and Mørk, K. J. (1994) "Probablistic Assessment of Dented and Corroded Pipeline" International Conference on Offshore and Polar Engineering, Osaka, Japan.
4. Bai, Y., Xu, T. and Bea, R., (1997) "Reliability-based Design and Re-qualification Criteria for Longitudinally Corroded Pipes", ISOPE'97.
5. BSI (1991) "PD6493 – Guidance on Methods for Assessing the Acceptability of Flaws in Fusion Welded Structures".
6. De Waard, C., Lotz, U., and Milliams, D.E. (1991) " Predictive model for CO2 corrosion engineering in wet natural gas pipelines", Corrosion, pp.976
7. Dugstad, A., Lunde, L., and Videm, K.(1994) "Parametric study of CO2 corrosion of carbon steel" Corrosion.
8. Folias, E. S., (1965) "An Axial Crack in a Pressurised Cylindrical Shell", Int. J. of Fracture Mechanics, Vol. 1 (1), pp.64-113.
9. Galambos, T.V. (1988) "Guide to Stability Design Criteria for Metal Structures", John Wiley & Sons, Inc. pp. 502-508.
10. Gunaltun, Y.M.(1996) "Combine Research and Field Data for Corrosion Rate Prediction," Corrosion, Paper No. 27
11. Hauch, S. and Bai, Y. (2000) "Bending Moment Capacity of Grove Corroded Pipes", ISOPE'2000.
12. Hopkins, P. and Jones, D. G., (1992) "A Study of the Behaviour of Long and Complex-shaped Corrosion in Transmission Pipelines", Proceedings of OMAE'92.
13. Jones, D. G., Turner T. and Ritchie, D. (1992) "Failure Behaviour of Internally Corroded Linepipe", OMAE'92.
14. Kastner, E., Roehrich, E., Schmitt, W. and Steinbuch, E. (1981) "Critical Crack Sizes in Ductile Piping", Int. J. Pres. Ves. and Pipeing, Vol. 9, pp. 197-219.
15. Kiefner, J. F. (1974) "Corroded Pipe Strength and Repair Methods", Symposium on Line Pipe Research, Pipeline Research Committee, American Gas Association.
16. Kiefner, J. F. and Vieth, P. H., (1989) "A Modified Criterion for Evaluating the Remaining Strength of Corroded Pipe, RSTRENG", Project PR 3-805 Pipeline Research Committee, American Gas Association.
17. Kiefner, J. F. and Vieth, P. H., (1990) "New Method Corrects Criterion for Evaluating Corroded Pipe", Oil & Gas Journal.

18. Kiefner, J. F. and Vieth, P. H., (1993) "RSTRENG Users Manual", Project PR 218-9205 Pipeline Research Committee, American Gas Association.
19. Klever, F. J., (1992) "Burst Strength of Corroded Pipe: 'Flow Stress' Revisited", Proceedings of Offshore Technology Conference, OTC 7029.
20. Mandke, J. S. (1990) "Corrosion Causes Most Pipeline Failure in the Gulf of Mexico" Oil and Gas Journal, Oct.29,1990.
21. Maxey, W. A., Kiefner, J. F., Eiber, R. J. and Duffy, A. R. (1971) "Ductile Fracture Initiation, Propagation and Arrest in Cylindrical Vessels" Fracture Toughness, Proceedings of the 1971 National Symposium on Fracture Mechanics, Part II, ASTM STP 514, American Society for Testing and Materials, pp.70-81.
22. Mok, D. H. B., Pick, R. J., and Glover, A. G., (1990) "Behaviour of Line Pipe with Long External Corrosion", Material Performance, Vol. 29 (5), pp. 75-79.
23. Mok, D. H. B., Pick, R. J., Glover, A. G. and Hoff, R., (1991) "Bursting of Line Pipe with Long External Corrosion", International Journal of Pressure Vessel and Piping, Vol. 46, pp. 159-216.
24. Sotberg, T. and Leira, B. J., (1994) "Reliability-based Pipeline Design and Code Calibration", Proceedings of 13th International Conference on Offshore Mechanics and Arctic Engineering.
25. Stewart, G., Klever, F. and Ritchie, D., (1994) "An Analytical Model to Predict the Burst Capacity of Pipelines", Proceedings of 13th International Conference on Offshore Mechanics and Arctic Engineering, Vol. 4.

Part VI

Integrity Management

Chapter 39 Residual Strength of Dented Pipes with Cracks

39.1 Introduction

With the increased use of pressure vessels, pipelines and piping systems, more and more pipes are being put into use. Mechanical damages in the form of dents and cracks occur frequently. These damages are mainly caused by 3rd party or operation activities, fabrication errors etc. Leakage of gas and oil from pipes due to structural failure may lead to reduced operating pressure or stopped production, human and environmental hazards and the significant economic loss consequently. Since the existence of dents especially at weld seams is one of the causes of leakage, it is important to arrive at a basis for assessing the structural integrity of dented pipe with cracks.

The first part of the chapter deals with the burst strength criteria of dented pipes with longitudinal and circumferential cracks. Subsequently, fracture assessment of damaged pipes is studied. Uncertainties involved in loading, strength and modeling are assessed. In the third part of the chapter, fracture reliability model of dented pipes with cracks is developed. Reliability-based calibration of safety factor and uncertainty modeling is performed. Conclusions and recommendations are also outlined.

39.2 Limit-state based Criteria for Dented Pipe

39.2.1 General

Pipeline Systems are more and more are used for transportation of gas and oil. They are usually unprotected when resting on seabed, and exposed to cyclic loading and corrosive fluids/gasses. Typical defects that may found in the pipeline systems include:

- Dents due to impact or local buckling
- Corrosion
- Cracks

Failure modes for the pipeline systems in service may be categorized as:

- Serviceability Limit State (Out of roundness)
- Ultimate Limit State

The remaining strength evaluation based on a limit state approach includes assessment against the following failure modes:

- Out of roundness
- Fracture
- Fatigue
- Bursting
- Collapse
- Local buckling

When damages have been found in the pipeline systems, possible actions are:

- None
- Seabed intervention
- Reduced operating pressure and pressure fluctuation
- Repair
- Replacement of pipe sections

In this section, a set of limit state design equations are developed for dented pipe, including bursting, buckling/collapse, fatigue and fracture for

- deepwater pipelines under combined loads
- metallic risers in deepwater
- high strength steel
- different safety levels

In particular, buckling/collapse equations for dented pipe have been newly developed for combined pressure, axial force, bending and torsional moment. Some of these limit-states are discussed in more detail as the following.

39.2.2 Serviceability Limit-state (Out of Roundness)

Out of roundness is a serviceability limit state mainly dictated by the operation of the pigging tools. The maximum allowable out of roundness is between 2.5 % and 5 % while fabrication tolerance is between 0.5 % and 1.5 %.

A detected out of roundness significantly higher than what is assumed in the design will influence the predicted collapse pressure and the bending moment capacity for external pressure load conditions.

39.2.3 Bursting Criterion for Dented Pipes

The bursting criterion is an ultimate limit state, which for design normally is given in form of a maximum allowable internal overpressure. The bursting limit state is often used in wall thickness design for pipes where internal pressure is the dominating load. The detection of cracks, corrosion, their size, number and orientation have high influence on the capacity.

The bursting criterion for dented pipe may be summarized as below:

- The bursting strength of a dented pipe is close to that for a new pipe if there is no crack in the dented area.
- A crack in the dented area may reduce bursting strength due to stress concentration at the crack tip.
- Bursting of dented pipe with crack is actually a fracture failure mode.
- The assessment of a dented pipe with cracks.

39.2.4 Fracture Criterion for Dented Pipes with Cracks

The fracture criterion is an ultimate limit state normally in form of a maximum allowable tensile strain. In pipeline design the fracture limit state is based on wall thickness and minimum detectable crack size given by pipe dimensions and measuring method.

A detected crack larger than assumed in the design or a corrosion defect deeper than an included corrosion allowance will change the predicted fracture strength and the fracture calculations shall be updated in accordance with the measured values.

39.2.5 Fatigue Criterion for Dented Pipes

The fatigue criterion is a limit state normally given in form of a maximum allowable stress range derived based on the S-N curve approach or fracture mechanics approach. The fatigue limit state is based on the design life and assumed load cycles during the life cycle to the pipeline. High plastic deformation of the pipe caused by a dent may reduce the fatigue life dramatically. A severe reduction in cross sectional area due to cracks or corrosion will also reduce the fatigue life.

The fatigue criterion may be divided into fatigue due to cyclic internal pressure and fatigue due to cyclic longitudinal forces and bending. The fatigue damage is influenced by detection of dents, cracks and corrosion.

Fatigue dented pipes due to cyclic pressure loads:

- The fatigue off dented pipe without a crack is to be calculated using S-N curve approach where stress concentration due to dent is to be included. The design is to estimate allowable cycles of stress range.
- The fatigue of dented pipe with crack is to be based on fracture mechanics approach that account for crack growth and final fracture. It is important to define correct input data on material and defects in the fracture mechanics assessment.

Fatigue of dented pipes due to cyclic longitudinal force and bending:

- The fatigue strength of dented pipe without crack is close to the fatigue strength of new pipes.
- Stress concentration due to dents should be accounted, and this is similar to design of new pipes.

39.2.6 Moment Criterion for Buckling and Collapse of Dented Pipes

Collapse Criterion

The Collapse criterion is an ultimate limit state, which for design normally is given in form of a maximum allowable external overpressure, see Section 4.4. The collapse limit state is often used in wall thickness design for pipes where external pressure is the dominating load. Cracks, corrosion, their size, number and orientation have high influence on the capacity.

Buckling/Collapse of Dented Pipe

The capacity equations are based on mechanism:

- The stress distribution is a fully plastic yielding problem.
- Integrating stress over the cross-section to get fully plastic interaction equation for pipe under combined internal pressure, axial force and bending, as well as torsion.
- The derived interaction equations are to be validated using finite element models, see Hauch and Bai (OMAE'98,'99).
- The effect of torsion may be included using plastic interaction equations derived by Fujikubo, Bai and Ueda (ISOPE'91).

When the dent angle is less than the angle to the plastic neutral axis, moment capacity of dented pipes may be expressed as below:

Dent in tensile side of cross section

$$M_{Dented} = M_{\text{Allowable}(F,p)} - 0.5\frac{\eta_{RM}}{\gamma_C} M_l(\sin(\beta) - \beta\cos(\beta))\left(\alpha\frac{p}{\eta_{RP} p_l} + \sqrt{1-(1-\alpha^2)\left(\frac{p}{\eta_{RP} p_l}\right)^2}\right)$$

Dent in compressive side of cross section

$$M_{Dented} = M_{\text{Allowable}(F,p)} - 0.5\frac{\eta_{RM}}{\gamma_C} M_l(\beta\cos(\beta) - \sin(\beta))\left(\alpha\frac{p}{\eta_{RP} p_l} - \sqrt{1-(1-\alpha^2)\left(\frac{p}{\eta_{RP} p_l}\right)^2}\right)$$

where:

M_C (F, p) : Hauch & Bai, OMAE'99;

M_l : Limit moment (Plastic moment capacity as for pipe without damage);

F_l : Longitudinal limit force (Longitudinal force capacity as for pipe without damage);

p_l : Limit pressure (The external limit pressure is given by Thimoshenko or Haagsma including high out of roundness and reduction in cross sectional area due to cracks, see Bai & Hauch, ISOPE'98.

For dented pipe under external pressure, the collapse strength of a dented pipe may be obtained by a curve between:

- collapse pressure for pipes without dent (e.g. Timoshenko's equations).
- collapse pressure for pipe with very deep dent or buckle propagation pressure.

The alpha factor is an important factor that affects the accuracy of the derived equations. Ideally its value should be calibrated against experimental and numerical tests. Tentatively it is suggested that: $\alpha = 0.25\frac{p_l}{F_l}$.

It may be concluded that:

- A new set of equations for calculating the bending moment strength are newly suggested for metallic pipes with corrosion defects or dent damage (and cracks).
- The Finite Element Method provides good predictions of the load versus deflection behaviour of pipes.
- Results from the developed analytical equations and finite element analyses are in good agreement for the range of variables studied
- Load and usage factors have been suggested for safety class low, normal and high, see Section 4.4.
- Full set of design criteria for corroded pipe, dented pipe, as an extension to Hauch and Bai, OMAE'99.

39.3 Fracture of Pipes with Longitudinal Cracks

The following assumptions are made for the analysis:

- Elastic-Plastic Fracture Mechanics is applied.
- The dent is assumed to be continuous and to have a constant length.
- The stress-concentrator is considered to be a notch located at the deepest point of the dent (infinite length, constant depth). The notch is longitudinal of length, L=2c, and depth, a.

39.3.1 Failure Pressure of Pipes with Longitudinal Cracks

Longitudinal surface cracks can occur as isolated cracks or in colonies of numerous closely spaced and parallel cracks. A procedure based on Maxey et al. (1972) for calculating the failure stress of longitudinal flaws is as follows:

Folias factor M_T is determined from Kiefner and Vieth (1989):

$$M_T = \sqrt{1 + 0.6275x^2 - 0.003375x^4} \quad \text{for } x \leq 7.07 \tag{39.1}$$

$$M_T = 0.032x^2 + 3.3 \quad \text{for } x > 7.07 \tag{39.2}$$

where:

- x : $L/(Dt)^{1/2}$
- L : total length of the crack (L=2c)
- D : pipe nominal outside diameter

T : pipe wall-thickness.

The failure pressure of pipes with longitudinal flaws is calculated as:

$$P_c = \frac{4t\,\sigma_{flow}}{\pi\,D\,M_S}\cos^{-1}\left(\exp(-B)\right) \tag{39.3}$$

where, σ_{flow} is the material flow stress and auxiliary parameters M_S and B are given as follows:

$$M_S = \frac{M_T\,t - a}{M_T\,(t - a)} \tag{39.4}$$

$$B = \frac{\pi}{4L}\left(\frac{K_{mat}}{\sigma_{flow}}\right)^2 \tag{39.5}$$

where:

a : crack depth

K_{mat} : material toughness, estimated from Charpy impact energy tests, as shown later.

By applying a safety factor γ, the allowable pressure can be calculated from:

$$P = P_c / \gamma \tag{39.6}$$

Safety factor γ can be calibrated by reliability methods as discussed in the following section. If no calibration is conducted, it is suggested that γ=2.0.

39.3.2 Burst Pressure of Pipes Containing Combined Dent and Longitudinal Notch

The fracture condition for the Bilby-Cottrell-Swinden dislocation model (Bilby, Cottrell and Swinden, 1963) is given as, (Heald et al., 1971)

$$\sigma = \frac{2\,\sigma_p}{\pi}\cos^{-1}\left(\exp\left(-\frac{\pi\,K_{mat}^2}{8a\,\sigma_p^2}\right)\right) \tag{39.7}$$

where:

σ : stress at failure (bursting)

σ_p : collapse stress for a pipe with an infinitely long defect notch of depth a.

This model has been used successfully to describe the failure of part-wall defects in pipes, but modifications are needed before it can be used for dented pipes with defects, as discussed below.

Toughness modification

Pipe toughness is measured in terms of the Charpy energy, C_v. This measure has been shown to be a good qualitative measure for pipe toughness but has no theoretical relation with the fracture toughness parameter, K_{mat}. It is, therefore, necessary to use an empirical relationship between K_{mat} and C_v.

The Battelle K_{mat}-C_v relationship has been derived based on non-linear regression on full-scale tests of mechanical damaged pipes. But the deterioration of the fracture toughness caused by the material deformation as a result of denting has not been taken into account. The K_{mat}-C_v relationship has been modified in Nederlanse Gasunie as:

$$K_{mat}^2 = 1000\frac{E}{A}(C_v - 17.6) \tag{39.8}$$

where:

K_{mat} : material toughness ($N/mm^{3/2}$)

C_v : Charpy energy (J)

E : Young's modulus (N/mm^2)

A : section area for Charpy test (mm^2), normally A=80 mm^2.

Compliance modification

The Bilby-Cottrell-Swinden Dislocation Model is for an embedded crack in an infinite body. For other geometry and crack shapes, it is necessary to introduce the elastic compliance factor, Y (or called geometry function Y). Rearranging the equation and introducing Y as described by Heald et al. (1971), stress intensity factor (SIF) K can be written as:

$$K = Y\sigma_p\left(\frac{8a}{\pi}\ln\left(\sec\left(\frac{\sigma}{\sigma_p}\frac{\pi}{2}\right)\right)\right)^{1/2} \tag{39.9}$$

In this chapter, geometry functions for a surface crack in plates by Newman and Raju (1981) are used. For the wide plate under combined tension and bending, the stress intensity factor K is the sum of tension and bending terms:

$$K = \frac{F}{\sqrt{Q}}\sigma\sqrt{\pi a} + H\frac{F}{\sqrt{Q}}\frac{6M}{t^2}\sqrt{\pi a} \tag{39.10}$$

where factors F, Q and bending correction factor H are given by Newman and Raju (1981).

Geometry function

Geometry correction factors Q, F and H are given by the following:

$$Q = 1 + 1.464\left(\frac{a}{c}\right)^{1.65} \quad \text{for } \frac{a}{c} \le 1$$

where

c : Half length of dent

$$F = \left[M_1 + M_2\left(\frac{a}{t}\right)^2 + M_3\left(\frac{a}{t}\right)^4\right]f_\phi g f_w$$

where

$$M_1 = 1.13 - 0.09\frac{a}{c}$$

$$M_2 = -0.54 + \frac{0.89}{0.2 + (a/c)}$$

$$M_3 = 0.5 - \frac{1}{0.65 + (a/c)} + 14\left(1 - \frac{a}{c}\right)^{24}$$

$$g = 1 + \left[0.1 + 0.35\left(\frac{a}{t}\right)^2\right](1 - \sin\phi)^2$$

where

ϕ : parametric angle of the elliptical-crack

The function f_ϕ, an angular function from the embedded elliptical-crack solution is:

$$f_\phi = \left[\left(\frac{a}{c}\right)^2 \cos^2\phi + \sin^2\phi\right]^{1/4}$$

The function f_w, a finite-width correction factor is:

$$f_w = \left[\sec\left(\frac{c}{D}\cdot\sqrt{\frac{a}{t}}\right)\right]^{1/2}$$

The function H has the form:

$$H = H_1 + (H_2 - H_1)\sin^p\phi$$

where

$$p = 0.2 + \left(\frac{a}{c}\right) + 0.6\left(\frac{a}{t}\right)$$

$$H_1 = 1 - 0.34\left(\frac{a}{t}\right) - 0.11\left(\frac{a}{c}\right)\left(\frac{a}{t}\right)$$

$$H_2 = 1 + G_1\frac{a}{t} + G_2\left(\frac{a}{t}\right)^2$$

and

$$G_1 = -1.22 - 0.12\left(\frac{a}{c}\right)$$

$$G_2 = 0.55 - 1.05\left(\frac{a}{c}\right)^{0.75} + 0.47\left(\frac{a}{c}\right)^{1.5}$$

Bending moment M and uniaxial tensile stress σ in a dented

Solutions for bending moment M and uniaxial tensile stress σ in a dented pipe are given by Shannon (1973). These complex functions can be approximately represented by the following relationships:

$$\sigma = \sigma_H\left(1 - 1.8\frac{D_d}{D}\right) \tag{39.11}$$

$$M = 0.85\,\sigma_H\, t\, D_d \tag{39.12}$$

where:

σ_H : nominal hoop stress

D_d : dent depth.

Substituting σ and M into Equation (39.10), we get:

$$K = \frac{F}{\sqrt{Q}}\left(1 - 1.8\left(\frac{D_d}{D}\right) + 5.1\,H\left(\frac{D_d}{t}\right)\right)\sigma_H\sqrt{\pi a} \tag{39.13}$$

Therefore, the geometry function, Y, can be expressed as:

$$Y = \frac{F}{\sqrt{Q}}\left(1 - 1.8\left(\frac{D_d}{D}\right) + 5.1\,H\left(\frac{D_d}{t}\right)\right) \tag{39.14}$$

The material fails when the following critical condition is satisfied:

$$K = K_{mat} \tag{39.15}$$

in which K_{mat} is related to the Charpy energy C_v.

Flow stress modification

A more accurate measure of the plastic failure stress would be the collapse stress with a defect present. Following the B31G, collapse stress for a rectangular defect in a pipe is:

$$\sigma_p = \sigma_f \frac{t - a}{t - a\,M_T^{-1}} \tag{39.16}$$

in which σ_f is the flow stress for intact pipe and can be estimated from API as:

$$\sigma_f = \alpha \cdot \sigma_y \tag{39.17}$$

where σ_Y is the pipe yield strength and parameter α is around 1.25, α decreases when σ_y increases.

39.3.3 Burst Strength Criteria

The critical stress at failure is obtained from Equations (39.9 and 39.15) as:

$$\sigma = \frac{2\,\sigma_p}{\pi}\cos^{-1}\left(\exp\left(-\frac{\pi\,K_{mat}^2}{Y^2\,8a\,\sigma_p^2}\right)\right) \tag{39.18}$$

Burst strength is given by:

$$P = 2\sigma\frac{t}{D} \tag{39.19}$$

Based on Failure Assessment Diagram (FAD), the aforementioned burst strength can also be obtained by use of the procedure presented in PD6493, in which iteratively solving the equation of assessment will be involved including safety factors, as described for the case for circumferential cracks, Section 39.3.

39.4 Fracture of Pipes with Circumferential Cracks

It is assumed that the stress-concentrator is a notch located at the deepest point of the dent, it is continuous (infinite length, constant depth) and has circumferential length 2c and depth, a.

39.4.1 Fracture Condition and Critical Stress

Based on PD6493, the equation of the fracture failure assessment curve is given by:

$$K_r = S_r\left(\frac{8}{\pi^2}\ln\left(\sec\left(\frac{\pi}{2}S_r\right)\right)\right)^{-1/2} \tag{39.20}$$

in which:

$$K_r = \frac{K_I}{K_{mat}} + \rho \tag{39.21}$$

where:

ρ : plasticity correction factor

K_I : Stress intensity factor, determined from the following equation:

$$K_I = Y\sigma\sqrt{\pi\,a} \tag{39.22}$$

where $Y\sigma$ is divided to primary stress term and secondary stress term as:

$$Y\sigma = (Y\sigma)_p + (Y\sigma)_s \tag{39.23}$$

The stress ratio S_r is defined as the ratio of net section stress σ_n to flow stress σ_{flow}:

$$S_r = \frac{\sigma_n}{\sigma_{flow}} \tag{39.24}$$

39.4.2 Material Toughness, K_{mat}

Several statistical correlation exists between standard full-size C_v (the Charpy V-notch) and K_{mat}. Rolfe and Novak (1970) developed the following correlation for upper shelf toughness in steels:

$$K_{mat} = \sigma_y \sqrt{\frac{0.6459\, C_v}{\sigma_y} - 0.25} \tag{39.25}$$

with K_{mat} is in $MPa(mm)^{1/2}$, C_v is in mm-N, and σ_y is in MPa.

39.4.3 Net Section Stress, σ_n

Following PD6493, the net section stress for pipes with surface flaw is:

$$\sigma_n = \frac{\sigma_b + \sqrt{\sigma_b^2 + 9\sigma_m^2(1-\alpha)^2}}{3(1-\alpha)^2} \tag{39.26}$$

where:

σ_b : bending stress

σ_m : membrane stress

$$\alpha = (2a/t)/(1+t/c) \tag{39.27}$$

$$\sigma_b = \frac{M}{t^2/6} \tag{39.28}$$

where M is given by Equation (39.12) substituting σ_H by nominal axial stress σ_{AX}.

39.4.4 Maximum Allowable Axial Stress

The critical stress at failure is obtained by iteratively solving the Level-2 FAD of PD6493 (Equation 39.20) including safety factors.

39.5 Reliability-based Assessment

39.5.1 Design Formats vs. LSF

Design format

If only internal pressure is considered, the partial safety factor approach given by Equation (39.6) leads to the design format as:

$$P_C \geq \gamma \cdot P_L \tag{39.29}$$

where:

P_C : characteristic strength of the pipe according to a criterion

P_L : characteristic load (internal pressure)

γ : safety factor.

The new design equation for dented pipes with cracks in operation with respect to fracture criterion can be formulated by substituting Equations (39.19) and (39.18) into Equation (39.29) as:

$$P_L \le \frac{1}{\gamma} \cdot 2\frac{t}{D} \cdot \frac{2\sigma_p}{\pi} \cos^{-1}\left(\exp\left(-\frac{\pi K_{mat}^2}{Y^2 8a \sigma_p^2} \right) \right) \tag{39.30}$$

All the parameters in the new design format can be referred to the aforementioned sections. It should be noted that characteristic values of those parameters will be used to estimate the design pressure.

Limit state function

LSF can be formed based on failure criteria for the specified case. Bursting of a pipe will happen at the uncontrolled tearing point in case the equivalent stress exceeds the flow stress. The bursting failure will lead to the pipe rupture. The LSF based on new fracture criterion can be formulated as:

$$g(Z) = 2\frac{t}{D} \cdot \frac{2\sigma_p}{\pi} \cos^{-1}\left(\exp\left(-\frac{\pi K_{mat}^2}{Y^2 8a \sigma_p^2} \right) \right) - P_L \tag{39.31}$$

where Z is the set of random variables involved in the new design format.

By introducing the normalized random variables including model error, as discussed in details below, the new LSF is given by:

$$g(Z) = \frac{4 t_c \sigma_{fc}}{\pi D_c} X_M X_t X_f M_S^{-1} \cos^{-1}\left(\exp\left(\frac{-\pi M_S^2 K_{mat}^2}{\sigma_{fc}^2 Y^2 8a X_Y^2 X_f^2} \right) \right) - X_P \cdot P_d \tag{39.32}$$

where P_d is the design pressure which can be estimated from new design Equation (39.30), parameters M_S and K_{mat} are given by Equations (39.4) and (39.8) respectively by introducing uncertainties into the corresponding random variables and the subscript c indicates the characteristic values of corresponding variables.

39.3.2 Uncertainty Measure

Considering uncertainties involved in the design format, each random variable X_i can be specified as:

$$X_i = B_X \cdot X_C \tag{39.33}$$

where X_C is the characteristic value of X_i, and B_X is a normalized variable reflecting the uncertainty in X_i. The statistical values for the above biases are given in Table 39.2 as below.

39.6 Design Examples

Cited from a practical evaluation of an existing dented pipe, an example is given to verify the presented model and demonstrate its application in assessing structural integrity of damaged pipes.

39.6.1 Case Description

The analysis is based on the following data given in Table 39.1 from an existing pipe.

Table 39.1 Basic input data of pipe.

Pipe outside diameter, D	:	1066.8 mm
Pipe yield strength, σ_y	:	413.7 N/mm^2
Material	:	API 5L60
Pipe wall-thickness, t	:	14.3 mm
Design pressure, P	:	1.913 MPa
Dent depth, D_d	:	45 mm
Hydrostatic test pressure	:	30 kg/cm^2

39.6.2 Parameter Measurements

A complete list of uncertainties parameters for reliability analysis are given in Table 39.2.

Table 39.2 Basic probabilistic parameters descriptions.

Random Variable	Distribution	Mean	COV
Wall-thickness factor, X_t	Normal	1.04	0.02
Flow stress factor, X_f	Normal	1.14	0.06
Flow stress model, X_M	Normal	0.92	0.11
Max. pressure factor, X_P	Gumbel	1.05	0.02
Crack length factor, X_L	Normal	1.00	0.10
Crack depth, a	Exponential	0.10	1.00
Dent depth factor, X_D	Normal	0.90	0.05
Y function factor, X_Y	Log-normal	1.00	0.10
Charpy energy, C_V	Log-normal	63.0	0.10
Young's modulus, E	Normal	210	0.03

39.6.3 Reliability Assessments

Fracture reliability assessment is performed by use of STRUREL and PROBAN respectively.

The influence of dent depth on fracture reliability is given in Figure 39.1, from which it is seen that no obvious changes can be observed if the dent depth is not serious. But failure probability increases dramatically with the increase of dent depth.

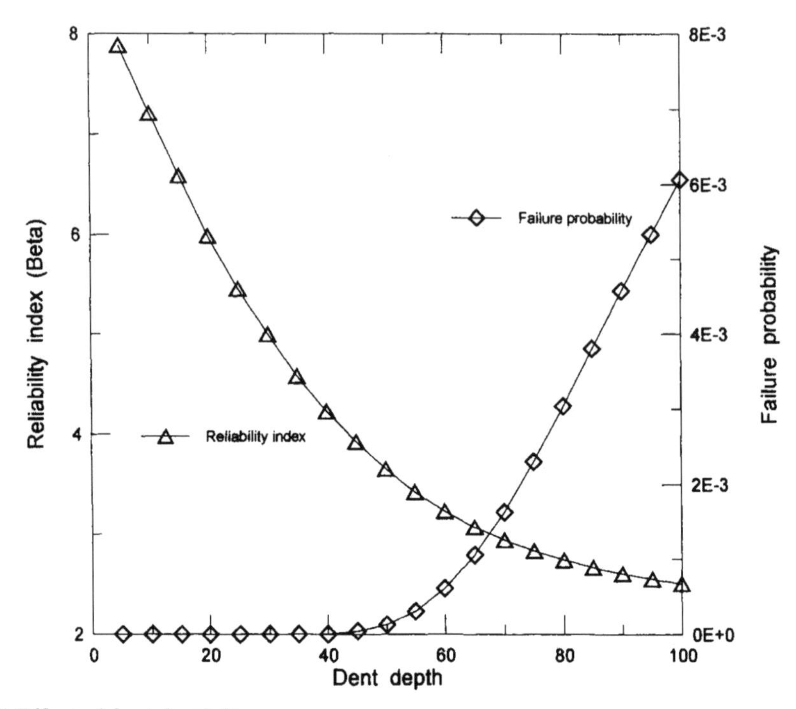

Figure 39. 1 Effect of dent depth D_d.

Figure 39.2 gives the results of the changes of failure probability and reliability index versus dent depth to wall-thickness ratio D_d/t. It is interesting to note that this ratio is a key factor affecting pipe fracture strength, since the stress concentration in the bottom of the dent is proportional to the dent depth.

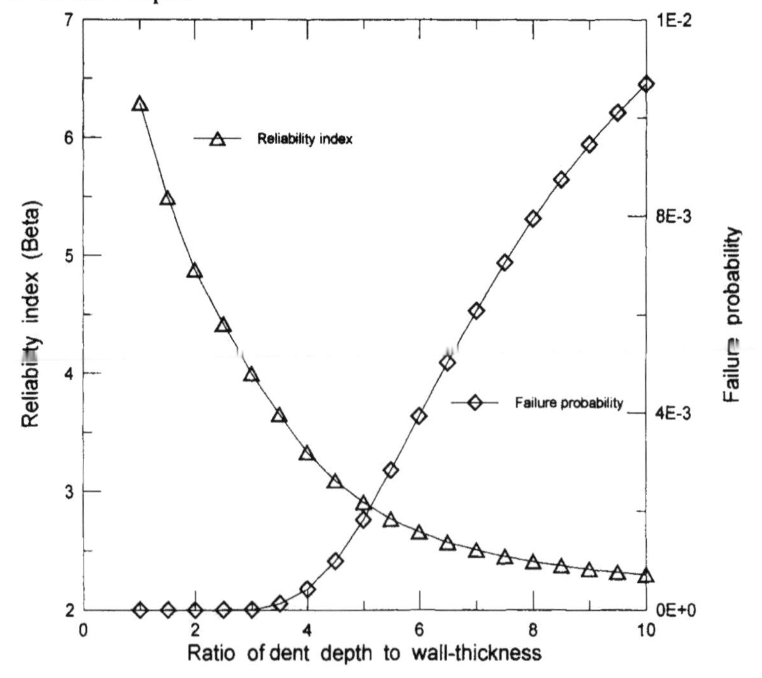

Figure 39.2 Effect of dent depth to thickness (D_d/t).

Parametric study results of dent depth to outside diameter D_d/D is shown in Figure 39.3, from which it is observed that failure probability increases rapidly when the ratio of D_d/D exceeds a certain value, say 4%. Care should be taken for the case of large D_d/D.

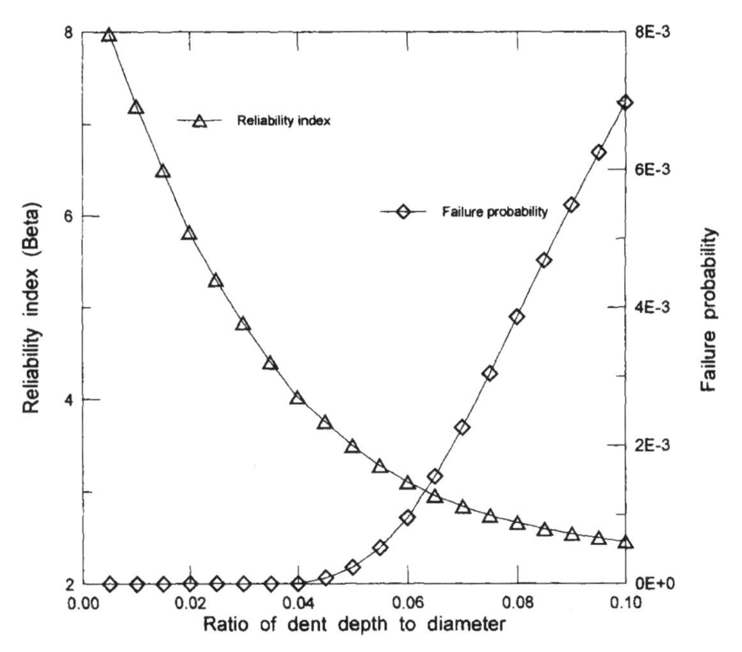

Figure 39.3 Effect of dent depth to diameter ratio (D_d/D).

The effect of crack depth to pipe wall-thickness ratio, a/t, on fracture reliability is studied and shown in Figure 39.4. From which it is observed that the ratio a/t is quite influential to fracture reliability. As the crack depth increase, the reliability decreases rapidly.

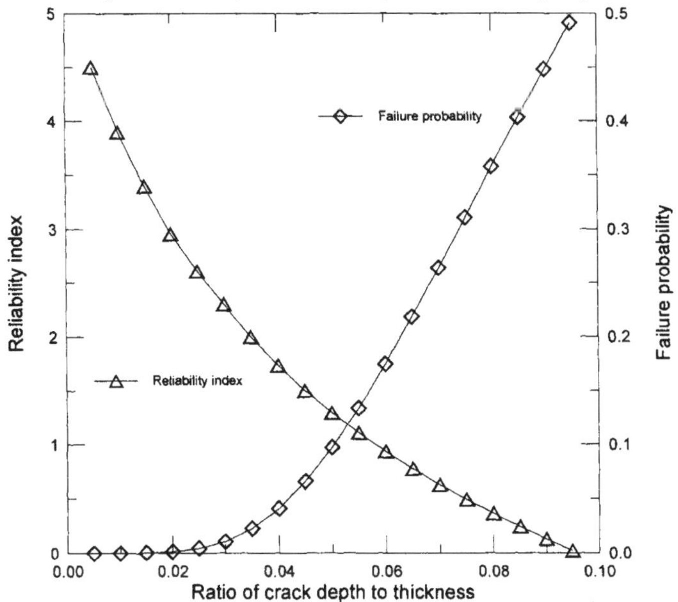

Figure 39.4 Effect of crack depth to thickness ratio (a/t).

39.6.4 Sensitivity Study

From Figure 39.1, it is seen that some dominating factors are very influential to the reliability index. Their effect on different target safety levels are studied and the results are shown in Table 39.3. Besides those parameters discussed above, other major parametric study results are listed in this table, in which the variation of safety factor are set to γ=1.6~2.2 and the investigation is performed based on the basic input parameters given in Table 39.3. The different parameter between investigated case and basic case is indicated in the table with given distribution type, mean and COV. A clearer picture about the parametric studies can be obtained from Table 39.3. It is important to note from Table 39.3 that crack depth, a, is very influential to reliability index. In the practical engineering, crack depth varies from case to case due to the measurability of the pressure vessels. For different crack size, there is a corresponding calibrated safety factor. Also, log-normal distribution may be applied to fit crack size (Kirkemo (1988)). In this case, it is noted from the comparison in Table 39.3 that the reliability index increases a great deal. So that it is essential to choose a suitable crack depth based on a practical considered case in order to have a rational results. It is observed from Table 39.3 that estimated reliability index is very sensitive to model uncertainty. In the interpretation of this result, it is important to be aware of that the results depend heavily on the chosen uncertainty model. Even a small change of X_M will lead to a big change in reliability index. So that, further study including tests and additional information from inspection is needed. It is also noted from this table that the uncertainty of pipe wall-thickness is also quite influential to reliability index. This is just as expected since wall-thickness is an important design parameter of pipes.

Table 39.3 Parameter studies.

Parametric studies		γ=1.6		γ=1.8		γ=2.0		γ=2.2	
		β	P_F	β	P_F	β	P_F	β	P_F
X_M	N(0.92, 0.11)	3.048	.115E-02	3.516	.219E-03	3.926	.432E-04	4.293	.882E-05
	N(1.0, 0.1)	3.298	.487E-03	3.768	.824E-04	4.183	.144E-04	4.557	.260E-05
	N(1.0, 0.2)	2.140	.162E-01	2.457	.700E-2	2.712	.334E-02	2.921	.174E-02
a	EXP(0.1)	3.048	.115E-02	3.516	.219E-03	3.926	.432E-04	4.293	.882E-05
	EXP(0.18)	2.436	.742E-02	2.861	.211E-02	3.241	.595E-03	3.588	.167E-03
	LN(0.09,1.0)	3.664	.124E-03	4.297	.868E-05	4.812	.747E-06	5.240	.806E-07
X_d	N(0.9, 0.05)	3.048	.115E-02	3.516	.219E-03	3.926	.432E-04	4.293	.882E-05
	N(0.9, 0.10)	2.990	.139E-02	3.440	.291E-03	3.833	.632E-04	4.186	.142E-04
	N(0.9, 0.15)	2.909	.185E-02	3.334	.428E-03	3.708	.105E-03	4.043	.264E-04
H_t	N(1.04, 0.02)	3.048	.115E-02	3.516	.219E-03	3.926	.432E-04	4.293	.882E-05
	N(1.04, 0.05)	2.910	.181E-02	3.355	.397E-03	3.742	.912E-04	4.088	.218E-04
	N(1.04, 0.10)	2.508	.608E-02	2.881	.198E-02	3.197	.694E-03	3.474	.257E-03
X_Y	LN(1.0, 0.10)	3.048	.115E-02	3.516	.219E-03	3.926	.432E-04	4.293	.882E-05
	LN(1.0, 0.20)	2.808	.249E-02	3.197	.695E-03	3.536	.204E-03	3.838	.621E-04
	LN(1.0, 0.30)	2.624	.435E-02	2.991	.139E-02	3.317	.454E-03	3.613	.151E-03
Cv	LN(63.0, 0.1)	3.048	.115E-02	3.516	.219E-03	3.926	.432E-04	4.293	.882E-05
	LN(63.0, 0.2)	2.843	.223E-02	3.251	.574E-03	3.605	.156E-03	3.917	.448E-04
	LN(63.0, 0.3)	2.530	.570E-02	2.845	.222E-02	3.102	.962E-03	3.313	.462E-03

Note: Distribution types used in the table include: N-Normal, LN-Log-normal, EXP-Exponential.

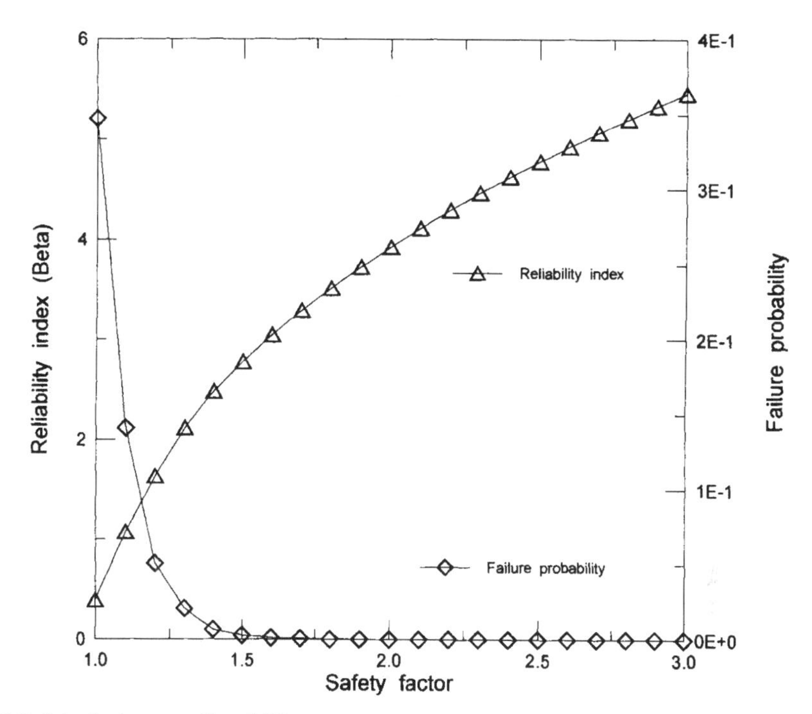

Figure 39. 5 Safety factor γ vs. β and Pf.

39.7 References

1. API: 5L Specifications, American Petroleum Institutes, (1993).
2. Bai. Y. and Song, R., (1997) "Fracture Assessment of Dented Pipes with Cracks and Reliability-based Calibration of Safety Factors", Int. Jour. Pressure Vessels and Piping, Vol. 74, (1997), pp. 221-229.
3. Bilby B.A., Cottrell A.H. and Swinden K.H., The spread of plastic yield from a notch, Proc. Roy. Soc. (A272), (1963) 304.
4. BSI: PD6493, Guidance on methods of assessing the acceptability of flaws in fusion welded structures, British Standards Institute, (1991).
5. Heald, P.T. et al., (1971) "Fracture initiation toughness measurement methods", Mat. Sci. and Eng., 10, 129.
6. Kiefner, J.F. and Vieth, P.H., "A modified criterion for evaluation the remaining strength of corroded pipe", RSTRENG, Project PR 3-805 Pipeline Research Committee, American Gas Association, Dec. 22, 1989.
7. Kirkemo, F., (1988) "Application of probabilistic fracture mechanics of offshore structures", Prof. of OMAE, Houston, USA.
8. Maxey, W.A., et al., (1972) "Ductile fracture initiation, propagation, and arrest in cylindrical pressure vessels", ASTM STP 514.
9. Newman, J.C. and Raju, I.S., "An empirical stress-intensity factor equation for the surface crack", Engineering Fracture Mechanics, 15 (1-2), (1981) 85-191.

10. PROBAN, (1996) General purpose probabilistic analysis program, DNV.
11. Rolfe, S.T. and Novak, S.T., (1970) "Slow bend KIC testing of medium strength high toughness steels", ASTM STP 463, American Society of Testing and Materials, Philadelphia.
12. Shannon, R.W., (1973) "The mechanics of low stress failure which occur as a result of severe mechanical interference - a preliminary hypothesis", ERS R.571.
13. STRUREL, (1996) A structural reliability analysis program system, users manual, RCP Consult, Munchen, Germany.

Part VI

Integrity Management

Chapter 40 Integrity Management of Subsea Systems

40.1 Introduction

40.1.1 General

In recent years risk analysis has become increasingly recognized as an effective tool for the management of safety, environmental pollution and financial risks in the pipeline industry. The purpose of this Chapter is to apply risk-based inspection planning methodologies to pipeline systems, by developing a set of methods and tools for the estimation of risks using structural reliability approach and incidental databases, and to illustrate our risk based inspection and management approach through a few examples.

After outlining the constituent steps of a complete risk analysis methodology, it is intended to give detailed information about each step of the methodology such that a complete risk analysis can be achieved (Sørheim and Bai, 1999). Willcocks and Bai (2000) gave a detailed guidance on evaluation of failure frequency, consequence, risk and risk-based inspection and integrity management of pipeline systems.

40.1.2 Risk Analysis Objectives

The objectives of risk analysis are:

- To identify and assess in terms of likelihood and consequence all reasonably expected hazards to Health, Safety and the Environment in the design, construction and installation of a pipeline;
- To ensure adherence to the appropriate international, national and organizational acceptance criteria.

40.1.3 Risk Analysis Concepts

After completing an investigation of initiating events, cause-analysis should then follow; the final stage would be an analysis of consequences. An outline of the methodology is given in Figure 40.1.

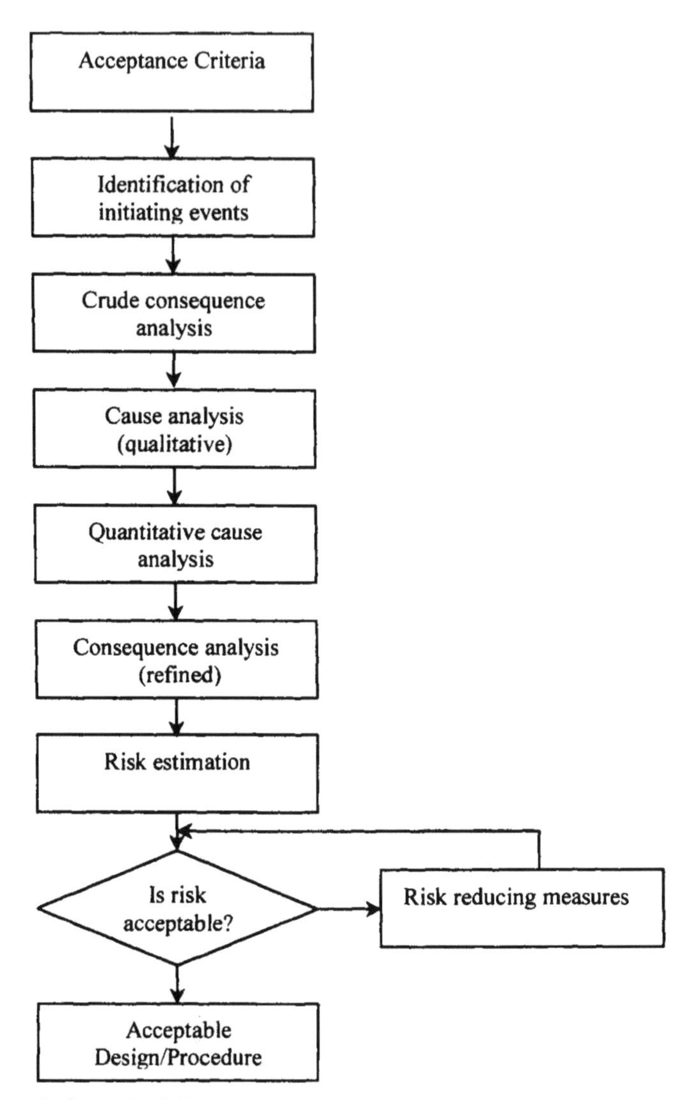

Figure 40.1 Risk analysis methodology.

40.1.4 Risk Based Inspection and Integrity Management (RBIM)

This is a means of focusing and optimizing the use of resources to 'high risk' areas in order to minimize costs, ensure effective and efficient asset management by ensuring the required confidence in the assets integrity and availability. It is employed in pipeline systems due to the high costs for a pipeline modeling, inspection and maintenance, but credible risks of failure.

RBIM is essentially the determination of required Structural Reliability Analysis (SRA), inspections (type, frequency/time and extent), maintenance tasks (e.g. repair to pipeline intervention, coatings, corrosion inhibitor etc.) to maintain the risk of failure via credible/potentially high risk modes below an 'acceptable level'. The establishment of failure patterns and failure rates (FCA) which also identifies failure warnings.

These together define the failure risks. Depending on the level of risk of failure for each mode and pattern of failure the required analysis, inspections, maintenance and repair tasks are selected. For example a review of historical failure databases e.g. PARLOC'00 indicates that the major failure modes are internal corrosion and external impact. Thus the main efforts (in terms of design, structural modeling, inspections etc) should be focused on these failure modes.

40.2 Acceptance Criteria

40.2.1 General

The acceptance criteria are distinctive, normative formulations against which the risk estimation can be compared. Most regulatory bodies give acceptance criteria either qualitatively or quantitatively. The NPD regulation states the following:

- In order to avoid or withstand accidental events, the operator shall define safety objectives to manage the activities.
- The operator shall define acceptance criteria before risk analysis is carried out.
- Risk analysis shall be carried out in order to identify the accidental events that may occur in the activities and the consequences of such accidental events for people, for the environment and for assets and financial interest.
- Probability reducing measures shall, to the extent possible be given priority over consequence reducing measures.
- Subsea pipeline systems shall be to a reasonable extent, be protected to prevent mechanical damage to the pipeline due to other activities along the route, including fishing and shipping activities.

Individual corporations may choose to implement internal acceptance criteria. These acceptance criteria may be based on the relative cost between implementing a risk reducing measure and the potential loss. Also many projects specify a pipeline availability requirement. Thus total losses must be such to ensure required availability.

If the risk estimation arrived at is not within the acceptable risk, then it is necessary to implement alterations. This new system should then be analyzed and compared with the risk acceptance to ensure adequate risk levels. This is an iterative process, which will eventually lead to a system/ design, which is acceptable.

40.2.2 Risk of Individuals

The FAR (Fatal Accident Rate) associated with post commissioning activities (the installation and retrieval of pigging equipment) has been evaluated. The FAR acceptance criteria are defined to be 10 fatalities per 10^8 working hours. The maximum FAR (Fatal Accident rate, No. of fatal accidents per 10^8 hours worked) for the operational phase should be ≤ 10. The maximum FAR for the installation phase should be ≤ 20.

40.2.3 Societal Risk

The society risk is 3rd Party (Societal) Risks posed to passing fishing vessels and merchant shipping. Acceptance of 3rd party risks posed by pipeline should be on the basis of the F-N curves shown in Figure 40.2 below.

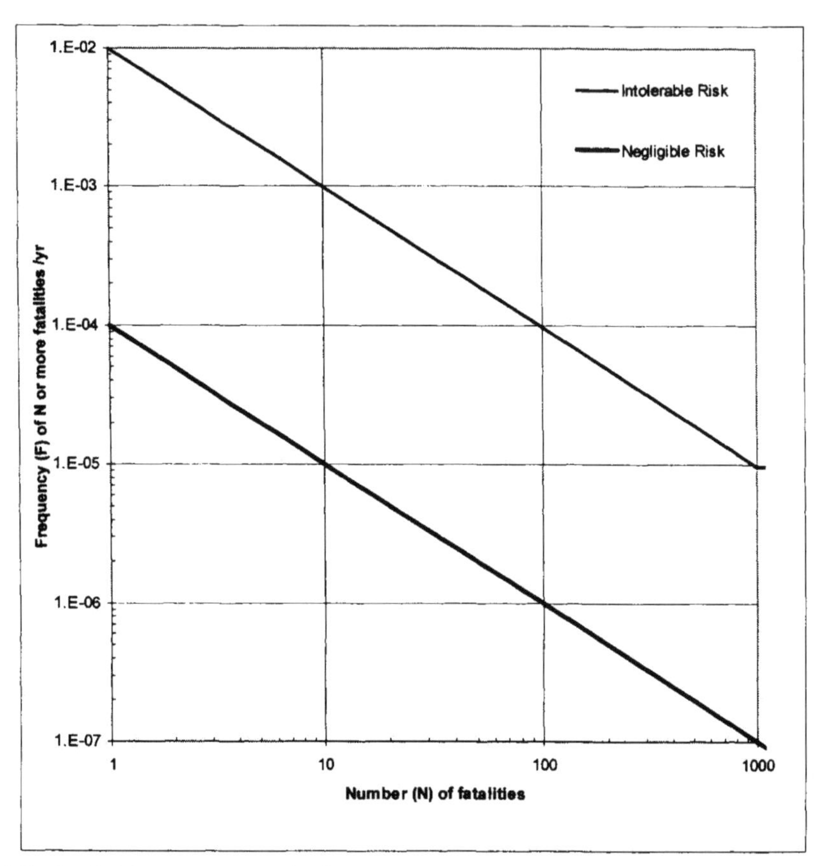

Figure 40.2 Societal risk acceptance criteria.

40.2.4 Environmental Risk

All incidents considered as initiating in the assessment of individual and societal risks during the operational phase are considered to be initiating for the purposes of determining the Environmental Risks. Loss of containment incidents during operation of pipeline will have minor local environmental effects. The environmental consequences of loss of containment incidents are therefore classified as being Category 1 (Table 40.1), i.e. the recovery period will be less than 1 year.

In addition any incidents having the potential to result in the release of corrosion inhibitors during commissioning of the pipeline are considered to be initiating with respect to Environmental Risks. Acceptance of the environmental risks associated with the construction and operation is normally based on the operator's criteria, which is established based on economical and political considerations.

Table 40.1 Acceptance criteria for environmental risk.

Category	Recovery period	Operational phase probability per year	Installation phase probability per operation
1	< 1 year	$< 1 \times 10^{-2}$	$< 1 \times 10^{-3}$
2	< 3 years	$< 2.5 \times 10^{-3}$	$< 2.5 \times 10^{-4}$
3	< 10 years	$< 1 \times 10^{-3}$	$< 1 \times 10^{-4}$
4	> 10 years	$< 5 \times 10^{-4}$	$< 5 \times 10^{-5}$

Causes of Loss of Containment incidents considered during the operational phase are:

- External impact (sinking vessels, dropped objects, trawl impact);
- Corrosion (external and internal);
- Material defect.

40.2.5 Financial Risks

All incidents considered as initiating in the assessment of individual and societal risks are considered to be initiating for the purposes of determining the Risks of Material Loss. In addition any incidents occurring during construction and installation and having the potential to result in damage to and/or delay in the construction of the pipeline are considered to be initiating with respect to Risks of Material Loss. The costs of incidents have been considered as being made up from:

- notional cost of fatalities;
- cost of repair;
- cost of deferred production.

The expected (average) number of loss of containment incidents and associated fatalities have been used to derive an expected annual cost incorporating each of the quantities given above. The acceptability of risks of material loss will be determined using cost benefit analysis. Risk reduction measures should be implemented if cost benefit analysis shows a net benefit over the full life cycle.

To summarize, the acceptance criteria shall be based upon a cost benefit evaluation, where the expected benefits must be much greater than the costs of implementing and operating with the risk reducing measure, i.e.:

$$C_{IMPL} + C_{OP} << C_{RED} \quad (40.1)$$

where:

C_{IMPL} : cost of implementing the risk reducing measure.

C_{OP} : net present value of operational cost related to the measure.

C_{RED} : net present value of expected benefits as a result of the risk reducing measure.

40.3 Identification of Initiating Events

Identification of initial events is regularly referred to as hazard identification, in the offshore industry. The main techniques that exist are:

- Check Lists- Review of possible accidents using lists which are developed by experts
- Accident and Failure Statistics- Similar to the checklists but are derived from failure events.
- Hazard and Operability Study- Used to detect sequences of failures and conditions that may exist in order to cause an initiating event.
- Comparison with detailed studies- Use of studies, which broadly match the situation being studied.

After the completion of this investigation it is necessary to examine the hazards and identify the significant hazards, which need to be analyzed further.

40.4 Cause Analysis

40.4.1 General

There are two purposes of cause analysis; firstly, it is necessary for the identification of the combinations of events that may lead to initiating events. Secondly, it is the assessment of the probability of the initiating event occurring. The first one is a qualitative assessment of the system and the latter is quantitative.

The qualitative analyses aim to; detect all causes and conditions that could result in an initiating event and develop the foundation for possible quantitative analysis. The aim of the quantitative analyses is to determine a probability value for the occurrence of an initiating event. The analysis tools that are available are stated below. This chapter will discuss only the first two approaches.

- Fault Tree Analysis;
- Event Tree Analysis;
- Synthesis Models;
- Monte Carlo Simulations;
- Equipment Failure Rate Databases.

40.4.2 Fault Tree Analysis

The fault tree is a graphical diagram of logical connections between events and conditions, which must be present if an initiating event should occur. A fault tree for a system can be regarded as a model showing how the system may fail or a model showing the system in an unwanted situation. The qualitative analysis maps systematically all possible combinations of causes for a defined unwanted event in the system. If available data can be supplied for the frequencies of the different failure causes, quantitative analysis may be performed. The quantitative analysis may give numerical estimates of the time between each time the unwanted event occurs, the probability of the event etc.

The Fault Tree Analysis (FTA) has three major phases:

1. Construction of the Fault Tree: this is the identification of combinations of failures and circumstances that may cause failures or accidents to occur.
2. Evaluation of the Fault Tree: this is the identification of particular sets of causes that separately will cause system failure or accident.
3. Quantification of the Fault Tree: this is overall failure probability assessment from the sets of causes as defined above.

40.4.3 Event Tree Analysis

An event tree is a visual model for description of possible event chains, which may develop from a hazardous situation. Top events are defined and associated probabilities of occurrence are estimated. Possible outcomes from the event are determined by a list of questions where each question is answered yes or no. The questions will often correspond to safety barriers in a system such as "isolation failed?" and the method reflects the designers' way of thinking. The events are partitioned for each question, and a probability is given for each branching point. The end events (terminal events) can be gathered in groups according to their consequence to give a risk picture.

40.5 Probability of Initiating Events

40.5.1 General

The methods stated above gives a methodology, which can be applied to any scenario such that it is possible to determine the conditions, which will result in an initiating event. However, it is necessary to determine how the probability value is to be assigned, when using the FTA and ETA.

Reliability analysis is used as the main method of determining the probability of failure caused by physical aspects of a pipeline i.e. corrosion, trawling impact, vortex-induced-vibrations etc.

Failure events that are not caused by physical failure of the pipeline may not be compatible with the reliability method of analysis; an example of this is the probability of human error. This type of failure requires deeper analysis using techniques such as historical data analysis or using comparable circumstances from other industries.

40.5.2 HOE Frequency

Human/organization error (HOE) probability is an area of pipeline risk analysis that is rarely quantified with reasonable accuracy, this is primarily due to physical and mental distance placed between individuals designing, constructing and operating the pipeline. A justifiable basis for a risk evaluation can be established by implementing an assessment of HOE. The purpose of a HOE evaluation is not to predict failure events, rather it is to identify the potentially critical flaws. The limitation of this is that one cannot analyze what one cannot predict.

There is little definitive information on the rates and effects of human errors and their interactions with organizations, environments, hardware and software. There is even less definitive information on how contributing factors influence the rates of human errors.

Lack of dependable quantitative data that is currently available on HOE in design and construction of pipeline structures can be compensated for using the following four primary sources of information, presented in work by Bea (1994).

1. Use of judgment based on expert evaluations;
2. Simulations of conditions in a laboratory, office or on sites;
3. Sampling general conditions that exist on site, laboratory and office;
4. Process reviews, accident and near miss databases.

Considering the quantity of conclusive data, which is available, the principle mode by which to quantify assessments is judgment method. As investigations into pipeline failures should eventually lead to comprehensive and reliable databases of HOE, these databases will compliment judgments and allow a more justifiable quantification to be arrived at.

It is necessary that any results that are deemed to be meaningful are qualified and unbiased. Investigations by Bea (1994) gives a number of biases that can distort the actual causes of HOE, these are listed in Table 40.2. It is important for the evaluator to try to minimize these biases, as it is impossible for them to be eliminated entirely.

Table 40.2 Influence on bias (Bea, 1994).

Type of Bias	Influence on Judgment
Availability	Probability of easily recalled events are distorted
Selective perception	Expectations distort observations of variables relevant to strategy
Illusory correlation	Encourages the belief that unrelated variables are correlated
Conservatism	Failure to sufficiently revise forecasts based on new information
Small samples	Over estimation of the degree to which small samples are representative of a population
Wishful thinking	Probability of desired outcomes judged to be inappropriately high
Illusions of Control	Over estimation of the personal control over outcomes
Logical construction	Logical construction of events which cannot be accurately controlled
Hindsight	Over estimation of the predictability of past events

Following research by Williams (1988), Swain and Guttman (1981) and Edmondson (1993), quantified data for HOE has been developed. This is based on experience gained in the nuclear power industry in the U.S.A. Experiments and simulations led to information regarding human task reliability.

Work undertaken by Swain and Guttman (1981) presents general error rates depending on the familiarity of the task being undertaken by the individual, included is a range of limitations or circumstances that the individual may be experiencing, this is shown in Figure 40.3. By assessing the intensity of these limitations or circumstances it is possible to adjust the value assigned to certain tasks. Other investigations (Williams, 1988) appear to correlate with this information. However, a multitude of influences impact upon these values and have potentially dramatic effects on the normal rates of errors (i.e. factors of 1E-3 or more). These influences include organizations, procedures, environments, hardware and interfaces. Information regarding these influences can be found in Bea (1994) and others.

It is important to establish the significance of any error that may occur as this is not established in the information developed. An error can be either major/significant or minor/not significant. Studies performed by Swain and Guttman (1981) and Dougherty and Frangola (1988) indicates that minor or not significant errors are often noticed and rectified, thus reducing their importance in human reliability. Further quantification of human reliability has been corroborated for a number of tasks relating specifically to structural design; the necessary information is investigated by Bea (1994).

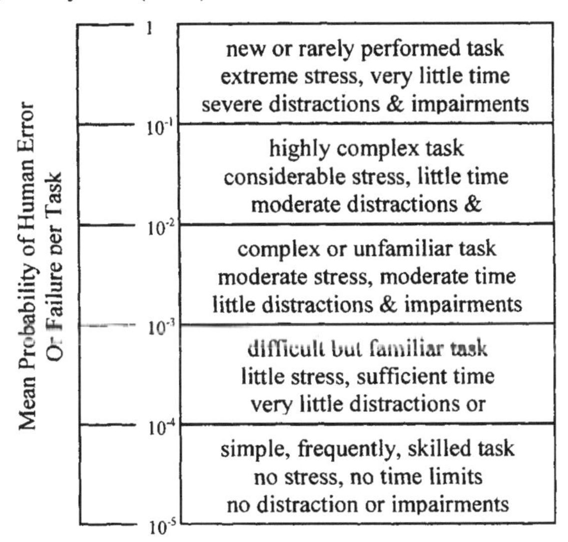

Figure 40.3 Human error rates.

40.6 Causes of Risks

40.6.1 General

This section will outline some common causes for the four different risk scenarios that were outlined in the introduction.

40.6.2 1st Party Individual Risk

The scope of this type of risk is limited to a consideration of the potential for ignited releases as a result of dropped object impact associated with maintenance/workover activities taking place after commissioning or random failure of the pipeline (discussed in next section).

The sources of the potential dropped objects are assumed to be the vessels employed for maintenance/workover. The assumptions made in order to determine the probability of loss of containment is as follows:

- Objects are assumed to fall in a 30° cone centered at a point directly above the pipeline;
- Objects are assumed to fall with equal probability at any point within the circle on the seabed defined by the drop cone. It is assumed that all dropped objects enter the sea, rather than landing on part of the vessel.
- The probability that the hazard zone, resulting from a loss of containment, coincides with the dropping vessel, is assumed to be 0.5.

Details of such operations are unlikely to be known during design, thus judgments are often required (based on previous experience) and the analysis updated later. During design this analysis necessary since decisions about protective requirements need to be considered.

40.6.3 Societal, Environmental and Material Loss Risk

Risks associated with construction, installation and commissioning of the pipeline do not impact on members of the general public. Only incidents that occur during the operation of the pipeline are therefore considered to be initiating with respect to Societal Risk.

The hazards giving rise to societal risks will also contribute to the environmental and material loss risks. These hazards include the following:

1. Fishing Interaction

Movement of fishing vessels around the location of subsea pipelines pose a risk. The frequency of such an event can be derived from existing databases (PARLOC).

2. Merchant Vessels

Incidents caused by passing merchant ships include emergency anchoring, dropped containers and sinking ships. Databases can again be used to determine the density of merchant vessels and the probability of the above incidents occurring.

3. Construction Vessels

Loss of containment incident frequencies as a result of construction vessel activities may be estimated based on databases. However, while it is accepted that construction activities contribute to the overall loss of containment frequency for pipelines, it is not considered to be appropriate to treat such incidents as initiating for Societal Risk calculations. This is because the presence of construction vessels will in itself exclude the presence of merchant shipping.

4. Random Failures

This may be due to any material failure of the pipeline and can usually be determined using reliability analysis.

40.7 Failure Probability Estimation Based on Qualitative Review and Databases

A risk review 'largely at a qualitative level' of the pipeline segments to generic failure and degradation modes is performed to establish the failure modes that may pose a threat at various locations along the pipeline. The risk review is based on:

- Generic/historical failure rates from relevant pipeline databases;
- Pipeline design and history of operating conditions;
- Known condition and incidents affecting pipeline and basic/high level structural damage and probability predictions.

The review is incorporated within a spreadsheet, in which as much design and operational data as possible regarding the pipeline is first input. Where relevant the pipeline is divided into suitable segments e.g. where other components in the line exist such as Tees, Risers, Riser bases, spools etc., crossing of shipping lanes, trawling areas and inshore areas. In this way the specific failure modes relevant to a particular location or item/component can be established.

A high level (basic) structural reliability assessment considering the specific failure mode is made, mainly considering the normal/accepted uncertainties. Failure predictions based on generic/historical failure data are also made. These represent the accidental cases where the pipeline will have been subject to conditions outside its design conditions or required design conditions were not achieved. The occurrence of accidental events are generally random, often result in immediate failure or within a very short time period such that inspecting the pipeline for accidental conditions/damage generally provides no benefit. The main means in dealing with accidental/unplanned conditions is to eliminate or reduce their likelihood to acceptable levels. Thus the identification of potential accidental events and their elimination is critical to the effective risk management of pipeline systems.

<u>Generic Hazard/Pipeline damage list</u>

Extreme Environmental Loads

- Earthquakes
- Severe wave and current loading
- Seabed movement & instability

Process Deviations

- Over-pressure
- Under-pressure

- Over & Under temperature
- Process upset – Offspec product into line

Excessive Internal Corrosion

Excessive Internal Erosion

Excessive External Corrosion

External Interference

Commercial Marine Traffic

- Dropped anchors
- Dragged anchors
- Sinking vessels
- Grounding vessels

Fishing/Trawling

- Impact loading
- Pull over loads
- Hooking
- Trawl pull-over combined with thermal buckling

Munitions

Falling/Rolling Boulders

Example of Risk Review - Considering Internal Corrosion

A sample of such a review is presented here considering internal corrosion of a new pipeline transporting normally dry gas but containing CO_2. The gas is dried to a high quality prior to export by a glycol drier. Thus the potential for internal corrosion (and hence failure) from CO_2 in wet service exists, it cannot be readily discounted and further assessment is required.

The general issues are; what are the required reliabilities of the gas drier, gas monitoring and process upset detection requirements, drying of the line after an upset, extent of corrosion allowance, if any and how often the line should be cleaned and inspected.

Part of the input data to the workbook is the product mass balance; the user is to ensure that all potential corrosive products are entered. For the case in question the cause of internal corrosion is CO_2 in wet service as a result of a) minute water content during normal production, b) process upset c) accidental operation (e.g. accidental water ingress into line during subsea pigging operations).

In this case the corrosion rates for each condition are calculated based on de Waard et.al. '93 (which is included in the workbook) and expected duration of wet service. For the process upset and accidental cases, estimates of the duration of wet service need to be established. These are based on:

- Probability/frequency of process upsets (based on driers reliability);
- Probability of detection (are there alarms, process trips on drier, monitoring of gas quality etc.);
- Probability of drying line within certain period after incident;
- Probability of accidental ingress (based on historical failure probability of gas pipelines from internal corrosion (PARLOC '96).

The uncertainties (potential variation) in these estimates are also input, high estimates are generally used. A basic conservative estimate of the pipelines structural reliability over time is then made based on the predicted safe operating pressure (accounting for corrosion damage) according to Bai et al '97 (– *Strength/Resistance model*), design pressure and uncertainties in the above estimates.

With respect to the estimates of accidental corrosive/wet service, the frequency of such an occurrence is based on the historical corrosion failure rate from PARLOC'96. Accidental internal corrosion conditions can result in very significant corrosion rates, but should rarely occur. Thus accidental corrosive service is not included in the 'normal/accepted' estimate of yearly corrosion rate over the life of the pipeline.

The causes of potential accidental events and resulting extreme conditions to the line are to be identified and estimated as far as reasonably practical e.g. undetected water ingress during subsea pigging or other activities in which water or other/increased corrosive products may be introduced (undetected). Eliminating or reducing the likelihood of these events is the main means of managing this risk. The potential time to failure in the event of such accidental operation is also predicted. For the case in question it was considered that water ingress from subsea pigging was the only potential accidental condition. If it occurred the service limit state acceptance criteria would be exceeded relatively quickly (within the year), though actual pipeline failure would be expected to take a few years. Thus normal cleaning pigging on a yearly base should protect against failure from such an accidental event, though if it occurs significant corrosion damage is inevitable.

It was found that the corrosion was very dependent on the upset frequency and incidental duration such that reliable means of detection and limitation of incidental duration is required. With such means in place the assessment predicts negligible corrosion such that no corrosion allowance is recommended nor intelligent pigging operations. However it is considered that an inspection should be made on a medium term basis (e.g. every 3-5 years), particularly in the early phase of operation to verify the corrosion prediction models and ensure no damage during RFO. Only if it is 100% certain that no significant upsets and accidental operations has occurred should such an inspection be omitted. With such means in place the assessment predicts negligible corrosion such that no corrosion allowance is recommended nor intelligent pigging operations.

40.8 Failure Probability Estimation Based on Structural Reliability Methods

40.8.1 General

Where the failure mode is identified as being significant and/or more specific details of the structural damage (defect data) to the pipeline are known (i.e. for pipelines that have been in operation for a number of years) then a more detailed SRA is possible and justified. Such analyses is performed using simple SRA spreadsheet based tools, based on simplistic 'boot strap' probabilistic methods (API 2A-LRFD), that do not require propriety software. For such analysis the following base data is required:

(1) Measured defect data from survey

If such data is available, then it needs to be correlated into useful input data e.g. the nature of the defects (type of corrosion – pitting, grove, girth welds), mean defect depth, length and area, along with the variance/standard deviation of such parameters. Else nominal defects are assumed based (where possible) on experience from similar lines.

(2) Pipeline Material/Strength Properties

The pipeline SMYS, SMTS and flow Stress are required. The variance in these mean values need to be established and is generally obtainable from published and manufactures data.

(3) Pipeline geometric properties

The following geometric properties in terms of mean value and variance need to be developed for each relevant section of the pipeline:

- Nominal diameter, it is not expected that this will vary along pipeline
- Wall thickness, the design/mean wall thickness and variation along the pipeline length
- Ovality, again such factor may vary along the pipeline length – though a design limit will be specified.

(4) Pipeline Loading Characteristics

Pressure loading at relevant segments along the pipeline is to be defined in terms of mean operating pressure, standard deviation and variance. Extreme high pressures should be accounted for, based on the reliabilities of pressure regulating and protection systems.

Temperature profile, this is not a strict loading, but affects the corrosion rates and axial force in pipeline. Over operating life the temperature profile may vary and may need to be accounted for.

40.8.2 Simplified Calculations of Reliability Index and Failure Probability

The probability of failure is dependent on the likelihood of the loading exceeding the pipelines strength/resistance as illustrated in Figure 40.4.

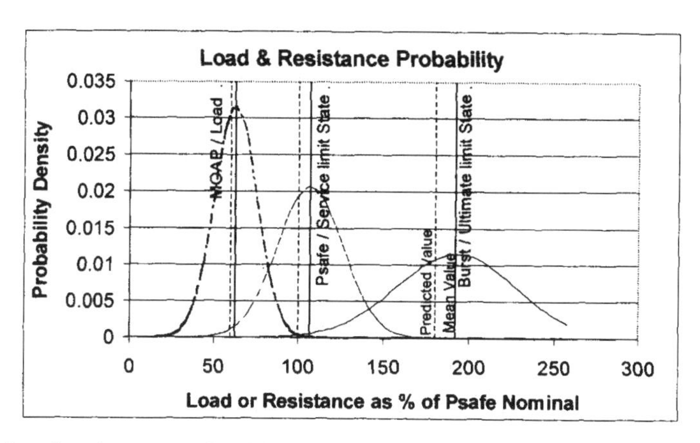

Figure 40.4. Load and resistance probability densities.

The Safety Index (β) is defined as (API 2A-LRFD):

$$\beta = \frac{\text{MeanSafetyMargin}}{\text{Uncertainty}} \equiv \frac{R_m - S_m}{\sigma_{RS}} = \frac{P_{safe.mean} - P_{op.mean}}{\sigma_{Psafe.meanPop.mean}} \qquad (40.2)$$

and the probability of failure P_f is calculated from:

$$Pf = 1 - \Phi(\beta)$$

By establishing and accounting for the main uncertainties in the:

- Pipelines nominal strength (load resistance against specific loading) dependent on the type of damage/degradation to the pipeline
- Pipelines nominal operating loads

The Reliability (Safety) Index (β) and probability of failure is calculated for a single defect as presented in the following text and illustration. These probabilities for all defects are then combined to give the safety index and failure probability of a pipeline segment and pipeline as a whole.

40.8.3 Strength/Resistance Models

An example model for the pipeline mean resistance (strength) against structural damage is presented here for damage from internal corrosion. The model was developed by Bai et al 1997, though many other models are available e.g. Shell '92 (D.Ritchie et al).

40.8.4 Evaluation of Strength Uncertainties

The uncertainty of pipeline strength is dependent on:

- Material strength uncertainty
- Defect measurement, detection and prediction uncertainty
- Pipeline parameter/geometry uncertainty.

- Strength model uncertainty

The uncertainties are measured in terms of standard deviation and variance from mean values and combine to give an uncertainty in the predicted pipeline safe operating pressure. The mean bias (B) and COV of the burst prediction model is (Bai et al 1997):

Model Bias, $B_M = \dfrac{P_{burst.actual}}{P_{burst.predicted}} \equiv 1.07$ with COV of 0.18.

and mean bias and variances of the equation parameters are:

$$B_{X_A} = \frac{X_{A.actua}}{X_{A.predicted}} \equiv 0.8 \text{ with a COV of } 0.08;$$

$$B_{X_f} = \frac{\sigma_{f.actual}}{\sigma_{f.predicted}} \equiv 1.14 \text{ with a COV of } 0.06;$$

$$B_{X_f} = \frac{X_{L.actual}}{X_{L.predicted}} \equiv 0.9 \text{ with a COV of } 0.05.$$

Multiplying the mean bias by the 'predicted value' gives the mean 'actual value': $B_{mean} \cdot X_{predicted} = X_{mean.actual}$

Thus the nominal/design $P_{safe.design}$ value is calculated using the predicted values. The mean P_{safe} value is calculated by substituting the measured or assumed $X_{mean.actual}$ values into the above equation and multiplying by the model bias B_M giving $P_{safe.mean}$.

$$P_{mean\ busrt\ value} = P_{mean\ safe\ value} \cdot \gamma$$

In the Bai et al '97 criterion P_{safe} is determined by applying a safety factor to the predicted burst pressure (Ultimate limit State), this is about 1.8 to give the desired/required reliability. P_{safe} is the Service Limit State.

P_{safe} is thus a factored value of P_{burst} to account for the P_{burst} bias and variance. The mean value and variance of P_{safe} is thus only dependent on the corrosion parameter predictions (length and depth) and not the model bias and variance, as the latter is accounted for by the safety factor.

40.9 Consequence Analysis

40.9.1 Consequence Modeling

The consequence model attempts to model the sequence of events that occur after a failure event. The sequence for consequence Modeling is shown in Figure 40.5. It should be noted that this method of consequence Modeling is only suitable for failures relating to the pipeline releasing some type of fluid or gas. The following steps for the Modeling of a release event gives only a general outline of the sequence of events that ultimately leads to a calculation of the various losses. Many different models exist for modeling these release characteristics (from simple to sophisticated/complex). However, there has not been extensive research/experimentation into Modeling of subsea releases so generally there is a high degree of uncertainty in this Modeling and conservatism is often used. One specific suite of computer Modeling programs available is the HGSystem written by Thornton Research Center.

Discharge

In order to determine dispersion, information is required for the discharge, this includes; hole size, duration, rate and quantity.

Dispersion of Gas

Leakage of a gas pipeline under water will result in a plume, which rises and exits from the surface of the water in the shape of a circle.

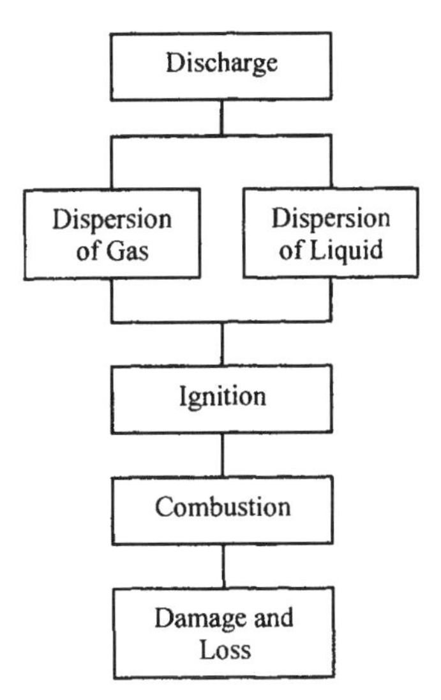

Figure 40.5 Modeling of consequence.

Dispersion of Liquid

The dispersion is dependent on the fluid released. Unstable condensate tends to be modeled as gas release (though a sound qualitative discussion about hydrate formation in water is required). Stable condensates will eventually rise to the surface to form a liquid pool at the surface. However, much of the dispersion is very complex and difficult to model.

Ignition

A leakage which does not ignite (i.e. not toxic, H_2S) will not present a risk to humans. A risk of ignition is developed using the following equation:

$$f_{fire}=f_{leakage} \times p_{ignition} \text{ (per year)} \tag{40.3}$$

$p_{ignition}$ is probability of ignition occurring, given a leak of a flammable substance. This can be determined using an ignition model, which considers all possible methods by which ignition could take place.

Subsea releases can usually be considered to be delayed hence, ignition will result in an explosion or flash fire (few unconfined flammable gas clouds will develop into an explosion) for gas leakage. Fire pool could arise from an oil leak. However, in the case of a shallow water release a low momentum jet fire may develop if ignition occurs before a significant cloud can develop. Such an ignition will result in a jet flame.

Combustion

Jet fire-There are a number models establishing jetfire characteristics e.g. Shell Thornton. A jetfire is characterized by flame length and radiated heat flux.

Pool fire- the height of the flame is highly dependent on the depth of the slick, the rate of combustion of the liquid and the wind speed.

Explosion-clouds of flammable gas can explode when ignited this is termed an unconfined vapour cloud explosion. (UVCE). This type of explosion is relatively mild, and has two effects; heat and force. The force effects can be modeled using the multi-energy method. For humans exposed to an explosion heat is the critical factor in determining bodily harm. Force can also act indirectly on persons exposed to the explosion, injury or death can result from flying debris or glass splinters. For structures it is the effect of force, which is critical.

Damage and Loss

It is also necessary to model the potential damage and loss that can occur to the following (Olshausen, 1998):

1. Humans
   - Heat from explosions or fires
     - The injury is dependent on the dose, which is D= time × $(kW/m^2)^{4/3}$
     - 50% death rate is likely when exposed to D_{50}= 2000 sec × (kW/m^2)
   - Force/missiles from explosions
     - There is a 50% chance of lung injury at 1.4 barg
     - There is a 50% chance of perforated eardrum at 0.5 barg
   - Toxic effects
     - For a majority of substances the D_{50} dose is known, that is a product of the time exposed and the $(\text{concentration})^n$ which results in a 50% likelihood of death.
2. Material loss
   - Repair of pipeline
   - Loss of Production
     - This is cost of lost income due to incapacity to provide a product to sell, this is a function of the time it takes to restore the pipeline to a functioning state.
3. Environmental damage

Uncertainty

All of the models in the sequence of analysis contain a significant degree of uncertainty. If taking a pessimistic approach and use factors of safety in the magnitude of 1.5 for each stage of calculation this will result in a total factor of safety of (1.5^4=) 5. This might be an unrealistic overestimate of the total value so it is necessary to adjust this figure to suit the situation.

Another difficulty with the consequence Modeling technique is that it is necessary to assume an initial discharge condition (i.e. the size of hole). This has a large influence over the models used, for a more comprehensive analysis a sample of likely release conditions could be evaluated. However, generalizations can be made regarding hole size based on failure rate data and type of failure, e.g. corrosion is likely to lead to small/pin pricks, where as third party interference tends to cause large diameter holes.

40.9.2 Estimation of Failure Consequence

The consequences of failure are:

- Consequential production losses
- Contract penalties (these can be extremely severe)
- Cost of repairing the pipeline
- Cost of repairing any damage to adjacent installations and environment
- Potential fatalities
- Cost of negative publicity

The potential consequences are very dependent on the operating pressure, pipeline length, diameter and content and size of the failure/release. The latter has been based on historical failure rates, for Subsea North Sea pipelines presented in PARLOC'96.

Potential fatalities, damage to adjacent installations and environment are assessed using standard consequence assessment methods incorporated within a spreadsheet suite of tools based on numerous published methods. The potential development of loss of containment is illustrated in the event tree, Figure 40.6. Consequence analysis techniques are generally well established within the Oil and Gas industry, though in certain areas better models are still required. The specific consequence models used for a subsea gas release are; pipeline time dependent gas release primarily based on Fannelop *et. al* '81, though other models are also used for comparison and Subsea plume modeling primarily based the methods for dispersion of Subsea Release reviewed by Rew, P.J. *et. al* '95 and hydrodynamics of underwater blowouts (Fanneløp, T.K. *et.al* '80). The potential surface gas cloud size and dispersion extent is modeled based on the methods reviewed by Rew, P.J. *et al* '95 and using HGSYSTEM suite. The potential explosion, flash fire extent and effects at the sea surface are calculated based on methods presented by AIChemE.

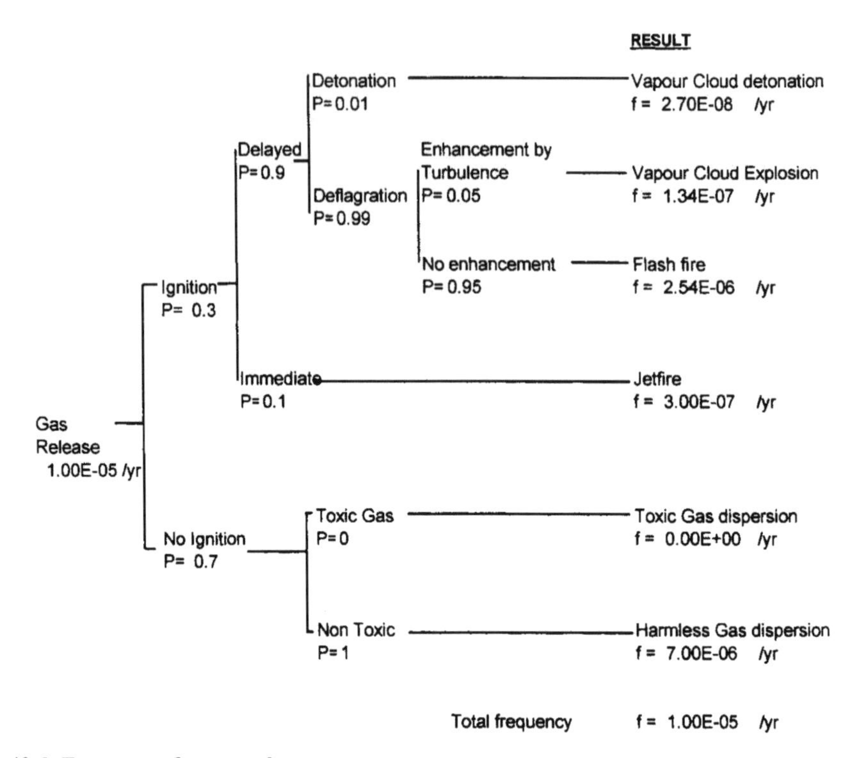

Figure 40.6. Event tree for gas release.

The calculated risks are compared against acceptance criteria, where these are not met further design or operational measures must be introduced to reduce the risks to within acceptable limits. The risk cost can calculated by adding all of the above consequence cost elements and multiplying it by the predicted frequency of pipeline failure and accident probabilities as presented by Bai et al '99 and Goldsmith et al. As the probability of failure increases with time (i.e. due time dependent structural degradation) the risk cost from last inspection can be plotted against the inspection and maintenance costs for increasing intervals, as illustrated in Figure 40.7 below. In this way the optimum value inspection interval can be selected.

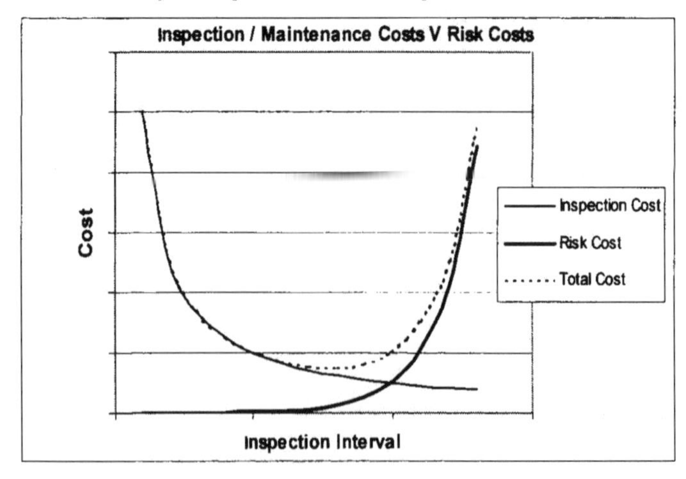

Figure 40.7 Maintenance costs verses risk costs.

40.10 Example 1: Risk Analysis for a Subsea Gas Pipeline

40.10.1 General

This risk analysis example will evaluate the risk acceptance and risk estimation of a North Sea pipeline transporting dry gas. This example will cover all aspects of the risk methodology developed in the chapter. By firstly determining the gas release for different hole sizes it is then possible to determine the potential effects on each type of risk.

40.10.2 Gas Releases

In order to provide an analysis that can be considered representative for the entire pipeline, the release rates have been estimated (conservatively) on the assumption that the water depth is 300m. This leads to a differential pressure at the site of loss of containment of ≈ 250 bar.

Representative hole sizes

Potential hole sizes will be modeled through the use of three representative hole sizes with diameters of 20mm, 80mm, and 200mm. The 20mm and 80mm hole sizes have been selected to provide ease of comparison with the hole sizes considered in the PARLOC database. The largest hole size considered is 200mm. This is considered to be a conservative upper bound to the equivalent hole size caused by major structural damage to the pipeline.

Discharge

Release rates have been estimated using SPILL. This is part of the HGSystem suite of programmer. The rates predicted for these hole sizes are given below. Indicative duration's for these releases are also shown below. These durations are based on the time required to blow down the pipeline through the hole and it is assumed that the mass release rates decrease linearly with time.

20mm hole	14.6 kg/sec	6000 hours
80mm hole	233.2 kg/sec	375 hours
200mm hole	1457.1 kg/sec	60 hours

The durations given above do not take into account emergency response actions initiated following the detection of a loss of containment. Hazard durations have therefore been assumed based on the time that it is expected to take for the existence of a release to be detected. These durations have been assumed to be 168, 48 and 6 hours, respectively. It should be noted that these times represent hazard duration's rather than leak duration's, i.e. they are estimates of the time required for the detection and location of a leak and for the imposition of measures to exclude shipping traffic from the affected locality. It should also be noted that the risk analysis results are not sensitive to the value assumed for the hazard duration for 20mm holes, since these do not result in flammable releases.

Subsea plume

The effect of a subsea gas release may be modeled as an inverted conical plume with a half cone angle of between 11 and 14 degrees in a zero current velocity situation. Assuming the most conservative case, this results for a 150m diameter release zone at the sea surface for the assumed 300m water depth.

Airborne dispersion

Airborne dispersion will be modeled using the program HEGADAS-S, part of the HGSystem suite. This program assumes that the gas evolves as a momentumless release from a rectangular pool. The pool has been taken to be 150m by 150m, so as to reflect the release into the atmosphere of the subsea plume.

Effect of water depth

Releases from greater depths will result in somewhat reduced mass flow rates. This is due to the increased seawater pressure at the site of loss of containment. Subsea dispersion over a greater depth will result in a larger gas evolution zone at the surface. These effects mean that the surface concentrations, and hence the dispersion distances and hazard zone dimensions will reduce with increasing release depth. The assumption of a 300m release depth for all loss of containment incidents is therefore conservative.

Stability

Pasquill stability classes define meteorological conditions from very unstable, A, to moderately stable conditions, F. These parameters are used in the Modeling of airborne dispersion. Two values of the Pasquill Stability Class have been used; these are Class D (Neutral Stability) and Class F (Moderately Stable Conditions). Class D is appropriate for night time and overcast day time, and has therefore been assumed to be representative of 75% of the time, with Class F being representative of the remaining 25%.

Wind speeds

Since there are no fixed installations at hazard as a result of subsea releases from the pipeline, wind direction is not required as an input to the risk assessments. Wind speeds are however required, since they determine the extent of the flammable gas clouds that may be generated by a release. The wind speeds and relative frequencies used to determine the hazard ranges associated with various releases are summarized in Table 40.3.

Table 40.3 Relative frequency of representative wind speeds.

Wind Speed Range (m/s)	Representative Wind Speed (m/s)	Relative Frequency
0 to 5	2	0.26
5 to 11	8	0.49
11 to 17	14	0.21
over 17	20	0.05

Hazard ranges

Hazard ranges are calculated in terms of the extent of the lower flammability limit (LFL) for different release rates, wind speeds and water depths. A concentration of 5% by volume has been used to represent the LFL. A total of eighteen gas dispersion analyses have been undertaken. These results are combined, using the data for relative frequency of Pasquill Class and wind speed, to provide an estimate of the hazard area associated with each of the three hole sizes. These are shown in Table 40.4.

Table 40.4 Average hazard areas for different hole sizes.

Hole Size	Hazard Area (m^2)
20 mm	0
80 mm	4900
200 mm	18650

40.10.3 Individual Risk

Acceptance criteria

The risks to which workers will be exposed are compared with the maximum operational FAR of 10 fatalities per 10^8 hours worked.

Cause analysis

Statistics of dropped object frequencies have been obtained from the 1992 Offshore Reliability Data Book, OREDA-92. This data source records a total of 7 dropped objects against a total calendar time of 648,200 hours or an operational time of 22,800 hours. Assuming an average lift duration of 5 minutes this is equivalent to 0.42 lifts per hour with a probability of a dropped object of 2.56×10^{-5} per lift.

Two lifting operations have been assumed at each work location, corresponding to one lift for installation of structures and one lift for pigging operations.

Assumptions

The following assumptions are made in addition to those stated earlier in the chapter.

1. Water depth has been assumed to be 300m.
2. The probability that the hazard zone resulting from a loss of containment coincides with the dropping vessel is assumed to be 0.5.
3. The probability of ignition has been taken as 0.3.
4. It is assumed that 50% of the persons on the vessel are working at any one time.

Consequence analysis

It is assumed that all persons on the vessel are at risk, the FAR is then a function of the proportion of persons on the vessel who are working, not of the total number of persons on the vessel.

Risk Estimation

The number of ignited releases per working location is given by:

$$f_{lift} \times p_{drop} \times p_{imp} \times p_{haz} \times p_{ign} = 2 \times 2.56 \cdot 10^{-5} \times 0.016 \times 0.5 \times 0.3 = 1.23 \cdot 10^{-7}$$

If the vessel remains on location for 48 hours and has n persons on board then this would result in x fatalities, as a result of 24n hours worked. The FAR is therefore equal to 0.51 x 10^{-8} (1.23 $\times 10^{-7}$ divided by 24). This is far less than the acceptance criteria established.

40.10.4 Societal Risk

Acceptance criteria

The acceptance criterion is 10^{-3} deaths per year.

Initiating incidents

Fishing interaction

Damage frequencies due to trawl gear interaction have been extracted from the PARLOC database. These are considered to be conservative, since the failure frequencies given in the PARLOC report are where no failures have been experienced. This is based on a theoretical analysis that does not take into account the robustness of the pipeline.

Merchant vessels

Because the minimum water depth for the pipeline is approximately 275m, emergency anchoring has not been considered. Incidents initiated by passing merchant vessels have therefore been restricted to dropped containers and sinking vessels. The initiating incident frequency data adopted is given in Table 40.5.

Table 40.5 Initiating incident frequencies.

Incident	Frequency	Hazard Distance
Dropped Container	5.15×10^{-6} per hour	15m
Sinking Vessel	2.11×10^{-7} per hour	150m

Construction vessels

Loss of containment incident frequencies, as a result of construction vessel activities, is given in PARLOC. However, while it is accepted that construction activities contribute to the overall loss of containment frequency for pipelines it is not considered to be appropriate to treat such incidents as initiating for societal risk calculations. This is because the presence of construction vessels will of itself exclude the presence of merchant shipping.

Random failures

Material and corrosion defect failure rates have been taken from PARLOC. Once again this data is considered to be conservative, particularly with respect to corrosion failure rates for export gas pipelines with a diameter > 10". It should, however, be understood that the corrosion defect failure rates used here can only be considered to be conservative provided that the pipeline is operated under the design conditions (i.e. dry). If the pipeline is to be frequently or continuously operated under wet conditions then the corrosion related failure rates would be significantly higher. The failure rates obtained from PARLOC are appropriate for the localized spot corrosion which may be experienced (often in association with a pre-existing defect) in a normally dry gas line in which corrosion is actively controlled and monitored on an ongoing basis.

Cause and consequence analysis

The total number of trawler crossings of the pipeline per year has been determined. It has been assumed that 50% of the trawlers will have a crew of 5 persons and 50% will have crews of 10 persons. It has been assumed that 15 people will on average be at risk per merchant vessel. This value is based on a population at risk of 10 people for 95% of vessels and 100 people for 5% of vessels.

In the absence of knowledge concerning the intensity of future 3rd Party construction activity it is not possible to predict the Societal Risks that will be associated with those activities. These risks will be subject to control by the 3rd Party concerned, and will contribute to the individual risks (the FAR) for those specific activities. In the absence of detailed information concerning the density of merchant vessel shipping, it has been assumed to be high. A merchant vessel crossing frequency of 29 per km year has been assumed.

The assumptions made with respect to the relative frequency of holes of different sizes are shown in Table 40.6.

Table 40.6 Calculated trawl impact frequencies.

Trawl Impact Frequency	Total Area	Pipeline
$f_{imp}/(\text{year} \times \text{km})$	2.63	0.42

Risk estimation

The expected number of 3rd party fatalities per year is 9.75×10^{-6} for the various scenarios considered. In view of the conservative nature of the calculations undertaken it is considered that the societal risks associated with the pipeline are acceptable.

40.10.5 Environmental Risk

No risk is posed since the material being transported is dry gas.

40.10.6 Risk of Material Loss

Initiating incidents

All incidents considered as initiating in the assessment of individual and societal risks are considered to be initiating for the purposes of determining the risks of material loss posed by the pipeline.

In addition any incidents occurring during construction and installation and having the potential to result in damage to and/or delay in the construction of the pipeline are considered to be initiating with respect to Risks of Material Loss.

Consequence analysis

Both repair cost and lost production cost have been assumed to be linearly related to the time taken for repair. Material costs for repairs have been neglected. Costs assumed are as follows:

- lost production 20 MNOK per day
- cost of repair spread 1 MNOK per day
- cost per fatality 100 MNOK

Time required for the repair of small or medium damage is assumed to be 16 days (clamp repair), time required for repair of large damage (new spoolpiece installed using mechanical connectors) is assumed to be 30 days. 3 days vessel mobilization has been assumed in each case. The costs (based on the above assumptions) incurred as the result of different sizes of damage are shown in Table 40.7. A discount factor of 7% is used to determine Net Present Values (1998 NOK) of future costs. The frequencies of incidents resulting in loss of containment are summarized in Table 40.8.

Table 40.7 Contributions to overall loss of containment rate.

	Small	Medium	Large	Total
Trawlers (Sinking)	0	0	$5.7x10^{-10}$	$5.7x10^{-10}$
Merchant (Sinking)	$1.3x10^{-8}$	$3.7 x10^{-9}$	$4.51x10^{-8}$	$6.18x10^{-8}$
Material Defect	$4.92 x10^{-7}$	$4.92 x10^{-7}$	$4.92x10^{-7}$	$1.48x10^{-6}$
Corrosion	$3.14 x10^{-6}$	0	0	$3.14x10^{-6}$
Trawl Impact	$1.16 x10^{-6}$	$2.91x10^{-7}$	0	$1.45x10^{-6}$
Subtotal (per km year)	$4.80 x10^{-6}$	$7.86x10^{-7}$	$5.60x10^{-7}$	$6.13x10^{-6}$
Maintenance/ Workover (per year)	$5.37 x10^{-7}$	$5.37x10^{-7}$	$5.37x10^{-7}$	$1.61x10^{-6}$
Total	$6 x10^{-4}$	$9.9 x10^{-5}$	$7.1 x10^{-5}$	$7.7 x10^{-4}$

Table 40.8 Costs of repairs.

Hole Size	Small	Medium	Large
Cost of repair (MNOK)	19	19	33
Cost of lost production (MNOK)	380	380	660

40.10.7 Risk Estimation

The expected discounted lifetime cost of incidents is deemed acceptable, as it is only a small percentage of the steel cost of the pipeline.

40.11 Example 2: Dropped Object Risk Analysis

40.11.1 General

This calculation is used to present an assessment of the risk posed by dropped objects hitting spools, umbilical and flowline sections around a template. This example will concentrate on the determination of the probability of dropped objects hitting subsea installations.

40.11.2 Acceptable Risk Levels

There is a need to distinguish SLS (Serviceability Limit State) and ULS (Ultimate Limit State). For this example, SLS is assumed as a dent damage larger than 3.5% of the pipe diameter, while ULS corresponds to bursting due to internal over pressure and combined dent and crack defects. The pipeline will not burst unless a large dent and a certain depth of cracks exist simultaneously.

The principle used in establishing the acceptance criteria is that the recovery time (for the most sensitive population) after an environmental damage incident should be insignificant relative to the frequency of occurrence of environmental damage. For this example, marine (pelagic) seabirds have been identified as the most sensitive resources during all seasons.

The damage category has been defined as minor for the field. The acceptance criterion is therefore a frequency $<2\cdot10^{-2}$ for the field as a whole. This can broadly be grouped into three main risk areas; pipelines, templates and topside and risers.

The acceptance criterion for pipelines alone is therefore assumed to be 1/3 of the field specific criterion, namely a frequency $< 7\cdot10^{-3}$.

40.11.3 Quantitative Cause Analysis

Probability cones

An object dropped at the sea surface is assumed to land within an area on the seabed which is swept out by a cone starting at the drop point. This area is determined by a cone with angle, ϕ.

It is further assumed that the probability of an object hitting a point within the cone follows a normal distribution and can be described as a function of distance x from the cone centerline.

$$p(x) = \frac{1}{\sigma\cdot\sqrt{2\pi}}\cdot\exp\left[-\frac{1}{2}\cdot\left(\frac{x-\mu}{\sigma}\right)^2\right] \tag{40.4}$$

where:

- $p(x)$: probability of hitting a point a distance x from the cone centerline
- σ : standard deviation
- x : distance from the cone centerline
- μ : mean value of x (here = 0)

The cone will sweep out an area with 99% cumulative probability of hit from a dropped object when:

$$X = d \cdot \tan\phi \tag{40.5}$$

where:

X : distance from the cone centerline giving 99% cumulative probability of hit

d : water depth

ϕ : cone angle

σ in Equation (40.4) can thus be determined by solving:

$$\int_{-a}^{a} p(X)dX = 0.99 \tag{40.6}$$

where: $a = X / \sigma$

Probability of flowline or spool hit

The probability of hitting parts of the pipeline or spool consists of three parts:

- Probability of dropping an object
- Probability of object landing within a cone area containing the flowline or spool.
- Probability of object hitting the spool or flowline (inside the cone area).

This is expressed in Equation (40.7).

$$P(hit) = P(drop) \cdot P(A_c) \cdot \frac{A_f}{A_c} \tag{40.7}$$

where:

P(hit): Probability of a dropped object hitting flowline, spool and/or umbilical

P(drop):Probability of an object being dropped

$P(A_c)$: Probability of a dropped object hitting the cone area A_c

A_f : Area of flowline, spool and/or umbilical within A_c, assumed = length x 1 m.

Energy absorbed by steel pipe

The energy required for a knife edge indentor to produce a dent in a pipeline may be calculated as follows:

$$E_d = 25 \cdot SMYS \cdot t^2 \cdot \sqrt{\frac{\Delta^3}{OD}} \tag{40.8}$$

where: t : wall thickness;

SMYS : Specified Minimum Yield Strength;

Δ : dent depth, assumed max. 3.5% of OD based on serviceability;

OD : outside diameter.

The effect of coatings and surface area of the falling object is conservatively neglected in Equation (40.9).

Basic data and assumptions for risk analysis

This example will consider the hit probabilities for a generalized L-spool. Table 40.9 presents the basic data for these calculations.

A 100m section of rockdump is assumed to follow directly after each spool. The hit probabilities are calculated for two areas:

- Probability of hitting the spool between the template and the start of the rockdump;
- Probability of hitting the pipeline outside the rockdump, but inside the 99% cone area.

The probability of the line being hit outside the 99% cone is considered negligible. Two flowlines and one umbilical are assumed for each template. The probability calculated considers a hit on any of these three items, for simplicity it is modeled as a total hit area of 3 x (one generalized spool length) x (a 1m corridor around each item).

The assessment is based on objects being dropped through the moon pool of the drill rig. Although objects may be dropped from the cranes, drops through the moon pool are assumed to be the worst case, as these will normally happen closest to the spools. A drill rig will be present on the field for the whole lifetime of the field (20yrs). A total of 17 templates has been assumed. This means that the time spent on one template will be 20yrs/17 ≈ 425 days. 75 days is added to this to account for increased drilling activities in the pre- and early production phase, after the lines are installed, giving a total of 500 days of drilling operations. There will be an average of 20 lifts/day during these 500 days, giving a total of 10000 lifts/20 years.

Table 40.9 Basic data and assumptions.

Item	Unit	Value
Water depth	m	300
Cone angle	°	30
P(drop)	-	$3 \cdot 10^{-5}$
Rig activity:	rig days/template/20 years	500
	Number of lifts/rig day	20
Design life	Years	20
Pipeline Outside Diameter	mm	259.8
Pipeline Wall thickness	mm	15.6

40.11.4 Results

Probabilities

Cone radii are found using simple geometric principles.

Cone radius, end spools: $(30^2 + 30^2)^{1/2} = 42$

Cone radius, end rockdump: $(130^2 + 30^2)^{1/2} = 133$

$X = 300\text{ m} \cdot \tan 30 = 173.2\text{ m}$

From Equation (40.5) and Table of the standard normal distribution:

$\sigma = X/2.575 = 67.2$ m (In a normal distribution; $P(-2.575<x<2.575) = 0.99$)

The cone area of the cone section encompassing the spools is:

$A_c = \pi \cdot (42)^2 = 5542 \text{ m}^2$

The spool area within this cone area is:

$A_f = 60\text{m} \cdot 3 \cdot 1 \text{ m} = 180 \text{ m}^2$ (length of pipe & umbilical within A_c with a 1m corridor)

Probability of hit within A_c:

$42 \text{ m}/67.2 \text{ m} = 0.625 \Rightarrow P(-0.625<x<0.625) = 0.468$

$$
\begin{aligned}
P(\text{hit}) &= 3 \cdot 10^5 \cdot 180/5542 \cdot 0.468 \\
&= 4.6 \cdot 10^7/\text{lift} \\
&= 4.6 \cdot 10^7/\text{lift} \cdot 20 \text{ lifts/rig day} \cdot 500 \text{ rig days/ } 20 \text{ years/template} \\
&= 4.6 \cdot 10^3/20 \text{ year/template} \\
&= 2.3 \cdot 10^4/\text{year/template} \cdot 17 \text{ templates} \\
&= 3.9 \cdot 10^3/\text{year}
\end{aligned}
$$

To calculate the probability of a dropped object hitting the flowlines outside the rockdumped area, the procedure above is repeated considering the cone section between the end of the rockdump and the end of the 99% cone area, giving:

$A'_c = \pi \cdot (173.2^2 - 133^2)^{1/2} = 38670 \text{ m}^2$

$A'_f = 3 \cdot (173.2\text{-}133) \cdot 1\text{m} = 120.6 \text{ m}^2$ (length of pipe & umbilical within A_c with a 1m corridor)

Probability of hit within A'_c:

$133 \text{ m}/67.2 \text{ m} = 1.979 \Rightarrow P(-1.979<x<1.979) = 0.952$

$P(\text{hit within } A'_c) = 0.99\text{-}0.952 = 0.038$

$$
\begin{aligned}
P'(\text{hit}) &= 3 \cdot 10^{-5} \cdot 120.6/38670 \cdot 0.038 = 3.6 \cdot 10^{-9}/\text{lift} \\
&= 3.6 \cdot 10^{-9}/\text{lift} \cdot 20 \text{ lifts/rig day} \cdot 500 \text{ rig days/20 years/template} \\
&= 3.6 \cdot 10^{-5}/20 \text{ years/template} \\
&= 1.8 \cdot 10^{-6}/\text{year/template} \cdot 17 \text{ templates} \\
&= 3.0 \cdot 10^{-5}/\text{year}
\end{aligned}
$$

Energy absorbed by steel pipe

The energy required to produce a dent of 3.5% of OD is found to be 5.2 kJ. Only items of approx. 1 tonne will have an impact energy less than 5.2 kJ. It is assumed that most dropped objects will be heavier than this, and consequently also assumed that all dropped objects will damage the spool/flowline enough for repair to be required.

This assumption is conservative because the falling object area (the object will not necessarily indent the pipe in a "knife edge" fashion) and the protection offered by the flowline coating is neglected.

40.11.5 Consequence Analysis

As stated earlier this example analysis pays little attention to the consequence of pipeline failure. The only consequence which is considered is the environmental damage that could be suffered. The damage category which the environment is likely to suffer is 'minor'.

40.12 Example 3: Example Use of RBIM to Reduce Operation Costs

40.12.1 General

The above RBIM approach can be used in the following aspects:

1. To optimise the intervals between planned shut-downs and the amount of inspections: the optimisation can be conducted through us of a cost-benefit analysis and/or structural target reliability levels, particularly where all costs cannot be accounted for.
2. To select inspection methods: An inspection method that yields most of the return for the dollars spent for safety and business is to be selected.
3. To prioritise the areas where risks are highest: for safety/business critical elements, it is necessary to accept additional inspection costs.
4. To prevent un-planned shut-downs: the cost associated with loss of production and transportation as a result of un-planned shut-down can be reduced by focusing inspection effort on safety/business critical elements.
5. To maintain the capacity of oil and gas transportation. Most of business risk is due to reduced value of maximum allowable operating pressure.

These targets are achieved through the establishment of inspection programs in which basic questions like what to inspect, when to inspect and how to inspect are answered.

The cost saving through the use of RBIM needs to be balanced with the costs of applying the RBIM. Much of the inspection expenditure is to satisfy prescriptive legislative requirements and many operators are concerned as to the value derived from such frequent inspection regimes. Risk Based Inspection (RBI) is increasingly becoming an interesting and profitable alternative to traditional, frequently performed inspections, which may bring little added value. An optimum interval of inspection may be obtained by minimising the total costs. The selected interval of inspection should, however, be less than that determined by the requirements of regulatory and company's safety and business criteria.

Use of RBI also allows operating expenditure to be focused on a few "critical elements" that will give the greatest return on expenditure.

40.12.2 Inspection Frequency for Corroded Pipelines

The spreadsheet tool receives processed defect data similar to the format below. The safe operating pressure, safety index and failure probability is calculated for each defect over the remaining life of the line and presented graphically in the spreadsheet as shown overleaf. The future corrosion damage in this case is predicted based on de Waard et al.'93 and considered process operating conditions. The individual defect failure probabilities are combined to give the failure probability for each segment and the pipeline as a whole.

The nominal P_{safe} is calculated from the equation below, using nominal calculated / measured values of X_A and X_L:

$$P_{safe} \equiv \frac{1}{\gamma} .2.\sigma_f.\frac{t}{D}.\frac{1 - X_A}{1 - \left(1 + 0.6275 \ .XL + 0.003275 \ .XL^2\right)^{-0.5}}$$

$= 137.6$ barg

where,

$$X_A = \frac{0.66\ L.d}{L.t} \equiv 0.33 \quad \& \quad X_L = \frac{L^2}{Dt} \equiv 0.4$$

P_{safe} is the operating pressure that gives an acceptable/desirable safety index (γ) i.e. probability of burst for the individual defect considered ($P_{burst} = \gamma \cdot P_{safe}$).

Input Data

Pipeline Section Properties			
Section Diameter	D	(m)	1
Section Nominal Wall thickness	Wt	(m)	2.50E-02
	δ_t	(m)	0.0005
Factor of Safety (New Criteria)	γ		2
Usage Factor	F		0.72
	SMYS	(MN/m^2)	445
Ultimate Tensile Strength	UTS	(MN/m^2)	553
MAOP	P	(MN/m^2)	16.02
		Bar	160.2
	P_{yield}	Bar	222.5
Pipeline Section Corrosion damage Parameters			
Type (Spiral, Pit, Groove, Circum Weld)			Groove
Measured max defect depth.	d_o	(m)	5.E-03
Stand Dev.	σ_d	(m)	5.E-04
Average Corrosion Rate	r	(m/yr)	4.00E-04
Stand Dev.	σ_r	m/yr)	4.0E-05
Measured Width			0.05
Spiral Angle			90.00
Measured Corrosion Length	Lm	(m)	0.05

The mean P_{burst} is calculated by substituting into the nominal P_{burst} equation above, the mean X_A and X_L values, which are obtained by multiplying the measured value by its bias i.e. X_A

$_{mean} = X_{A\ nominal} \cdot B_{XA}$. The bias being obtained from analysis of experimental data and for the case in question given above. The equation is further multiplied by the P_{burst} model bias X_M and normalized by dividing through by the SMYS. Thus the normalized mean P_{burst} (R_m) is given by:

$$P_{safe} \equiv .2 \cdot \sigma_f \cdot \frac{t}{D} \cdot \frac{1 - X_A}{1 - \left(1 + 0.6275 \cdot XL + 0.003275 \cdot XL^2\right)^{-0.5}} . B_{XF} \cdot B_{XM}$$

= 336 barg. The mean load is taken to be 137. barg multiplied by the load bias 1.05 giving a mean Load of 144.5 barg.

The variance of the mean resistance R_m is estimated from the variances of X_A, X_L X_F X_M, values of which are given above, thus $V_{Rm} \sim (V_A^2 + V_L^2 + V_F^2 + V_M^2)^{0.5} \sim 0.212$. The variance for the load (S_m) is taken from Bai '99 as 0.02. The Safety index β is calculated as $[\ln(R_m/S_m)/\sigma_{\ln RS}]$ where:

$$\sigma_{\ln RS} = ((\ln((R_m + V_{Rm}.R_m)/R_m))^2 + (\ln((S_m - V_{Sm}.S_m)/S_m))^2)^{0.5} = 0.19$$

Thus $\beta = \ln(336.0 / (137.6 \times 1.05)) / (0.19) = 4.56$ and the probability of failure $P_f = 1 - \Phi[\beta] = 2.33\text{x}10^{-6}$. This is well below the Ultimate Limit State acceptance criteria of $1\text{x}10^{-4}$. Thus if only this defect exists the safety factor of 2 can be reduced. For year 0, the following Safety levels are calculated for lower safety factors:

Safety factor	P_{safe}	Safety index	P_f
1,8	152.2	4	$3.1\text{x}10^{-5}$
1,6	171.2	3.75	$8.61\text{x}10^{-5}$

Note: for $\gamma = 1.6$ $P_{operating} = 160$ barg.

P_{burst}, P_{safe}, and safety index are predicted for the service life of the pipeline as illustrated in Figures 40.8 and 40.8 below. This analysis is repeated for every defect considered and an overall failure probability established.

If a safety factor greater than 1.6 is required then the cost of repairing the defect(s) verses reducing the operating pressure needs to be evaluated. However the defect(s) may be located near the export end of the pipeline such that the local operating pressure is much less than P_{safe}.. Thus depending on the relative costs, pressure protection systems may be put in place to prevent the local pressure exceeding P_{safe}, without reducing the inlet pressure and thus transport rates. If a safety factor of 1.6 is adequate (e.g. few significant defects), then initially no pressure derating is required. After approximately 5 years the ULS acceptance criteria is exceeded. At which point, either an intelligent inspection is performed to verify the predicted corrosion damage or the pressure is reduced (to a level that accounts for the uncertainty in predicted corrosion damage), depending on the relative costs. If many defects exist, the particular defects and segments can be ranked in terms of contribution to failure probability. Alternatively the line could be inspected when the predicted failure probability falls below $1\text{x}10^{-4}$ to establish whether the predicted corrosion rates are correct.

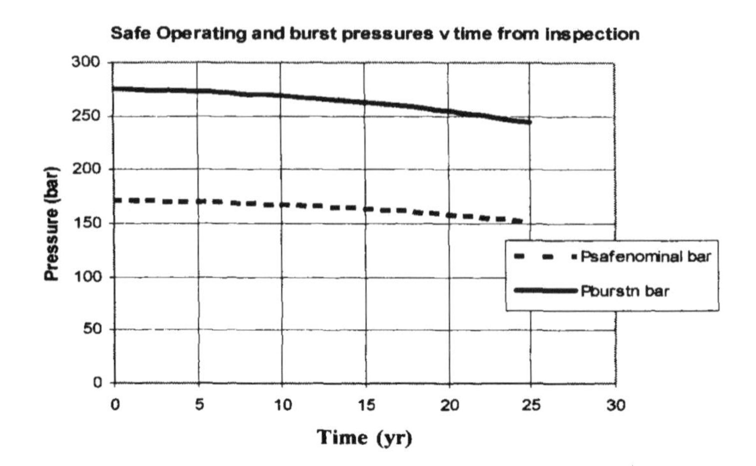

Figure 40.8 Operating and burst pressure vs. time from inspection.

Case B represents the situation where the operating pressure is reduced to P_{safe}.

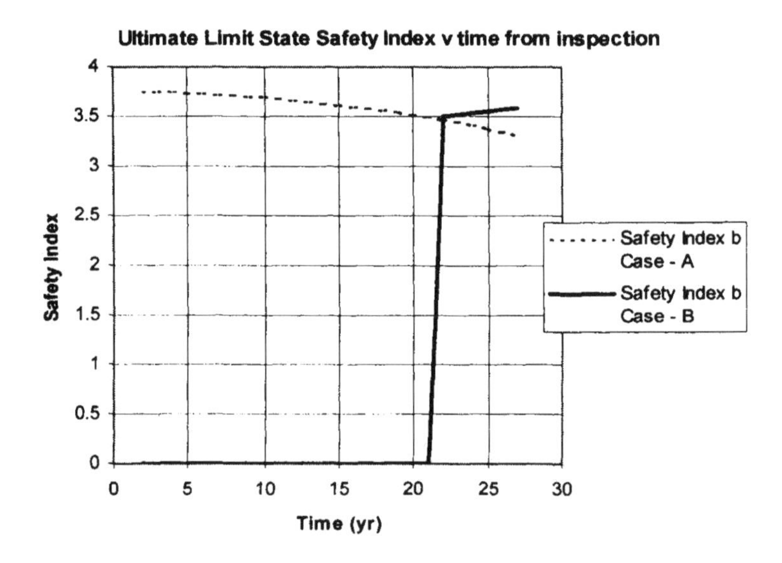

Figure 40.9 ULS safety index vs. time form inspection.

40.12.3 Examples of Prioritising Tasks

The RBIM tools can also be used to prioritise which areas to inspect and repair or other corrective actions to take. For a pipeline system all pressure containing parts that cannot be readily isolated (so that their failure does not affect the overall system) are generally equally critical and cannot be prioritised by this means. However supporting structures and equipment such as rock supports, protection structures and riser supports etc. will have varying criticality levels which can also be used to rank their inspection and maintenance requirements.

From the previous example, the failure predictions for each defect location are plotted verse location and time and compared to required reliability targets, see Figure 40.10.

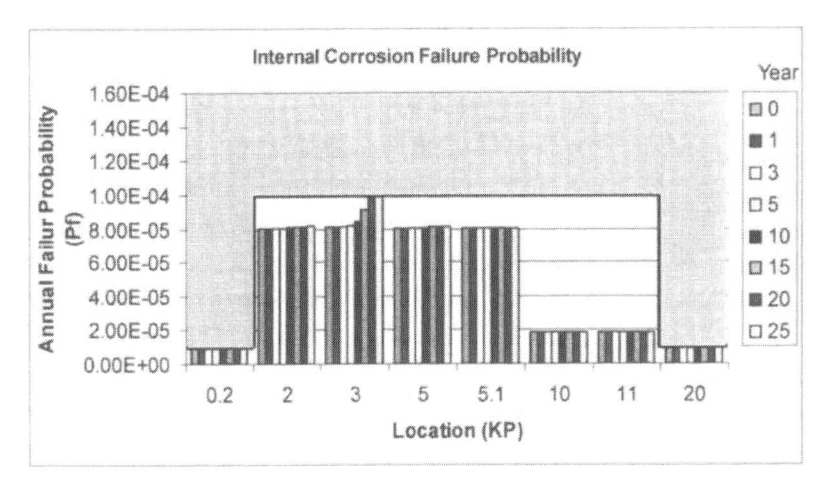

Figure 40.10 Specific defect/location failure probability with time.

Due to the high safety risks within the platform safety zone, the target ULS reliability is $1\text{x}10^{-5}$/yr compared to $1\text{x}10^{-4}$/yr for the midline. Thus it can be seen that the defects at KP 0.2 need to be repaired soon and that the export pressure needs to be reduced until the repair has been carried out. For the defect/segment at KP 20 also within a safety zone, it also needs to be repaired shortly unless the local operating pressure can be kept below P_{safe}.

For the midline section, all defects are within the acceptance criteria, though that at KP 3 indicates that it could develop into a concern. Also overall these defects exceed the acceptance criteria, thus those that pose the greatest likelihood of failure can be repaired first e.g. KP 2–5.1.

The local failure predictions can also be converted into local fatality, material loss and environmental damage risks and compared against risk acceptance criteria. This paper has presented a number of examples and approaches of applying a pipeline Risk Based Inspection and Integrity Management (RBIM).

40.13 References

1. A.IChemE, Guidelines for Evaluating the Characteristics of Vapour Cloud Explosions, flash fires and BLEVE's.
2. API 2A-LRFD: Planning, Designing and Construction of Fixed Offshore Platforms – Load and Resistance Factor Design.
3. Bea, R. (1994) "The Role of Human Error in the Design, Construction and Reliability of Marine Structures", Ship Structure Committee, USA, SSC-378.
4. Bea, R. (1997) in Journal of Reliability Engineering and System Safety, Vol. 52, Elsevier Science Limited.
5. Dougherty, E.M. and Fragola, J.R. (1986) "Human Reliability Analysis", John Wiley and Sons, New York, 1986.
6. Edmondson, J.N. (1993) "Human Reliability Estimates within Offshore Safety Cases", Proc. of symposium on Human Factors in Offshore Design Cases, Aberdeen, Scotland.

7. Fanneløp, T.E. and Lryhming, I. (1981) 'Massive Release of Gas from Long Pipelines' Jan. '81 J.Energy.
8. Fanneløp, T.K. & Sjøon (1980), 'Hydrodynamics of Underwater Blowouts', published by The Ship Research Institute of Norway R-97.80, Jan 1980.
9. Goldsmith, R. et al; Lifetime Risk-Adjusted Cost Comparison for Deepwater Well Riser Systems, OTC 10976.
10. HGSystem 3.0, Edited by Post, I., Shell Research Limited, Thornton Research Center, PO Box 1, Chester, United Kingdom, TNER 94.059.
11. Norwegian Petroleum Directorate, (1992) "Regulations Concerning Implementation and use of Risk Analysis in the Petroleum Industry", YA-049.
12. Offshore Engineer, November 1993, Total Maintenance takes hold on Forties field.
13. Olshausen, K. D. (1998) "Consequence Modeling", Seminar on Pipeline Safety at Statoil by Scandpower A.S., Stavanger.
14. OREDA (1992) "Offshore Reliability Data Book", Veritec.
15. PARLOC'94 "The Update of Loss of Containment Data for Offshore Pipelines", OTH 95 468, HSE Books, 1996.
16. PARLOC 96: The Update of the Loss of Containment Data for Offshore Pipelines, OTH 551.
17. Rew, P.J.,Gallagher,P. & Deaves, D.M. (1995), 'Dispersion of Subsea Release, Review of Prediction Methodologies',prepared for the Health and Safety Executive, OTH 95 465, 1995.
18. Ritchie, D., Voermans, C., Larsen,M.H., Vrankx, W.R.; Planning Repair and Inspection of Ageing Corroded Lines Using Probabilistic Methods, Risk Based and Limit State Design and operations of pipelines –Aberdeen '98 IBC UK Conferences Ltd.,
19. Sotberg, T. et al.(1997) 'The SUPERB Project: Recommended Target Safety Levels for Limit Based Design of Offshore Pipelines' Proc. of OMAE'97.
20. Sørheim, M. and Bai, Y. (1999) "Risk Analysis Applied to Subsea Pipeline Engineering", Proc. of OMAE'99.
21. Swain, A. D. and Guttman, H.E. (1981) "Handbook of Human Reliability Analysis with Emphasis on Nuclear Power Plant Applications", NUREG/CR-1278, Washington D.C., US Nuclear Regulatory Commission.
22. Willcocks, J. and Bai, Y. (2000) "Risk Based Inspection and Integrity Management of Pipeline Systems", Proc. of ISOPE'2000.
23. Williams, J.C. (1988) "A Data based Method for Assessing and Reducing Human Error to Improve Operational Experience", Proc. of IEEE, 4th Conference on Human Factors in Power Plants, California.
24. de Waard,C. and Lotz, U. (1993), 'Prediction of CO2 Corrosion of Carbon Steel ' CORROSION/93, paper no. 69.

Part VI

Integrity Management

Chapter 41 LCC Modeling as a Decision Making Tool in Pipeline Design

41.1 Introduction

41.1.1 General

Pipeline engineering projects can be divided into specific stages, each a separate source of cost, these stages include; Pre-engineering, conceptual engineering, detailed engineering, fabrication, construction, operation and abandonment. Although all of the different cost aspects are considered, this occurs in a segmented manner. The costs related to activities such as conceptual engineering, fabrication and installation are considered as isolated and addressed at different points in the pipeline life cycle and not viewed on an integrated basis.

It is necessary to assess these costs as interdependent entities. Thus in addressing the economic aspects of pipelines, one must look at the total cost in the context of the overall life cycle, especially in the early stages of conceptual design. Life-cycle Cost, when included as a variable in the pipeline development process, provides opportunity to design economically optimized pipelines.

The benefit of the Life-cycle Cost model of decision is that it is very flexible (Fabrycky and Blanchard, 1991). It is possible to analyze any aspect of the system being designed. In the case of pipeline engineering this type of analysis can be used at all levels of design and management, it can be used as a management tool in assessment of which training programs to implement, such that workforce efficiency is increased. Alternatively, it could be used by the engineer to work out the most economic method of preventing failure due to corrosion (i.e. inhibitors, corrosion allowance or high quality materials).

This chapter will present a generic model that is used in most industries such that it can be implemented into pipeline engineering, see Bai et al. (1999). The Life-cycle Cost (LCC) method will be discussed and a step-by-step procedure will be developed. Each of the steps will be discussed in terms of pipeline engineering, such that it can be used for future reference in the determination of LCC.

41.1.2 Probabilistic vs. Deterministic LCC Models

By using the LCC it is possible to express the total cost of a design alternative in terms of a mathematical expression, which can be generically described as follows:

$$\text{TOTAL(NPV)} = \text{CAPEX(NPV)} + \text{OPEX(NPV)} + \text{RISKEX(NPV)} \tag{41.1}$$

where:

CAPEX	: capital expenditure or initial investment;
OPEX	: operational costs, this includes planned (regular maintenance) and unplanned costs (repair of failures);
RISKEX	: risk expenditure;
NPV	: net present value.

A deterministic method of solving this expression would involve identifying and estimating any foreseeable costs based on historical data and past events. There are several different methods of estimating cost in this way, these include: engineering judgment, analogy and parametric method (see Fabrycky and Blanchard, 1991).

A probabilistic method of solving this expression would involve identifying costs and developing a probability distribution, which would best approximate the cost. There are various statistical methods that exist for developing a probability distribution based on historical data.

41.1.3 Economic Value Analysis

The paper written by Cui et al. (1998), introduces the idea of Economic Value Analysis (EVA), this analysis is based on the LCC model. It uses the idea that there exists a trade-off between the quality and cost. Quality is defined as the ability to satisfy requirements. In pipeline engineering these requirements include serviceability, safety, compatibility and durability (Bea 1998). Good quality in the design and construction of a pipeline can increase the safety and thus reduce the maintenance cost. However, introducing strict quality controls, the capital costs will increase and may not be recovered from the revenue generated in the operational phase. So Economic Value analysis develops the LCC model into a method by which it is possible to minimize the total Life-cycle Cost of a structure. The chapter developed a methodology for the Economic Value Analysis:

1. Identify the structure/system to be considered;

2. Identify the quality item(s) to be considered for the system;

3. Identify the principal failure modes for the structure/ system to be considered. In general, there may be several failure modes to be considered for a complex structure such as a pipeline or part of it (buckling, fatigue, on-bottom stability etc).

4. Write the limit state equations for each failure mode for the structure/system. This equation describes failure condition. The main point of this step is that in limit state equations, the quality item identified in step 2 must be explicitly considered.

5. Collect all of the statistical data for each parameter in the limit state equations. This will consist in the limit state equations. This can be in terms of probabilistic methods (statistical distributions) or deterministic values.

6. Compute the probability of failure, P_{fs}, as a function of the quality measure.

7. Define the consequences of failure and the related costs of these consequences for the structural system, C_f.

8. Compute the expected cost of failure E(C) of the system during service life as a function of the quality measure.

9. Define the initial costs of construction (C_o) as a function of the quality measure.

10. Perform the EVA, computing the quality measure or tolerance that will minimize total expected Life-cycle Costs, E(C).

$$\text{Min. } E(C) = \text{Min. } (C_o + C_f \cdot P_{fs}) \tag{41.2}$$

It should be noted that Equation 41.2 can be related to Equation 41.1. C_o corresponds to either an initial investment or planned costs. The second part, $C_f \cdot P_{fs}$, corresponds to unplanned costs that may occur during the pipeline lifetime.

This chapter identifies several quality aspects that can be modified. By then introducing basic financial risk theory it is then possible to complete an LCC analysis.

41.2 Initial Cost

41.2.1 General

When making a decision at any level it is always beneficial to identify the possible alternatives. In business situations the alternatives have nearly always related initial costs. This initial cost is always a function of some quality aspect of the alternative. Quality can be defined as a measurement of the extent to which the alternative covers the requirements of the situation. In engineering businesses these requirements include those of serviceability, safety, compatibility and durability.

- Serviceability is suitability for the proposed purposes, i.e. functionality. Serviceability is intended to guarantee the use of the system for the agreed purpose and under the agreed conditions of use.
- Safety is the freedom from excessive danger to human life, the environment and property damage. Safety is the state of being free of undesirable and hazardous

situations. The capacity of a structure to withstand its loading and other hazards is directly related to and most often associated with safety.

- Compatibility assures that the system does not have unnecessary or excessive negative impacts on the environment and society during its life-cycle. Compatibility is also the ability to meet economic and time requirements.
- Durability assures that serviceability, safety and environment compatibility are maintained during the intended life of the marine system. Durability is freedom from unanticipated maintenance problems and costs.

The alternatives available must fulfill the minimum criteria for each of these requirements, which is set forward by those that own, operate, design, construct and regulate pipelines. Any additional quality that is attained from an alternative will have financial implications over the lifetime of the product, as explained earlier.

This section will define the different types of quality aspects that exist in pipeline engineering. These different types include Management, Design/Engineering Services, Material and Fabrication, Marine Operations and Operation. It is important to recognize that the quality aspect to be analyzed will possibly lead to a failure, and that the calculation of risk of failure can be found using the techniques discussed in the risk section of this chapter.

41.2.2 Management

Management can be defined as the co-ordination and control of individuals and systems. The activity of management is present throughout the entire pipeline development process. By implementing different strategies or plans it is possible to influence the quality of performance of the individuals and systems. Research carried out by Bea (1994) implies that the quality of performance of individuals and systems in the design, construction and reliability of marine structures is a function of the frequency of Human/Organizational Errors (HOE).

Factors that contribute to HOE can be categorized into individual, organizational and systems (hardware, software) errors. Individual or human errors are those that are made by a single person which can contribute to an accident.

The sources of organizational errors can be placed into three general categories. The first is upper level management. The lack of appropriate resources and commitments to achieve reliability and the provision of conflicting goals and incentives (e.g. maintain production when it needs to be decreased to allow maintenance to be performed on the system) are examples of upper level management errors. The second is front line management. Information filtering (make it look better than it really is, tell the boss what he wants to hear- good news), and redirection of resources to achieve production at the expense of safety are examples of front line management errors.

The third category is the design, construction, or operating team. Team work in which there is an inherent and thorough process of checking and verification have proven to be particularly important: "if you find a problem, you own it until it is either solved or you find someone to solve it".

Errors can also be observed with human- system (equipment, structure, software or instructions manuals) interfacing. These are described as system (hardware errors) and procedure (software errors). System errors can be attributed to design errors and result in an operator making improper decisions. Similarly the procedures and guidelines provided to design, construct or operate a system could be seriously flawed.

The effects of management errors should be included in the risk in a quantitative manner. Very often the largest risk is that associated with intrinsic human errors. The influence of human errors on LCC should be accounted for through use of quantitative risk analysis in which failure probability and consequence are estimated.

Through the above subdivisions of HOE it is possible to specify quality aspects that can be varied. One example of the numerous ways in which this could be implemented could be when deciding on the recruitment of new engineers. By recruiting an experienced engineer the likelihood of design error is low and salary high, however, if a graduate engineer is hired the likelihood of design errors taking place is quite high and salary is low. This can then be assessed using Life-cycle Cost analysis and the most economically viable solution may be reached.

41.2.3 Design/Engineering Services

The scope of the quality aspects that this category covers is conceptual engineering and preliminary engineering. The detailed engineering of a pipeline structure allows very little scope for the alteration of quality aspects of the pipeline and is hence not discussed. The limits of each of these areas are outlined as follows, based on work by Langford and Kelly (1990).

1. Conceptual Engineering
   - To establish technical feasibility and constraints on the system design and construction
   - To eliminate non-viable options
   - To identify the required information for the forthcoming design and construction
   - To identify interfaces with other systems planned or currently in existence.

2. Preliminary Engineering
   - Perform pipeline so that system concept is fixed. This includes:
   - To verify the sizing of the pipeline
   - Determining the pipeline grade (included in material section) and wall thickness
   - Verifying the pipeline against design and code requirements for installation, commissioning and operation.

The level of engineering is sometimes specified as being sufficient to detail the design for inclusion into an 'Engineering, Procurement and Construction' (EPC) tender. The EPC contractor should then be able to perform the detailed design with the minimum number of variations as detailed in their design.

41.2.4 Materials and Fabrication

This category of quality aspects is probably the one in which most experience has been gained in terms of financial analysis of the options available. This category covers the quality of all materials that are used in the pipeline development and the quality of fabrication of these materials.

41.2.5 Marine Operations

This category of quality aspects covers all marine operations that are required prior to the operation of the pipeline and also the extraordinary marine operations that are required to maintain operation of the pipeline (i.e. repair). An example of the application of LCC would be when deciding on the type of Lay barge to use, a balance between day rates and days down would be analyzed.

41.2.6 Operation

The operation of a pipeline includes all activities that are performed after the installation of the pipeline. This primarily involves the inspection of the pipeline, but not, as stated above, the repair of the pipeline.

41.3 Financial Risk

41.3.1 General

Through the use of risk analysis it is possible to arrive at financial values which represent the financial losses that are likely to occur in a pipeline. This method relies on the Quantitative Risk Analysis approach.

Risk analysis can be determined by the following generic expression:

$$\text{Risk} = \text{Probability of Failure} \times \text{Consequence of Failure} \tag{41.3}$$

The two elements that are used to calculate risk can be separated into the Probability of Failure, which is equivalent to the Frequency of Failure, and the Consequence of Failure. In using risk for financial analysis it is necessary to determine the consequence of failure in monetary terms.

In the paper of Sørheim and Bai (1999) a detailed development of the risk analysis is given. This chapter discusses the different methods of determining the Probability of Failure and Consequence Modeling.

41.3.2 Probability of Failure

In determining the probability of failure two different levels of failure causes can be identified, direct failure and indirect failure. Direct failures are related to physical aspects of the pipeline failing, such as corrosion, fatigue or on-bottom stability. Indirect failures pertain to system or human errors which may eventually lead to a direct failure. The direct failures can be determined using structural reliability analysis. For reliability analysis to be considered a probability of failure, it is necessary to incorporate a deterministic value for human error (usually a factor between 5 and 10). The indirect failures can be modeled using a number of quantitative risk analysis techniques including event tree analysis. Both the structural reliability analysis and quantitative risk analysis techniques are developed fully in Sørheim and Bai (1999).

41.3.3 Consequence

41.3.3.1 General

Consequence is the determination of the possible outcome(s) of a failure event. Two methods are available to measure the consequences of a release event, these are consequence modeling and the interval method.

- Consequence modeling - This is an analytical method which assess the sequence of events after a failure has occurred. The different stages that occur after a release include; discharge, dispersion, ignition, combustion and damage and loss. This method is discussed further in Sørheim and Bai (1999).

- Interval method - The second, Interval method is an approximate method. By using engineering judgment and historical data it is possible to give estimated upper and lower bound consequence scenarios. This allows a scope of different consequence scenarios to be evaluated, thus a decision can be reached on which scenario best suits the philosophy of the decision-maker, (optimist- low consequences, pessimist- high consequences or other).

The different types of consequences that are likely to occur as a result of a release event are:

- Cost associated with averting fatalities and injuries
- Environmental damage
- Production Loss
- Material Repair

41.3.3.2 Cost Associated with Averting Fatalities and Injuries

Although any human loss is unacceptable, it is necessary to account for all possible scenarios. Cost associated with averting fatalities and injuries (another wording for "human loss") would place a financial burden on the Owner. There are currently two main methods used for determining the economic value of a human life. It must be noted that this is a 'statistical' life,' not an identifiable individual. Society has always been ready to spend much more to save an individual in a specific situation- trapped coal miner, for instance. The statistical life reflects the amount that society is willing to spend to reduce the statistical risk of accidental death by one individual.

The first method is the human capital approach in which the value is based upon the economic loss of future contributions to society by an individual. The second approximation willingness to pay, identifies how much an organization is willing to pay (in terms of other goods and services given up) to gain a reduction in the probability of accidental death. Each method has drawbacks and benefits.

Injuries frequently cost more than fatalities. This cost should also be included in consequence modeling.

1. Material Repair

Material repair is a function of the extent of damage that the pipeline has experienced. There are three ways in which a breach of containment is likely to be repaired; hyperbaric weld repair, spoolpiece installation and bolted sleeve installation. Information regarding the cost of these repairs are available from most operating companies.

2. Production Loss

The production loss calculates the financial loss due to the time which is lost due to the damage of the pipeline, this is a function of the time it takes to repair the pipeline. This can be calculated from the value of the product being transported per unit and the volume of product that could potentially be transported during repair.

The cost that will arise from inconvenience caused to the receivers of the transported goods must also be included. By assessing contractual agreements between operator and purchaser, it is possible to identify potential costs.

3. Environmental Damage

It is necessary to assess each case on its own merits. The following factors will be the most influential in determining any cost.

- Volume and type of product lost
- Probable currents and exposed coastline.
- Topography and location of 'sensitive areas' (nature reserves, farming, recreational areas, potable water sources etc.)
- Existing emergency response capacity

A useful source of information for an estimated value of the financial loss suffered is the use of risk matrices available from most operators of offshore installations. A typical risk matrix would include information that could be correlated to the circumstances of a pipeline failure.

41.4 Time Value of Money

The time value of money in the form of an interest rate is an important element in most decision situations involving the flow of money over time. The reason for this is that money earns interest through its investment over a period of time, a dollar to be received at some future date is not worth as much as a dollar in the hand at present.

Money also has a time value due to the purchasing power of a dollar through time. During periods of inflation the amount of goods that can be bought for a particular amount of money decreases, as the time of purchase occurs further in the future. Therefore, when considering the time value of money it is important to recognize both the earning power of money and the purchasing power of money.

In analyzing the time value of money for a LCC model it is necessary to evaluate all costs on a common basis, this is usually when an initial investment is made, therefore, all costs must be evaluated in terms of the initial investment cost. At this stage it is necessary to assess the types of costs that are likely to be encountered, single payment, annual payments or varying annual payments.

When calculating the cost of risk it is necessary to recognize, that different types of probabilities exist; immediate, time independent and time dependent. Immediate failure is a failure which occurs immediately upon installation of the pipeline (e.g. hydrostatic collapse or hoop stress criterion). Since the failure occurs immediately the cost does not have to be adjusted to account for time value of money principles.

$$\text{Risk} = \text{Consequence cost } (t=0) \times P_f \tag{41.4}$$

The second type, time independent, is a failure that can occur at any point during the lifetime of the pipeline (e.g. trawl impact or dropped objects). It is therefore necessary calculate the present value of the consequence on the basis of a failure occurring at the midpoint of its life. This gives an equal assessment of the failure occurring at any given point in time.

$$\text{Risk} = \text{NPV (Consequence cost } (t= \tfrac{1}{2} \text{ Total life time})) \times P_f \tag{41.5}$$

The time dependent failure will result in the most complex assessment of the cost assessment. These types of failures include fatigue and corrosion. Failure probability increases per year, hence it is necessary to adjust the consequence cost for each year and multiply by the failure rate of the same year, this can then be cumulated to give a total risk cost.

$$\text{Total Risk Cost} = \Sigma(\text{NPV (Consequence cost } (t)) \times P_f(t)) \tag{41.6}$$

41.5 Fabrication Tolerance Example Using the Life-cycle Cost Model

41.5.1 General

The purpose of this calculation example is to demonstrate the validity of Life-cycle Cost modeling as a method by which to justify choices between design alternatives. This example will look at the practicalities of assessing the failure probability, the cost of consequence, the implementation of economic theory and the utilization of interval method. This calculation example I inspired by Nødland et al. (1997), will follow the steps outlined in the introduction.

41.5.2 Background

Pipeline fabrication quality is one particular aspect of pipeline design that could give potential cost savings over the life cycle. Good quality in the fabrication of pipeline can increase the safety and thus reduce cost of unplanned maintenance and cost of consequences. However, too stringent quality requirements can drive up fabrication costs and this increase of initial cost may not compare favorably to lesser quality options. This design example will compare the Life-cycle Cost of two different fabrication qualities, in terms of the probability of failure due to corrosion, and thus arrive at a judgement as to which fabricator is more economically viable.

41.5.3 Step 1- Definition of Structure

The structure to be considered is a subsea pipeline.

41.5.4 Step 2- Quality Aspect Considered

The quality aspect that is to be considered in this example is the fabrication tolerance that is to be used. This calculation example will consider two fabricators each of which produce a different quality of pipe. The different qualities of pipeline will be implemented into the problem through a random variable, modeling the uncertainty in wall thickness. The exact nature of this variable is described fully in Step 5- Definition of Parameters and Variables.

41.5.5 Step 3- Failure Modes Considered

In order to simplify the scope of the example, the only design aspect which is to be considered is the design criteria for corrosion allowance, from this only two failure modes are likely, these are hoop stress and hydrostatic collapse.

41.5.6 Step 4- Limit State Equations

1. General

By considering corrosion depth as the load and wall thickness as the resistance, it is possible to apply Load-Resistance Factored Design (LRFD) methodology to pipeline corrosion allowance design. This introduces a welcomed opportunity to take into account the range of uncertainties inherent in corrosion rate calculations and residual strength of corroded pipelines in the design.

2. Corrosion rate and defect length

Corrosion rate (CR) is based on De Waard '93 formula. This gives corrosion rate as a function of temperature, pressure and CO_2 content. In addition, effects due to pH, saturation of corrosion products, glycol content and scale formation may be accounted for.

The following assumptions are made:

- The corrosion rate during normal operation is negligible;
- Corrosion will only occur following an upset where water or wet gas is introduced into the line. The time during which corrosion can occur is labeled t_w (= total duration of wet service). The determination of t_w is described in Step 5,
- Glycol will be present in the line as a continuous film on the pipe wall because of carry-over from the glycol drying unit. The water ingress in the line will thus result in an increased water content in the glycol film. The water content in the glycol film is assumed to increase to a maximum of 50% following an upset.

The depth of an attack is modeled as shown in Equation 41.7:

$$d = CR \cdot t_w \tag{41.7}$$

where:

CR : corrosion rate;

t_w : total duration of wet service.

3. Allowable Corrosion Depth Based on Hoop Stress

ASME B31G (1993) defines a safe operating pressure, P', for a corroded pipe with a short defect (i.e. A ≤ 4, see Equation 41.9):

$$\Delta P' = 1.1 \cdot \Delta P \cdot \left(\frac{1 - \frac{2}{3} \cdot \frac{d}{t}}{1 - \frac{2}{3} \cdot \left(\frac{d}{t \cdot \sqrt{A^2 + 1}} \right)} \right) \tag{41.8}$$

where:

ΔP : Design pressure (internal – external);

D : Maximum allowable depth of corroded area;

T : Nominal wall thickness of pipe;

A : Constant $= 0.893 \cdot \frac{L}{\sqrt{D \cdot t}}$ (41.9)

L : Axial extent of the defect;

D : Nominal outside diameter of pipe.

Pressure is related to wall thickness as shown in Equation 41.10.

$$\Delta P = SMYS \cdot \frac{2 \cdot t}{D - t} \cdot \eta \tag{41.10}$$

where:

η : Usage factor;

Equations 41.8, 41.9 and 41.10 are combined to give Equation 41.11 for short corrosion defects ($L \le 4.48 \cdot \sqrt{D \cdot t}$) (from Nødland, Bai and Damsleth, 1997). The increased strength of the pipe wall in uncorroded sections of the pipe (due to the remaining corrosion allowance) has thus been taken into account. These calculations allow for no reduction in design pressure during the lifetime.

$$d_h = \frac{1.5 \cdot t \cdot \left(\frac{1.1 \cdot t}{t_{-c}} \cdot \frac{D - t_{-c}}{D - t} - 1 \right)}{\frac{1.1 \cdot t}{t_{-c}} \cdot \frac{D - t_{-c}}{D - t} - \frac{1}{\sqrt{1 + \frac{0.8 \cdot L^2}{D \cdot t}}}} \tag{41.11}$$

where:

d_h : Allowable corrosion depth based on hoop stress;

T : Wall thickness incl. corrosion allowance;

t_{-c} : Wall thickness excl. corrosion allowance.

4. Allowable Corrosion Depth Based on Collapse

A corrosion defect may reduce the hoop buckling capacity of a pipe. The allowable corrosion depth based on collapse may easily be derived based on the formulation in Chapter 3.

5. Limit State Function

The limit state function, $g(\overline{X})$, forms the basis for the reliability calculations. This function expresses 'Resistance' - 'Load' as a function of $\overline{X}$, where $\overline{X}$ is a vector containing all the basic uncertainty variables describing the 'loads' and 'resistance's'. Deterministic values may also be included in g. The criterion for non-acceptance (or failure) is consequently defined as $g(\overline{X}) < 0$, with the corresponding probability:

$$P(g(\overline{X}) < 0) = \int_V f_{\overline{X}}(\overline{x}) d\overline{x} \tag{41.12}$$

where:

V : failure domain = $\{\overline{x} | g(\overline{X}) < 0\}$

$f_{\overline{X}}(\overline{x})$: joint density function for $\overline{X}$

$\overline{x}$: realization of $\overline{X}$ in the basic variable space.

Since two failure modes are investigated (i.e. hoop stress and local collapse for load and displacement control), two limit state functions are needed to describe the system. The system probability of failure may therefore be approximated by:

$$P_{system} = P(g_1(\overline{X}) < 0) + P(g_2(\overline{X}) < 0) \tag{41.13}$$

The calculation of the probability of failure is done by the proprietary software SYSREL. Second Order Reliability Method (SORM) is used.

The limit state functions used for the corrosion allowance calculations are shown in Equations 41.14 and 41.15.

$$g_1(\overline{X}) = \frac{1.5 \cdot x_t \cdot t \cdot \left(\frac{1.1 \cdot x_t \cdot t}{x_t \cdot t_{-c}} \cdot \frac{D - x_t \cdot t_{-c}}{D - x_t \cdot t} - 1 \right)}{\frac{1.1 \cdot x_t \cdot t}{x_t \cdot t_{-c}} \cdot \frac{D - x_t \cdot t_{-c}}{D - x_t \cdot t} - \frac{1}{\sqrt{1 + \frac{0.8 \cdot L^2}{D \cdot x_t \cdot t}}}} - CR \cdot x_m \cdot \frac{t_w}{365} \tag{41.14}$$

where:

x_m : corrosion rate model uncertainty factor;

x_t : wall thickness uncertainty factor (from manufacturing process).

$$g_2(\overline{X}) = p_c(eq.\,4.8) - \frac{\gamma_R \cdot p_e}{\sqrt{1 - \left[\frac{M_{F,c} \cdot \gamma_R \cdot \gamma_F \cdot \gamma_c}{M_c} \right]^2}} \tag{41.15}$$

Note that the expression for g_2 is somewhat simplified for clarity. In the analysis, the Equation 2.20 is solved for p_c with h as given in Equation 41.16.

$$h = t - CR \cdot x_m \cdot \frac{t_w}{365} \tag{41.16}$$

In addition, the expression for wall thickness, t, in the limit state function, is always multiplied with it's uncertainty factor, x_t, and M_c is calculated from Equation 3.48.

41.5.7 Step 5- Definition of Parameters and Variables

1. Pipeline, Operational and Environmental Data

The pipeline data presented in Table 41.1 has been used. The required wall thickness for hoop stress is 13.0 mm. However, a wall thickness of 15.9 mm is chosen because high moments and strains are expected in the line due to the uneven seabed. The expected functional strains and moments are shown in Table 41.2. The benefit from the higher wall thickness will be increased local buckling capacity and hence reduced need for seabed intervention.

Table 41.1 Pipeline and environmental data.

PARAMETER	UNIT	VALUE
Internal diameter	mm	425.2
Wall thickness	mm	15.9
Wall thickness, req'd for hoop stress only	mm	13.0
Ovality	-	1.5%
External pressure	MPa	3.65

The calculation of the allowable collapse pressure is based on the parameters given in Table 41.2.

Table 41.2 Functional moments and force.

PARAMETER	VALUE	REFERENCE
M_F	0.6 MPa	preliminary in-place analysis
F	250 kN	preliminary in-place analysis

The operational data presented in Table 41.3 has been used. The temperature drop in the bundle is estimated, based on calculated temperature profiles, an inlet temperature of 70°C, and a bundle length of 400 m.

Table 41.3 Operational data.

PARAMETER	UNIT	VALUE
Design Pressure	Bar	225
Temperature @end bundle/start pipeline	°C	45
CO_2 content	Mole %	3

2. Defect Length

Intelligent pig inspections from the Emden gas pipeline show defect sizes after approx. 20 years of service which can be illustrated by the distribution function shown in Figure 41.1. (Nødland, Bai and Damsleth, 1998). This function has been used to describe the expected defect length in the line.

It should be noted that the defects in the Emden line have occurred following a history of operational difficulties. After these have been sorted out, the defect growth and occurrences of new defects have decreased significantly.

3. Wall Thickness Uncertainty

In this calculation example the parameter which is being investigated is the fabrication quality of the pipeline. As additional complexity would be introduced into the limit state equations it has been chosen to represent this difference in fabrication quality through the wall thickness uncertainty variable. This parameter is represented by the following.

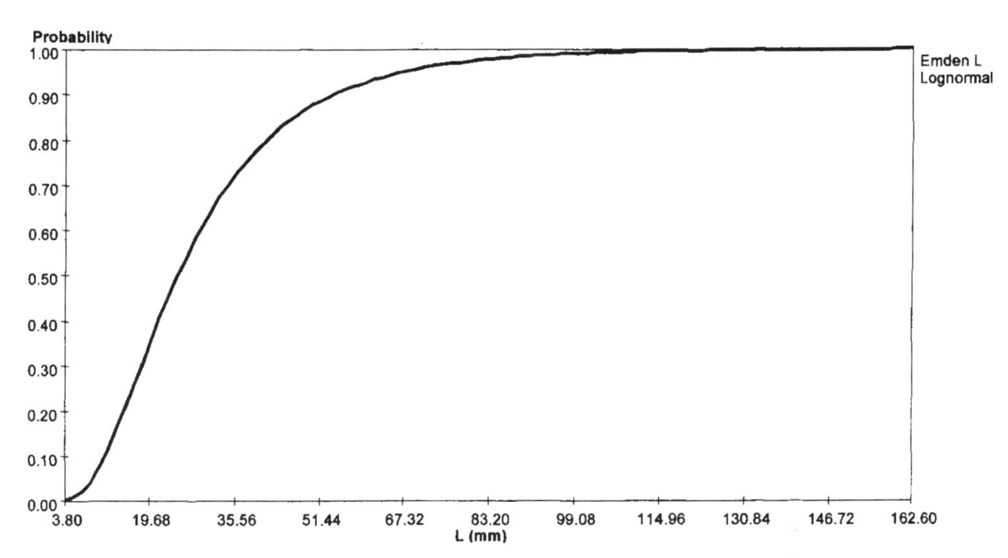

Figure 41.1 Cumulative probability distribution of defect lengths, L, found in Emden gas line; distribution function = lognormal, mean = 30, std. dev. = 20.

Table 41.4 Wall thickness parameter.

Variable	Description	Distribution	Mean	Variance	Upper Bound	Lower Bound
xt	Wall thickness uncertainty	Beta	1.02	High Quality = 0.005 Low Quality = 0.035	1.1	0.95

It is important to note that the difference between the qualities of pipeline depends on the variance of the parameter; the greater the variance the less likely the fabricator is of manufacturing to the specified size. The smaller the variation the more constant the fabricator is in producing the pipe. These values were chosen using engineering judgment such that reality is simulated to a reasonable extent.

4. Common Input Parameters

The basic parameters are summarized in Table 41.5.

For the purpose of this example an upset is defined as water or wet gas ingress into the pipeline. Detection of an upset is assumed to lead to immediate shutdown. It is assumed that the upsets are independent occurrences and that the line will be dried after the previous upset before a new upset occurs. The total time of wet operation (t_w) is a product of the number of upsets per year, the duration of each upset and the number of years operated.

Table 41.5 Summary of common input parameters.

Parameter	Comment	Distribution	Basic value
CR	Corrosion rate, mm/yr.	Constant	2.0 mm/yr.
x_m	Model uncertainty	Gumbel, max	Mean: 1.5 Std. dev.: 0.5
U	Number of upsets per year	Normal	Mean: 3 Std. dev.: 2
t (single upset)	Duration of wet line operation, single upset, days	Lognormal	Mean: 3 Std. dev.: 2
L	Length of defect, mm	Lognormal	Mean: 30 Std. dev.: 20
Wt	Wall thickness	Constant	15.9
ID	Internal diameter	Constant	425.2 mm

41.5.8 Step 6- Reliability Analysis

Through the use of SYSREL, a reliability analysis program, it was possible to determine the cumulative failure probability of each year of operation. The annual failure probability was found using the equation give below.

$$P_f(n) = CP_f(n) - CP_f(n-1) \qquad (41.17)$$

where:

P_f : annual probability of failure;

CP_f : cumulative probability of failure;

n : year .

Figure 41.2 gives the distribution of annual failure probabilities.

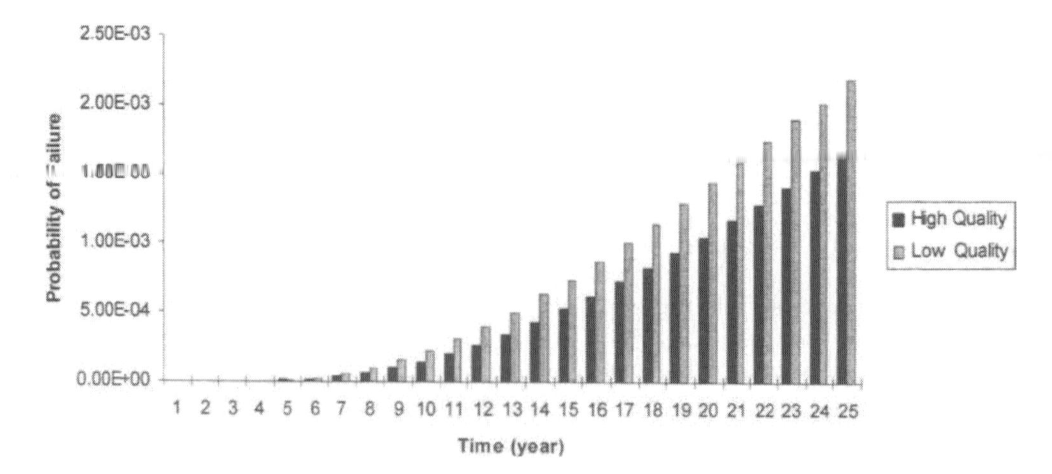

Figure 41.2 Annual failure probabilities.

41.5.9 Step 7- Cost of Consequence

In evaluating the cost of consequences of failure of pipeline, it is necessary to consider the mode of failure and the potential out comes of those modes of failure. In developing the consequence costs, four separate costs, can be considered unplanned maintenance, environmental damage and clean-up, loss of production and loss of human life.

In order to calculate these costs accurately, it is necessary to carry out a thorough analysis of the possible consequences of pipeline failure. This can be simplified to the extreme boundaries of these consequences. As in our case the following consequences can be envisaged:

Table 41.6 Cost of consequences.

Cost	Variable	Boundary	Description	Cost (NOK)
Unplanned Maintenance	C_{UM}^{U}	Upper	Spool Replacement	20,000,000
	C_{UM}^{L}	Lower	Sleeve Clamp	9,000,000
Environmental Cost	C_{E}^{U}	Upper	>2500m^3	5,000,000
	C_{E}^{L}	Lower	<100m^3	250,000
Loss of Production	C_{LP}^{U}	Upper	See note 1	0
	C_{LP}^{L}	Lower	See note 1	0
Human Loss	C_{HL}^{U}	Upper	See note 2	0
	C_{HL}^{L}	Lower	See note 2	0

Note 1: Loss of production is a function of the time to repair; it is assumed that the time to repair of either case is approximately equal.

Note 2: it is assumed that no human life is lost as a result of failure.

41.5.10 Step 8- Calculation of Expected Costs

In this calculation example the development of corrosion increases with time, this has already been accounted for in the calculation of the probability of failure and is shown in Figure 41.2. In order to calculate the expected cost of these failure probabilities it is necessary to consider time value of money principles, such that the cost of consequences are projected to reflect the year in which the failure occurs. By multiplying this future expected value by the probability of failure for that year it is possible to calculate the expected cost. Finally, this value must then be represented in present value form, such that it is possible to evaluate all costs equally. Summation over all of the years being evaluated gives an Expected Cost in terms of present value:

$$EC = \Sigma\, NPV\{rate, n, [FV(inflation, n, C)] \times P_{fn}\} \qquad (41.18)$$

where:

EC : expected cost;

NPV (...) : Economic expression for deriving a present value based on a future value;

rate : economic return that can be expected from an alternative investment;

n : year;

FV(...): economic expression for deriving a future value based on a present value;

Inflation: the amount by which the relevant is expected to rise by each year;

C : cost being evaluated;

P_{fn} : probability of failure in year n.

In this calculation example the inflation rate used is 2% and the "interest rate" used is 6%. The Expected Costs are given in Table 41.7.

Table 41.7 Expected costs.

Cost Type	Variable	Low Quality Expected Cost	High Quality Expected Cost
Maintenance Unplanned	C_{UM}^U	NOK 517,451	NOK 374,392
	C_{UM}^L	NOK 232,853	NOK 168,477
Environmental	C_E^U	NOK 129,363	NOK 94,490
	C_E^L	NOK 6,468	NOK 4,680

41.5.11 Step 9- Initial Cost

The initial cost of the low quality pipeline is assumed to be NOK6,500 per tone, from this a total cost for the pipeline can be calculated, it is assumed that the high quality pipeline will cost 5% over the cost of the low quality pipeline. The initial costs are outlined in Table 41.8

41.5.12 Step 10- Comparison of Life-cycle Costs

In this final step it is possible to compare the two different Life-cycle Costs which are generated. In order to do this it is necessary to consider all the combinations of the expected costs, thus giving the decision maker a full set of information which can be used to justify the final decision. Using the following equation the final Life-cycle Costs were found:

$$LCC = C_I + C_F \tag{41.19}$$

where:

C_I : initial cost;

C_F : sum of all costs associated with the failure/loss of performance of the pipeline.

In this case this involves the following:

C_{UM} : interval cost of unplanned maintenance, $[C_{UM}^U, C_{UM}^L]$;

C_E : interval cost of environmental damage, $[C_E^U, C_E^L]$.

A graphical representation found in Figure 41.3 shows that for the different combinations of consequence that were used, the optimal pipeline fabrication varies between high and low quality. This provides a basis from which a decision about the fabrication tolerance can be selected.

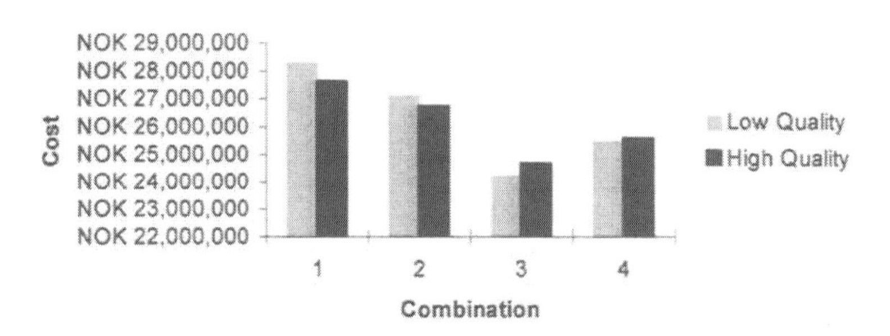

Figure 41.3 Comparison of LCC for alternative combinations.

41.6 On-Bottom Stability Example

41.6.1 Introduction

This example will outline the use of LCC modeling when deciding the method by which to stabilize a pipeline. This problem will only be discussed in general terms from which it would be possible to complete a more detailed assessment.

41.6.2 Step 1- Definition of System

The system to be considered is a pipeline.

41.6.3 Step 2– Quality Aspects Considered

The quality aspect to be considered is on-bottom stability. From this it is possible to identify different mechanisms by which the pipeline may be stabilized. These include:

- Concrete Coating;
- Rock Dumping;
- Dynamic Stability.

41.6.4 Step 3- Failure Modes

The failure modes, which exist for on-bottom stability design criteria, are:

- Sliding Lateral Stability;
- Uplift Vertical Stability.

Given the high degree of correlation between the uplift and sliding modes of failure, the probability of stability failure is equal to the maximum of either the probability of sliding failure or the probability of uplift failure.

41.6.5 Step 4- Limit State Equations

Two limit state equations, one for each of the failure modes, these can be expressed as follows.

Uplift Failure:

$$g_1(X) = W - F_L \tag{41.20}$$

Lateral Failure:

$$g_2(X) = Ru - F_D \tag{41.21}$$

where:

- W : submerged weight of pipe;
- F_L : hydrostatic uplift force;
- Ru : resistance of soil (friction);
- F_D : hydrostatic drag force.

41.6.6 Step 5- Definition of Variables and Parameters

As can be noticed from the Equations 41.20 and 41.21, there are few variables to be considered. However, a greater amount of complexity can be added to the model by introducing probabilistic variables.

41.6.7 Step 6- Reliability Analysis

The reliability analysis could be performed using SYSREL (as in the previous example). It is important to note the type of probability of failure that is determined in this procedure. For this example the failure would be a time independent failure, since the forces causing failure (currents and wave action) are random in nature.

41.6.8 Step 7- Cost of Consequence

Movement of the pipeline could result in buckling, this could result in similar consequence scenarios as those presented in the previous example. Alternatively, the consequence may be to stabilize the pipeline further. This is a very case-specific matter, which would require further details.

In determining the cost consequence it is necessary to use the time value of money principles to determine the NPV of cost of consequences.

41.6.9 Step 8- Expected Cost

By multiplying the cost of consequence and the risk found, it is possible to determine the expected cost of failure.

41.6.10 Step 9- Initial Cost

The initial cost of the method by which the pipeline is stabilized can be found universally among pipeline design consultants and operators.

Table 41.8 Initial costs.

	Low Quality	High Quality
Initial Cost (C_I)	NOK 21,859,125	NOK 22,952,081

41.6.11 Step 10- Comparison of Life-cycle Cost

The final product of this analysis will render a range of on-bottom stability methods along with their potential Life-cycle Costs, allowing an informed decision to be reached.

Table 41.9 Combination of consequence outcomes.

Combination	Unplanned Maintenance	Environment
1	Upper	Upper
2	Upper	Lower
3	Lower	Lower
4	Lower	Upper

41.7 References

1. ASME B31G (1993) "Manual for Determining the Remaining Strength of Corroded Pipes", American Society of Mechanical Engineers
2. Bai, Y., Sørheim, M., Nødland, S. and Damsleth, P.A. (1999) "LCC Modeling as a Decision Making Tool in Pipeline Design". OMAE'99.
3. Bea, R. (1994) "The role of human error in the design, construction and reliability of marine structures", Ship Structure Committee, USA.
4. Bea, R. (1998) "Human and organization factors in the safety of offshore structures", in Risk and reliability in Marine Technology, edited by C. Guedes Soares, Published by A.A.Balkema.
5. Bea, R. et al. (1996) "Life-cycle reliability characteristic of minimum structures" OMAE'96.
6. Cui, W., Mansour, A.E, Elsayed, T. and Wirsching, W. (1998) "Reliability based quality and cost optimisation of unstiffened plates in ship structures", Proc. of PRADS '98, Edited by M.W. C. Oosterveld and S. G. Tan, Elsevier Science B.V.
7. deWaard, Lotz U. (1993) "Prediction of CO_2 Corrosion of Carbon Steel", CORROSION '93, paper no. 69.
8. Fabrycky, W.J. and Blanchard, B.S. (1991) "Life-cycle Cost and Economic Analysis", Prentice-Hall
9. Langford, G. and Kelly, P.G. (1990) "Design, Installation and Tie-in of Flowlines", JPK Report No. 4680.1

10. Nødland, S., Bai, Y. and Damsleth, P., (1997) "Reliability approach to optimise corrosion allowance", IBC Conference on risk based & limit state design & operation of pipelines.
11. Sørheim, M. and Bai, Y. (1999) "Risk Analysis Applied to Subsea Pipeline Engineering" OMAE'99.
12. SYSREL (1996), A Structural System Reliability program, RCP Consulting, Munich, rev. 9.10.

SUBJECT INDEX